Advanced Textbooks in Control and Signal Processing

Series Editors

Professor Michael J. Grimble, Professor of Industrial Systems and Director
Professor Michael A. Johnson, Professor Emeritus of Control Systems and Deputy Director
Industrial Control Centre, Department of Electronic and Electrical Engineering,
University of Strathclyde, Graham Hills Building, 50 George Street, Glasgow G1 1QE, UK

Other titles published in this series:

Genetic Algorithms
K.F. Man, K.S. Tang and S. Kwong

Introduction to Optimal Estimation
E.W. Kamen and J.K. Su

Discrete-time Signal Processing
D. Williamson

Neural Networks for Modelling and Control of Dynamic Systems
M. Nørgaard, O. Ravn, N.K. Poulsen and L.K. Hansen

Fault Detection and Diagnosis in Industrial Systems
L.H. Chiang, E.L. Russell and R.D. Braatz

Soft Computing
L. Fortuna, G. Rizzotto, M. Lavorgna, G. Nunnari, M.G. Xibilia and R. Caponetto

Statistical Signal Processing
T. Chonavel

Discrete-time Stochastic Processes (2nd Edition)
T. Söderström

Parallel Computing for Real-time Signal Processing and Control
M.O. Tokhi, M.A. Hossain and M.H. Shaheed

Multivariable Control Systems
P. Albertos and A. Sala

Control Systems with Input and Output Constraints
A.H. Glattfelder and W. Schaufelberger

Analysis and Control of Non-linear Process Systems
K.M. Hangos, J. Bokor and G. Szederkényi

Model Predictive Control (2nd Edition)
E.F. Camacho and C. Bordons

Principles of Adaptive Filters and Self-learning Systems
A. Zaknich

Digital Self-tuning Controllers
V. Bobál, J. Böhm, J. Fessl and J. Macháček

Control of Robot Manipulators in Joint Space
R. Kelly, V. Santibáñez and A. Loría

Receding Horizon Control
W.H. Kwon and S. Han

Robust Control Design with MATLAB®
D.-W. Gu, P.H. Petkov and M.M. Konstantinov

Control of Dead-time Processes
J.E. Normey-Rico and E.F. Camacho

Modeling and Control of Discrete-event Dynamic Systems
B. Hrúz and M.C. Zhou

Bruno Siciliano • Lorenzo Sciavicco
Luigi Villani • Giuseppe Oriolo

Robotics

Modelling, Planning and Control

Springer

Bruno Siciliano, PhD
Dipartimento di Informatica e Sistemistica
Università di Napoli Federico II
Via Claudio 21
80125 Napoli
Italy

Lorenzo Sciavicco, DrEng
Dipartimento di Informatica e Automazione
Università di Roma Tre
Via della Vasca Navale 79
00146 Roma
Italy

Luigi Villani, PhD
Dipartimento di Informatica e Sistemistica
Università di Napoli Federico II
Via Claudio 21
80125 Napoli
Italy

Giuseppe Oriolo, PhD
Dipartimento di Informatica e Sistemistica
Università di Roma "La Sapienza"
Via Ariosto 25
00185 Roma
Italy

ISBN 978-1-84996-634-4 e-ISBN 978-1-84628-642-1

DOI 10.1007/978-1-84628-642-1

Advanced Textbooks in Control and Signal Processing series ISSN 1439-2232

A catalogue record for this book is available from the British Library

Cover design: eStudio Calamar S.L., Girona, Spain

Printed on acid-free paper

9 8 7 6 5 4 3 2 1

springer.com

to our families

Series Editors' Foreword

The topics of control engineering and signal processing continue to flourish and develop. In common with general scientific investigation, new ideas, concepts and interpretations emerge quite spontaneously and these are then discussed, used, discarded or subsumed into the prevailing subject paradigm. Sometimes these innovative concepts coalesce into a new sub-discipline within the broad subject tapestry of control and signal processing. This preliminary battle between old and new usually takes place at conferences, through the Internet and in the journals of the discipline. After a little more maturity has been acquired by the new concepts then archival publication as a scientific or engineering monograph may occur.

A new concept in control and signal processing is known to have arrived when sufficient material has evolved for the topic to be taught as a specialised tutorial workshop or as a course to undergraduate, graduate or industrial engineers. *Advanced Textbooks in Control and Signal Processing* are designed as a vehicle for the systematic presentation of course material for both popular and innovative topics in the discipline. It is hoped that prospective authors will welcome the opportunity to publish a structured and systematic presentation of some of the newer emerging control and signal processing technologies in the textbook series.

Robots have appeared extensively in the artistic field of science fiction writing. The actual name robot arose from its use by the playwright Karel Čapek in the play *Rossum's Universal Robots* (1920). Not surprisingly, the artistic focus has been on mechanical bipeds with anthropomorphic personalities often termed androids. This focus has been the theme of such cinematic productions as, *I, Robot* (based on Isaac Asimov's stories) and Stanley Kubrick's film, *A.I.*; however, this book demonstrates that robot technology is already widely used in industry and that there is some robot technology which is at prototype stage rapidly approaching introduction to commercial use. Currently, robots may be classified according to their mobility attributes as shown in the figure.

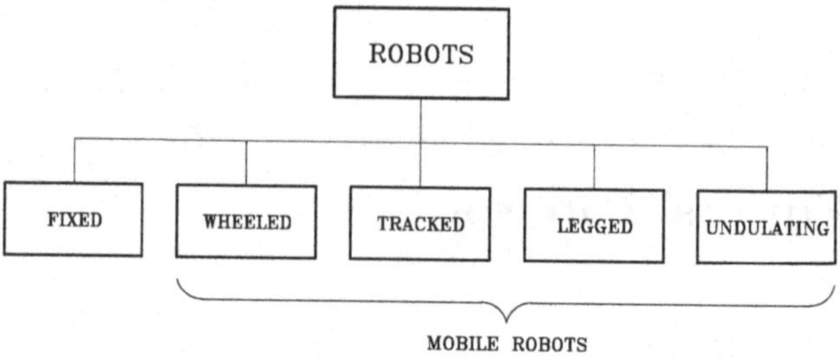

The largest class of robots extant today is that of the fixed robot which does repetitive but often precise mechanical and physical tasks. These robots pervade many areas of modern industrial automation and are mainly concerned with tasks performed in a structured environment. It seems highly likely that as the technology develops the number of mobile robots will significantly increase and become far more visible as more applications and tasks in an unstructured environment are serviced by robotic technology.

What then is robotics? A succinct definition is given in *The Chamber's Dictionary* (2003): *the branch of technology dealing with the design, construction and use of robots*. This definition certainly captures the spirit of this volume in the *Advanced Textbooks in Control and Signal Processing* series entitled *Robotics* and written by Bruno Siciliano, Lorenzo Sciavicco, Luigi Villani and Giuseppe Oriolo. This book is a greatly extended and revised version of an earlier book in the series, *Modelling and Control of Robot Manipulators* (2000, ISBN: 978-1-85233-221-1). As can be seen from the figure above, robots cover a wide variety of types and the new book seeks to present a unified approach to robotics whilst focusing on the two leading classes of robots, the fixed and the wheeled types. The textbook series publishes volumes in support of new disciplines that are emerging with their own novel identity, and robotics as a subject certainly falls into this category. The full scope of robotics lies at the intersection of mechanics, electronics, signal processing, control engineering, computing and mathematical modelling. However, within this very broad framework the authors have pursued the themes of *modelling, planning and control*. These are, and will remain, fundamental aspects of robot design and operation for years to come. Some interesting innovations in this text include material on wheeled robots and on vision as used in the control of robots. Thus, the book provides a thorough theoretical grounding in an area where the technologies are evolving and developing in new applications.

The series is one of textbooks for advanced courses, and volumes in the series have useful pedagogical features. This volume has twelve chapters covering both fundamental and specialist topics, and there is a *Problems* section at the end of each chapter. Five appendices have been included to give more depth to some of the advanced methods used in the text. There are over twelve pages of references and nine pages of index. The details of the citations and index should also facilitate the use of the volume as a source of reference as well as a course study text. We expect that the student, the researcher, the lecturer and the engineer will find this volume of great value for the study of robotics.

Glasgow
August 2008

Michael J. Grimble
Michael A. Johnson

Preface

In the last 25 years, the field of *robotics* has stimulated an increasing interest in a wide number of scholars, and thus literature has been conspicuous, both in terms of textbooks and monographs, and in terms of specialized journals dedicated to robotics. This strong interest is also to be attributed to the interdisciplinary character of robotics, which is a science having roots in different areas. Cybernetics, mechanics, controls, computers, bioengineering, electronics — to mention the most important ones — are all cultural domains which undoubtedly have boosted the development of this science.

Despite robotics representing as yet a relatively young discipline, its foundations are to be considered well-assessed in the classical textbook literature. Among these, *modelling*, *planning* and *control* play a basic role, not only in the traditional context of industrial robotics, but also for the advanced scenarios of field and service robots, which have attracted an increasing interest from the research community in the last 15 years.

This book is the natural evolution of the previous text *Modelling and Control of Robot Manipulators* by the first two co-authors, published in 1995, and in 2000 with its second edition. The cut of the original textbook has been confirmed with the educational goal of blending the fundamental and technological aspects with those advanced aspects, on a uniform track as regards a rigorous formalism.

The fundamental and technological aspects are mainly concentrated in the first six chapters of the book and concern the theory of manipulator structures, including kinematics, statics and trajectory planning, and the technology of robot actuators, sensors and control units.

The advanced aspects are dealt with in the subsequent six chapters and concern dynamics and motion control of robot manipulators, interaction with the environment using exteroceptive sensory data (force and vision), mobile robots and motion planning.

The book contents are organized in 12 chapters and 5 appendices.

In Chap. 1, the differences between *industrial* and *advanced* applications are enlightened in the general robotics context. The most common mechanical structures of robot manipulators and wheeled mobile robots are presented. Topics are also introduced which are developed in the subsequent chapters.

In Chap. 2 *kinematics* is presented with a systematic and general approach which refers to the Denavit-Hartenberg convention. The *direct kinematics equation* is formulated which relates joint space variables to operational space variables. This equation is utilized to find manipulator workspace as well as to derive a kinematic calibration technique. The *inverse kinematics problem* is also analyzed and closed-form solutions are found for typical manipulation structures.

Differential kinematics is presented in Chap. 3. The relationship between joint velocities and end-effector linear and angular velocities is described by the geometric *Jacobian*. The difference between the geometric Jacobian and the analytical Jacobian is pointed out. The Jacobian constitutes a fundamental tool to characterize a manipulator, since it allows the determination of singular configurations, an analysis of redundancy and the expression of the relationship between forces and moments applied to the end-effector and the resulting joint torques at equilibrium configurations (*statics*). Moreover, the Jacobian allows the formulation of inverse kinematics algorithms that solve the inverse kinematics problem even for manipulators not having a closed-form solution.

In Chap. 4, *trajectory planning* techniques are illustrated which deal with the computation of interpolating polynomials through a sequence of desired points. Both the case of point-to-point motion and that of motion through a sequence of points are treated. Techniques are developed for generating trajectories both in the *joint space* and in the *operational space*, with a special concern to orientation for the latter.

Chapter 5 is devoted to the presentation of *actuators* and *sensors*. After an illustration of the general features of an actuating system, methods to control electric and hydraulic *drives* are presented. The most common proprioceptive and exteroceptive sensors in robotics are described.

In Chap. 6, the functional architecture of a robot *control* system is illustrated. The characteristics of programming environments are presented with an emphasis on teaching-by-showing and robot-oriented programming. A general model for the hardware architecture of an industrial robot control system is finally discussed.

Chapter 7 deals with the derivation of manipulator *dynamics*, which plays a fundamental role in motion simulation, manipulation structure analysis and control algorithm synthesis. The dynamic model is obtained by explicitly taking into account the presence of actuators. Two approaches are considered, namely, one based on *Lagrange* formulation, and the other based on *Newton–Euler* formulation. The former is conceptually simpler and systematic, whereas the latter allows computation of a dynamic model in a recursive form. Notable properties of the dynamic model are presented, including linearity in the parameters which is utilized to develop a model identification technique. Finally, the transformations needed to express the dynamic model in the operational space are illustrated.

In Chap. 8 the problem of *motion control* in free space is treated. The distinction between joint space *decentralized* and *centralized* control strategies is pointed out. With reference to the former, the *independent joint control* technique is presented which is typically used for industrial robot control. As a premise to centralized control, the computed torque feedforward control technique is introduced. Advanced schemes are then introduced including PD control with gravity compensation, *inverse dynamics control*, robust control, and adaptive control. Centralized techniques are extended to *operational space control*.

Force control of a manipulator in contact with the working environment is tackled in Chap. 9. The concepts of mechanical *compliance* and *impedance* are defined as a natural extension of operational space control schemes to the constrained motion case. Force control schemes are then presented which are obtained by the addition of an outer force feedback loop to a motion control scheme. The *hybrid force/motion control* strategy is finally presented with reference to the formulation of natural and artificial constraints describing an interaction task.

In Chap. 10, *visual control* is introduced which allows the use of information on the environment surrounding the robotic system. The problems of camera *position and orientation estimate* with respect to the objects in the scene are solved by resorting to both analytical and numerical techniques. After presenting the advantages to be gained with *stereo vision* and a suitable camera *calibration*, the two main visual control strategies are illustrated, namely in the *operational space* and in the *image space*, whose advantages can be effectively combined in the *hybrid visual control* scheme.

Wheeled *mobile robots* are dealt with in Chap. 11, which extends some modelling, planning and control aspects of the previous chapters. As far as modelling is concerned, it is worth distinguishing between the *kinematic model*, strongly characterized by the type of constraint imposed by wheel rolling, and the *dynamic model* which accounts for the forces acting on the robot. The peculiar structure of the kinematic model is keenly exploited to develop both *path* and *trajectory planning* techniques. The *control* problem is tackled with reference to two main motion tasks: *trajectory tracking* and *configuration regulation*. Further, it is evidenced how the implementation of the control schemes utilizes *odometric localization* methods.

Chapter 12 reprises the planning problems treated in Chaps. 4 and 11 for robot manipulators and mobile robots respectively, in the case when obstacles are present in the workspace. In this framework, *motion planning* is referred to, which is effectively formulated in the *configuration space*. Several planning techniques for *mobile robots* are then presented: retraction, cell decomposition, probabilistic, artificial potential. The extension to the case of *robot manipulators* is finally discussed.

This chapter concludes the presentation of the topical contents of the textbook; five appendices follow which have been included to recall background methodologies.

Appendix A is devoted to *linear algebra* and presents the fundamental notions on matrices, vectors and related operations.

Appendix B presents those basic concepts of *rigid body mechanics* which are preliminary to the study of manipulator kinematics, statics and dynamics.

Appendix C illustrates the principles of *feedback control* of linear systems and presents a general method based on Lyapunov theory for control of nonlinear systems.

Appendix D deals with some concepts of *differential geometry* needed for control of mechanical systems subject to nonholonomic constraints.

Appendix E is focused on *graph search algorithms* and their complexity in view of application to motion planning methods.

The organization of the contents according to the above illustrated scheme allows the adoption of the book as a reference text for a senior undergraduate or graduate course in automation, computer, electrical, electronics, or mechanical engineering with strong robotics content.

From a pedagogical viewpoint, the various topics are presented in an instrumental manner and are developed with a gradually increasing level of difficulty. Problems are raised and proper tools are established to find engineering-oriented solutions. Each chapter is introduced by a brief preamble providing the rationale and the objectives of the subject matter. The topics needed for a proficient study of the text are presented in the five appendices, whose purpose is to provide students of different extraction with a homogeneous background.

The book contains more than 310 illustrations and more than 60 worked-out examples and case studies spread throughout the text with frequent resort to simulation. The results of computer implementations of inverse kinematics algorithms, trajectory planning techniques, inverse dynamics computation, motion, force and visual control algorithms for robot manipulators, and motion control for mobile robots are presented in considerable detail in order to facilitate the comprehension of the theoretical development, as well as to increase sensitivity of application in practical problems. In addition, nearly 150 end-of-chapter problems are proposed, some of which contain further study matter of the contents, and the book is accompanied by an electronic solutions manual (downloadable from `www.springer.com/978-1-84628-641-4`) containing the MATLAB® code for computer problems; this is available free of charge to those adopting this volume as a text for courses. Special care has been devoted to the selection of bibliographical references (more than 250) which are cited at the end of each chapter in relation to the historical development of the field.

Finally, the authors wish to acknowledge all those who have been helpful in the preparation of this book.

With reference to the original work, as the basis of the present textbook, devoted thanks go to Pasquale Chiacchio and Stefano Chiaverini for their contributions to the writing of the chapters on trajectory planning and force control, respectively. Fabrizio Caccavale and Ciro Natale have been of great help in the revision of the contents for the second edition.

A special note of thanks goes to Alessandro De Luca for his punctual and critical reading of large portions of the text, as well as to Vincenzo Lippiello, Agostino De Santis, Marilena Vendittelli and Luigi Freda for their contributions and comments on some sections.

Naples and Rome
July 2008

Bruno Siciliano
Lorenzo Sciavicco
Luigi Villani
Giuseppe Oriolo

Contents

1 **Introduction** 1
1.1 Robotics 1
1.2 Robot Mechanical Structure 3
1.2.1 Robot Manipulators 4
1.2.2 Mobile Robots 10
1.3 Industrial Robotics 15
1.4 Advanced Robotics 25
1.4.1 Field Robots 26
1.4.2 Service Robots 27
1.5 Robot Modelling, Planning and Control 29
1.5.1 Modelling 30
1.5.2 Planning 32
1.5.3 Control 32
Bibliography 33

2 **Kinematics** 39
2.1 Pose of a Rigid Body 39
2.2 Rotation Matrix 40
2.2.1 Elementary Rotations 41
2.2.2 Representation of a Vector 42
2.2.3 Rotation of a Vector 44
2.3 Composition of Rotation Matrices 45
2.4 Euler Angles 48
2.4.1 ZYZ Angles 49
2.4.2 RPY Angles 51
2.5 Angle and Axis 52
2.6 Unit Quaternion 54
2.7 Homogeneous Transformations 56
2.8 Direct Kinematics 58
2.8.1 Open Chain 60
2.8.2 Denavit–Hartenberg Convention 61

2.8.3 Closed Chain ... 65
2.9 Kinematics of Typical Manipulator Structures ... 68
2.9.1 Three-link Planar Arm ... 69
2.9.2 Parallelogram Arm ... 70
2.9.3 Spherical Arm ... 72
2.9.4 Anthropomorphic Arm ... 73
2.9.5 Spherical Wrist ... 75
2.9.6 Stanford Manipulator ... 76
2.9.7 Anthropomorphic Arm with Spherical Wrist ... 77
2.9.8 DLR Manipulator ... 79
2.9.9 Humanoid Manipulator ... 81
2.10 Joint Space and Operational Space ... 83
2.10.1 Workspace ... 85
2.10.2 Kinematic Redundancy ... 87
2.11 Kinematic Calibration ... 88
2.12 Inverse Kinematics Problem ... 90
2.12.1 Solution of Three-link Planar Arm ... 91
2.12.2 Solution of Manipulators with Spherical Wrist ... 94
2.12.3 Solution of Spherical Arm ... 95
2.12.4 Solution of Anthropomorphic Arm ... 96
2.12.5 Solution of Spherical Wrist ... 99
Bibliography ... 100
Problems ... 100

3 Differential Kinematics and Statics ... 105
3.1 Geometric Jacobian ... 105
3.1.1 Derivative of a Rotation Matrix ... 106
3.1.2 Link Velocities ... 108
3.1.3 Jacobian Computation ... 111
3.2 Jacobian of Typical Manipulator Structures ... 113
3.2.1 Three-link Planar Arm ... 113
3.2.2 Anthropomorphic Arm ... 114
3.2.3 Stanford Manipulator ... 115
3.3 Kinematic Singularities ... 116
3.3.1 Singularity Decoupling ... 117
3.3.2 Wrist Singularities ... 119
3.3.3 Arm Singularities ... 119
3.4 Analysis of Redundancy ... 121
3.5 Inverse Differential Kinematics ... 123
3.5.1 Redundant Manipulators ... 124
3.5.2 Kinematic Singularities ... 127
3.6 Analytical Jacobian ... 128
3.7 Inverse Kinematics Algorithms ... 132
3.7.1 Jacobian (Pseudo-)inverse ... 133
3.7.2 Jacobian Transpose ... 134

3.7.3 Orientation Error 137
3.7.4 Second-order Algorithms 141
3.7.5 Comparison Among Inverse Kinematics Algorithms 143
3.8 Statics 147
3.8.1 Kineto-Statics Duality 148
3.8.2 Velocity and Force Transformation 149
3.8.3 Closed Chain 151
3.9 Manipulability Ellipsoids 152
Bibliography 158
Problems 159

4 Trajectory Planning 161
4.1 Path and Trajectory 161
4.2 Joint Space Trajectories 162
4.2.1 Point-to-Point Motion 163
4.2.2 Motion Through a Sequence of Points 168
4.3 Operational Space Trajectories 179
4.3.1 Path Primitives 181
4.3.2 Position 184
4.3.3 Orientation 187
Bibliography 188
Problems 189

5 Actuators and Sensors 191
5.1 Joint Actuating System 191
5.1.1 Transmissions 192
5.1.2 Servomotors 193
5.1.3 Power Amplifiers 197
5.1.4 Power Supply 198
5.2 Drives 198
5.2.1 Electric Drives 198
5.2.2 Hydraulic Drives 202
5.2.3 Transmission Effects 204
5.2.4 Position Control 206
5.3 Proprioceptive Sensors 209
5.3.1 Position Transducers 210
5.3.2 Velocity Transducers 214
5.4 Exteroceptive Sensors 215
5.4.1 Force Sensors 215
5.4.2 Range Sensors 219
5.4.3 Vision Sensors 225
Bibliography 230
Problems 230

6 Control Architecture ... 233
6.1 Functional Architecture ... 233
6.2 Programming Environment ... 238
6.2.1 Teaching-by-Showing ... 240
6.2.2 Robot-oriented Programming ... 241
6.3 Hardware Architecture ... 242
Bibliography ... 245
Problems ... 245

7 Dynamics ... 247
7.1 Lagrange Formulation ... 247
7.1.1 Computation of Kinetic Energy ... 249
7.1.2 Computation of Potential Energy ... 255
7.1.3 Equations of Motion ... 255
7.2 Notable Properties of Dynamic Model ... 257
7.2.1 Skew-symmetry of Matrix $\dot{\boldsymbol{B}} - 2\boldsymbol{C}$ ... 257
7.2.2 Linearity in the Dynamic Parameters ... 259
7.3 Dynamic Model of Simple Manipulator Structures ... 264
7.3.1 Two-link Cartesian Arm ... 264
7.3.2 Two-link Planar Arm ... 265
7.3.3 Parallelogram Arm ... 277
7.4 Dynamic Parameter Identification ... 280
7.5 Newton–Euler Formulation ... 282
7.5.1 Link Accelerations ... 285
7.5.2 Recursive Algorithm ... 286
7.5.3 Example ... 289
7.6 Direct Dynamics and Inverse Dynamics ... 292
7.7 Dynamic Scaling of Trajectories ... 294
7.8 Operational Space Dynamic Model ... 296
7.9 Dynamic Manipulability Ellipsoid ... 299
Bibliography ... 301
Problems ... 301

8 Motion Control ... 303
8.1 The Control Problem ... 303
8.2 Joint Space Control ... 305
8.3 Decentralized Control ... 309
8.3.1 Independent Joint Control ... 311
8.3.2 Decentralized Feedforward Compensation ... 319
8.4 Computed Torque Feedforward Control ... 324
8.5 Centralized Control ... 327
8.5.1 PD Control with Gravity Compensation ... 328
8.5.2 Inverse Dynamics Control ... 330
8.5.3 Robust Control ... 333
8.5.4 Adaptive Control ... 338

8.6 Operational Space Control 343
8.6.1 General Schemes 344
8.6.2 PD Control with Gravity Compensation 345
8.6.3 Inverse Dynamics Control 347
8.7 Comparison Among Various Control Schemes 349
Bibliography 359
Problems 360

9 Force Control 363
9.1 Manipulator Interaction with Environment 363
9.2 Compliance Control 364
9.2.1 Passive Compliance 366
9.2.2 Active Compliance 367
9.3 Impedance Control 372
9.4 Force Control 378
9.4.1 Force Control with Inner Position Loop 379
9.4.2 Force Control with Inner Velocity Loop 380
9.4.3 Parallel Force/Position Control 381
9.5 Constrained Motion 384
9.5.1 Rigid Environment 385
9.5.2 Compliant Environment 389
9.6 Natural and Artificial Constraints 391
9.6.1 Analysis of Tasks 392
9.7 Hybrid Force/Motion Control 396
9.7.1 Compliant Environment 397
9.7.2 Rigid Environment 401
Bibliography 403
Problems 404

10 Visual Servoing 407
10.1 Vision for Control 407
10.1.1 Configuration of the Visual System 409
10.2 Image Processing 410
10.2.1 Image Segmentation 411
10.2.2 Image Interpretation 416
10.3 Pose Estimation 418
10.3.1 Analytic Solution 419
10.3.2 Interaction Matrix 424
10.3.3 Algorithmic Solution 427
10.4 Stereo Vision 433
10.4.1 Epipolar Geometry 433
10.4.2 Triangulation 435
10.4.3 Absolute Orientation 436
10.4.4 3D Reconstruction from Planar Homography 438
10.5 Camera Calibration 440

10.6 The Visual Servoing Problem 443
10.7 Position-based Visual Servoing 445
10.7.1 PD Control with Gravity Compensation 446
10.7.2 Resolved-velocity Control 447
10.8 Image-based Visual Servoing 449
10.8.1 PD Control with Gravity Compensation 449
10.8.2 Resolved-velocity Control 451
10.9 Comparison Among Various Control Schemes 453
10.10 Hybrid Visual Servoing 460
Bibliography 465
Problems 466

11 Mobile Robots 469
11.1 Nonholonomic Constraints 469
11.1.1 Integrability Conditions 473
11.2 Kinematic Model 476
11.2.1 Unicycle 478
11.2.2 Bicycle 479
11.3 Chained Form 482
11.4 Dynamic Model 485
11.5 Planning 489
11.5.1 Path and Timing Law 489
11.5.2 Flat Outputs 491
11.5.3 Path Planning 492
11.5.4 Trajectory Planning 498
11.5.5 Optimal Trajectories 499
11.6 Motion Control 502
11.6.1 Trajectory Tracking 503
11.6.2 Regulation 510
11.7 Odometric Localization 514
Bibliography 518
Problems 518

12 Motion Planning 523
12.1 The Canonical Problem 523
12.2 Configuration Space 525
12.2.1 Distance 527
12.2.2 Obstacles 527
12.2.3 Examples of Obstacles 528
12.3 Planning via Retraction 532
12.4 Planning via Cell Decomposition 536
12.4.1 Exact Decomposition 536
12.4.2 Approximate Decomposition 539
12.5 Probabilistic Planning 541
12.5.1 PRM Method 541

12.5.2 Bidirectional RRT Method 543
12.6 Planning via Artificial Potentials 546
12.6.1 Attractive Potential 546
12.6.2 Repulsive Potential 547
12.6.3 Total Potential 549
12.6.4 Planning Techniques 550
12.6.5 The Local Minima Problem 551
12.7 The Robot Manipulator Case 554
Bibliography 557
Problems 557

Appendices

A Linear Algebra 563
A.1 Definitions 563
A.2 Matrix Operations 565
A.3 Vector Operations 569
A.4 Linear Transformation 572
A.5 Eigenvalues and Eigenvectors 573
A.6 Bilinear Forms and Quadratic Forms 574
A.7 Pseudo-inverse 575
A.8 Singular Value Decomposition 577
Bibliography 578

B Rigid-body Mechanics 579
B.1 Kinematics 579
B.2 Dynamics 581
B.3 Work and Energy 584
B.4 Constrained Systems 585
Bibliography 588

C Feedback Control 589
C.1 Control of Single-input/Single-output Linear Systems 589
C.2 Control of Nonlinear Mechanical Systems 594
C.3 Lyapunov Direct Method 596
Bibliography 598

D Differential Geometry 599
D.1 Vector Fields and Lie Brackets 599
D.2 Nonlinear Controllability 603
Bibliography 604

E Graph Search Algorithms ... 605
E.1 Complexity ... 605
E.2 Breadth-first and Depth-first Search ... 606
E.3 $A^{\star}$ Algorithm ... 607
Bibliography ... 608

References ... 609

Index ... 623

1

Introduction

Robotics is concerned with the study of those machines that can replace human beings in the execution of a task, as regards both physical activity and decision making. The goal of the introductory chapter is to point out the problems related to the use of *robots* in *industrial* applications, as well as the perspectives offered by *advanced robotics*. A classification of the most common mechanical structures of *robot manipulators* and *mobile robots* is presented. Topics of *modelling*, *planning* and *control* are introduced which will be examined in the following chapters. The chapter ends with a list of references dealing with subjects both of specific interest and of related interest to those covered by this textbook.

1.1 Robotics

Robotics has profound cultural roots. Over the course of centuries, human beings have constantly attempted to seek substitutes that would be able to mimic their behaviour in the various instances of interaction with the surrounding environment. Several motivations have inspired this continuous search referring to philosophical, economic, social and scientific principles.

One of human beings' greatest ambitions has been to give life to their artifacts. The legend of the Titan Prometheus, who molded humankind from clay, as well as that of the giant Talus, the bronze slave forged by Hephaestus, testify how Greek mythology was influenced by that ambition, which has been revisited in the tale of Frankenstein in modern times.

Just as the giant Talus was entrusted with the task of protecting the island of Crete from invaders, in the Industrial Age a mechanical creature (*automaton*) has been entrusted with the task of substituting a human being in subordinate labor duties. This concept was introduced by the Czech playwright Karel Čapek who wrote the play *Rossum's Universal Robots (R.U.R.)* in 1920. On that occasion he coined the term *robot* — derived from the term *robota* that means executive labour in Slav languages — to denote the automaton built by Rossum who ends up by rising up against humankind in the science fiction tale.

In the subsequent years, in view of the development of science fiction, the behaviour conceived for the robot has often been conditioned by feelings. This has contributed to rendering the robot more and more similar to its creator.

It is worth noticing how Rossum's robots were represented as creatures made with organic material. The image of the robot as a mechanical artifact starts in the 1940s when the Russian Isaac Asimov, the well-known science fiction writer, conceived the robot as an automaton of human appearance but devoid of feelings. Its behaviour was dictated by a "positronic" brain programmed by a human being in such a way as to satisfy certain rules of ethical conduct. The term *robotics* was then introduced by Asimov as the science devoted to the study of robots which was based on the *three fundamental laws*:

1. A robot may not injure a human being or, through inaction, allow a human being to come to harm.
2. A robot must obey the orders given by human beings, except when such orders would conflict with the first law.
3. A robot must protect its own existence, as long as such protection does not conflict with the first or second law.

These laws established rules of behaviour to consider as specifications for the design of a robot, which since then has attained the connotation of an industrial product designed by engineers or specialized technicians.

Science fiction has influenced the man and the woman in the street that continue to imagine the robot as a humanoid who can speak, walk, see, and hear, with an appearance very much like that presented by the robots of the movie *Metropolis*, a precursor of modern cinematography on robots, with *Star Wars* and more recently with *I, Robot* inspired by Asimov's novels.

According to a scientific interpretation of the science-fiction scenario, the robot is seen as a machine that, independently of its exterior, is able to modify the environment in which it operates. This is accomplished by carrying out actions that are conditioned by certain rules of behaviour intrinsic in the machine as well as by some data the robot acquires on its status and on the environment. In fact, *robotics* is commonly defined as the science studying the *intelligent connection between perception and action*.

With reference to this definition, a *robotic system* is in reality a complex system, functionally represented by multiple subsystems (Fig. 1.1).

The essential component of a robot is the *mechanical system* endowed, in general, with a locomotion apparatus (wheels, crawlers, mechanical legs) and a manipulation apparatus (mechanical arms, end-effectors, artificial hands). As an example, the mechanical system in Fig. 1.1 consists of two mechanical arms (manipulation apparatus), each of which is carried by a mobile vehicle

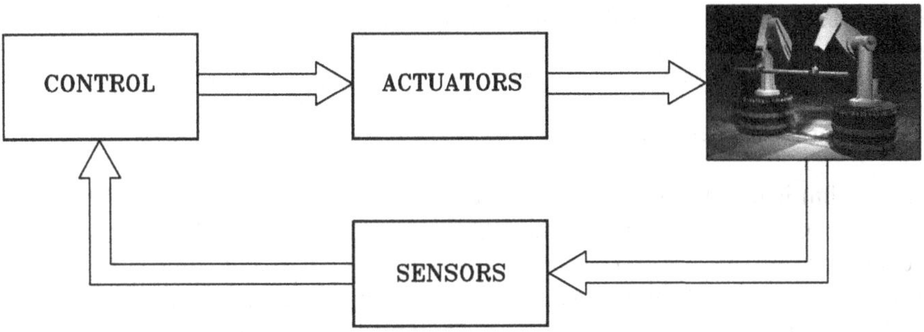

Fig. 1.1. Components of a robotic system

(locomotion apparatus). The realization of such a system refers to the context of design of articulated mechanical systems and choice of materials.

The capability to exert an action, both locomotion and manipulation, is provided by an *actuation system* which animates the mechanical components of the robot. The concept of such a system refers to the context of *motion control*, dealing with *servomotors*, *drives* and *transmissions*.

The capability for perception is entrusted to a *sensory system* which can acquire data on the internal status of the mechanical system (*proprioceptive sensors*, such as position transducers) as well as on the external status of the environment (*exteroceptive sensors*, such as force sensors and cameras). The realization of such a system refers to the context of materials properties, signal conditioning, data processing, and information retrieval.

The capability for connecting action to perception in an intelligent fashion is provided by a *control system* which can command the execution of the action in respect to the goals set by a task *planning* technique, as well as of the constraints imposed by the robot and the environment. The realization of such a system follows the same feedback principle devoted to *control* of human body functions, possibly exploiting the description of the robotic system's components (*modelling*). The context is that of cybernetics, dealing with control and supervision of robot motions, artificial intelligence and expert systems, the computational architecture and programming environment.

Therefore, it can be recognized that robotics is an interdisciplinary subject concerning the cultural areas of *mechanics*, *control*, *computers*, and *electronics*.

1.2 Robot Mechanical Structure

The key feature of a robot is its mechanical structure. Robots can be classified as those with a fixed base, *robot manipulators*, and those with a mobile base, *mobile robots*. In the following, the geometrical features of the two classes are presented.

1.2.1 Robot Manipulators

The mechanical structure of a *robot manipulator* consists of a sequence of rigid bodies (*links*) interconnected by means of articulations (*joints*); a manipulator is characterized by an *arm* that ensures mobility, a *wrist* that confers dexterity, and an *end-effector* that performs the task required of the robot.

The fundamental structure of a manipulator is the serial or *open kinematic chain*. From a topological viewpoint, a kinematic chain is termed open when there is only one sequence of links connecting the two ends of the chain. Alternatively, a manipulator contains a *closed kinematic chain* when a sequence of links forms a loop.

A manipulator's mobility is ensured by the presence of joints. The articulation between two consecutive links can be realized by means of either a *prismatic* or a *revolute* joint. In an open kinematic chain, each prismatic or revolute joint provides the structure with a single degree of freedom (DOF). A prismatic joint creates a relative translational motion between the two links, whereas a revolute joint creates a relative rotational motion between the two links. Revolute joints are usually preferred to prismatic joints in view of their compactness and reliability. On the other hand, in a closed kinematic chain, the number of DOFs is less than the number of joints in view of the constraints imposed by the loop.

The *degrees of freedom* should be properly distributed along the mechanical structure in order to have a sufficient number to execute a given task. In the most general case of a task consisting of arbitrarily positioning and orienting an object in three-dimensional (3D) space, *six* DOFs are required, three for positioning a point on the object and three for orienting the object with respect to a reference coordinate frame. If more DOFs than task variables are available, the manipulator is said to be *redundant* from a kinematic viewpoint.

The *workspace* represents that portion of the environment the manipulator's end-effector can access. Its shape and volume depend on the manipulator structure as well as on the presence of mechanical joint limits.

The task required of the arm is to position the wrist which then is required to orient the end-effector. The type and sequence of the arm's DOFs, starting from the base joint, allows a classification of manipulators as *Cartesian*, *cylindrical*, *spherical*, *SCARA*, and *anthropomorphic*.

Cartesian geometry is realized by three prismatic joints whose axes typically are mutually orthogonal (Fig. 1.2). In view of the simple geometry, each DOF corresponds to a Cartesian space variable and thus it is natural to perform straight motions in space. The Cartesian structure offers very good mechanical stiffness. Wrist positioning accuracy is constant everywhere in the workspace. This is the volume enclosed by a rectangular parallel-piped

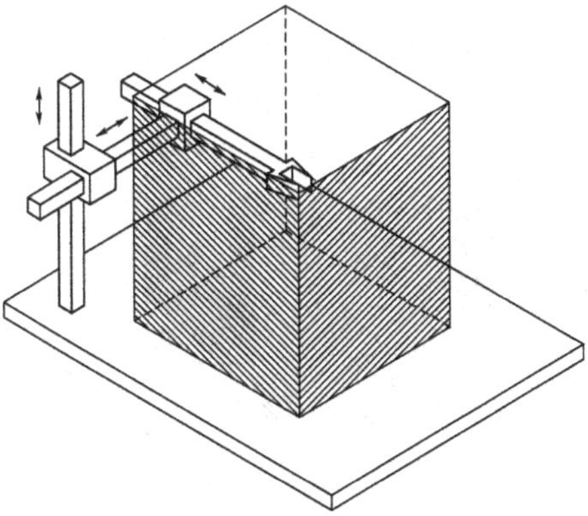

Fig. 1.2. Cartesian manipulator and its workspace

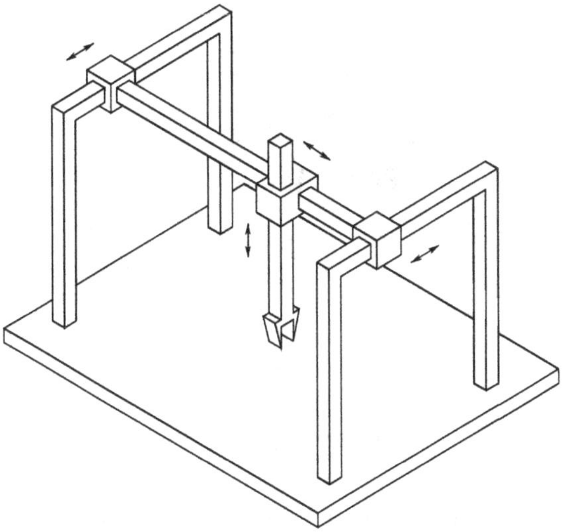

Fig. 1.3. Gantry manipulator

(Fig. 1.2). As opposed to high accuracy, the structure has low dexterity since all the joints are prismatic. The direction of approach in order to manipulate an object is from the side. On the other hand, if it is desired to approach an object from the top, the Cartesian manipulator can be realized by a *gantry* structure as illustrated in Fig. 1.3. Such a structure makes available a workspace with a large volume and enables the manipulation of objects of large dimensions and heavy weight. Cartesian manipulators are employed for material handling and assembly. The motors actuating the joints of a Cartesian manipulator are typically electric and occasionally pneumatic.

Cylindrical geometry differs from Cartesian in that the first prismatic joint is replaced with a revolute joint (Fig. 1.4). If the task is described in cylindri-

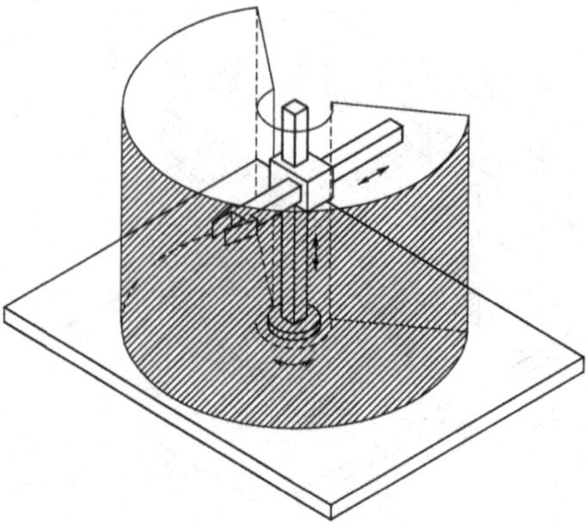

Fig. 1.4. Cylindrical manipulator and its workspace

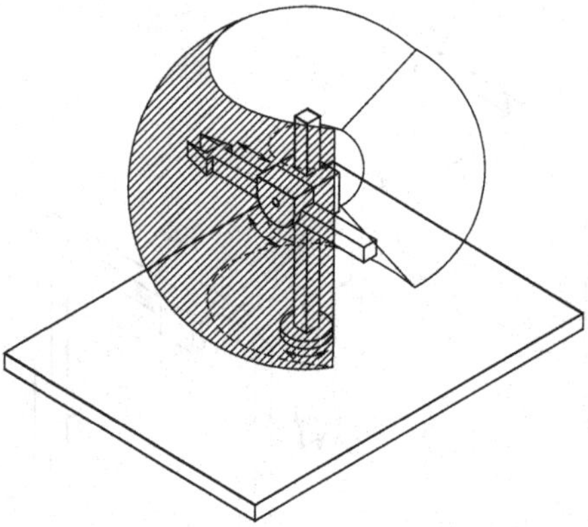

Fig. 1.5. Spherical manipulator and its workspace

cal coordinates, in this case each DOF also corresponds to a Cartesian space variable. The cylindrical structure offers good mechanical stiffness. Wrist positioning accuracy decreases as the horizontal stroke increases. The workspace is a portion of a hollow cylinder (Fig. 1.4). The horizontal prismatic joint makes the wrist of a cylindrical manipulator suitable to access horizontal cavities. Cylindrical manipulators are mainly employed for carrying objects even of large dimensions; in such a case the use of hydraulic motors is to be preferred to that of electric motors.

Spherical geometry differs from cylindrical in that the second prismatic joint is replaced with a revolute joint (Fig. 1.5). Each DOF corresponds to a Cartesian space variable provided that the task is described in spherical coordinates. Mechanical stiffness is lower than the above two geometries and mechanical construction is more complex. Wrist positioning accuracy decreases as the radial stroke increases. The workspace is a portion of a hollow sphere (Fig. 1.5); it can also include the supporting base of the manipulator and thus

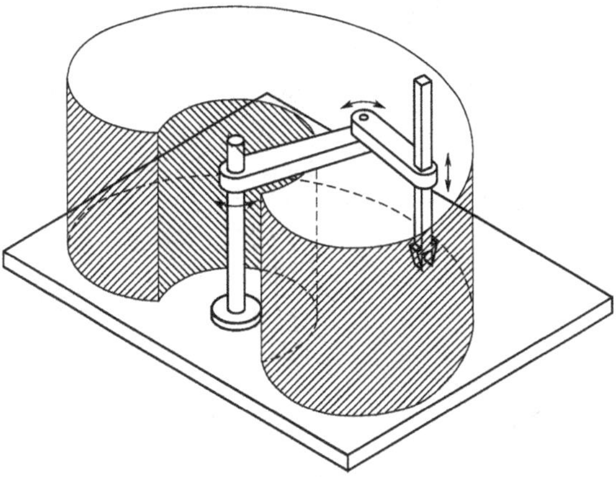

Fig. 1.6. SCARA manipulator and its workspace

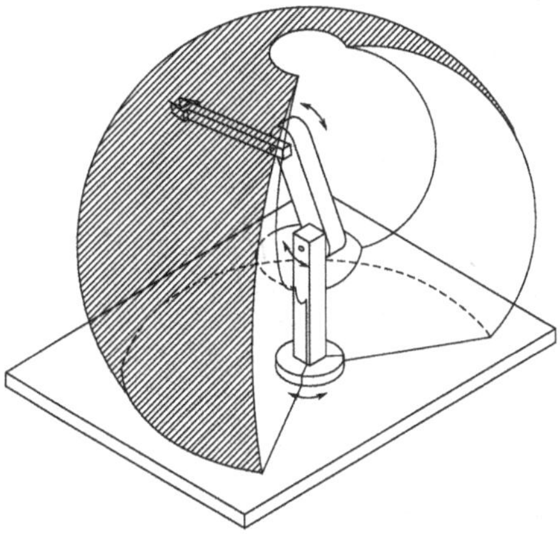

Fig. 1.7. Anthropomorphic manipulator and its workspace

it can allow manipulation of objects on the floor. Spherical manipulators are mainly employed for machining. Electric motors are typically used to actuate the joints.

A special geometry is *SCARA* geometry that can be realized by disposing two revolute joints and one prismatic joint in such a way that all the axes of motion are parallel (Fig. 1.6). The acronym SCARA stands for *Selective Compliance Assembly Robot Arm* and characterizes the mechanical features of a structure offering high stiffness to vertical loads and compliance to horizontal loads. As such, the SCARA structure is well-suited to vertical assembly tasks. The correspondence between the DOFs and Cartesian space variables is maintained only for the vertical component of a task described in Cartesian coordinates. Wrist positioning accuracy decreases as the distance of the wrist from the first joint axis increases. The typical workspace is illustrated

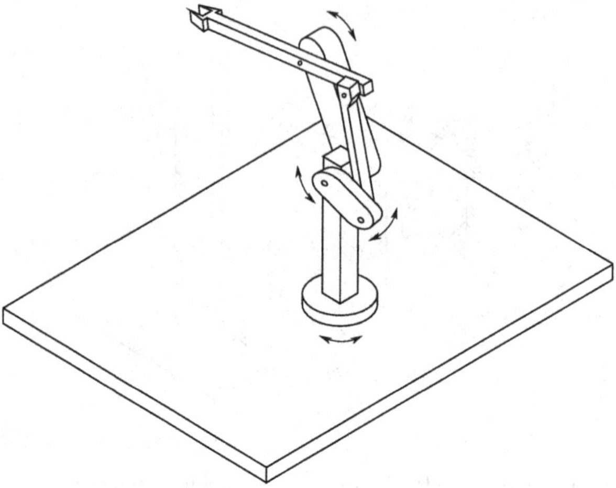

Fig. 1.8. Manipulator with parallelogram

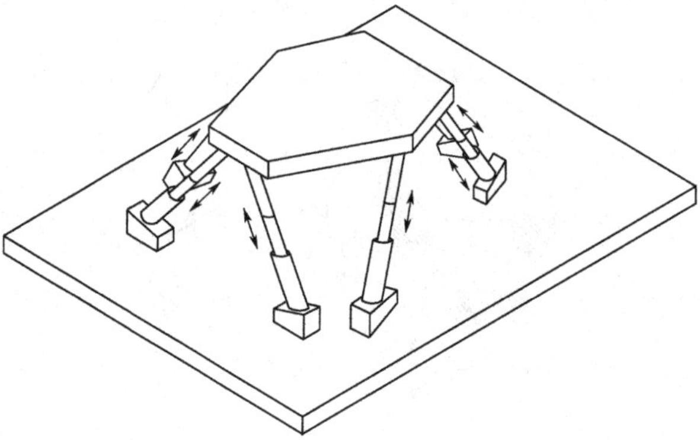

Fig. 1.9. Parallel manipulator

in Fig. 1.6. The SCARA manipulator is suitable for manipulation of small objects; joints are actuated by electric motors.

Anthropomorphic geometry is realized by three revolute joints; the revolute axis of the first joint is orthogonal to the axes of the other two which are parallel (Fig. 1.7). By virtue of its similarity with the human arm, the second joint is called the shoulder joint and the third joint the elbow joint since it connects the "arm" with the "forearm." The anthropomorphic structure is the most dexterous one, since all the joints are revolute. On the other hand, the correspondence between the DOFs and the Cartesian space variables is lost, and wrist positioning accuracy varies inside the workspace. This is approximately a portion of a sphere (Fig. 1.7) and its volume is large compared to manipulator encumbrance. Joints are typically actuated by electric motors. The range of industrial applications of anthropomorphic manipulators is wide.

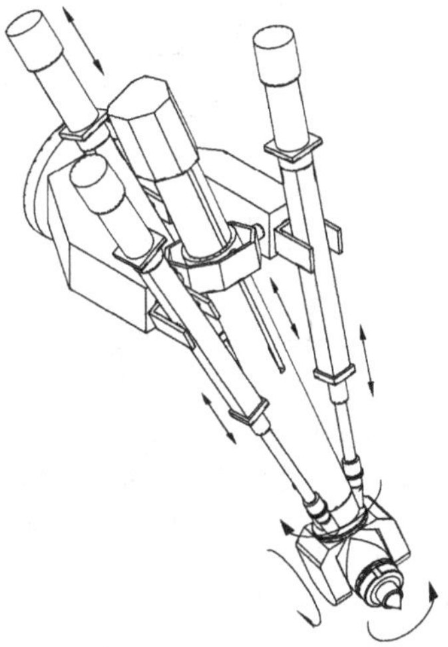

Fig. 1.10. Hybrid parallel-serial manipulator

According to the latest report by the *International Federation of Robotics* (IFR), up to 2005, 59% of installed robot manipulators worldwide has anthropomorphic geometry, 20% has Cartesian geometry, 12% has cylindrical geometry, and 8% has SCARA geometry.

All the previous manipulators have an open kinematic chain. Whenever larger payloads are required, the mechanical structure will have higher stiffness to guarantee comparable positioning accuracy. In such a case, resorting to a closed kinematic chain is advised. For instance, for an anthropomorphic structure, parallelogram geometry between the shoulder and elbow joints can be adopted, so as to create a closed kinematic chain (Fig. 1.8).

An interesting closed-chain geometry is *parallel* geometry (Fig. 1.9) which has multiple kinematic chains connecting the base to the end-effector. The fundamental advantage is seen in the high structural stiffness, with respect to open-chain manipulators, and thus the possibility to achieve high operational speeds; the drawback is that of having a reduced workspace.

The geometry illustrated in Fig. 1.10 is of hybrid type, since it consists of a parallel arm and a serial kinematic chain. This structure is suitable for the execution of manipulation tasks requiring large values of force along the vertical direction.

The manipulator structures presented above are required to position the wrist which is then required to orient the manipulator's end-effector. If arbitrary orientation in 3D space is desired, the wrist must possess at least three DOFs provided by revolute joints. Since the wrist constitutes the terminal part of the manipulator, it has to be compact; this often complicates its mechanical design. Without entering into construction details, the realization endowing the wrist with the highest dexterity is one where the three revolute

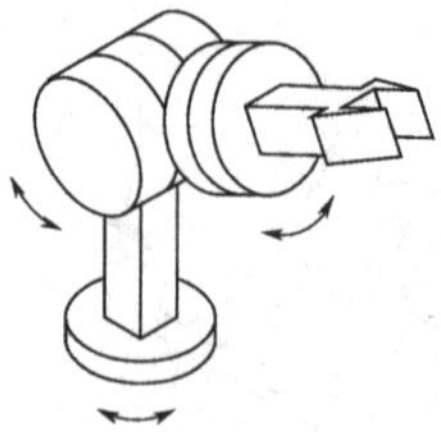

Fig. 1.11. Spherical wrist

axes intersect at a single point. In such a case, the wrist is called a *spherical wrist*, as represented in Fig. 1.11. The key feature of a spherical wrist is the decoupling between position and orientation of the end-effector; the arm is entrusted with the task of positioning the above point of intersection, whereas the wrist determines the end-effector orientation. Those realizations where the wrist is not spherical are simpler from a mechanical viewpoint, but position and orientation are coupled, and this complicates the coordination between the motion of the arm and that of the wrist to perform a given task.

The *end-effector* is specified according to the task the robot should execute. For material handling tasks, the end-effector consists of a gripper of proper shape and dimensions determined by the object to be grasped (Fig. 1.11). For machining and assembly tasks, the end-effector is a tool or a specialized device, e.g., a welding torch, a spray gun, a mill, a drill, or a screwdriver.

The versatility and flexibility of a robot manipulator should not induce the conviction that all mechanical structures are equivalent for the execution of a given task. The choice of a robot is indeed conditioned by the application which sets constraints on the workspace dimensions and shape, the maximum payload, positioning accuracy, and dynamic performance of the manipulator.

1.2.2 Mobile Robots

The main feature of *mobile robots* is the presence of a mobile base which allows the robot to move freely in the environment. Unlike manipulators, such robots are mostly used in service applications, where extensive, autonomous motion capabilities are required. From a mechanical viewpoint, a mobile robot consists of one or more rigid bodies equipped with a *locomotion* system. This description includes the following two main classes of mobile robots:[1]

- *Wheeled* mobile robots typically consist of a rigid body (*base* or *chassis*) and a system of wheels which provide motion with respect to the ground.

[1] Other types of mechanical locomotion systems are not considered here. Among these, it is worth mentioning *tracked locomotion*, very effective on uneven terrain, and *undulatory locomotion*, inspired by snake gaits, which can be achieved without specific devices. There also exist types of locomotion that are not constrained to the ground, such as flying and navigation.

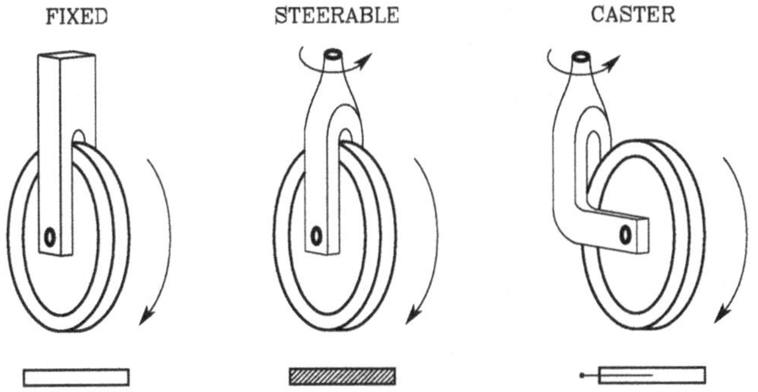

Fig. 1.12. The three types of conventional wheels with their respective icons

Other rigid bodies (*trailers*), also equipped with wheels, may be connected to the base by means of revolute joints.

- *Legged* mobile robots are made of multiple rigid bodies, interconnected by prismatic joints or, more often, by revolute joints. Some of these bodies form lower limbs, whose extremities (*feet*) periodically come in contact with the ground to realize locomotion. There is a large variety of mechanical structures in this class, whose design is often inspired by the study of living organisms (*biomimetic robotics*): they range from biped humanoids to hexapod robots aimed at replicating the biomechanical efficiency of insects.

Only wheeled vehicles are considered in the following, as they represent the vast majority of mobile robots actually used in applications. The basic mechanical element of such robots is indeed the wheel. Three types of conventional wheels exist, which are shown in Fig. 1.12 together with the icons that will be used to represent them:

- The *fixed wheel* can rotate about an axis that goes through the center of the wheel and is orthogonal to the wheel plane. The wheel is rigidly attached to the chassis, whose orientation with respect to the wheel is therefore constant.
- The *steerable wheel* has two axes of rotation. The first is the same as a fixed wheel, while the second is vertical and goes through the center of the wheel. This allows the wheel to change its orientation with respect to the chassis.
- The *caster wheel* has two axes of rotation, but the vertical axis does not pass through the center of the wheel, from which it is displaced by a constant *offset*. Such an arrangement causes the wheel to swivel automatically, rapidly aligning with the direction of motion of the chassis. This type of wheel is therefore introduced to provide a supporting point for static balance without affecting the mobility of the base; for instance, caster wheels are commonly used in shopping carts as well as in chairs with wheels.

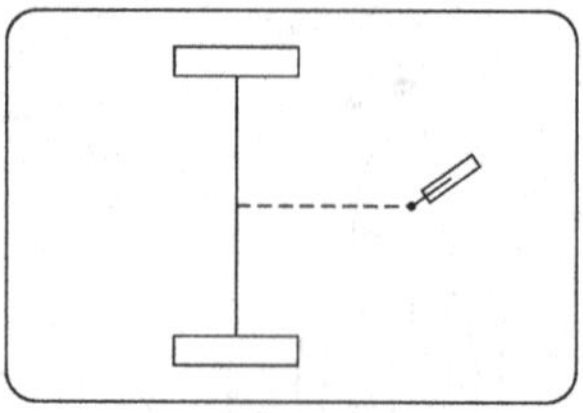

Fig. 1.13. A differential-drive mobile robot

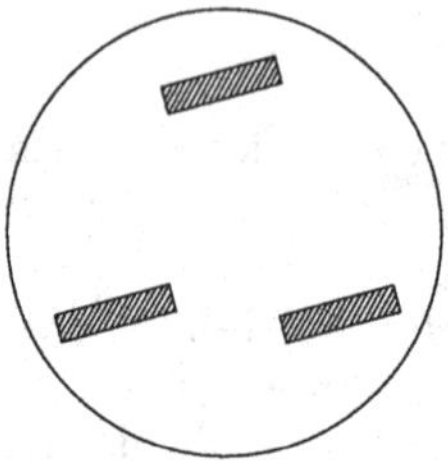

Fig. 1.14. A synchro-drive mobile robot

The variety of kinematic structures that can be obtained by combining the three conventional wheels is wide. In the following, the most relevant arrangements are briefly examined.

In a *differential-drive* vehicle there are two fixed wheels with a common axis of rotation, and one or more caster wheels, typically smaller, whose function is to keep the robot statically balanced (Fig. 1.13). The two fixed wheels are separately controlled, in that different values of angular velocity may be arbitrarily imposed, while the caster wheel is passive. Such a robot can rotate on the spot (i.e., without moving the midpoint between the wheels), provided that the angular velocities of the two wheels are equal and opposite.

A vehicle with similar mobility is obtained using a *synchro-drive* kinematic arrangement (Fig. 1.14). This robot has three aligned steerable wheels which are synchronously driven by only two motors through a mechanical coupling, e.g., a chain or a transmission belt. The first motor controls the rotation of the wheels around the horizontal axis, thus providing the driving force (traction) to the vehicle. The second motor controls the rotation of the wheels around the vertical axis, hence affecting their orientation. Note that the heading of the chassis does not change during the motion. Often, a third motor is used in this type of robot to rotate independently the upper part of the chassis (a turret) with respect to the lower part. This may be useful to orient arbitrarily a directional sensor (e.g., a camera) or in any case to recover an orientation error.

In a *tricycle* vehicle (Fig. 1.15) there are two fixed wheels mounted on a rear axle and a steerable wheel in front. The fixed wheels are driven by a single

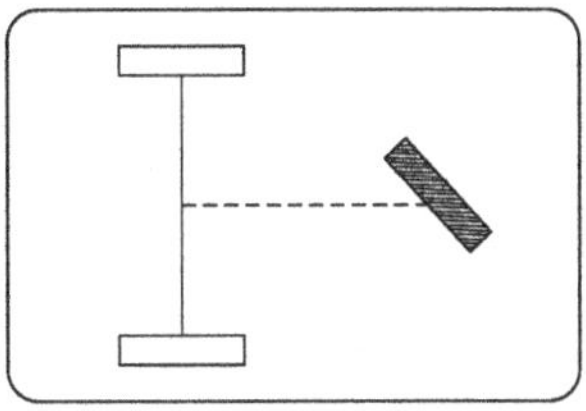

Fig. 1.15. A tricycle mobile robot

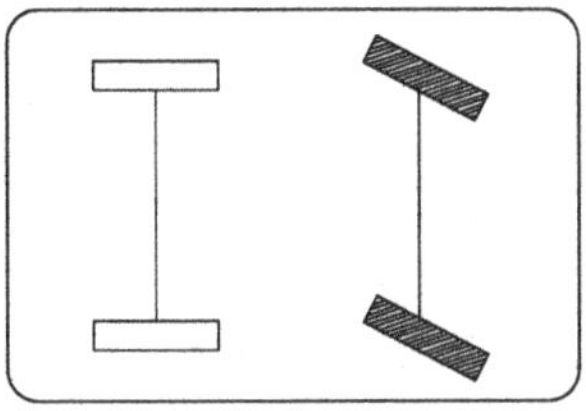

Fig. 1.16. A car-like mobile robot

motor which controls their traction,[2] while the steerable wheel is driven by another motor which changes its orientation, acting then as a steering device. Alternatively, the two rear wheels may be passive and the front wheel may provide traction as well as steering.

A *car-like* vehicle has two fixed wheels mounted on a rear axle and two steerable wheels mounted on a front axle, as shown in Fig. 1.16. As in the previous case, one motor provides (front or rear) traction while the other changes the orientation of the front wheels with respect to the vehicle. It is worth pointing out that, to avoid slippage, the two front wheels must have a different orientation when the vehicle moves along a curve; in particular, the internal wheel is slightly more steered with respect to the external one. This is guaranteed by the use of a specific device called *Ackermann steering*.

Finally, consider the robot in Fig. 1.17, which has three caster wheels usually arranged in a symmetric pattern. The traction velocities of the three wheels are independently driven. Unlike the previous cases, this vehicle is *omnidirectional*: in fact, it can move instantaneously in any Cartesian direction, as well as re-orient itself on the spot.

In addition to the above conventional wheels, there exist other special types of wheels, among which is notably the *Mecanum* (or *Swedish*) *wheel*, shown in Fig. 1.18. This is a fixed wheel with passive rollers placed along the external rim; the axis of rotation of each roller is typically inclined by 45° with respect to the plane of the wheel. A vehicle equipped with four such wheels mounted in pairs on two parallel axles is also omnidirectional.

[2] The distribution of the traction torque on the two wheels must take into account the fact that in general they move with different speeds. The mechanism which equally distributes traction is the *differential*.

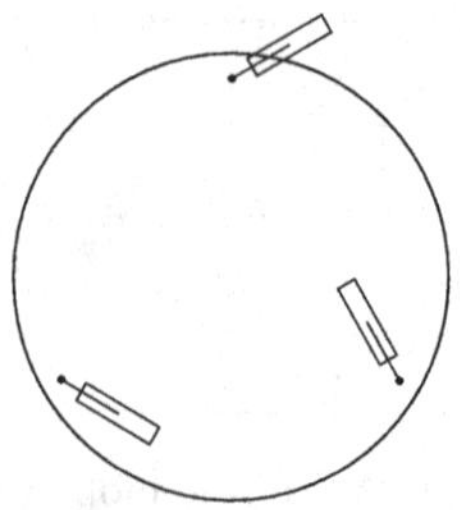

Fig. 1.17. An omnidirectional mobile robot with three independently driven caster wheels

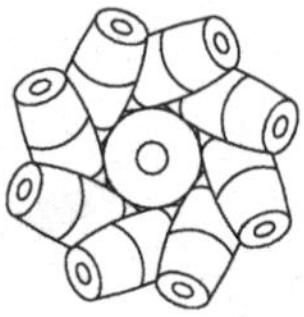

Fig. 1.18. A Mecanum (or Swedish) wheel

In the design of a wheeled robot, the mechanical balance of the structure does not represent a problem in general. In particular, a three-wheel robot is statically balanced as long as its center of mass falls inside the *support triangle*, which is defined by the contact points between the wheels and ground. Robots with more than three wheels have a support *polygon*, and thus it is typically easier to guarantee the above balance condition. It should be noted, however, that when the robot moves on uneven terrain a suspension system is needed to maintain the contact between each wheel and the ground.

Unlike the case of manipulators, the *workspace* of a mobile robot (defined as the portion of the surrounding environment that the robot can access) is potentially unlimited. Nevertheless, the local mobility of a non-omnidirectional mobile robot is always reduced; for instance, the tricycle robot in Fig. 1.15 cannot move instantaneously in a direction parallel to the rear wheel axle. Despite this fact, the tricycle can be manoeuvered so as to obtain, at the end of the motion, a net displacement in that direction. In other words, many mobile robots are subject to constraints on the admissible instantaneous motions, without actually preventing the possibility of attaining any position and orientation in the workspace. This also implies that the number of DOFs of the robot (meant as the number of admissible instantaneous motions) is lower than the number of its configuration variables.

It is obviously possible to merge the mechanical structure of a manipulator with that of a mobile vehicle by mounting the former on the latter. Such a robot is called a *mobile manipulator* and combines the dexterity of the articulated arm with the unlimited mobility of the base. An example of such a mechanical structure is shown in Fig. 1.19. However, the design of a mobile manipulator involves additional difficulties related, for instance, to the static

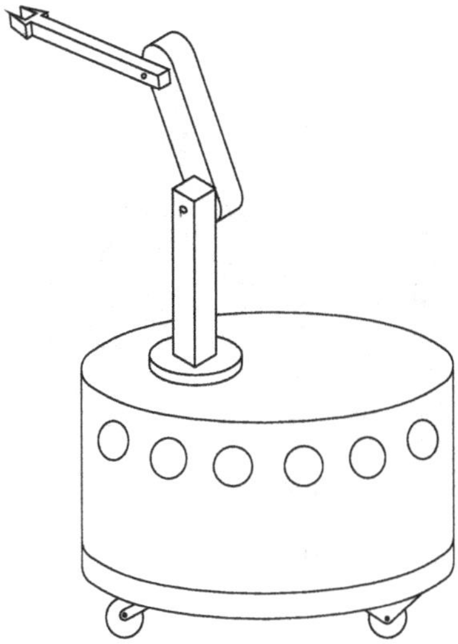

Fig. 1.19. A mobile manipulator obtained by mounting an anthropomorphic arm on a differential-drive vehicle

and dynamic mechanical balance of the robot, as well as to the actuation of the two systems.

1.3 Industrial Robotics

Industrial robotics is the discipline concerning robot design, control and applications in industry, and its products have by now reached the level of a mature technology. The connotation of a robot for industrial applications is that of operating in a *structured environment* whose geometrical or physical characteristics are mostly known a priori. Hence, limited autonomy is required.

The early industrial robots were developed in the 1960s, at the confluence of two technologies: numerical control machines for precise manufacturing, and teleoperators for remote radioactive material handling. Compared to its precursors, the first robot manipulators were characterized by:

- versatility, in view of the employment of different end-effectors at the tip of the manipulator,
- adaptability to a priori unknown situations, in view of the use of sensors,
- positioning accuracy, in view of the adoption of feedback control techniques,
- execution repeatability, in view of the programmability of various operations.

During the subsequent decades, industrial robots have gained a wide popularity as essential components for the realization of automated manufacturing

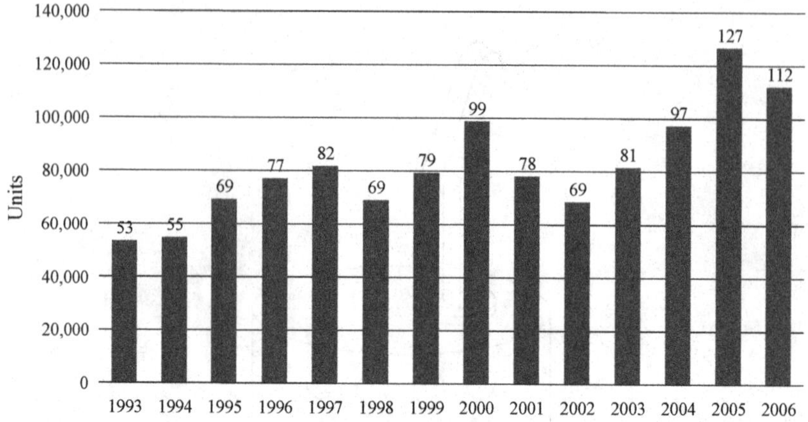

Fig. 1.20. Yearly installations of industrial robots worldwide

systems. The main factors having determined the spread of robotics technology in an increasingly wider range of applications in the manufacturing industry are reduction of manufacturing costs, increase of productivity, improvement of product quality standards and, last but not least, the possibility of eliminating harmful or off-putting tasks for the human operator in a manufacturing system.

By its usual meaning, the term *automation* denotes a technology aimed at replacing human beings with machines in a manufacturing process, as regards not only the execution of physical operations but also the intelligent processing of information on the status of the process. Automation is then the synthesis of industrial technologies typical of the manufacturing process and computer technology allowing information management. The three levels of automation one may refer to are rigid automation, programmable automation, and flexible automation.

Rigid automation deals with a factory context oriented to the mass manufacture of products of the same type. The need to manufacture large numbers of parts with high productivity and quality standards demands the use of fixed operational sequences to be executed on the workpiece by special purpose machines.

Programmable automation deals with a factory context oriented to the manufacture of low-to-medium batches of products of different types. A programmable automated system permits changing easy the sequence of operations to be executed on the workpieces in order to vary the range of products. The machines employed are more versatile and are capable of manufacturing different objects belonging to the same group technology. The majority of the products available on the market today are manufactured by programmable automated systems.

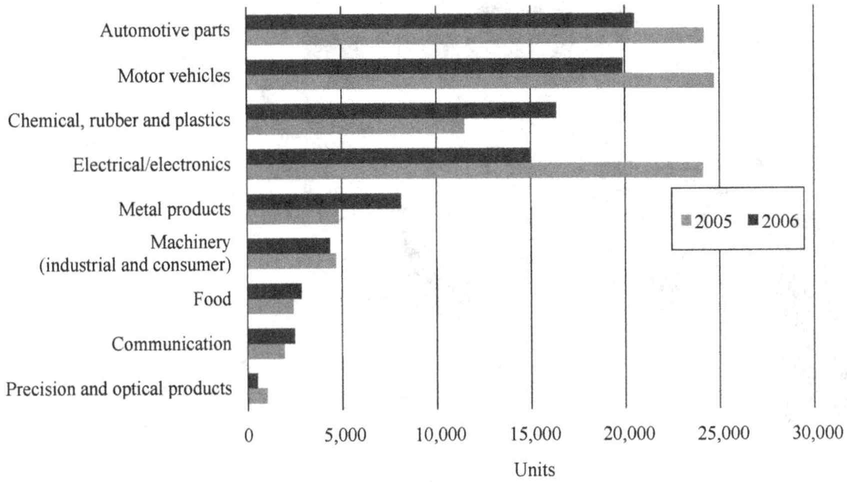

Fig. 1.21. Yearly supply of industrial robots by main industries

Flexible automation represents the evolution of programmable automation. Its goal is to allow manufacturing of variable batches of different products by minimizing the time lost for reprogramming the sequence of operations and the machines employed to pass from one batch to the next. The realization of a flexible manufacturing system (FMS) demands strong integration of computer technology with industrial technology.

The *industrial robot* is a machine with significant characteristics of versatility and flexibility. According to the widely accepted definition of the Robot Institute of America, *a robot is a reprogrammable multifunctional manipulator designed to move materials, parts, tools or specialized devices through variable programmed motions for the performance of a variety of tasks*. Such a definition, dating back to 1980, reflects the current status of robotics technology.

By virtue of its programmability, the industrial robot is a typical component of programmable automated systems. Nonetheless, robots can be entrusted with tasks in both rigid and flexible automated systems.

According to the above-mentioned IFR report, up to 2006 nearly one million industrial robots are in use worldwide, half of which are in Asia, one third in Europe, and 16% in North America. The four countries with the largest number of robots are Japan, Germany, United States and Italy. The figures for robot installations in the last 15 years are summarized in the graph in Fig. 1.20; by the end of 2007, an increase of 10% in sales with respect to the previous year is foreseen, with milder increase rates in the following years, reaching a worldwide figure of 1,200,000 units at work by the end of 2010.

In the same report it is shown how the average service life of an industrial robot is about 12 years, which may increase to 15 in a few years from now. An interesting statistic is robot density based on the total number of persons employed: this ranges from 349 robots in operation per 10,000 workers to

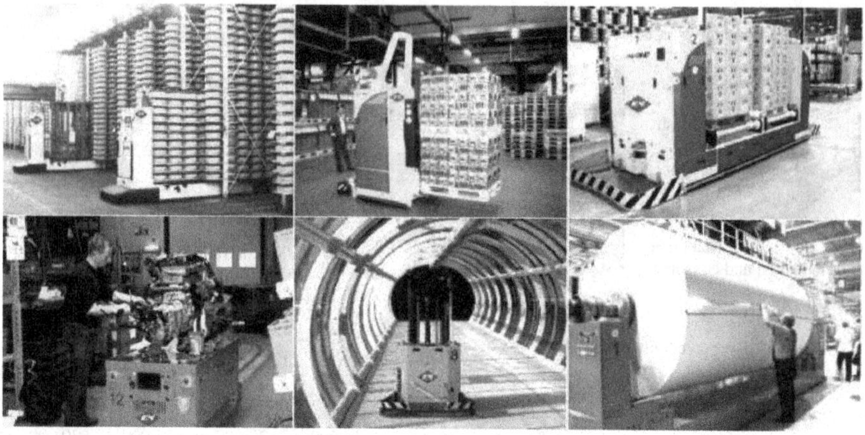

Fig. 1.22. Examples of AGVs for material handling (courtesy of E&K Automation GmbH)

187 in Korea, 186 in Germany, and 13 in Italy. The United States has just 99 robots per 10,000 workers. The average cost of a 6-axis industrial robot, including the control unit and development software, ranges from 20,000 to 60,000 euros, depending on the size and applications.

The automotive industry is still the predominant user of industrial robots. The graph in Fig. 1.21 referring to 2005 and 2006, however, reveals how both the chemical industry and the electrical/electronics industry are gaining in importance, and new industrial applications, such as metal products, constitute an area with a high potential investment.

Industrial robots present three fundamental capacities that make them useful for a manufacturing process: *material handling*, *manipulation*, and *measurement*.

In a manufacturing process, each object has to be transferred from one location in the factory to another in order to be stored, manufactured, assembled, and packed. During transfer, the physical characteristics of the object do not undergo any alteration. The robot's capability to pick up an object, move it in space on predefined paths and release it makes the robot itself an ideal candidate for material handling operations. Typical applications include:

- palletizing (placing objects on a pallet in an ordered way),
- warehouse loading and unloading,
- mill and machine tool tending,
- part sorting,
- packaging.

In these applications, besides robots, *Automated Guided Vehicles* (AGV) are utilized which ensure handling of parts and tools around the shop floor

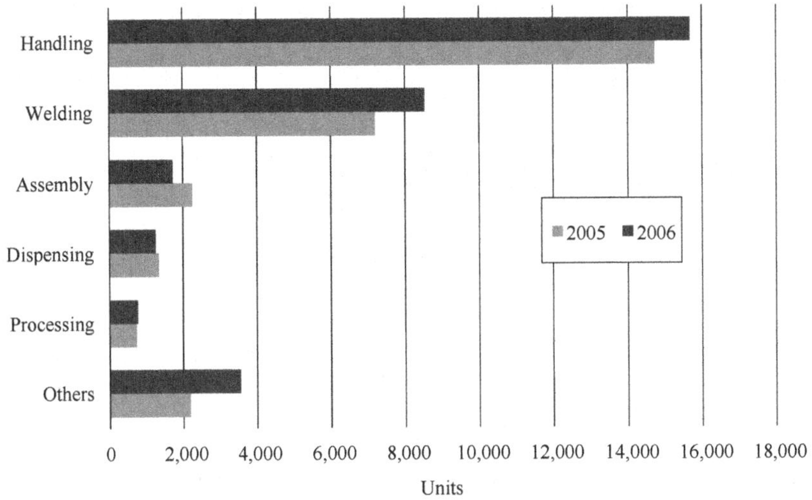

Fig. 1.23. Yearly supply of industrial robots in Europe for manufacturing operations

from one manufacturing cell to the next (Fig. 1.22). As compared to the traditional fixed guide paths for vehicles (inductive guide wire, magnetic tape, or optical visible line), modern AGVs utilize high-tech systems with onboard microprocessors and sensors (laser, odometry, GPS) which allow their localization within the plant layout, and manage their work flow and functions, allowing their complete integration in the FMS. The mobile robots employed in advanced applications can be considered as the natural evolution of the AGV systems, as far as enhanced autonomy is concerned.

Manufacturing consists of transforming objects from raw material into finished products; during this process, the part either changes its own physical characteristics as a result of machining, or loses its identity as a result of an assembly of more parts. The robot's capability to manipulate both objects and tools make it suitable to be employed in manufacturing. Typical applications include:

- arc and spot welding,
- painting and coating,
- gluing and sealing,
- laser and water jet cutting,
- milling and drilling,
- casting and die spraying,
- deburring and grinding,
- screwing, wiring and fastening,
- assembly of mechanical and electrical groups,
- assembly of electronic boards.

Fig. 1.24. The AdeptOne XL robot (courtesy of Adept Technology Inc)

Besides material handling and manipulation, in a manufacturing process it is necessary to perform measurements to test product quality. The robot's capability to explore 3D space together with the availability of measurements on the manipulator's status allow a robot to be used as a measuring device. Typical applications include:

- object inspection,
- contour finding,
- detection of manufacturing imperfections.

The graph in Fig. 1.23 reports the number of robots employed in Europe in 2005 and 2006 for various operations, which reveals how material handling requires twice as many robots employed for welding, whereas a limited number of robots is still employed for assembly.

In the following some industrial robots are illustrated in terms of their features and application fields.

The AdeptOne XL robot in Fig. 1.24 has a four-joint SCARA structure. Direct drive motors are employed. The maximum reach is 800 mm, with a repeatability of 0.025 mm horizontally and 0.038 mm vertically. Maximum speeds are 1200 mm/s for the prismatic joint, while they range from to 650 to 3300 deg/s for the three revolute joints. The maximum payload[3] is 12 kg. Typical industrial applications include small-parts material handling, assembly and packaging.

[3] Repeatability and payload are classical parameters found in industrial robot data sheets. The former gives a measure of the manipulator's ability to return to a previously reached position, while the latter indicates the average load to be carried at the robot's end-effector.

Fig. 1.25. The COMAU Smart NS robot (courtesy of COMAU SpA Robotica)

Fig. 1.26. The ABB IRB 4400 robot (courtesy of ABB Robotics)

The Comau SMART NS robot in Fig. 1.25 has a six-joint anthropomorphic structure with spherical wrist. In its four versions, the outreach ranges from 1650 and 1850 mm horizontally, with a repeatability of 0.05 mm. Maximum speeds range from 155 to 170 deg/s for the inner three joints, and from 350 to 550 deg/s for the outer three joints. The maximum payload is 16 kg. Both floor and ceiling mounting positions are allowed. Typical industrial applications include arc welding, light handling, assembly and technological processes.

The ABB IRB 4400 robot in Fig. 1.26 also has a six-joint anthropomorphic structure, but unlike the previous open-chain structure, it possesses a closed chain of parallelogram type between the shoulder and elbow joints. The outreach ranges from 1960 to 2550 mm for the various versions, with a

Fig. 1.27. The KUKA KR 60 Jet robot (courtesy of KUKA Roboter GmbH)

repeatability from 0.07 to 0.1 mm. The maximum speed at the end-effector is 2200 mm/s. The maximum payload is 60 kg. Floor or shelf-mounting is available. Typical industrial applications include material handling, machine tending, grinding, gluing, casting, die spraying and assembly.

The KUKA KR 60 Jet robot in Fig. 1.27 is composed of a five-axis structure, mounted on a sliding track with a gantry-type installation; the upright installation is also available. The linear unit has a stroke from a minimum of 400 mm to a maximum of 20 m (depending on customer's request), and a maximum speed of 3200 mm/s. On the other hand, the robot has a payload of 60 kg, an outreach of 820 mm and a repeatability of 0.15 mm. Maximum speeds are 120 deg/s and 166 deg/s for the first two joints, while they range from 260 to 322 deg/s for the outer three joints. Typical industrial applications include machine tending, arc welding, deburring, coating, sealing, plasma and waterjet cutting.

The ABB IRB340 FlexPicker robot in Fig. 1.28 adopts a parallel geometry with four axes; in view of its reduced weight and floor mounting, the robot can transport 150 objects a minute (cycle time of just 0.4 s), reaching record speeds of 10 m/s and accelerations of 100 m/s^2, for a payload of 1 kg, with a repeatability of 0.1 mm. In its 'clean' aluminum version, it is particularly suitable for packaging in the food and pharmaceutical industries.

The Fanuc M-16iB robot in Fig. 1.29 has a six-joint anthropomorphic structure with a spherical wrist. In its two versions, the outreach varies from 1667 to 1885 mm horizontally, with a repeatability of 0.1 mm. Maximum speeds range from 165 to 175 deg/s for the inner three joints, and from 340 to 520 deg/s for the outer three joints. Payload varies from 10 to 20 kg. The peculiarity of this robot consists of the integrated sensors in the control unit, including a servoing system based on 3D vision and a six-axis force sensor.

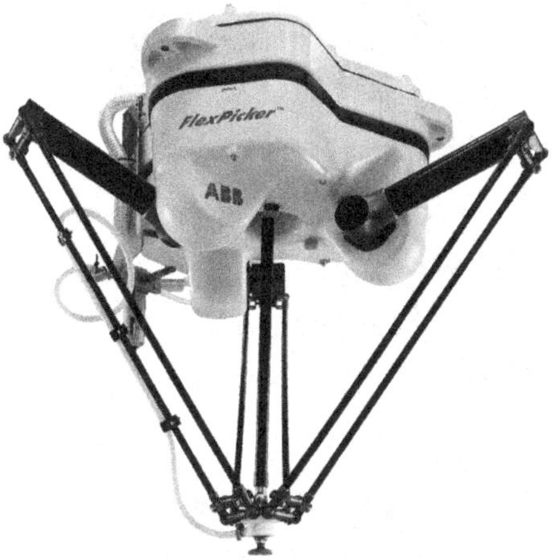

Fig. 1.28. The ABB IRB 340 FlexPicker robot (courtesy of ABB Robotics)

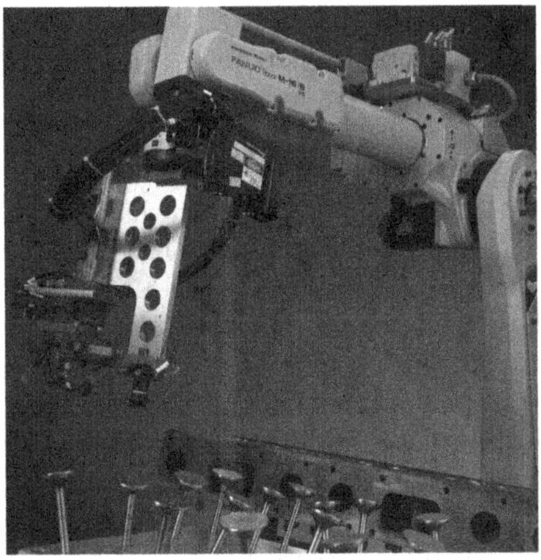

Fig. 1.29. The Fanuc M-16iB robot (courtesy of Fanuc Ltd)

The robot is utilized for handling arbitrarily located objects, deburring, sealing and waterjet cutting.

The Light Weight Robot (LWR) in Fig. 1.30 with a seven-axis structure was introduced in 2006 as the outcome of technology transfer from DLR (the German Aerospace Agency) to KUKA. In view of the adoption of lightweight materials, as well as the adoption of torque sensors at the joints, the robot can manipulate a payload of 7 to 14 kg, in the face of a weight of the structure of just 15 kg. The horizontal outreach is 868 mm, with joint speeds ranging from 110 to 210 deg/s. On the other hand, the presence of the seventh axis of motion confers kinematic redundancy to the robot, which can then be reconfigured into more dexterous postures for the execution of given tasks. Such

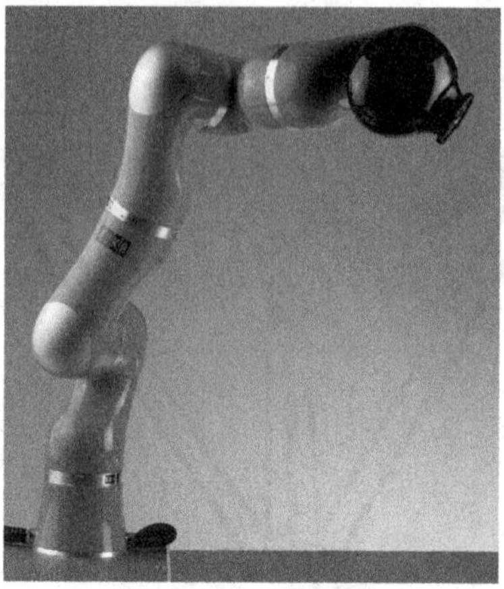

Fig. 1.30. The KUKA LWR robot (courtesy of KUKA Roboter GmbH)

a manipulator represents one of the most advanced industrial products and, in view of its lightweight feature, it offers interesting performance for interaction with the environment, ensuring an inherent safety in case of contact with human beings.

In most industrial applications requiring object manipulation, typical grippers are utilized as end-effectors. Nevertheless, whenever enhanced manipulability and dexterity is desired, multifingered robot hands are available.

The BarrettHand (Fig. 1.31), endowed with a fixed finger and two mobile fingers around the base of the palm, allows the manipulation of objects of different dimension, shape and orientation.

The SCHUNK Antropomorphic Hand (SAH) in Fig. 1.32 is the outcome of technology transfer from DLR and Harbin Institute of Technology (China) to SCHUNK. Characterized by three independent aligned fingers and an opposing finger which is analogous to the human thumb. The finger joints are endowed with magnetic angular sensors and torque sensors. This hand offers good dexterity and approaches the characteristics of the human hand.

LWR technology has been employed for the realization of the two arms of Justin, a humanoid manipulator made by DLR, composed of a three-joint torso with an anthropomorphic structure, two seven-axis arms and a sensorized head. The robot is illustrated in Fig. 1.33 in the execution of a bimanual manipulation task; the hands employed are previous versions of the SAH anthropomorphic hand.

The applications listed describe the current employment of robots as components of industrial automation systems. They all refer to strongly structured working environments and thus do not exhaust all the possible utilizations of robots for industrial applications. Whenever it is desired to tackle problems requiring the adaptation of the robot to a changeable working environment, the fall-out of advanced robotics products are of concern. In this regard, the

Fig. 1.31. The BarrettHand (courtesy of Barrett Technology Inc)

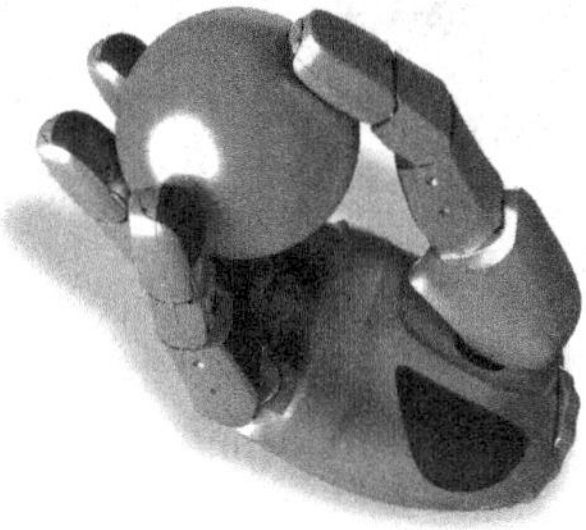

Fig. 1.32. The SCHUNK Anthropomorphic Hand (courtesy of SCHUNK Intec Ltd)

lightweight robot, the hands and the humanoid manipulator presented above are to be considered at the transition from traditional industrial robotics systems toward those innovative systems of advanced robotics.

1.4 Advanced Robotics

The expression *advanced robotics* usually refers to the science studying robots with marked characteristics of *autonomy*, operating in scarcely structured or *unstructured environments*, whose geometrical or physical characteristics would not be known a priori.

Nowadays, advanced robotics is still in its youth. It has indeed featured the realization of prototypes only, because the associated technology is not yet mature. There are many motivations which strongly encourage advances in knowledge within this field. They range from the need for automata whenever human operators are not available or are not safe (*field robots*), to the opportunity of developing products for potentially wide markets which are aimed at improving quality of life (*service robots*).

The graph in Fig. 1.34 reports the number of robots in stock for non-industrial applications at the end of 2006 and the forecast to 2010. Such applications are characterized by the complexity level, the uncertainty and variability of the environment with which the robot interacts, as shown in the following examples.

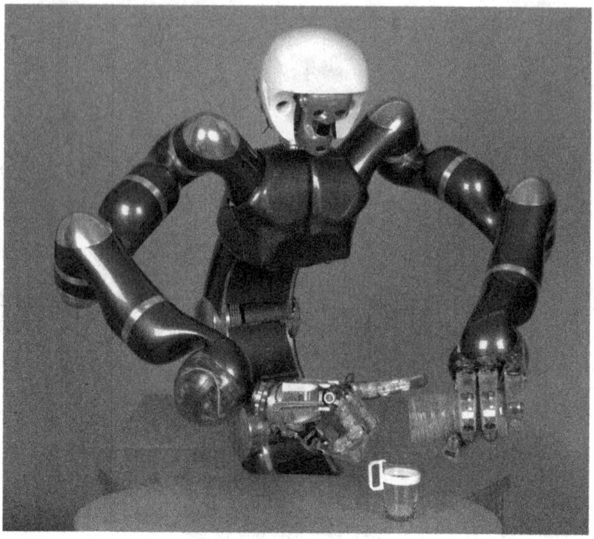

Fig. 1.33. The Justin humanoid robot manipulator (courtesy of DLR)

1.4.1 Field Robots

The context is that of deploying robots in areas where human beings could not survive or be exposed to unsustainable risks. Such robots should carry out exploration tasks and report useful data on the environment to a remote operator, using suitable onboard sensors. Typical scenarios are the exploration of a volcano, the intervention in areas contaminated by poisonous gas or radiation, or the exploration of the deep ocean or space. As is well known, NASA succeeded in delivering some mobile robots (rovers) to Mars (Fig. 1.35) which navigated on the Martian soil, across rocks, hills and crevasses. Such rovers were partially teleoperated from earth and have successfully explored the environment with sufficient autonomy. Some mini-robots were deployed on September 11, 2001 at Ground Zero after the collapse of the Twin Towers in New York, to penetrate the debris in the search for survivors.

A similar scenario is that of disasters caused by fires in tunnels or earthquakes; in such occurrences, there is a danger of further explosions, escape of harmful gases or collapse, and thus human rescue teams may cooperate with robot rescue teams. Also in the military field, unmanned autonomous aircrafts and missiles are utilized, as well as teleoperated robots with onboard cameras to explore buildings. The 'Grand Challenge' of October 2005 (Fig. 1.36) was financially supported by the US Department of Defense (DARPA) with the goal of developing autonomous vehicles to carry weapons and sensors, thus reducing soldier employment.

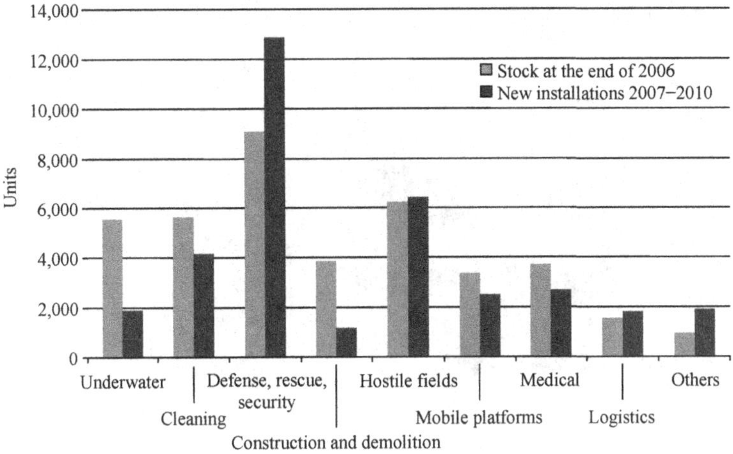

Fig. 1.34. Robots on stock for non-industrial applications

Fig. 1.35. The Sojourner rover was deployed by the Pathfinder lander and explored 250 m^2 of Martian soil in 1997 (courtesy of NASA)

1.4.2 Service Robots

Autonomous vehicles are also employed for civil applications, i.e., for mass transit systems (Fig. 1.37), thus contributing to the reduction of pollution levels. Such vehicles are part of the so-called Intelligent Transportation Systems (ITS) devoted to traffic management in urban areas. Another feasible application where the adoption of mobile robots offers potential advantages is museum guided tours (Fig. 1.38).

Many countries are investing in establishing the new market of service robots which will co-habitat with human beings in everyday life. According to the above-mentioned IFR report, up to 2005 1.9 million service robots for domestic applications (Fig. 1.39) and 1 million toy robots have been sold.

Technology is ready to transform into commercial products the prototypes of robotic aids to enhance elderly and impaired people's autonomy in everyday life; autonomous wheelchairs, mobility aid lifters, feeding aids and rehabilitation robots allowing tetraplegics to perform manual labor tasks are examples of such service devices. In perspective, other than an all-purpose robot waiter,

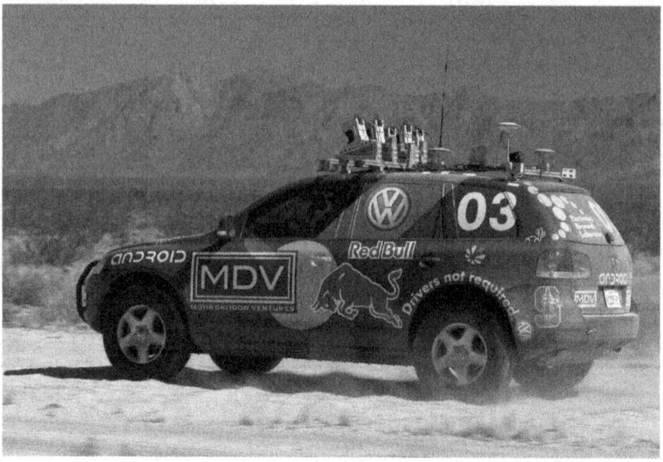

Fig. 1.36. The unmanned car Stanley autonomously completed a path of 132 miles in the record time of 6 h and 53 min (courtesy of DARPA)

Fig. 1.37. The Cycab is an electrically-driven vehicle for autonomous transportation in urban environments (courtesy of INRIA)

assistance, and healthcare systems integrating robotic and telematic modules will be developed for home service management (domotics).

Several robotic systems are employed for medical applications. Surgery assistance systems exploit a robot's high accuracy to position a tool, i.e., for hip prosthesis implant. Yet, in minimally-invasive surgery, i.e., cardiac surgery, the surgeon operates while seated comfortably at a console viewing a 3D image of the surgical field, and operating the surgical instruments remotely by means of a haptic interface (Fig. 1.40).

Further, in diagnostic and endoscopic surgery systems, small teleoperated robots travels through the cavities of human body, i.e., in the gastrointestinal system, bringing live images or intervening in situ for biopsy, dispensing drugs or removing neoplasms.

Fig. 1.38. Rhino, employing the synchro-drive mobile base B21 by Real World Interface, was one of the first robots for museum guided tours (courtesy of Deutsches Museum Bonn)

Fig. 1.39. The vacuum robot Roomba, employing a differential-drive kinematics, autonomously sweeps and cleans floors (courtesy of I-Robot Corp)

Finally, in motor rehabilitation systems, a hemiplegic patient wears an exoskeleton, which actively interacts, sustains and corrects the movements according to the physiotherapist's programmed plan.

Another wide market segment comes from entertainment, where robots are used as toy companions for children, and life companions for the elderly, such as humanoid robots (Fig. 1.41) and the pet robots (Fig. 1.42) being developed in Japan. It is reasonable to predict that service robots will be naturally integrated into our society. Tomorrow, robots will be as pervasive and personal as today's personal computers, or just as TV sets in the homes of 20 years ago. Robotics will then become ubiquitous, a challenge under discussion within the scientific community.

1.5 Robot Modelling, Planning and Control

In all robot applications, completion of a generic task requires the execution of a specific motion prescribed to the robot. The correct execution of such

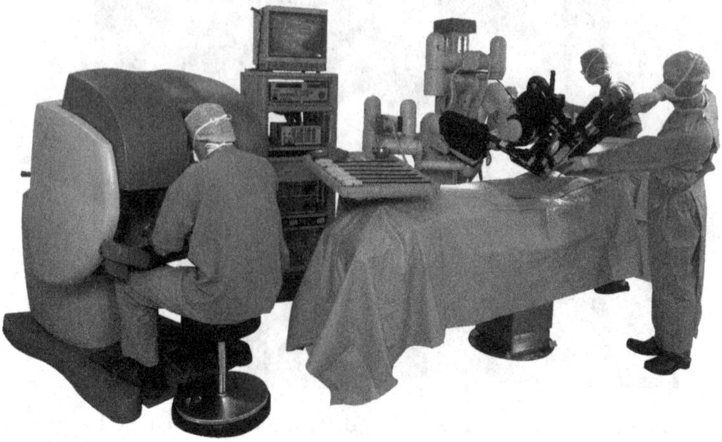

Fig. 1.40. The da Vinci robotic system for laparoscopic surgery (courtesy of Intuitive Surgical Inc)

motion is entrusted to the control system which should provide the robot's actuators with the commands consistent with the desired motion. Motion control demands an accurate analysis of the characteristics of the mechanical structure, actuators, and sensors. The goal of such analysis is the derivation of the mathematical models describing the input/output relationship characterizing the robot components. Modelling a robot manipulator is therefore a necessary premise to finding motion control strategies.

Significant topics in the study of modelling, planning and control of robots which constitute the subject of subsequent chapters are illustrated below.

1.5.1 Modelling

Kinematic analysis of the mechanical structure of a robot concerns the description of the motion with respect to a fixed reference Cartesian frame by ignoring the forces and moments that cause motion of the structure. It is meaningful to distinguish between kinematics and differential kinematics. With reference to a robot manipulator, *kinematics* describes the analytical relationship between the joint positions and the end-effector position and orientation. *Differential kinematics* describes the analytical relationship between the joint motion and the end-effector motion in terms of velocities, through the manipulator Jacobiann.

The formulation of the kinematics relationship allows the study of two key problems of robotics, namely, the direct kinematics problem and the inverse kinematics problem. The former concerns the determination of a systematic, general method to describe the end-effector motion as a function of the joint motion by means of linear algebra tools. The latter concerns the

Fig. 1.41. The Asimo humanoid robot, launched in 1996, has been endowed with even more natural locomotion and human-robot interaction skills (courtesy of Honda Motor Company Ltd)

Fig. 1.42. The AIBO dog had been the most widely diffused entertainment robot in the recent years (courtesy of Sony Corp)

inverse problem; its solution is of fundamental importance to transform the desired motion, naturally prescribed to the end-effector in the workspace, into the corresponding joint motion.

The availability of a manipulator's kinematic model is also useful to determine the relationship between the forces and torques applied to the joints and the forces and moments applied to the end-effector in *static* equilibrium configurations.

Chapter 2 is dedicated to the study of kinematics. Chapter 3 is dedicated to the study of differential kinematics and statics, whereas Appendix A provides a useful brush-up on *linear algebra*.

Kinematics of a manipulator represents the basis of a systematic, general derivation of its *dynamics*, i.e., the equations of motion of the manipulator as a function of the forces and moments acting on it. The availability of the dynamic model is very useful for mechanical design of the structure, choice of actuators, determination of control strategies, and computer simulation of manipulator motion. Chapter 7 is dedicated to the study of dynamics, whereas Appendix B recalls some fundamentals on *rigid body mechanics*.

Modelling of *mobile robots* requires a preliminary analysis of the kinematic constraints imposed by the presence of wheels. Depending on the mechanical structure, such constraints can be integrable or not; this has direct consequence on a robot's mobility. The *kinematic model* of a mobile robot is essentially the description of the admissible instantaneous motions in respect of the constraints. On the other hand, the *dynamic model* accounts for the reaction forces and describes the relationship between the above motions and the generalized forces acting on the robot. These models can be expressed in a canonical form which is convenient for design of planning and control techniques. Kinematic and dynamic analysis of mobile robots is developed in Chap. 11, while Appendix D contains some useful concepts of *differential geometry*.

1.5.2 Planning

With reference to the tasks assigned to a manipulator, the issue is whether to specify the motion at the joints or directly at the end-effector. In material handling tasks, it is sufficient to assign only the pick-up and release locations of an object (point-to-point motion), whereas, in machining tasks, the end-effector has to follow a desired trajectory (path motion). The goal of *trajectory planning* is to generate the timing laws for the relevant variables (joint or end-effector) starting from a concise description of the desired motion. Chapter 4 is dedicated to trajectory planning for robot manipulators.

The motion planning problem for a mobile robot concerns the generation of trajectories to take the vehicle from a given initial configuration to a desired final configuration. Such a problem is more complex than that of robot manipulators, since trajectories have to be generated in respect of the kinematic constraints imposed by the wheels. Some solution techniques are presented in Chap. 11, which exploit the specific differential structure of the mobile robots' kinematic models.

Whenever obstacles are present in a mobile robot's workspace, the planned motions must be safe, so as to avoid collisions. Such a problem, known as *motion planning*, can be formulated in an effective fashion for both robot manipulators and mobile robots utilizing the configuration space concept. The solution techniques are essentially of algorithmic nature and include exact, probabilistic and heuristic methods. Chapter 12 is dedicated to motion planning problem, while Appendix E provides some basic concepts on *graph search algorithms*.

1.5.3 Control

Realization of the motion specified by the control law requires the employment of *actuators* and *sensors*. The functional characteristics of the most commonly used actuators and sensors for robots are described in Chap. 5.

Chapter 6 is concerned with the hardware/software *architecture* of a robot's *control system* which is in charge of implementation of control laws as well as of interface with the operator.

The trajectories generated constitute the reference inputs to the *motion control* system of the mechanical structure. The problem of *robot manipulator* control is to find the time behaviour of the forces and torques to be delivered by the joint actuators so as to ensure the execution of the reference trajectories. This problem is quite complex, since a manipulator is an articulated system and, as such, the motion of one link influences the motion of the others. Manipulator equations of motion indeed reveal the presence of coupling dynamic effects among the joints, except in the case of a Cartesian structure with mutually orthogonal axes. The synthesis of the joint forces and torques cannot be made on the basis of the sole knowledge of the dynamic model, since this does not completely describe the real structure. Therefore, manipulator control is entrusted to the closure of feedback loops; by computing the deviation between the reference inputs and the data provided by the proprioceptive sensors, a feedback control system is capable of satisfying accuracy requirements on the execution of the prescribed trajectories.

Chapter 8 is dedicated to the presentation of motion control techniques, whereas Appendix C illustrates the basic principles of *feedback control*.

Control of a *mobile robot* substantially differs from the analogous problem for robot manipulators. This is due, in turn, to the availability of fewer control inputs than the robot has configuration variables. An important consequence is that the structure of a controller allowing a robot to follow a trajectory (tracking problem) is unavoidably different from that of a controller aimed at taking the robot to a given configuration (regulation problem). Further, since a mobile robot's proprioceptive sensors do not yield any data on the vehicle's configuration, it is necessary to develop localization methods for the robot in the environment. The control design problem for wheeled mobile robots is treated in Chap. 11.

If a manipulation task requires interaction between the robot and the environment, the control problem should account for the data provided by the exteroceptive sensors; the forces exchanged at the contact with the environment, and the objects' position as detected by suitable cameras. Chapter 9 is dedicated to *force control* techniques for robot manipulators, while Chap. 10 presents *visual control* techniques.

Bibliography

In the last 30 years, the robotics field has stimulated the interest of an increasing number of scholars. A truly respectable international research community has been established. Literature production has been conspicuous, both in terms of textbooks and scientific monographs and in terms of journals dedicated to robotics. Therefore, it seems appropriate to close this introduction by offering a selection of *bibliographical reference sources* to those readers who wish to make a thorough study of robotics.

Besides indicating those basic textbooks sharing an affinity of contents with this one, the following lists include specialized books on related subjects, collections of contributions on the state of the art of research, scientific journals, and series of international conferences.

Basic textbooks

- J. Angeles, *Fundamentals of Robotic Mechanical Systems: Theory, Methods, and Algorithms*, Springer-Verlag, New York, 1997.
- H. Asada, J.-J.E. Slotine, *Robot Analysis and Control*, Wiley, New York, 1986.
- G.A. Bekey, *Autonomous Robots*, MIT Press, Cambridge, MA, 2005.
- C. Canudas de Wit, B. Siciliano, G. Bastin, (Eds.), *Theory of Robot Control*, Springer-Verlag, London, 1996.
- J.J. Craig, *Introduction to Robotics: Mechanics and Control*, 3rd ed., Pearson Prentice Hall, Upper Saddle River, NJ, 2004.
- A.J. Critchlow, *Introduction to Robotics*, Macmillan, New York, 1985.
- J.F. Engelberger, *Robotics in Practice*, Amacom, New York, 1980.
- J.F. Engelberger, *Robotics in Service*, MIT Press, Cambridge, MA, 1989.
- K.S. Fu, R.C. Gonzalez, C.S.G. Lee, *Robotics: Control, Sensing, Vision, and Intelligence*, McGraw-Hill, New York, 1987.
- W. Khalil, E. Dombre, *Modeling, Identification and Control of Robots*, Hermes Penton Ltd, London, 2002.
- A.J. Koivo, *Fundamentals for Control of Robotic Manipulators*, Wiley, New York, 1989.
- Y. Koren, *Robotics for Engineers*, McGraw-Hill, New York, 1985.
- F.L. Lewis, C.T. Abdallah, D.M. Dawson, *Control of Robot Manipulators*, Macmillan, New York, 1993.
- P.J. McKerrow, *Introduction to Robotics*, Addison-Wesley, Sydney, Australia, 1991.
- R.M. Murray, Z. Li, S.S. Sastry, *A Mathematical Introduction to Robotic Manipulation*, CRC Press, Boca Raton, FL, 1994.
- S.B. Niku, *Introduction to Robotics: Analysis, Systems, Applications*, Prentice-Hall, Upper Saddle River, NJ, 2001.
- R.P. Paul, *Robot Manipulators: Mathematics, Programming, and Control* MIT Press, Cambridge, MA, 1981.
- R.J. Schilling, *Fundamentals of Robotics: Analysis and Control*, Prentice-Hall, Englewood Cliffs, NJ, 1990.
- L. Sciavicco, B. Siciliano, *Modelling and Control of Robot Manipulators*, 2nd ed., Springer, London, UK, 2000.
- W.E. Snyder, *Industrial Robots: Computer Interfacing and Control*, Prentice-Hall, Englewood Cliffs, NJ, 1985.

- M.W. Spong, S. Hutchinson, M. Vidyasagar, *Robot Modeling and Control*, Wiley, New York, 2006.
- M. Vukobratović, *Introduction to Robotics*, Springer-Verlag, Berlin, Germany, 1989.
- T. Yoshikawa, *Foundations of Robotics*, MIT Press, Boston, MA, 1990.

Specialized books

Topics of related interest to robot modelling, planning and control are:

- manipulator mechanical design,
- manipulation tools,
- manipulators with elastic members,
- parallel robots,
- locomotion apparatus,
- mobile robots,
- underwater and space robots,
- control architectures
- motion and force control,
- robot vision,
- multisensory data fusion,
- telerobotics,
- human-robot interaction.

The following books are dedicated to these topics:

- G. Antonelli, *Underwater Robots: Motion and Force Control of Vehicle-Manipulator Systems*, 2nd ed., Springer, Heidelberg, Germany, 2006.
- R.C. Arkin, *Behavior-Based Robotics*, MIT Press, Cambridge, MA, 1998.
- J. Baeten, J. De Schutter, *Integrated Visual Servoing and Force Control: The Task Frame Approach*, Springer, Heidelberg, Germany, 2003.
- M. Buehler, K. Iagnemma, S. Singh, (Eds.), *The 2005 DARPA Grand Challenge: The Great Robot Race*, Springer, Heidelberg, Germany, 2007.
- J.F. Canny, *The Complexity of Robot Motion Planning*, MIT Press, Cambridge, MA, 1988.
- H. Choset, K.M. Lynch, S. Hutchinson, G. Kantor, W. Burgard, L.E. Kavraki, S. Thrun, *Principles of Robot Motion: Theory, Algorithms, and Implementations*, MIT Press, Cambridge, MA, 2005.
- P.I. Corke, *Visual Control of Robots: High-Performance Visual Servoing*, Research Studies Press, Taunton, UK, 1996.
- M.R. Cutkosky, *Robotic Grasping and Fine Manipulation*, Kluwer, Boston, MA, 1985.
- H.F. Durrant-Whyte, *Integration, Coordination and Control of Multi-Sensor Robot Systems*, Kluwer, Boston, MA, 1988.
- A. Ellery, *An Introduction to Space Robotics*, Springer-Verlag, London, UK, 2000.

- A.R. Fraser, R.W. Daniel, *Perturbation Techniques for Flexible Manipulators*, Kluwer, Boston, MA, 1991.
- B.K. Ghosh, N. Xi, T.-J. Tarn, (Eds.), *Control in Robotics and Automation: Sensor-Based Integration*, Academic Press, San Diego, CA, 1999.
- K. Goldberg, (Ed.), *The Robot in the Garden: Telerobotics and Telepistemology in the Age of the Internet*, MIT Press, Cambridge, MA, 2000.
- S. Hirose, *Biologically Inspired Robots*, Oxford University Press, Oxford, UK, 1993.
- B.K.P. Horn, *Robot Vision*, McGraw-Hill, New York, 1986.
- K. Iagnemma, S. Dubowsky, *Mobile Robots in Rough Terrain Estimation: Motion Planning, and Control with Application to Planetary Rovers Series*, Springer, Heidelberg, Germany, 2004.
- R. Kelly, V. Santibañez, A. Loría, *Control of Robot Manipulators in Joint Space*, Springer-Verlag, London, UK, 2005.
- J.-C. Latombe, *Robot Motion Planning*, Kluwer, Boston, MA, 1991.
- M.T. Mason, *Mechanics of Robotic Manipulation*, MIT Press, Cambridge, MA, 2001.
- M.T. Mason, J.K. Salisbury, *Robot Hands and the Mechanics of Manipulation*, MIT Press, Cambridge, MA, 1985.
- J.-P. Merlet, *Parallel Robots*, 2nd ed., Springer, Dordrecht, The Netherlands, 2006.
- R.R. Murphy, *Introduction to AI Robotics*, MIT Press, Cambridge, MA, 2000.
- C. Natale, *Interaction Control of Robot Manipulators: Six-degrees-of-freedom Tasks*, Springer, Heidelberg, Germany, 2003.
- M. Raibert, *Legged Robots that Balance*, MIT Press, Cambridge, MA, 1985.
- E.I. Rivin, *Mechanical Design of Robots*, McGraw-Hill, New York, 1987.
- B. Siciliano, L. Villani, *Robot Force Control*, Kluwer, Boston, MA, 2000.
- R. Siegwart, *Introduction to Autonomous Mobile Robots*, MIT Press, Cambridge, MA, 2004.
- S. Thrun, W. Burgard, D. Fox, *Probabilistic Robotics*, MIT Press, Cambridge, MA, 2005.
- D.J. Todd, *Walking Machines, an Introduction to Legged Robots*, Chapman Hall, London, UK, 1985.
- L.-W. Tsai, *Robot Analysis: The Mechanics of Serial and Parallel Manipulators*, Wiley, New York, 1999.

Edited collections on the state of the art of research

- M. Brady, (Ed.), *Robotics Science*, MIT Press, Cambridge, MA, 1989.
- M. Brady, J.M. Hollerbach, T.L. Johnson, T. Lozano-Pérez, M.T. Mason, (Eds.), *Robot Motion: Planning and Control*, MIT Press, Cambridge, MA, 1982.
- R.C. Dorf, *International Encyclopedia of Robotics*, Wiley, New York, 1988.

- V.D. Hunt, *Industrial Robotics Handbook*, Industrial Press, New York, 1983.
- O. Khatib, J.J. Craig, T. Lozano-Pérez, (Eds.), *The Robotics Review 1*, MIT Press, Cambridge, MA, 1989.
- O. Khatib, J.J. Craig, T. Lozano-Pérez, (Eds.), *The Robotics Review 2*, MIT Press, Cambridge, MA., 1992.
- T.R. Kurfess, (Ed.), *Robotics and Automation Handbook*, CRC Press, Boca Raton, FL, 2005.
- B. Siciliano, O. Khatib, (Eds.), *Springer Handbook of Robotics*, Springer, Heidelberg, Germany, 2008.
- C.S.G. Lee, R.C. Gonzalez, K.S. Fu, (Eds.), *Tutorial on Robotics*, 2nd ed., IEEE Computer Society Press, Silver Spring, MD, 1986.
- M.W. Spong, F.L. Lewis, C.T. Abdallah, (Eds.), *Robot Control: Dynamics, Motion Planning, and Analysis*, IEEE Press, New York, 1993.

Scientific journals

- Advanced Robotics
- Autonomous Robots
- IEEE Robotics and Automation Magazine
- IEEE Transactions on Robotics
- International Journal of Robotics Research
- Journal of Field Robotics
- Journal of Intelligent and Robotic Systems
- Robotica
- Robotics and Autonomous Systems

Series of international scientific conferences

- IEEE International Conference on Robotics and Automation
- IEEE/RSJ International Conference on Intelligent Robots and Systems
- International Conference on Advanced Robotics
- International Symposium of Robotics Research
- International Symposium on Experimental Robotics
- Robotics: Science and Systems

The above journals and conferences represent the reference sources for the international scientific community. Many other robotics journals and conferences exist which are devoted to specific topics, such as kinematics, control, vision, algorithms, haptics, industrial applications, space and underwater exploration, humanoid robotics, and human-robot interaction. On the other hand, several journals and prestigious conferences in other fields, such as mechanics, control, sensors, and artificial intelligence, offer generous space to robotics topics.

2

Kinematics

A *manipulator* can be schematically represented from a mechanical viewpoint as a kinematic chain of rigid bodies (*links*) connected by means of revolute or prismatic *joints*. One end of the chain is constrained to a base, while an *end-effector* is mounted to the other end. The resulting motion of the structure is obtained by composition of the elementary motions of each link with respect to the previous one. Therefore, in order to manipulate an object in space, it is necessary to describe the end-effector position and orientation. This chapter is dedicated to the derivation of the *direct kinematics equation* through a systematic, general approach based on linear algebra. This allows the end-effector position and orientation (*pose*) to be expressed as a function of the joint variables of the mechanical structure with respect to a reference frame. Both open-chain and closed-chain kinematic structures are considered. With reference to a *minimal representation of orientation*, the concept of *operational space* is introduced and its relationship with the *joint space* is established. Furthermore, a *calibration* technique of the manipulator kinematic parameters is presented. The chapter ends with the derivation of solutions to the *inverse kinematics problem*, which consists of the determination of the joint variables corresponding to a given end-effector pose.

2.1 Pose of a Rigid Body

A *rigid body* is completely described in space by its *position* and *orientation* (in brief *pose*) with respect to a reference frame. As shown in Fig. 2.1, let O–xyz be the orthonormal reference frame and $\boldsymbol{x}$, $\boldsymbol{y}$, $\boldsymbol{z}$ be the unit vectors of the frame axes.

The position of a point O' on the rigid body with respect to the coordinate frame O–xyz is expressed by the relation

$$\boldsymbol{o}' = o'_x\boldsymbol{x} + o'_y\boldsymbol{y} + o'_z\boldsymbol{z},$$

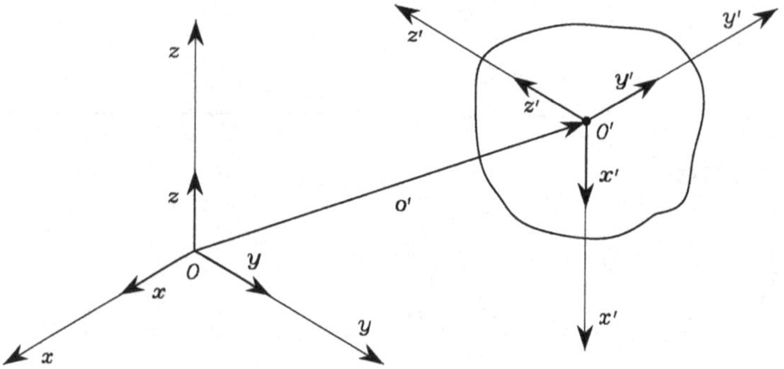

Fig. 2.1. Position and orientation of a rigid body

where o'_x, o'_y, o'_z denote the components of the vector $\boldsymbol{o}' \in \mathbb{R}^3$ along the frame axes; the position of O' can be compactly written as the (3×1) vector

$$\boldsymbol{o}' = \begin{bmatrix} o'_x \\ o'_y \\ o'_z \end{bmatrix}. \tag{2.1}$$

Vector $\boldsymbol{o}'$ is a bound vector since its line of application and point of application are both prescribed, in addition to its direction and norm.

In order to describe the rigid body orientation, it is convenient to consider an orthonormal frame attached to the body and express its unit vectors with respect to the reference frame. Let then O'–$x'y'z'$ be such a frame with origin in O' and $\boldsymbol{x}'$, $\boldsymbol{y}'$, $\boldsymbol{z}'$ be the unit vectors of the frame axes. These vectors are expressed with respect to the reference frame O–xyz by the equations:

$$\begin{aligned} \boldsymbol{x}' &= x'_x \boldsymbol{x} + x'_y \boldsymbol{y} + x'_z \boldsymbol{z} \\ \boldsymbol{y}' &= y'_x \boldsymbol{x} + y'_y \boldsymbol{y} + y'_z \boldsymbol{z} \\ \boldsymbol{z}' &= z'_x \boldsymbol{x} + z'_y \boldsymbol{y} + z'_z \boldsymbol{z}. \end{aligned} \tag{2.2}$$

The components of each unit vector are the direction cosines of the axes of frame O'–$x'y'z'$ with respect to the reference frame O–xyz.

2.2 Rotation Matrix

By adopting a compact notation, the three unit vectors in (2.2) describing the body orientation with respect to the reference frame can be combined in the (3×3) matrix

$$\boldsymbol{R} = \begin{bmatrix} \boldsymbol{x}' & \boldsymbol{y}' & \boldsymbol{z}' \end{bmatrix} = \begin{bmatrix} x'_x & y'_x & z'_x \\ x'_y & y'_y & z'_y \\ x'_z & y'_z & z'_z \end{bmatrix} = \begin{bmatrix} \boldsymbol{x}'^T\boldsymbol{x} & \boldsymbol{y}'^T\boldsymbol{x} & \boldsymbol{z}'^T\boldsymbol{x} \\ \boldsymbol{x}'^T\boldsymbol{y} & \boldsymbol{y}'^T\boldsymbol{y} & \boldsymbol{z}'^T\boldsymbol{y} \\ \boldsymbol{x}'^T\boldsymbol{z} & \boldsymbol{y}'^T\boldsymbol{z} & \boldsymbol{z}'^T\boldsymbol{z} \end{bmatrix}, \tag{2.3}$$

which is termed *rotation matrix*.

It is worth noting that the column vectors of matrix $\boldsymbol{R}$ are mutually orthogonal since they represent the unit vectors of an orthonormal frame, i.e.,

$$\boldsymbol{x}'^T\boldsymbol{y}' = 0 \qquad \boldsymbol{y}'^T\boldsymbol{z}' = 0 \qquad \boldsymbol{z}'^T\boldsymbol{x}' = 0.$$

Also, they have unit norm

$$\boldsymbol{x}'^T\boldsymbol{x}' = 1 \qquad \boldsymbol{y}'^T\boldsymbol{y}' = 1 \qquad \boldsymbol{z}'^T\boldsymbol{z}' = 1.$$

As a consequence, $\boldsymbol{R}$ is an *orthogonal* matrix meaning that

$$\boldsymbol{R}^T\boldsymbol{R} = \boldsymbol{I}_3 \tag{2.4}$$

where $\boldsymbol{I}_3$ denotes the (3×3) identity matrix.

If both sides of (2.4) are postmultiplied by the inverse matrix $\boldsymbol{R}^{-1}$, the useful result is obtained:

$$\boldsymbol{R}^T = \boldsymbol{R}^{-1}, \tag{2.5}$$

that is, the transpose of the rotation matrix is equal to its inverse. Further, observe that $\det(\boldsymbol{R}) = 1$ if the frame is right-handed, while $\det(\boldsymbol{R}) = -1$ if the frame is left-handed.

The above-defined rotation matrix belongs to the *special orthonormal group* $SO(m)$ of the real $(m \times m)$ matrices with othonormal columns and determinant equal to 1; in the case of spatial rotations it is $m = 3$, whereas in the case of planar rotations it is $m = 2$.

2.2.1 Elementary Rotations

Consider the frames that can be obtained via *elementary rotations* of the reference frame about one of the coordinate axes. These rotations are positive if they are made counter-clockwise about the relative axis.

Suppose that the reference frame O–xyz is rotated by an angle α about axis z (Fig. 2.2), and let O–$x'y'z'$ be the rotated frame. The unit vectors of the new frame can be described in terms of their components with respect to the reference frame. Consider the frames that can be obtained via *elementary rotations* of the reference frame about one of the coordinate axes. These rotations are positive if they are made counter-clockwise about the relative axis.

Suppose that the reference frame O–xyz is rotated by an angle α about axis z (Fig. 2.2), and let O–$x'y'z'$ be the rotated frame. The unit vectors of the new frame can be described in terms of their components with respect to the reference frame, i.e.,

$$\boldsymbol{x}' = \begin{bmatrix} \cos\alpha \\ \sin\alpha \\ 0 \end{bmatrix} \qquad \boldsymbol{y}' = \begin{bmatrix} -\sin\alpha \\ \cos\alpha \\ 0 \end{bmatrix} \qquad \boldsymbol{z}' = \begin{bmatrix} 0 \\ 0 \\ 1 \end{bmatrix}.$$

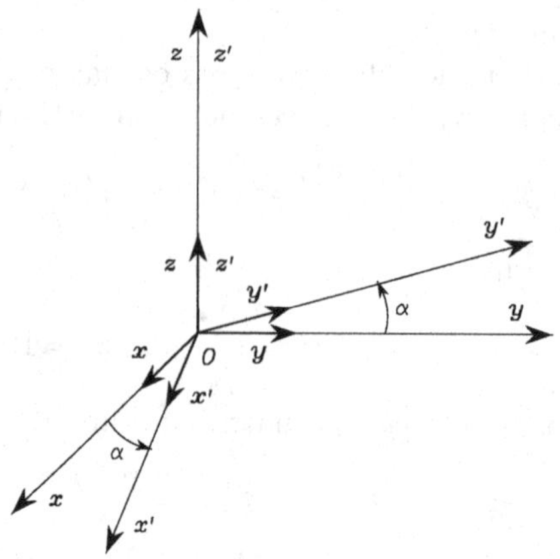

Fig. 2.2. Rotation of frame O–xyz by an angle α about axis z

Hence, the rotation matrix of frame O–$x'y'z'$ with respect to frame O–xyz is

$$\boldsymbol{R}_z(\alpha) = \begin{bmatrix} \cos\alpha & -\sin\alpha & 0 \\ \sin\alpha & \cos\alpha & 0 \\ 0 & 0 & 1 \end{bmatrix}. \tag{2.6}$$

In a similar manner, it can be shown that the rotations by an angle β about axis y and by an angle γ about axis x are respectively given by

$$\boldsymbol{R}_y(\beta) = \begin{bmatrix} \cos\beta & 0 & \sin\beta \\ 0 & 1 & 0 \\ -\sin\beta & 0 & \cos\beta \end{bmatrix} \tag{2.7}$$

$$\boldsymbol{R}_x(\gamma) = \begin{bmatrix} 1 & 0 & 0 \\ 0 & \cos\gamma & -\sin\gamma \\ 0 & \sin\gamma & \cos\gamma \end{bmatrix}. \tag{2.8}$$

These matrices will be useful to describe rotations about an arbitrary axis in space.

It is easy to verify that for the elementary rotation matrices in (2.6)–(2.8) the following property holds:

$$\boldsymbol{R}_k(-\vartheta) = \boldsymbol{R}_k^T(\vartheta) \qquad k = x, y, z. \tag{2.9}$$

In view of (2.6)–(2.8), the rotation matrix can be attributed a geometrical meaning; namely, the matrix $\boldsymbol{R}$ describes the rotation about an axis in space needed to align the axes of the reference frame with the corresponding axes of the body frame.

2.2.2 Representation of a Vector

In order to understand a further geometrical meaning of a rotation matrix, consider the case when the origin of the body frame coincides with the origin

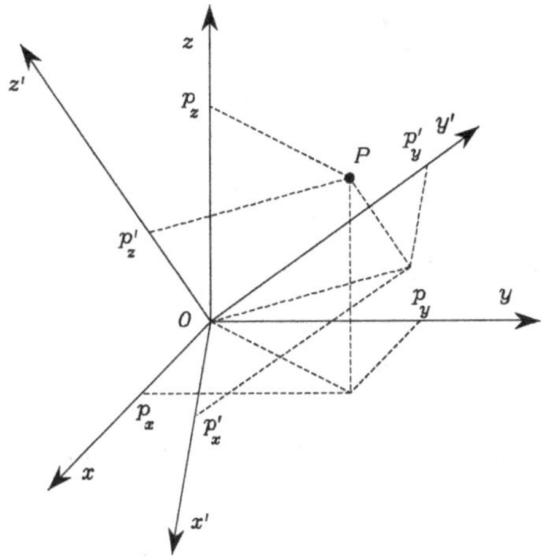

Fig. 2.3. Representation of a point P in two different coordinate frames

of the reference frame (Fig. 2.3); it follows that $\boldsymbol{o}' = \boldsymbol{0}$, where $\boldsymbol{0}$ denotes the (3×1) null vector. A point P in space can be represented either as

$$\boldsymbol{p} = \begin{bmatrix} p_x \\ p_y \\ p_z \end{bmatrix}$$

with respect to frame O–xyz, or as

$$\boldsymbol{p}' = \begin{bmatrix} p'_x \\ p'_y \\ p'_z \end{bmatrix}$$

with respect to frame O–$x'y'z'$.

Since $\boldsymbol{p}$ and $\boldsymbol{p}'$ are representations of the same point P, it is

$$\boldsymbol{p} = p'_x\boldsymbol{x}' + p'_y\boldsymbol{y}' + p'_z\boldsymbol{z}' = \begin{bmatrix} \boldsymbol{x}' & \boldsymbol{y}' & \boldsymbol{z}' \end{bmatrix} \boldsymbol{p}'$$

and, accounting for (2.3), it is

$$\boldsymbol{p} = \boldsymbol{R}\boldsymbol{p}'. \tag{2.10}$$

The rotation matrix $\boldsymbol{R}$ represents the *transformation matrix* of the vector coordinates in frame O–$x'y'z'$ into the coordinates of the same vector in frame O–xyz. In view of the orthogonality property (2.4), the inverse transformation is simply given by

$$\boldsymbol{p}' = \boldsymbol{R}^T\boldsymbol{p}. \tag{2.11}$$

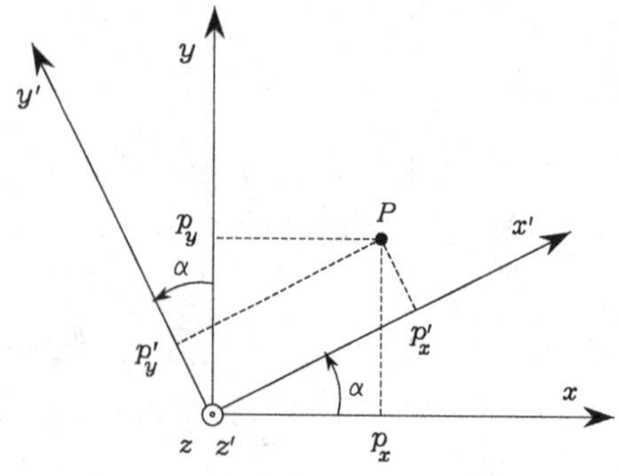

Fig. 2.4. Representation of a point P in rotated frames

Example 2.1

Consider two frames with common origin mutually rotated by an angle α about the axis z. Let $\boldsymbol{p}$ and $\boldsymbol{p}'$ be the vectors of the coordinates of a point P, expressed in the frames O–xyz and O–$x'y'z'$, respectively (Fig. 2.4). On the basis of simple geometry, the relationship between the coordinates of P in the two frames is

$$\begin{aligned}p_x &= p'_x \cos\alpha - p'_y \sin\alpha\\ p_y &= p'_x \sin\alpha + p'_y \cos\alpha\\ p_z &= p'_z.\end{aligned}$$

Therefore, the matrix (2.6) represents not only the orientation of a frame with respect to another frame, but it also describes the transformation of a vector from a frame to another frame with the same origin.

2.2.3 Rotation of a Vector

A rotation matrix can be also interpreted as the matrix operator allowing rotation of a vector by a given angle about an arbitrary axis in space. In fact, let $\boldsymbol{p}'$ be a vector in the reference frame O–xyz; in view of orthogonality of the matrix $\boldsymbol{R}$, the product $\boldsymbol{R}\boldsymbol{p}'$ yields a vector $\boldsymbol{p}$ with the same norm as that of $\boldsymbol{p}'$ but rotated with respect to $\boldsymbol{p}'$ according to the matrix $\boldsymbol{R}$. The norm equality can be proved by observing that $\boldsymbol{p}^T\boldsymbol{p} = \boldsymbol{p}'^T\boldsymbol{R}^T\boldsymbol{R}\boldsymbol{p}'$ and applying (2.4). This interpretation of the rotation matrix will be revisited later.

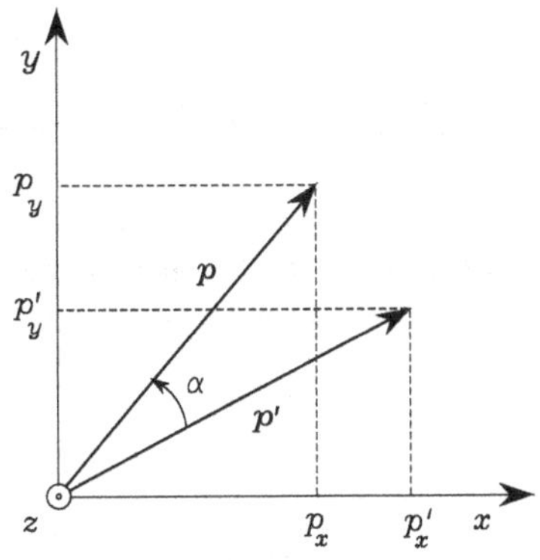

Fig. 2.5. Rotation of a vector

Example 2.2

Consider the vector $\boldsymbol{p}$ which is obtained by rotating a vector $\boldsymbol{p}'$ in the plane xy by an angle α about axis z of the reference frame (Fig. 2.5). Let (p'_x, p'_y, p'_z) be the coordinates of the vector $\boldsymbol{p}'$. The vector $\boldsymbol{p}$ has components

$$\begin{aligned} p_x &= p'_x \cos\alpha - p'_y \sin\alpha \\ p_y &= p'_x \sin\alpha + p'_y \cos\alpha \\ p_z &= p'_z. \end{aligned}$$

It is easy to recognize that $\boldsymbol{p}$ can be expressed as

$$\boldsymbol{p} = \boldsymbol{R}_z(\alpha)\boldsymbol{p}',$$

where $\boldsymbol{R}_z(\alpha)$ is the same rotation matrix as in (2.6).

In sum, a rotation matrix attains three *equivalent geometrical meanings*:

- It describes the mutual orientation between two coordinate frames; its column vectors are the direction cosines of the axes of the rotated frame with respect to the original frame.
- It represents the coordinate transformation between the coordinates of a point expressed in two different frames (with common origin).
- It is the operator that allows the rotation of a vector in the same coordinate frame.

2.3 Composition of Rotation Matrices

In order to derive composition rules of rotation matrices, it is useful to consider the expression of a vector in two different reference frames. Let then $O\text{–}x_0y_0z_0$,

O–$x_1y_1z_1$, O–$x_2y_2z_2$ be three frames with common origin O. The vector $\boldsymbol{p}$ describing the position of a generic point in space can be expressed in each of the above frames; let $\boldsymbol{p}^0$, $\boldsymbol{p}^1$, $\boldsymbol{p}^2$ denote the expressions of $\boldsymbol{p}$ in the three frames.[1]

At first, consider the relationship between the expression $\boldsymbol{p}^2$ of the vector $\boldsymbol{p}$ in Frame 2 and the expression $\boldsymbol{p}^1$ of the same vector in Frame 1. If $\boldsymbol{R}_i^j$ denotes the rotation matrix of Frame i with respect to Frame j, it is

$$\boldsymbol{p}^1 = \boldsymbol{R}_2^1 \boldsymbol{p}^2. \tag{2.12}$$

Similarly, it turns out that

$$\boldsymbol{p}^0 = \boldsymbol{R}_1^0 \boldsymbol{p}^1 \tag{2.13}$$

$$\boldsymbol{p}^0 = \boldsymbol{R}_2^0 \boldsymbol{p}^2. \tag{2.14}$$

On the other hand, substituting (2.12) in (2.13) and using (2.14) gives

$$\boldsymbol{R}_2^0 = \boldsymbol{R}_1^0 \boldsymbol{R}_2^1. \tag{2.15}$$

The relationship in (2.15) can be interpreted as the composition of successive rotations. Consider a frame initially aligned with the frame O–$x_0y_0z_0$. The rotation expressed by matrix $\boldsymbol{R}_2^0$ can be regarded as obtained in two steps:

- First rotate the given frame according to $\boldsymbol{R}_1^0$, so as to align it with frame O–$x_1y_1z_1$.
- Then rotate the frame, now aligned with frame O–$x_1y_1z_1$, according to $\boldsymbol{R}_2^1$, so as to align it with frame O–$x_2y_2z_2$.

Notice that the overall rotation can be expressed as a sequence of partial rotations; each rotation is defined with respect to the preceding one. The frame with respect to which the rotation occurs is termed *current frame*. Composition of successive rotations is then obtained by postmultiplication of the rotation matrices following the given order of rotations, as in (2.15). With the adopted notation, in view of (2.5), it is

$$\boldsymbol{R}_i^j = (\boldsymbol{R}_j^i)^{-1} = (\boldsymbol{R}_j^i)^T. \tag{2.16}$$

Successive rotations can be also specified by constantly referring them to the initial frame; in this case, the rotations are made with respect to a *fixed frame*. Let $\boldsymbol{R}_1^0$ be the rotation matrix of frame O–$x_1y_1z_1$ with respect to the fixed frame O–$x_0y_0z_0$. Let then $\bar{\boldsymbol{R}}_2^0$ denote the matrix characterizing frame O–$x_2y_2z_2$ with respect to Frame 0, which is obtained as a rotation of Frame 1 according to the matrix $\bar{\boldsymbol{R}}_2^1$. Since (2.15) gives a composition rule of successive rotations about the axes of the current frame, the overall rotation can be regarded as obtained in the following steps:

[1] Hereafter, the superscript of a vector or a matrix denotes the frame in which its components are expressed.

- First realign Frame 1 with Frame 0 by means of rotation $\boldsymbol{R}_0^1$.
- Then make the rotation expressed by $\bar{\boldsymbol{R}}_2^1$ with respect to the current frame.
- Finally compensate for the rotation made for the realignment by means of the inverse rotation $\boldsymbol{R}_1^0$.

Since the above rotations are described with respect to the current frame, the application of the composition rule (2.15) yields

$$\bar{\boldsymbol{R}}_2^0 = \boldsymbol{R}_1^0 \boldsymbol{R}_0^1 \bar{\boldsymbol{R}}_2^1 \boldsymbol{R}_1^0.$$

In view of (2.16), it is

$$\bar{\boldsymbol{R}}_2^0 = \bar{\boldsymbol{R}}_2^1 \boldsymbol{R}_1^0 \qquad (2.17)$$

where the resulting $\bar{\boldsymbol{R}}_2^0$ is different from the matrix $\boldsymbol{R}_2^0$ in (2.15). Hence, it can be stated that composition of successive rotations with respect to a fixed frame is obtained by premultiplication of the single rotation matrices in the order of the given sequence of rotations.

By recalling the meaning of a rotation matrix in terms of the orientation of a current frame with respect to a fixed frame, it can be recognized that its columns are the direction cosines of the axes of the current frame with respect to the fixed frame, while its rows (columns of its transpose and inverse) are the direction cosines of the axes of the fixed frame with respect to the current frame.

An important issue of composition of rotations is that the matrix product is not commutative. In view of this, it can be concluded that two rotations in general do not commute and its composition depends on the order of the single rotations.

Example 2.3

Consider an object and a frame attached to it. Figure 2.6 shows the effects of two successive rotations of the object with respect to the current frame by changing the order of rotations. It is evident that the final object orientation is different in the two cases. Also in the case of rotations made with respect to the current frame, the final orientations differ (Fig. 2.7). It is interesting to note that the effects of the sequence of rotations with respect to the fixed frame are interchanged with the effects of the sequence of rotations with respect to the current frame. This can be explained by observing that the order of rotations in the fixed frame commutes with respect to the order of rotations in the current frame.

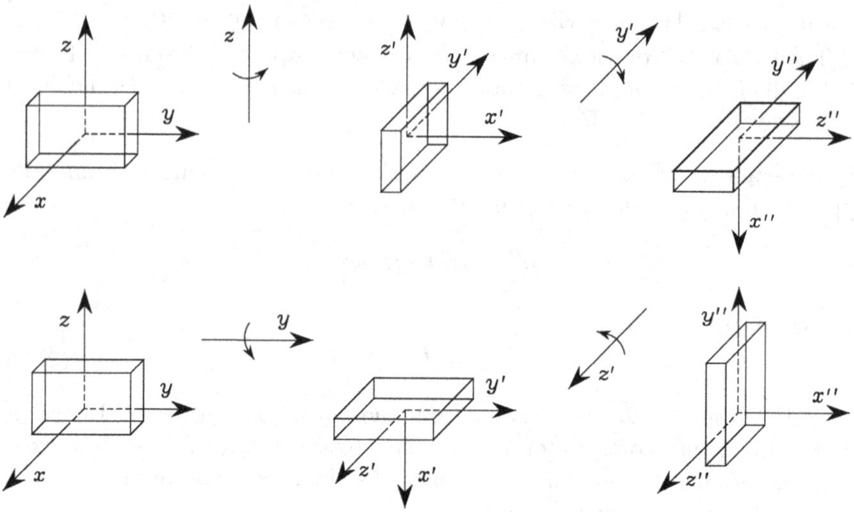

Fig. 2.6. Successive rotations of an object about axes of current frame

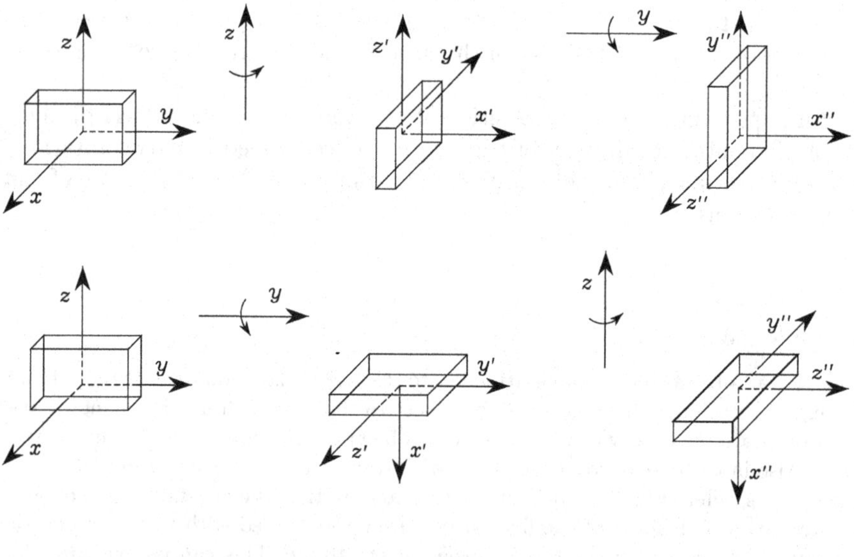

Fig. 2.7. Successive rotations of an object about axes of fixed frame

2.4 Euler Angles

Rotation matrices give a redundant description of frame orientation; in fact, they are characterized by nine elements which are not independent but related by six constraints due to the orthogonality conditions given in (2.4). This implies that *three parameters* are sufficient to describe orientation of a rigid body

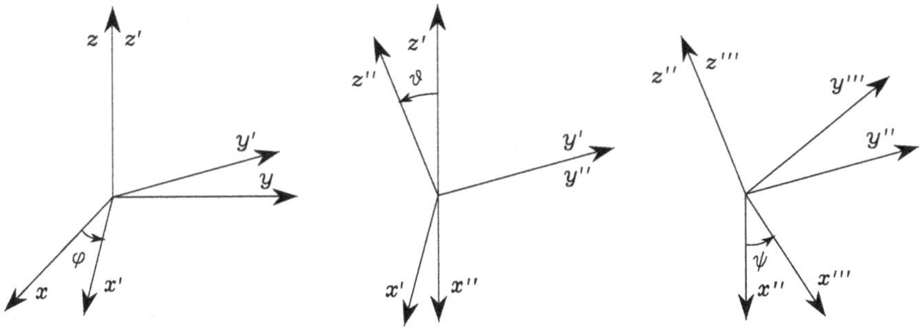

Fig. 2.8. Representation of Euler angles ZYZ

in space. A representation of orientation in terms of three independent parameters constitutes a *minimal representation*. In fact, a minimal representation of the special orthonormal group $SO(m)$ requires $m(m-1)/2$ parameters; thus, three parameters are needed to parameterize $SO(3)$, whereas only one parameter is needed for a planar rotation $SO(2)$.

A minimal representation of orientation can be obtained by using a set of three angles $\boldsymbol{\phi} = [\varphi \quad \vartheta \quad \psi]^T$. Consider the rotation matrix expressing the elementary rotation about one of the coordinate axes as a function of a single angle. Then, a generic rotation matrix can be obtained by composing a suitable sequence of three elementary rotations while guaranteeing that two successive rotations are not made about parallel axes. This implies that 12 distinct sets of angles are allowed out of all 27 possible combinations; each set represents a triplet of *Euler angles*. In the following, two sets of Euler angles are analyzed; namely, the ZYZ angles and the ZYX (or Roll–Pitch–Yaw) angles.

2.4.1 ZYZ Angles

The rotation described by *ZYZ angles* is obtained as composition of the following elementary rotations (Fig. 2.8):

- Rotate the reference frame by the angle φ about axis z; this rotation is described by the matrix $\boldsymbol{R}_z(\varphi)$ which is formally defined in (2.6).
- Rotate the current frame by the angle ϑ about axis y'; this rotation is described by the matrix $\boldsymbol{R}_{y'}(\vartheta)$ which is formally defined in (2.7).
- Rotate the current frame by the angle ψ about axis z''; this rotation is described by the matrix $\boldsymbol{R}_{z''}(\psi)$ which is again formally defined in (2.6).

The resulting frame orientation is obtained by composition of rotations with respect to *current frames*, and then it can be computed via postmultiplication of the matrices of elementary rotation, i.e.,[2]

$$\boldsymbol{R}(\boldsymbol{\phi}) = \boldsymbol{R}_z(\varphi)\boldsymbol{R}_{y'}(\vartheta)\boldsymbol{R}_{z''}(\psi) \tag{2.18}$$
$$= \begin{bmatrix} c_\varphi c_\vartheta c_\psi - s_\varphi s_\psi & -c_\varphi c_\vartheta s_\psi - s_\varphi c_\psi & c_\varphi s_\vartheta \\ s_\varphi c_\vartheta c_\psi + c_\varphi s_\psi & -s_\varphi c_\vartheta s_\psi + c_\varphi c_\psi & s_\varphi s_\vartheta \\ -s_\vartheta c_\psi & s_\vartheta s_\psi & c_\vartheta \end{bmatrix}.$$

It is useful to solve the *inverse problem*, that is to determine the set of Euler angles corresponding to a given rotation matrix

$$\boldsymbol{R} = \begin{bmatrix} r_{11} & r_{12} & r_{13} \\ r_{21} & r_{22} & r_{23} \\ r_{31} & r_{32} & r_{33} \end{bmatrix}.$$

Compare this expression with that of $\boldsymbol{R}(\boldsymbol{\phi})$ in (2.18). By considering the elements $[1,3]$ and $[2,3]$, under the assumption that $r_{13} \neq 0$ and $r_{23} \neq 0$, it follows that

$$\varphi = \text{Atan2}(r_{23}, r_{13})$$

where $\text{Atan2}(y,x)$ is the arctangent function of two arguments[3]. Then, squaring and summing the elements $[1,3]$ and $[2,3]$ and using the element $[3,3]$ yields

$$\vartheta = \text{Atan2}\left(\sqrt{r_{13}^2 + r_{23}^2}, r_{33}\right).$$

The choice of the positive sign for the term $\sqrt{r_{13}^2 + r_{23}^2}$ limits the range of feasible values of ϑ to $(0,\pi)$. On this assumption, considering the elements $[3,1]$ and $[3,2]$ gives

$$\psi = \text{Atan2}(r_{32}, -r_{31}).$$

In sum, the requested solution is

$$\begin{aligned} \varphi &= \text{Atan2}(r_{23}, r_{13}) \\ \vartheta &= \text{Atan2}\left(\sqrt{r_{13}^2 + r_{23}^2}, r_{33}\right) \\ \psi &= \text{Atan2}(r_{32}, -r_{31}). \end{aligned} \tag{2.19}$$

It is possible to derive another solution which produces the same effects as solution (2.19). Choosing ϑ in the range $(-\pi, 0)$ leads to

$$\varphi = \text{Atan2}(-r_{23}, -r_{13})$$

---

[2] The notations c_ϕ and s_ϕ are the abbreviations for $\cos\phi$ and $\sin\phi$, respectively; short-hand notations of this kind will be adopted often throughout the text.

[3] The function $\text{Atan2}(y,x)$ computes the arctangent of the ratio y/x but utilizes the sign of each argument to determine which quadrant the resulting angle belongs to; this allows the correct determination of an angle in a range of 2π.

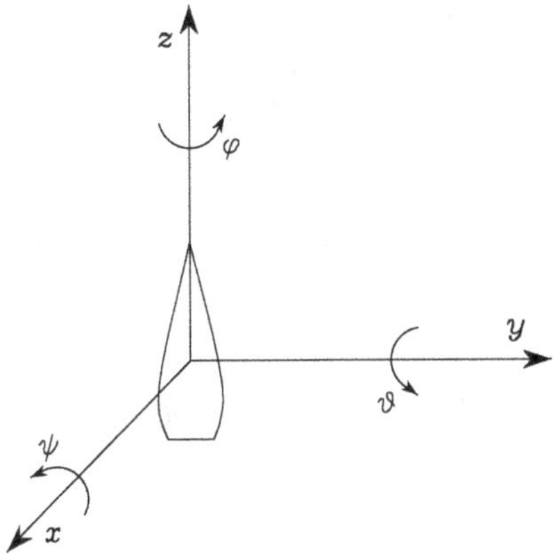

Fig. 2.9. Representation of Roll–Pitch–Yaw angles

$$\vartheta = \text{Atan2}\Big(-\sqrt{r_{13}^2 + r_{23}^2}, r_{33}\Big) \tag{2.20}$$
$$\psi = \text{Atan2}(-r_{32}, r_{31}).$$

Solutions (2.19), (2.20) degenerate when $s_\vartheta = 0$; in this case, it is possible to determine only the sum or difference of φ and ψ. In fact, if $\vartheta = 0, \pi$, the successive rotations of φ and ψ are made about axes of current frames which are parallel, thus giving equivalent contributions to the rotation; see Problem 2.2.[4]

2.4.2 RPY Angles

Another set of Euler angles originates from a representation of orientation in the (aero)nautical field. These are the ZYX angles, also called *Roll–Pitch–Yaw angles*, to denote the typical changes of attitude of an (air)craft. In this case, the angles $\boldsymbol{\phi} = [\varphi \quad \vartheta \quad \psi]^T$ represent rotations defined with respect to a fixed frame attached to the centre of mass of the craft (Fig. 2.9).

The rotation resulting from Roll–Pitch–Yaw angles can be obtained as follows:

- Rotate the reference frame by the angle ψ about axis x (yaw); this rotation is described by the matrix $\boldsymbol{R}_x(\psi)$ which is formally defined in (2.8).
- Rotate the reference frame by the angle ϑ about axis y (pitch); this rotation is described by the matrix $\boldsymbol{R}_y(\vartheta)$ which is formally defined in (2.7).
- Rotate the reference frame by the angle φ about axis z (roll); this rotation is described by the matrix $\boldsymbol{R}_z(\varphi)$ which is formally defined in (2.6).

[4] In the following chapter, it will be seen that these configurations characterize the so-called representation *singularities* of the Euler angles.

The resulting frame orientation is obtained by composition of rotations with respect to the *fixed frame*, and then it can be computed via premultiplication of the matrices of elementary rotation, i.e.,[5]

$$\begin{aligned}\boldsymbol{R}(\boldsymbol{\phi}) &= \boldsymbol{R}_z(\varphi)\boldsymbol{R}_y(\vartheta)\boldsymbol{R}_x(\psi) \\ &= \begin{bmatrix} c_\varphi c_\vartheta & c_\varphi s_\vartheta s_\psi - s_\varphi c_\psi & c_\varphi s_\vartheta c_\psi + s_\varphi s_\psi \\ s_\varphi c_\vartheta & s_\varphi s_\vartheta s_\psi + c_\varphi c_\psi & s_\varphi s_\vartheta c_\psi - c_\varphi s_\psi \\ -s_\vartheta & c_\vartheta s_\psi & c_\vartheta c_\psi \end{bmatrix}.\end{aligned} \tag{2.21}$$

As for the Euler angles ZYZ, the *inverse solution* to a given rotation matrix

$$\boldsymbol{R} = \begin{bmatrix} r_{11} & r_{12} & r_{13} \\ r_{21} & r_{22} & r_{23} \\ r_{31} & r_{32} & r_{33} \end{bmatrix},$$

can be obtained by comparing it with the expression of $\boldsymbol{R}(\boldsymbol{\phi})$ in (2.21). The solution for ϑ in the range $(-\pi/2, \pi/2)$ is

$$\begin{aligned}\varphi &= \text{Atan2}(r_{21}, r_{11}) \\ \vartheta &= \text{Atan2}\left(-r_{31}, \sqrt{r_{32}^2 + r_{33}^2}\right) \\ \psi &= \text{Atan2}(r_{32}, r_{33}).\end{aligned} \tag{2.22}$$

The other equivalent solution for ϑ in the range $(\pi/2, 3\pi/2)$ is

$$\begin{aligned}\varphi &= \text{Atan2}(-r_{21}, -r_{11}) \\ \vartheta &= \text{Atan2}\left(-r_{31}, -\sqrt{r_{32}^2 + r_{33}^2}\right) \\ \psi &= \text{Atan2}(-r_{32}, -r_{33}).\end{aligned} \tag{2.23}$$

Solutions (2.22), (2.23) degenerate when $c_\vartheta = 0$; in this case, it is possible to determine only the sum or difference of φ and ψ.

2.5 Angle and Axis

A nonminimal representation of orientation can be obtained by resorting to *four parameters* expressing a rotation of a given angle about an axis in space. This can be advantageous in the problem of trajectory planning for a manipulator's end-effector orientation.

Let $\boldsymbol{r} = [\, r_x \quad r_y \quad r_z \,]^T$ be the unit vector of a rotation axis with respect to the reference frame O–xyz. In order to derive the rotation matrix $\boldsymbol{R}(\vartheta, \boldsymbol{r})$ expressing the rotation of an *angle* ϑ about *axis* $\boldsymbol{r}$, it is convenient to compose

[5] The ordered sequence of rotations XYZ about axes of the fixed frame is equivalent to the sequence ZYX about axes of the current frame.

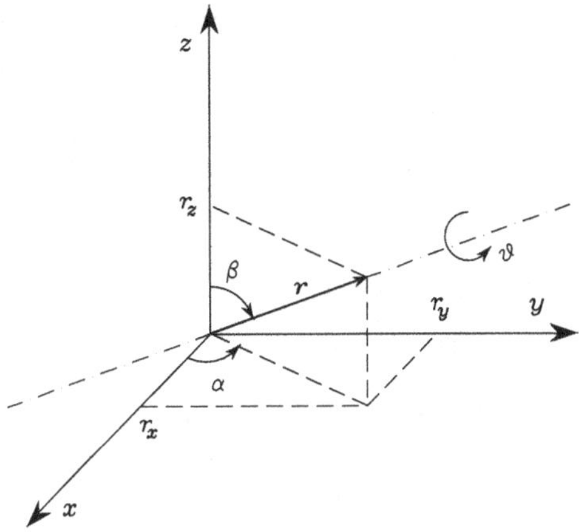

Fig. 2.10. Rotation of an angle about an axis

elementary rotations about the coordinate axes of the reference frame. The angle is taken to be positive if the rotation is made counter-clockwise about axis $\boldsymbol{r}$.

As shown in Fig. 2.10, a possible solution is to rotate first $\boldsymbol{r}$ by the angles necessary to align it with axis z, then to rotate by ϑ about z and finally to rotate by the angles necessary to align the unit vector with the initial direction. In detail, the sequence of rotations, to be made always with respect to axes of fixed frame, is the following:

- Align $\boldsymbol{r}$ with z, which is obtained as the sequence of a rotation by $-\alpha$ about z and a rotation by $-\beta$ about y.
- Rotate by ϑ about z.
- Realign with the initial direction of $\boldsymbol{r}$, which is obtained as the sequence of a rotation by β about y and a rotation by α about z.

In sum, the resulting rotation matrix is

$$\boldsymbol{R}(\vartheta, \boldsymbol{r}) = \boldsymbol{R}_z(\alpha)\boldsymbol{R}_y(\beta)\boldsymbol{R}_z(\vartheta)\boldsymbol{R}_y(-\beta)\boldsymbol{R}_z(-\alpha). \tag{2.24}$$

From the components of the unit vector $\boldsymbol{r}$ it is possible to extract the transcendental functions needed to compute the rotation matrix in (2.24), so as to eliminate the dependence from α and β; in fact, it is

$$\sin\alpha = \frac{r_y}{\sqrt{r_x^2 + r_y^2}} \qquad \cos\alpha = \frac{r_x}{\sqrt{r_x^2 + r_y^2}}$$

$$\sin\beta = \sqrt{r_x^2 + r_y^2} \qquad \cos\beta = r_z.$$

Then, it can be found that the rotation matrix corresponding to a given angle and axis is — see Problem 2.4 —

$$\boldsymbol{R}(\vartheta,\boldsymbol{r})=\begin{bmatrix} r_x^2(1-c_\vartheta)+c_\vartheta & r_xr_y(1-c_\vartheta)-r_zs_\vartheta & r_xr_z(1-c_\vartheta)+r_ys_\vartheta \\ r_xr_y(1-c_\vartheta)+r_zs_\vartheta & r_y^2(1-c_\vartheta)+c_\vartheta & r_yr_z(1-c_\vartheta)-r_xs_\vartheta \\ r_xr_z(1-c_\vartheta)-r_ys_\vartheta & r_yr_z(1-c_\vartheta)+r_xs_\vartheta & r_z^2(1-c_\vartheta)+c_\vartheta \end{bmatrix}. \tag{2.25}$$

For this matrix, the following property holds:

$$\boldsymbol{R}(-\vartheta,-\boldsymbol{r})=\boldsymbol{R}(\vartheta,\boldsymbol{r}), \tag{2.26}$$

i.e., a rotation by $-\vartheta$ about $-\boldsymbol{r}$ cannot be distinguished from a rotation by ϑ about $\boldsymbol{r}$; hence, such representation is not unique.

If it is desired to solve the *inverse problem* to compute the axis and angle corresponding to a given rotation matrix

$$\boldsymbol{R}=\begin{bmatrix} r_{11} & r_{12} & r_{13} \\ r_{21} & r_{22} & r_{23} \\ r_{31} & r_{32} & r_{33} \end{bmatrix},$$

the following result is useful:

$$\vartheta=\cos^{-1}\left(\frac{r_{11}+r_{22}+r_{33}-1}{2}\right) \tag{2.27}$$

$$\boldsymbol{r}=\frac{1}{2\sin\vartheta}\begin{bmatrix} r_{32}-r_{23} \\ r_{13}-r_{31} \\ r_{21}-r_{12} \end{bmatrix}, \tag{2.28}$$

for $\sin\vartheta \neq 0$. Notice that the expressions (2.27), (2.28) describe the rotation in terms of four parameters; namely, the angle and the three components of the axis unit vector. However, it can be observed that the three components of $\boldsymbol{r}$ are not independent but are constrained by the condition

$$r_x^2+r_y^2+r_z^2=1. \tag{2.29}$$

If $\sin\vartheta = 0$, the expressions (2.27), (2.28) become meaningless. To solve the inverse problem, it is necessary to directly refer to the particular expressions attained by the rotation matrix $\boldsymbol{R}$ and find the solving formulae in the two cases $\vartheta = 0$ and $\vartheta = \pi$. Notice that, when $\vartheta = 0$ (null rotation), the unit vector $\boldsymbol{r}$ is arbitrary (singularity). See also Problem 2.5.

2.6 Unit Quaternion

The drawbacks of the angle/axis representation can be overcome by a different four-parameter representation; namely, the unit *quaternion*, viz. Euler parameters, defined as $\mathcal{Q}=\{\eta,\boldsymbol{\epsilon}\}$ where:

$$\eta=\cos\frac{\vartheta}{2} \tag{2.30}$$

$$\epsilon = \sin\frac{\vartheta}{2}\boldsymbol{r}; \tag{2.31}$$

η is called the scalar part of the quaternion while $\epsilon = [\,\epsilon_x \quad \epsilon_y \quad \epsilon_z\,]^T$ is called the vector part of the quaternion. They are constrained by the condition

$$\eta^2 + \epsilon_x^2 + \epsilon_y^2 + \epsilon_z^2 = 1, \tag{2.32}$$

hence, the name *unit* quaternion. It is worth remarking that, unlike the angle/axis representation, a rotation by $-\vartheta$ about $-\boldsymbol{r}$ gives the same quaternion as that associated with a rotation by ϑ about $\boldsymbol{r}$; this solves the above nonuniqueness problem. In view of (2.25), (2.30), (2.31), (2.32), the rotation matrix corresponding to a given quaternion takes on the form — see Problem 2.6 —

$$\boldsymbol{R}(\eta,\epsilon) = \begin{bmatrix} 2(\eta^2+\epsilon_x^2)-1 & 2(\epsilon_x\epsilon_y-\eta\epsilon_z) & 2(\epsilon_x\epsilon_z+\eta\epsilon_y) \\ 2(\epsilon_x\epsilon_y+\eta\epsilon_z) & 2(\eta^2+\epsilon_y^2)-1 & 2(\epsilon_y\epsilon_z-\eta\epsilon_x) \\ 2(\epsilon_x\epsilon_z-\eta\epsilon_y) & 2(\epsilon_y\epsilon_z+\eta\epsilon_x) & 2(\eta^2+\epsilon_z^2)-1 \end{bmatrix}. \tag{2.33}$$

If it is desired to solve the *inverse problem* to compute the quaternion corresponding to a given rotation matrix

$$\boldsymbol{R} = \begin{bmatrix} r_{11} & r_{12} & r_{13} \\ r_{21} & r_{22} & r_{23} \\ r_{31} & r_{32} & r_{33} \end{bmatrix},$$

the following result is useful:

$$\eta = \frac{1}{2}\sqrt{r_{11}+r_{22}+r_{33}+1} \tag{2.34}$$

$$\epsilon = \frac{1}{2}\begin{bmatrix} \operatorname{sgn}(r_{32}-r_{23})\sqrt{r_{11}-r_{22}-r_{33}+1} \\ \operatorname{sgn}(r_{13}-r_{31})\sqrt{r_{22}-r_{33}-r_{11}+1} \\ \operatorname{sgn}(r_{21}-r_{12})\sqrt{r_{33}-r_{11}-r_{22}+1} \end{bmatrix}, \tag{2.35}$$

where conventionally $\operatorname{sgn}(x) = 1$ for $x \geq 0$ and $\operatorname{sgn}(x) = -1$ for $x < 0$. Notice that in (2.34) it has been implicitly assumed $\eta \geq 0$; this corresponds to an angle $\vartheta \in [-\pi, \pi]$, and thus any rotation can be described. Also, compared to the inverse solution in (2.27), (2.28) for the angle and axis representation, no singularity occurs for (2.34), (2.35). See also Problem 2.8.

The quaternion extracted from $\boldsymbol{R}^{-1} = \boldsymbol{R}^T$ is denoted as $\mathcal{Q}^{-1}$, and can be computed as

$$\mathcal{Q}^{-1} = \{\eta, -\epsilon\}. \tag{2.36}$$

Let $\mathcal{Q}_1 = \{\eta_1, \epsilon_1\}$ and $\mathcal{Q}_2 = \{\eta_2, \epsilon_2\}$ denote the quaternions corresponding to the rotation matrices $\boldsymbol{R}_1$ and $\boldsymbol{R}_2$, respectively. The quaternion corresponding to the product $\boldsymbol{R}_1\boldsymbol{R}_2$ is given by

$$\mathcal{Q}_1 * \mathcal{Q}_2 = \{\eta_1\eta_2 - \epsilon_1^T\epsilon_2, \eta_1\epsilon_2 + \eta_2\epsilon_1 + \epsilon_1 \times \epsilon_2\} \tag{2.37}$$

where the quaternion product operator "$*$" has been formally introduced. It is easy to see that if $\mathcal{Q}_2 = \mathcal{Q}_1^{-1}$ then the quaternion $\{1, \boldsymbol{0}\}$ is obtained from (2.37) which is the identity element for the product. See also Problem 2.9.

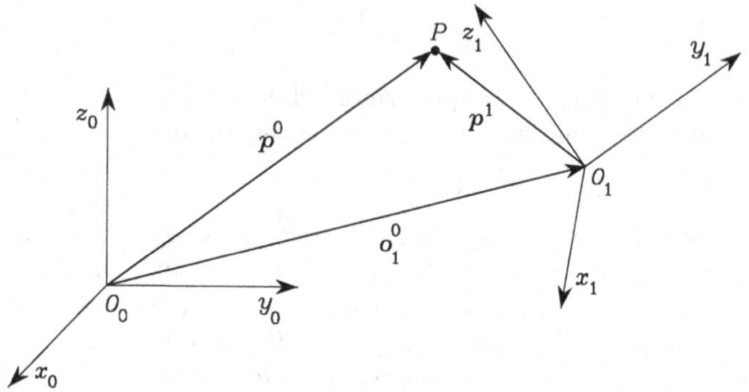

Fig. 2.11. Representation of a point P in different coordinate frames

2.7 Homogeneous Transformations

As illustrated at the beginning of the chapter, the position of a rigid body in space is expressed in terms of the position of a suitable point on the body with respect to a reference frame (translation), while its orientation is expressed in terms of the components of the unit vectors of a frame attached to the body — with origin in the above point — with respect to the same reference frame (rotation).

As shown in Fig. 2.11, consider an arbitrary point P in space. Let $\boldsymbol{p}^0$ be the vector of coordinates of P with respect to the reference frame O_0–$x_0y_0z_0$. Consider then another frame in space O_1–$x_1y_1z_1$. Let $\boldsymbol{o}_1^0$ be the vector describing the origin of Frame 1 with respect to Frame 0, and $\boldsymbol{R}_1^0$ be the rotation matrix of Frame 1 with respect to Frame 0. Let also $\boldsymbol{p}^1$ be the vector of coordinates of P with respect to Frame 1. On the basis of simple geometry, the position of point P with respect to the reference frame can be expressed as

$$\boldsymbol{p}^0 = \boldsymbol{o}_1^0 + \boldsymbol{R}_1^0\boldsymbol{p}^1. \tag{2.38}$$

Hence, (2.38) represents the *coordinate transformation* (*translation + rotation*) of a bound vector between two frames.

The inverse transformation can be obtained by premultiplying both sides of (2.38) by $\boldsymbol{R}_1^{0T}$; in view of(2.4), it follows that

$$\boldsymbol{p}^1 = -\boldsymbol{R}_1^{0T}\boldsymbol{o}_1^0 + \boldsymbol{R}_1^{0T}\boldsymbol{p}^0 \tag{2.39}$$

which, via (2.16), can be written as

$$\boldsymbol{p}^1 = -\boldsymbol{R}_0^1\boldsymbol{o}_1^0 + \boldsymbol{R}_0^1\boldsymbol{p}^0. \tag{2.40}$$

In order to achieve a compact representation of the relationship between the coordinates of the same point in two different frames, the *homogeneous representation* of a generic vector $\boldsymbol{p}$ can be introduced as the vector $\widetilde{\boldsymbol{p}}$ formed by adding a fourth unit component, i.e.,

$$\widetilde{\boldsymbol{p}} = \begin{bmatrix} \boldsymbol{p} \\ 1 \end{bmatrix}. \tag{2.41}$$

By adopting this representation for the vectors $\boldsymbol{p}^0$ and $\boldsymbol{p}^1$ in (2.38), the coordinate transformation can be written in terms of the (4×4) matrix

$$\boldsymbol{A}_1^0 = \begin{bmatrix} \boldsymbol{R}_1^0 & \boldsymbol{o}_1^0 \\ \boldsymbol{0}^T & 1 \end{bmatrix} \tag{2.42}$$

which, according to (2.41), is termed *homogeneous transformation matrix*. Since $\boldsymbol{o}_1^0 \in \mathbb{R}^3$ e $\boldsymbol{R}_1^0 \in SO(3)$, this matrix belongs to the *special Euclidean group* $SE(3) = \mathbb{R}^3 \times SO(3)$.

As can be easily seen from (2.42), the transformation of a vector from Frame 1 to Frame 0 is expressed by a single matrix containing the rotation matrix of Frame 1 with respect to Frame 0 and the translation vector from the origin of Frame 0 to the origin of Frame 1.[6] Therefore, the coordinate transformation (2.38) can be compactly rewritten as

$$\widetilde{\boldsymbol{p}}^0 = \boldsymbol{A}_1^0 \widetilde{\boldsymbol{p}}^1. \tag{2.43}$$

The coordinate transformation between Frame 0 and Frame 1 is described by the homogeneous transformation matrix $\boldsymbol{A}_0^1$ which satisfies the equation

$$\widetilde{\boldsymbol{p}}^1 = \boldsymbol{A}_0^1 \widetilde{\boldsymbol{p}}^0 = \left(\boldsymbol{A}_1^0\right)^{-1} \widetilde{\boldsymbol{p}}^0. \tag{2.44}$$

This matrix is expressed in a block-partitioned form as

$$\boldsymbol{A}_0^1 = \begin{bmatrix} \boldsymbol{R}_1^{0T} & -\boldsymbol{R}_1^{0T} \boldsymbol{o}_1^0 \\ \boldsymbol{0}^T & 1 \end{bmatrix} = \begin{bmatrix} \boldsymbol{R}_0^1 & -\boldsymbol{R}_0^1 \boldsymbol{o}_1^0 \\ \boldsymbol{0}^T & 1 \end{bmatrix}, \tag{2.45}$$

which gives the homogeneous representation form of the result already established by (2.39), (2.40) — see Problem 2.10.

Notice that for the homogeneous transformation matrix the orthogonality property does not hold; hence, in general,

$$\boldsymbol{A}^{-1} \neq \boldsymbol{A}^T. \tag{2.46}$$

In sum, a homogeneous transformation matrix expresses the coordinate transformation between two frames in a compact form. If the frames have the

[6] It can be shown that in (2.42) non-null values of the first three elements of the fourth row of $\boldsymbol{A}$ produce a perspective effect, while values other than unity for the fourth element give a scaling effect.

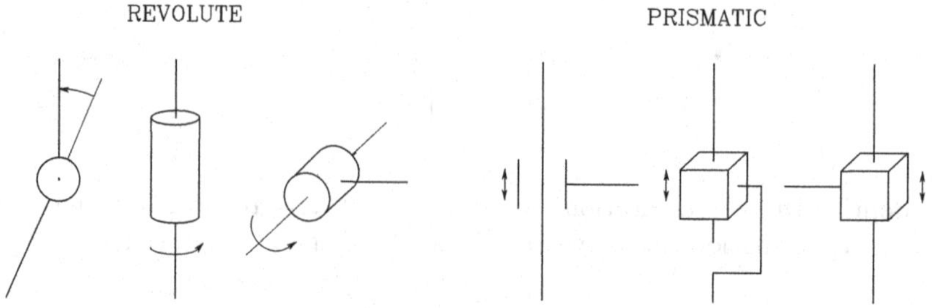

Fig. 2.12. Conventional representations of joints

same origin, it reduces to the rotation matrix previously defined. Instead, if the frames have distinct origins, it allows the notation with superscripts and subscripts to be kept which directly characterize the current frame and the fixed frame.

Analogously to what presented for the rotation matrices, it is easy to verify that a sequence of coordinate transformations can be composed by the product

$$\widetilde{\boldsymbol{p}}^0 = \boldsymbol{A}_1^0 \boldsymbol{A}_2^1 \dots \boldsymbol{A}_n^{n-1} \widetilde{\boldsymbol{p}}^n \tag{2.47}$$

where $\boldsymbol{A}_i^{i-1}$ denotes the homogeneous transformation relating the description of a point in Frame i to the description of the same point in Frame $i-1$.

2.8 Direct Kinematics

A manipulator consists of a series of rigid bodies (*links*) connected by means of kinematic pairs or *joints*. Joints can be essentially of two types: *revolute* and *prismatic*; conventional representations of the two types of joints are sketched in Fig. 2.12. The whole structure forms a *kinematic chain*. One end of the chain is constrained to a base. An *end-effector* (gripper, tool) is connected to the other end allowing manipulation of objects in space.

From a topological viewpoint, the kinematic chain is termed *open* when there is only one sequence of links connecting the two ends of the chain. Alternatively, a manipulator contains a *closed* kinematic chain when a sequence of links forms a loop.

The mechanical structure of a manipulator is characterized by a number of degrees of freedom (DOFs) which uniquely determine its *posture*.[7] Each DOF is typically associated with a joint articulation and constitutes a *joint variable*. The aim of *direct kinematics* is to compute the pose of the end-effector as a function of the joint variables.

[7] The term *posture* of a kinematic chain denotes the pose of all the rigid bodies composing the chain. Whenever the kinematic chain reduces to a single rigid body, then the posture coincides with the pose of the body.

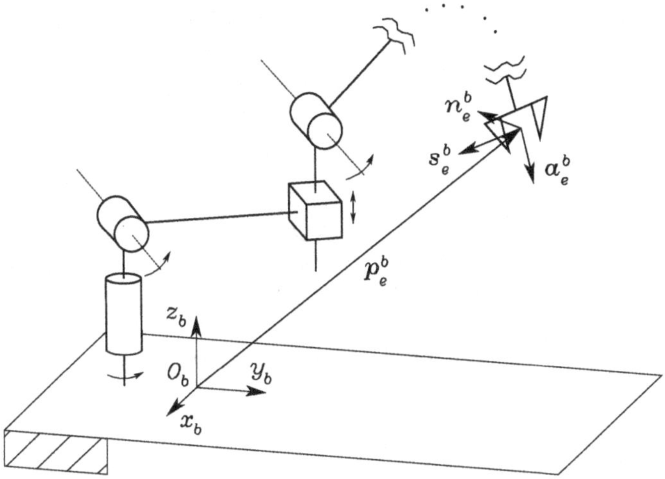

Fig. 2.13. Description of the position and orientation of the end-effector frame

It was previously illustrated that the pose of a body with respect to a reference frame is described by the position vector of the origin and the unit vectors of a frame attached to the body. Hence, with respect to a reference frame O_b–$x_b y_b z_b$, the direct kinematics function is expressed by the homogeneous transformation matrix

$$\boldsymbol{T}_e^b(\boldsymbol{q}) = \begin{bmatrix} \boldsymbol{n}_e^b(\boldsymbol{q}) & \boldsymbol{s}_e^b(\boldsymbol{q}) & \boldsymbol{a}_e^b(\boldsymbol{q}) & \boldsymbol{p}_e^b(\boldsymbol{q}) \\ 0 & 0 & 0 & 1 \end{bmatrix}, \tag{2.48}$$

where $\boldsymbol{q}$ is the $(n \times 1)$ vector of joint variables, $\boldsymbol{n}_e$, $\boldsymbol{s}_e$, $\boldsymbol{a}_e$ are the unit vectors of a frame attached to the end-effector, and $\boldsymbol{p}_e$ is the position vector of the origin of such a frame with respect to the origin of the base frame O_b–$x_b y_b z_b$ (Fig. 2.13). Note that $\boldsymbol{n}_e$, $\boldsymbol{s}_e$, $\boldsymbol{a}_e$ and $\boldsymbol{p}_e$ are a function of $\boldsymbol{q}$.

The frame O_b–$x_b y_b z_b$ is termed *base frame*. The frame attached to the end-effector is termed *end-effector frame* and is conveniently chosen according to the particular task geometry. If the end-effector is a gripper, the origin of the end-effector frame is located at the centre of the gripper, the unit vector $\boldsymbol{a}_e$ is chosen in the *approach* direction to the object, the unit vector $\boldsymbol{s}_e$ is chosen normal to $\boldsymbol{a}_e$ in the *sliding* plane of the jaws, and the unit vector $\boldsymbol{n}_e$ is chosen *normal* to the other two so that the frame $(\boldsymbol{n}_e, \boldsymbol{s}_e, \boldsymbol{a}_e)$ is right-handed.

A first way to compute direct kinematics is offered by a geometric analysis of the structure of the given manipulator.

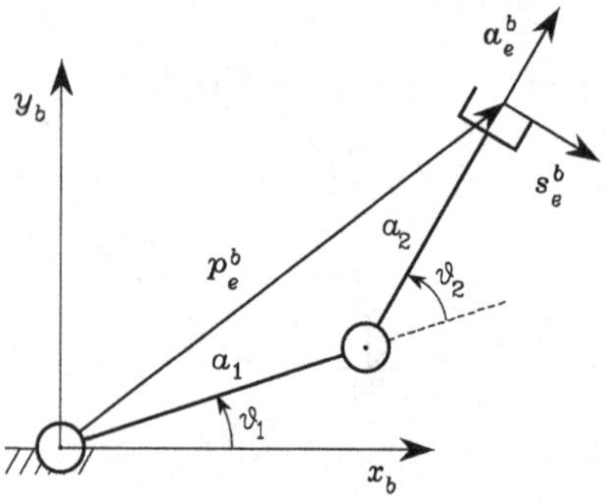

Fig. 2.14. Two-link planar arm

Example 2.4

Consider the two-link planar arm in Fig. 2.14. On the basis of simple trigonometry, the choice of the joint variables, the base frame, and the end-effector frame leads to[8]

$$\boldsymbol{T}_e^b(\boldsymbol{q}) = \begin{bmatrix} \boldsymbol{n}_e^b & \boldsymbol{s}_e^b & \boldsymbol{a}_e^b & \boldsymbol{p}_e^b \\ 0 & 0 & 0 & 1 \end{bmatrix} = \begin{bmatrix} 0 & s_{12} & c_{12} & a_1c_1 + a_2c_{12} \\ 0 & -c_{12} & s_{12} & a_1s_1 + a_2s_{12} \\ 1 & 0 & 0 & 0 \\ 0 & 0 & 0 & 1 \end{bmatrix}. \tag{2.49}$$

It is not difficult to infer that the effectiveness of a geometric approach to the direct kinematics problem is based first on a convenient choice of the relevant quantities and then on the ability and geometric intuition of the problem solver. Whenever the manipulator structure is complex and the number of joints increases, it is preferable to adopt a less direct solution, which, though, is based on a systematic, general procedure. The problem becomes even more complex when the manipulator contains one or more closed kinematic chains. In such a case, as it will be discussed later, there is no guarantee to obtain an analytical expression for the direct kinematics function in (2.48).

2.8.1 Open Chain

Consider an *open-chain* manipulator constituted by $n+1$ links connected by n joints, where Link 0 is conventionally fixed to the ground. It is assumed that each joint provides the mechanical structure with a single DOF, corresponding to the joint variable.

The construction of an operating procedure for the computation of direct kinematics is naturally derived from the typical open kinematic chain of the manipulator structure. In fact, since each joint connects two consecutive

[8] The notations $s_{i\ldots j}$, $c_{i\ldots j}$ denote respectively $\sin(q_i + \ldots + q_j)$, $\cos(q_i + \ldots + q_j)$.

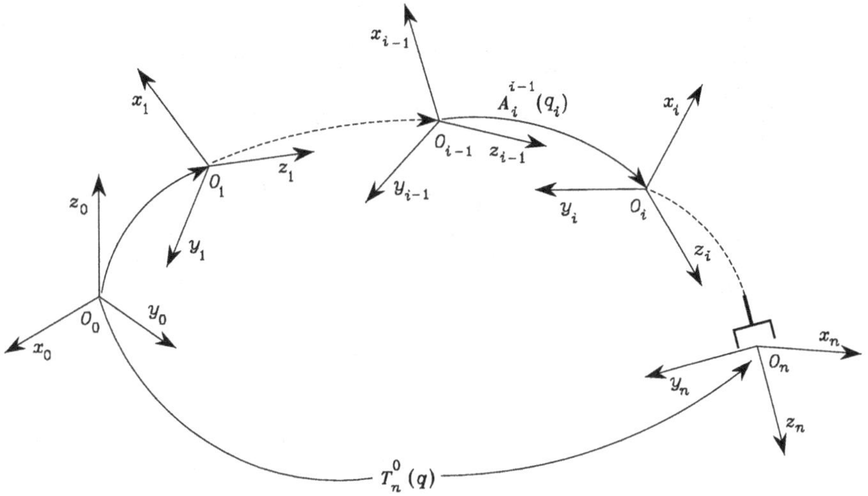

Fig. 2.15. Coordinate transformations in an open kinematic chain

links, it is reasonable to consider first the description of kinematic relationship between consecutive links and then to obtain the overall description of manipulator kinematics in a recursive fashion. To this purpose, it is worth defining a coordinate frame attached to each link, from Link 0 to Link n. Then, the coordinate transformation describing the position and orientation of Frame n with respect to Frame 0 (Fig. 2.15) is given by

$$\boldsymbol{T}_n^0(\boldsymbol{q}) = \boldsymbol{A}_1^0(q_1)\boldsymbol{A}_2^1(q_2)\dots\boldsymbol{A}_n^{n-1}(q_n). \tag{2.50}$$

As requested, the computation of the direct kinematics function is recursive and is obtained in a systematic manner by simple products of the homogeneous transformation matrices $\boldsymbol{A}_i^{i-1}(q_i)$ (for $i = 1, \dots, n$), each of which is a function of a single joint variable.

With reference to the direct kinematics equation in (2.49), the actual coordinate transformation describing the position and orientation of the end-effector frame with respect to the base frame can be obtained as

$$\boldsymbol{T}_e^b(\boldsymbol{q}) = \boldsymbol{T}_0^b\boldsymbol{T}_n^0(\boldsymbol{q})\boldsymbol{T}_e^n \tag{2.51}$$

where $\boldsymbol{T}_0^b$ and $\boldsymbol{T}_e^n$ are two (typically) constant homogeneous transformations describing the position and orientation of Frame 0 with respect to the base frame, and of the end-effector frame with respect to Frame n, respectively.

2.8.2 Denavit–Hartenberg Convention

In order to compute the direct kinematics equation for an open-chain manipulator according to the recursive expression in (2.50), a systematic, general

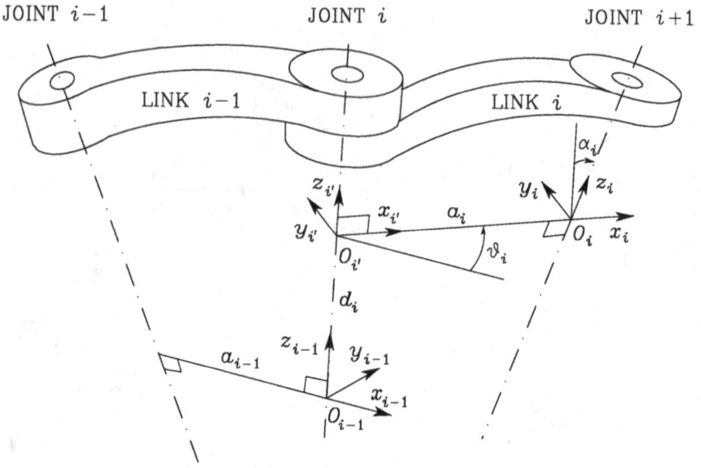

Fig. 2.16. Denavit–Hartenberg kinematic parameters

method is to be derived to define the relative position and orientation of two consecutive links; the problem is that to determine two frames attached to the two links and compute the coordinate transformations between them. In general, the frames can be arbitrarily chosen as long as they are attached to the link they are referred to. Nevertheless, it is convenient to set some rules also for the definition of the link frames.

With reference to Fig. 2.16, let Axis i denote the axis of the joint connecting Link $i-1$ to Link i; the so-called *Denavit–Hartenberg convention* (DH) is adopted to define link Frame i:

- Choose axis z_i along the axis of Joint $i+1$.
- Locate the origin O_i at the intersection of axis z_i with the common normal[9] to axes z_{i-1} and z_i. Also, locate $O_{i'}$ at the intersection of the common normal with axis z_{i-1}.
- Choose axis x_i along the common normal to axes z_{i-1} and z_i with direction from Joint i to Joint $i+1$.
- Choose axis y_i so as to complete a right-handed frame.

The Denavit–Hartenberg convention gives a nonunique definition of the link frame in the following cases:

- For Frame 0, only the direction of axis z_0 is specified; then O_0 and x_0 can be arbitrarily chosen.
- For Frame n, since there is no Joint $n+1$, z_n is not uniquely defined while x_n has to be normal to axis z_{n-1}. Typically, Joint n is revolute, and thus z_n is to be aligned with the direction of z_{n-1}.

[9] The common normal between two lines is the line containing the minimum distance segment between the two lines.

- When two consecutive axes are parallel, the common normal between them is not uniquely defined.
- When two consecutive axes intersect, the direction of x_i is arbitrary.
- When Joint i is prismatic, the direction of z_{i-1} is arbitrary.

In all such cases, the indeterminacy can be exploited to simplify the procedure; for instance, the axes of consecutive frames can be made parallel.

Once the link frames have been established, the position and orientation of Frame i with respect to Frame $i-1$ are completely specified by the following *parameters*:

a_i distance between O_i and $O_{i'}$,
d_i coordinate of $O_{i'}$ along z_{i-1},
α_i angle between axes z_{i-1} and z_i about axis x_i to be taken positive when rotation is made counter-clockwise,
ϑ_i angle between axes x_{i-1} and x_i about axis z_{i-1} to be taken positive when rotation is made counter-clockwise.

Two of the four parameters (a_i and α_i) are always constant and depend only on the geometry of connection between consecutive joints established by Link i. Of the remaining two parameters, only one is variable depending on the type of joint that connects Link $i-1$ to Link i. In particular:

- if Joint i is *revolute* the variable is ϑ_i,
- if Joint i is *prismatic* the variable is d_i.

At this point, it is possible to express the coordinate transformation between Frame i and Frame $i-1$ according to the following steps:

- Choose a frame aligned with Frame $i-1$.
- Translate the chosen frame by d_i along axis z_{i-1} and rotate it by ϑ_i about axis z_{i-1}; this sequence aligns the current frame with Frame i' and is described by the homogeneous transformation matrix

$$\boldsymbol{A}_{i'}^{i-1} = \begin{bmatrix} c_{\vartheta_i} & -s_{\vartheta_i} & 0 & 0 \\ s_{\vartheta_i} & c_{\vartheta_i} & 0 & 0 \\ 0 & 0 & 1 & d_i \\ 0 & 0 & 0 & 1 \end{bmatrix}.$$

- Translate the frame aligned with Frame i' by a_i along axis $x_{i'}$ and rotate it by α_i about axis $x_{i'}$; this sequence aligns the current frame with Frame i and is described by the homogeneous transformation matrix

$$\boldsymbol{A}_{i}^{i'} = \begin{bmatrix} 1 & 0 & 0 & a_i \\ 0 & c_{\alpha_i} & -s_{\alpha_i} & 0 \\ 0 & s_{\alpha_i} & c_{\alpha_i} & 0 \\ 0 & 0 & 0 & 1 \end{bmatrix}.$$

- The resulting coordinate transformation is obtained by postmultiplication of the single transformations as

$$\boldsymbol{A}_i^{i-1}(q_i) = \boldsymbol{A}_{i'}^{i-1}\boldsymbol{A}_i^{i'} = \begin{bmatrix} c_{\vartheta_i} & -s_{\vartheta_i}c_{\alpha_i} & s_{\vartheta_i}s_{\alpha_i} & a_i c_{\vartheta_i} \\ s_{\vartheta_i} & c_{\vartheta_i}c_{\alpha_i} & -c_{\vartheta_i}s_{\alpha_i} & a_i s_{\vartheta_i} \\ 0 & s_{\alpha_i} & c_{\alpha_i} & d_i \\ 0 & 0 & 0 & 1 \end{bmatrix}. \tag{2.52}$$

Notice that the transformation matrix from Frame i to Frame $i-1$ is a function only of the joint variable q_i, that is, ϑ_i for a revolute joint or d_i for a prismatic joint.

To summarize, the Denavit–Hartenberg convention allows the construction of the direct kinematics function by composition of the individual coordinate transformations expressed by (2.52) into one homogeneous transformation matrix as in (2.50). The procedure can be applied to any open kinematic chain and can be easily rewritten in an operating form as follows.

1. Find and number consecutively the joint axes; set the directions of axes $z_0, \ldots, z_{n-1}$.
2. Choose Frame 0 by locating the origin on axis z_0; axes x_0 and y_0 are chosen so as to obtain a right-handed frame. If feasible, it is worth choosing Frame 0 to coincide with the base frame.

Execute steps from **3** to **5** for $i = 1, \ldots, n-1$:

3. Locate the origin O_i at the intersection of z_i with the common normal to axes z_{i-1} and z_i. If axes z_{i-1} and z_i are parallel and Joint i is revolute, then locate O_i so that $d_i = 0$; if Joint i is prismatic, locate O_i at a reference position for the joint range, e.g., a mechanical limit.
4. Choose axis x_i along the common normal to axes z_{i-1} and z_i with direction from Joint i to Joint $i+1$.
5. Choose axis y_i so as to obtain a right-handed frame.

To complete:

6. Choose Frame n; if Joint n is revolute, then align z_n with z_{n-1}, otherwise, if Joint n is prismatic, then choose z_n arbitrarily. Axis x_n is set according to step **4**.
7. For $i = 1, \ldots, n$, form the table of parameters $a_i, d_i, \alpha_i, \vartheta_i$.
8. On the basis of the parameters in **7**, compute the homogeneous transformation matrices $\boldsymbol{A}_i^{i-1}(q_i)$ for $i = 1, \ldots, n$.
9. Compute the homogeneous transformation $\boldsymbol{T}_n^0(\boldsymbol{q}) = \boldsymbol{A}_1^0 \ldots \boldsymbol{A}_n^{n-1}$ that yields the position and orientation of Frame n with respect to Frame 0.
10. Given $\boldsymbol{T}_0^b$ and $\boldsymbol{T}_e^n$, compute the direct kinematics function as $\boldsymbol{T}_e^b(\boldsymbol{q}) = \boldsymbol{T}_0^b\boldsymbol{T}_n^0\boldsymbol{T}_e^n$ that yields the position and orientation of the end-effector frame with respect to the base frame.

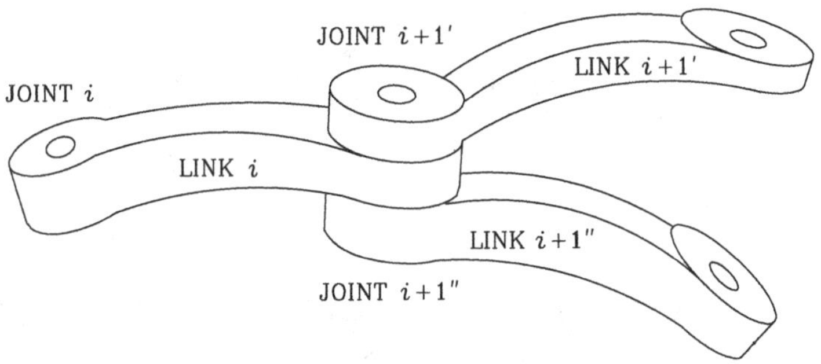

Fig. 2.17. Connection of a single link in the chain with two links

For what concerns the computational aspects of direct kinematics, it can be recognized that the heaviest load derives from the evaluation of transcendental functions. On the other hand, by suitably factorizing the transformation equations and introducing local variables, the number of flops (additions + multiplications) can be reduced. Finally, for computation of orientation it is convenient to evaluate the two unit vectors of the end-effector frame of simplest expression and derive the third one by vector product of the first two.

2.8.3 Closed Chain

The above direct kinematics method based on the DH convention exploits the inherently recursive feature of an open-chain manipulator. Nevertheless, the method can be extended to the case of manipulators containing closed kinematic chains according to the technique illustrated below.

Consider a *closed-chain* manipulator constituted by $n+1$ links. Because of the presence of a loop, the number of joints l must be greater than n; in particular, it can be understood that the number of closed loops is equal to $l-n$.

With reference to Fig. 2.17, Links 0 through i are connected successively through the first i joints as in an open kinematic chain. Then, Joint $i+1'$ connects Link i with Link $i+1'$ while Joint $i+1''$ connects Link i with Link $i+1''$; the axes of Joints $i+1'$ and $i+1''$ are assumed to be aligned. Although not represented in the figure, Links $i+1'$ and $i+1''$ are members of the closed kinematic chain. In particular, Link $i+1'$ is further connected to Link $i+2'$ via Joint $i+2'$ and so forth, until Link j via Joint j. Likewise, Link $i+1''$ is further connected to Link $i+2''$ via Joint $i+2''$ and so forth, until Link k via Joint k. Finally, Links j and k are connected together at Joint $j+1$ to form a closed chain. In general, $j \neq k$.

In order to attach frames to the various links and apply DH convention, one closed kinematic chain is taken into account. The closed chain can be virtually cut open at Joint $j+1$, i.e., the joint between Link j and Link k. An equivalent tree-structured open kinematic chain is obtained, and thus link

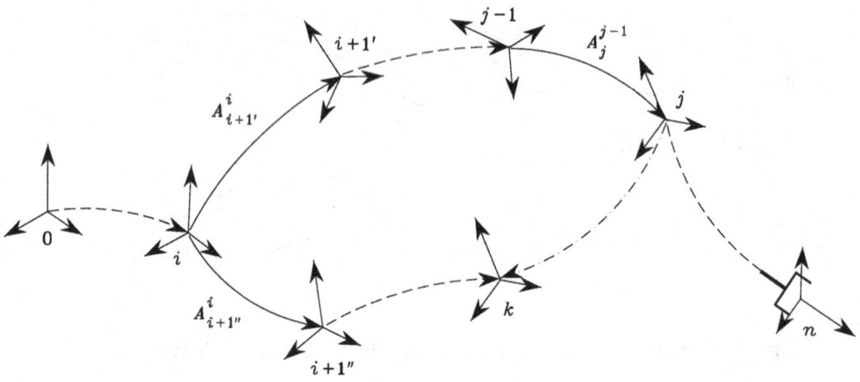

Fig. 2.18. Coordinate transformations in a closed kinematic chain

frames can be defined as in Fig. 2.18. Since Links 0 through i occur before the two branches of the tree, they are left out of the analysis. For the same reason, Links $j+1$ through n are left out as well. Notice that Frame i is to be chosen with axis z_i aligned with the axes of Joints $i+1'$ and $i+1''$.

It follows that the position and orientation of Frame j with respect to Frame i can be expressed by composing the homogeneous transformations as

$$\boldsymbol{A}^i_j(\boldsymbol{q}') = \boldsymbol{A}^i_{i+1'}(q_{i+1'}) \dots \boldsymbol{A}^{j-1}_j(q_j) \tag{2.53}$$

where $\boldsymbol{q}' = [\, q_{i+1'} \quad \dots \quad q_j \,]^T$. Likewise, the position and orientation of Frame k with respect to Frame i is given by

$$\boldsymbol{A}^i_k(\boldsymbol{q}'') = \boldsymbol{A}^i_{i+1''}(q_{i+1''}) \dots \boldsymbol{A}^{k-1}_k(q_k) \tag{2.54}$$

where $\boldsymbol{q}'' = [\, q_{i+1''} \quad \dots \quad q_k \,]^T$.

Since Links j and k are connected to each other through Joint $j+1$, it is worth analyzing the mutual position and orientation between Frames j and k, as illustrated in Fig. 2.19. Notice that, since Links j and k are connected to form a closed chain, axes z_j and z_k are aligned. Therefore, the following orientation constraint has to be imposed between Frames j and k:

$$\boldsymbol{z}^i_j(\boldsymbol{q}') = \boldsymbol{z}^i_k(\boldsymbol{q}''), \tag{2.55}$$

where the unit vectors of the two axes have been conveniently referred to Frame i.

Moreover, if Joint $j+1$ is prismatic, the angle ϑ_{jk} between axes x_j and x_k is fixed; hence, in addition to (2.55), the following constraint is obtained:

$$\boldsymbol{x}^{iT}_j(\boldsymbol{q}')\boldsymbol{x}^i_k(\boldsymbol{q}'') = \cos\vartheta_{jk}. \tag{2.56}$$

Obviously, there is no need to impose a similar constraint on axes y_j and y_k since that would be redundant.

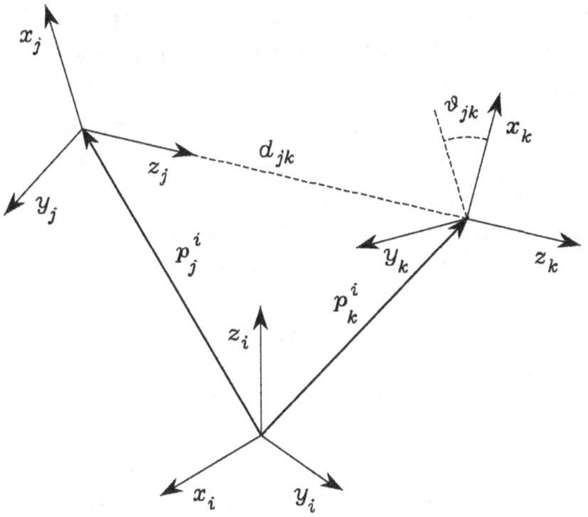

Fig. 2.19. Coordinate transformation at the cut joint

Regarding the position constraint between Frames j and k, let $\boldsymbol{p}^i_j$ and $\boldsymbol{p}^i_k$ respectively denote the positions of the origins of Frames j and k, when referred to Frame i. By projecting on Frame j the distance vector of the origin of Frame k from Frame j, the following constraint has to be imposed:

$$\boldsymbol{R}^j_i(\boldsymbol{q}')\left(\boldsymbol{p}^i_j(\boldsymbol{q}') - \boldsymbol{p}^i_k(\boldsymbol{q}'')\right) = [\,0 \quad 0 \quad d_{jk}\,]^T \tag{2.57}$$

where $\boldsymbol{R}^j_i = \boldsymbol{R}^{iT}_j$ denotes the orientation of Frame i with respect to Frame j. At this point, if Joint $j+1$ is revolute, then d_{jk} is a fixed offset along axis z_j; hence, the three equalities of (2.57) fully describe the position constraint. If, however, Joint $j+1$ is prismatic, then d_{jk} varies. Consequently, only the first two equalities of (2.57) describe the position constraint, i.e.,

$$\begin{bmatrix} \boldsymbol{x}^{iT}_j(\boldsymbol{q}') \\ \boldsymbol{y}^{iT}_j(\boldsymbol{q}') \end{bmatrix} \left(\boldsymbol{p}^i_j(\boldsymbol{q}') - \boldsymbol{p}^i_k(\boldsymbol{q}'')\right) = \begin{bmatrix} 0 \\ 0 \end{bmatrix} \tag{2.58}$$

where $\boldsymbol{R}^i_j = [\,\boldsymbol{x}^i_j \quad \boldsymbol{y}^i_j \quad \boldsymbol{z}^i_j\,]$.

In summary, if Joint $j+1$ is *revolute* the constraints are

$$\begin{cases} \boldsymbol{R}^j_i(\boldsymbol{q}')\left(\boldsymbol{p}^i_j(\boldsymbol{q}') - \boldsymbol{p}^i_k(\boldsymbol{q}'')\right) = [\,0 \quad 0 \quad d_{jk}\,]^T \\ \boldsymbol{z}^i_j(\boldsymbol{q}') = \boldsymbol{z}^i_k(\boldsymbol{q}''), \end{cases} \tag{2.59}$$

whereas if Joint $j+1$ is *prismatic* the constraints are

$$\begin{cases} \begin{bmatrix} \boldsymbol{x}^{iT}_j(\boldsymbol{q}') \\ \boldsymbol{y}^{iT}_j(\boldsymbol{q}') \end{bmatrix} \left(\boldsymbol{p}^i_j(\boldsymbol{q}') - \boldsymbol{p}^i_k(\boldsymbol{q}'')\right) = \begin{bmatrix} 0 \\ 0 \end{bmatrix} \\ \boldsymbol{z}^i_j(\boldsymbol{q}') = \boldsymbol{z}^i_k(\boldsymbol{q}'') \\ \boldsymbol{x}^{iT}_j(\boldsymbol{q}')\boldsymbol{x}^i_k(\boldsymbol{q}'') = \cos\vartheta_{jk}. \end{cases} \tag{2.60}$$

In either case, there are six equalities that must be satisfied. Those should be solved for a reduced number of independent joint variables to be keenly chosen among the components of $\boldsymbol{q}'$ and $\boldsymbol{q}''$ which characterize the DOFs of the closed chain. These are the natural candidates to be the actuated joints, while the other joints in the chain (including the cut joint) are typically not actuated. Such independent variables, together with the remaining joint variables not involved in the above analysis, constitute the joint vector $\boldsymbol{q}$ that allows the direct kinematics equation to be computed as

$$\boldsymbol{T}_n^0(\boldsymbol{q}) = \boldsymbol{A}_i^0 \boldsymbol{A}_j^i \boldsymbol{A}_n^j, \tag{2.61}$$

where the sequence of successive transformations after the closure of the chain has been conventionally resumed from Frame j.

In general, there is no guarantee to solve the constraints in closed form unless the manipulator has a simple kinematic structure. In other words, for a given manipulator with a specific geometry, e.g., a planar structure, some of the above equalities may become dependent. Hence, the number of independent equalities is less than six and it should likely be easier to solve them.

To conclude, it is worth sketching the operating form of the procedure to compute the direct kinematics function for a closed-chain manipulator using the Denavit–Hartenberg convention.

1. In the closed chain, select one joint that is not actuated. Assume that the joint is cut open so as to obtain an open chain in a tree structure.
2. Compute the homogeneous transformations according to DH convention.
3. Find the equality constraints for the two frames connected by the cut joint.
4. Solve the constraints for a reduced number of joint variables.
5. Express the homogeneous transformations in terms of the above joint variables and compute the direct kinematics function by composing the various transformations from the base frame to the end-effector frame.

2.9 Kinematics of Typical Manipulator Structures

This section contains several examples of computation of the direct kinematics function for typical manipulator structures that are often encountered in industrial robots.

With reference to the schematic representation of the kinematic chain, manipulators are usually illustrated in postures where the joint variables, defined according to the DH convention, are different from zero; such values might differ from the null references utilized for robot manipulator programming. Hence, it will be necessary to sum constant contributions (offsets) to the values of the joint variables measured by the robot sensory system, so as to match the references.

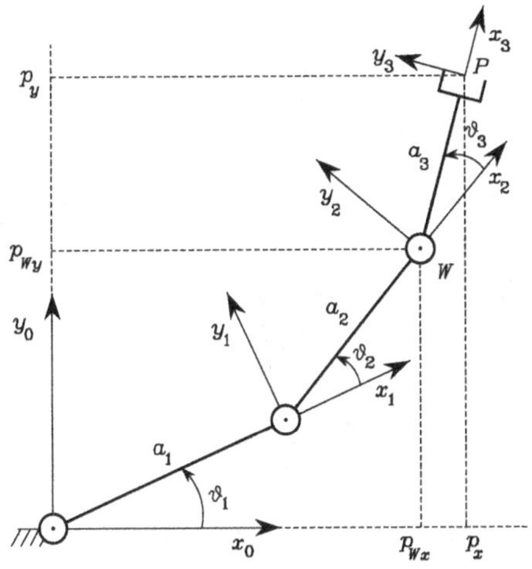

Fig. 2.20. Three-link planar arm

2.9.1 Three-link Planar Arm

Consider the three-link planar arm in Fig. 2.20, where the link frames have been illustrated. Since the revolute axes are all parallel, the simplest choice was made for all axes x_i along the direction of the relative links (the direction of x_0 is arbitrary) and all lying in the plane (x_0, y_0). In this way, all the parameters d_i are null and the angles between the axes x_i directly provide the joint variables. The DH parameters are specified in Table 2.1.

Table 2.1. DH parameters for the three-link planar arm

Link	a_i	α_i	d_i	ϑ_i
1	a_1	0	0	ϑ_1
2	a_2	0	0	ϑ_2
3	a_3	0	0	ϑ_3

Since all joints are revolute, the homogeneous transformation matrix defined in (2.52) has the same structure for each joint, i.e.,

$$\boldsymbol{A}_i^{i-1}(\vartheta_i) = \begin{bmatrix} c_i & -s_i & 0 & a_i c_i \\ s_i & c_i & 0 & a_i s_i \\ 0 & 0 & 1 & 0 \\ 0 & 0 & 0 & 1 \end{bmatrix} \qquad i = 1, 2, 3. \tag{2.62}$$

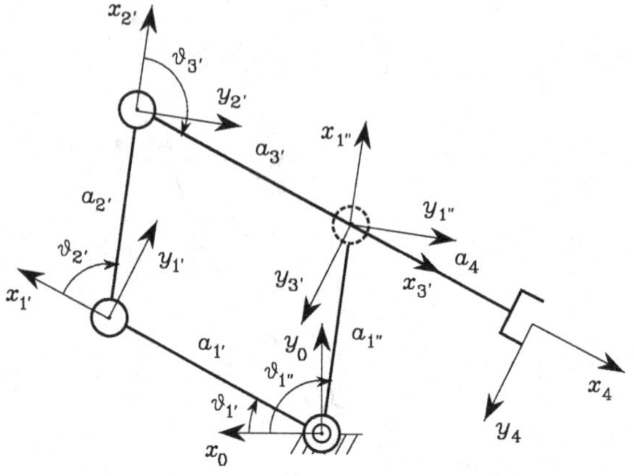

Fig. 2.21. Parallelogram arm

Computation of the direct kinematics function as in (2.50) yields

$$T_3^0(q) = A_1^0 A_2^1 A_3^2 = \begin{bmatrix} c_{123} & -s_{123} & 0 & a_1c_1 + a_2c_{12} + a_3c_{123} \\ s_{123} & c_{123} & 0 & a_1s_1 + a_2s_{12} + a_3s_{123} \\ 0 & 0 & 1 & 0 \\ 0 & 0 & 0 & 1 \end{bmatrix} \tag{2.63}$$

where $q = [\,\vartheta_1 \quad \vartheta_2 \quad \vartheta_3\,]^T$. Notice that the unit vector z_3^0 of Frame 3 is aligned with $z_0 = [\,0 \quad 0 \quad 1\,]^T$, in view of the fact that all revolute joints are parallel to axis z_0. Obviously, $p_z = 0$ and all three joints concur to determine the end-effector position in the plane of the structure. It is worth pointing out that Frame 3 does not coincide with the end-effector frame (Fig. 2.13), since the resulting approach unit vector is aligned with x_3^0 and not with z_3^0. Thus, assuming that the two frames have the same origin, the constant transformation

$$T_e^3 = \begin{bmatrix} 0 & 0 & 1 & 0 \\ 0 & 1 & 0 & 0 \\ -1 & 0 & 0 & 0 \\ 0 & 0 & 0 & 1 \end{bmatrix}.$$

is needed, having taken n aligned with z_0.

2.9.2 Parallelogram Arm

Consider the parallelogram arm in Fig. 2.21. A closed chain occurs where the first two joints connect Link 1′ and Link 1″ to Link 0, respectively. Joint 4 was selected as the cut joint, and the link frames have been established accordingly. The DH parameters are specified in Table 2.2, where $a_{1'} = a_{3'}$ and $a_{2'} = a_{1''}$ in view of the parallelogram structure.

Notice that the parameters for Link 4 are all constant. Since the joints are revolute, the homogeneous transformation matrix defined in (2.52) has

Table 2.2. DH parameters for the parallelogram arm

Link	a_i	α_i	d_i	ϑ_i
$1'$	$a_{1'}$	0	0	$\vartheta_{1'}$
$2'$	$a_{2'}$	0	0	$\vartheta_{2'}$
$3'$	$a_{3'}$	0	0	$\vartheta_{3'}$
$1''$	$a_{1''}$	0	0	$\vartheta_{1''}$
4	a_4	0	0	0

the same structure for each joint, i.e., as in (2.62) for Joints $1'$, $2'$, $3'$ and $1''$. Therefore, the coordinate transformations for the two branches of the tree are respectively:

$$\boldsymbol{A}_{3'}^{0}(\boldsymbol{q}') = \boldsymbol{A}_{1'}^{0}\boldsymbol{A}_{2'}^{1'}\boldsymbol{A}_{3'}^{2'} = \begin{bmatrix} c_{1'2'3'} & -s_{1'2'3'} & 0 & a_{1'}c_{1'} + a_{2'}c_{1'2'} + a_{3'}c_{1'2'3'} \\ s_{1'2'3'} & c_{1'2'3'} & 0 & a_{1'}s_{1'} + a_{2'}s_{1'2'} + a_{3'}s_{1'2'3'} \\ 0 & 0 & 1 & 0 \\ 0 & 0 & 0 & 1 \end{bmatrix}$$

where $\boldsymbol{q}' = [\,\vartheta_{1'} \quad \vartheta_{2'} \quad \vartheta_{3'}\,]^T$, and

$$\boldsymbol{A}_{1''}^{0}(q'') = \begin{bmatrix} c_{1''} & -s_{1''} & 0 & a_{1''}c_{1''} \\ s_{1''} & c_{1''} & 0 & a_{1''}s_{1''} \\ 0 & 0 & 1 & 0 \\ 0 & 0 & 0 & 1 \end{bmatrix}$$

where $q'' = \vartheta_{1''}$. To complete, the constant homogeneous transformation for the last link is

$$\boldsymbol{A}_{4}^{3'} = \begin{bmatrix} 1 & 0 & 0 & a_4 \\ 0 & 1 & 0 & 0 \\ 0 & 0 & 1 & 0 \\ 0 & 0 & 0 & 1 \end{bmatrix}.$$

With reference to (2.59), the position constraints are ($d_{3'1''} = 0$)

$$\boldsymbol{R}_{0}^{3'}(\boldsymbol{q}')\left(\boldsymbol{p}_{3'}^{0}(\boldsymbol{q}') - \boldsymbol{p}_{1''}^{0}(q'')\right) = \begin{bmatrix} 0 \\ 0 \\ 0 \end{bmatrix}$$

while the orientation constraints are satisfied independently of $\boldsymbol{q}'$ and q''. Since $a_{1'} = a_{3'}$ and $a_{2'} = a_{1''}$, two independent constraints can be extracted, i.e.,

$$a_{1'}(c_{1'} + c_{1'2'3'}) + a_{1''}(c_{1'2'} - c_{1''}) = 0$$
$$a_{1'}(s_{1'} + s_{1'2'3'}) + a_{1''}(s_{1'2'} - s_{1''}) = 0.$$

In order to satisfy them for any choice of $a_{1'}$ and $a_{1''}$, it must be

$$\vartheta_{2'} = \vartheta_{1''} - \vartheta_{1'}$$
$$\vartheta_{3'} = \pi - \vartheta_{2'} = \pi - \vartheta_{1''} + \vartheta_{1'}$$

Therefore, the vector of joint variables is $\boldsymbol{q} = [\vartheta_{1'} \quad \vartheta_{1''}]^T$. These joints are natural candidates to be the actuated joints.[10] Substituting the expressions of $\vartheta_{2'}$ and $\vartheta_{3'}$ into the homogeneous transformation $\boldsymbol{A}^0_{3'}$ and computing the direct kinematics function as in (2.61) yields

$$\boldsymbol{T}^0_4(\boldsymbol{q}) = \boldsymbol{A}^0_{3'}(\boldsymbol{q})\boldsymbol{A}^{3'}_4 = \begin{bmatrix} -c_{1'} & s_{1'} & 0 & a_{1''}c_{1''} - a_4 c_{1'} \\ -s_{1'} & -c_{1'} & 0 & a_{1''}s_{1''} - a_4 s_{1'} \\ 0 & 0 & 1 & 0 \\ 0 & 0 & 0 & 1 \end{bmatrix}. \tag{2.64}$$

A comparison between (2.64) and (2.49) reveals that the parallelogram arm is kinematically equivalent to a two-link planar arm. The noticeable difference, though, is that the two actuated joints — providing the DOFs of the structure — are located at the base. This will greatly simplify the dynamic model of the structure, as will be seen in Sect. 7.3.3.

2.9.3 Spherical Arm

Consider the spherical arm in Fig. 2.22, where the link frames have been illustrated. Notice that the origin of Frame 0 was located at the intersection of z_0 with z_1 so that $d_1 = 0$; analogously, the origin of Frame 2 was located at the intersection between z_1 and z_2. The DH parameters are specified in Table 2.3.

Table 2.3. DH parameters for the spherical arm

Link	a_i	α_i	d_i	ϑ_i
1	0	$-\pi/2$	0	ϑ_1
2	0	$\pi/2$	d_2	ϑ_2
3	0	0	d_3	0

The homogeneous transformation matrices defined in (2.52) are for the single joints:

$$\boldsymbol{A}^0_1(\vartheta_1) = \begin{bmatrix} c_1 & 0 & -s_1 & 0 \\ s_1 & 0 & c_1 & 0 \\ 0 & -1 & 0 & 0 \\ 0 & 0 & 0 & 1 \end{bmatrix} \qquad \boldsymbol{A}^1_2(\vartheta_2) = \begin{bmatrix} c_2 & 0 & s_2 & 0 \\ s_2 & 0 & -c_2 & 0 \\ 0 & 1 & 0 & d_2 \\ 0 & 0 & 0 & 1 \end{bmatrix}$$

$$\boldsymbol{A}^2_3(d_3) = \begin{bmatrix} 1 & 0 & 0 & 0 \\ 0 & 1 & 0 & 0 \\ 0 & 0 & 1 & d_3 \\ 0 & 0 & 0 & 1 \end{bmatrix}.$$

[10] Notice that it is not possible to solve (2.64) for $\vartheta_{2'}$ and $\vartheta_{3'}$ since they are constrained by the condition $\vartheta_{2'} + \vartheta_{3'} = \pi$.

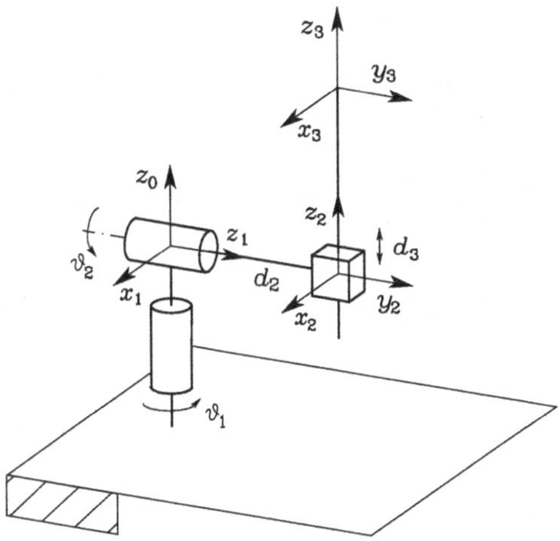

Fig. 2.22. Spherical arm

Computation of the direct kinematics function as in (2.50) yields

$$T_3^0(q) = A_1^0 A_2^1 A_3^2 = \begin{bmatrix} c_1c_2 & -s_1 & c_1s_2 & c_1s_2d_3 - s_1d_2 \\ s_1c_2 & c_1 & s_1s_2 & s_1s_2d_3 + c_1d_2 \\ -s_2 & 0 & c_2 & c_2d_3 \\ 0 & 0 & 0 & 1 \end{bmatrix} \tag{2.65}$$

where $q = [\vartheta_1 \quad \vartheta_2 \quad d_3]^T$. Notice that the third joint does not obviously influence the rotation matrix. Further, the orientation of the unit vector y_3^0 is uniquely determined by the first joint, since the revolute axis of the second joint z_1 is parallel to axis y_3. Different from the previous structures, in this case Frame 3 can represent an end-effector frame of unit vectors (n_e, s_e, a_e), i.e., $T_e^3 = I_4$.

2.9.4 Anthropomorphic Arm

Consider the anthropomorphic arm in Fig. 2.23. Notice how this arm corresponds to a two-link planar arm with an additional rotation about an axis of the plane. In this respect, the parallelogram arm could be used in lieu of the two-link planar arm, as found in some industrial robots with an anthropomorphic structure.

The link frames have been illustrated in the figure. As for the previous structure, the origin of Frame 0 was chosen at the intersection of z_0 with z_1 ($d_1 = 0$); further, z_1 and z_2 are parallel and the choice of axes x_1 and x_2 was made as for the two-link planar arm. The DH parameters are specified in Table 2.4.

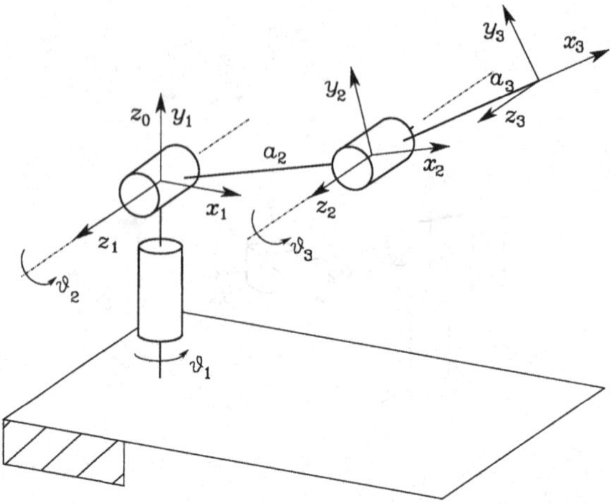

Fig. 2.23. Anthropomorphic arm

Table 2.4. DH parameters for the anthropomorphic arm

Link	a_i	α_i	d_i	ϑ_i
1	0	$\pi/2$	0	ϑ_1
2	a_2	0	0	ϑ_2
3	a_3	0	0	ϑ_3

The homogeneous transformation matrices defined in (2.52) are for the single joints:

$$\boldsymbol{A}_1^0(\vartheta_1) = \begin{bmatrix} c_1 & 0 & s_1 & 0 \\ s_1 & 0 & -c_1 & 0 \\ 0 & 1 & 0 & 0 \\ 0 & 0 & 0 & 1 \end{bmatrix}$$

$$\boldsymbol{A}_i^{i-1}(\vartheta_i) = \begin{bmatrix} c_i & -s_i & 0 & a_i c_i \\ s_i & c_i & 0 & a_i s_i \\ 0 & 0 & 1 & 0 \\ 0 & 0 & 0 & 1 \end{bmatrix} \qquad i = 2, 3.$$

Computation of the direct kinematics function as in (2.50) yields

$$\boldsymbol{T}_3^0(\boldsymbol{q}) = \boldsymbol{A}_1^0 \boldsymbol{A}_2^1 \boldsymbol{A}_3^2 = \begin{bmatrix} c_1 c_{23} & -c_1 s_{23} & s_1 & c_1(a_2 c_2 + a_3 c_{23}) \\ s_1 c_{23} & -s_1 s_{23} & -c_1 & s_1(a_2 c_2 + a_3 c_{23}) \\ s_{23} & c_{23} & 0 & a_2 s_2 + a_3 s_{23} \\ 0 & 0 & 0 & 1 \end{bmatrix} \tag{2.66}$$

where $\boldsymbol{q} = [\,\vartheta_1 \quad \vartheta_2 \quad \vartheta_3\,]^T$. Since z_3 is aligned with z_2, Frame 3 does not coincide with a possible end-effector frame as in Fig. 2.13, and a proper constant transformation would be needed.

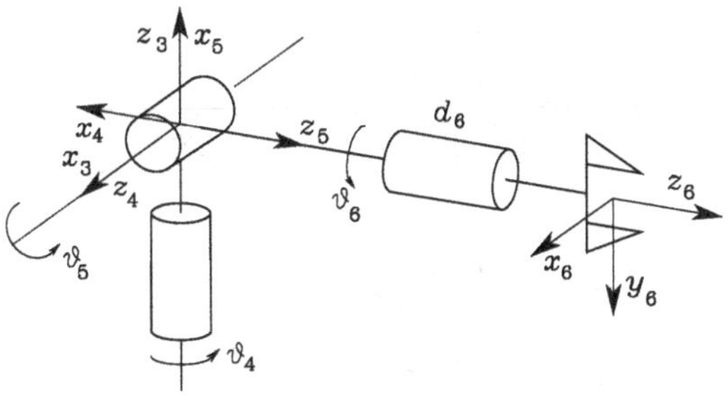

Fig. 2.24. Spherical wrist

2.9.5 Spherical Wrist

Consider a particular type of structure consisting just of the wrist of Fig. 2.24. Joint variables were numbered progressively starting from 4, since such a wrist is typically thought of as mounted on a three-DOF arm of a six-DOF manipulator. It is worth noticing that the wrist is spherical since all revolute axes intersect at a single point. Once z_3, z_4, z_5 have been established, and x_3 has been chosen, there is an indeterminacy on the directions of x_4 and x_5. With reference to the frames indicated in Fig. 2.24, the DH parameters are specified in Table 2.5.

Table 2.5. DH parameters for the spherical wrist

Link	a_i	α_i	d_i	ϑ_i
4	0	$-\pi/2$	0	ϑ_4
5	0	$\pi/2$	0	ϑ_5
6	0	0	d_6	ϑ_6

The homogeneous transformation matrices defined in (2.52) are for the single joints:

$$\boldsymbol{A}_4^3(\vartheta_4) = \begin{bmatrix} c_4 & 0 & -s_4 & 0 \\ s_4 & 0 & c_4 & 0 \\ 0 & -1 & 0 & 0 \\ 0 & 0 & 0 & 1 \end{bmatrix} \qquad \boldsymbol{A}_5^4(\vartheta_5) = \begin{bmatrix} c_5 & 0 & s_5 & 0 \\ s_5 & 0 & -c_5 & 0 \\ 0 & 1 & 0 & 0 \\ 0 & 0 & 0 & 1 \end{bmatrix}$$

$$\boldsymbol{A}_6^5(\vartheta_6) = \begin{bmatrix} c_6 & -s_6 & 0 & 0 \\ s_6 & c_6 & 0 & 0 \\ 0 & 0 & 1 & d_6 \\ 0 & 0 & 0 & 1 \end{bmatrix}.$$

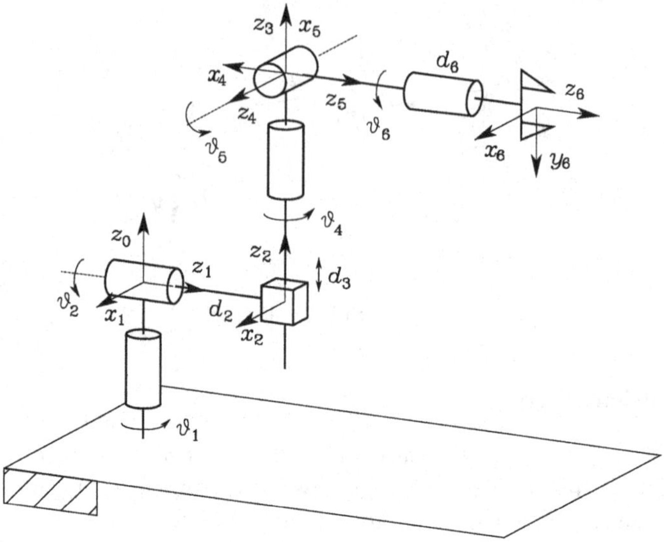

Fig. 2.25. Stanford manipulator

Computation of the direct kinematics function as in (2.50) yields

$$T_6^3(q) = A_4^3 A_5^4 A_6^5 = \begin{bmatrix} c_4c_5c_6 - s_4s_6 & -c_4c_5s_6 - s_4c_6 & c_4s_5 & c_4s_5d_6 \\ s_4c_5c_6 + c_4s_6 & -s_4c_5s_6 + c_4c_6 & s_4s_5 & s_4s_5d_6 \\ -s_5c_6 & s_5s_6 & c_5 & c_5d_6 \\ 0 & 0 & 0 & 1 \end{bmatrix} \tag{2.67}$$

where $q = [\,\vartheta_4 \quad \vartheta_5 \quad \vartheta_6\,]^T$. Notice that, as a consequence of the choice made for the coordinate frames, the block matrix R_6^3 that can be extracted from T_6^3 coincides with the rotation matrix of Euler angles (2.18) previously derived, that is, $\vartheta_4, \vartheta_5, \vartheta_6$ constitute the set of ZYZ angles with respect to the reference frame O_3–$x_3y_3z_3$. Moreover, the unit vectors of Frame 6 coincide with the unit vectors of a possible end-effector frame according to Fig. 2.13.

2.9.6 Stanford Manipulator

The so-called Stanford manipulator is composed of a spherical arm and a spherical wrist (Fig. 2.25). Since Frame 3 of the spherical arm coincides with Frame 3 of the spherical wrist, the direct kinematics function can be obtained via simple composition of the transformation matrices (2.65), (2.67) of the previous examples, i.e.,

$$T_6^0 = T_3^0 T_6^3 = \begin{bmatrix} n^0 & s^0 & a^0 & p^0 \\ 0 & 0 & 0 & 1 \end{bmatrix}.$$

Carrying out the products yields

$$p_6^0 = \begin{bmatrix} c_1 s_2 d_3 - s_1 d_2 + \big(c_1(c_2 c_4 s_5 + s_2 c_5) - s_1 s_4 s_5\big) d_6 \\ s_1 s_2 d_3 + c_1 d_2 + \big(s_1(c_2 c_4 s_5 + s_2 c_5) + c_1 s_4 s_5\big) d_6 \\ c_2 d_3 + (-s_2 c_4 s_5 + c_2 c_5) d_6 \end{bmatrix} \tag{2.68}$$

for the end-effector position, and

$$\begin{aligned} n_6^0 &= \begin{bmatrix} c_1\big(c_2(c_4 c_5 c_6 - s_4 s_6) - s_2 s_5 c_6\big) - s_1(s_4 c_5 c_6 + c_4 s_6) \\ s_1\big(c_2(c_4 c_5 c_6 - s_4 s_6) - s_2 s_5 c_6\big) + c_1(s_4 c_5 c_6 + c_4 s_6) \\ -s_2(c_4 c_5 c_6 - s_4 s_6) - c_2 s_5 c_6 \end{bmatrix} \\ s_6^0 &= \begin{bmatrix} c_1\big(-c_2(c_4 c_5 s_6 + s_4 c_6) + s_2 s_5 s_6\big) - s_1(-s_4 c_5 s_6 + c_4 c_6) \\ s_1\big(-c_2(c_4 c_5 s_6 + s_4 c_6) + s_2 s_5 s_6\big) + c_1(-s_4 c_5 s_6 + c_4 c_6) \\ s_2(c_4 c_5 s_6 + s_4 c_6) + c_2 s_5 s_6 \end{bmatrix} \\ a_6^0 &= \begin{bmatrix} c_1(c_2 c_4 s_5 + s_2 c_5) - s_1 s_4 s_5 \\ s_1(c_2 c_4 s_5 + s_2 c_5) + c_1 s_4 s_5 \\ -s_2 c_4 s_5 + c_2 c_5 \end{bmatrix} \end{aligned} \tag{2.69}$$

for the end-effector orientation.

A comparison of the vector p_6^0 in (2.68) with the vector p_3^0 in (2.65) relative to the sole spherical arm reveals the presence of additional contributions due to the choice of the origin of the end-effector frame at a distance d_6 from the origin of Frame 3 along the direction of a_6^0. In other words, if it were $d_6 = 0$, the position vector would be the same. This feature is of fundamental importance for the solution of the inverse kinematics for this manipulator, as will be seen later.

2.9.7 Anthropomorphic Arm with Spherical Wrist

A comparison between Fig. 2.23 and Fig. 2.24 reveals that the direct kinematics function cannot be obtained by multiplying the transformation matrices T_3^0 and T_6^3, since Frame 3 of the anthropomorphic arm cannot coincide with Frame 3 of the spherical wrist.

Direct kinematics of the entire structure can be obtained in two ways. One consists of interposing a constant transformation matrix between T_3^0 and T_6^3 which allows the alignment of the two frames. The other refers to the Denavit–Hartenberg operating procedure with the frame assignment for the entire structure illustrated in Fig. 2.26. The DH parameters are specified in Table 2.6.

Since Rows 3 and 4 differ from the corresponding rows of the tables for the two single structures, the relative homogeneous transformation matrices A_3^2 and A_4^3 have to be modified into

$$A_3^2(\vartheta_3) = \begin{bmatrix} c_3 & 0 & s_3 & 0 \\ s_3 & 0 & -c_3 & 0 \\ 0 & 1 & 0 & 0 \\ 0 & 0 & 0 & 1 \end{bmatrix} \qquad A_4^3(\vartheta_4) = \begin{bmatrix} c_4 & 0 & -s_4 & 0 \\ s_4 & 0 & c_4 & 0 \\ 0 & -1 & 0 & d_4 \\ 0 & 0 & 0 & 1 \end{bmatrix}$$

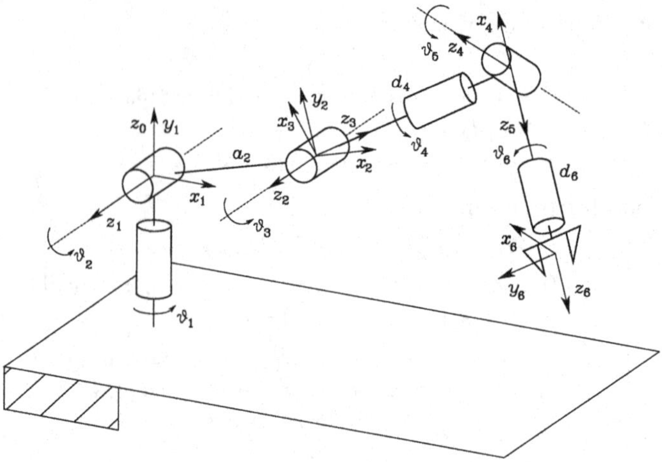

Fig. 2.26. Anthropomorphic arm with spherical wrist

Table 2.6. DH parameters for the anthropomorphic arm with spherical wrist

Link	a_i	α_i	d_i	ϑ_i
1	0	$\pi/2$	0	ϑ_1
2	a_2	0	0	ϑ_2
3	0	$\pi/2$	0	ϑ_3
4	0	$-\pi/2$	d_4	ϑ_4
5	0	$\pi/2$	0	ϑ_5
6	0	0	d_6	ϑ_6

while the other transformation matrices remain the same. Computation of the direct kinematics function leads to expressing the position and orientation of the end-effector frame as:

$$\boldsymbol{p}_6^0 = \begin{bmatrix} a_2c_1c_2 + d_4c_1s_{23} + d_6\big(c_1(c_{23}c_4s_5 + s_{23}c_5) + s_1s_4s_5\big) \\ a_2s_1c_2 + d_4s_1s_{23} + d_6\big(s_1(c_{23}c_4s_5 + s_{23}c_5) - c_1s_4s_5\big) \\ a_2s_2 - d_4c_{23} + d_6(s_{23}c_4s_5 - c_{23}c_5) \end{bmatrix} \tag{2.70}$$

and

$$\begin{aligned} \boldsymbol{n}_6^0 &= \begin{bmatrix} c_1\big(c_{23}(c_4c_5c_6 - s_4s_6) - s_{23}s_5c_6\big) + s_1(s_4c_5c_6 + c_4s_6) \\ s_1\big(c_{23}(c_4c_5c_6 - s_4s_6) - s_{23}s_5c_6\big) - c_1(s_4c_5c_6 + c_4s_6) \\ s_{23}(c_4c_5c_6 - s_4s_6) + c_{23}s_5c_6 \end{bmatrix} \\ \boldsymbol{s}_6^0 &= \begin{bmatrix} c_1\big(-c_{23}(c_4c_5s_6 + s_4c_6) + s_{23}s_5s_6\big) + s_1(-s_4c_5s_6 + c_4c_6) \\ s_1\big(-c_{23}(c_4c_5s_6 + s_4c_6) + s_{23}s_5s_6\big) - c_1(-s_4c_5s_6 + c_4c_6) \\ -s_{23}(c_4c_5s_6 + s_4c_6) - c_{23}s_5s_6 \end{bmatrix} \\ \boldsymbol{a}_6^0 &= \begin{bmatrix} c_1(c_{23}c_4s_5 + s_{23}c_5) + s_1s_4s_5 \\ s_1(c_{23}c_4s_5 + s_{23}c_5) - c_1s_4s_5 \\ s_{23}c_4s_5 - c_{23}c_5 \end{bmatrix}. \end{aligned} \tag{2.71}$$

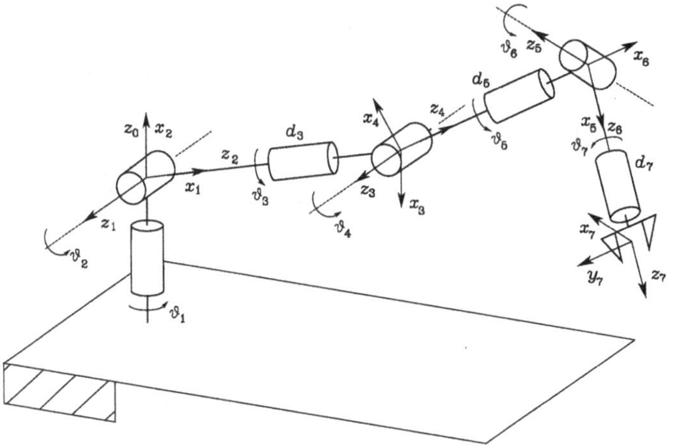

Fig. 2.27. DLR manipulator

By setting $d_6 = 0$, the position of the wrist axes intersection is obtained. In that case, the vector $\boldsymbol{p}^0$ in (2.70) corresponds to the vector $\boldsymbol{p}_3^0$ for the sole anthropomorphic arm in (2.66), because d_4 gives the length of the forearm (a_3) and axis x_3 in Fig. 2.26 is rotated by $\pi/2$ with respect to axis x_3 in Fig. 2.23.

2.9.8 DLR Manipulator

Consider the DLR manipulator, whose development is at the basis of the realization of the robot in Fig. 1.30; it is characterized by seven DOFs and as such it is inherently redundant. This manipulator has two possible configurations for the outer three joints (wrist). With reference to a spherical wrist similar to that introduced in Sect. 2.9.5, the resulting kinematic structure is illustrated in Fig. 2.27, where the frames attached to the links are evidenced.

As in the case of the spherical arm, notice that the origin of Frame 0 has been chosen so as to zero d_1. The DH parameters are specified in Table 2.7.

Table 2.7. DH parameters for the DLR manipulator

Link	a_i	α_i	d_i	ϑ_i
1	0	$\pi/2$	0	ϑ_1
2	0	$\pi/2$	0	ϑ_2
3	0	$\pi/2$	d_3	ϑ_3
4	0	$\pi/2$	0	ϑ_4
5	0	$\pi/2$	d_5	ϑ_5
6	0	$\pi/2$	0	ϑ_6
7	0	0	d_7	ϑ_7

The generic homogeneous transformation matrix defined in (2.52) is ($\alpha_i = \pi/2$)

$$\boldsymbol{A}_i^{i-1} = \begin{bmatrix} c_i & 0 & s_i & 0 \\ s_i & 0 & -c_i & 0 \\ 0 & 1 & 0 & d_i \\ 0 & 0 & 0 & 1 \end{bmatrix} \qquad i = 1, \dots, 6 \tag{2.72}$$

while, since $\alpha_7 = 0$, it is

$$\boldsymbol{A}_7^6 = \begin{bmatrix} c_7 & -s_7 & 0 & 0 \\ s_7 & c_7 & 0 & 0 \\ 0 & 0 & 1 & d_7 \\ 0 & 0 & 0 & 1 \end{bmatrix}. \tag{2.73}$$

The direct kinematics function, computed as in (2.50), leads to the following expressions for the end-effector frame

$$\boldsymbol{p}_7^0 = \begin{bmatrix} d_3 x_{d_3} + d_5 x_{d_5} + d_7 x_{d_7} \\ d_3 y_{d_3} + d_5 y_{d_5} + d_7 y_{d_7} \\ d_3 z_{d_3} + d_5 z_{d_5} + d_7 z_{d_7} \end{bmatrix} \tag{2.74}$$

with

$$\begin{aligned} x_{d_3} &= c_1 s_2 \\ x_{d_5} &= c_1(c_2 c_3 s_4 - s_2 c_4) + s_1 s_3 s_4 \\ x_{d_7} &= c_1(c_2 k_1 + s_2 k_2) + s_1 k_3 \\ y_{d_3} &= s_1 s_2 \\ y_{d_5} &= s_1(c_2 c_3 s_4 - s_2 c_4) - c_1 s_3 s_4 \\ y_{d_7} &= s_1(c_2 k_1 + s_2 k_2) - c_1 k_3 \\ z_{d_3} &= -c_2 \\ z_{d_5} &= c_2 c_4 + s_2 c_3 s_4 \\ z_{d_7} &= s_2(c_3(c_4 c_5 s_6 - s_4 c_6) + s_3 s_5 s_6) - c_2 k_2, \end{aligned}$$

where

$$\begin{aligned} k_1 &= c_3(c_4 c_5 s_6 - s_4 c_6) + s_3 s_5 s_6 \\ k_2 &= s_4 c_5 s_6 + c_4 c_6 \\ k_3 &= s_3(c_4 c_5 s_6 - s_4 c_6) - c_3 s_5 s_6. \end{aligned}$$

Furthermore, the end-effector frame orientation can be derived as

$$\boldsymbol{n}_7^0 = \begin{bmatrix} ((x_a c_5 + x_c s_5) c_6 + x_b s_6) c_7 + (x_a s_5 - x_c c_5) s_7 \\ ((y_a c_5 + y_c s_5) c_6 + y_b s_6) c_7 + (y_a s_5 - y_c c_5) s_7 \\ (z_a c_6 + z_c s_6) c_7 + z_b s_7 \end{bmatrix}$$

$$
\boldsymbol{s}_7^0 = \begin{bmatrix} -((x_a c_5 + x_c s_5)c_6 + x_b s_6)s_7 + (x_a s_5 - x_c c_5)c_7 \\ -((y_a c_5 + y_c s_5)c_6 + y_b s_6)s_7 + (y_a s_5 - y_c c_5)c_7 \\ -(z_a c_6 + z_c s_6)s_7 + z_b c_7 \end{bmatrix} \tag{2.75}
$$

$$
\boldsymbol{a}_7^0 = \begin{bmatrix} (x_a c_5 + x_c s_5)s_6 - x_b c_6 \\ (y_a c_5 + y_c s_5)s_6 - y_b c_6 \\ z_a s_6 - z_c c_6 \end{bmatrix},
$$

where

$$
\begin{aligned}
x_a &= (c_1 c_2 c_3 + s_1 s_3)c_4 + c_1 s_2 s_4 \\
x_b &= (c_1 c_2 c_3 + s_1 s_3)s_4 - c_1 s_2 c_4 \\
x_c &= c_1 c_2 s_3 - s_1 c_3 \\
y_a &= (s_1 c_2 c_3 - c_1 s_3)c_4 + s_1 s_2 s_4 \\
y_b &= (s_1 c_2 c_3 - c_1 s_3)s_4 - s_1 s_2 c_4 \\
y_c &= s_1 c_2 s_3 + c_1 c_3 \\
z_a &= (s_2 c_3 c_4 - c_2 s_4)c_5 + s_2 s_3 s_5 \\
z_b &= (s_2 c_3 s_4 + c_2 c_4)s_5 - s_2 s_3 c_5 \\
z_c &= s_2 c_3 s_4 + c_2 c_4.
\end{aligned} \tag{2.76}
$$

As in the case of the anthropomorphic arm with spherical wrist, it occurs that Frame 4 cannot coincide with the base frame of the wrist.

Finally, consider the possibility to mount a different type of spherical wrist, where Joint 7 is so that $\alpha_7 = \pi/2$. In such a case, the computation of the direct kinematics function changes, since the seventh row of the kinematic parameters table changes. In particular, notice that, since $d_7 = 0$, $a_7 \neq 0$, then

$$
\boldsymbol{A}_7^6 = \begin{bmatrix} c_7 & 0 & s_7 & a_7 c_7 \\ s_7 & 0 & -c_7 & a_7 s_7 \\ 0 & 0 & 1 & 0 \\ 0 & 0 & 0 & 1 \end{bmatrix}. \tag{2.77}
$$

It follows, however, that Frame 7 does not coincide with the end-effector frame, as already discussed for the three-link planar arm, since the approach unit vector $\boldsymbol{a}_7^0$ is aligned with x_7.

2.9.9 Humanoid Manipulator

The term humanoid refers to a robot showing a kinematic structure similar to that of the human body. It is commonly thought that the most relevant feature of humanoid robots is biped locomotion. However, in detail, a humanoid manipulator refers to an articulated structure with a kinematics analogous to

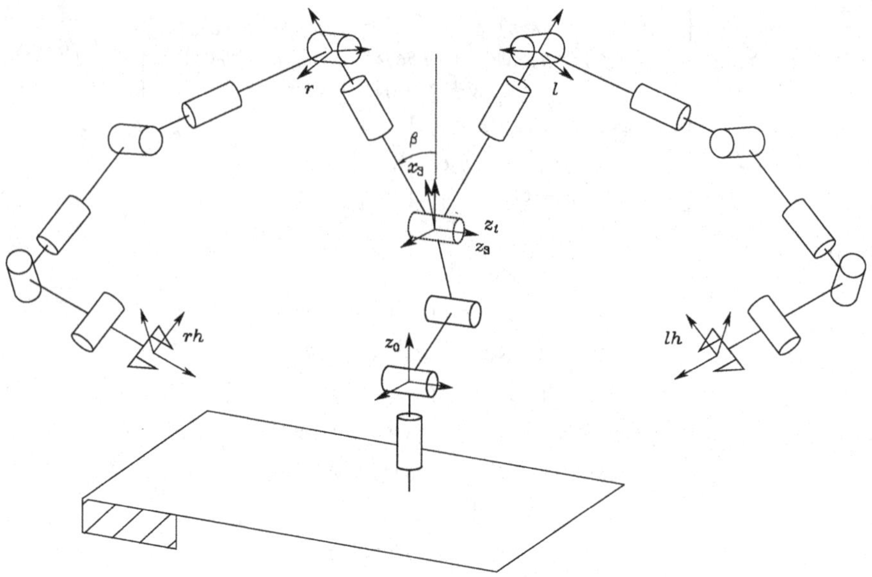

Fig. 2.28. Humanoid manipulator

that of the human body upper part: torso, arms, end-effectors similar to human hands and a 'head' which, eventually, includes an artificial vision system — see Chap. 10.

For the humanoid manipulator in Fig. 1.33, it is worth noticing the presence of two end-effectors (where the 'hands' are mounted), while the arms consist of two DLR manipulators, introduced in the previous section, each with seven DOFs. In particular, consider the configuration where the last joint is so that $\alpha_7 = \pi/2$.

To simplify, the kinematic structure allowing the articulation of the robot's head in Fig. 1.33. The torso can be modelled as an anthropomorphic arm (three DOFs), for a total of seventeen DOFs.

Further, a connecting device exists between the end-effector of the anthropomorphic torso and the base frames of the two manipulators. Such device permits keeping the 'chest' of the humanoid manipulator always orthogonal to the ground. With reference to Fig. 2.28, this device is represented by a further joint, located at the end of the torso. Hence, the corresponding parameter ϑ_4 does not constitute a DOF, yet it varies so as to compensate Joints 2 and 3 rotations of the anthropomorphic torso.

To compute the direct kinematics function, it is possible to resort to a DH parameters table for each of the two tree kinematic structures, which can be identified from the base of the manipulator to each of the two end-effectors. Similarly to the case of mounting a spherical wrist onto an anthropomorphic arm, this implies the change of some rows of the transformation matrices of those manipulators, described in the previous sections, constituting the torso and the arms.

Alternatively, it is possible to consider intermediate transformation matrices between the relevant structures. In detail, as illustrated in Fig. 2.28, if t denotes the frame attached to the torso, r and l the base frames, respectively, of the right arm and the left arm, and rh and lh the frames attached to the two hands (end-effectors), it is possible to compute for the right arm and the left arm, respectively:

$$\boldsymbol{T}_{rh}^{0} = \boldsymbol{T}_{3}^{0}\,\boldsymbol{T}_{t}^{3}\,\boldsymbol{T}_{r}^{t}\boldsymbol{T}_{rh}^{r} \tag{2.78}$$

$$\boldsymbol{T}_{lh}^{0} = \boldsymbol{T}_{3}^{0}\,\boldsymbol{T}_{t}^{3}\,\boldsymbol{T}_{l}^{t}\boldsymbol{T}_{lh}^{l} \tag{2.79}$$

where the matrix $\boldsymbol{T}_t^3$ describes the transformation imposed by the motion of Joint 4 (dashed line in Fig. 2.28), located at the end-effector of the torso. Frame 4 coincides with Frame t in Fig. 2.27. In view of the property of parameter ϑ_4, it is $\vartheta_4 = -\vartheta_2 - \vartheta_3$, and thus

$$\boldsymbol{T}_t^3 = \begin{bmatrix} c_{23} & s_{23} & 0 & 0 \\ -s_{23} & c_{23} & 0 & 0 \\ 0 & 0 & 1 & 0 \\ 0 & 0 & 0 & 1 \end{bmatrix}.$$

The matrix $\boldsymbol{T}_3^0$ is given by (2.66), whereas the matrices $\boldsymbol{T}_r^t$ and $\boldsymbol{T}_l^t$ relating the torso end-effector frame to the base frames of the two manipulators have constant values. With reference to Fig. 2.28, the elements of these matrices depend on the angle β and on the distances between the origin of Frame t and the origins of Frames r and l. Finally, the expressions of the matrices $\boldsymbol{T}_{rh}^r$ and $\boldsymbol{T}_{lh}^l$ must be computed by considering the change in the seventh row of the DH parameters table of the DLR manipulator, so as to account for the different kinematic structure of the wrist (see Problem 2.14).

2.10 Joint Space and Operational Space

As described in the previous sections, the direct kinematics equation of a manipulator allows the position and orientation of the end-effector frame to be expressed as a function of the joint variables with respect to the base frame.

If a task is to be specified for the end-effector, it is necessary to assign the end-effector position and orientation, eventually as a function of time (trajectory). This is quite easy for the position. On the other hand, specifying the orientation through the unit vector triplet $(\boldsymbol{n}_e, \boldsymbol{s}_e, \boldsymbol{a}_e)$[11] is quite difficult, since their nine components must be guaranteed to satisfy the orthonormality constraints imposed by (2.4) at each time instant. This problem will be resumed in Chap. 4.

[11] To simplify, the indication of the reference frame in the superscript is omitted.

The problem of describing end-effector orientation admits a natural solution if one of the above minimal representations is adopted. In this case, indeed, a motion trajectory can be assigned to the set of angles chosen to represent orientation.

Therefore, the position can be given by a minimal number of coordinates with regard to the geometry of the structure, and the orientation can be specified in terms of a minimal representation (Euler angles) describing the rotation of the end-effector frame with respect to the base frame. In this way, it is possible to describe the end-effector pose by means of the $(m \times 1)$ vector, with $m \leq n$,

$$\boldsymbol{x}_e = \begin{bmatrix} \boldsymbol{p}_e \\ \boldsymbol{\phi}_e \end{bmatrix} \tag{2.80}$$

where $\boldsymbol{p}_e$ describes the end-effector position and $\boldsymbol{\phi}_e$ its orientation.

This representation of position and orientation allows the description of an end-effector task in terms of a number of inherently independent parameters. The vector $\boldsymbol{x}_e$ is defined in the space in which the manipulator task is specified; hence, this space is typically called *operational space*. On the other hand, the *joint space* (configuration space) denotes the space in which the $(n \times 1)$ vector of joint variables

$$\boldsymbol{q} = \begin{bmatrix} q_1 \\ \vdots \\ q_n \end{bmatrix}, \tag{2.81}$$

is defined; it is $q_i = \vartheta_i$ for a revolute joint and $q_i = d_i$ for a prismatic joint. Accounting for the dependence of position and orientation from the joint variables, the direct kinematics equation can be written in a form other than (2.50), i.e.,

$$\boldsymbol{x}_e = \boldsymbol{k}(\boldsymbol{q}). \tag{2.82}$$

The $(m \times 1)$ vector function $\boldsymbol{k}(\cdot)$ — nonlinear in general — allows computation of the operational space variables from the knowledge of the joint space variables.

It is worth noticing that the dependence of the orientation components of the function $\boldsymbol{k}(\boldsymbol{q})$ in (2.82) on the joint variables is not easy to express except for simple cases. In fact, in the most general case of a six-dimensional operational space ($m = 6$), the computation of the three components of the function $\boldsymbol{\phi}_e(\boldsymbol{q})$ cannot be performed in closed form but goes through the computation of the elements of the rotation matrix, i.e., $\boldsymbol{n}_e(\boldsymbol{q})$, $\boldsymbol{s}_e(\boldsymbol{q})$, $\boldsymbol{a}_e(\boldsymbol{q})$. The equations that allow the determination of the Euler angles from the triplet of unit vectors $\boldsymbol{n}_e$, $\boldsymbol{s}_e$, $\boldsymbol{a}_e$ were given in Sect. 2.4.

Example 2.5

Consider again the three-link planar arm in Fig. 2.20. The geometry of the structure suggests that the end-effector position is determined by the two coordinates p_x and p_y, while its orientation is determined by the angle ϕ formed by the end-effector with the axis x_0. Expressing these operational variables as a function of the joint variables, the two position coordinates are given by the first two elements of the fourth column of the homogeneous transformation matrix (2.63), while the orientation angle is simply given by the sum of joint variables. In sum, the direct kinematics equation can be written in the form

$$\boldsymbol{x}_e = \begin{bmatrix} p_x \\ p_y \\ \phi \end{bmatrix} = \boldsymbol{k}(\boldsymbol{q}) = \begin{bmatrix} a_1 c_1 + a_2 c_{12} + a_3 c_{123} \\ a_1 s_1 + a_2 s_{12} + a_3 s_{123} \\ \vartheta_1 + \vartheta_2 + \vartheta_3 \end{bmatrix}. \tag{2.83}$$

This expression shows that three joint space variables allow specification of at most three independent operational space variables. On the other hand, if orientation is of no concern, it is $\boldsymbol{x}_e = [\,p_x \quad p_y\,]^T$ and there is *kinematic redundancy* of DOFs with respect to a pure positioning end-effector task; this concept will be dealt with in detail afterwards.

2.10.1 Workspace

With reference to the operational space, an index of robot performance is the so-called *workspace*; this is the region described by the origin of the end-effector frame when all the manipulator joints execute all possible motions. It is often customary to distinguish between *reachable* workspace and *dexterous* workspace. The latter is the region that the origin of the end-effector frame can describe while attaining different orientations, while the former is the region that the origin of the end-effector frame can reach with at least one orientation. Obviously, the dexterous workspace is a subspace of the reachable workspace. A manipulator with less than six DOFs cannot take any arbitrary position and orientation in space.

The workspace is characterized by the manipulator geometry and the mechanical joint limits. For an n-DOF manipulator, the reachable workspace is the geometric locus of the points that can be achieved by considering the direct kinematics equation for the sole position part, i.e.,

$$\boldsymbol{p}_e = \boldsymbol{p}_e(\boldsymbol{q}) \qquad q_{im} \le q_i \le q_{iM} \quad i = 1, \dots, n,$$

where q_{im} (q_{iM}) denotes the minimum (maximum) limit at Joint i. This volume is finite, closed, connected — $\boldsymbol{p}_e(\boldsymbol{q})$ is a continuous function — and thus is defined by its bordering surface. Since the joints are revolute or prismatic, it is easy to recognize that this surface is constituted by surface elements of planar, spherical, toroidal and cylindrical type. The manipulator workspace

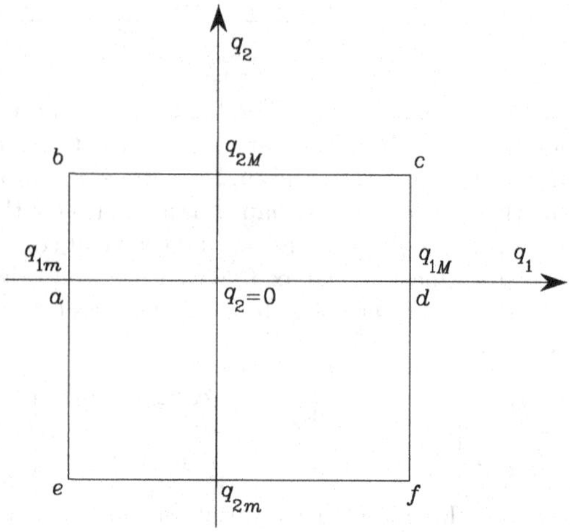

Fig. 2.29. Region of admissible configurations for a two-link arm

(without end-effector) is reported in the data sheet given by the robot manufacturer in terms of a top view and a side view. It represents a basic element to evaluate robot performance for a desired application.

Example 2.6

Consider the simple two-link planar arm. If the mechanical joint limits are known, the arm can attain all the joint space configurations corresponding to the points in the rectangle in Fig. 2.29.

The reachable workspace can be derived via a graphical construction of the image of the rectangle perimeter in the plane of the arm. To this purpose, it is worth considering the images of the segments ab, bc, cd, da, ae, ef, fd. Along the segments ab, bc, cd, ae, ef, fd a loss of mobility occurs due to a joint limit; a loss of mobility occurs also along the segment ad because the arm and forearm are aligned.[12] Further, a change of the arm posture occurs at points a and d: for $q_2 > 0$ the *elbow-down* posture is obtained, while for $q_2 < 0$ the arm is in the *elbow-up* posture.

In the plane of the arm, start drawing the arm in configuration A corresponding to q_{1m} and $q_2 = 0$ (a); then, the segment ab describing motion from $q_2 = 0$ to q_{2M} generates the arc AB; the subsequent arcs BC, CD, DA, AE, EF, FD are generated in a similar way (Fig. 2.30). The external contour of the area $CDAEFHC$ delimits the requested workspace. Further, the area $BCDAB$ is relative to elbow-down postures while the area $DAEFD$ is relative to elbow-up postures; hence, the points in the area $BADHB$ are reachable by the end-effector with both postures.

---

[12] In the following chapter, it will be seen that this configuration characterizes a kinematic *singularity* of the arm.

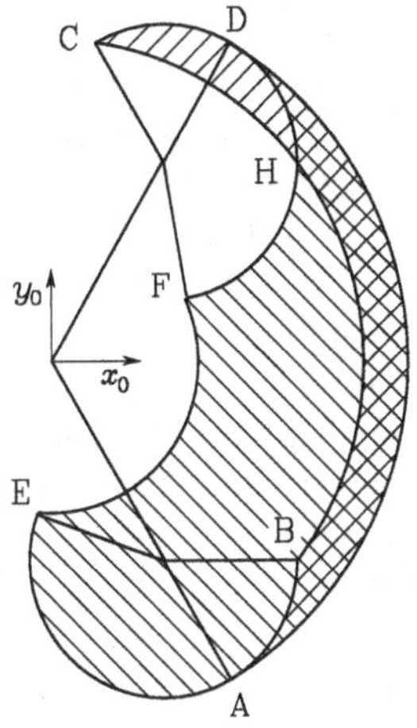

Fig. 2.30. Workspace of a two-link planar arm

In a real manipulator, for a given set of joint variables, the actual values of the operational space variables deviate from those computed via direct kinematics. The direct kinematics equation has indeed a dependence from the DH parameters which is not explicit in (2.82). If the mechanical dimensions of the structure differ from the corresponding parameter of the table because of mechanical tolerances, a deviation arises between the position reached in the assigned posture and the position computed via direct kinematics. Such a deviation is defined *accuracy*; this parameter attains typical values below one millimeter and depends on the structure as well as on manipulator dimensions. Accuracy varies with the end-effector position in the workspace and it is a relevant parameter when robot programming oriented environments are adopted, as will be seen in the last chapter.

Another parameter that is usually listed in the performance data sheet of an industrial robot is *repeatability* which gives a measure of the manipulator's ability to return to a previously reached position; this parameter is relevant for programming an industrial robot by the teaching–by–showing technique which will be presented in Chap. 6. Repeatability depends not only on the characteristics of the mechanical structure but also on the transducers and controller; it is expressed in metric units and is typically smaller than accuracy. For instance, for a manipulator with a maximum reach of 1.5 m, accuracy varies from 0.2 to 1 mm in the workspace, while repeatability varies from 0.02 to 0.2 mm.

2.10.2 Kinematic Redundancy

A manipulator is termed *kinematically redundant* when it has a number of DOFs which is greater than the number of variables that are necessary to describe a given task. With reference to the above-defined spaces, a manipulator is intrinsically redundant when the dimension of the operational space is smaller than the dimension of the joint space ($m < n$). Redundancy is, anyhow, a concept *relative* to the task assigned to the manipulator; a manipulator can be redundant with respect to a task and nonredundant with respect to another. Even in the case of $m = n$, a manipulator can be functionally redundant when only a number of r components of operational space are of concern for the specific task, with $r < m$.

Consider again the three-DOF planar arm of Sect. 2.9.1. If only the end-effector position (in the plane) is specified, that structure presents a functional redundancy ($n = m = 3$, $r = 2$); this is lost when also the end-effector orientation in the plane is specified ($n = m = r = 3$). On the other hand, a four-DOF planar arm is intrinsically redundant ($n = 4$, $m = 3$).

Yet, take the typical industrial robot with six DOFs; such manipulator is not intrinsically redundant ($n = m = 6$), but it can become functionally redundant with regard to the task to execute. Thus, for instance, in a laser-cutting task a functional redundancy will occur since the end-effector rotation about the approach direction is irrelevant to completion of the task ($r = 5$).

At this point, a question should arise spontaneously: Why to intentionally utilize a redundant manipulator? The answer is to recognize that redundancy can provide the manipulator with dexterity and versatility in its motion. The typical example is constituted by the human arm that has *seven* DOFs: three in the shoulder, one in the elbow and three in the wrist, without considering the DOFs in the fingers. This manipulator is intrinsically redundant; in fact, if the base and the hand position and orientation are both fixed — requiring six DOFs — the elbow can be moved, thanks to the additional available DOF. Then, for instance, it is possible to avoid obstacles in the workspace. Further, if a joint of a redundant manipulator reaches its mechanical limit, there might be other joints that allow execution of the prescribed end-effector motion.

A formal treatment of redundancy will be presented in the following chapter.

2.11 Kinematic Calibration

The Denavit–Hartenberg parameters for direct kinematics need to be computed as precisely as possible in order to improve manipulator accuracy. *Kinematic calibration* techniques are devoted to finding accurate estimates of DH parameters from a series of measurements on the manipulator's end-effector pose. Hence, they do not allow direct measurement of the geometric parameters of the structure.

Consider the direct kinematics equation in (2.82) which can be rewritten by emphasizing the dependence of the operational space variables on the fixed DH parameters, besides the joint variables. Let $\boldsymbol{a} = [a_1 \quad \dots \quad a_n]^T$, $\boldsymbol{\alpha} =$

$[\alpha_1 \;\; \ldots \;\; \alpha_n]^T$, $\boldsymbol{d} = [d_1 \;\; \ldots \;\; d_n]^T$, and $\boldsymbol{\vartheta} = [\theta_1 \;\; \ldots \;\; \theta_n]^T$ denote the vectors of DH parameters for the whole structure; then (2.82) becomes

$$\boldsymbol{x}_e = \boldsymbol{k}(\boldsymbol{a}, \boldsymbol{\alpha}, \boldsymbol{d}, \boldsymbol{\vartheta}). \tag{2.84}$$

The manipulator's end-effector pose should be measured with high precision for the effectiveness of the kinematic calibration procedure. To this purpose a mechanical apparatus can be used that allows the end-effector to be constrained at given poses with a priori known precision. Alternatively, direct measurement systems of object position and orientation in the Cartesian space can be used which employ triangulation techniques.

Let $\boldsymbol{x}_m$ be the measured pose and $\boldsymbol{x}_n$ the nominal pose that can be computed via (2.84) with the nominal values of the parameters $\boldsymbol{a}$, $\boldsymbol{\alpha}$, $\boldsymbol{d}$, $\boldsymbol{\vartheta}$. The nominal values of the fixed parameters are set equal to the design data of the mechanical structure, whereas the nominal values of the joint variables are set equal to the data provided by the position transducers at the given manipulator posture. The deviation $\Delta\boldsymbol{x} = \boldsymbol{x}_m - \boldsymbol{x}_n$ gives a measure of accuracy at the given posture. On the assumption of small deviations, at first approximation, it is possible to derive the following relation from (2.84):

$$\Delta\boldsymbol{x} = \frac{\partial \boldsymbol{k}}{\partial \boldsymbol{a}}\Delta\boldsymbol{a} + \frac{\partial \boldsymbol{k}}{\partial \boldsymbol{\alpha}}\Delta\boldsymbol{\alpha} + \frac{\partial \boldsymbol{k}}{\partial \boldsymbol{d}}\Delta\boldsymbol{d} + \frac{\partial \boldsymbol{k}}{\partial \boldsymbol{\vartheta}}\Delta\boldsymbol{\vartheta} \tag{2.85}$$

where $\Delta\boldsymbol{a}$, $\Delta\boldsymbol{\alpha}$, $\Delta\boldsymbol{d}$, $\Delta\boldsymbol{\vartheta}$ denote the deviations between the values of the parameters of the real structure and the nominal ones. Moreover, $\partial\boldsymbol{k}/\partial\boldsymbol{a}$, $\partial\boldsymbol{k}/\partial\boldsymbol{\alpha}$, $\partial\boldsymbol{k}/\partial\boldsymbol{d}$, $\partial\boldsymbol{k}/\partial\boldsymbol{\vartheta}$ denote the $(m \times n)$ matrices whose elements are the partial derivatives of the components of the direct kinematics function with respect to the single parameters.[13]

Group the parameters in the $(4n \times 1)$ vector $\boldsymbol{\zeta} = [\boldsymbol{a}^T \;\; \boldsymbol{\alpha}^T \;\; \boldsymbol{d}^T \;\; \boldsymbol{\vartheta}^T]^T$. Let $\Delta\boldsymbol{\zeta} = \boldsymbol{\zeta}_m - \boldsymbol{\zeta}_n$ denote the parameter variations with respect to the nominal values, and $\boldsymbol{\Phi} = [\partial\boldsymbol{k}/\partial\boldsymbol{a} \;\; \partial\boldsymbol{k}/\partial\boldsymbol{\alpha} \;\; \partial\boldsymbol{k}/\partial\boldsymbol{d} \;\; \partial\boldsymbol{k}/\partial\boldsymbol{\vartheta}]$ the $(m \times 4n)$ *kinematic calibration matrix* computed for the nominal values of the parameters $\boldsymbol{\zeta}_n$. Then (2.85) can be compactly rewritten as

$$\Delta\boldsymbol{x} = \boldsymbol{\Phi}(\boldsymbol{\zeta}_n)\Delta\boldsymbol{\zeta}. \tag{2.86}$$

It is desired to compute $\Delta\boldsymbol{\zeta}$ starting from the knowledge of $\boldsymbol{\zeta}_n, \boldsymbol{x}_n$ and the measurement of $\boldsymbol{x}_m$. Since (2.86) constitutes a system of m equations into $4n$ unknowns with $m < 4n$, a sufficient number of end-effector pose measurements has to be performed so as to obtain a system of at least $4n$ equations. Therefore, if measurements are made for a number of l poses, (2.86) yields

$$\Delta\bar{\boldsymbol{x}} = \begin{bmatrix} \Delta\boldsymbol{x}_1 \\ \vdots \\ \Delta\boldsymbol{x}_l \end{bmatrix} = \begin{bmatrix} \boldsymbol{\Phi}_1 \\ \vdots \\ \boldsymbol{\Phi}_l \end{bmatrix} \Delta\boldsymbol{\zeta} = \bar{\boldsymbol{\Phi}}\Delta\boldsymbol{\zeta}. \tag{2.87}$$

[13] These matrices are the Jacobians of the transformations between the parameter space and the operational space.

As regards the nominal values of the parameters needed for the computation of the matrices $\boldsymbol{\Phi}_i$, it should be observed that the geometric parameters are constant whereas the joint variables depend on the manipulator configuration at pose i.

In order to avoid ill-conditioning of matrix $\bar{\boldsymbol{\Phi}}$, it is advisable to choose l so that $lm \gg 4n$ and then solve (2.87) with a least-squares technique; in this case the solution is of the form

$$\Delta\boldsymbol{\zeta} = (\bar{\boldsymbol{\Phi}}^T\bar{\boldsymbol{\Phi}})^{-1}\bar{\boldsymbol{\Phi}}^T\Delta\bar{\boldsymbol{x}} \tag{2.88}$$

where $(\bar{\boldsymbol{\Phi}}^T\bar{\boldsymbol{\Phi}})^{-1}\bar{\boldsymbol{\Phi}}^T$ is the *left pseudo-inverse* matrix of $\bar{\boldsymbol{\Phi}}$.[14] By computing $\bar{\boldsymbol{\Phi}}$ with the nominal values of the parameters $\boldsymbol{\zeta}_n$, the first parameter *estimate* is given by

$$\boldsymbol{\zeta}' = \boldsymbol{\zeta}_n + \Delta\boldsymbol{\zeta}. \tag{2.89}$$

This is a nonlinear parameter estimate problem and, as such, the procedure should be iterated until $\Delta\boldsymbol{\zeta}$ converges within a given threshold. At each iteration, the calibration matrix $\bar{\boldsymbol{\Phi}}$ is to be updated with the parameter estimates $\boldsymbol{\zeta}'$ obtained via (2.89) at the previous iteration. In a similar manner, the deviation $\Delta\bar{\boldsymbol{x}}$ is to be computed as the difference between the measured values for the l end-effector poses and the corresponding poses computed by the direct kinematics function with the values of the parameters at the previous iteration. As a result of the kinematic calibration procedure, more accurate estimates of the real manipulator geometric parameters as well as possible corrections to make on the joint transducers measurements are obtained.

Kinematic calibration is an operation that is performed by the robot manufacturer to guarantee the accuracy reported in the data sheet. There is another kind of calibration that is performed by the robot user which is needed for the measurement system *start-up* to guarantee that the position transducers data are consistent with the attained manipulator posture. For instance, in the case of incremental (nonabsolute) position transducers, such calibration consists of taking the mechanical structure into a given reference posture (*home*) and initializing the position transducers with the values at that posture.

2.12 Inverse Kinematics Problem

The direct kinematics equation, either in the form (2.50) or in the form (2.82), establishes the functional relationship between the joint variables and the end-effector position and orientation. The *inverse kinematics problem* consists of the determination of the joint variables corresponding to a given end-effector position and orientation. The solution to this problem is of fundamental importance in order to transform the motion specifications, assigned to the end-effector in the operational space, into the corresponding joint space motions that allow execution of the desired motion.

[14] See Sect. A.7 for the definition of the pseudo-inverse of a matrix.

As regards the direct kinematics equation in (2.50), the end-effector position and rotation matrix are computed in a unique manner, once the joint variables are known[15]. On the other hand, the inverse kinematics problem is much more complex for the following reasons:

- The equations to solve are in general nonlinear, and thus it is not always possible to find a *closed-form solution*.
- *Multiple solutions* may exist.
- *Infinite solutions* may exist, e.g., in the case of a kinematically redundant manipulator.
- There might be no *admissible* solutions, in view of the manipulator kinematic structure.

The existence of solutions is guaranteed only if the given end-effector position and orientation belong to the manipulator dexterous workspace.

On the other hand, the problem of multiple solutions depends not only on the number of DOFs but also on the number of non-null DH parameters; in general, the greater the number of non-null parameters, the greater the number of admissible solutions. For a six-DOF manipulator without mechanical joint limits, there are in general up to 16 admissible solutions. Such occurrence demands some criterion to choose among admissible solutions (e.g., the elbow-up/elbow-down case of Example 2.6). The existence of mechanical joint limits may eventually reduce the number of admissible multiple solutions for the real structure.

Computation of closed-form solutions requires either *algebraic intuition* to find those significant equations containing the unknowns or *geometric intuition* to find those significant points on the structure with respect to which it is convenient to express position and/or orientation as a function of a reduced number of unknowns. The following examples will point out the ability required to an inverse kinematics problem solver. On the other hand, in all those cases when there are no — or it is difficult to find — closed-form solutions, it might be appropriate to resort to *numerical solution techniques*; these clearly have the advantage of being applicable to any kinematic structure, but in general they do not allow computation of all admissible solutions. In the following chapter, it will be shown how suitable algorithms utilizing the manipulator Jacobian can be employed to solve the inverse kinematics problem.

2.12.1 Solution of Three-link Planar Arm

Consider the arm shown in Fig. 2.20 whose direct kinematics was given in (2.63). It is desired to find the joint variables ϑ_1, ϑ_2, ϑ_3 corresponding to a given end-effector position and orientation.

[15] In general, this cannot be said for (2.82) too, since the Euler angles are not uniquely defined.

As already pointed out, it is convenient to specify position and orientation in terms of a minimal number of parameters: the two coordinates p_x, p_y and the angle ϕ with axis x_0, in this case. Hence, it is possible to refer to the direct kinematics equation in the form (2.83).

A first *algebraic solution* technique is illustrated below. Having specified the orientation, the relation

$$\phi = \vartheta_1 + \vartheta_2 + \vartheta_3 \tag{2.90}$$

is one of the equations of the system to solve[16]. From (2.63) the following equations can be obtained:

$$p_{Wx} = p_x - a_3 c_\phi = a_1 c_1 + a_2 c_{12} \tag{2.91}$$

$$p_{Wy} = p_y - a_3 s_\phi = a_1 s_1 + a_2 s_{12} \tag{2.92}$$

which describe the position of point W, i.e., the origin of Frame 2; this depends only on the first two angles ϑ_1 and ϑ_2. Squaring and summing (2.91), (2.92) yields

$$p_{Wx}^2 + p_{Wy}^2 = a_1^2 + a_2^2 + 2a_1 a_2 c_2$$

from which

$$c_2 = \frac{p_{Wx}^2 + p_{Wy}^2 - a_1^2 - a_2^2}{2a_1 a_2}.$$

The existence of a solution obviously imposes that $-1 \le c_2 \le 1$, otherwise the given point would be outside the arm reachable workspace. Then, set

$$s_2 = \pm\sqrt{1 - c_2^2},$$

where the positive sign is relative to the elbow-down posture and the negative sign to the elbow-up posture. Hence, the angle ϑ_2 can be computed as

$$\vartheta_2 = \text{Atan2}(s_2, c_2).$$

Having determined ϑ_2, the angle ϑ_1 can be found as follows. Substituting ϑ_2 into (2.91), (2.92) yields an algebraic system of two equations in the two unknowns s_1 and c_1, whose solution is

$$s_1 = \frac{(a_1 + a_2 c_2) p_{Wy} - a_2 s_2 p_{Wx}}{p_{Wx}^2 + p_{Wy}^2}$$

$$c_1 = \frac{(a_1 + a_2 c_2) p_{Wx} + a_2 s_2 p_{Wy}}{p_{Wx}^2 + p_{Wy}^2}.$$

In analogy to the above, it is

$$\vartheta_1 = \text{Atan2}(s_1, c_1).$$

[16] If ϕ is not specified, then the arm is redundant and there exist infinite solutions to the inverse kinematics problem.

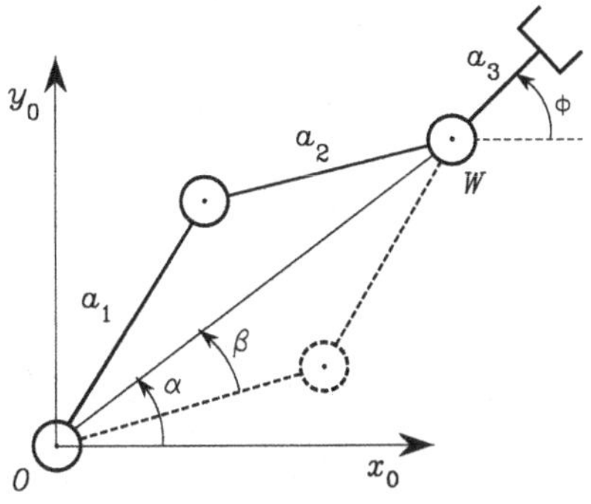

Fig. 2.31. Admissible postures for a two-link planar arm

In the case when $s_2 = 0$, it is obviously $\vartheta_2 = 0, \pi$; as will be shown in the following, in such a posture the manipulator is at a kinematic *singularity*. Yet, the angle ϑ_1 can be determined uniquely, unless $a_1 = a_2$ and it is required $p_{Wx} = p_{Wy} = 0$.

Finally, the angle ϑ_3 is found from (2.90) as

$$\vartheta_3 = \phi - \vartheta_1 - \vartheta_2.$$

An alternative *geometric solution* technique is presented below. As above, the orientation angle is given as in (2.90) and the coordinates of the origin of Frame 2 are computed as in (2.91), (2.92). The application of the cosine theorem to the triangle formed by links a_1, a_2 and the segment connecting points W and O gives

$$p_{Wx}^2 + p_{Wy}^2 = a_1^2 + a_2^2 - 2a_1a_2\cos(\pi - \vartheta_2);$$

the two admissible configurations of the triangle are shown in Fig. 2.31. Observing that $\cos(\pi - \vartheta_2) = -\cos\vartheta_2$ leads to

$$c_2 = \frac{p_{Wx}^2 + p_{Wy}^2 - a_1^2 - a_2^2}{2a_1a_2}.$$

For the existence of the triangle, it must be $\sqrt{p_{Wx}^2 + p_{Wy}^2} \leq a_1 + a_2$. This condition is not satisfied when the given point is outside the arm reachable workspace. Then, under the assumption of admissible solutions, it is

$$\vartheta_2 = \pm\cos^{-1}(c_2);$$

the elbow-up posture is obtained for $\vartheta_2 \in (-\pi, 0)$ while the elbow-down posture is obtained for $\vartheta_2 \in (0, \pi)$.

To find ϑ_1 consider the angles α and β in Fig. 2.31. Notice that the determination of α depends on the sign of p_{Wx} and p_{Wy}; then, it is necessary to compute α as

$$\alpha = \text{Atan2}(p_{Wy}, p_{Wx}).$$

To compute β, applying again the cosine theorem yields

$$c_\beta \sqrt{p_{Wx}^2 + p_{Wy}^2} = a_1 + a_2 c_2$$

and resorting to the expression of c_2 given above leads to

$$\beta = \cos^{-1}\left(\frac{p_{Wx}^2 + p_{Wy}^2 + a_1^2 - a_2^2}{2a_1\sqrt{p_{Wx}^2 + p_{Wy}^2}}\right)$$

with $\beta \in (0, \pi)$ so as to preserve the existence of triangles. Then, it is

$$\vartheta_1 = \alpha \pm \beta,$$

where the positive sign holds for $\vartheta_2 < 0$ and the negative sign for $\vartheta_2 > 0$. Finally, ϑ_3 is computed from (2.90).

It is worth noticing that, in view of the substantial equivalence between the two-link planar arm and the parallelogram arm, the above techniques can be formally applied to solve the inverse kinematics of the arm in Sect. 2.9.2.

2.12.2 Solution of Manipulators with Spherical Wrist

Most of the existing manipulators are kinematically simple, since they are typically formed by an arm, of the kind presented above, and a spherical wrist; see the manipulators in Sects. 2.9.6–2.9.8. This choice is partly motivated by the difficulty to find solutions to the inverse kinematics problem in the general case. In particular, a *six*-DOF kinematic structure has closed-form inverse kinematics solutions if:

- three consecutive revolute joint axes intersect at a common point, like for the spherical wrist;
- three consecutive revolute joint axes are parallel.

In any case, algebraic or geometric intuition is required to obtain closed-form solutions.

Inspired by the previous solution to a three-link planar arm, a suitable point along the structure can be found whose position can be expressed both as a function of the given end-effector position and orientation and as a function of a reduced number of joint variables. This is equivalent to articulating the inverse kinematics problem into two subproblems, since the solution for the *position* is *decoupled* from that for the *orientation*.

For a manipulator with spherical wrist, the natural choice is to locate such point W at the intersection of the three terminal revolute axes (Fig. 2.32). In fact, once the end-effector position and orientation are specified in terms of $\boldsymbol{p}_e$ and $\boldsymbol{R}_e = [\boldsymbol{n}_e \quad \boldsymbol{s}_e \quad \boldsymbol{a}_e]$, the wrist position can be found as

$$\boldsymbol{p}_W = \boldsymbol{p}_e - d_6 \boldsymbol{a}_e \tag{2.93}$$

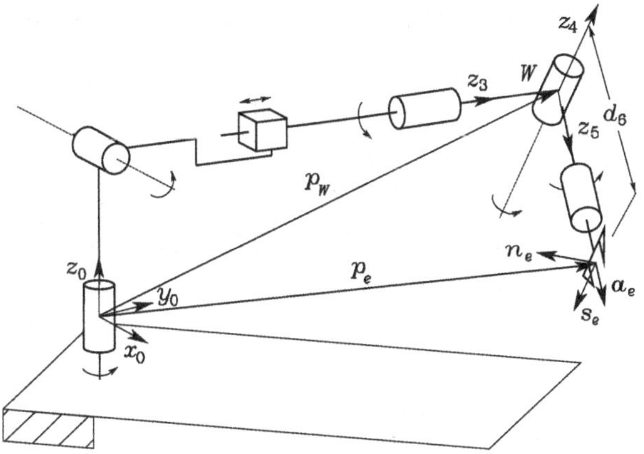

Fig. 2.32. Manipulator with spherical wrist

which is a function of the sole joint variables that determine the arm position[17]. Hence, in the case of a (nonredundant) three-DOF arm, the inverse kinematics can be solved according to the following steps:

- Compute the wrist position $\boldsymbol{p}_W(q_1, q_2, q_3)$ as in (2.93).
- Solve inverse kinematics for (q_1, q_2, q_3).
- Compute $\boldsymbol{R}_3^0(q_1, q_2, q_3)$.
- Compute $\boldsymbol{R}_6^3(\vartheta_4, \vartheta_5, \vartheta_6) = \boldsymbol{R}_3^{0T}\boldsymbol{R}$.
- Solve inverse kinematics for orientation $(\vartheta_4, \vartheta_5, \vartheta_6)$.

Therefore, on the basis of this kinematic decoupling, it is possible to solve the inverse kinematics for the arm separately from the inverse kinematics for the spherical wrist. Below are presented the solutions for two typical arms (spherical and anthropomorphic) as well as the solution for the spherical wrist.

2.12.3 Solution of Spherical Arm

Consider the spherical arm shown in Fig. 2.22, whose direct kinematics was given in (2.65). It is desired to find the joint variables ϑ_1, ϑ_2, d_3 corresponding to a given end-effector position $\boldsymbol{p}_W$.

In order to separate the variables on which $\boldsymbol{p}_W$ depends, it is convenient to express the position of $\boldsymbol{p}_W$ with respect to Frame 1; then, consider the matrix equation

$$(\boldsymbol{A}_1^0)^{-1}\boldsymbol{T}_3^0 = \boldsymbol{A}_2^1\boldsymbol{A}_3^2.$$

[17] Note that the same reasoning was implicitly adopted in Sect. 2.12.1 for the three-link planar arm; $\boldsymbol{p}_W$ described the one-DOF wrist position for the two-DOF arm obtained by considering only the first two links.

Equating the first three elements of the fourth columns of the matrices on both sides yields

$$\boldsymbol{p}_W^1 = \begin{bmatrix} p_{Wx}c_1 + p_{Wy}s_1 \\ -p_{Wz} \\ -p_{Wx}s_1 + p_{Wy}c_1 \end{bmatrix} = \begin{bmatrix} d_3 s_2 \\ -d_3 c_2 \\ d_2 \end{bmatrix} \tag{2.94}$$

which depends only on ϑ_2 and d_3. To solve this equation, set

$$t = \tan\frac{\vartheta_1}{2}$$

so that

$$c_1 = \frac{1-t^2}{1+t^2} \qquad s_1 = \frac{2t}{1+t^2}.$$

Substituting this equation in the third component on the left-hand side of (2.94) gives

$$(d_2 + p_{Wy})t^2 + 2p_{Wx}t + d_2 - p_{Wy} = 0,$$

whose solution is

$$t = \frac{-p_{Wx} \pm \sqrt{p_{Wx}^2 + p_{Wy}^2 - d_2^2}}{d_2 + p_{Wy}}.$$

The two solutions correspond to two different postures. Hence, it is

$$\vartheta_1 = 2\text{Atan2}\Big(-p_{Wx} \pm \sqrt{p_{Wx}^2 + p_{Wy}^2 - d_2^2},\, d_2 + p_{Wy}\Big).$$

Once ϑ_1 is known, squaring and summing the first two components of (2.94) yields

$$d_3 = \sqrt{(p_{Wx}c_1 + p_{Wy}s_1)^2 + p_{Wz}^2},$$

where only the solution with $d_3 \geq 0$ has been considered. Note that the same value of d_3 corresponds to both solutions for ϑ_1. Finally, if $d_3 \neq 0$, from the first two components of (2.94) it is

$$\frac{p_{Wx}c_1 + p_{Wy}s_1}{-p_{Wz}} = \frac{d_3 s_2}{-d_3 c_2},$$

from which

$$\vartheta_2 = \text{Atan2}(p_{Wx}c_1 + p_{Wy}s_1, p_{Wz}).$$

Notice that, if $d_3 = 0$, then ϑ_2 cannot be uniquely determined.

2.12.4 Solution of Anthropomorphic Arm

Consider the anthropomorphic arm shown in Fig. 2.23. It is desired to find the joint variables ϑ_1, ϑ_2, ϑ_3 corresponding to a given end-effector position $\boldsymbol{p}_W$. Notice that the direct kinematics for $\boldsymbol{p}_W$ is expressed by (2.66) which can be obtained from (2.70) by setting $d_6 = 0$, $d_4 = a_3$ and replacing ϑ_3 with the angle $\vartheta_3 + \pi/2$ because of the misalignment of the Frames 3 for the structures in Fig. 2.23 and in Fig. 2.26, respectively. Hence, it follows

$$p_{Wx} = c_1(a_2c_2 + a_3c_{23}) \tag{2.95}$$
$$p_{Wy} = s_1(a_2c_2 + a_3c_{23}) \tag{2.96}$$
$$p_{Wz} = a_2s_2 + a_3s_{23}. \tag{2.97}$$

Proceeding as in the case of the two-link planar arm, it is worth squaring and summing (2.95)–(2.97) yielding

$$p_{Wx}^2 + p_{Wy}^2 + p_{Wz}^2 = a_2^2 + a_3^2 + 2a_2a_3c_3$$

from which

$$c_3 = \frac{p_{Wx}^2 + p_{Wy}^2 + p_{Wz}^2 - a_2^2 - a_3^2}{2a_2a_3} \tag{2.98}$$

where the admissibility of the solution obviously requires that $-1 \le c_3 \le 1$, or equivalently $|a_2 - a_3| \le \sqrt{p_{Wx}^2 + p_{Wy}^2 + p_{Wz}^2} \le a_2 + a_3$, otherwise the wrist point is outside the reachable workspace of the manipulator. Hence it is

$$s_3 = \pm\sqrt{1 - c_3^2} \tag{2.99}$$

and thus

$$\vartheta_3 = \text{Atan2}(s_3, c_3)$$

giving the two solutions, according to the sign of s_3,

$$\vartheta_{3,\text{I}} \in [-\pi, \pi] \tag{2.100}$$
$$\vartheta_{3,\text{II}} = -\vartheta_{3,I}. \tag{2.101}$$

Having determined ϑ_3, it is possible to compute ϑ_2 as follows. Squaring and summing (2.95), (2.96) gives

$$p_{Wx}^2 + p_{Wy}^2 = (a_2c_2 + a_3c_{23})^2$$

from which

$$a_2c_2 + a_3c_{23} = \pm\sqrt{p_{Wx}^2 + p_{Wy}^2}. \tag{2.102}$$

The system of the two Eqs. (2.102), (2.97), for each of the solutions (2.100), (2.101), admits the solutions:

$$c_2 = \frac{\pm\sqrt{p_{Wx}^2 + p_{Wy}^2}(a_2 + a_3c_3) + p_{Wz}a_3s_3}{a_2^2 + a_3^2 + 2a_2a_3c_3} \tag{2.103}$$

$$s_2 = \frac{p_{Wz}(a_2 + a_3c_3) \mp \sqrt{p_{Wx}^2 + p_{Wy}^2}a_3s_3}{a_2^2 + a_3^2 + 2a_2a_3c_3}. \tag{2.104}$$

From (2.103), (2.104) it follows

$$\vartheta_2 = \text{Atan2}(s_2, c_2)$$

which gives the four solutions for ϑ_2, according to the sign of s_3 in (2.99):

$$\begin{aligned}\vartheta_{2,\text{I}} = \text{Atan2}\Big((a_2 + a_3c_3)p_{Wz} - a_3s_3^+\sqrt{p_{Wx}^2 + p_{Wy}^2},\\ (a_2 + a_3c_3)\sqrt{p_{Wx}^2 + p_{Wy}^2} + a_3s_3^+p_{Wz}\Big)\end{aligned} \tag{2.105}$$

$$\begin{aligned}\vartheta_{2,\text{II}} = \text{Atan2}\Big((a_2 + a_3c_3)p_{Wz} + a_3s_3^+\sqrt{p_{Wx}^2 + p_{Wy}^2},\\ -(a_2 + a_3c_3)\sqrt{p_{Wx}^2 + p_{Wy}^2} + a_3s_3^+p_{Wz}\Big)\end{aligned} \tag{2.106}$$

corresponding to $s_3^+ = \sqrt{1 - c_3^2}$, and

$$\begin{aligned}\vartheta_{2,\text{III}} = \text{Atan2}\Big((a_2 + a_3c_3)p_{Wz} - a_3s_3^-\sqrt{p_{Wx}^2 + p_{Wy}^2},\\ (a_2 + a_3c_3)\sqrt{p_{Wx}^2 + p_{Wy}^2} + a_3s_3^-p_{Wz}\Big)\end{aligned} \tag{2.107}$$

$$\begin{aligned}\vartheta_{2,\text{IV}} = \text{Atan2}\Big((a_2 + a_3c_3)p_{Wz} + a_3s_3^-\sqrt{p_{Wx}^2 + p_{Wy}^2},\\ -(a_2 + a_3c_3)\sqrt{p_{Wx}^2 + p_{Wy}^2} + a_3s_3^-p_{Wz}\Big)\end{aligned} \tag{2.108}$$

corresponding to $s_3^- = -\sqrt{1 - c_3^2}$.

Finally, to compute ϑ_1, it is sufficient to rewrite (2.95), (2.96), using (2.102), as

$$p_{Wx} = \pm c_1\sqrt{p_{Wx}^2 + p_{Wy}^2}$$

$$p_{Wy} = \pm s_1\sqrt{p_{Wx}^2 + p_{Wy}^2}$$

which, once solved, gives the two solutions:

$$\vartheta_{1,\text{I}} = \text{Atan2}(p_{Wy}, p_{Wx}) \tag{2.109}$$

$$\vartheta_{1,\text{II}} = \text{Atan2}(-p_{Wy}, -p_{Wx}). \tag{2.110}$$

Notice that (2.110) gives[18]

$$\vartheta_{1,\text{II}} = \begin{cases} \text{Atan2}(p_{Wy}, p_{Wx}) - \pi & p_{Wy} \geq 0 \\ \text{Atan2}(p_{Wy}, p_{Wx}) + \pi & p_{Wy} < 0. \end{cases}$$

---

[18] It is easy to show that $\text{Atan2}(-y, -x) = -\text{Atan2}(y, -x)$ and

$$\text{Atan2}(y, -x) = \begin{cases} \pi - \text{Atan2}(y, x) & y \geq 0 \\ -\pi - \text{Atan2}(y, x) & y < 0. \end{cases}$$

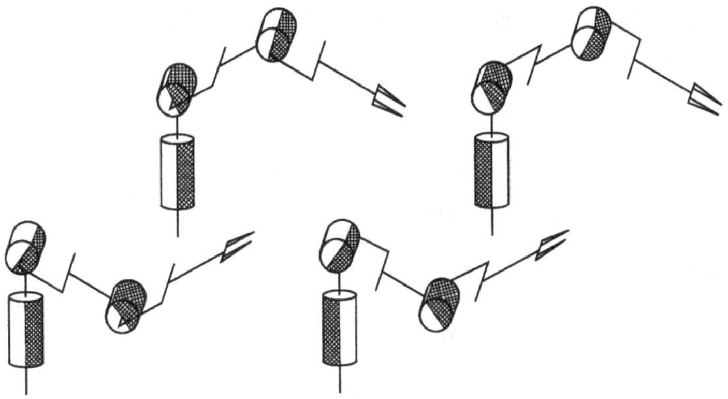

Fig. 2.33. The four configurations of an anthropomorphic arm compatible with a given wrist position

As can be recognized, there exist four solutions according to the values of ϑ_3 in (2.100), (2.101), ϑ_2 in (2.105)–(2.108) and ϑ_1 in (2.109), (2.110):

$$(\vartheta_{1,\mathrm{I}}, \vartheta_{2,\mathrm{I}}, \vartheta_{3,\mathrm{I}}) \quad (\vartheta_{1,\mathrm{I}}, \vartheta_{2,\mathrm{III}}, \vartheta_{3,\mathrm{II}}) \quad (\vartheta_{1,\mathrm{II}}, \vartheta_{2,\mathrm{II}}, \vartheta_{3,\mathrm{I}}) \quad (\vartheta_{1,\mathrm{II}}, \vartheta_{2,\mathrm{IV}}, \vartheta_{3,\mathrm{II}}),$$

which are illustrated in Fig. 2.33: shoulder–right/elbow–up, shoulder–left/elbow–up, shoulder–right/elbow–down, shoulder–left/elbow–down; obviously, the forearm orientation is different for the two pairs of solutions.

Notice finally how it is possible to find the solutions only if at least

$$p_{Wx} \neq 0 \qquad \text{or} \qquad p_{Wy} \neq 0.$$

In the case $p_{Wx} = p_{Wy} = 0$, an infinity of solutions is obtained, since it is possible to determine the joint variables ϑ_2 and ϑ_3 independently of the value of ϑ_1; in the following, it will be seen that the arm in such configuration is kinematically *singular* (see Problem 2.18).

2.12.5 Solution of Spherical Wrist

Consider the spherical wrist shown in Fig. 2.24, whose direct kinematics was given in (2.67). It is desired to find the joint variables ϑ_4, ϑ_5, ϑ_6 corresponding to a given end-effector orientation $\boldsymbol{R}_6^3$. As previously pointed out, these angles constitute a set of Euler angles ZYZ with respect to Frame 3. Hence, having computed the rotation matrix

$$\boldsymbol{R}_6^3 = \begin{bmatrix} n_x^3 & s_x^3 & a_x^3 \\ n_y^3 & s_y^3 & a_y^3 \\ n_z^3 & s_z^3 & a_z^3 \end{bmatrix},$$

from its expression in terms of the joint variables in (2.67), it is possible to compute the solutions directly as in (2.19), (2.20), i.e.,

$$\begin{aligned}\vartheta_4 &= \text{Atan2}(a_y^3, a_x^3)\\ \vartheta_5 &= \text{Atan2}\Big(\sqrt{(a_x^3)^2 + (a_y^3)^2}, a_z^3\Big)\\ \vartheta_6 &= \text{Atan2}(s_z^3, -n_z^3)\end{aligned} \tag{2.111}$$

for $\vartheta_5 \in (0, \pi)$, and

$$\begin{aligned}\vartheta_4 &= \text{Atan2}(-a_y^3, -a_x^3)\\ \vartheta_5 &= \text{Atan2}\Big(-\sqrt{(a_x^3)^2 + (a_y^3)^2}, a_z^3\Big)\\ \vartheta_6 &= \text{Atan2}(-s_z^3, n_z^3)\end{aligned} \tag{2.112}$$

for $\vartheta_5 \in (-\pi, 0)$.

Bibliography

The treatment of kinematics of robot manipulators can be found in several classical robotics texts, such as [180, 10, 200, 217]. Specific texts are [23, 6, 151].

For the descriptions of the orientation of a rigid body, see [187]. Quaternion algebra can be found in [46]; see [204] for the extraction of quaternions from rotation matrices.

The Denavit–Hartenberg convention was first introduced in [60]. A modified version is utilized in [53, 248, 111]. The use of homogeneous transformation matrices for the computation of open-chain manipulator direct kinematics is presented in [181], while in [183] sufficient conditions are given for the closed-form computation of the inverse kinematics problem. For kinematics of closed chains see [144, 111]. The design of the Stanford manipulator is due to [196].

The problem of kinematic calibration is considered in [188, 98]. Methods which do not require the use of external sensors for direct measurement of end-effector position and orientation are proposed in [68].

The kinematic decoupling deriving from the spherical wrist is utilized in [76, 99, 182]. Numerical methods for the solution of the inverse kinematics problem based on iterative algorithms are proposed in [232, 86].

Problems

2.1. Find the rotation matrix corresponding to the set of Euler angles ZXZ.

2.2. Discuss the inverse solution for the Euler angles ZYZ in the case $s_\vartheta = 0$.

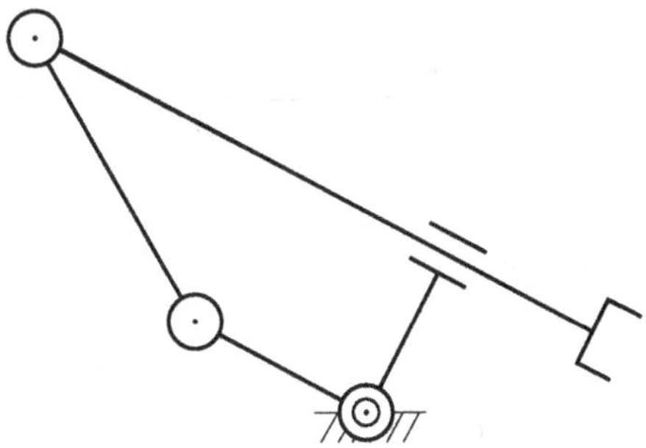

Fig. 2.34. Four-link closed-chain planar arm with prismatic joint

2.3. Discuss the inverse solution for the Roll–Pitch–Yaw angles in the case $c_\vartheta = 0$.

2.4. Verify that the rotation matrix corresponding to the rotation by an angle about an arbitrary axis is given by (2.25).

2.5. Prove that the angle and the unit vector of the axis corresponding to a rotation matrix are given by (2.27), (2.28). Find inverse formulae in the case of $\sin\vartheta = 0$.

2.6. Verify that the rotation matrix corresponding to the unit quaternion is given by (2.33).

2.7. Prove that the unit quaternion is invariant with respect to the rotation matrix and its transpose, i.e., $\boldsymbol{R}(\eta,\boldsymbol{\epsilon})\boldsymbol{\epsilon} = \boldsymbol{R}^T(\eta,\boldsymbol{\epsilon})\boldsymbol{\epsilon} = \boldsymbol{\epsilon}$.

2.8. Prove that the unit quaternion corresponding to a rotation matrix is given by (2.34), (2.35).

2.9. Prove that the quaternion product is expressed by (2.37).

2.10. By applying the rules for inverting a block-partitioned matrix, prove that matrix $\boldsymbol{A}_0^1$ is given by (2.45).

2.11. Find the direct kinematics equation of the four-link closed-chain planar arm in Fig. 2.34, where the two links connected by the prismatic joint are orthogonal to each other

2.12. Find the direct kinematics equation for the cylindrical arm in Fig. 2.35.

2.13. Find the direct kinematics equation for the SCARA manipulator in Fig. 2.36.

2.14. Find the complete direct kinematics equation for the humanoid manipulator in Fig. 2.28.

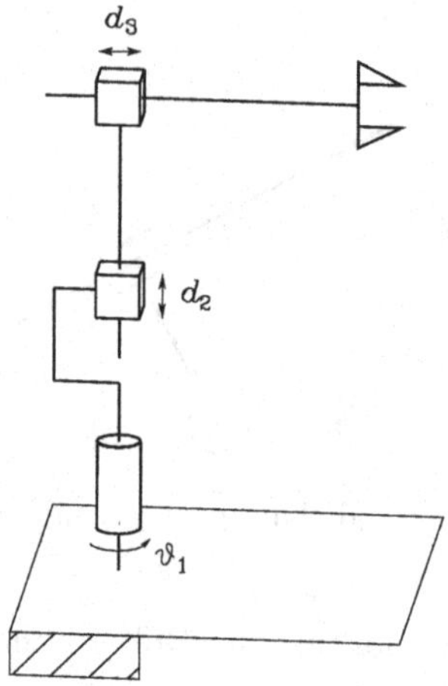

Fig. 2.35. Cylindrical arm

2.15. For the set of minimal representations of orientation ϕ, define the sum operation in terms of the composition of rotations. By means of an example, show that the commutative property does not hold for that operation.

2.16. Consider the elementary rotations about coordinate axes given by infinitesimal angles. Show that the rotation resulting from any two elementary rotations does not depend on the order of rotations. [*Hint*: for an infinitesimal angle $d\phi$, approximate $\cos(d\phi) \approx 1$ and $\sin(d\phi) \approx d\phi \ldots$]. Further, define $\boldsymbol{R}(d\phi_x, d\phi_y, d\phi_z) = \boldsymbol{R}_x(d\phi_x)\boldsymbol{R}_y(d\phi_y)\boldsymbol{R}_z(d\phi_z)$; show that

$$\boldsymbol{R}(d\phi_x, d\phi_y, d\phi_z)\boldsymbol{R}(d\phi'_x, d\phi'_y, d\phi'_z) = \boldsymbol{R}(d\phi_x + d\phi'_x, d\phi_y + d\phi'_y, d\phi_z + d\phi'_z).$$

2.17. Draw the workspace of the three-link planar arm in Fig. 2.20 with the data:

$$a_1 = 0.5 \qquad a_2 = 0.3 \qquad a_3 = 0.2$$

$$-\pi/3 \le q_1 \le \pi/3 \qquad -2\pi/3 \le q_2 \le 2\pi/3 \qquad -\pi/2 \le q_3 \le \pi/2.$$

2.18. With reference to the inverse kinematics of the anthropomorphic arm in Sect. 2.12.4, discuss the number of solutions in the singular cases of $s_3 = 0$ and $p_{Wx} = p_{Wy} = 0$.

2.19. Solve the inverse kinematics for the cylindrical arm in Fig. 2.35.

2.20. Solve the inverse kinematics for the SCARA manipulator in Fig. 2.36.

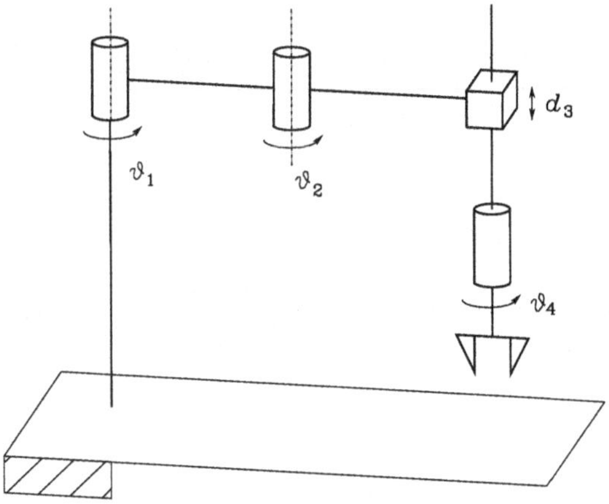

Fig. 2.36. SCARA manipulator

3

Differential Kinematics and Statics

In the previous chapter, direct and inverse kinematics equations establishing the relationship between the joint variables and the end-effector pose were derived. In this chapter, *differential kinematics* is presented which gives the relationship between the joint velocities and the corresponding end-effector linear and angular velocity. This mapping is described by a matrix, termed *geometric Jacobian*, which depends on the manipulator configuration. Alternatively, if the end-effector pose is expressed with reference to a minimal representation in the operational space, it is possible to compute the Jacobian matrix via differentiation of the direct kinematics function with respect to the joint variables. The resulting Jacobian, termed *analytical Jacobian*, in general differs from the geometric one. The Jacobian constitutes one of the most important tools for manipulator characterization; in fact, it is useful for finding *singularities*, analyzing *redundancy*, determining *inverse kinematics algorithms*, describing the mapping between forces applied to the end-effector and resulting torques at the joints (*statics*) and, as will be seen in the following chapters, deriving dynamic equations of motion and designing operational space control schemes. Finally, the *kineto-statics duality* concept is illustrated, which is at the basis of the definition of velocity and force *manipulability ellipsoids*.

3.1 Geometric Jacobian

Consider an n-DOF manipulator. The direct kinematics equation can be written in the form

$$\boldsymbol{T}_e(\boldsymbol{q}) = \begin{bmatrix} \boldsymbol{R}_e(\boldsymbol{q}) & \boldsymbol{p}_e(\boldsymbol{q}) \\ \boldsymbol{0}^T & 1 \end{bmatrix} \tag{3.1}$$

where $\boldsymbol{q} = [\, q_1 \quad \dots \quad q_n \,]^T$ is the vector of joint variables. Both end-effector position and orientation vary as $\boldsymbol{q}$ varies.

The goal of the differential kinematics is to find the relationship between the joint velocities and the end-effector linear and angular velocities. In other words, it is desired to express the end-effector linear velocity $\dot{\boldsymbol{p}}_e$ and angular velocity $\boldsymbol{\omega}_e$ as a function of the joint velocities $\dot{\boldsymbol{q}}$. As will be seen afterwards, the sought relations are both linear in the joint velocities, i.e.,

$$\dot{\boldsymbol{p}}_e = \boldsymbol{J}_P(\boldsymbol{q})\dot{\boldsymbol{q}} \tag{3.2}$$

$$\boldsymbol{\omega}_e = \boldsymbol{J}_O(\boldsymbol{q})\dot{\boldsymbol{q}}. \tag{3.3}$$

In (3.2) $\boldsymbol{J}_P$ is the $(3 \times n)$ matrix relating the contribution of the joint velocities $\dot{\boldsymbol{q}}$ to the end-effector *linear* velocity $\dot{\boldsymbol{p}}_e$, while in (3.3) $\boldsymbol{J}_O$ is the $(3 \times n)$ matrix relating the contribution of the joint velocities $\dot{\boldsymbol{q}}$ to the end-effector *angular* velocity $\boldsymbol{\omega}_e$. In compact form, (3.2), (3.3) can be written as

$$\boldsymbol{v}_e = \begin{bmatrix} \dot{\boldsymbol{p}}_e \\ \boldsymbol{\omega}_e \end{bmatrix} = \boldsymbol{J}(\boldsymbol{q})\dot{\boldsymbol{q}} \tag{3.4}$$

which represents the manipulator *differential kinematics equation*. The $(6 \times n)$ matrix $\boldsymbol{J}$ is the manipulator *geometric Jacobian*

$$\boldsymbol{J} = \begin{bmatrix} \boldsymbol{J}_P \\ \boldsymbol{J}_O \end{bmatrix}, \tag{3.5}$$

which in general is a function of the joint variables.

In order to compute the geometric Jacobian, it is worth recalling a number of properties of rotation matrices and some important results of rigid body kinematics.

3.1.1 Derivative of a Rotation Matrix

The manipulator direct kinematics equation in (3.1) describes the end-effector pose, as a function of the joint variables, in terms of a position vector and a rotation matrix. Since the aim is to characterize the end-effector linear and angular velocities, it is worth considering first the *derivative of a rotation matrix* with respect to time.

Consider a time-varying rotation matrix $\boldsymbol{R} = \boldsymbol{R}(t)$. In view of the orthogonality of $\boldsymbol{R}$, one has the relation

$$\boldsymbol{R}(t)\boldsymbol{R}^T(t) = \boldsymbol{I}$$

which, differentiated with respect to time, gives the identity

$$\dot{\boldsymbol{R}}(t)\boldsymbol{R}^T(t) + \boldsymbol{R}(t)\dot{\boldsymbol{R}}^T(t) = \boldsymbol{O}.$$

Set

$$\boldsymbol{S}(t) = \dot{\boldsymbol{R}}(t)\boldsymbol{R}^T(t); \tag{3.6}$$

the (3×3) matrix $\boldsymbol{S}$ is *skew-symmetric* since

$$\boldsymbol{S}(t) + \boldsymbol{S}^T(t) = \boldsymbol{O}. \tag{3.7}$$

Postmultiplying both sides of (3.6) by $\boldsymbol{R}(t)$ gives

$$\dot{\boldsymbol{R}}(t) = \boldsymbol{S}(t)\boldsymbol{R}(t) \tag{3.8}$$

that allows the time derivative of $\boldsymbol{R}(t)$ to be expressed as a function of $\boldsymbol{R}(t)$ itself.

Equation (3.8) relates the rotation matrix $\boldsymbol{R}$ to its derivative by means of the skew-symmetric operator $\boldsymbol{S}$ and has a meaningful physical interpretation. Consider a constant vector $\boldsymbol{p}'$ and the vector $\boldsymbol{p}(t) = \boldsymbol{R}(t)\boldsymbol{p}'$. The time derivative of $\boldsymbol{p}(t)$ is

$$\dot{\boldsymbol{p}}(t) = \dot{\boldsymbol{R}}(t)\boldsymbol{p}',$$

which, in view of (3.8), can be written as

$$\dot{\boldsymbol{p}}(t) = \boldsymbol{S}(t)\boldsymbol{R}(t)\boldsymbol{p}'.$$

If the vector $\boldsymbol{\omega}(t)$ denotes the *angular velocity* of frame $\boldsymbol{R}(t)$ with respect to the reference frame at time t, it is known from mechanics that

$$\dot{\boldsymbol{p}}(t) = \boldsymbol{\omega}(t) \times \boldsymbol{R}(t)\boldsymbol{p}'.$$

Therefore, the matrix operator $\boldsymbol{S}(t)$ describes the vector product between the vector $\boldsymbol{\omega}$ and the vector $\boldsymbol{R}(t)\boldsymbol{p}'$. The matrix $\boldsymbol{S}(t)$ is so that its symmetric elements with respect to the main diagonal represent the components of the vector $\boldsymbol{\omega}(t) = [\,\omega_x \quad \omega_y \quad \omega_z\,]^T$ in the form

$$\boldsymbol{S} = \begin{bmatrix} 0 & -\omega_z & \omega_y \\ \omega_z & 0 & -\omega_x \\ -\omega_y & \omega_x & 0 \end{bmatrix}, \tag{3.9}$$

which justifies the expression $\boldsymbol{S}(t) = \boldsymbol{S}(\boldsymbol{\omega}(t))$. Hence, (3.8) can be rewritten as

$$\dot{\boldsymbol{R}} = \boldsymbol{S}(\boldsymbol{\omega})\boldsymbol{R}. \tag{3.10}$$

Furthermore, if $\boldsymbol{R}$ denotes a rotation matrix, it can be shown that the following relation holds:

$$\boldsymbol{R}\boldsymbol{S}(\boldsymbol{\omega})\boldsymbol{R}^T = \boldsymbol{S}(\boldsymbol{R}\boldsymbol{\omega}) \tag{3.11}$$

which will be useful later (see Problem 3.1).

Example 3.1

Consider the elementary rotation matrix about axis z given in (2.6). If α is a function of time, by computing the time derivative of $\boldsymbol{R}_z(\alpha(t))$, (3.6) becomes

$$\boldsymbol{S}(t) = \begin{bmatrix} -\dot{\alpha}\sin\alpha & -\dot{\alpha}\cos\alpha & 0 \\ \dot{\alpha}\cos\alpha & -\dot{\alpha}\sin\alpha & 0 \\ 0 & 0 & 0 \end{bmatrix} \begin{bmatrix} \cos\alpha & \sin\alpha & 0 \\ -\sin\alpha & \cos\alpha & 0 \\ 0 & 0 & 1 \end{bmatrix}$$

$$= \begin{bmatrix} 0 & -\dot{\alpha} & 0 \\ \dot{\alpha} & 0 & 0 \\ 0 & 0 & 0 \end{bmatrix} = \boldsymbol{S}(\boldsymbol{\omega}(t)).$$

According to (3.9), it is

$$\boldsymbol{\omega} = [\,0 \quad 0 \quad \dot{\alpha}\,]^T$$

that expresses the angular velocity of the frame about axis z.

With reference to Fig. 2.11, consider the coordinate transformation of a point P from Frame 1 to Frame 0; in view of (2.38), this is given by

$$\boldsymbol{p}^0 = \boldsymbol{o}_1^0 + \boldsymbol{R}_1^0\boldsymbol{p}^1. \tag{3.12}$$

Differentiating (3.12) with respect to time gives

$$\dot{\boldsymbol{p}}^0 = \dot{\boldsymbol{o}}_1^0 + \boldsymbol{R}_1^0\dot{\boldsymbol{p}}^1 + \dot{\boldsymbol{R}}_1^0\boldsymbol{p}^1; \tag{3.13}$$

utilizing the expression of the derivative of a rotation matrix (3.8) and specifying the dependence on the angular velocity gives

$$\dot{\boldsymbol{p}}^0 = \dot{\boldsymbol{o}}_1^0 + \boldsymbol{R}_1^0\dot{\boldsymbol{p}}^1 + \boldsymbol{S}(\boldsymbol{\omega}_1^0)\boldsymbol{R}_1^0\boldsymbol{p}^1.$$

Further, denoting the vector $\boldsymbol{R}_1^0\boldsymbol{p}^1$ by $\boldsymbol{r}_1^0$, it is

$$\dot{\boldsymbol{p}}^0 = \dot{\boldsymbol{o}}_1^0 + \boldsymbol{R}_1^0\dot{\boldsymbol{p}}^1 + \boldsymbol{\omega}_1^0 \times \boldsymbol{r}_1^0 \tag{3.14}$$

which is the known form of the velocity composition rule.

Notice that, if $\boldsymbol{p}^1$ is *fixed* in Frame 1, then it is

$$\dot{\boldsymbol{p}}^0 = \dot{\boldsymbol{o}}_1^0 + \boldsymbol{\omega}_1^0 \times \boldsymbol{r}_1^0 \tag{3.15}$$

since $\dot{\boldsymbol{p}}^1 = \boldsymbol{0}$.

3.1.2 Link Velocities

Consider the generic Link i of a manipulator with an open kinematic chain. According to the Denavit–Hartenberg convention adopted in the previous chapter, Link i connects Joints i and $i+1$; Frame i is attached to Link i

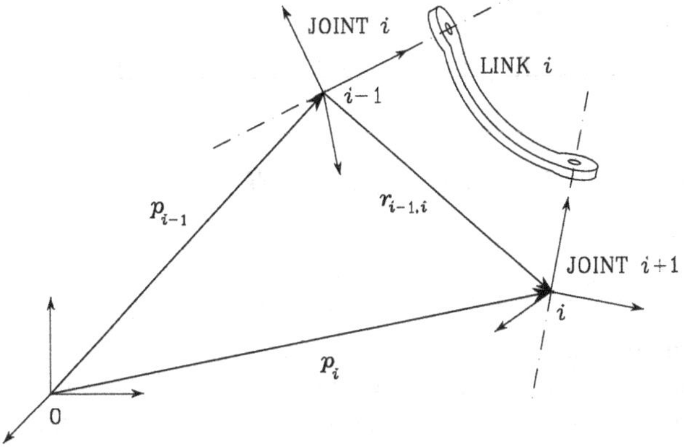

Fig. 3.1. Characterization of generic Link i of a manipulator

and has origin along Joint $i+1$ axis, while Frame $i-1$ has origin along Joint i axis (Fig. 3.1).

Let $\boldsymbol{p}_{i-1}$ and $\boldsymbol{p}_i$ be the position vectors of the origins of Frames $i-1$ and i, respectively. Also, let $\boldsymbol{r}_{i-1,i}^{i-1}$ denote the position of the origin of Frame i with respect to Frame $i-1$ expressed in Frame $i-1$. According to the coordinate transformation (3.10), one can write[1]

$$\boldsymbol{p}_i = \boldsymbol{p}_{i-1} + \boldsymbol{R}_{i-1}\boldsymbol{r}_{i-1,i}^{i-1}.$$

Then, by virtue of (3.14), it is

$$\dot{\boldsymbol{p}}_i = \dot{\boldsymbol{p}}_{i-1} + \boldsymbol{R}_{i-1}\dot{\boldsymbol{r}}_{i-1,i}^{i-1} + \boldsymbol{\omega}_{i-1} \times \boldsymbol{R}_{i-1}\boldsymbol{r}_{i-1,i}^{i-1} = \dot{\boldsymbol{p}}_{i-1} + \boldsymbol{v}_{i-1,i} + \boldsymbol{\omega}_{i-1} \times \boldsymbol{r}_{i-1,i} \tag{3.16}$$

which gives the expression of the linear velocity of Link i as a function of the translational and rotational velocities of Link $i-1$. Note that $\boldsymbol{v}_{i-1,i}$ denotes the velocity of the origin of Frame i with respect to the origin of Frame $i-1$.

Concerning link angular velocity, it is worth starting from the rotation composition

$$\boldsymbol{R}_i = \boldsymbol{R}_{i-1}\boldsymbol{R}_i^{i-1};$$

from (3.8), its time derivative can be written as

$$\boldsymbol{S}(\boldsymbol{\omega}_i)\boldsymbol{R}_i = \boldsymbol{S}(\boldsymbol{\omega}_{i-1})\boldsymbol{R}_i + \boldsymbol{R}_{i-1}\boldsymbol{S}(\boldsymbol{\omega}_{i-1,i}^{i-1})\boldsymbol{R}_i^{i-1} \tag{3.17}$$

where $\boldsymbol{\omega}_{i-1,i}^{i-1}$ denotes the angular velocity of Frame i with respect to Frame $i-1$ expressed in Frame $i-1$. From (2.4), the second term on the right-hand side of (3.17) can be rewritten as

$$\boldsymbol{R}_{i-1}\boldsymbol{S}(\boldsymbol{\omega}_{i-1,i}^{i-1})\boldsymbol{R}_i^{i-1} = \boldsymbol{R}_{i-1}\boldsymbol{S}(\boldsymbol{\omega}_{i-1,i}^{i-1})\boldsymbol{R}_{i-1}^T\boldsymbol{R}_{i-1}\boldsymbol{R}_i^{i-1};$$

[1] Hereafter, the indication of superscript '0' is omitted for quantities referred to Frame 0. Also, without loss of generality, Frame 0 and Frame n are taken as the base frame and the end-effector frame, respectively.

in view of property (3.11), it is

$$\boldsymbol{R}_{i-1}\boldsymbol{S}(\boldsymbol{\omega}_{i-1,i}^{i-1})\boldsymbol{R}_i^{i-1} = \boldsymbol{S}(\boldsymbol{R}_{i-1}\boldsymbol{\omega}_{i-1,i}^{i-1})\boldsymbol{R}_i.$$

Then, (3.17) becomes

$$\boldsymbol{S}(\boldsymbol{\omega}_i)\boldsymbol{R}_i = \boldsymbol{S}(\boldsymbol{\omega}_{i-1})\boldsymbol{R}_i + \boldsymbol{S}(\boldsymbol{R}_{i-1}\boldsymbol{\omega}_{i-1,i}^{i-1})\boldsymbol{R}_i$$

leading to the result

$$\boldsymbol{\omega}_i = \boldsymbol{\omega}_{i-1} + \boldsymbol{R}_{i-1}\boldsymbol{\omega}_{i-1,i}^{i-1} = \boldsymbol{\omega}_{i-1} + \boldsymbol{\omega}_{i-1,i}, \tag{3.18}$$

which gives the expression of the angular velocity of Link i as a function of the angular velocities of Link $i-1$ and of Link i with respect to Link $i-1$.

The relations (3.16), (3.18) attain different expressions depending on the type of Joint i (*prismatic* or *revolute*).

Prismatic joint

Since orientation of Frame i with respect to Frame $i-1$ does not vary by moving Joint i, it is

$$\boldsymbol{\omega}_{i-1,i} = \boldsymbol{0}. \tag{3.19}$$

Further, the linear velocity is

$$\boldsymbol{v}_{i-1,i} = \dot{d}_i\boldsymbol{z}_{i-1} \tag{3.20}$$

where $\boldsymbol{z}_{i-1}$ is the unit vector of Joint i axis. Hence, the expressions of angular velocity (3.18) and linear velocity (3.16) respectively become

$$\boldsymbol{\omega}_i = \boldsymbol{\omega}_{i-1} \tag{3.21}$$

$$\dot{\boldsymbol{p}}_i = \dot{\boldsymbol{p}}_{i-1} + \dot{d}_i\boldsymbol{z}_{i-1} + \boldsymbol{\omega}_i \times \boldsymbol{r}_{i-1,i}, \tag{3.22}$$

where the relation $\boldsymbol{\omega}_i = \boldsymbol{\omega}_{i-1}$ has been exploited to derive (3.22).

Revolute joint

For the angular velocity it is obviously

$$\boldsymbol{\omega}_{i-1,i} = \dot{\vartheta}_i\boldsymbol{z}_{i-1}, \tag{3.23}$$

while for the linear velocity it is

$$\boldsymbol{v}_{i-1,i} = \boldsymbol{\omega}_{i-1,i} \times \boldsymbol{r}_{i-1,i} \tag{3.24}$$

due to the rotation of Frame i with respect to Frame $i-1$ induced by the motion of Joint i. Hence, the expressions of angular velocity (3.18) and linear velocity (3.16) respectively become

$$\boldsymbol{\omega}_i = \boldsymbol{\omega}_{i-1} + \dot{\vartheta}_i\boldsymbol{z}_{i-1} \tag{3.25}$$

$$\dot{\boldsymbol{p}}_i = \dot{\boldsymbol{p}}_{i-1} + \boldsymbol{\omega}_i \times \boldsymbol{r}_{i-1,i}, \tag{3.26}$$

where (3.18) has been exploited to derive (3.26).

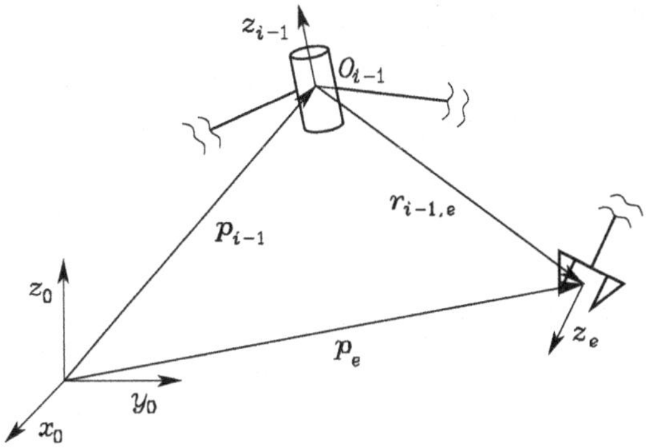

Fig. 3.2. Representation of vectors needed for the computation of the velocity contribution of a revolute joint to the end-effector linear velocity

3.1.3 Jacobian Computation

In order to compute the Jacobian, it is convenient to proceed separately for the linear velocity and the angular velocity.

For the contribution to the *linear velocity*, the time derivative of $\boldsymbol{p}_e(\boldsymbol{q})$ can be written as

$$\dot{\boldsymbol{p}}_e = \sum_{i=1}^{n} \frac{\partial \boldsymbol{p}_e}{\partial q_i} \dot{q}_i = \sum_{i=1}^{n} \boldsymbol{\jmath}_{Pi} \dot{q}_i. \tag{3.27}$$

This expression shows how $\dot{\boldsymbol{p}}_e$ can be obtained as the sum of the terms $\dot{q}_i \boldsymbol{\jmath}_{Pi}$. Each term represents the contribution of the velocity of single Joint i to the end-effector linear velocity when all the other joints are still.

Therefore, by distinguishing the case of a *prismatic* joint ($q_i = d_i$) from the case of a *revolute* joint ($q_i = \vartheta_i$), it is:

- If Joint i is *prismatic*, from (3.20) it is

$$\dot{q}_i \boldsymbol{\jmath}_{Pi} = \dot{d}_i \boldsymbol{z}_{i-1}$$

  and then

$$\boldsymbol{\jmath}_{Pi} = \boldsymbol{z}_{i-1}.$$

- If Joint i is *revolute*, observing that the contribution to the linear velocity is to be computed with reference to the origin of the end-effector frame (Fig. 3.2), it is

$$\dot{q}_i \boldsymbol{\jmath}_{Pi} = \boldsymbol{\omega}_{i-1,i} \times \boldsymbol{r}_{i-1,e} = \dot{\vartheta}_i \boldsymbol{z}_{i-1} \times (\boldsymbol{p}_e - \boldsymbol{p}_{i-1})$$

  and then

$$\boldsymbol{\jmath}_{Pi} = \boldsymbol{z}_{i-1} \times (\boldsymbol{p}_e - \boldsymbol{p}_{i-1}).$$

For the contribution to the *angular velocity*, in view of (3.18), it is

$$\boldsymbol{\omega}_e = \boldsymbol{\omega}_n = \sum_{i=1}^{n} \boldsymbol{\omega}_{i-1,i} = \sum_{i=1}^{n} \boldsymbol{\jmath}_{Oi}\dot{q}_i, \tag{3.28}$$

where (3.19) and (3.23) have been utilized to characterize the terms $\dot{q}_i\boldsymbol{\jmath}_{Oi}$, and thus in detail:

- If Joint i is *prismatic*, from (3.19) it is

$$\dot{q}_i\boldsymbol{\jmath}_{Oi} = \mathbf{0}$$

  and then

$$\boldsymbol{\jmath}_{Oi} = \mathbf{0}.$$

- If Joint i is *revolute*, from (3.23) it is

$$\dot{q}_i\boldsymbol{\jmath}_{Oi} = \dot{\vartheta}_i\boldsymbol{z}_{i-1}$$

  and then

$$\boldsymbol{\jmath}_{Oi} = \boldsymbol{z}_{i-1}.$$

In summary, the Jacobian in (3.5) can be partitioned into the (3×1) column vectors $\boldsymbol{\jmath}_{Pi}$ and $\boldsymbol{\jmath}_{Oi}$ as

$$\boldsymbol{J} = \begin{bmatrix} \boldsymbol{\jmath}_{P1} & & \boldsymbol{\jmath}_{Pn} \\ & \dots & \\ \boldsymbol{\jmath}_{O1} & & \boldsymbol{\jmath}_{On} \end{bmatrix}, \tag{3.29}$$

where

$$\begin{bmatrix} \boldsymbol{\jmath}_{Pi} \\ \boldsymbol{\jmath}_{Oi} \end{bmatrix} = \begin{cases} \begin{bmatrix} \boldsymbol{z}_{i-1} \\ \mathbf{0} \end{bmatrix} & \text{for a } \textit{prismatic} \text{ joint} \\ \begin{bmatrix} \boldsymbol{z}_{i-1} \times (\boldsymbol{p}_e - \boldsymbol{p}_{i-1}) \\ \boldsymbol{z}_{i-1} \end{bmatrix} & \text{for a } \textit{revolute} \text{ joint.} \end{cases} \tag{3.30}$$

The expressions in (3.30) allow Jacobian computation in a simple, systematic way on the basis of direct kinematics relations. In fact, the vectors $\boldsymbol{z}_{i-1}$, $\boldsymbol{p}_e$ and $\boldsymbol{p}_{i-1}$ are all functions of the joint variables. In particular:

- $\boldsymbol{z}_{i-1}$ is given by the third column of the rotation matrix $\boldsymbol{R}_{i-1}^0$, i.e.,

$$\boldsymbol{z}_{i-1} = \boldsymbol{R}_1^0(q_1)\dots\boldsymbol{R}_{i-1}^{i-2}(q_{i-1})\boldsymbol{z}_0 \tag{3.31}$$

  where $\boldsymbol{z}_0 = [\,0 \quad 0 \quad 1\,]^T$ allows the selection of the third column.
- $\boldsymbol{p}_e$ is given by the first three elements of the fourth column of the transformation matrix $\boldsymbol{T}_e^0$, i.e., by expressing $\widetilde{\boldsymbol{p}}_e$ in the (4×1) homogeneous form

$$\widetilde{\boldsymbol{p}}_e = \boldsymbol{A}_1^0(q_1)\dots\boldsymbol{A}_n^{n-1}(q_n)\widetilde{\boldsymbol{p}}_0 \tag{3.32}$$

  where $\widetilde{\boldsymbol{p}}_0 = [\,0 \quad 0 \quad 0 \quad 1\,]^T$ allows the selection of the fourth column.

- $\boldsymbol{p}_{i-1}$ is given by the first three elements of the fourth column of the transformation matrix $\boldsymbol{T}_{i-1}^{0}$, i.e., it can be extracted from

$$\widetilde{\boldsymbol{p}}_{i-1} = \boldsymbol{A}_1^0(q_1)\dots\boldsymbol{A}_{i-1}^{i-2}(q_{i-1})\widetilde{\boldsymbol{p}}_0. \tag{3.33}$$

The above equations can be conveniently used to compute the translational and rotational velocities of any point along the manipulator structure, as long as the direct kinematics functions relative to that point are known.

Finally, notice that the Jacobian matrix depends on the frame in which the end-effector velocity is expressed. The above equations allow computation of the geometric Jacobian with respect to the base frame. If it is desired to represent the Jacobian in a different Frame u, it is sufficient to know the relative rotation matrix $\boldsymbol{R}^u$. The relationship between velocities in the two frames is

$$\begin{bmatrix} \dot{\boldsymbol{p}}_e^u \\ \boldsymbol{\omega}_e^u \end{bmatrix} = \begin{bmatrix} \boldsymbol{R}^u & \boldsymbol{O} \\ \boldsymbol{O} & \boldsymbol{R}^u \end{bmatrix} \begin{bmatrix} \dot{\boldsymbol{p}}_e \\ \boldsymbol{\omega}_e \end{bmatrix},$$

which, substituted in (3.4), gives

$$\begin{bmatrix} \dot{\boldsymbol{p}}_e^u \\ \boldsymbol{\omega}_e^u \end{bmatrix} = \begin{bmatrix} \boldsymbol{R}^u & \boldsymbol{O} \\ \boldsymbol{O} & \boldsymbol{R}^u \end{bmatrix} \boldsymbol{J}\dot{\boldsymbol{q}}$$

and then

$$\boldsymbol{J}^u = \begin{bmatrix} \boldsymbol{R}^u & \boldsymbol{O} \\ \boldsymbol{O} & \boldsymbol{R}^u \end{bmatrix} \boldsymbol{J}, \tag{3.34}$$

where $\boldsymbol{J}^u$ denotes the geometric Jacobian in Frame u, which has been assumed to be time-invariant.

3.2 Jacobian of Typical Manipulator Structures

In the following, the Jacobian is computed for some of the typical manipulator structures presented in the previous chapter.

3.2.1 Three-link Planar Arm

In this case, from (3.30) the Jacobian is

$$\boldsymbol{J}(\boldsymbol{q}) = \begin{bmatrix} \boldsymbol{z}_0 \times (\boldsymbol{p}_3 - \boldsymbol{p}_0) & \boldsymbol{z}_1 \times (\boldsymbol{p}_3 - \boldsymbol{p}_1) & \boldsymbol{z}_2 \times (\boldsymbol{p}_3 - \boldsymbol{p}_2) \\ \boldsymbol{z}_0 & \boldsymbol{z}_1 & \boldsymbol{z}_2 \end{bmatrix}.$$

Computation of the position vectors of the various links gives

$$\boldsymbol{p}_0 = \begin{bmatrix} 0 \\ 0 \\ 0 \end{bmatrix} \quad \boldsymbol{p}_1 = \begin{bmatrix} a_1 c_1 \\ a_1 s_1 \\ 0 \end{bmatrix} \quad \boldsymbol{p}_2 = \begin{bmatrix} a_1 c_1 + a_2 c_{12} \\ a_1 s_1 + a_2 s_{12} \\ 0 \end{bmatrix}$$

$$p_3 = \begin{bmatrix} a_1c_1 + a_2c_{12} + a_3c_{123} \\ a_1s_1 + a_2s_{12} + a_3s_{123} \\ 0 \end{bmatrix}$$

while computation of the unit vectors of revolute joint axes gives

$$z_0 = z_1 = z_2 = \begin{bmatrix} 0 \\ 0 \\ 1 \end{bmatrix}$$

since they are all parallel to axis z_0. From (3.29) it is

$$J = \begin{bmatrix} -a_1s_1 - a_2s_{12} - a_3s_{123} & -a_2s_{12} - a_3s_{123} & -a_3s_{123} \\ a_1c_1 + a_2c_{12} + a_3c_{123} & a_2c_{12} + a_3c_{123} & a_3c_{123} \\ 0 & 0 & 0 \\ 0 & 0 & 0 \\ 0 & 0 & 0 \\ 1 & 1 & 1 \end{bmatrix}. \tag{3.35}$$

In the Jacobian (3.35), only the three non-null rows are relevant (the rank of the matrix is at most 3); these refer to the two components of linear velocity along axes x_0, y_0 and the component of angular velocity about axis z_0. This result can be derived by observing that three DOFs allow specification of at most three end-effector variables; v_z, ω_x, ω_y are always null for this kinematic structure. If orientation is of no concern, the (2×3) Jacobian for the positional part can be derived by considering just the first two rows, i.e.,

$$J_P = \begin{bmatrix} -a_1s_1 - a_2s_{12} - a_3s_{123} & -a_2s_{12} - a_3s_{123} & -a_3s_{123} \\ a_1c_1 + a_2c_{12} + a_3c_{123} & a_2c_{12} + a_3c_{123} & a_3c_{123} \end{bmatrix}. \tag{3.36}$$

3.2.2 Anthropomorphic Arm

In this case, from (3.30) the Jacobian is

$$J = \begin{bmatrix} z_0 \times (p_3 - p_0) & z_1 \times (p_3 - p_1) & z_2 \times (p_3 - p_2) \\ z_0 & z_1 & z_2 \end{bmatrix}.$$

Computation of the position vectors of the various links gives

$$p_0 = p_1 = \begin{bmatrix} 0 \\ 0 \\ 0 \end{bmatrix} \qquad p_2 = \begin{bmatrix} a_2c_1c_2 \\ a_2s_1c_2 \\ a_2s_2 \end{bmatrix}$$

$$p_3 = \begin{bmatrix} c_1(a_2c_2 + a_3c_{23}) \\ s_1(a_2c_2 + a_3c_{23}) \\ a_2s_2 + a_3s_{23} \end{bmatrix}$$

while computation of the unit vectors of revolute joint axes gives

$$z_0 = \begin{bmatrix} 0 \\ 0 \\ 1 \end{bmatrix} \qquad z_1 = z_2 = \begin{bmatrix} s_1 \\ -c_1 \\ 0 \end{bmatrix}.$$

From (3.29) it is

$$\boldsymbol{J} = \begin{bmatrix} -s_1(a_2c_2 + a_3c_{23}) & -c_1(a_2s_2 + a_3s_{23}) & -a_3c_1s_{23} \\ c_1(a_2c_2 + a_3c_{23}) & -s_1(a_2s_2 + a_3s_{23}) & -a_3s_1s_{23} \\ 0 & a_2c_2 + a_3c_{23} & a_3c_{23} \\ 0 & s_1 & s_1 \\ 0 & -c_1 & -c_1 \\ 1 & 0 & 0 \end{bmatrix}. \tag{3.37}$$

Only three of the six rows of the Jacobian (3.37) are linearly independent. Having 3 DOFs only, it is worth considering the upper (3×3) block of the Jacobian

$$\boldsymbol{J}_P = \begin{bmatrix} -s_1(a_2c_2 + a_3c_{23}) & -c_1(a_2s_2 + a_3s_{23}) & -a_3c_1s_{23} \\ c_1(a_2c_2 + a_3c_{23}) & -s_1(a_2s_2 + a_3s_{23}) & -a_3s_1s_{23} \\ 0 & a_2c_2 + a_3c_{23} & a_3c_{23} \end{bmatrix} \tag{3.38}$$

that describes the relationship between the joint velocities and the end-effector linear velocity. This structure does not allow an arbitrary angular velocity $\boldsymbol{\omega}$ to be obtained; in fact, the two components ω_x and ω_y are not independent $(s_1\omega_y = -c_1\omega_x)$.

3.2.3 Stanford Manipulator

In this case, from (3.30) it is

$$\boldsymbol{J} = \begin{bmatrix} z_0 \times (p_6 - p_0) & z_1 \times (p_6 - p_1) & z_2 \\ z_0 & z_1 & \boldsymbol{0} \end{bmatrix.$$
$$\begin{matrix} z_3 \times (p_6 - p_3) & z_4 \times (p_6 - p_4) & z_5 \times (p_6 - p_5) \\ z_3 & z_4 & z_5 \end{matrix}\Bigg].$$

Computation of the position vectors of the various links gives

$$p_0 = p_1 = \begin{bmatrix} 0 \\ 0 \\ 0 \end{bmatrix} \qquad p_3 = p_4 = p_5 = \begin{bmatrix} c_1s_2d_3 - s_1d_2 \\ s_1s_2d_3 + c_1d_2 \\ c_2d_3 \end{bmatrix}$$

$$p_6 = \begin{bmatrix} c_1s_2d_3 - s_1d_2 + \big(c_1(c_2c_4s_5 + s_2c_5) - s_1s_4s_5\big)d_6 \\ s_1s_2d_3 + c_1d_2 + \big(s_1(c_2c_4s_5 + s_2c_5) + c_1s_4s_5\big)d_6 \\ c_2d_3 + (-s_2c_4s_5 + c_2c_5)d_6 \end{bmatrix},$$

while computation of the unit vectors of joint axes gives

$$\boldsymbol{z}_0 = \begin{bmatrix} 0 \\ 0 \\ 1 \end{bmatrix} \quad \boldsymbol{z}_1 = \begin{bmatrix} -s_1 \\ c_1 \\ 0 \end{bmatrix} \quad \boldsymbol{z}_2 = \boldsymbol{z}_3 = \begin{bmatrix} c_1 s_2 \\ s_1 s_2 \\ c_2 \end{bmatrix}$$

$$\boldsymbol{z}_4 = \begin{bmatrix} -c_1 c_2 s_4 - s_1 c_4 \\ -s_1 c_2 s_4 + c_1 c_4 \\ s_2 s_4 \end{bmatrix} \quad \boldsymbol{z}_5 = \begin{bmatrix} c_1(c_2 c_4 s_5 + s_2 c_5) - s_1 s_4 s_5 \\ s_1(c_2 c_4 s_5 + s_2 c_5) + c_1 s_4 s_5 \\ -s_2 c_4 s_5 + c_2 c_5 \end{bmatrix}.$$

The sought Jacobian can be obtained by developing the computations as in (3.29), leading to expressing end-effector linear and angular velocity as a function of joint velocities.

3.3 Kinematic Singularities

The Jacobian in the differential kinematics equation of a manipulator defines a linear mapping

$$\boldsymbol{v}_e = \boldsymbol{J}(\boldsymbol{q})\dot{\boldsymbol{q}} \tag{3.39}$$

between the vector $\dot{\boldsymbol{q}}$ of joint velocities and the vector $\boldsymbol{v}_e = [\,\dot{\boldsymbol{p}}_e^T \;\; \boldsymbol{\omega}_e^T\,]^T$ of end-effector velocity. The Jacobian is, in general, a function of the configuration $\boldsymbol{q}$; those configurations at which $\boldsymbol{J}$ is rank-deficient are termed *kinematic singularities*. To find the singularities of a manipulator is of great interest for the following reasons:

a) Singularities represent configurations at which mobility of the structure is reduced, i.e., it is not possible to impose an arbitrary motion to the end-effector.
b) When the structure is at a singularity, infinite solutions to the inverse kinematics problem may exist.
c) In the neighbourhood of a singularity, small velocities in the operational space may cause large velocities in the joint space.

Singularities can be classified into:

- *Boundary* singularities that occur when the manipulator is either out-stretched or retracted. It may be understood that these singularities do not represent a true drawback, since they can be avoided on condition that the manipulator is not driven to the boundaries of its reachable workspace.
- *Internal* singularities that occur inside the reachable workspace and are generally caused by the alignment of two or more axes of motion, or else by the attainment of particular end-effector configurations. Unlike the above, these singularities constitute a serious problem, as they can be encountered anywhere in the reachable workspace for a planned path in the operational space.

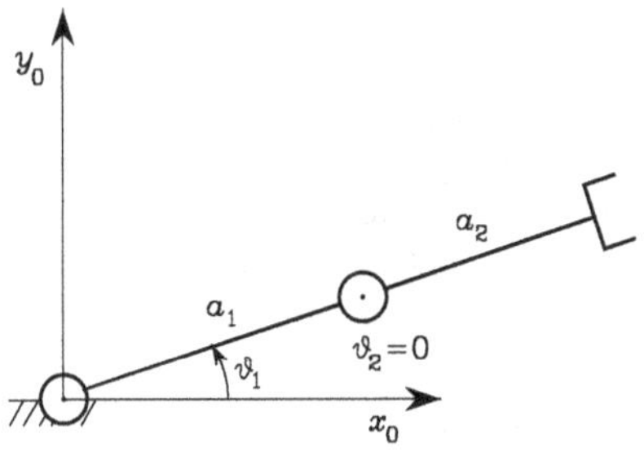

Fig. 3.3. Two-link planar arm at a boundary singularity

Example 3.2

To illustrate the behaviour of a manipulator at a singularity, consider a two-link planar arm. In this case, it is worth considering only the components $\dot{p}_x$ and $\dot{p}_y$ of the linear velocity in the plane. Thus, the Jacobian is the (2×2) matrix

$$\boldsymbol{J} = \begin{bmatrix} -a_1 s_1 - a_2 s_{12} & -a_2 s_{12} \\ a_1 c_1 + a_2 c_{12} & a_2 c_{12} \end{bmatrix}. \tag{3.40}$$

To analyze matrix rank, consider its determinant given by

$$\det(\boldsymbol{J}) = a_1 a_2 s_2. \tag{3.41}$$

For $a_1, a_2 \neq 0$, it is easy to find that the determinant in (3.41) vanishes whenever

$$\vartheta_2 = 0 \qquad \vartheta_2 = \pi,$$

ϑ_1 being irrelevant for the determination of singular configurations. These occur when the arm tip is located either on the outer ($\vartheta_2 = 0$) or on the inner ($\vartheta_2 = \pi$) boundary of the reachable workspace. Figure 3.3 illustrates the arm posture for $\vartheta_2 = 0$.

By analyzing the differential motion of the structure in such configuration, it can be observed that the two column vectors $[\,-(a_1+a_2)s_1 \quad (a_1+a_2)c_1\,]^T$ and $[\,-a_2 s_1 \quad a_2 c_1\,]^T$ of the Jacobian become parallel, and thus the Jacobian rank becomes one; this means that the tip velocity components are not independent (see point **a)** above).

3.3.1 Singularity Decoupling

Computation of internal singularities via the Jacobian determinant may be tedious and of no easy solution for complex structures. For manipulators having a spherical wrist, by analogy with what has already been seen for inverse kinematics, it is possible to split the problem of singularity computation into two separate problems:

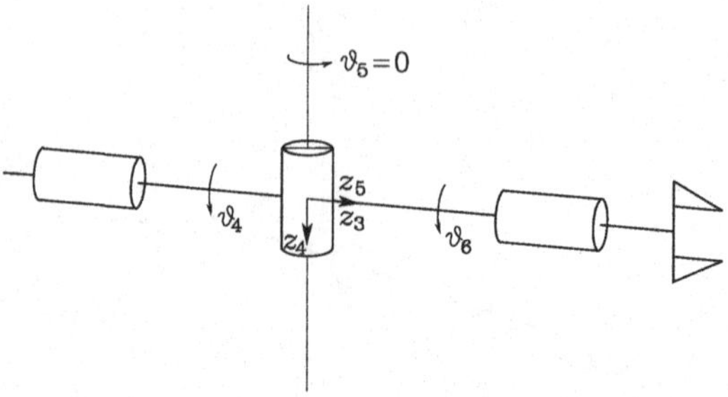

Fig. 3.4. Spherical wrist at a singularity

- computation of *arm singularities* resulting from the motion of the first 3 or more links,
- computation of *wrist singularities* resulting from the motion of the wrist joints.

For the sake of simplicity, consider the case $n = 6$; the Jacobian can be partitioned into (3×3) blocks as follows:

$$\boldsymbol{J} = \begin{bmatrix} \boldsymbol{J}_{11} & \boldsymbol{J}_{12} \\ \boldsymbol{J}_{21} & \boldsymbol{J}_{22} \end{bmatrix} \tag{3.42}$$

where, since the outer 3 joints are all revolute, the expressions of the two right blocks are respectively

$$\boldsymbol{J}_{12} = \begin{bmatrix} \boldsymbol{z}_3 \times (\boldsymbol{p}_e - \boldsymbol{p}_3) & \boldsymbol{z}_4 \times (\boldsymbol{p}_e - \boldsymbol{p}_4) & \boldsymbol{z}_5 \times (\boldsymbol{p}_e - \boldsymbol{p}_5) \end{bmatrix}$$

$$\boldsymbol{J}_{22} = \begin{bmatrix} \boldsymbol{z}_3 & \boldsymbol{z}_4 & \boldsymbol{z}_5 \end{bmatrix}. \tag{3.43}$$

As singularities are typical of the mechanical structure and do not depend on the frames chosen to describe kinematics, it is convenient to choose the origin of the end-effector frame at the intersection of the wrist axes (see Fig. 2.32). The choice $\boldsymbol{p} = \boldsymbol{p}_W$ leads to

$$\boldsymbol{J}_{12} = \begin{bmatrix} \boldsymbol{0} & \boldsymbol{0} & \boldsymbol{0} \end{bmatrix},$$

since all vectors $\boldsymbol{p}_W - \boldsymbol{p}_i$ are parallel to the unit vectors $\boldsymbol{z}_i$, for $i = 3, 4, 5$, no matter how Frames $3, 4, 5$ are chosen according to DH convention. In view of this choice, the overall Jacobian becomes a block lower-triangular matrix. In this case, computation of the determinant is greatly simplified, as this is given by the product of the determinants of the two blocks on the diagonal, i.e.,

$$\det(\boldsymbol{J}) = \det(\boldsymbol{J}_{11})\det(\boldsymbol{J}_{22}). \tag{3.44}$$

In turn, a true *singularity decoupling* has been achieved; the condition

$$\det(\boldsymbol{J}_{11}) = 0$$

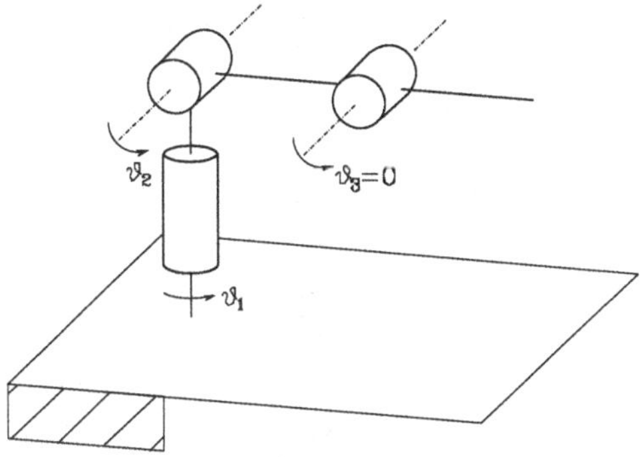

Fig. 3.5. Anthropomorphic arm at an elbow singularity

leads to determining the *arm singularities*, while the condition

$$\det(\boldsymbol{J}_{22}) = 0$$

leads to determining the *wrist singularities*.

Notice, however, that this form of Jacobian does not provide the relationship between the joint velocities and the end-effector velocity, but it leads to simplifying singularity computation. Below the two types of singularities are analyzed in detail.

3.3.2 Wrist Singularities

On the basis of the above singularity decoupling, wrist singularities can be determined by inspecting the block $\boldsymbol{J}_{22}$ in (3.43). It can be recognized that the wrist is at a singular configuration whenever the unit vectors $\boldsymbol{z}_3$, $\boldsymbol{z}_4$, $\boldsymbol{z}_5$ are linearly dependent. The wrist kinematic structure reveals that a singularity occurs when $\boldsymbol{z}_3$ and $\boldsymbol{z}_5$ are aligned, i.e., whenever

$$\vartheta_5 = 0 \qquad \vartheta_5 = \pi.$$

Taking into consideration only the first configuration (Fig. 3.4), the loss of mobility is caused by the fact that rotations of equal magnitude about opposite directions on ϑ_4 and ϑ_6 do not produce any end-effector rotation. Further, the wrist is not allowed to rotate about the axis orthogonal to $\boldsymbol{z}_4$ and $\boldsymbol{z}_3$, (see point **a)** above). This singularity is naturally described in the joint space and can be encountered anywhere inside the manipulator reachable workspace; as a consequence, special care is to be taken in programming an end-effector motion.

3.3.3 Arm Singularities

Arm singularities are characteristic of a specific manipulator structure; to illustrate their determination, consider the anthropomorphic arm (Fig. 2.23),

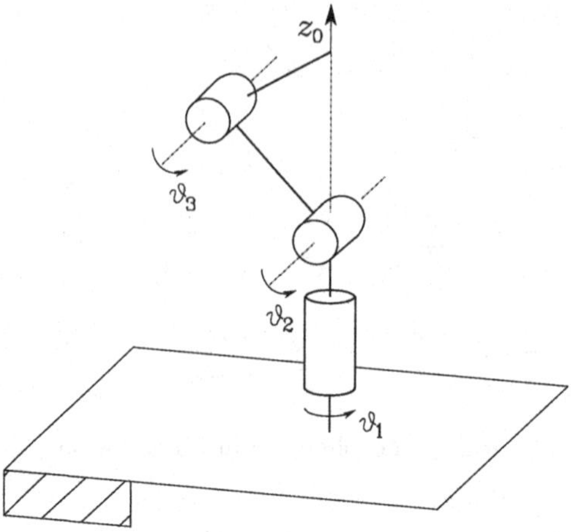

Fig. 3.6. Anthropomorphic arm at a shoulder singularity

whose Jacobian for the linear velocity part is given by (3.38). Its determinant is

$$\det(\boldsymbol{J}_P) = -a_2 a_3 s_3 (a_2 c_2 + a_3 c_{23}).$$

Like in the case of the planar arm of Example 3.2, the determinant does not depend on the first joint variable.

For $a_2, a_3 \neq 0$, the determinant vanishes if $s_3 = 0$ and/or $(a_2 c_2 + a_3 c_{23}) = 0$. The first situation occurs whenever

$$\vartheta_3 = 0 \qquad \vartheta_3 = \pi$$

meaning that the elbow is outstretched (Fig. 3.5) or retracted, and is termed *elbow singularity*. Notice that this type of singularity is conceptually equivalent to the singularity found for the two-link planar arm.

By recalling the direct kinematics equation in (2.66), it can be observed that the second situation occurs when the wrist point lies on axis z_0 (Fig. 3.6); it is thus characterized by

$$p_x = p_y = 0$$

and is termed *shoulder singularity*.

Notice that the whole axis z_0 describes a continuum of singular configurations; a rotation of ϑ_1 does not cause any translation of the wrist position (the first column of $\boldsymbol{J}_P$ is always null at a shoulder singularity), and then the kinematics equation admits infinite solutions; moreover, motions starting from the singular configuration that take the wrist along the z_1 direction are not allowed (see point **b)** above).

If a spherical wrist is connected to an anthropomorphic arm (Fig. 2.26), the arm direct kinematics is different. In this case the Jacobian to consider represents the block $\boldsymbol{J}_{11}$ of the Jacobian in (3.42) with $\boldsymbol{p} = \boldsymbol{p}_W$. Analyzing its determinant leads to finding the same singular configurations, which are relative to different values of the third joint variables, though — compare (2.66) and (2.70).

Finally, it is important to remark that, unlike the wrist singularities, the arm singularities are well identified in the operational space, and thus they can be suitably avoided in the end-effector trajectory planning stage.

3.4 Analysis of Redundancy

The concept of *kinematic redundancy* has been introduced in Sect. 2.10.2; redundancy is related to the number n of DOFs of the structure, the number m of operational space variables, and the number r of operational space variables necessary to specify a given task.

In order to perform a systematic analysis of redundancy, it is worth considering differential kinematics in lieu of direct kinematics (2.82). To this end, (3.39) is to be interpreted as the differential kinematics mapping relating the n components of the joint velocity vector to the $r \leq m$ components of the velocity vector $\boldsymbol{v}_e$ of concern for the specific task. To clarify this point, consider the case of a 3-link planar arm; that is not intrinsically redundant ($n = m = 3$) and its Jacobian (3.35) has 3 null rows accordingly. If the task does not specify ω_z ($r = 2$), the arm becomes functionally redundant and the Jacobian to consider for redundancy analysis is the one in (3.36).

A different case is that of the anthropomorphic arm for which only position variables are of concern ($n = m = 3$). The relevant Jacobian is the one in (3.38). The arm is neither intrinsically redundant nor can become functionally redundant if it is assigned a planar task; in that case, indeed, the task would set constraints on the 3 components of end-effector linear velocity.

Therefore, the differential kinematics equation to consider can be formally written as in (3.39), i.e.,

$$\boldsymbol{v}_e = \boldsymbol{J}(\boldsymbol{q})\dot{\boldsymbol{q}}, \tag{3.45}$$

where now $\boldsymbol{v}_e$ is meant to be the $(r \times 1)$ vector of end-effector velocity of concern for the specific task and $\boldsymbol{J}$ is the corresponding $(r \times n)$ Jacobian matrix that can be extracted from the geometric Jacobian; $\dot{\boldsymbol{q}}$ is the $(n \times 1)$ vector of joint velocities. If $r < n$, the manipulator is kinematically redundant and there exist $(n - r)$ *redundant DOFs*.

The Jacobian describes the linear mapping from the joint velocity space to the end-effector velocity space. In general, it is a function of the configuration. In the context of differential kinematics, however, the Jacobian has to be regarded as a constant matrix, since the instantaneous velocity mapping is of interest for a given posture. The mapping is schematically illustrated in Fig. 3.7 with a typical notation from set theory.

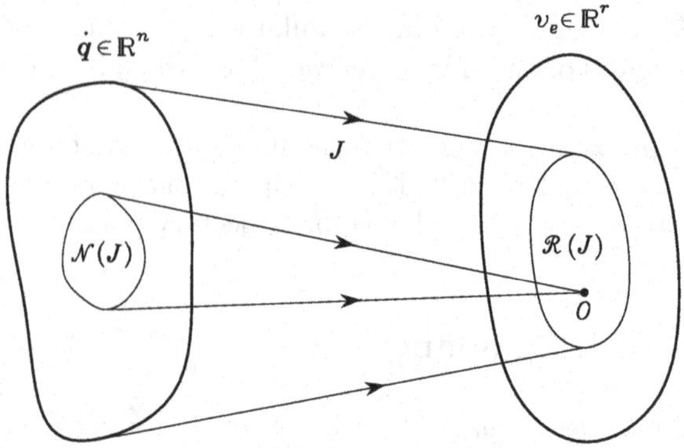

Fig. 3.7. Mapping between the joint velocity space and the end-effector velocity space

The differential kinematics equation in (3.45) can be characterized in terms of the *range* and *null* spaces of the mapping;[2] specifically, one has that:

- The *range* space of $\boldsymbol{J}$ is the subspace $\mathcal{R}(\boldsymbol{J})$ in $\mathbb{R}^r$ of the end-effector velocities that can be generated by the joint velocities, in the given manipulator posture.
- The *null* space of $\boldsymbol{J}$ is the subspace $\mathcal{N}(\boldsymbol{J})$ in $\mathbb{R}^n$ of joint velocities that do not produce any end-effector velocity, in the given manipulator posture.

If the Jacobian has *full rank*, one has

$$\dim\big(\mathcal{R}(\boldsymbol{J})\big) = r \qquad \dim\big(\mathcal{N}(\boldsymbol{J})\big) = n - r$$

and the range of $\boldsymbol{J}$ spans the entire space $\mathbb{R}^r$. Instead, if the Jacobian degenerates at a *singularity*, the dimension of the range space decreases while the dimension of the null space increases, since the following relation holds:

$$\dim\big(\mathcal{R}(\boldsymbol{J})\big) + \dim\big(\mathcal{N}(\boldsymbol{J})\big) = n$$

independently of the rank of the matrix $\boldsymbol{J}$.

The existence of a subspace $\mathcal{N}(\boldsymbol{J}) \neq \emptyset$ for a redundant manipulator allows determination of systematic techniques for handling redundant DOFs. To this end, if $\dot{\boldsymbol{q}}^*$ denotes a solution to (3.45) and $\boldsymbol{P}$ is an $(n \times n)$ matrix so that

$$\mathcal{R}(\boldsymbol{P}) \equiv \mathcal{N}(\boldsymbol{J}),$$

the joint velocity vector

$$\dot{\boldsymbol{q}} = \dot{\boldsymbol{q}}^* + \boldsymbol{P}\dot{\boldsymbol{q}}_0, \tag{3.46}$$

with arbitrary $\dot{\boldsymbol{q}}_0$, is also a solution to (3.45). In fact, premultiplying both sides of (3.46) by $\boldsymbol{J}$ yields

$$\boldsymbol{J}\dot{\boldsymbol{q}} = \boldsymbol{J}\dot{\boldsymbol{q}}^* + \boldsymbol{J}\boldsymbol{P}\dot{\boldsymbol{q}}_0 = \boldsymbol{J}\dot{\boldsymbol{q}}^* = \boldsymbol{v}_e$$

[2] See Sect. A.4 for the linear mappings.

since $\boldsymbol{J}\boldsymbol{P}\dot{\boldsymbol{q}}_0 = \boldsymbol{0}$ for any $\dot{\boldsymbol{q}}_0$. This result is of fundamental importance for redundancy resolution; a solution of the kind (3.46) points out the possibility of choosing the vector of arbitrary joint velocities $\dot{\boldsymbol{q}}_0$ so as to exploit advantageously the redundant DOFs. In fact, the effect of $\dot{\boldsymbol{q}}_0$ is to generate *internal motions* of the structure that do not change the end-effector position and orientation but may allow, for instance, manipulator reconfiguration into more dexterous postures for execution of a given task.

3.5 Inverse Differential Kinematics

In Sect. 2.12 it was shown how the inverse kinematics problem admits closed-form solutions only for manipulators having a simple kinematic structure. Problems arise whenever the end-effector attains a particular position and/or orientation in the operational space, or the structure is complex and it is not possible to relate the end-effector pose to different sets of joint variables, or else the manipulator is redundant. These limitations are caused by the highly nonlinear relationship between joint space variables and operational space variables.

On the other hand, the differential kinematics equation represents a linear mapping between the joint velocity space and the operational velocity space, although it varies with the current configuration. This fact suggests the possibility to utilize the differential kinematics equation to tackle the inverse kinematics problem.

Suppose that a motion trajectory is assigned to the end-effector in terms of $\boldsymbol{v}_e$ and the initial conditions on position and orientation. The aim is to determine a feasible joint trajectory $(\boldsymbol{q}(t), \dot{\boldsymbol{q}}(t))$ that reproduces the given trajectory.

By considering (3.45) with $n = r$, the joint velocities can be obtained via simple inversion of the Jacobian matrix

$$\dot{\boldsymbol{q}} = \boldsymbol{J}^{-1}(\boldsymbol{q})\boldsymbol{v}_e. \tag{3.47}$$

If the initial manipulator posture $\boldsymbol{q}(0)$ is known, joint positions can be computed by integrating velocities over time, i.e.,

$$\boldsymbol{q}(t) = \int_0^t \dot{\boldsymbol{q}}(\varsigma)d\varsigma + \boldsymbol{q}(0).$$

The integration can be performed in discrete time by resorting to numerical techniques. The simplest technique is based on the Euler integration method; given an integration interval Δt, if the joint positions and velocities at time t_k are known, the joint positions at time $t_{k+1} = t_k + \Delta t$ can be computed as

$$\boldsymbol{q}(t_{k+1}) = \boldsymbol{q}(t_k) + \dot{\boldsymbol{q}}(t_k)\Delta t. \tag{3.48}$$

This technique for inverting kinematics is independent of the solvability of the kinematic structure. Nonetheless, it is necessary that the *Jacobian* be *square* and of *full rank*; this demands further insight into the cases of *redundant* manipulators and kinematic *singularity* occurrence.

3.5.1 Redundant Manipulators

When the manipulator is *redundant* ($r < n$), the Jacobian matrix has more columns than rows and infinite solutions exist to (3.45). A viable solution method is to formulate the problem as a constrained linear optimization problem.

In detail, once the end-effector velocity $\boldsymbol{v}_e$ and Jacobian $\boldsymbol{J}$ are given (for a given configuration $\boldsymbol{q}$), it is desired to find the solutions $\dot{\boldsymbol{q}}$ that satisfy the linear equation in (3.45) and *minimize* the quadratic cost functional of joint velocities[3]

$$g(\dot{\boldsymbol{q}}) = \frac{1}{2}\dot{\boldsymbol{q}}^T\boldsymbol{W}\dot{\boldsymbol{q}}$$

where $\boldsymbol{W}$ is a suitable $(n \times n)$ symmetric positive definite weighting matrix.

This problem can be solved with the *method of Lagrange multipliers*. Consider the modified cost functional

$$g(\dot{\boldsymbol{q}}, \boldsymbol{\lambda}) = \frac{1}{2}\dot{\boldsymbol{q}}^T\boldsymbol{W}\dot{\boldsymbol{q}} + \boldsymbol{\lambda}^T(\boldsymbol{v}_e - \boldsymbol{J}\dot{\boldsymbol{q}}),$$

where $\boldsymbol{\lambda}$ is an $(r \times 1)$ vector of unknown multipliers that allows the incorporation of the constraint (3.45) in the functional to minimize. The requested solution has to satisfy the necessary conditions:

$$\left(\frac{\partial g}{\partial \dot{\boldsymbol{q}}}\right)^T = \boldsymbol{0} \qquad \left(\frac{\partial g}{\partial \boldsymbol{\lambda}}\right)^T = \boldsymbol{0}.$$

From the first one, it is $\boldsymbol{W}\dot{\boldsymbol{q}} - \boldsymbol{J}^T\boldsymbol{\lambda} = \boldsymbol{0}$ and thus

$$\dot{\boldsymbol{q}} = \boldsymbol{W}^{-1}\boldsymbol{J}^T\boldsymbol{\lambda} \tag{3.49}$$

where the inverse of $\boldsymbol{W}$ exists. Notice that the solution (3.49) is a minimum, since $\partial^2 g/\partial\dot{\boldsymbol{q}}^2 = \boldsymbol{W}$ is positive definite. From the second condition above, the constraint

$$\boldsymbol{v}_e = \boldsymbol{J}\dot{\boldsymbol{q}}$$

is recovered. Combining the two conditions gives

$$\boldsymbol{v}_e = \boldsymbol{J}\boldsymbol{W}^{-1}\boldsymbol{J}^T\boldsymbol{\lambda};$$

under the assumption that $\boldsymbol{J}$ has full rank, $\boldsymbol{J}\boldsymbol{W}^{-1}\boldsymbol{J}^T$ is an $(r \times r)$ square matrix of rank r and thus can be inverted. Solving for $\boldsymbol{\lambda}$ yields

$$\boldsymbol{\lambda} = (\boldsymbol{J}\boldsymbol{W}^{-1}\boldsymbol{J}^T)^{-1}\boldsymbol{v}_e$$

[3] Quadratic forms and the relative operations are recalled in Sect. A.6.

which, substituted into (3.49), gives the sought optimal solution

$$\dot{\boldsymbol{q}} = \boldsymbol{W}^{-1}\boldsymbol{J}^T(\boldsymbol{J}\boldsymbol{W}^{-1}\boldsymbol{J}^T)^{-1}\boldsymbol{v}_e. \tag{3.50}$$

Premultiplying both sides of (3.50) by $\boldsymbol{J}$, it is easy to verify that this solution satisfies the differential kinematics equation in (3.45).

A particular case occurs when the weighting matrix $\boldsymbol{W}$ is the identity matrix $\boldsymbol{I}$ and the solution simplifies into

$$\dot{\boldsymbol{q}} = \boldsymbol{J}^{\dagger}\boldsymbol{v}_e; \tag{3.51}$$

the matrix

$$\boldsymbol{J}^{\dagger} = \boldsymbol{J}^T(\boldsymbol{J}\boldsymbol{J}^T)^{-1} \tag{3.52}$$

is the *right pseudo-inverse* of $\boldsymbol{J}$.[4] The obtained solution locally minimizes the norm of joint velocities.

It was pointed out above that if $\dot{\boldsymbol{q}}^*$ is a solution to (3.45), $\dot{\boldsymbol{q}}^* + \boldsymbol{P}\dot{\boldsymbol{q}}_0$ is also a solution, where $\dot{\boldsymbol{q}}_0$ is a vector of arbitrary joint velocities and $\boldsymbol{P}$ is a projector in the null space of $\boldsymbol{J}$. Therefore, in view of the presence of redundant DOFs, the solution (3.51) can be modified by the introduction of another term of the kind $\boldsymbol{P}\dot{\boldsymbol{q}}_0$. In particular, $\dot{\boldsymbol{q}}_0$ can be specified so as to satisfy an additional constraint to the problem.

In that case, it is necessary to consider a new cost functional in the form

$$g'(\dot{\boldsymbol{q}}) = \frac{1}{2}(\dot{\boldsymbol{q}} - \dot{\boldsymbol{q}}_0)^T(\dot{\boldsymbol{q}} - \dot{\boldsymbol{q}}_0);$$

this choice is aimed at minimizing the norm of vector $\dot{\boldsymbol{q}} - \dot{\boldsymbol{q}}_0$; in other words, solutions are sought which satisfy the constraint (3.45) and are as close as possible to $\dot{\boldsymbol{q}}_0$. In this way, the objective specified through $\dot{\boldsymbol{q}}_0$ becomes unavoidably a secondary objective to satisfy with respect to the primary objective specified by the constraint (3.45).

Proceeding in a way similar to the above yields

$$g'(\dot{\boldsymbol{q}}, \boldsymbol{\lambda}) = \frac{1}{2}(\dot{\boldsymbol{q}} - \dot{\boldsymbol{q}}_0)^T(\dot{\boldsymbol{q}} - \dot{\boldsymbol{q}}_0) + \boldsymbol{\lambda}^T(\boldsymbol{v}_e - \boldsymbol{J}\dot{\boldsymbol{q}});$$

from the first necessary condition it is

$$\dot{\boldsymbol{q}} = \boldsymbol{J}^T\boldsymbol{\lambda} + \dot{\boldsymbol{q}}_0 \tag{3.53}$$

which, substituted into (3.45), gives

$$\boldsymbol{\lambda} = (\boldsymbol{J}\boldsymbol{J}^T)^{-1}(\boldsymbol{v}_e - \boldsymbol{J}\dot{\boldsymbol{q}}_0).$$

Finally, substituting $\boldsymbol{\lambda}$ back in (3.53) gives

$$\dot{\boldsymbol{q}} = \boldsymbol{J}^{\dagger}\boldsymbol{v}_e + (\boldsymbol{I}_n - \boldsymbol{J}^{\dagger}\boldsymbol{J})\dot{\boldsymbol{q}}_0. \tag{3.54}$$

---

[4] See Sect. A.7 for the definition of the pseudo-inverse of a matrix.

As can be easily recognized, the obtained solution is composed of two terms. The first is relative to minimum norm joint velocities. The second, termed *homogeneous solution*, attempts to satisfy the additional constraint to specify via $\dot{\boldsymbol{q}}_0$;[5] the matrix $(\boldsymbol{I} - \boldsymbol{J}^\dagger \boldsymbol{J})$ is one of those matrices $\boldsymbol{P}$ introduced in (3.46) which allows the projection of the vector $\dot{\boldsymbol{q}}_0$ in the null space of $\boldsymbol{J}$, so as not to violate the constraint (3.45). A direct consequence is that, in the case $\boldsymbol{v}_e = \boldsymbol{0}$, is is possible to generate *internal motions* described by $(\boldsymbol{I} - \boldsymbol{J}^\dagger \boldsymbol{J})\dot{\boldsymbol{q}}_0$ that reconfigure the manipulator structure without changing the end-effector position and orientation.

Finally, it is worth discussing the way to specify the vector $\dot{\boldsymbol{q}}_0$ for a convenient utilization of redundant DOFs. A typical choice is

$$\dot{\boldsymbol{q}}_0 = k_0 \left(\frac{\partial w(\boldsymbol{q})}{\partial \boldsymbol{q}} \right)^T \tag{3.55}$$

where $k_0 > 0$ and $w(\boldsymbol{q})$ is a (secondary) objective function of the joint variables. Since the solution moves along the direction of the gradient of the objective function, it attempts to *maximize* it *locally* compatible to the primary objective (kinematic constraint). Typical objective functions are:

- The *manipulability measure*, defined as

$$w(\boldsymbol{q}) = \sqrt{\det\big(\boldsymbol{J}(\boldsymbol{q})\boldsymbol{J}^T(\boldsymbol{q})\big)} \tag{3.56}$$

  which vanishes at a singular configuration; thus, by maximizing this measure, redundancy is exploited to move away from singularities.[6]
- The *distance from mechanical joint limits*, defined as

$$w(\boldsymbol{q}) = -\frac{1}{2n} \sum_{i=1}^{n} \left(\frac{q_i - \bar{q}_i}{q_{iM} - q_{im}} \right)^2 \tag{3.57}$$

  where q_{iM} (q_{im}) denotes the maximum (minimum) joint limit and $\bar{q}_i$ the middle value of the joint range; thus, by maximizing this distance, redundancy is exploited to keep the joint variables as close as possible to the centre of their ranges.
- The *distance from an obstacle*, defined as

$$w(\boldsymbol{q}) = \min_{\boldsymbol{p},\boldsymbol{o}} \|\boldsymbol{p}(\boldsymbol{q}) - \boldsymbol{o}\| \tag{3.58}$$

  where $\boldsymbol{o}$ is the position vector of a suitable point on the obstacle (its centre, for instance, if the obstacle is modelled as a sphere) and $\boldsymbol{p}$ is the

[5] It should be recalled that the additional constraint has secondary priority with respect to the primary kinematic constraint.

[6] The manipulability measure is given by the product of the singular values of the Jacobian (see Problem 3.8).

position vector of a generic point along the structure; thus, by maximizing this distance, redundancy is exploited to avoid collision of the manipulator with an obstacle (see also Problem 3.9).[7]

3.5.2 Kinematic Singularities

Both solutions (3.47) and (3.51) can be computed only when the Jacobian has full rank. Hence, they become meaningless when the manipulator is at a singular configuration; in such a case, the system $\boldsymbol{v}_e = \boldsymbol{J}\dot{\boldsymbol{q}}$ contains linearly dependent equations.

It is possible to find a solution $\dot{\boldsymbol{q}}$ by extracting all the linearly independent equations only if $\boldsymbol{v}_e \in \mathcal{R}(\boldsymbol{J})$. The occurrence of this situation means that the assigned path is physically executable by the manipulator, even though it is at a singular configuration. If instead $\boldsymbol{v}_e \notin \mathcal{R}(\boldsymbol{J})$, the system of equations has no solution; this means that the operational space path cannot be executed by the manipulator at the given posture.

It is important to underline that the inversion of the Jacobian can represent a serious inconvenience not only at a singularity but also in the neighbourhood of a singularity. For instance, for the Jacobian inverse it is well known that its computation requires the computation of the determinant; in the neighbourhood of a singularity, the determinant takes on a relatively small value which can cause large joint velocities (see point **c)** in Sect. 3.3). Consider again the above example of the shoulder singularity for the anthropomorphic arm. If a path is assigned to the end-effector which passes nearby the base rotation axis (geometric locus of singular configurations), the base joint is forced to make a rotation of about π in a relatively short time to allow the end-effector to keep tracking the imposed trajectory.

A more rigorous analysis of the solution features in the neighbourhood of singular configurations can be developed by resorting to the singular value decomposition (SVD) of matrix $\boldsymbol{J}$.[8]

An alternative solution overcoming the problem of inverting differential kinematics in the neighbourhood of a singularity is provided by the so-called *damped least-squares (DLS) inverse*

$$\boldsymbol{J}^{\star} = \boldsymbol{J}^T(\boldsymbol{J}\boldsymbol{J}^T + k^2\boldsymbol{I})^{-1} \tag{3.59}$$

where k is a damping factor that renders the inversion better conditioned from a numerical viewpoint. It can be shown that such a solution can be

[7] If an obstacle occurs along the end-effector path, it is opportune to invert the order of priority between the kinematic constraint and the additional constraint; in this way the obstacle may be avoided, but one gives up tracking the desired path.

[8] See Sect. A.8.

obtained by reformulating the problem in terms of the minimization of the cost functional

$$g''(\dot{\boldsymbol{q}}) = \frac{1}{2}(\boldsymbol{v}_e - \boldsymbol{J}\dot{\boldsymbol{q}})^T(\boldsymbol{v}_e - \boldsymbol{J}\dot{\boldsymbol{q}}) + \frac{1}{2}k^2\dot{\boldsymbol{q}}^T\dot{\boldsymbol{q}},$$

where the introduction of the first term allows a finite inversion error to be tolerated, with the advantage of norm-bounded velocities. The factor k establishes the relative weight between the two objectives, and there exist techniques for selecting optimal values for the damping factor (see Problem 3.10).

3.6 Analytical Jacobian

The above sections have shown the way to compute the end-effector velocity in terms of the velocity of the end-effector frame. The Jacobian is computed according to a *geometric technique* in which the contributions of each joint velocity to the components of end-effector linear and angular velocity are determined.

If the end-effector pose is specified in terms of a minimal number of parameters in the operational space as in (2.80), it is natural to ask whether it is possible to compute the Jacobian via differentiation of the direct kinematics function with respect to the joint variables. To this end, an *analytical technique* is presented below to compute the Jacobian, and the existing relationship between the two Jacobians is found.

The translational velocity of the end-effector frame can be expressed as the time derivative of vector $\boldsymbol{p}_e$, representing the origin of the end-effector frame with respect to the base frame, i.e.,

$$\dot{\boldsymbol{p}}_e = \frac{\partial \boldsymbol{p}_e}{\partial \boldsymbol{q}}\dot{\boldsymbol{q}} = \boldsymbol{J}_P(\boldsymbol{q})\dot{\boldsymbol{q}}. \tag{3.60}$$

For what concerns the rotational velocity of the end-effector frame, the minimal representation of orientation in terms of three variables $\boldsymbol{\phi}_e$ can be considered. Its time derivative $\dot{\boldsymbol{\phi}}_e$ in general differs from the angular velocity vector defined above. In any case, once the function $\boldsymbol{\phi}_e(\boldsymbol{q})$ is known, it is formally correct to consider the Jacobian obtained as

$$\dot{\boldsymbol{\phi}}_e = \frac{\partial \boldsymbol{\phi}_e}{\partial \boldsymbol{q}}\dot{\boldsymbol{q}} = \boldsymbol{J}_\phi(\boldsymbol{q})\dot{\boldsymbol{q}}. \tag{3.61}$$

Computing the Jacobian $\boldsymbol{J}_\phi(\boldsymbol{q})$ as $\partial\boldsymbol{\phi}_e/\partial\boldsymbol{q}$ is not straightforward, since the function $\boldsymbol{\phi}_e(\boldsymbol{q})$ is not usually available in direct form, but requires computation of the elements of the relative rotation matrix.

Upon these premises, the differential kinematics equation can be obtained as the time derivative of the direct kinematics equation in (2.82), i.e.,

$$\dot{\boldsymbol{x}}_e = \begin{bmatrix} \dot{\boldsymbol{p}}_e \\ \dot{\boldsymbol{\phi}}_e \end{bmatrix} = \begin{bmatrix} \boldsymbol{J}_P(\boldsymbol{q}) \\ \boldsymbol{J}_\phi(\boldsymbol{q}) \end{bmatrix}\dot{\boldsymbol{q}} = \boldsymbol{J}_A(\boldsymbol{q})\dot{\boldsymbol{q}} \tag{3.62}$$

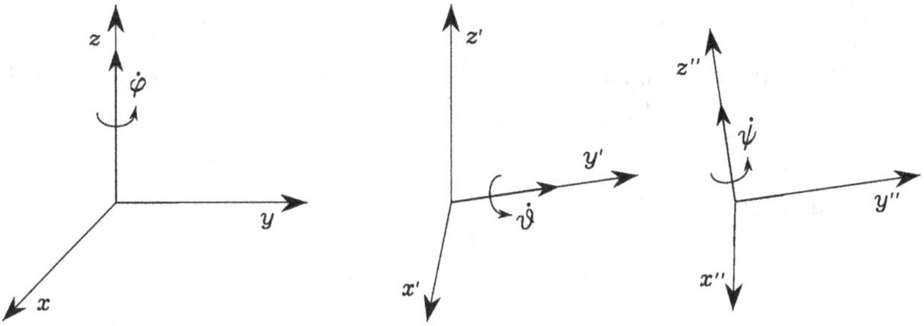

Fig. 3.8. Rotational velocities of Euler angles ZYZ in current frame

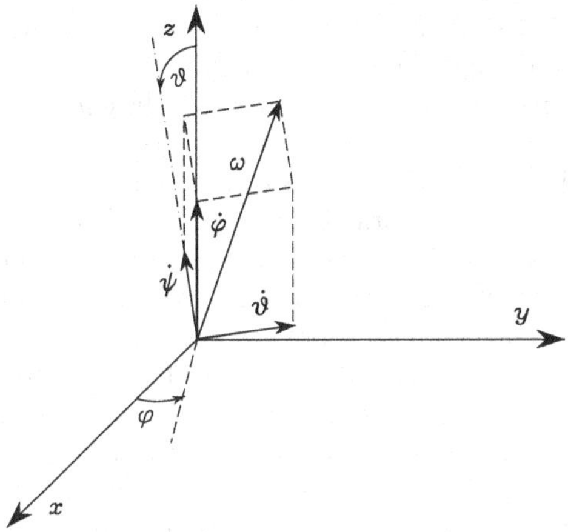

Fig. 3.9. Composition of elementary rotational velocities for computing angular velocity

where the *analytical Jacobian*

$$\boldsymbol{J}_A(\boldsymbol{q}) = \frac{\partial \boldsymbol{k}(\boldsymbol{q})}{\partial \boldsymbol{q}} \tag{3.63}$$

is different from the geometric Jacobian $\boldsymbol{J}$, since the end-effector angular velocity $\boldsymbol{\omega}_e$ with respect to the base frame is not given by $\dot{\boldsymbol{\phi}}_e$.

It is possible to find the relationship between the angular velocity $\boldsymbol{\omega}_e$ and the rotational velocity $\dot{\boldsymbol{\phi}}_e$ for a given set of orientation angles. For instance, consider the Euler angles ZYZ defined in Sect. 2.4.1; in Fig. 3.8, the vectors corresponding to the rotational velocities $\dot{\varphi}$, $\dot{\vartheta}$, $\dot{\psi}$ have been represented with reference to the current frame. Figure 3.9 illustrates how to compute the contributions of each rotational velocity to the components of angular velocity about the axes of the reference frame:

- as a result of $\dot{\varphi}$: $[\omega_x \quad \omega_y \quad \omega_z]^T = \dot{\varphi}\,[0 \quad 0 \quad 1]^T$
- as a result of $\dot{\vartheta}$: $[\omega_x \quad \omega_y \quad \omega_z]^T = \dot{\vartheta}\,[-s_\varphi \quad c_\varphi \quad 0]^T$

- as a result of $\dot{\psi}$: $[\,\omega_x \quad \omega_y \quad \omega_z\,]^T = \dot{\psi}\,[\,c_\varphi s_\vartheta \quad s_\varphi s_\vartheta \quad c_\vartheta\,]^T$,

and then the equation relating the angular velocity $\boldsymbol{\omega}_e$ to the time derivative of the Euler angles $\dot{\boldsymbol{\phi}}_e$ is[9]

$$\boldsymbol{\omega}_e = \boldsymbol{T}(\boldsymbol{\phi}_e)\dot{\boldsymbol{\phi}}_e, \tag{3.64}$$

where, in this case,

$$\boldsymbol{T} = \begin{bmatrix} 0 & -s_\varphi & c_\varphi s_\vartheta \\ 0 & c_\varphi & s_\varphi s_\vartheta \\ 1 & 0 & c_\vartheta \end{bmatrix}.$$

The determinant of matrix $\boldsymbol{T}$ is $-s_\vartheta$, which implies that the relationship cannot be inverted for $\vartheta = 0, \pi$. This means that, even though all rotational velocities of the end-effector frame can be expressed by means of a suitable angular velocity vector $\boldsymbol{\omega}_e$, there exist angular velocities which cannot be expressed by means of $\dot{\boldsymbol{\phi}}_e$ when the orientation of the end-effector frame causes $s_\vartheta = 0$.[10] In fact, in this situation, the angular velocities that can be described by $\dot{\boldsymbol{\phi}}_e$ should have linearly dependent components in the directions orthogonal to axis z ($\omega_x^2 + \omega_y^2 = \dot{\vartheta}^2$). An orientation for which the determinant of the transformation matrix vanishes is termed *representation singularity* of $\boldsymbol{\phi}_e$.

From a physical viewpoint, the meaning of $\boldsymbol{\omega}_e$ is more intuitive than that of $\dot{\boldsymbol{\phi}}_e$. The three components of $\boldsymbol{\omega}_e$ represent the components of angular velocity with respect to the base frame. Instead, the three elements of $\dot{\boldsymbol{\phi}}_e$ represent nonorthogonal components of angular velocity defined with respect to the axes of a frame that varies as the end-effector orientation varies. On the other hand, while the integral of $\dot{\boldsymbol{\phi}}_e$ over time gives $\boldsymbol{\phi}_e$, the integral of $\boldsymbol{\omega}_e$ does not admit a clear physical interpretation, as can be seen in the following example.

Example 3.3

Consider an object whose orientation with respect to a reference frame is known at time $t = 0$. Assign the following time profiles to $\boldsymbol{\omega}$:
- $\boldsymbol{\omega} = [\,\pi/2 \quad 0 \quad 0\,]^T \quad 0 \le t \le 1 \qquad \boldsymbol{\omega} = [\,0 \quad \pi/2 \quad 0\,]^T \quad 1 < t \le 2$,
- $\boldsymbol{\omega} = [\,0 \quad \pi/2 \quad 0\,]^T \quad 0 \le t \le 1 \qquad \boldsymbol{\omega} = [\,\pi/2 \quad 0 \quad 0\,]^T \quad 1 < t \le 2$.

The integral of $\boldsymbol{\omega}$ gives the same result in the two cases

$$\int_0^2 \boldsymbol{\omega} dt = [\,\pi/2 \quad \pi/2 \quad 0\,]^T$$

but the final object orientation corresponding to the second timing law is clearly different from the one obtained with the first timing law (Fig. 3.10).

---

[9] This relation can also be obtained from the rotation matrix associated with the three angles (see Problem 3.11).

[10] In Sect. 2.4.1, it was shown that for this orientation the inverse solution of the Euler angles degenerates.

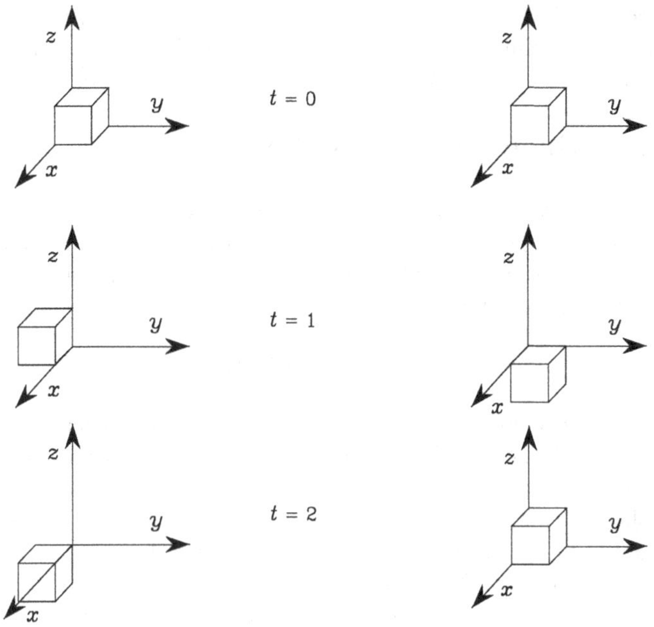

Fig. 3.10. Nonuniqueness of orientation computed as the integral of angular velocity

Once the transformation $\boldsymbol{T}$ between $\boldsymbol{\omega}_e$ and $\dot{\boldsymbol{\phi}}_e$ is given, the analytical Jacobian can be related to the geometric Jacobian as

$$\boldsymbol{v}_e = \begin{bmatrix} \boldsymbol{I} & \boldsymbol{O} \\ \boldsymbol{O} & \boldsymbol{T}(\boldsymbol{\phi}_e) \end{bmatrix} \dot{\boldsymbol{x}}_e = \boldsymbol{T}_A(\boldsymbol{\phi}_e)\dot{\boldsymbol{x}}_e \tag{3.65}$$

which, in view of (3.4), (3.62), yields

$$\boldsymbol{J} = \boldsymbol{T}_A(\boldsymbol{\phi})\boldsymbol{J}_A. \tag{3.66}$$

This relationship shows that $\boldsymbol{J}$ and $\boldsymbol{J}_A$, in general, differ. Regarding the use of either one or the other in all those problems where the influence of the Jacobian matters, it is anticipated that the geometric Jacobian will be adopted whenever it is necessary to refer to quantities of clear physical meaning, while the analytical Jacobian will be adopted whenever it is necessary to refer to differential quantities of variables defined in the operational space.

For certain manipulator geometries, it is possible to establish a substantial equivalence between $\boldsymbol{J}$ and $\boldsymbol{J}_A$. In fact, when the DOFs cause rotations of the end-effector all about the same fixed axis in space, the two Jacobians are essentially the same. This is the case of the above three-link planar arm. Its geometric Jacobian (3.35) reveals that only rotations about axis z_0 are permitted. The (3×3) analytical Jacobian that can be derived by considering the end-effector position components in the plane of the structure and defining the end-effector orientation as $\phi = \vartheta_1 + \vartheta_2 + \vartheta_3$ coincides with the matrix that is obtained by eliminating the three null rows of the geometric Jacobian.

3.7 Inverse Kinematics Algorithms

In Sect. 3.5 it was shown how to invert kinematics by using the differential kinematics equation. In the numerical implementation of (3.48), computation of joint velocities is obtained by using the inverse of the Jacobian evaluated with the joint variables at the previous instant of time

$$\boldsymbol{q}(t_{k+1}) = \boldsymbol{q}(t_k) + \boldsymbol{J}^{-1}(\boldsymbol{q}(t_k))\boldsymbol{v}_e(t_k)\Delta t.$$

It follows that the computed joint velocities $\dot{\boldsymbol{q}}$ do not coincide with those satisfying (3.47) in the continuous time. Therefore, reconstruction of joint variables $\boldsymbol{q}$ is entrusted to a numerical integration which involves *drift* phenomena of the solution; as a consequence, the end-effector pose corresponding to the computed joint variables differs from the desired one.

This inconvenience can be overcome by resorting to a solution scheme that accounts for the *operational space error* between the desired and the actual end-effector position and orientation. Let

$$\boldsymbol{e} = \boldsymbol{x}_d - \boldsymbol{x}_e \tag{3.67}$$

be the expression of such error.

Consider the time derivative of (3.67), i.e.,

$$\dot{\boldsymbol{e}} = \dot{\boldsymbol{x}}_d - \dot{\boldsymbol{x}}_e \tag{3.68}$$

which, according to differential kinematics (3.62), can be written as

$$\dot{\boldsymbol{e}} = \dot{\boldsymbol{x}}_d - \boldsymbol{J}_A(\boldsymbol{q})\dot{\boldsymbol{q}}. \tag{3.69}$$

Notice in (3.69) that the use of operational space quantities has naturally lead to using the analytical Jacobian in lieu of the geometric Jacobian. For this equation to lead to an *inverse kinematics algorithm*, it is worth relating the computed joint velocity vector $\dot{\boldsymbol{q}}$ to the error $\boldsymbol{e}$ so that (3.69) gives a differential equation describing error evolution over time. Nonetheless, it is necessary to choose a relationship between $\dot{\boldsymbol{q}}$ and $\boldsymbol{e}$ that ensures convergence of the error to zero.

Having formulated inverse kinematics in algorithmic terms implies that the joint variables $\boldsymbol{q}$ corresponding to a given end-effector pose $\boldsymbol{x}_d$ are accurately computed only when the error $\boldsymbol{x}_d - \boldsymbol{k}(\boldsymbol{q})$ is reduced within a given threshold; such settling time depends on the dynamic characteristics of the error differential equation. The choice of $\dot{\boldsymbol{q}}$ as a function of $\boldsymbol{e}$ permits finding inverse kinematics algorithms with different features.

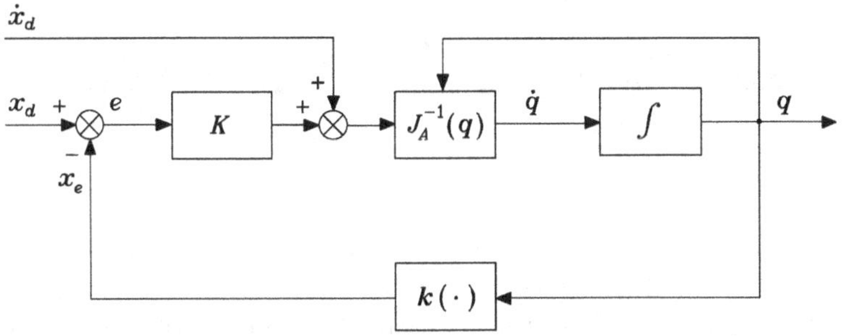

Fig. 3.11. Inverse kinematics algorithm with Jacobian inverse

3.7.1 Jacobian (Pseudo-)inverse

On the assumption that matrix $\boldsymbol{J}_A$ is square and nonsingular, the choice

$$\dot{\boldsymbol{q}} = \boldsymbol{J}_A^{-1}(\boldsymbol{q})(\dot{\boldsymbol{x}}_d + \boldsymbol{K}\boldsymbol{e}) \tag{3.70}$$

leads to the equivalent linear system

$$\dot{\boldsymbol{e}} + \boldsymbol{K}\boldsymbol{e} = \boldsymbol{0}. \tag{3.71}$$

If $\boldsymbol{K}$ is a positive definite (usually diagonal) matrix, the system (3.71) is *asymptotically stable*. The error tends to zero along the trajectory with a convergence rate that depends on the eigenvalues of matrix $\boldsymbol{K}$;[11] the larger the eigenvalues, the faster the convergence. Since the scheme is practically implemented as a discrete-time system, it is reasonable to predict that an upper bound exists on the eigenvalues; depending on the sampling time, there will be a limit for the maximum eigenvalue of $\boldsymbol{K}$ under which asymptotic stability of the error system is guaranteed.

The block scheme corresponding to the inverse kinematics algorithm in (3.70) is illustrated in Fig. 3.11, where $\boldsymbol{k}(\cdot)$ indicates the direct kinematics function in (2.82). This scheme can be revisited in terms of the usual feedback control schemes. Specifically, it can observed that the nonlinear block $\boldsymbol{k}(\cdot)$ is needed to compute $\boldsymbol{x}$ and thus the tracking error $\boldsymbol{e}$, while the block $\boldsymbol{J}_A^{-1}(\boldsymbol{q})$ has been introduced to compensate for $\boldsymbol{J}_A(\boldsymbol{q})$ and making the system linear. The block scheme shows the presence of a string of integrators on the forward loop and then, for a constant reference ($\dot{\boldsymbol{x}}_d = \boldsymbol{0}$), guarantees a null steady-state error. Further, the *feedforward* action provided by $\dot{\boldsymbol{x}}_d$ for a time-varying reference ensures that the error is kept to zero (in the case $\boldsymbol{e}(0) = \boldsymbol{0}$) along the whole trajectory, independently of the type of desired reference $\boldsymbol{x}_d(t)$.

Finally, notice that (3.70), for $\dot{\boldsymbol{x}}_d = \boldsymbol{0}$, corresponds to the Newton method for solving a system of nonlinear equations. Given a constant end-effector pose $\boldsymbol{x}_d$, the algorithm can be keenly applied to compute one of the admissible

[11] See Sect. A.5.

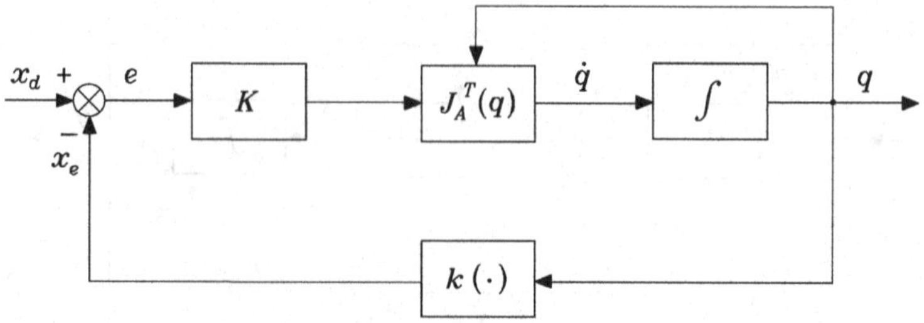

Fig. 3.12. Block scheme of the inverse kinematics algorithm with Jacobian transpose

solutions to the inverse kinematics problem, whenever that does not admit closed-form solutions, as discussed in Sect. 2.12. Such a method is also useful in practice at the start-up of the manipulator for a given task, to compute the corresponding joint configuration.

In the case of a *redundant manipulator*, solution (3.70) can be generalized into

$$\dot{\boldsymbol{q}} = \boldsymbol{J}_A^{\dagger}(\dot{\boldsymbol{x}}_d + \boldsymbol{K}\boldsymbol{e}) + (\boldsymbol{I}_n - \boldsymbol{J}_A^{\dagger}\boldsymbol{J}_A)\dot{\boldsymbol{q}}_0, \tag{3.72}$$

which represents the algorithmic version of solution (3.54).

The structure of the inverse kinematics algorithm can be conceptually adopted for a simple robot control technique, known under the name of *kinematic control*. As will be seen in Chap. 7, a manipulator is actually an electromechanical system actuated by motor torques, while in Chaps. 8–10 dynamic control techniques will be presented which will properly account for the nonlinear and coupling effects of the dynamic model.

At first approximation, however, it is possible to consider a kinematic command as system input, typically a velocity. This is possible in view of the presence of a low-level control loop, which 'ideally' imposes any specified reference velocity. On the other hand, such a loop already exists in a 'closed' control unit, where the user can also intervene with kinematic commands. In other words, the scheme in Fig. 3.11 can implement a kinematic control, provided that the integrator is regarded as a simplified model of the robot, thanks to the presence of single joint local servos, which ensure a more or less accurate reproduction of the velocity commands. Nevertheless, it is worth underlining that such a kinematic control technique yields satisfactory performance only when one does not require too fast motions or rapid accelerations. The performance of the independent joint control will be analyzed in Sect. 8.3.

3.7.2 Jacobian Transpose

A computationally simpler algorithm can be derived by finding a relationship between $\dot{\boldsymbol{q}}$ and $\boldsymbol{e}$ that ensures error convergence to zero, without requiring linearization of (3.69). As a consequence, the error dynamics is governed by a nonlinear differential equation. The Lyapunov direct method can be utilized to determine a dependence $\dot{\boldsymbol{q}}(\boldsymbol{e})$ that ensures asymptotic stability of the error system. Choose as Lyapunov function candidate the positive definite quadratic form[12]

$$V(\boldsymbol{e}) = \frac{1}{2}\boldsymbol{e}^T\boldsymbol{K}\boldsymbol{e}, \tag{3.73}$$

where $\boldsymbol{K}$ is a symmetric positive definite matrix. This function is so that

$$V(\boldsymbol{e}) > 0 \quad \forall \boldsymbol{e} \neq \boldsymbol{0}, \qquad V(\boldsymbol{0}) = 0.$$

Differentiating (3.73) with respect to time and accounting for (3.68) gives

$$\dot{V} = \boldsymbol{e}^T\boldsymbol{K}\dot{\boldsymbol{x}}_d - \boldsymbol{e}^T\boldsymbol{K}\dot{\boldsymbol{x}}_e. \tag{3.74}$$

In view of (3.62), it is

$$\dot{V} = \boldsymbol{e}^T\boldsymbol{K}\dot{\boldsymbol{x}}_d - \boldsymbol{e}^T\boldsymbol{K}\boldsymbol{J}_A(\boldsymbol{q})\dot{\boldsymbol{q}}. \tag{3.75}$$

At this point, the choice of joint velocities as

$$\dot{\boldsymbol{q}} = \boldsymbol{J}_A^T(\boldsymbol{q})\boldsymbol{K}\boldsymbol{e} \tag{3.76}$$

leads to

$$\dot{V} = \boldsymbol{e}^T\boldsymbol{K}\dot{\boldsymbol{x}}_d - \boldsymbol{e}^T\boldsymbol{K}\boldsymbol{J}_A(\boldsymbol{q})\boldsymbol{J}_A^T(\boldsymbol{q})\boldsymbol{K}\boldsymbol{e}. \tag{3.77}$$

Consider the case of a constant reference ($\dot{\boldsymbol{x}}_d = \boldsymbol{0}$). The function in (3.77) is negative definite, under the assumption of full rank for $\boldsymbol{J}_A(\boldsymbol{q})$. The condition $\dot{V} < 0$ with $V > 0$ implies that the system trajectories uniformly converge to $\boldsymbol{e} = \boldsymbol{0}$, i.e., the system is *asymptotically stable*. When $\mathcal{N}(\boldsymbol{J}_A^T) \neq \emptyset$, the function in (3.77) is only negative semi-definite, since $\dot{V} = 0$ for $\boldsymbol{e} \neq \boldsymbol{0}$ with $\boldsymbol{K}\boldsymbol{e} \in \mathcal{N}(\boldsymbol{J}_A^T)$. In this case, the algorithm can get stuck at $\dot{\boldsymbol{q}} = \boldsymbol{0}$ with $\boldsymbol{e} \neq \boldsymbol{0}$. However, the example that follows will show that this situation occurs only if the assigned end-effector position is not actually reachable from the current configuration.

The resulting block scheme is illustrated in Fig. 3.12, which shows the notable feature of the algorithm to require *computation only of direct kinematics functions* $\boldsymbol{k}(\boldsymbol{q})$, $\boldsymbol{J}_A^T(\boldsymbol{q})$.

It can be recognized that (3.76) corresponds to the gradient method for the solution of a system on nonlinear equations. As in the case of the Jacobian inverse solution, for a given constant end-effector pose $\boldsymbol{x}_d$, the Jacobian transpose algorithm can be keenly employed to solve the inverse kinematics problem, or more simply to initialize the values of the manipulator joint variables.

The case when $\boldsymbol{x}_d$ is a time-varying function ($\dot{\boldsymbol{x}}_d \neq \boldsymbol{0}$) deserves a separate analysis. In order to obtain $\dot{V} < 0$ also in this case, it would be sufficient to choose a $\dot{\boldsymbol{q}}$ that depends on the (pseudo-)inverse of the Jacobian as in (3.70),

[12] See Sect. C.3 for the presentation of the Lyapunov direct method.

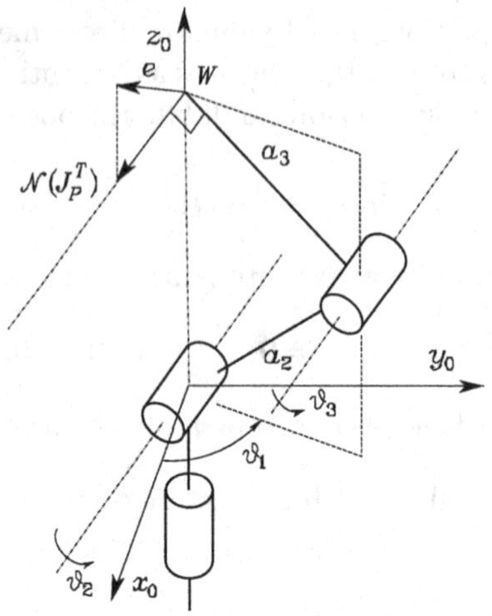

Fig. 3.13. Characterization of the anthropomorphic arm at a shoulder singularity for the admissible solutions of the Jacobian transpose algorithm

recovering the asymptotic stability result derived above.[13] For the inversion scheme based on the transpose, the first term on the right-hand side of (3.77) is not cancelled any more and nothing can be said about its sign. This implies that asymptotic stability along the trajectory cannot be achieved. The tracking error $\boldsymbol{e}(t)$ is, anyhow, norm-bounded; the larger the norm of $\boldsymbol{K}$, the smaller the norm of $\boldsymbol{e}$.[14] In practice, since the inversion scheme is to be implemented in discrete-time, there is an upper bound on the norm of $\boldsymbol{K}$ with reference to the adopted sampling time.

Example 3.4

Consider the anthropomorphic arm; a shoulder singularity occurs whenever $a_2c_2 + a_3c_{23} = 0$ (Fig. 3.6). In this configuration, the transpose of the Jacobian in (3.38) is

$$\boldsymbol{J}_P^T = \begin{bmatrix} 0 & 0 & 0 \\ -c_1(a_2s_2 + a_3s_{23}) & -s_1(a_2s_2 + a_3s_{23}) & 0 \\ -a_3c_1s_{23} & -a_3s_1s_{23} & a_3c_{23} \end{bmatrix}.$$

By computing the null space of $\boldsymbol{J}_P^T$, if ν_x, ν_y and ν_z denote the components of vector $\boldsymbol{\nu}$ along the axes of the base frame, one has the result

$$\frac{\nu_y}{\nu_x} = -\frac{1}{\tan\vartheta_1} \qquad \nu_z = 0,$$

13 Notice that, anyhow, in case of kinematic singularities, it is necessary to resort to an inverse kinematics scheme that does not require inversion of the Jacobian.

14 Notice that the negative definite term is a quadratic function of the error, while the other term is a linear function of the error. Therefore, for an error of very small norm, the linear term prevails over the quadratic term, and the norm of $\boldsymbol{K}$ should be increased to reduce the norm of $\boldsymbol{e}$ as much as possible.

implying that the direction of $\mathcal{N}(\boldsymbol{J}_P^T)$ coincides with the direction orthogonal to the plane of the structure (Fig. 3.13). The Jacobian transpose algorithm gets stuck if, with $\boldsymbol{K}$ diagonal and having all equal elements, the desired position is along the line normal to the plane of the structure at the intersection with the wrist point. On the other hand, the end-effector cannot physically move from the singular configuration along such a line. Instead, if the prescribed path has a non-null component in the plane of the structure at the singularity, algorithm convergence is ensured, since in that case $\boldsymbol{K}\boldsymbol{e} \notin \mathcal{N}(\boldsymbol{J}_P^T)$.

In summary, the algorithm based on the computation of the Jacobian transpose provides a computationally efficient inverse kinematics method that can be utilized also for paths crossing kinematic singularities.

3.7.3 Orientation Error

The inverse kinematics algorithms presented in the above sections utilize the analytical Jacobian since they operate on error variables (position and orientation) that are defined in the operational space.

For what concerns the position error, it is obvious that its expression is given by

$$\boldsymbol{e}_P = \boldsymbol{p}_d - \boldsymbol{p}_e(\boldsymbol{q}) \tag{3.78}$$

where $\boldsymbol{p}_d$ and $\boldsymbol{p}_e$ denote respectively the desired and computed end-effector positions. Further, its time derivative is

$$\dot{\boldsymbol{e}}_P = \dot{\boldsymbol{p}}_d - \dot{\boldsymbol{p}}_e. \tag{3.79}$$

On the other hand, for what concerns the *orientation error*, its expression depends on the particular representation of end-effector orientation, namely, Euler angles, angle and axis, and unit quaternion.

Euler angles

The orientation error is chosen according to an expression formally analogous to (3.78), i.e.,

$$\boldsymbol{e}_O = \boldsymbol{\phi}_d - \boldsymbol{\phi}_e(\boldsymbol{q}) \tag{3.80}$$

where $\boldsymbol{\phi}_d$ and $\boldsymbol{\phi}_e$ denote respectively the desired and computed set of Euler angles. Further, its time derivative is

$$\dot{\boldsymbol{e}}_O = \dot{\boldsymbol{\phi}}_d - \dot{\boldsymbol{\phi}}_e. \tag{3.81}$$

Therefore, assuming that neither kinematic nor representation singularities occur, the Jacobian inverse solution for a nonredundant manipulator is derived from (3.70), i.e.,

$$\dot{\boldsymbol{q}} = \boldsymbol{J}_A^{-1}(\boldsymbol{q}) \begin{bmatrix} \dot{\boldsymbol{p}}_d + \boldsymbol{K}_P \boldsymbol{e}_P \\ \dot{\boldsymbol{\phi}}_d + \boldsymbol{K}_O \boldsymbol{e}_O \end{bmatrix} \tag{3.82}$$

where $\boldsymbol{K}_P$ and $\boldsymbol{K}_O$ are positive definite matrices.

As already pointed out in Sect. 2.10 for computation of the direct kinematics function in the form (2.82), the determination of the orientation variables from the joint variables is not easy except for simple cases (see Example 2.5). To this end, it is worth recalling that computation of the angles $\boldsymbol{\phi}_e$, in a minimal representation of orientation, requires computation of the rotation matrix $\boldsymbol{R}_e = [\,\boldsymbol{n}_e \quad \boldsymbol{s}_e \quad \boldsymbol{a}_e\,]$; in fact, only the dependence of $\boldsymbol{R}_e$ on $\boldsymbol{q}$ is known in closed form, but not that of $\boldsymbol{\phi}_e$ on $\boldsymbol{q}$. Further, the use of inverse functions (Atan2) in (2.19), (2.22) involves a non-negligible complexity in the computation of the analytical Jacobian, and the occurrence of representation singularities constitutes another drawback for the orientation error based on Euler angles.

Different kinds of remarks are to be made about the way to assign a time profile for the reference variables $\boldsymbol{\phi}_d$ chosen to represent end-effector orientation. The most intuitive way to specify end-effector orientation is to refer to the orientation of the end-effector frame ($\boldsymbol{n}_d$, $\boldsymbol{s}_d$, $\boldsymbol{a}_d$) with respect to the base frame. Given the limitations pointed out in Sect. 2.10 about guaranteeing orthonormality of the unit vectors along time, it is necessary first to compute the Euler angles corresponding to the initial and final orientation of the end-effector frame via (2.19), (2.22); only then a time evolution can be generated. Such solutions will be presented in Chap. 4.

A radical simplification of the problem at issue can be obtained for manipulators having a spherical wrist. Section 2.12.2 pointed out the possibility to solve the inverse kinematics problem for the position part separately from that for the orientation part. This result also has an impact at algorithmic level. In fact, the implementation of an inverse kinematics algorithm for determining the joint variables influencing the wrist position allows the computation of the time evolution of the wrist frame $\boldsymbol{R}_W(t)$. Hence, once the desired time evolution of the end-effector frame $\boldsymbol{R}_d(t)$ is given, it is sufficient to compute the Euler angles ZYZ from the matrix $\boldsymbol{R}_W^T \boldsymbol{R}_d$ by applying (2.19). As shown in Sect. 2.12.5, these angles are directly the joint variables of the spherical wrist. See also Problem 3.14.

The above considerations show that the inverse kinematics algorithms based on the analytical Jacobian are effective for kinematic structures having a spherical wrist which are of significant interest. For manipulator structures which cannot be reduced to that class, it may be appropriate to reformulate the inverse kinematics problem on the basis of a different definition of the orientation error.

Angle and axis

If $\boldsymbol{R}_d = [\boldsymbol{n}_d \quad \boldsymbol{s}_d \quad \boldsymbol{a}_d]$ denotes the desired rotation matrix of the end-effector frame and $\boldsymbol{R}_e = [\boldsymbol{n}_e \quad \boldsymbol{s}_e \quad \boldsymbol{a}_e]$ the rotation matrix that can be computed from the joint variables, the orientation error between the two frames can be expressed as

$$\boldsymbol{e}_O = \boldsymbol{r}\sin\vartheta \tag{3.83}$$

where ϑ and $\boldsymbol{r}$ identify the *angle and axis* of the equivalent rotation that can be deduced from the matrix

$$\boldsymbol{R}(\vartheta, \boldsymbol{r}) = \boldsymbol{R}_d \boldsymbol{R}_e^T(\boldsymbol{q}), \tag{3.84}$$

describing the rotation needed to align $\boldsymbol{R}$ with $\boldsymbol{R}_d$. Notice that (3.83) gives a unique relationship for $-\pi/2 < \vartheta < \pi/2$. The angle ϑ represents the magnitude of an orientation error, and thus the above limitation is not restrictive since the tracking error is typically small for an inverse kinematics algorithm.

By comparing the off-diagonal terms of the expression of $\boldsymbol{R}(\vartheta, \boldsymbol{r})$ in (2.25) with the corresponding terms resulting on the right-hand side of (3.84), it can be found that a functional expression of the orientation error in (3.83) is (see Problem 3.16)

$$\boldsymbol{e}_O = \frac{1}{2}(\boldsymbol{n}_e(\boldsymbol{q}) \times \boldsymbol{n}_d + \boldsymbol{s}_e(\boldsymbol{q}) \times \boldsymbol{s}_d + \boldsymbol{a}_e(\boldsymbol{q}) \times \boldsymbol{a}_d); \tag{3.85}$$

the limitation on ϑ is transformed in the condition $\boldsymbol{n}_e^T \boldsymbol{n}_d \geq 0$, $\boldsymbol{s}_e^T \boldsymbol{s}_d \geq 0$, $\boldsymbol{a}_e^T \boldsymbol{a}_d \geq 0$.

Differentiating (3.85) with respect to time and accounting for the expression of the columns of the derivative of a rotation matrix in (3.8) gives (see Problem 3.19)

$$\dot{\boldsymbol{e}}_O = \boldsymbol{L}^T \boldsymbol{\omega}_d - \boldsymbol{L}\boldsymbol{\omega}_e \tag{3.86}$$

where

$$\boldsymbol{L} = -\frac{1}{2}\big(\boldsymbol{S}(\boldsymbol{n}_d)\boldsymbol{S}(\boldsymbol{n}_e) + \boldsymbol{S}(\boldsymbol{s}_d)\boldsymbol{S}(\boldsymbol{s}_e) + \boldsymbol{S}(\boldsymbol{a}_d)\boldsymbol{S}(\boldsymbol{a}_e)\big). \tag{3.87}$$

At this point, by exploiting the relations (3.2), (3.3) of the geometric Jacobian expressing $\dot{\boldsymbol{p}}_e$ and $\boldsymbol{\omega}_e$ as a function of $\dot{\boldsymbol{q}}$, (3.79), (3.86) become

$$\dot{\boldsymbol{e}} = \begin{bmatrix} \dot{\boldsymbol{e}}_P \\ \dot{\boldsymbol{e}}_O \end{bmatrix} = \begin{bmatrix} \dot{\boldsymbol{p}}_d - \boldsymbol{J}_P(\boldsymbol{q})\dot{\boldsymbol{q}} \\ \boldsymbol{L}^T\boldsymbol{\omega}_d - \boldsymbol{L}\boldsymbol{J}_O(\boldsymbol{q})\dot{\boldsymbol{q}} \end{bmatrix} = \begin{bmatrix} \dot{\boldsymbol{p}}_d \\ \boldsymbol{L}^T\boldsymbol{\omega}_d \end{bmatrix} - \begin{bmatrix} \boldsymbol{I} & \boldsymbol{O} \\ \boldsymbol{O} & \boldsymbol{L} \end{bmatrix} \boldsymbol{J}\dot{\boldsymbol{q}}. \tag{3.88}$$

The expression in (3.88) suggests the possibility of devising inverse kinematics algorithms analogous to the ones derived above, but using the geometric Jacobian in place of the analytical Jacobian. For instance, the Jacobian inverse solution for a nonredundant nonsingular manipulator is

$$\dot{\boldsymbol{q}} = \boldsymbol{J}^{-1}(\boldsymbol{q}) \begin{bmatrix} \dot{\boldsymbol{p}}_d + \boldsymbol{K}_P\boldsymbol{e}_P \\ \boldsymbol{L}^{-1}\left(\boldsymbol{L}^T\boldsymbol{\omega}_d + \boldsymbol{K}_O\boldsymbol{e}_O\right) \end{bmatrix}. \tag{3.89}$$

It is worth remarking that the inverse kinematics solution based on (3.89) is expected to perform better than the solution based on (3.82) since it uses the geometric Jacobian in lieu of the analytical Jacobian, thus avoiding the occurrence of representation singularities.

Unit quaternion

In order to devise an inverse kinematics algorithm based on the *unit quaternion*, a suitable orientation error should be defined. Let $\mathcal{Q}_d = \{\eta_d, \boldsymbol{\epsilon}_d\}$ and $\mathcal{Q}_e = \{\eta_e, \boldsymbol{\epsilon}_e\}$ represent the quaternions associated with $\boldsymbol{R}_d$ and $\boldsymbol{R}_e$, respectively. The orientation error can be described by the rotation matrix $\boldsymbol{R}_d\boldsymbol{R}_e^T$ and, in view of (2.37), can be expressed in terms of the quaternion $\Delta\mathcal{Q} = \{\Delta\eta, \Delta\boldsymbol{\epsilon}\}$ where

$$\Delta\mathcal{Q} = \mathcal{Q}_d * \mathcal{Q}_e^{-1}. \tag{3.90}$$

It can be recognized that $\Delta\mathcal{Q} = \{1, \boldsymbol{0}\}$ if and only if $\boldsymbol{R}_e$ and $\boldsymbol{R}_d$ are aligned. Hence, it is sufficient to define the orientation error as

$$\boldsymbol{e}_O = \Delta\boldsymbol{\epsilon} = \eta_e(\boldsymbol{q})\boldsymbol{\epsilon}_d - \eta_d\boldsymbol{\epsilon}_e(\boldsymbol{q}) - \boldsymbol{S}(\boldsymbol{\epsilon}_d)\boldsymbol{\epsilon}_e(\boldsymbol{q}), \tag{3.91}$$

where the skew-symmetric operator $\boldsymbol{S}(\cdot)$ has been used. Notice, however, that the explicit computation of η_e and $\boldsymbol{\epsilon}_e$ from the joint variables is not possible but it requires the intermediate computation of the rotation matrix $\boldsymbol{R}_e$ that is available from the manipulator direct kinematics; then, the quaternion can be extracted using (2.34).

At this point, a Jacobian inverse solution can be computed as

$$\dot{\boldsymbol{q}} = \boldsymbol{J}^{-1}(\boldsymbol{q}) \begin{bmatrix} \dot{\boldsymbol{p}}_d + \boldsymbol{K}_P\boldsymbol{e}_P \\ \boldsymbol{\omega}_d + \boldsymbol{K}_O\boldsymbol{e}_O \end{bmatrix} \tag{3.92}$$

where noticeably the geometric Jacobian has been used. Substituting (3.92) into (3.4) gives (3.79) and

$$\boldsymbol{\omega}_d - \boldsymbol{\omega}_e + \boldsymbol{K}_O\boldsymbol{e}_O = \boldsymbol{0}. \tag{3.93}$$

It should be observed that now the orientation error equation is nonlinear in $\boldsymbol{e}_O$ since it contains the end-effector angular velocity error instead of the time derivative of the orientation error. To this end, it is worth considering the relationship between the time derivative of the quaternion $\mathcal{Q}_e$ and the angular velocity $\boldsymbol{\omega}_e$. This can be found to be (see Problem 3.19)

$$\dot{\eta}_e = -\frac{1}{2}\boldsymbol{\epsilon}_e^T\boldsymbol{\omega}_e \tag{3.94}$$

$$\dot{\boldsymbol{\epsilon}}_e = \frac{1}{2}\left(\eta_e\boldsymbol{I}_3 - \boldsymbol{S}(\boldsymbol{\epsilon}_e)\right)\boldsymbol{\omega}_e \tag{3.95}$$

which is the so-called *quaternion propagation*. A similar relationship holds between the time derivative of $\mathcal{Q}_d$ and $\boldsymbol{\omega}_d$.

To study stability of system (3.93), consider the positive definite Lyapunov function candidate

$$V = (\eta_d - \eta_e)^2 + (\boldsymbol{\epsilon}_d - \boldsymbol{\epsilon}_e)^T(\boldsymbol{\epsilon}_d - \boldsymbol{\epsilon}_e). \tag{3.96}$$

In view of (3.94), (3.95), differentiating (3.96) with respect to time and accounting for (3.93) yields (see Problem 3.20)

$$\dot{V} = -\boldsymbol{e}_O^T \boldsymbol{K}_O \boldsymbol{e}_O \tag{3.97}$$

which is negative definite, implying that $\boldsymbol{e}_O$ converges to zero.

In summary, the inverse kinematics solution based on (3.92) uses the geometric Jacobian as the solution based on (3.89) but is computationally lighter.

3.7.4 Second-order Algorithms

The above inverse kinematics algorithms can be defined as *first-order* algorithms, in that they allow the inversion of a motion trajectory, specified at the end-effector in terms of of position and orientation, into the equivalent joint positions and velocities.

Nevertheless, as will be seen in Chap. 8, for control purposes it may be necessary to invert a motion trajectory specified in terms of position, velocity and acceleration. On the other hand, the manipulator is inherently a *second-order* mechanical system, as will be revealed by the dynamic model to be derived in Chap. 7.

The time differentiation of the differential kinematics equation (3.62) leads to

$$\ddot{\boldsymbol{x}}_e = \boldsymbol{J}_A(\boldsymbol{q})\ddot{\boldsymbol{q}} + \dot{\boldsymbol{J}}_A(\boldsymbol{q}, \dot{\boldsymbol{q}})\dot{\boldsymbol{q}} \tag{3.98}$$

which gives the relationship between the joint space accelerations and the operational space accelerations.

Under the assumption of a square and non-singular matrix $\boldsymbol{J}_A$, the second-order differential kinematics (3.98) can be inverted in terms of the joint accelerations

$$\ddot{\boldsymbol{q}} = \boldsymbol{J}_A^{-1}(\boldsymbol{q})\left(\ddot{\boldsymbol{x}}_e - \dot{\boldsymbol{J}}_A(\boldsymbol{q}, \dot{\boldsymbol{q}})\dot{\boldsymbol{q}}\right). \tag{3.99}$$

The numerical integration of (3.99) to reconstruct the joint velocities and positions would unavoidably lead to a drift of the solution; therefore, similarly to the inverse kinematics algorithm with the Jacobian inverse, it is worth considering the error defined in (3.68) along with its derivative

$$\ddot{\boldsymbol{e}} = \ddot{\boldsymbol{x}}_d - \ddot{\boldsymbol{x}}_e \tag{3.100}$$

which, in view of (3.98), yields

$$\ddot{\boldsymbol{e}} = \ddot{\boldsymbol{x}}_d - \boldsymbol{J}_A(\boldsymbol{q})\ddot{\boldsymbol{q}} - \dot{\boldsymbol{J}}_A(\boldsymbol{q}, \dot{\boldsymbol{q}})\dot{\boldsymbol{q}}. \tag{3.101}$$

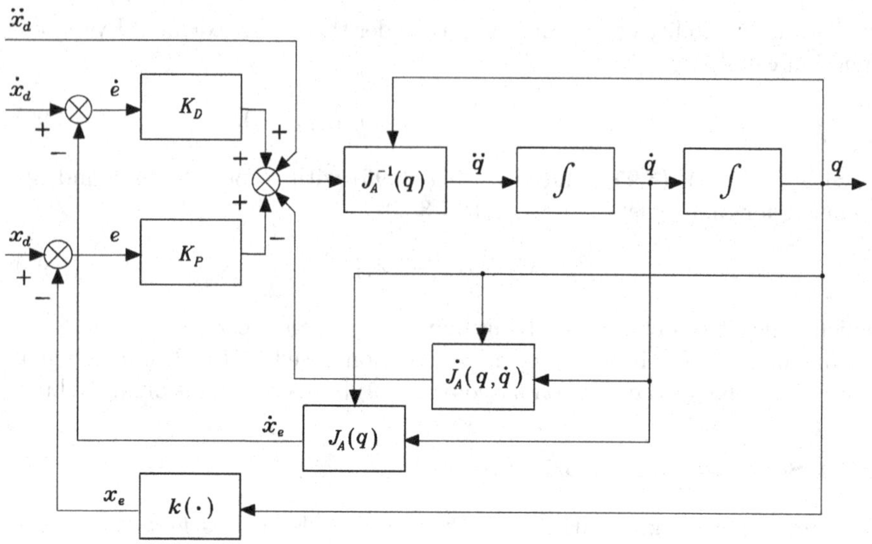

Fig. 3.14. Block scheme of the second-order inverse kinematics algorithm with Jacobian inverse

At this point, it is advisable to choose the joint acceleration vector as

$$\ddot{q} = \boldsymbol{J}_A^{-1}(\boldsymbol{q})\left(\ddot{\boldsymbol{x}}_d + \boldsymbol{K}_D\dot{\boldsymbol{e}} + \boldsymbol{K}_P\boldsymbol{e} - \dot{\boldsymbol{J}}_A(\boldsymbol{q},\dot{\boldsymbol{q}})\dot{\boldsymbol{q}}\right) \tag{3.102}$$

where $\boldsymbol{K}_D$ and $\boldsymbol{K}_P$ are positive definite (typically diagonal) matrices. Substituting (3.102) into (3.101) leads to the equivalent linear error system

$$\ddot{e} + \boldsymbol{K}_D\dot{\boldsymbol{e}} + \boldsymbol{K}_P\boldsymbol{e} = \boldsymbol{0} \tag{3.103}$$

which is *asymptotically stable*: the error tends to zero along the trajectory with a convergence speed depending on the choice of the matrices $\boldsymbol{K}_P$ e $\boldsymbol{K}_D$. The second-order inverse kinematics algorithm is illustrated in the block scheme of Fig. 3.14.

In the case of a *redundant manipulator*, the generalization of (3.102) leads to an algorithmic solution based on the Jacobian pseudo-inverse of the kind

$$\ddot{q} = \boldsymbol{J}_A^{\dagger}\left(\ddot{\boldsymbol{x}}_d + \boldsymbol{K}_D\dot{\boldsymbol{e}} + \boldsymbol{K}_P\boldsymbol{e} - \dot{\boldsymbol{J}}_A(\boldsymbol{q},\dot{\boldsymbol{q}})\dot{\boldsymbol{q}}\right) + (\boldsymbol{I}_n - \boldsymbol{J}_A^{\dagger}\boldsymbol{J}_A)\ddot{\boldsymbol{q}}_0 \tag{3.104}$$

where the vector $\ddot{\boldsymbol{q}}_0$ represents arbitrary joint accelerations which can be chosen so as to (locally) optimize an objective function like those considered in Sect. 3.5.1.

As for the first-order inverse kinematics algorithms, it is possible to consider other expressions for the orientation error which, unlike the Euler angles, refer to an angle and axis description, else to the unit quaternion.

3.7.5 Comparison Among Inverse Kinematics Algorithms

In order to make a comparison of performance among the inverse kinematics algorithms presented above, consider the 3-link planar arm in Fig. 2.20 whose link lengths are $a_1 = a_2 = a_3 = 0.5\,\text{m}$. The direct kinematics for this arm is given by (2.83), while its Jacobian can be found from (3.35) by considering the 3 non-null rows of interest for the operational space.

Let the arm be at the initial posture $\boldsymbol{q} = [\,\pi \quad -\pi/2 \quad -\pi/2\,]^T$ rad, corresponding to the end-effector pose: $\boldsymbol{p} = [\,0 \quad 0.5\,]^T$ m, $\phi = 0$ rad. A circular path of radius 0.25 m and centre at $(0.25, 0.5)$ m is assigned to the end-effector. Let the motion trajectory be

$$\boldsymbol{p}_d(t) = \begin{bmatrix} 0.25(1 - \cos \pi t) \\ 0.25(2 + \sin \pi t) \end{bmatrix} \qquad 0 \le t \le 4;$$

i.e., the end-effector has to make two complete circles in a time of 2 s per circle. As regards end-effector orientation, initially it is required to follow the trajectory

$$\phi_d(t) = \sin \frac{\pi}{24} t \qquad 0 \le t \le 4;$$

i.e., the end-effector has to attain a different orientation ($\phi_d = 0.5$ rad) at the end of the two circles.

The inverse kinematics algorithms were implemented on a computer by adopting the Euler numerical integration scheme (3.48) with an integration time $\Delta t = 1$ ms.

At first, the inverse kinematics along the given trajectory has been performed by using (3.47). The results obtained in Fig. 3.15 show that the norm of the position error along the whole trajectory is bounded; at steady state, after $t = 4$, the error sets to a constant value in view of the typical *drift* of *open-loop* schemes. A similar drift can be observed for the orientation error.

Next, the inverse kinematics algorithm based on (3.70) using the Jacobian *inverse* has been used, with the matrix gain $\boldsymbol{K} = \text{diag}\{500, 500, 100\}$. The resulting joint positions and velocities as well as the tracking errors are shown in Fig. 3.16. The norm of the position error is radically decreased and converges to zero at steady state, thanks to the *closed-loop* feature of the scheme; the orientation error, too, is decreased and tends to zero at steady state.

On the other hand, if the end-effector orientation is not constrained, the operational space becomes two-dimensional and is characterized by the first two rows of the direct kinematics in (2.83) as well as by the Jacobian in (3.36); a *redundant* DOF is then available. Hence, the inverse kinematics algorithm based on (3.72) using the Jacobian *pseudo-inverse* has been used with $\boldsymbol{K} = \text{diag}\{500, 500\}$. If redundancy is not exploited ($\dot{\boldsymbol{q}}_0 = \boldsymbol{0}$), the results in Fig. 3.17 reveal that position tracking remains satisfactory and, of course, the end-effector orientation freely varies along the given trajectory.

With reference to the previous situation, the use of the Jacobian *transpose* algorithm based on (3.76) with $\boldsymbol{K} = \text{diag}\{500, 500\}$ gives rise to a tracking

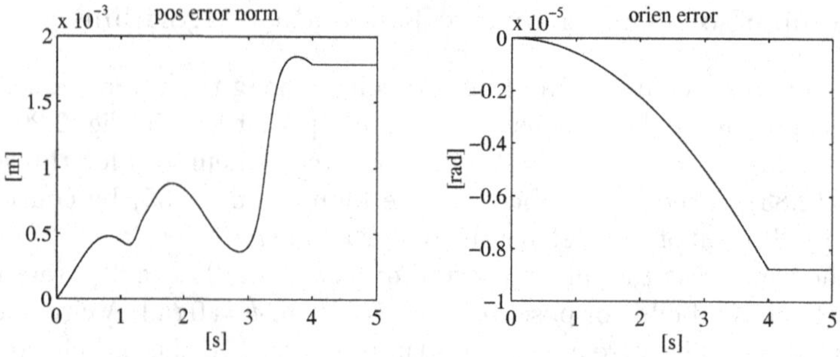

Fig. 3.15. Time history of the norm of end-effector position error and orientation error with the open-loop inverse Jacobian algorithm

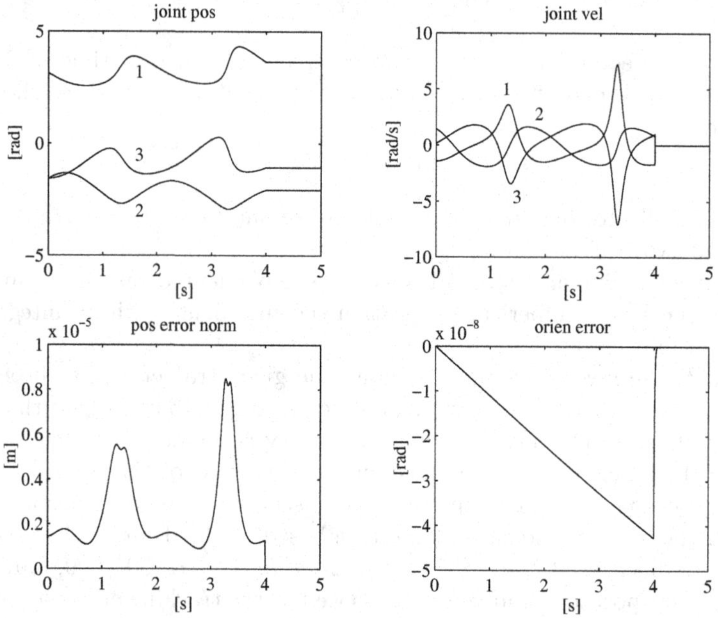

Fig. 3.16. Time history of the joint positions and velocities, and of the norm of end-effector position error and orientation error with the closed-loop inverse Jacobian algorithm

error (Fig. 3.18) which is anyhow bounded and rapidly tends to zero at steady state.

In order to show the capability of handling the degree of redundancy, the algorithm based on (3.72) with $\dot{\boldsymbol{q}}_0 \neq \boldsymbol{0}$ has been used; two types of constraints

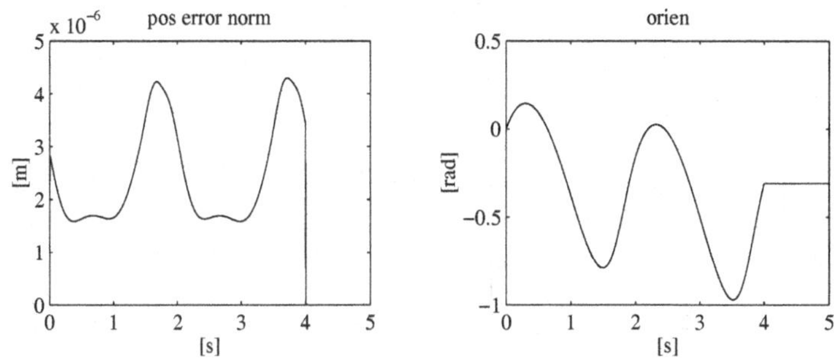

Fig. 3.17. Time history of the norm of end-effector position error and orientation with the Jacobian pseudo-inverse algorithm

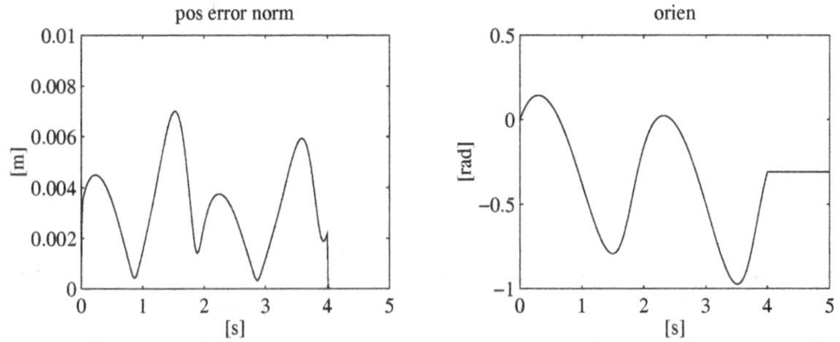

Fig. 3.18. Time history of the norm of end-effector position error and orientation with the Jacobian transpose algorithm

have been considered concerning an objective function to locally maximize according to the choice (3.55). The first function is

$$w(\vartheta_2, \vartheta_3) = \frac{1}{2}(s_2^2 + s_3^2)$$

that provides a *manipulability measure*. Notice that such a function is computationally simpler than the function in (3.56), but it still describes a distance from kinematic singularities in an effective way. The gain in (3.55)) has been set to $k_0 = 50$. In Fig. 3.19, the joint trajectories are reported for the two cases with and without ($k_0 = 0$) constraint. The addition of the constraint leads to having coincident trajectories for Joints 2 and 3. The manipulability measure in the constrained case (*continuous line*) attains larger values along the trajectory compared to the unconstrained case (*dashed line*). It is worth underlining that the tracking position error is practically the same in the two cases (Fig. 3.17), since the additional joint velocity contribution is projected in the null space of the Jacobian so as not to alter the performance of the end-effector position task.

Finally, it is worth noticing that in the constrained case the resulting joint trajectories are *cyclic*, i.e., they take on the same values after a period of

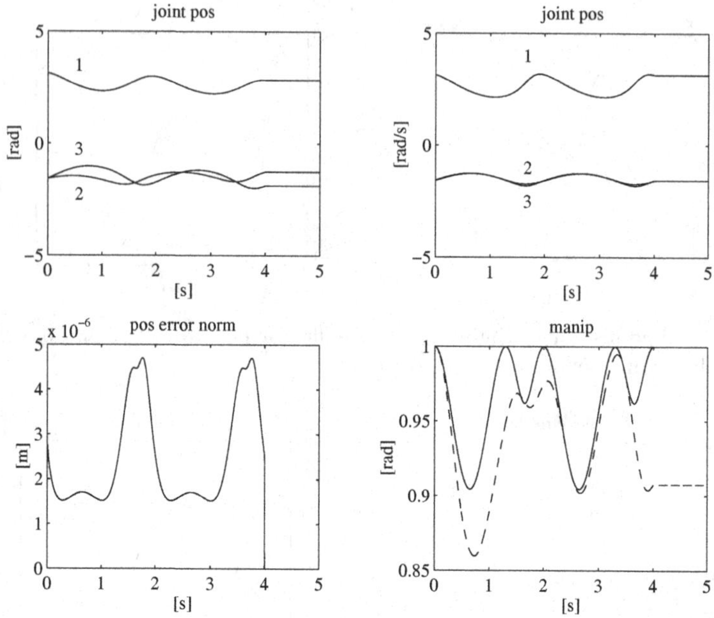

Fig. 3.19. Time history of the joint positions, the norm of end-effector position error, and the manipulability measure with the Jacobian pseudo-inverse algorithm and manipulability constraint; *upper left*: with the unconstrained solution, *upper right*: with the constrained solution

the circular path. This does not happen for the unconstrained case, since the internal motion of the structure causes the arm to be in a different posture after one circle.

The second objective function considered is the *distance from mechanical joint limits* in (3.57). Specifically, it is assumed what follows: the first joint does not have limits ($q_{1m} = -2\pi$, $q_{1M} = 2\pi$), the second joint has limits $q_{2m} = -\pi/2$, $q_{2M} = \pi/2$, and the third joint has limits $q_{3m} = -3\pi/2$, $q_{3M} = -\pi/2$. It is not difficult to verify that, in the unconstrained case, the trajectories of Joints 2 and 3 in Fig. 3.19 violate the respective limits. The gain in (3.55) has been set to $k_0 = 250$. The results in Fig. 3.20 show the effectiveness of the technique with utilization of redundancy, since both Joints 2 and 3 tend to invert their motion — with respect to the unconstrained trajectories in Fig. 3.19 — and keep far from the minimum limit for Joint 2 and the maximum limit for Joint 3, respectively. Such an effort does not appreciably affect the position tracking error, whose norm is bounded anyhow within acceptable values.

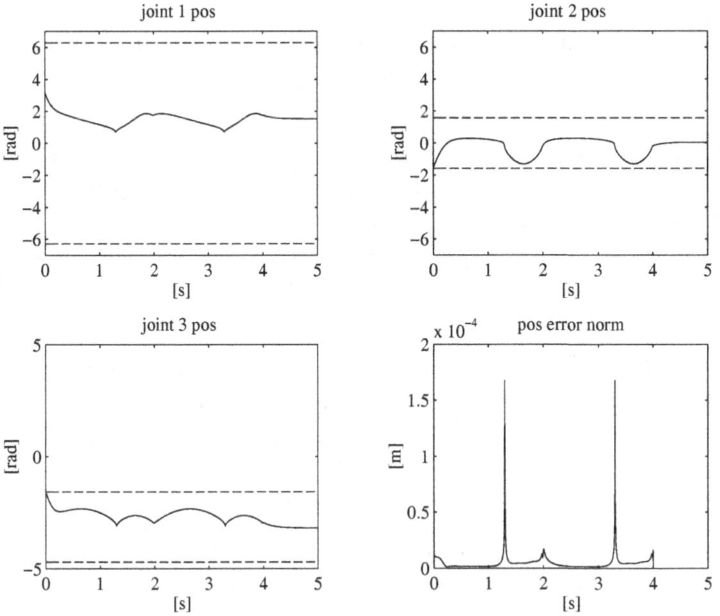

Fig. 3.20. Time history of the joint positions and the norm of end-effector position error with the Jacobian pseudo-inverse algorithm and joint limit constraint (joint limits are denoted by *dashed lines*)

3.8 Statics

The goal of *statics* is to determine the relationship between the generalized forces applied to the end-effector and the generalized forces applied to the joints — forces for prismatic joints, torques for revolute joints — with the manipulator at an equilibrium configuration.

Let $\boldsymbol{\tau}$ denote the $(n \times 1)$ vector of joint torques and $\boldsymbol{\gamma}$ the $(r \times 1)$ vector of end-effector forces[15] where r is the dimension of the operational space of interest.

The application of the *principle of virtual work* allows the determination of the required relationship. The mechanical manipulators considered are systems with time-invariant, holonomic constraints, and thus their configurations depend only on the joint variables $\boldsymbol{q}$ and not explicitly on time. This implies that virtual displacements coincide with elementary displacements.

Consider the elementary works performed by the two force systems. As for the joint torques, the elementary work associated with them is

$$dW_{\tau} = \boldsymbol{\tau}^T d\boldsymbol{q}. \tag{3.105}$$

[15] Hereafter, generalized forces at the joints are often called *torques*, while generalized forces at the end-effector are often called *forces*.

As for the end-effector forces $\boldsymbol{\gamma}$, if the force contributions $\boldsymbol{f}_e$ are separated by the moment contributions $\boldsymbol{\mu}_e$, the elementary work associated with them is

$$dW_\gamma = \boldsymbol{f}_e^T d\boldsymbol{p}_e + \boldsymbol{\mu}_e^T \boldsymbol{\omega}_e dt, \tag{3.106}$$

where $d\boldsymbol{p}_e$ is the linear displacement and $\boldsymbol{\omega}_e dt$ is the angular displacement[16]

By accounting for the differential kinematics relationship in (3.4), (3.5), the relation (3.106) can be rewritten as

$$\begin{aligned} dW_\gamma &= \boldsymbol{f}_e^T \boldsymbol{J}_P(\boldsymbol{q})d\boldsymbol{q} + \boldsymbol{\mu}_e^T \boldsymbol{J}_O(\boldsymbol{q})d\boldsymbol{q} \\ &= \boldsymbol{\gamma}_e^T \boldsymbol{J}(\boldsymbol{q})d\boldsymbol{q} \end{aligned} \tag{3.107}$$

where $\boldsymbol{\gamma}_e = [\, \boldsymbol{f}_e^T \quad \boldsymbol{\mu}_e^T \,]^T$. Since virtual and elementary displacements coincide, the virtual works associated with the two force systems are

$$\delta W_\tau = \boldsymbol{\tau}^T \delta \boldsymbol{q} \tag{3.108}$$

$$\delta W_\gamma = \boldsymbol{\gamma}_e^T \boldsymbol{J}(\boldsymbol{q})\delta\boldsymbol{q}, \tag{3.109}$$

where δ is the usual symbol to indicate virtual quantities.

According to the principle of virtual work, the manipulator is at *static equilibrium* if and only if

$$\delta W_\tau = \delta W_\gamma \qquad \forall \delta\boldsymbol{q}, \tag{3.110}$$

i.e., the difference between the virtual work of the joint torques and the virtual work of the end-effector forces must be null for all joint displacements.

From (3.109), notice that the virtual work of the end-effector forces is null for any displacement in the null space of $\boldsymbol{J}$. This implies that the joint torques associated with such displacements must be null at static equilibrium. Substituting (3.108), (3.109) into (3.110) leads to the notable result

$$\boldsymbol{\tau} = \boldsymbol{J}^T(\boldsymbol{q})\boldsymbol{\gamma}_e \tag{3.111}$$

stating that the relationship between the end-effector forces and the joint torques is established by the transpose of the manipulator geometric Jacobian.

3.8.1 Kineto-Statics Duality

The statics relationship in (3.111), combined with the differential kinematics equation in (3.45), points out a property of *kineto-statics duality*. In fact, by adopting a representation similar to that of Fig. 3.7 for differential kinematics, one has that (Fig. 3.21):

- The range space of $\boldsymbol{J}^T$ is the subspace $\mathcal{R}(\boldsymbol{J}^T)$ in $\mathbb{R}^n$ of the joint torques that can balance the end-effector forces, in the given manipulator posture.

[16] The angular displacement has been indicated by $\boldsymbol{\omega}_e dt$ in view of the problems of integrability of $\boldsymbol{\omega}_e$ discussed in Sect. 3.6.

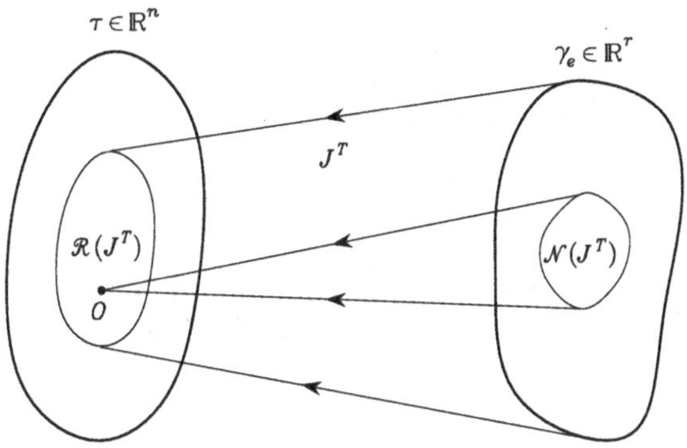

Fig. 3.21. Mapping between the end-effector force space and the joint torque space

- The null space of $\boldsymbol{J}^T$ is the subspace $\mathcal{N}(\boldsymbol{J}^T)$ in $\mathbb{R}^r$ of the end-effector forces that do not require any balancing joint torques, in the given manipulator posture.

It is worth remarking that the end-effector forces $\boldsymbol{\gamma}_e \in \mathcal{N}(\boldsymbol{J}^T)$ are entirely absorbed by the structure in that the mechanical constraint reaction forces can balance them exactly. Hence, a manipulator at a singular configuration remains in the given posture whatever end-effector force $\boldsymbol{\gamma}_e$ is applied so that $\boldsymbol{\gamma}_e \in \mathcal{N}(\boldsymbol{J}^T)$.

The relations between the two subspaces are established by

$$\mathcal{N}(\boldsymbol{J}) \equiv \mathcal{R}^{\perp}(\boldsymbol{J}^T) \qquad\qquad \mathcal{R}(\boldsymbol{J}) \equiv \mathcal{N}^{\perp}(\boldsymbol{J}^T)$$

and then, once the manipulator Jacobian is known, it is possible to characterize completely differential kinematics and statics in terms of the range and null spaces of the Jacobian and its transpose.

On the basis of the above duality, the inverse kinematics scheme with the Jacobian transpose in Fig. 3.12 admits an interesting physical interpretation. Consider a manipulator with ideal dynamics $\boldsymbol{\tau} = \dot{\boldsymbol{q}}$ (null masses and unit viscous friction coefficients); the algorithm update law $\dot{\boldsymbol{q}} = \boldsymbol{J}^T\boldsymbol{K}\boldsymbol{e}$ plays the role of a generalized spring of stiffness constant $\boldsymbol{K}$ generating a force $\boldsymbol{K}\boldsymbol{e}$ that pulls the end-effector towards the desired posture in the operational space. If this manipulator is allowed to move, e.g., in the case $\boldsymbol{K}\boldsymbol{e} \notin \mathcal{N}(\boldsymbol{J}^T)$, the end-effector attains the desired posture and the corresponding joint variables are determined.

3.8.2 Velocity and Force Transformation

The kineto-statics duality concept presented above can be useful to characterize the transformation of velocities and forces between two coordinate frames.

Consider a reference coordinate frame O_0–$x_0y_0z_0$ and a rigid body moving with respect to such a frame. Then let O_1–$x_1y_1z_1$ and O_2–$x_2y_2z_2$ be two

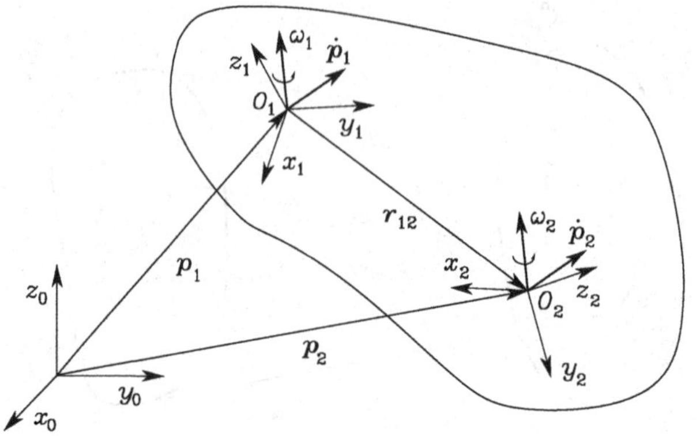

Fig. 3.22. Representation of linear and angular velocities in different coordinate frames on the same rigid body

coordinate frames attached to the body (Fig. 3.22). The relationships between translational and rotational velocities of the two frames with respect to the reference frame are given by

$$\begin{aligned}\boldsymbol{\omega}_2 &= \boldsymbol{\omega}_1\\ \dot{\boldsymbol{p}}_2 &= \dot{\boldsymbol{p}}_1 + \boldsymbol{\omega}_1 \times \boldsymbol{r}_{12}.\end{aligned}$$

By exploiting the skew-symmetric operator $\boldsymbol{S}(\cdot)$ in (3.9), the above relations can be compactly written as

$$\begin{bmatrix}\dot{\boldsymbol{p}}_2\\ \boldsymbol{\omega}_2\end{bmatrix} = \begin{bmatrix}\boldsymbol{I} & -\boldsymbol{S}(\boldsymbol{r}_{12})\\ \boldsymbol{O} & \boldsymbol{I}\end{bmatrix}\begin{bmatrix}\dot{\boldsymbol{p}}_1\\ \boldsymbol{\omega}_1\end{bmatrix}. \tag{3.112}$$

All vectors in (3.112) are meant to be referred to the reference frame O_0–$x_0y_0z_0$. On the other hand, if vectors are referred to their own frames, it is

$$\boldsymbol{r}_{12} = \boldsymbol{R}_1\boldsymbol{r}_{12}^1$$

and also

$$\begin{aligned}\dot{\boldsymbol{p}}_1 &= \boldsymbol{R}_1\dot{\boldsymbol{p}}_1^1 & \dot{\boldsymbol{p}}_2 &= \boldsymbol{R}_2\dot{\boldsymbol{p}}_2^2 = \boldsymbol{R}_1\boldsymbol{R}_2^1\dot{\boldsymbol{p}}_2^2\\ \boldsymbol{\omega}_1 &= \boldsymbol{R}_1\boldsymbol{\omega}_1^1 & \boldsymbol{\omega}_2 &= \boldsymbol{R}_2\boldsymbol{\omega}_2^2 = \boldsymbol{R}_1\boldsymbol{R}_2^1\boldsymbol{\omega}_2^2.\end{aligned}$$

Accounting for (3.112) and (3.11) gives

$$\begin{aligned}\boldsymbol{R}_1\boldsymbol{R}_2^1\dot{\boldsymbol{p}}_2^2 &= \boldsymbol{R}_1\dot{\boldsymbol{p}}_1^1 - \boldsymbol{R}_1\boldsymbol{S}(\boldsymbol{r}_{12}^1)\boldsymbol{R}_1^T\boldsymbol{R}_1\boldsymbol{\omega}_1^1\\ \boldsymbol{R}_1\boldsymbol{R}_2^1\boldsymbol{\omega}_2^2 &= \boldsymbol{R}_1\boldsymbol{\omega}_1^1.\end{aligned}$$

Eliminating the dependence on $\boldsymbol{R}_1$, which is premultiplied to each term on both sides of the previous relations, yields[17]

$$\begin{bmatrix}\dot{\boldsymbol{p}}_2^2\\ \boldsymbol{\omega}_2^2\end{bmatrix} = \begin{bmatrix}\boldsymbol{R}_1^2 & -\boldsymbol{R}_1^2\boldsymbol{S}(\boldsymbol{r}_{12}^1)\\ \boldsymbol{O} & \boldsymbol{R}_1^2\end{bmatrix}\begin{bmatrix}\dot{\boldsymbol{p}}_1^1\\ \boldsymbol{\omega}_1^1\end{bmatrix} \tag{3.113}$$

[17] Recall that $\boldsymbol{R}^T\boldsymbol{R} = \boldsymbol{I}$, as in (2.4).

giving the sought general relationship of *velocity transformation* between two frames.

It may be observed that the transformation matrix in (3.113) plays the role of a true Jacobian, since it characterizes a velocity transformation, and thus (3.113) may be shortly written as

$$\boldsymbol{v}_2^2 = \boldsymbol{J}_1^2 \boldsymbol{v}_1^1. \tag{3.114}$$

At this point, by virtue of the kineto-statics duality, the *force transformation* between two frames can be directly derived in the form

$$\boldsymbol{\gamma}_1^1 = \boldsymbol{J}_1^{2T} \boldsymbol{\gamma}_2^2 \tag{3.115}$$

which can be detailed into[18]

$$\begin{bmatrix} \boldsymbol{f}_1^1 \\ \boldsymbol{\mu}_1^1 \end{bmatrix} = \begin{bmatrix} \boldsymbol{R}_2^1 & \boldsymbol{O} \\ \boldsymbol{S}(\boldsymbol{r}_{12}^1)\boldsymbol{R}_2^1 & \boldsymbol{R}_2^1 \end{bmatrix} \begin{bmatrix} \boldsymbol{f}_2^2 \\ \boldsymbol{\mu}_2^2 \end{bmatrix}. \tag{3.116}$$

Finally, notice that the above analysis is instantaneous in that, if a coordinate frame varies with respect to the other, it is necessary to recompute the Jacobian of the transformation through the computation of the related rotation matrix of one frame with respect to the other.

3.8.3 Closed Chain

As discussed in Sect. 2.8.3, whenever the manipulator contains a closed chain, there is a functional relationship between the joint variables. In particular, the closed chain structure is transformed into a tree-structured open chain by virtually cutting the loop at a joint. It is worth choosing such a cut joint as one of the unactuated joints. Then, the constraints (2.59) or (2.60) should be solved for a reduced number of joint variables, corresponding to the DOFs of the chain. Therefore, it is reasonable to assume that at least such independent joints are actuated, while the others may or may not be actuated. Let $\boldsymbol{q}_o = [\,\boldsymbol{q}_a^T \quad \boldsymbol{q}_u^T\,]^T$ denote the vector of joint variables of the tree-structured open chain, where $\boldsymbol{q}_a$ and $\boldsymbol{q}_u$ are the vectors of *actuated* and *unactuated* joint variables, respectively. Assume that from the above constraints it is possible to determine a functional expression

$$\boldsymbol{q}_u = \boldsymbol{q}_u(\boldsymbol{q}_a). \tag{3.117}$$

Time differentiation of (3.117) gives the relationship between joint velocities in the form

$$\dot{\boldsymbol{q}}_o = \boldsymbol{\Upsilon} \dot{\boldsymbol{q}}_a \tag{3.118}$$

where

$$\boldsymbol{\Upsilon} = \begin{bmatrix} \boldsymbol{I} \\ \dfrac{\partial \boldsymbol{q}_u}{\partial \boldsymbol{q}_a} \end{bmatrix} \tag{3.119}$$

[18] The skew-symmetry property $\boldsymbol{S} + \boldsymbol{S}^T = \boldsymbol{O}$ is utilized.

is the transformation matrix between the two vectors of joint velocities, which in turn plays the role of a Jacobian.

At this point, according to an intuitive kineto-statics duality concept, it is possible to describe the transformation between the corresponding vectors of joint torques in the form

$$\boldsymbol{\tau}_a = \boldsymbol{\Upsilon}^T \boldsymbol{\tau}_o \tag{3.120}$$

where $\boldsymbol{\tau}_o = [\,\boldsymbol{\tau}_a^T \quad \boldsymbol{\tau}_u^T\,]^T$, with obvious meaning of the quantities.

Example 3.5

Consider the parallelogram arm of Sect. 2.9.2. On the assumption to actuate the two Joints 1′ and 1″ at the base, it is $\boldsymbol{q}_a = [\,\vartheta_{1'} \quad \vartheta_{1''}\,]^T$ and $\boldsymbol{q}_u = [\,\vartheta_{2'} \quad \vartheta_{3'}\,]^T$. Then, using (2.64), the transformation matrix in (3.119) is

$$\boldsymbol{\Upsilon} = \begin{bmatrix} 1 & 0 \\ 0 & 1 \\ -1 & 1 \\ 1 & -1 \end{bmatrix}.$$

Hence, in view of (3.120), the torque vector of the actuated joints is

$$\boldsymbol{\tau}_a = \begin{bmatrix} \tau_{1'} - \tau_{2'} + \tau_{3'} \\ \tau_{1''} + \tau_{2'} - \tau_{3'} \end{bmatrix} \tag{3.121}$$

while obviously $\boldsymbol{\tau}_u = [\,0 \quad 0\,]^T$ in agreement with the fact that both Joints 2′ and 3′ are unactuated.

3.9 Manipulability Ellipsoids

The differential kinematics equation in (3.45) and the statics equation in (3.111), together with the duality property, allow the definition of indices for the evaluation of manipulator performance. Such indices can be helpful both for mechanical manipulator design and for determining suitable manipulator postures to execute a given task in the current configuration.

First, it is desired to represent the attitude of a manipulator to arbitrarily change end-effector position and orientation. This capability is described in an effective manner by the *velocity manipulability ellipsoid*.

Consider the set of joint velocities of constant (unit) norm

$$\dot{\boldsymbol{q}}^T \dot{\boldsymbol{q}} = 1; \tag{3.122}$$

this equation describes the points on the surface of a sphere in the joint velocity space. It is desired to describe the operational space velocities that can be generated by the given set of joint velocities, with the manipulator in a given posture. To this end, one can utilize the differential kinematics equation in (3.45) solved for the joint velocities; in the general case of a redundant manipulator ($r < n$) at a nonsingular configuration, the minimum-norm solution $\dot{\boldsymbol{q}} = \boldsymbol{J}^{\dagger}(\boldsymbol{q})\boldsymbol{v}_e$ can be considered which, substituted into (3.122), yields

$$\boldsymbol{v}_e^T \big(\boldsymbol{J}^{\dagger T}(\boldsymbol{q})\boldsymbol{J}^{\dagger}(\boldsymbol{q})\big)\boldsymbol{v}_e = 1.$$

Accounting for the expression of the pseudo-inverse of $\boldsymbol{J}$ in (3.52) gives

$$\boldsymbol{v}_e^T \big(\boldsymbol{J}(\boldsymbol{q})\boldsymbol{J}^T(\boldsymbol{q})\big)^{-1}\boldsymbol{v}_e = 1, \tag{3.123}$$

which is the equation of the points on the surface of an ellipsoid in the end-effector velocity space.

The choice of the minimum-norm solution rules out the presence of internal motions for the redundant structure. If the general solution (3.54) is used for $\dot{\boldsymbol{q}}$, the points satisfying (3.122) are mapped into points inside the ellipsoid whose surface is described by (3.123).

For a nonredundant manipulator, the differential kinematics solution (3.47) is used to derive (3.123); in this case the points on the surface of the sphere in the joint velocity space are mapped into points on the surface of the ellipsoid in the end-effector velocity space.

Along the direction of the major axis of the ellipsoid, the end-effector can move at large velocity, while along the direction of the minor axis small end-effector velocities are obtained. Further, the closer the ellipsoid is to a sphere — unit eccentricity — the better the end-effector can move isotropically along all directions of the operational space. Hence, it can be understood why this ellipsoid is an index characterizing manipulation ability of the structure in terms of velocities.

As can be recognized from (3.123), the shape and orientation of the ellipsoid are determined by the core of its quadratic form and then by the matrix $\boldsymbol{J}\boldsymbol{J}^T$ which is in general a function of the manipulator configuration. The directions of the principal axes of the ellipsoid are determined by the eigenvectors $\boldsymbol{u}_i$, for $i = 1, \ldots, r$, of the matrix $\boldsymbol{J}\boldsymbol{J}^T$, while the dimensions of the axes are given by the singular values of $\boldsymbol{J}$, $\sigma_i = \sqrt{\lambda_i(\boldsymbol{J}\boldsymbol{J}^T)}$, for $i = 1, \ldots, r$, where $\lambda_i(\boldsymbol{J}\boldsymbol{J}^T)$ denotes the generic eigenvalue of $\boldsymbol{J}\boldsymbol{J}^T$.

A global representative measure of manipulation ability can be obtained by considering the volume of the ellipsoid. This volume is proportional to the quantity

$$w(\boldsymbol{q}) = \sqrt{\det\big(\boldsymbol{J}(\boldsymbol{q})\boldsymbol{J}^T(\boldsymbol{q})\big)}$$

which is the *manipulability measure* already introduced in (3.56). In the case of a nonredundant manipulator ($r = n$), w reduces to

$$w(\boldsymbol{q}) = \big|\det\big(\boldsymbol{J}(\boldsymbol{q})\big)\big|. \tag{3.124}$$

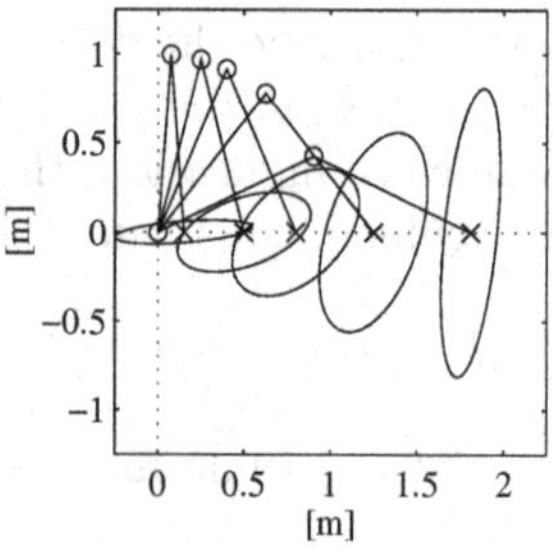

Fig. 3.23. Velocity manipulability ellipses for a two-link planar arm in different postures

It is easy to recognize that it is always $w > 0$, except for a manipulator at a singular configuration when $w = 0$. For this reason, this measure is usually adopted as a distance of the manipulator from singular configurations.

Example 3.6

Consider the two-link planar arm. From the expression in (3.41), the manipulability measure is in this case

$$w = |\det(\boldsymbol{J})| = a_1 a_2 |s_2|.$$

Therefore, as a function of the arm postures, the manipulability is maximum for $\vartheta_2 = \pm\pi/2$. On the other hand, for a given constant reach $a_1 + a_2$, the structure offering the maximum manipulability, independently of ϑ_1 and ϑ_2, is the one with $a_1 = a_2$.

These results have a biomimetic interpretation in the human arm, if that is regarded as a two-link arm (arm + forearm). The condition $a_1 = a_2$ is satisfied with good approximation. Further, the elbow angle ϑ_2 is usually in the neighbourhood of $\pi/2$ in the execution of several tasks, such as that of writing. Hence, the human being tends to dispose the arm in the most dexterous configuration from a manipulability viewpoint.

Figure 3.23 illustrates the velocity manipulability ellipses for a certain number of postures with the tip along the horizontal axis and $a_1 = a_2 = 1$. It can be seen that when the arm is outstretched the ellipsoid is very thin along the vertical direction. Hence, one recovers the result anticipated in the study of singularities that the arm in this posture can generate tip velocities preferably along the vertical direction. In Fig. 3.24, moreover, the behaviour of the minimum and maximum singular values of the matrix $\boldsymbol{J}$ is illustrated as a function of tip position along axis x; it can be verified that the minimum singular value is null when the manipulator is at a singularity (retracted or outstretched).

Therefore, with reference to the postures, manipulability has a maximum for $\vartheta_2 = \pm\pi/2$. On the other hand, for a given total extension $a_1 + a_2$, the structure which, independently of ϑ_1 and ϑ_2, offers the largest manipulability is that with $a_1 = a_2$.

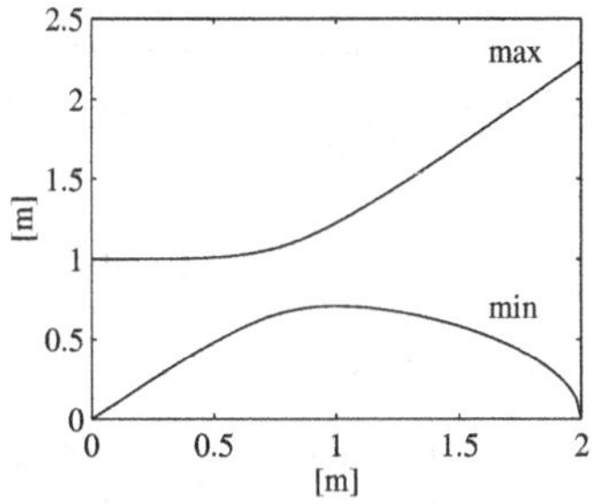

Fig. 3.24. Minimum and maximum singular values of $\boldsymbol{J}$ for a two-link planar arm as a function of the arm posture

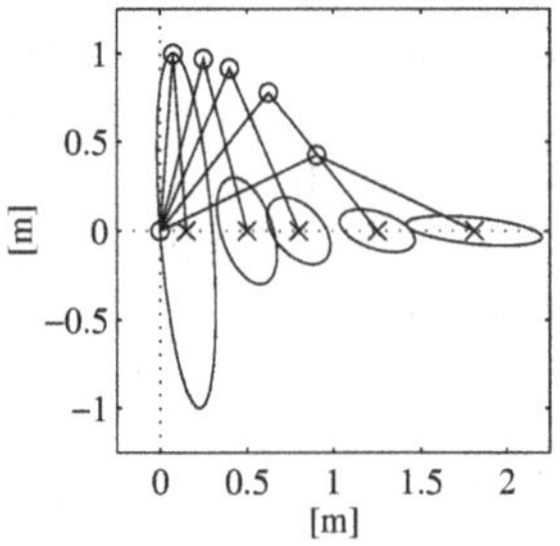

Fig. 3.25. Force manipulability ellipses for a two-link planar arm in different postures

The manipulability measure w has the advantage of being easy to compute, through the determinant of matrix $\boldsymbol{J}\boldsymbol{J}^T$. However, its numerical value does not constitute an absolute measure of the actual closeness of the manipulator to a singularity. It is enough to consider the above example and take two arms of identical structure, one with links of 1 m and the other with links of 1 cm. Two different values of manipulability are obtained which differ by four orders of magnitude. Hence, in that case it is convenient to consider only $|s_2|$ — eventually $|\vartheta_2|$ — as the manipulability measure. In more general cases when it is not easy to find a simple, meaningful index, one can consider the ratio between the minimum and maximum singular values of the Jacobian σ_r/σ_1 which is equivalent to the inverse of the condition number of matrix $\boldsymbol{J}$. This ratio gives not only a measure of the distance from a singularity ($\sigma_r = 0$), but also a direct measure of eccentricity of the ellipsoid. The disadvantage in utilizing this index is its computational complexity; it is practically impossible to compute it in symbolic form, i.e., as a function of the joint configuration, except for matrices of reduced dimension.

On the basis of the existing duality between differential kinematics and statics, it is possible to describe the manipulability of a structure not only with reference to velocities, but also with reference to forces. To be specific, one can consider the sphere in the space of joint torques

$$\boldsymbol{\tau}^T\boldsymbol{\tau} = 1 \tag{3.125}$$

which, accounting for (3.111), is mapped into the ellipsoid in the space of end-effector forces

$$\boldsymbol{\gamma}_e^T\big(\boldsymbol{J}(\boldsymbol{q})\boldsymbol{J}^T(\boldsymbol{q})\big)\boldsymbol{\gamma}_e = 1 \tag{3.126}$$

which is defined as the *force manipulability ellipsoid*. This ellipsoid characterizes the end-effector forces that can be generated with the given set of joint torques, with the manipulator in a given posture.

As can be easily recognized from (3.126), the core of the quadratic form is constituted by the inverse of the matrix core of the velocity ellipsoid in (3.123). This feature leads to the notable result that the principal axes of the force manipulability ellipsoid coincide with the principal axes of the velocity manipulability ellipsoid, while the dimensions of the respective axes are in inverse proportion. Therefore, according to the concept of force/velocity duality, a direction along which good velocity manipulability is obtained is a direction along which poor force manipulability is obtained, and vice versa.

In Fig. 3.25, the manipulability ellipses for the same postures as those of the example in Fig. 3.23 are illustrated. A comparison of the shape and orientation of the ellipses confirms the force/velocity duality effect on the manipulability along different directions.

It is worth pointing out that these manipulability ellipsoids can be represented geometrically in all cases of an operational space of dimension at most 3. Therefore, if it is desired to analyze manipulability in a space of greater dimension, it is worth separating the components of linear velocity (force) from those of angular velocity (moment), also avoiding problems due to non-homogeneous dimensions of the relevant quantities (e.g., m/s vs rad/s). For instance, for a manipulator with a spherical wrist, the manipulability analysis is naturally prone to a decoupling between arm and wrist.

An effective interpretation of the above results can be achieved by regarding the manipulator as a *mechanical transformer* of velocities and forces from the joint space to the operational space. Conservation of energy dictates that an amplification in the velocity transformation is necessarily accompanied by a reduction in the force transformation, and vice versa. The transformation ratio along a given direction is determined by the intersection of the vector along that direction with the surface of the ellipsoid. Once a unit vector $\boldsymbol{u}$ along a direction has been assigned, it is possible to compute the transformation ratio for the force manipulability ellipsoid as

$$\alpha(\boldsymbol{q}) = \Big(\boldsymbol{u}^T\boldsymbol{J}(\boldsymbol{q})\boldsymbol{J}^T(\boldsymbol{q})\boldsymbol{u}\Big)^{-1/2} \tag{3.127}$$

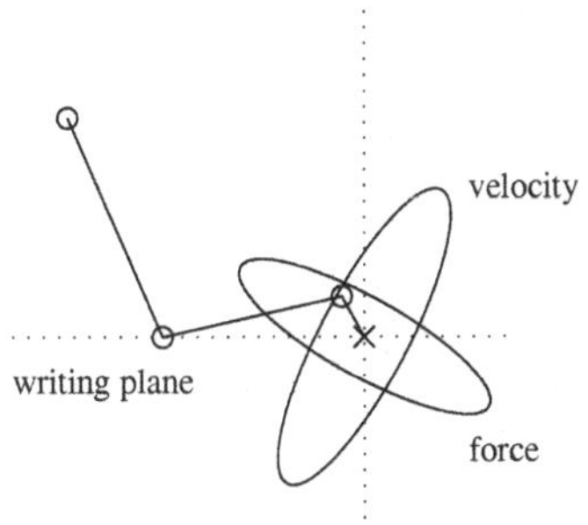

Fig. 3.26. Velocity and force manipulability ellipses for a 3-link planar arm in a typical configuration for a task of controlling force and velocity

and for the velocity manipulability ellipsoid as

$$\beta(\boldsymbol{q}) = \left(\boldsymbol{u}^T \left(\boldsymbol{J}(\boldsymbol{q})\boldsymbol{J}^T(\boldsymbol{q})\right)^{-1} \boldsymbol{u}\right)^{-1/2}. \tag{3.128}$$

The manipulability ellipsoids can be conveniently utilized not only for analyzing manipulability of the structure along different directions of the operational space, but also for determining compatibility of the structure to execute a task assigned along a direction. To this end, it is useful to distinguish between actuation tasks and control tasks of velocity and force. In terms of the relative ellipsoid, the task of actuating a velocity (force) requires preferably a large transformation ratio along the task direction, since for a given set of joint velocities (forces) at the joints it is possible to generate a large velocity (force) at the end-effector. On the other hand, for a control task it is important to have a small transformation ratio so as to gain good sensitivity to errors that may occur along the given direction.

Revisiting once again the duality between velocity manipulability ellipsoid and force manipulability ellipsoid, it can be found that an optimal direction to actuate a velocity is also an optimal direction to control a force. Analogously, a good direction to actuate a force is also a good direction to control a velocity.

To have a tangible example of the above concept, consider the typical task of writing on a horizontal surface for the human arm; this time, the arm is regarded as a 3-link planar arm: arm + forearm + hand. Restricting the analysis to a two-dimensional task space (the direction vertical to the surface and the direction of the line of writing), one has to achieve fine control of the vertical force (the pressure of the pen on the paper) and of the horizontal velocity (to write in good calligraphy). As a consequence, the force manipulability ellipse tends to be oriented horizontally for correct task execution. Correspondingly, the velocity manipulability ellipse tends to be oriented vertically in perfect agreement with the task requirement. In this case, from Fig. 3.26 the typical configuration of the human arm when writing can be recognized.

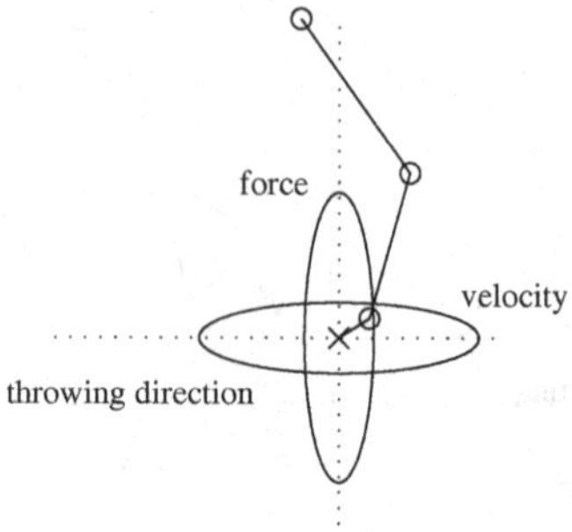

Fig. 3.27. Velocity and force manipulability ellipses for a 3-link planar arm in a typical configuration for a task of actuating force and velocity

An opposite example to the previous one is that of the human arm when throwing a weight in the horizontal direction. In fact, now it is necessary to actuate a large vertical force (to sustain the weight) and a large horizontal velocity (to throw the load for a considerable distance). Unlike the above, the force (velocity) manipulability ellipse tends to be oriented vertically (horizontally) to successfully execute the task. The relative configuration in Fig. 3.27 is representative of the typical attitude of the human arm when, for instance, releasing the ball in a bowling game.

In the above two examples, it is worth pointing out that the presence of a two-dimensional operational space is certainly advantageous to try reconfiguring the structure in the best configuration compatible with the given task. In fact, the transformation ratios defined in (3.127) and (3.128) are scalar functions of the manipulator configurations that can be optimized locally according to the technique for exploiting redundant DOFs previously illustrated.

Bibliography

The concept of geometric Jacobian was originally introduced in [240] and the problem of its computationally efficient determination is considered in [173]. The concept of analytical Jacobian is presented in [114] with reference to operational space control.

Inverse differential kinematics dates back to [240] under the name of resolved rate control. The use of the Jacobian pseudo-inverse is due to [118]. The adoption of the damped least-squares inverse has been independently proposed by [161] and [238]; a tutorial on the topic is [42]. The inverse kinematics algorithm based on the Jacobian transpose has been originally proposed in [198, 16]. Further details about the orientation error are found in [142, 250, 132, 41].

The utilization of the joint velocities in the null space of the Jacobian for redundancy resolution is proposed in [129] and further refined in [147] regarding the choice of the objective functions. The approach based on task priority is presented in [163]; other approaches based on the concept of augmented task space are presented in [14, 69, 199, 203, 194, 37]. For global redundancy resolutions see [162]. A complete treatment of redundant manipulators can be found in [160] while a tutorial is [206].

The extension of inverse kinematics to the second order has been proposed in [207], while the symbolic differentiation of the solutions in terms of joint velocities to obtain stable acceleration solutions can be found in [208]. Further details about redundancy resolution are in [59].

The concepts of kineto-statics duality are discussed in [191]. The manipulability ellipsoids are proposed in [245, 248] and employed in [44] for posture dexterity analysis with regard to manipulation tasks.

Problems

3.1. Prove (3.11).

3.2. Compute the Jacobian of the cylindrical arm in Fig. 2.35.

3.3. Compute the Jacobian of the SCARA manipulator in Fig. 2.36.

3.4. Find the singularities of the 3-link planar arm in Fig. 2.20.

3.5. Find the singularities of the spherical arm in Fig. 2.22.

3.6. Find the singularities of the cylindrical arm in Fig. 2.35.

3.7. Find the singularities of the SCARA manipulator in Fig. 2.36.

3.8. Show that the manipulability measure defined in (3.56) is given by the product of the singular values of the Jacobian matrix.

3.9. For the 3-link planar arm in Fig. 2.20, find an expression of the distance of the arm from a circular obstacle of given radius and coordinates.

3.10. Find the solution to the differential kinematics equation with the damped least-square inverse in (3.59).

3.11. Prove (3.64) in an alternative way, i.e., by computing $\boldsymbol{S}(\boldsymbol{\omega}_e)$ as in (3.6) starting from $\boldsymbol{R}(\boldsymbol{\phi})$ in (2.18).

3.12. With reference to (3.64), find the transformation matrix $\boldsymbol{T}(\boldsymbol{\phi}_e)$ in the case of RPY angles.

3.13. With reference to (3.64), find the triplet of Euler angles for which $\boldsymbol{T}(\boldsymbol{0}) = \boldsymbol{I}$.

3.14. Show how the inverse kinematics scheme of Fig. 3.11 can be simplified in the case of a manipulator having a spherical wrist.

3.15. Find an expression of the upper bound on the norm of $\boldsymbol{e}$ for the solution (3.76) in the case $\dot{\boldsymbol{x}}_d \neq \boldsymbol{0}$.

3.16. Prove (3.81).

3.17. Prove (3.86), (3.87).

3.18. Prove that the equation relating the angular velocity to the time derivative of the quaternion is given by

$$\boldsymbol{\omega} = 2\boldsymbol{S}(\boldsymbol{\epsilon})\dot{\boldsymbol{\epsilon}} + 2\eta\dot{\boldsymbol{\epsilon}} - 2\dot{\eta}\boldsymbol{\epsilon}.$$

[*Hint*: Start by showing that (2.33) can be rewritten as $\boldsymbol{R}(\eta, \boldsymbol{\epsilon}) = (2\eta^2 - 1)\boldsymbol{I} + 2\boldsymbol{\epsilon}\boldsymbol{\epsilon}^T + 2\eta\boldsymbol{S}(\boldsymbol{\epsilon})$].

3.19. Prove (3.94), (3.95).

3.20. Prove that the time derivative of the Lyapunov function in (3.96) is given by (3.97).

3.21. Consider the 3-link planar arm in Fig. 2.20, whose link lengths are respectively 0.5 m, 0.3 m, 0.3 m. Perform a computer implementation of the inverse kinematics algorithm using the Jacobian pseudo-inverse along the operational space path given by a straight line connecting the points of coordinates $(0.8, 0.2)$ m and $(0.8, -0.2)$ m. Add a constraint aimed at avoiding link collision with a circular object located at $\boldsymbol{o} = [\,0.3 \quad 0\,]^T$ m of radius 0.1 m. The initial arm configuration is chosen so that $\boldsymbol{p}_e(0) = \boldsymbol{p}_d(0)$. The final time is 2 s. Use sinusoidal motion timing laws. Adopt the Euler numerical integration scheme (3.48) with an integration time $\Delta t = 1$ ms.

3.22. Consider the SCARA manipulator in Fig. 2.36, whose links both have a length of 0.5 m and are located at a height of 1 m from the supporting plane. Perform a computer implementation of the inverse kinematics algorithms with both Jacobian inverse and Jacobian transpose along the operational space path whose position is given by a straight line connecting the points of coordinates $(0.7, 0, 0)$ m and $(0, 0.8, 0.5)$ m, and whose orientation is given by a rotation from 0 rad to $\pi/2$ rad. The initial arm configuration is chosen so that $\boldsymbol{x}_e(0) = \boldsymbol{x}_d(0)$. The final time is 2 s. Use sinusoidal motion timing laws. Adopt the Euler numerical integration scheme (3.48) with an integration time $\Delta t = 1$ ms.

3.23. Prove that the directions of the principal axes of the force and velocity manipulability ellipsoids coincide while their dimensions are in inverse proportion.

4

Trajectory Planning

For the execution of a specific robot task, it is worth considering the main features of motion planning algorithms. The goal of *trajectory planning* is to generate the reference inputs to the motion control system which ensures that the manipulator executes the planned trajectories. The user typically specifies a number of parameters to describe the desired trajectory. Planning consists of generating a time sequence of the values attained by an interpolating function (typically a polynomial) of the desired trajectory. This chapter presents some techniques for trajectory generation, both in the case when the initial and final point of the path are assigned (*point-to-point motion*), and in the case when a finite sequence of points are assigned along the path (*motion through a sequence of points*). First, the problem of trajectory planning in the *joint space* is considered, and then the basic concepts of trajectory planning in the *operational space* are illustrated. The treatment of the motion planning problem for mobile robots is deferred to Chap. 12.

4.1 Path and Trajectory

The minimal requirement for a manipulator is the capability to move from an initial posture to a final assigned posture. The transition should be characterized by motion laws requiring the actuators to exert joint generalized forces which do not violate the saturation limits and do not excite the typically modelled resonant modes of the structure. It is then necessary to devise planning algorithms that generate suitably smooth trajectories.

In order to avoid confusion between terms often used as synonyms, the difference between a path and a trajectory is to be explained. A *path* denotes the locus of points in the joint space, or in the operational space, which the manipulator has to follow in the execution of the assigned motion; a path is then a pure geometric description of motion. On the other hand, a *trajectory* is a path on which a timing law is specified, for instance in terms of velocities and/or accelerations at each point.

In principle, it can be conceived that the inputs to a *trajectory planning* algorithm are the path description, the path constraints, and the constraints imposed by manipulator dynamics, whereas the outputs are the end-effector trajectories in terms of a time sequence of the values attained by position, velocity and acceleration.

A geometric path cannot be fully specified by the user for obvious complexity reasons. Typically, a reduced number of parameters is specified such as extremal points, possible intermediate points, and geometric primitives interpolating the points. Also, the motion timing law is not typically specified at each point of the geometric path, but rather it regards the total trajectory time, the constraints on the maximum velocities and accelerations, and eventually the assignment of velocity and acceleration at points of particular interest. On the basis of the above information, the trajectory planning algorithm generates a time sequence of variables that describe end-effector position and orientation over time in respect of the imposed constraints. Since the control action on the manipulator is carried out in the joint space, a suitable inverse kinematics algorithm is to be used to reconstruct the time sequence of joint variables corresponding to the above sequence in the operational space.

Trajectory planning in the operational space naturally allows the presence of path constraints to be accounted; these are due to regions of workspace which are forbidden to the manipulator, e.g., due to the presence of obstacles. In fact, such constraints are typically better described in the operational space, since their corresponding points in the joint space are difficult to compute.

With regard to motion in the neighbourhood of singular configurations and presence of redundant DOFs, trajectory planning in the operational space may involve problems difficult to solve. In such cases, it may be advisable to specify the path in the joint space, still in terms of a reduced number of parameters. Hence, a time sequence of joint variables has to be generated which satisfy the constraints imposed on the trajectory.

For the sake of clarity, in the following, the case of joint space trajectory planning is treated first. The results will then be extended to the case of trajectories in the operational space.

4.2 Joint Space Trajectories

A manipulator motion is typically assigned in the operational space in terms of trajectory parameters such as the initial and final end-effector pose, possible intermediate poses, and travelling time along particular geometric paths. If it is desired to plan a trajectory in the *joint space*, the values of the joint variables have to be determined first from the end-effector position and orientation specified by the user. It is then necessary to resort to an inverse kinematics algorithm, if planning is done off-line, or to directly measure the above variables, if planning is done by the teaching-by-showing technique (see Chap. 6).

The planning algorithm generates a function $\boldsymbol{q}(t)$ interpolating the given vectors of joint variables at each point, in respect of the imposed constraints.

In general, a joint space trajectory planning algorithm is required to have the following features:

- the generated trajectories should be not very demanding from a computational viewpoint,
- the joint positions and velocities should be continuous functions of time (continuity of accelerations may be imposed, too),
- undesirable effects should be minimized, e.g., nonsmooth trajectories interpolating a sequence of points on a path.

At first, the case is examined when only the initial and final points on the path and the traveling time are specified (*point-to-point*); the results are then generalized to the case when also intermediate points along the path are specified (*motion through a sequence of points*). Without loss of generality, the single joint variable $q(t)$ is considered.

4.2.1 Point-to-Point Motion

In *point-to-point motion*, the manipulator has to move from an initial to a final joint configuration in a given time t_f. In this case, the actual end-effector path is of no concern. The algorithm should generate a trajectory which, in respect to the above general requirements, is also capable of optimizing some performance index when the joint is moved from one position to another.

A suggestion for choosing the motion primitive may stem from the analysis of an incremental motion problem. Let I be the moment of inertia of a rigid body about its rotation axis. It is required to take the angle q from an initial value q_i to a final value q_f in a time t_f. It is obvious that infinite solutions exist to this problem. Assumed that rotation is executed through a torque τ supplied by a motor, a solution can be found which minimizes the energy dissipated in the motor. This optimization problem can be formalized as follows. Having set $\dot{q} = \omega$, determine the solution to the differential equation

$$I\dot{\omega} = \tau$$

subject to the condition

$$\int_o^{t_f} \omega(t)dt = q_f - q_i$$

so as to minimize the performance index

$$\int_0^{t_f} \tau^2(t)dt.$$

It can be shown that the resulting solution is of the type

$$\omega(t) = at^2 + bt + c.$$

Even though the joint dynamics cannot be described in the above simple manner,[1] the choice of a third-order polynomial function to generate a joint trajectory represents a valid solution for the problem at issue.

Therefore, to determine a joint motion, the *cubic polynomial*

$$q(t) = a_3 t^3 + a_2 t^2 + a_1 t + a_0 \tag{4.1}$$

can be chosen, resulting into a parabolic velocity profile

$$\dot{q}(t) = 3a_3 t^2 + 2a_2 t + a_1$$

and a linear acceleration profile

$$\ddot{q}(t) = 6a_3 t + 2a_2.$$

Since four coefficients are available, it is possible to impose, besides the initial and final joint position values q_i and q_f, also the initial and final joint velocity values $\dot{q}_i$ and $\dot{q}_f$ which are usually set to zero. Determination of a specific trajectory is given by the solution to the following system of equations:

$$\begin{aligned} a_0 &= q_i \\ a_1 &= \dot{q}_i \\ a_3 t_f^3 + a_2 t_f^2 + a_1 t_f + a_0 &= q_f \\ 3a_3 t_f^2 + 2a_2 t_f + a_1 &= \dot{q}_f, \end{aligned}$$

that allows the computation of the coefficients of the polynomial in (4.1).[2] Figure 4.1 illustrates the timing law obtained with the following data: $q_i = 0$, $q_f = \pi$, $t_f = 1$, and $\dot{q}_i = \dot{q}_f = 0$. As anticipated, velocity has a parabolic profile, while acceleration has a linear profile with initial and final discontinuity.

If it is desired to assign also the initial and final values of acceleration, six constraints have to be satisfied and then a polynomial of at least *fifth* order is needed. The motion timing law for the generic joint is then given by

$$q(t) = a_5 t^5 + a_4 t^4 + a_3 t^3 + a_2 t^2 + a_1 t + a_0, \tag{4.2}$$

whose coefficients can be computed, as for the previous case, by imposing the conditions for $t = 0$ and $t = t_f$ on the joint variable $q(t)$ and on its first two derivatives. With the choice (4.2), one obviously gives up minimizing the above performance index.

An alternative approach with timing laws of blended polynomial type is frequently adopted in industrial practice, which allows a direct verification

---

[1] In fact, recall that the moment of inertia about the joint axis is a function of manipulator configuration.

[2] Notice that it is possible to normalize the computation of the coefficients, so as to be independent both on the final time t_f and on the path length $|q_f - q_i|$.

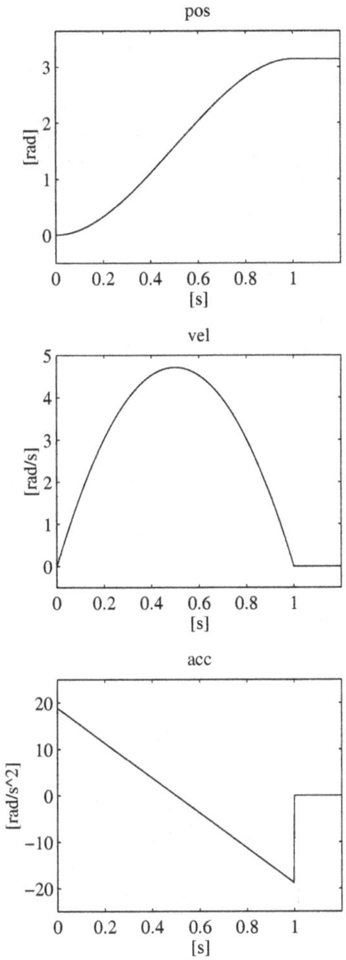

Fig. 4.1. Time history of position, velocity and acceleration with a cubic polynomial timing law

of whether the resulting velocities and accelerations can be supported by the physical mechanical manipulator.

In this case, a *trapezoidal velocity profile* is assigned, which imposes a constant acceleration in the start phase, a cruise velocity, and a constant deceleration in the arrival phase. The resulting trajectory is formed by a linear segment connected by two parabolic segments to the initial and final positions.

In the following, the problem is formulated by assuming that the final time of trajectory duration has been assigned. However, in industrial practice, the user is offered the option to specify the velocity percentage with respect to the maximum allowable velocity; this choice is aimed at avoiding occurrences when

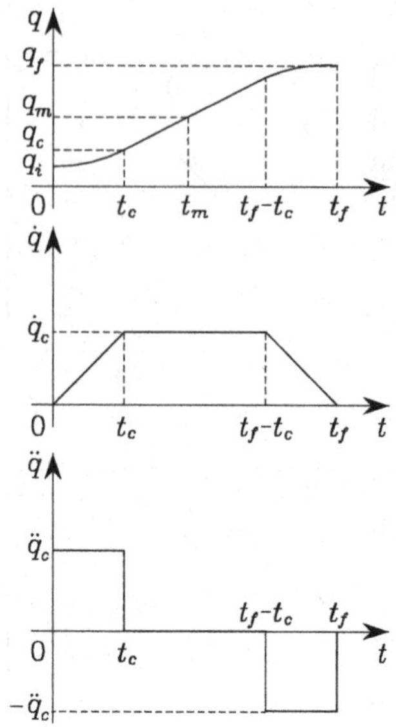

Fig. 4.2. Characterization of a timing law with trapezoidal velocity profile in terms of position, velocity and acceleration

the specification of a much too short motion duration would involve much too large values of velocities and/or accelerations, beyond those achievable by the manipulator.

As can be seen from the velocity profiles in Fig. 4.2, it is assumed that both initial and final velocities are null and the segments with constant accelerations have the same time duration; this implies an equal magnitude $\ddot{q}_c$ in the two segments. Notice also that the above choice leads to a symmetric trajectory with respect to the average point $q_m = (q_f + q_i)/2$ at $t_m = t_f/2$.

The trajectory has to satisfy some constraints to ensure the transition from q_i to q_f in a time t_f. The velocity at the end of the parabolic segment must be equal to the (constant) velocity of the linear segment, i.e.,

$$\ddot{q}_c t_c = \frac{q_m - q_c}{t_m - t_c} \tag{4.3}$$

where q_c is the value attained by the joint variable at the end of the parabolic segment at time t_c with constant acceleration $\ddot{q}_c$ (recall that $\dot{q}(0) = 0$). It is then

$$q_c = q_i + \frac{1}{2}\ddot{q}_c t_c^2. \tag{4.4}$$

Combining (4.3), (4.4) gives

$$\ddot{q}_c t_c^2 - \ddot{q}_c t_f t_c + q_f - q_i = 0. \tag{4.5}$$

Usually, $\ddot{q}_c$ is specified with the constraint that $\operatorname{sgn}\ddot{q}_c = \operatorname{sgn}(q_f - q_i)$; hence, for given t_f, q_i and q_f, the solution for t_c is computed from (4.5) as ($t_c \le t_f/2$)

$$t_c = \frac{t_f}{2} - \frac{1}{2}\sqrt{\frac{t_f^2\ddot{q}_c - 4(q_f - q_i)}{\ddot{q}_c}}. \tag{4.6}$$

Acceleration is then subject to the constraint

$$|\ddot{q}_c| \ge \frac{4|q_f - q_i|}{t_f^2}. \tag{4.7}$$

When the acceleration $\ddot{q}_c$ is chosen so as to satisfy (4.7) with the equality sign, the resulting trajectory does not feature the constant velocity segment any more and has only the acceleration and deceleration segments (*triangular* profile).

Given q_i, q_f and t_f, and thus also an average transition velocity, the constraint in (4.7) allows the imposition of a value of acceleration consistent with the trajectory. Then, t_c is computed from (4.6), and the following sequence of polynomials is generated:

$$q(t) = \begin{cases} q_i + \frac{1}{2}\ddot{q}_c t^2 & 0 \le t \le t_c \\ q_i + \ddot{q}_c t_c (t - t_c/2) & t_c < t \le t_f - t_c \\ q_f - \frac{1}{2}\ddot{q}_c (t_f - t)^2 & t_f - t_c < t \le t_f. \end{cases} \tag{4.8}$$

Figure 4.3 illustrates a representation of the motion timing law obtained by imposing the data: $q_i = 0$, $q_f = \pi$, $t_f = 1$, and $|\ddot{q}_c| = 6\pi$.

Specifying acceleration in the parabolic segment is not the only way to determine trajectories with trapezoidal velocity profile. Besides q_i, q_f and t_f, one can specify also the cruise velocity $\dot{q}_c$ which is subject to the constraint

$$\frac{|q_f - q_i|}{t_f} < |\dot{q}_c| \le \frac{2|q_f - q_i|}{t_f}. \tag{4.9}$$

By recognizing that $\dot{q}_c = \ddot{q}_c t_c$, (4.5) allows the computation of t_c as

$$t_c = \frac{q_i - q_f + \dot{q}_c t_f}{\dot{q}_c}, \tag{4.10}$$

and thus the resulting acceleration is

$$\ddot{q}_c = \frac{\dot{q}_c^2}{q_i - q_f + \dot{q}_c t_f}. \tag{4.11}$$

The computed values of t_c and $\ddot{q}_c$ as in (4.10), (4.11) allow the generation of the sequence of polynomials expressed by (4.8).

The adoption of a trapezoidal velocity profile results in a worse performance index compared to the cubic polynomial. The decrease is, however, limited; the term $\int_0^{t_f} \tau^2 dt$ increases by 12.5% with respect to the optimal case.

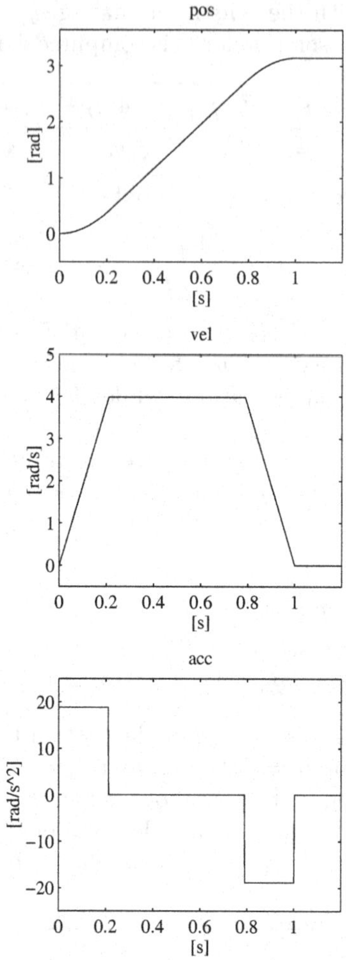

Fig. 4.3. Time history of position, velocity and acceleration with a trapezoidal velocity profile timing law

4.2.2 Motion Through a Sequence of Points

In several applications, the path is described in terms of a number of points greater than two. For instance, even for the simple point-to-point motion of a pick-and-place task, it may be worth assigning two intermediate points between the initial point and the final point; suitable positions can be set for lifting off and setting down the object, so that reduced velocities are obtained with respect to direct transfer of the object. For more complex applications, it may be convenient to assign a *sequence of points* so as to guarantee better monitoring on the executed trajectories; the points are to be specified more densely in those segments of the path where obstacles have to be avoided

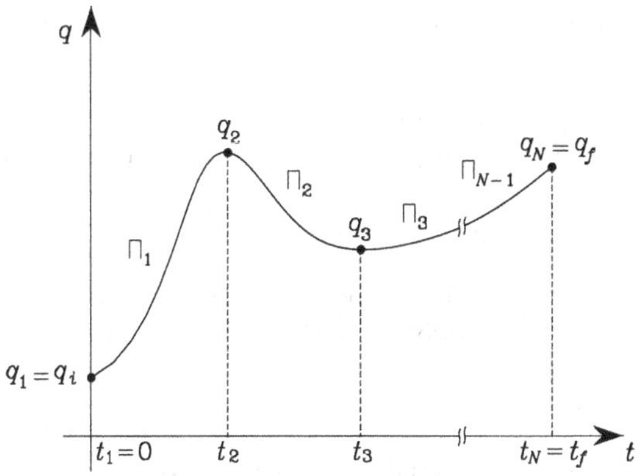

Fig. 4.4. Characterization of a trajectory on a given path obtained through interpolating polynomials

or a high path curvature is expected. It should not be forgotten that the corresponding joint variables have to be computed from the operational space poses.

Therefore, the problem is to generate a trajectory when N points, termed *path points*, are specified and have to be reached by the manipulator at certain instants of time. For each joint variable there are N constraints, and then one might want to use an $(N-1)$-order polynomial. This choice, however, has the following disadvantages:

- It is not possible to assign the initial and final velocities.
- As the order of a polynomial increases, its oscillatory behaviour increases, and this may lead to trajectories which are not natural for the manipulator.
- Numerical accuracy for computation of polynomial coefficients decreases as order increases.
- The resulting system of constraint equations is heavy to solve.
- Polynomial coefficients depend on all the assigned points; thus, if it is desired to change a point, all of them have to be recomputed.

These drawbacks can be overcome if a suitable number of low-order *interpolating polynomials*, continuous at the path points, are considered in place of a single high-order polynomial.

According to the previous section, the interpolating polynomial of lowest order is the *cubic polynomial*, since it allows the imposition of continuity of velocities at the path points. With reference to the single joint variable, a function $q(t)$ is sought, formed by a sequence of $N-1$ cubic polynomials $\Pi_k(t)$, for $k = 1, \ldots, N-1$, continuous with continuous first derivatives. The function $q(t)$ attains the values q_k for $t = t_k$ $(k = 1, \ldots, N)$, and $q_1 = q_i$, $t_1 = 0$, $q_N = q_f$, $t_N = t_f$; the q_k's represent the path points describing the desired trajectory at $t = t_k$ (Fig. 4.4). The following situations can be considered:

- Arbitrary values of $\dot{q}(t)$ are imposed at the path points.
- The values of $\dot{q}(t)$ at the path points are assigned according to a certain criterion.
- The acceleration $\ddot{q}(t)$ has to be continuous at the path points.

To simplify the problem, it is also possible to find interpolating polynomials of order less than three which determine trajectories passing nearby the path points at the given instants of time.

Interpolating polynomials with imposed velocities at path points

This solution requires the user to be able to specify the desired velocity at each path point; the solution does not possess any novelty with respect to the above concepts.

The system of equations allowing computation of the coefficients of the $N-1$ cubic polynomials interpolating the N path points is obtained by imposing the following conditions on the generic polynomial $\Pi_k(t)$ interpolating q_k and q_{k+1}, for $k = 1, \ldots, N-1$:

$$\begin{aligned}\Pi_k(t_k) &= q_k\\ \Pi_k(t_{k+1}) &= q_{k+1}\\ \dot{\Pi}_k(t_k) &= \dot{q}_k\\ \dot{\Pi}_k(t_{k+1}) &= \dot{q}_{k+1}.\end{aligned}$$

The result is $N-1$ systems of four equations in the four unknown coefficients of the generic polynomial; these can be solved one independently of the other. The initial and final velocities of the trajectory are typically set to zero ($\dot{q}_1 = \dot{q}_N = 0$) and continuity of velocity at the path points is ensured by setting

$$\dot{\Pi}_k(t_{k+1}) = \dot{\Pi}_{k+1}(t_{k+1})$$

for $k = 1, \ldots, N-2$.

Figure 4.5 illustrates the time history of position, velocity and acceleration obtained with the data: $q_1 = 0$, $q_2 = 2\pi$, $q_3 = \pi/2$, $q_4 = \pi$, $t_1 = 0$, $t_2 = 2$, $t_3 = 3$, $t_4 = 5$, $\dot{q}_1 = 0$, $\dot{q}_2 = \pi$, $\dot{q}_3 = -\pi$, $\dot{q}_4 = 0$. Notice the resulting discontinuity on the acceleration, since only continuity of velocity is guaranteed.

Interpolating polynomials with computed velocities at path points

In this case, the joint velocity at a path point has to be computed according to a certain criterion. By interpolating the path points with linear segments, the relative velocities can be computed according to the following rules:

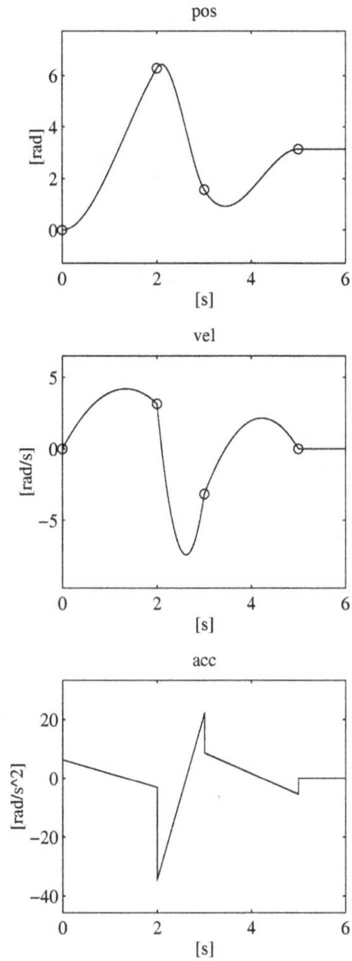

Fig. 4.5. Time history of position, velocity and acceleration with a timing law of interpolating polynomials with velocity constraints at path points

$$
\begin{aligned}
\dot{q}_1 &= 0 \\
\dot{q}_k &= \begin{cases} 0 & \operatorname{sgn}(v_k) \neq \operatorname{sgn}(v_{k+1}) \\ \frac{1}{2}(v_k + v_{k+1}) & \operatorname{sgn}(v_k) = \operatorname{sgn}(v_{k+1}) \end{cases} \\
\dot{q}_N &= 0,
\end{aligned}
\tag{4.12}
$$

where $v_k = (q_k - q_{k-1})/(t_k - t_{k-1})$ gives the slope of the segment in the time interval $[t_{k-1}, t_k]$. With the above settings, the determination of the interpolating polynomials is reduced to the previous case.

Figure 4.6 illustrates the time history of position, velocity and acceleration obtained with the following data: $q_1 = 0$, $q_2 = 2\pi$, $q_3 = \pi/2$, $q_4 = \pi$, $t_1 = 0$, $t_2 = 2$, $t_3 = 3$, $t_4 = 5$, $\dot{q}_1 = 0$, $\dot{q}_4 = 0$. It is easy to recognize that the imposed

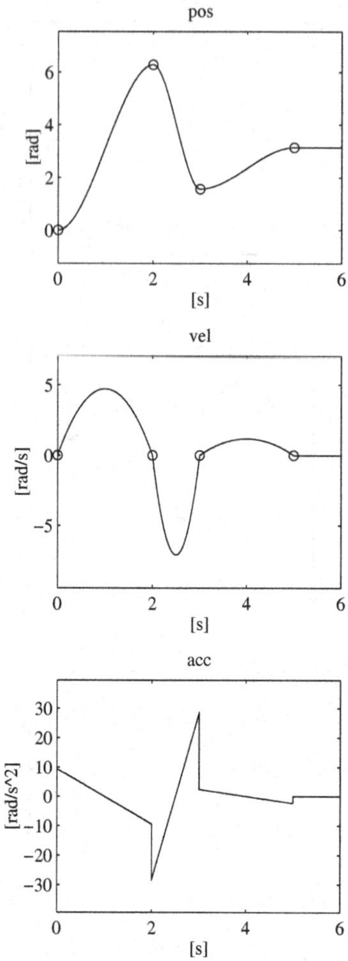

Fig. 4.6. Time history of position, velocity and acceleration with a timing law of interpolating polynomials with computed velocities at path points

sequence of path points leads to having zero velocity at the intermediate points.

Interpolating polynomials with continuous accelerations at path points (splines)

Both the above two solutions do not ensure continuity of accelerations at the path points. Given a sequence of N path points, the acceleration is also continuous at each t_k if four constraints are imposed, namely, two position constraints for each of the adjacent cubics and two constraints guaranteeing continuity of velocity and acceleration. The following equations have then to be satisfied:

$$\begin{aligned}\Pi_{k-1}(t_k) &= q_k \\ \Pi_{k-1}(t_k) &= \Pi_k(t_k) \\ \dot{\Pi}_{k-1}(t_k) &= \dot{\Pi}_k(t_k) \\ \ddot{\Pi}_{k-1}(t_k) &= \ddot{\Pi}_k(t_k).\end{aligned}$$

The resulting system for the N path points, including the initial and final points, cannot be solved. In fact, it is formed by $4(N-2)$ equations for the intermediate points and 6 equations for the extremal points; the position constraints for the polynomials $\Pi_0(t_1) = q_i$ and $\Pi_N(t_f) = q_f$ have to be excluded since they are not defined. Also, $\dot{\Pi}_0(t_1)$, $\ddot{\Pi}_0(t_1)$, $\dot{\Pi}_N(t_f)$, $\ddot{\Pi}_N(t_f)$ do not have to be counted as polynomials since they are just the imposed values of initial and final velocities and accelerations. In summary, one has $4N-2$ equations in $4(N-1)$ unknowns.

The system can be solved only if one eliminates the two equations which allow the arbitrary assignment of the initial and final acceleration values. Fourth-order polynomials should be used to include this possibility for the first and last segment.

On the other hand, if only third-order polynomials are to be used, the following deception can be operated. Two *virtual points* are introduced for which continuity constraints on position, velocity and acceleration can be imposed, without specifying the actual positions, though. It is worth remarking that the effective location of these points is irrelevant, since their position constraints regard continuity only. Hence, the introduction of two virtual points implies the determination of $N+1$ cubic polynomials.

Consider $N+2$ time instants t_k, where t_2 and t_{N+1} conventionally refer to the virtual points. The system of equations for determining the $N+1$ cubic polynomials can be found by taking the $4(N-2)$ equations:

$$\Pi_{k-1}(t_k) = q_k \tag{4.13}$$
$$\Pi_{k-1}(t_k) = \Pi_k(t_k) \tag{4.14}$$
$$\dot{\Pi}_{k-1}(t_k) = \dot{\Pi}_k(t_k) \tag{4.15}$$
$$\ddot{\Pi}_{k-1}(t_k) = \ddot{\Pi}_k(t_k) \tag{4.16}$$

for $k = 3, \ldots, N$, written for the $N-2$ intermediate path points, the 6 equations:

$$\Pi_1(t_1) = q_i \tag{4.17}$$
$$\dot{\Pi}_1(t_1) = \dot{q}_i \tag{4.18}$$
$$\ddot{\Pi}_1(t_1) = \ddot{q}_i, \tag{4.19}$$

$$\Pi_{N+1}(t_{N+2}) = q_f \tag{4.20}$$

$$\dot{\Pi}_{N+1}(t_{N+2}) = \dot{q}_f \tag{4.21}$$

$$\ddot{\Pi}_{N+1}(t_{N+2}) = \ddot{q}_f \tag{4.22}$$

written for the initial and final points, and the 6 equations:

$$\Pi_{k-1}(t_k) = \Pi_k(t_k) \tag{4.23}$$

$$\dot{\Pi}_{k-1}(t_k) = \dot{\Pi}_k(t_k) \tag{4.24}$$

$$\ddot{\Pi}_{k-1}(t_k) = \ddot{\Pi}_k(t_k) \tag{4.25}$$

for $k = 2, N+1$, written for the two virtual points. The resulting system has $4(N+1)$ equations in $4(N+1)$ unknowns, that are the coefficients of the $N+1$ cubic polynomials.

The solution to the system is computationally demanding, even for low values of N. Nonetheless, the problem can be cast in a suitable form so as to solve the resulting system of equations with a computationally efficient algorithm. Since the generic polynomial $\Pi_k(t)$ is a cubic, its second derivative must be a linear function of time which then can be written as

$$\ddot{\Pi}_k(t) = \frac{\ddot{\Pi}_k(t_k)}{\Delta t_k}(t_{k+1} - t) + \frac{\ddot{\Pi}_k(t_{k+1})}{\Delta t_k}(t - t_k) \qquad k = 1, \ldots, N+1, \tag{4.26}$$

where $\Delta t_k = t_{k+1} - t_k$ indicates the time interval to reach q_{k+1} from q_k. By integrating (4.26) twice over time, the generic polynomial can be written as

$$\begin{aligned}\Pi_k(t) = {} & \frac{\ddot{\Pi}_k(t_k)}{6\Delta t_k}(t_{k+1} - t)^3 + \frac{\ddot{\Pi}_k(t_{k+1})}{6\Delta t_k}(t - t_k)^3 \\ & + \left(\frac{\Pi_k(t_{k+1})}{\Delta t_k} - \frac{\Delta t_k \ddot{\Pi}_k(t_{k+1})}{6}\right)(t - t_k) \\ & + \left(\frac{\Pi_k(t_k)}{\Delta t_k} - \frac{\Delta t_k \ddot{\Pi}_k(t_k)}{6}\right)(t_{k+1} - t) \qquad k = 1, \ldots, N+1,\end{aligned} \tag{4.27}$$

which depends on the 4 unknowns: $\Pi_k(t_k)$, $\Pi_k(t_{k+1})$, $\ddot{\Pi}_k(t_k)$, $\ddot{\Pi}_k(t_{k+1})$.

Notice that the N variables q_k for $k \neq 2, N+1$ are given via (4.13), while continuity is imposed for q_2 and q_{N+1} via (4.23). By using (4.14), (4.17), (4.20), the unknowns in the $N+1$ equations in (4.27) reduce to $2(N+2)$. By observing that the equations in (4.18), (4.21) depend on q_2 and q_{N+1}, and that $\dot{q}_i$ and $\dot{q}_f$ are given, q_2 and q_{N+1} can be computed as a function of $\ddot{\Pi}_1(t_1)$ and $\ddot{\Pi}_{N+1}(t_{N+2})$, respectively. Thus, a number of $2(N+1)$ unknowns are left.

By accounting for (4.16), (4.25), and noticing that in ((4.19), (4.22) $\ddot{q}_i$ and $\ddot{q}_f$ are given, the unknowns reduce to N.

At this point, (4.15), (4.24) can be utilized to write the system of N equations in N unknowns:

$$\dot{\Pi}_1(t_2) = \dot{\Pi}_2(t_2)$$

$$\vdots$$

$$\dot{\Pi}_N(t_{N+1}) = \dot{\Pi}_{N+1}(t_{N+1}).$$

Time-differentiation of (4.27) gives both $\ddot{\Pi}_k(t_{k+1})$ and $\ddot{\Pi}_{k+1}(t_{k+1})$ for $k = 1, \ldots, N$, and thus it is possible to write a system of linear equations of the kind

$$\boldsymbol{A}\,[\,\ddot{\Pi}_2(t_2) \quad \ldots \quad \ddot{\Pi}_{N+1}(t_{N+1})\,]^T = \boldsymbol{b} \tag{4.28}$$

which presents a vector $\boldsymbol{b}$ of known terms and a nonsingular coefficient matrix $\boldsymbol{A}$; the solution to this system always exists and is unique. It can be shown that the matrix $\boldsymbol{A}$ has a tridiagonal band structure of the type

$$\boldsymbol{A} = \begin{bmatrix} a_{11} & a_{12} & \ldots & 0 & 0 \\ a_{21} & a_{22} & \ldots & 0 & 0 \\ \vdots & \vdots & \ddots & \vdots & \vdots \\ 0 & 0 & \ldots & a_{N-1,N-1} & a_{N-1,N} \\ 0 & 0 & \ldots & a_{N,N-1} & a_{NN} \end{bmatrix},$$

which simplifies the solution to the system (see Problem 4.4). This matrix is the same for all joints, since it depends only on the time intervals Δt_k specified.

An efficient solution algorithm exists for the above system which is given by a *forward* computation followed by a *backward* computation. From the first equation, $\ddot{\Pi}_2(t_2)$ can be computed as a function of $\ddot{\Pi}_3(t_3)$ and then substituted in the second equation, which then becomes an equation in the unknowns $\ddot{\Pi}_3(t_3)$ and $\ddot{\Pi}_4(t_4)$. This is carried out forward by transforming all the equations in equations with two unknowns, except the last one which will have $\ddot{\Pi}_{N+1}(t_{N+1})$ only as unknown. At this point, all the unknowns can be determined step by step through a backward computation.

The above sequence of cubic polynomials is termed *spline* to indicate smooth functions that interpolate a sequence of given points ensuring continuity of the function and its derivatives.

Figure 4.7 illustrates the time history of position, velocity and acceleration obtained with the data: $q_1 = 0$, $q_3 = 2\pi$, $q_4 = \pi/2$, $q_6 = \pi$, $t_1 = 0$, $t_3 = 2$, $t_4 = 3$, $t_6 = 5$, $\dot{q}_1 = 0$, $\dot{q}_6 = 0$. Two different pairs of virtual points were considered at the time instants: $t_2 = 0.5$, $t_5 = 4.5$ (solid line in the figure), and $t_2 = 1.5$, $t_5 = 3.5$ (dashed line in the figure), respectively. Notice the parabolic velocity profile and the linear acceleration profile. Further, for the second pair, larger values of acceleration are obtained, since the relative time instants are closer to those of the two intermediate points.

Interpolating linear polynomials with parabolic blends

A simplification in trajectory planning can be achieved as follows. Consider the case when it is desired to interpolate N path points $q_1, \ldots, q_N$ at time

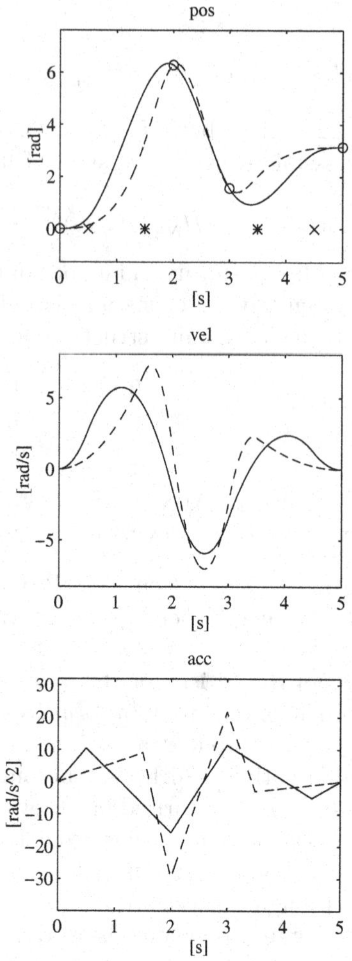

Fig. 4.7. Time history of position, velocity and acceleration with a timing law of cubic splines for two different pairs of virtual points

instants $t_1, \ldots, t_N$ with linear segments. To avoid discontinuity problems on the first derivative at the time instants t_k, the function $q(t)$ must have a parabolic profile (*blend*) around t_k; as a consequence, the entire trajectory is composed of a sequence of linear and quadratic polynomials, which in turn implies that a discontinuity on $\ddot{q}(t)$ is tolerated.

Then let $\Delta t_k = t_{k+1} - t_k$ be the time distance between q_k and q_{k+1}, and $\Delta t_{k,k+1}$ be the time interval during which the trajectory interpolating q_k and q_{k+1} is a linear function of time. Also let $\dot{q}_{k,k+1}$ be the constant velocity and $\ddot{q}_k$ be the acceleration in the parabolic blend whose duration is $\Delta t'_k$. The resulting trajectory is illustrated in Fig. 4.8. The values of q_k, Δt_k, and $\Delta t'_k$

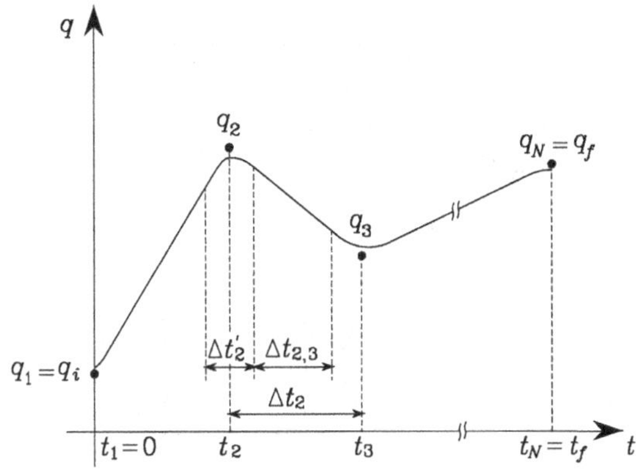

Fig. 4.8. Characterization of a trajectory with interpolating linear polynomials with parabolic blends

are assumed to be given. Velocity and acceleration for the intermediate points are computed as

$$\dot{q}_{k-1,k} = \frac{q_k - q_{k-1}}{\Delta t_{k-1}} \tag{4.29}$$

$$\ddot{q}_k = \frac{\dot{q}_{k,k+1} - \dot{q}_{k-1,k}}{\Delta t'_k}; \tag{4.30}$$

these equations are straightforward.

The first and last segments deserve special care. In fact, if it is desired to maintain the coincidence of the trajectory with the first and last segments, at least for a portion of time, the resulting trajectory has a longer duration given by $t_N - t_1 + (\Delta t'_1 + \Delta t'_N)/2$, where $\dot{q}_{0,1} = \dot{q}_{N,N+1} = 0$ has been imposed for computing initial and final accelerations.

Notice that $q(t)$ reaches none of the path points q_k but passes nearby (Fig. 4.8). In this situation, the path points are more appropriately termed *via points*; the larger the blending acceleration, the closer the passage to a via point.

On the basis of the given q_k, Δt_k and $\Delta t'_k$, the values of $\dot{q}_{k-1,k}$ and $\ddot{q}_k$ are computed via (4.29), (4.30) and a sequence of linear polynomials with parabolic blends is generated. Their expressions as a function of time are not derived here to avoid further loading of the analytic presentation.

Figure 4.9 illustrates the time history of position, velocity and acceleration obtained with the data: $q_1 = 0$, $q_2 = 2\pi$, $q_3 = \pi/2$, $q_4 = \pi$, $t_1 = 0$, $t_2 = 2$, $t_3 = 3$, $t_4 = 5$, $\dot{q}_1 = 0$, $\dot{q}_4 = 0$. Two different values for the blend times have been considered: $\Delta t'_k = 0.2$ (solid line in the figure) and $\Delta t'_k = 0.6$ (dashed line in the figure), for $k = 1, \ldots, 4$, respectively. Notice that in the first case the passage of $q(t)$ is closer to the via points, though at the expense of higher acceleration values.

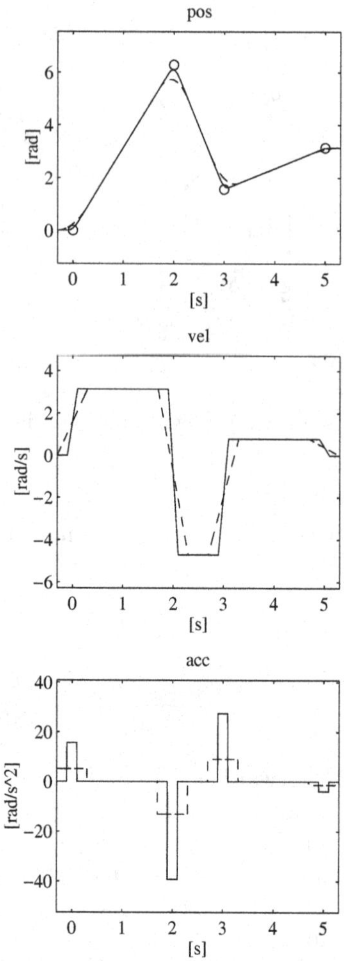

Fig. 4.9. Time history of position, velocity and acceleration with a timing law of interpolating linear polynomials with parabolic blends

The technique presented above turns out to be an application of the trapezoidal velocity profile law to the interpolation problem. If one gives up a trajectory passing near a via point at a prescribed instant of time, the use of trapezoidal velocity profiles allows the development of a trajectory planning algorithm which is attractive for its simplicity.

In particular, consider the case of one intermediate point only, and suppose that trapezoidal velocity profiles are considered as motion primitives with the possibility to specify the initial and final point and the duration of the motion only; it is assumed that $\dot{q}_i = \dot{q}_f = 0$. If two segments with trapezoidal velocity profiles were generated, the manipulator joint would certainly reach the intermediate point, but it would be forced to stop there, before continuing the motion towards the final point. A keen alternative is to start generating the second segment ahead of time with respect to the end of the first segment, using the sum of velocities (or positions) as a reference. In this way, the joint is guaranteed to reach the final position; crossing of the intermediate point at the specified instant of time is not guaranteed, though.

Figure 4.10 illustrates the time history of position, velocity and acceleration obtained with the data: $q_i = 0$, $q_f = 3\pi/2$, $t_i = 0$, $t_f = 2$. The intermediate point is located at $q = \pi$ with $t = 1$, the maximum acceleration values in the two segments are respectively $|\ddot{q}_c| = 6\pi$ and $|\ddot{q}_c| = 3\pi$, and the time anticipation is 0.18. As predicted, with time anticipation, the assigned intermediate position becomes a via point with the advantage of an overall shorter time duration. Notice, also, that velocity does not vanish at the intermediate point.

4.3 Operational Space Trajectories

A joint space trajectory planning algorithm generates a time sequence of values for the joint variables $\boldsymbol{q}(t)$ so that the manipulator is taken from the initial to the final configuration, eventually by moving through a sequence of intermediate configurations. The resulting end-effector motion is not easily predictable, in view of the nonlinear effects introduced by direct kinematics. Whenever it is desired that the end-effector motion follows a geometrically specified path in the *operational space*, it is necessary to plan trajectory execution directly in the same space. Planning can be done either by interpolating a sequence of prescribed path points or by generating the analytical motion primitive and the relative trajectory in a punctual way.

In both cases, the time sequence of the values attained by the operational space variables is utilized in real time to obtain the corresponding sequence of values of the joint space variables, via an inverse kinematics algorithm. In this regard, the computational complexity induced by trajectory generation in the operational space and related kinematic inversion sets an upper limit on the maximum sampling rate to generate the above sequences. Since these sequences constitute the reference inputs to the motion control system, a linear *microinterpolation* is typically carried out. In this way, the frequency at which reference inputs are updated is increased so as to enhance dynamic performance of the system.

Whenever the path is not to be followed exactly, its characterization can be performed through the assignment of N points specifying the values of the variables $\boldsymbol{x}_e$ chosen to describe the end-effector pose in the operational space at given time instants t_k, for $k = 1, \dots, N$. Similar to what was presented in the above sections, the trajectory is generated by determining a smooth interpolating vector function between the various path points. Such a function

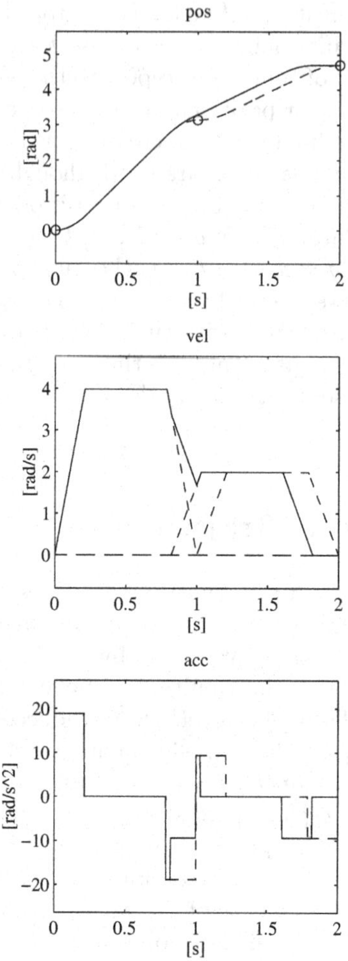

Fig. 4.10. Time history of position, velocity and acceleration with a timing law of interpolating linear polynomials with parabolic blends obtained by anticipating the generation of the second segment of trajectory

can be computed by applying to each component of $\boldsymbol{x}_e$ any of the interpolation techniques illustrated in Sect. 4.2.2 for the single joint variable.

Therefore, for given path (or via) points $\boldsymbol{x}_e(t_k)$, the corresponding components $x_{ei}(t_k)$, for $i = 1, \ldots r$ (where r is the dimension of the operational space of interest) can be interpolated with a sequence of cubic polynomials, a sequence of linear polynomials with parabolic blends, and so on.

On the other hand, if the end-effector motion has to follow a prescribed trajectory of motion, this must be expressed analytically. It is then necessary to refer to motion primitives defining the geometric features of the path and time primitives defining the timing law on the path itself.

4.3.1 Path Primitives

For the definition of *path primitives* it is convenient to refer to the parametric description of paths in space. Then let $\boldsymbol{p}$ be a (3×1) vector and $\boldsymbol{f}(\sigma)$ a continuous vector function defined in the interval $[\sigma_i, \sigma_f]$. Consider the equation

$$\boldsymbol{p} = \boldsymbol{f}(\sigma); \tag{4.31}$$

with reference to its geometric description, the sequence of values of $\boldsymbol{p}$ with σ varying in $[\sigma_i, \sigma_f]$ is termed *path* in space. The equation in (4.31) defines the *parametric representation* of the path Γ and the scalar σ is called parameter. As σ increases, the point $\boldsymbol{p}$ moves on the path in a given direction. This direction is said to be the direction induced on Γ by the parametric representation (4.31). A path is closed when $\boldsymbol{p}(\sigma_f) = \boldsymbol{p}(\sigma_i)$; otherwise it is open.

Let $\boldsymbol{p}_i$ be a point on the open path Γ on which a direction has been fixed. The *arc length* s of the generic point $\boldsymbol{p}$ is the length of the arc of Γ with extremes $\boldsymbol{p}$ and $\boldsymbol{p}_i$ if $\boldsymbol{p}$ follows $\boldsymbol{p}_i$, the opposite of this length if $\boldsymbol{p}$ precedes $\boldsymbol{p}_i$. The point $\boldsymbol{p}_i$ is said to be the origin of the arc length ($s = 0$).

From the above presentation it follows that to each value of s a well-determined path point corresponds, and then the arc length can be used as a parameter in a different parametric representation of the path Γ:

$$\boldsymbol{p} = \boldsymbol{f}(s); \tag{4.32}$$

the range of variation of the parameter s will be the sequence of arc lengths associated with the points of Γ.

Consider a path Γ represented by (4.32). Let $\boldsymbol{p}$ be a point corresponding to the arc length s. Except for special cases, $\boldsymbol{p}$ allows the definition of three unit vectors characterizing the path. The orientation of such vectors depends exclusively on the path geometry, while their direction depends also on the direction induced by (4.32) on the path.

The first of such unit vectors is the *tangent unit vector* denoted by $\boldsymbol{t}$. This vector is oriented along the direction induced on the path by s.

The second unit vector is the *normal unit vector* denoted by $\boldsymbol{n}$. This vector is oriented along the line intersecting $\boldsymbol{p}$ at a right angle with $\boldsymbol{t}$ and lies in the so-called *osculating plane* $\mathcal{O}$ (Fig. 4.11); such plane is the limit position of the plane containing the unit vector $\boldsymbol{t}$ and a point $\boldsymbol{p}' \in \Gamma$ when $\boldsymbol{p}'$ tends to $\boldsymbol{p}$ along the path. The direction of $\boldsymbol{n}$ is so that the path Γ, in the neighbourhood of $\boldsymbol{p}$ with respect to the plane containing $\boldsymbol{t}$ and normal to $\boldsymbol{n}$, lies on the same side of $\boldsymbol{n}$.

The third unit vector is the *binormal unit vector* denoted by $\boldsymbol{b}$. This vector is so that the frame $(\boldsymbol{t}, \boldsymbol{n}, \boldsymbol{b})$ is right-handed (Fig. 4.11). Notice that it is not always possible to define uniquely such a frame.

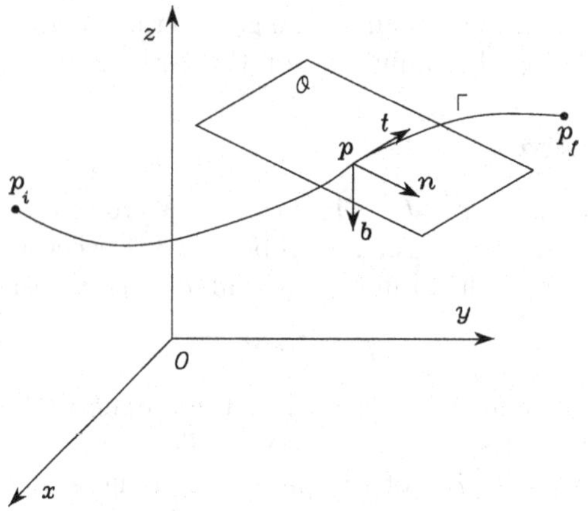

Fig. 4.11. Parametric representation of a path in space

It can be shown that the above three unit vectors are related by simple relations to the path representation Γ as a function of the arc length. In particular, it is

$$\begin{aligned} \boldsymbol{t} &= \frac{d\boldsymbol{p}}{ds} \\ \boldsymbol{n} &= \frac{1}{\left\|\frac{d^2\boldsymbol{p}}{ds^2}\right\|}\frac{d^2\boldsymbol{p}}{ds^2} \\ \boldsymbol{b} &= \boldsymbol{t} \times \boldsymbol{n}. \end{aligned} \tag{4.33}$$

Typical path parametric representations are reported below which are useful for trajectory generation in the operational space.

Rectilinear path

Consider the linear segment connecting point $\boldsymbol{p}_i$ to point $\boldsymbol{p}_f$. The parametric representation of this path is

$$\boldsymbol{p}(s) = \boldsymbol{p}_i + \frac{s}{\|\boldsymbol{p}_f - \boldsymbol{p}_i\|}(\boldsymbol{p}_f - \boldsymbol{p}_i). \tag{4.34}$$

Notice that $\boldsymbol{p}(0) = \boldsymbol{p}_i$ and $\boldsymbol{p}(\|\boldsymbol{p}_f - \boldsymbol{p}_i\|) = \boldsymbol{p}_f$. Hence, the direction induced on Γ by the parametric representation (4.34) is that going from $\boldsymbol{p}_i$ to $\boldsymbol{p}_f$. Differentiating (4.34) with respect to s gives

$$\frac{d\boldsymbol{p}}{ds} = \frac{1}{\|\boldsymbol{p}_f - \boldsymbol{p}_i\|}(\boldsymbol{p}_f - \boldsymbol{p}_i) \tag{4.35}$$

$$\frac{d^2\boldsymbol{p}}{ds^2} = \boldsymbol{0}. \tag{4.36}$$

In this case it is not possible to define the frame $(\boldsymbol{t}, \boldsymbol{n}, \boldsymbol{b})$ uniquely.

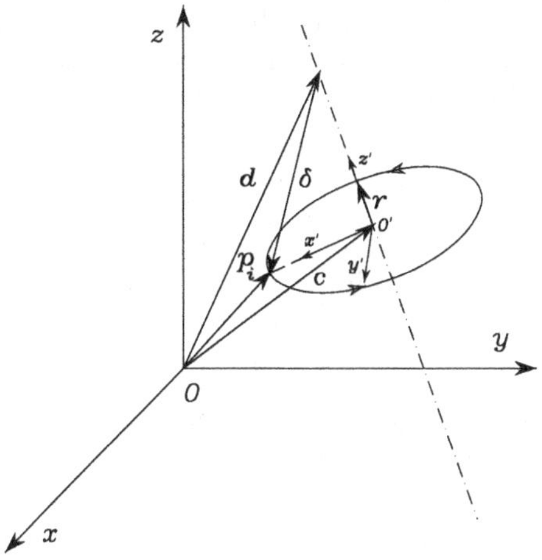

Fig. 4.12. Parametric representation of a circle in space

Circular path

Consider a circle Γ in space. Before deriving its parametric representation, it is necessary to introduce its significant parameters. Suppose that the circle is specified by assigning (Fig. 4.12):

- the unit vector of the circle axis $\boldsymbol{r}$,
- the position vector $\boldsymbol{d}$ of a point along the circle axis,
- the position vector $\boldsymbol{p}_i$ of a point on the circle.

With these parameters, the position vector $\boldsymbol{c}$ of the centre of the circle can be found. Let $\boldsymbol{\delta} = \boldsymbol{p}_i - \boldsymbol{d}$; for $\boldsymbol{p}_i$ not to be on the axis, i.e., for the circle not to degenerate into a point, it must be

$$|\boldsymbol{\delta}^T \boldsymbol{r}| < \|\boldsymbol{\delta}\|;$$

in this case it is

$$\boldsymbol{c} = \boldsymbol{d} + (\boldsymbol{\delta}^T \boldsymbol{r})\boldsymbol{r}. \tag{4.37}$$

It is now desired to find a parametric representation of the circle as a function of the arc length. Notice that this representation is very simple for a suitable choice of the reference frame. To see this, consider the frame O'–$x'y'z'$, where O' coincides with the centre of the circle, axis x' is oriented along the direction of the vector $\boldsymbol{p}_i - \boldsymbol{c}$, axis z' is oriented along $\boldsymbol{r}$ and axis y' is chosen so as to complete a right-handed frame. When expressed in this reference frame, the parametric representation of the circle is

$$\boldsymbol{p}'(s) = \begin{bmatrix} \rho\cos{(s/\rho)} \\ \rho\sin{(s/\rho)} \\ 0 \end{bmatrix}, \tag{4.38}$$

where $\rho = \|\boldsymbol{p}_i - \boldsymbol{c}\|$ is the radius of the circle and the point $\boldsymbol{p}_i$ has been assumed as the origin of the arc length. For a different reference frame, the path representation becomes

$$\boldsymbol{p}(s) = \boldsymbol{c} + \boldsymbol{R}\boldsymbol{p}'(s), \tag{4.39}$$

where $\boldsymbol{c}$ is expressed in the frame O–xyz and $\boldsymbol{R}$ is the rotation matrix of frame O'– $x'y'z'$ with respect to frame O–xyz which, in view of (2.3), can be written as

$$\boldsymbol{R} = [\,\boldsymbol{x}' \quad \boldsymbol{y}' \quad \boldsymbol{z}'\,];$$

$\boldsymbol{x}'$, $\boldsymbol{y}'$, $\boldsymbol{z}'$ indicate the unit vectors of the frame expressed in the frame O–xyz. Differentiating (4.39) with respect to s gives

$$\frac{d\boldsymbol{p}}{ds} = \boldsymbol{R}\begin{bmatrix} -\sin(s/\rho) \\ \cos(s/\rho) \\ 0 \end{bmatrix} \tag{4.40}$$

$$\frac{d^2\boldsymbol{p}}{ds^2} = \boldsymbol{R}\begin{bmatrix} -\cos(s/\rho)/\rho \\ -\sin(s/\rho)/\rho \\ 0 \end{bmatrix}. \tag{4.41}$$

4.3.2 Position

Let $\boldsymbol{x}_e$ be the vector of operational space variables expressing the *pose* of the manipulator's end-effector as in (2.80). Generating a trajectory in the operational space means to determine a function $\boldsymbol{x}_e(t)$ taking the end-effector frame from the initial to the final pose in a time t_f along a given path with a specific motion timing law. First, consider end-effector position. Orientation will follow.

Let $\boldsymbol{p}_e = \boldsymbol{f}(s)$ be the (3×1) vector of the parametric representation of the path Γ as a function of the arc length s; the origin of the end-effector frame moves from $\boldsymbol{p}_i$ to $\boldsymbol{p}_f$ in a time t_f. For simplicity, suppose that the origin of the arc length is at $\boldsymbol{p}_i$ and the direction induced on Γ is that going from $\boldsymbol{p}_i$ to $\boldsymbol{p}_f$. The arc length then goes from the value $s = 0$ at $t = 0$ to the value $s = s_f$ (path length) at $t = t_f$. The timing law along the path is described by the function $s(t)$.

In order to find an analytic expression for $s(t)$, any of the above techniques for joint trajectory generation can be employed. In particular, either a cubic polynomial or a sequence of linear segments with parabolic blends can be chosen for $s(t)$.

It is worth making some remarks on the time evolution of $\boldsymbol{p}_e$ on Γ, for a given timing law $s(t)$. The velocity of point $\boldsymbol{p}_e$ is given by the time derivative of $\boldsymbol{p}_e$

$$\dot{\boldsymbol{p}}_e = \dot{s}\frac{d\boldsymbol{p}_e}{ds} = \dot{s}\boldsymbol{t},$$

where $\boldsymbol{t}$ is the tangent vector to the path at point $\boldsymbol{p}$ in (4.33). Then, $\dot{s}$ represents the magnitude of the velocity vector relative to point $\boldsymbol{p}$, taken with the positive or negative sign depending on the direction of $\dot{\boldsymbol{p}}$ along $\boldsymbol{t}$. The magnitude of $\dot{\boldsymbol{p}}$ starts from zero at $t = 0$, then it varies with a parabolic or trapezoidal profile as per either of the above choices for $s(t)$, and finally it returns to zero at $t = t_f$.

As a first example, consider the segment connecting point $\boldsymbol{p}_i$ with point $\boldsymbol{p}_f$. The parametric representation of this path is given by (4.34). Velocity and acceleration of $\boldsymbol{p}_e$ can be easily computed by recalling the rule of differentiation of compound functions, i.e.,

$$\dot{\boldsymbol{p}}_e = \frac{\dot{s}}{\|\boldsymbol{p}_f - \boldsymbol{p}_i\|}(\boldsymbol{p}_f - \boldsymbol{p}_i) = \dot{s}\boldsymbol{t} \tag{4.42}$$

$$\ddot{\boldsymbol{p}}_e = \frac{\ddot{s}}{\|\boldsymbol{p}_f - \boldsymbol{p}_i\|}(\boldsymbol{p}_f - \boldsymbol{p}_i) = \ddot{s}\boldsymbol{t}. \tag{4.43}$$

As a further example, consider a circle Γ in space. From the parametric representation derived above, in view of (4.40), (4.41), velocity and acceleration of point $\boldsymbol{p}_e$ on the circle are

$$\dot{\boldsymbol{p}}_e = \boldsymbol{R}\begin{bmatrix} -\dot{s}\sin(s/\rho) \\ \dot{s}\cos(s/\rho) \\ 0 \end{bmatrix} \tag{4.44}$$

$$\ddot{\boldsymbol{p}}_e = \boldsymbol{R}\begin{bmatrix} -\dot{s}^2\cos(s/\rho)/\rho - \ddot{s}\sin(s/\rho) \\ -\dot{s}^2\sin(s/\rho)/\rho + \ddot{s}\cos(s/\rho) \\ 0 \end{bmatrix}. \tag{4.45}$$

Notice that the velocity vector is aligned with $\boldsymbol{t}$, and the acceleration vector is given by two contributions: the first is aligned with $\boldsymbol{n}$ and represents the centripetal acceleration, while the second is aligned with $\boldsymbol{t}$ and represents the tangential acceleration.

Finally, consider the path consisting of a sequence of $N + 1$ points, $\boldsymbol{p}_0, \boldsymbol{p}_1, \ldots, \boldsymbol{p}_N$, connected by N segments. A feasible parametric representation of the overall path is the following:

$$\boldsymbol{p}_e = \boldsymbol{p}_0 + \sum_{j=1}^{N} \frac{s_j}{\|\boldsymbol{p}_j - \boldsymbol{p}_{j-1}\|}(\boldsymbol{p}_j - \boldsymbol{p}_{j-1}), \tag{4.46}$$

with $j = 1, \ldots, N$. In (4.46) s_j is the arc length associated with the j-th segment of the path, connecting point $\boldsymbol{p}_{j-1}$ to point $\boldsymbol{p}_j$, defined as

$$s_j(t) = \begin{cases} 0 & 0 \le t \le t_{j-1} \\ s_j'(t) & t_{j-1} < t < t_j \\ \|\boldsymbol{p}_j - \boldsymbol{p}_{j-1}\| & t_j \le t \le t_f, \end{cases} \tag{4.47}$$

where $t_0 = 0$ and $t_N = t_f$ are respectively the initial and final time instants of the trajectory, t_j is the time instant corresponding to point $\boldsymbol{p}_j$ and $s'_j(t)$ can be an analytical function of cubic polynomial type, linear type with parabolic blends, and so forth, which varies continuously from the value $s'_j = 0$ at $t = t_{j-1}$ to the value $s'_j = \|\boldsymbol{p}_j - \boldsymbol{p}_{j-1}\|$ at $t = t_j$.

The velocity and acceleration of $\boldsymbol{p}_e$ can be easily found by differentiating (4.46):

$$\dot{\boldsymbol{p}}_e = \sum_{j=1}^{N} \frac{\dot{s}_j}{\|\boldsymbol{p}_j - \boldsymbol{p}_{j-1}\|}(\boldsymbol{p}_j - \boldsymbol{p}_{j-1}) = \sum_{j=1}^{N} \dot{s}_j \boldsymbol{t}_j \tag{4.48}$$

$$\ddot{\boldsymbol{p}}_e = \sum_{j=1}^{N} \frac{\ddot{s}_j}{\|\boldsymbol{p}_j - \boldsymbol{p}_{j-1}\|}(\boldsymbol{p}_j - \boldsymbol{p}_{j-1}) = \sum_{j=1}^{N} \ddot{s}_j \boldsymbol{t}_j, \tag{4.49}$$

where $\boldsymbol{t}_j$ is the tangent unit vector of the j-th segment.

Because of the discontinuity of the first derivative at the path points between two non-aligned segments, the manipulator will have to stop and then go along the direction of the following segment. Assumed a relaxation of the constraint to pass through the path points, it is possible to avoid a manipulator stop by connecting the segments near the above points, which will then be named *operational space via points* so as to guarantee, at least, continuity of the first derivative.

As already illustrated for planning of interpolating linear polynomials with parabolic blends passing by the via points in the joint space, the use of trapezoidal velocity profiles for the arc lengths allows the development of a rather simple planning algorithm

In detail, it will be sufficient to properly anticipate the generation of the single segments, before the preceding segment has been completed. This leads to modifying (4.47) as follows:

$$s_j(t) = \begin{cases} 0 & 0 \le t \le t_{j-1} - \Delta t_j \\ s'_j(t + \Delta t_j) & t_{j-1} - \Delta t_j < t < t_j - \Delta t_j \\ \|\boldsymbol{p}_j - \boldsymbol{p}_{j-1}\| & t_j - \Delta t_j \le t \le t_f - \Delta t_N, \end{cases} \tag{4.50}$$

where Δt_j is the time advance at which the j-th segment is generated, which can be recursively evaluated as

$$\Delta t_j = \Delta t_{j-1} + \delta t_j,$$

with $j = 1, \ldots, N$ e $\Delta t_0 = 0$. Notice that this time advance is given by the sum of two contributions: the former, Δt_{j-1}, accounts for the sum of the time advances at which the preceding segments have been generated, while the latter, δt_j, is the time advance at which the generation of the current segment starts.

4.3.3 Orientation

Consider now end-effector orientation. Typically, this is specified in terms of the rotation matrix of the (time-varying) end-effector frame with respect to the base frame. As is well known, the three columns of the rotation matrix represent the three unit vectors of the end-effector frame with respect to the base frame. To generate a trajectory, however, a linear interpolation on the unit vectors $\boldsymbol{n}_e$, $\boldsymbol{s}_e$, $\boldsymbol{a}_e$ describing the initial and final orientation does not guarantee orthonormality of the above vectors at each instant of time.

Euler angles

In view of the above difficulty, for trajectory generation purposes, orientation is often described in terms of the Euler angles triplet $\boldsymbol{\phi}_e = (\varphi, \vartheta, \psi)$ for which a timing law can be specified. Usually, $\boldsymbol{\phi}_e$ moves along the segment connecting its initial value $\boldsymbol{\phi}_i$ to its final value $\boldsymbol{\phi}_f$. Also in this case, it is convenient to choose a cubic polynomial or a linear segment with parabolic blends timing law. In this way, in fact, the angular velocity $\boldsymbol{\omega}_e$ of the time-varying frame, which is related to $\dot{\boldsymbol{\phi}}_e$ by the linear relationship (3.64), will have continuous magnitude.

Therefore, for given $\boldsymbol{\phi}_i$ and $\boldsymbol{\phi}_f$ and timing law, the position, velocity and acceleration profiles are

$$\begin{aligned}
\boldsymbol{\phi}_e &= \boldsymbol{\phi}_i + \frac{s}{\|\boldsymbol{\phi}_f - \boldsymbol{\phi}_i\|}(\boldsymbol{\phi}_f - \boldsymbol{\phi}_i) \\
\dot{\boldsymbol{\phi}}_e &= \frac{\dot{s}}{\|\boldsymbol{\phi}_f - \boldsymbol{\phi}_i\|}(\boldsymbol{\phi}_f - \boldsymbol{\phi}_i) \\
\ddot{\boldsymbol{\phi}}_e &= \frac{\ddot{s}}{\|\boldsymbol{\phi}_f - \boldsymbol{\phi}_i\|}(\boldsymbol{\phi}_f - \boldsymbol{\phi}_i);
\end{aligned} \tag{4.51}$$

where the timing law for $s(t)$ has to be specified. The three unit vectors of the end-effector frame can be computed — with reference to Euler angles ZYZ — as in (2.18), the end-effector frame angular velocity as in (3.64), and the angular acceleration by differentiating (3.64) itself.

Angle and axis

An alternative way to generate a trajectory for orientation of clearer interpretation in the Cartesian space can be derived by resorting to the the angle and axis description presented in Sect. 2.5. Given two coordinate frames in the Cartesian space with the same origin and different orientation, it is always possible to determine a unit vector so that the second frame can be obtained from the first frame by a rotation of a proper angle about the axis of such unit vector.

Let $\boldsymbol{R}_i$ and $\boldsymbol{R}_f$ denote respectively the rotation matrices of the initial frame O_i–$x_iy_iz_i$ and the final frame O_f–$x_fy_fz_f$, both with respect to the base frame. The rotation matrix between the two frames can be computed by recalling that $\boldsymbol{R}_f = \boldsymbol{R}_i\boldsymbol{R}^i_f$; the expression in (2.5) leads to

$$\boldsymbol{R}^i_f = \boldsymbol{R}^T_i\boldsymbol{R}_f = \begin{bmatrix} r_{11} & r_{12} & r_{13} \\ r_{21} & r_{22} & r_{23} \\ r_{31} & r_{32} & r_{33} \end{bmatrix}.$$

If the matrix $\boldsymbol{R}^i(t)$ is defined to describe the transition from $\boldsymbol{R}_i$ to $\boldsymbol{R}_f$, it must be $\boldsymbol{R}^i(0) = \boldsymbol{I}$ and $\boldsymbol{R}^i(t_f) = \boldsymbol{R}^i_f$. Hence, the matrix $\boldsymbol{R}^i_f$ can be expressed as the rotation matrix about a fixed axis in space; the unit vector $\boldsymbol{r}^i$ of the axis and the angle of rotation ϑ_f can be computed by using (2.27):

$$\vartheta_f = \cos^{-1}\left(\frac{r_{11}+r_{22}+r_{33}-1}{2}\right) \tag{4.52}$$

$$\boldsymbol{r} = \frac{1}{2\sin\vartheta_f}\begin{bmatrix} r_{32}-r_{23} \\ r_{13}-r_{31} \\ r_{21}-r_{12} \end{bmatrix} \tag{4.53}$$

for $\sin\vartheta_f \neq 0$.

The matrix $\boldsymbol{R}^i(t)$ can be interpreted as a matrix $\boldsymbol{R}^i(\vartheta(t), \boldsymbol{r}^i)$ and computed via (2.25); it is then sufficient to assign a timing law to ϑ, of the type of those presented for the single joint with $\vartheta(0) = 0$ and $\vartheta(t_f) = \vartheta_f$, and compute the components of $\boldsymbol{r}^i$ from (4.52). Since $\boldsymbol{r}^i$ is constant, the resulting velocity and acceleration are respectively

$$\boldsymbol{\omega}^i = \dot{\vartheta}\,\boldsymbol{r}^i \tag{4.54}$$

$$\dot{\boldsymbol{\omega}}^i = \ddot{\vartheta}\,\boldsymbol{r}^i. \tag{4.55}$$

Finally, in order to characterize the end-effector orientation trajectory with respect to the base frame, the following transformations are needed:

$$\boldsymbol{R}_e(t) = \boldsymbol{R}_i\boldsymbol{R}^i(t)$$

$$\boldsymbol{\omega}_e(t) = \boldsymbol{R}_i\boldsymbol{\omega}^i(t)$$

$$\dot{\boldsymbol{\omega}}_e(t) = \boldsymbol{R}_i\dot{\boldsymbol{\omega}}^i(t).$$

Once a path and a trajectory have been specified in the operational space in terms of $\boldsymbol{p}_e(t)$ and $\boldsymbol{\phi}_e(t)$ or $\boldsymbol{R}_e(t)$, inverse kinematics techniques can be used to find the corresponding trajectories in the joint space $\boldsymbol{q}(t)$.

Bibliography

Trajectory planning for robot manipulators has been addressed since the first works in the field of robotics [178]. The formulation of the interpolation problem of the path points by means of different classes of functions has been suggested in [26].

The generation of motion trajectories through sequences of points in the joint space using splines is due to [131]. Alternative formulations for this problem are found in [56]. For a complete treatment of splines, including geometric properties and computational aspects, see [54]. In [155] a survey on the functions employed for trajectory planning of a single motion axis is given, which accounts for performance indices and effects of unmodelled flexible dynamics.

Cartesian space trajectory planning and the associated motion control problem have been originally treated in [179]. The systematic management of the motion by the via points using interpolating linear polynomials with parabolic blends has been proposed in [229]. A detailed presentation of the general aspects of the geometric primitives that can be utilized in robotics to define Cartesian space paths can be found in the computer graphics text [73].

Problems

4.1. Compute the joint trajectory from $q(0) = 1$ to $q(2) = 4$ with null initial and final velocities and accelerations.

4.2. Compute the timing law $q(t)$ for a joint trajectory with velocity profile of the type $\dot{q}(t) = k(1 - \cos(at))$ from $q(0) = 0$ to $q(2) = 3$.

4.3. Given the values for the joint variable: $q(0) = 0$, $q(2) = 2$, and $q(4) = 3$, compute the two fifth-order interpolating polynomials with continuous velocities and accelerations.

4.4. Show that the matrix $\boldsymbol{A}$ in (4.28) has a tridiagonal band structure.

4.5. Given the values for the joint variable: $q(0) = 0$, $q(2) = 2$, and $q(4) = 3$, compute the cubic interpolating spline with null initial and final velocities and accelerations.

4.6. Given the values for the joint variable: $q(0) = 0$, $q(2) = 2$, and $q(4) = 3$, find the interpolating polynomial with linear segments and parabolic blends with null initial and final velocities.

4.7. Find the timing law $\boldsymbol{p}(t)$ for a Cartesian space rectilinear path with trapezoidal velocity profile from $\boldsymbol{p}(0) = [\,0 \quad 0.5 \quad 0\,]^T$ to $\boldsymbol{p}(2) = [\,1 \quad -0.5 \quad 0\,]^T$.

4.8. Find the timing law $\boldsymbol{p}(t)$ for a Cartesian space circular path with trapezoidal velocity profile from $\boldsymbol{p}(0) = [\,0 \quad 0.5 \quad 1\,]^T$ to $\boldsymbol{p}(2) = [\,0 \quad -0.5 \quad 1\,]^T$; the circle is located in the plane $x = 0$ with centre at $\boldsymbol{c} = [\,0 \quad 0 \quad 1\,]^T$ and radius $\rho = 0.5$, and is executed clockwise for an observer aligned with x.

5

Actuators and Sensors

In this chapter, two basic robot components are treated: *actuators* and *sensors*. In the first part, the features of an *actuating system* are presented in terms of the power supply, power amplifier, servomotor and transmission. In view of their control versatility, two types of servomotors are used, namely, *electric servomotors* for actuating the joints of small and medium size manipulators, and *hydraulic servomotors* for actuating the joints of large size manipulators. The models describing the input/output relationship for such servomotors are derived, together with the control schemes of the *drives*. The electric servomotors are also employed to actuate the wheels of the mobile robots, which will be dealt with in Chap. 11. Successively, *proprioceptive sensors* are presented which allow measurement of the quantities characterizing the internal state of the manipulator, namely, *encoders* and *resolvers* for joint position measurement, *tachometers* for joint velocity measurement; further, *exteroceptive sensors* are presented including *force sensors* for end-effector force measurement, *distance sensors* for detection of objects in the workspace, and *vision sensors* for the measurement of the characteristic parameters of such objects, whenever the manipulator interacts with the environment.

5.1 Joint Actuating System

The motion imposed to a manipulator's joint is realized by an *actuating system* which in general consists of:

- a *power supply*,
- a *power amplifier*,
- a *servomotor*,
- a *transmission*.

The connection between the various components is illustrated in Fig. 5.1 where the exchanged powers are shown. To this end, recall that power can

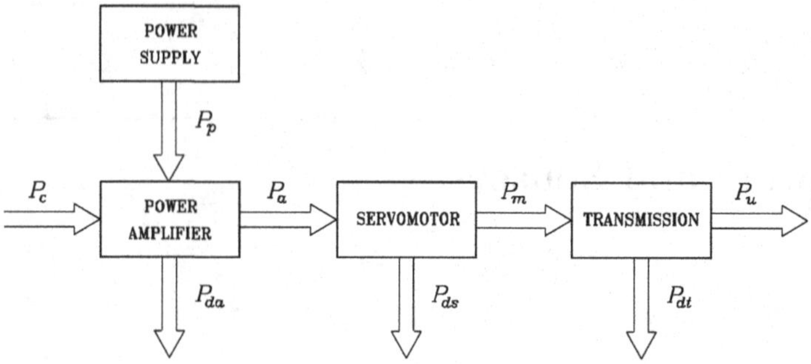

Fig. 5.1. Components of a joint actuating system

always be expressed as the product of a flow and a force quantity, whose physical context allows the specification of the nature of the power (mechanical, electric, hydraulic, or pneumatic).

In terms of a global input/output relationship, P_c denotes the (usually electric) power associated with the control law signal, whereas P_u represents the mechanical power required to the joint to actuate the motion. The intermediate connections characterize the supply power P_a of the motor (of electric, hydraulic, or pneumatic type), the power provided by the primary source P_p of the same physical nature as that of P_a, and the mechanical power P_m developed by the motor. Moreover, P_{da}, P_{ds} and P_{dt} denote the powers lost for dissipation in the conversions performed respectively by the amplifier, motor and transmission.

To choose the components of an actuating system, it is worth starting from the requirements imposed on the mechanical power P_u by the force and velocity that describe the joint motion.

5.1.1 Transmissions

The execution of joint motions of a manipulator demands *low speeds* with *high torques*. In general, such requirements do not allow an effective use of the mechanical features of servomotors, which typically provide high speeds with low torques in optimal operating conditions. It is then necessary to interpose a *transmission* (*gear*) to optimize the transfer of mechanical power from the motor (P_m) to the joint (P_u). During this transfer, the power P_{dt} is dissipated as a result of friction.

The choice of the transmission depends on the power requirements, the kind of desired motion, and the allocation of the motor with respect to the joint. In fact, the transmission allows the outputs of the motor to be transformed both quantitatively (velocity and torque) and qualitatively (a rotational motion about the motor axis into a translational motion of the joint). Also, it allows the static and dynamic performance of a manipulator to be optimized, by reducing the effective loads when the motor is located upstream of the joint; for instance, if some motors are mounted to the base of the robot, the total weight of the manipulator is decreased and the power-to-weight ratio is increased.

The following transmissions are typically used for industrial robots:

- *Spur gears* that modify the characteristics of the rotational motion of the motor by changing the axis of rotation and/or by translating the application point; spur gears are usually constructed with wide cross-section teeth and squat shafts.
- *Lead screws* that convert rotational motion of the motor into translational motion, as needed for actuation of prismatic joints; in order to reduce friction, ball screws are usually employed that are preloaded so as to increase stiffness and decrease backlash.
- *Timing belts* and *chains* which are equivalent from a kinematic viewpoint and are employed to locate the motor remotely from the axis of the actuated joint. The stress on timing belts may cause strain, and then these are used in applications requiring high speeds and low forces. On the other hand, chains are used in applications requiring low speeds, since their large mass may induce vibration at high speeds.

On the assumption of rigid transmissions with no backlash, the relationship between input forces (velocities) and output forces (velocities) is purely proportional.

The mechanical features of the motor used for an actuating system may sometimes allow a direct connection of the motor to the joint without the use of any transmission element (*direct drive*). The drawbacks due to transmission elasticity and backlash are thus eliminated, although more sophisticated control algorithms are required, since the absence of reduction gears does not allow the nonlinear coupling terms in the dynamic model to be neglected. The use of direct-drive actuating systems is not yet popular for industrial manipulators, in view of the cost and size of the motors as well as of control complexity.

5.1.2 Servomotors

Actuation of joint motions is entrusted to *motors* which allow the realization of a desired motion for the mechanical system. Concerning the kind of input power P_a, motors can be classified into three groups:

- *Pneumatic motors* which utilize the pneumatic energy provided by a compressor and transform it into mechanical energy by means of pistons or turbines.
- *Hydraulic motors* which transform the hydraulic energy stored in a reservoir into mechanical energy by means of suitable pumps.
- *Electric motors* whose primary supply is the electric energy available from the electric distribution system.

A portion of the input power P_a is converted to output as mechanical power P_m, and the rest (P_{ds}) is dissipated because of mechanical, electric, hydraulic, or pneumatic loss.

The motors employed in robotics are the evolution of the motors employed in industrial automation having powers ranging from about 10 W to about 10 kW. For the typical performance required, such motors should have the following requirements with respect to those employed in conventional applications:

- low inertia and high power-to-weight ratio,
- possibility of overload and delivery of impulse torques,
- capability to develop high accelerations,
- wide velocity range (from 1 to 1000 revolutes/min),
- high positioning accuracy (at least 1/1000 of a circle),
- low torque ripple so as to guarantee continuous rotation even at low speed.

These requirements are enhanced by the good trajectory tracking and positioning accuracy demanded for an actuating system for robots, and thus the motor must play the role of a *servomotor*. In this respect, pneumatic motors are difficult to control accurately, in view of the unavoidable fluid compressibility errors. Therefore, they are not widely employed, if not for the actuation of the typical opening and closing motions of the jaws in a gripper tool, then for the actuation of simple arms used in applications where continuous motion control is not of concern.

The most employed motors in robotics applications are *electric servomotors*. Among them, the most popular are permanent-magnet direct-current (DC) servomotors and brushless DC servomotors, in view of their good control flexibility.

The *permanent-magnet DC servomotor* consists of:

- A stator coil that generates magnetic flux; this generator is always a permanent magnet made by ferromagnetic ceramics or rare earths (high fields in contained space).
- An armature that includes the current-carrying winding that surrounds a rotary ferromagnetic core (rotor).
- A commutator that provides an electric connection by means of brushes between the rotating armature winding and the external feed winding, according to a commutation logic determined by the rotor motion.

The *brushless DC servomotor* consists of:

- A rotating coil (rotor) that generates magnetic flux; this generator is a permanent magnet made by ferromagnetic ceramics or rare earths.
- A stationary armature (stator) made by a polyphase winding.
- A static commutator that, on the basis of the signals provided by a position sensor located on the motor shaft, generates the feed sequence of the armature winding phases as a function of the rotor motion.

With reference to the above details of constructions, a comparison between the operating principle of a permanent-magnet DC and a brushless DC servomotor leads to the following considerations.

In the brushless DC motor, by means of the rotor position sensor, the winding orthogonal to the magnetic field of the coil is found; then, feeding the winding makes the rotor rotate. As a consequence of rotation, the electronic control module commutes the feeding on the winding of the various phases in such a way that the resulting field at the armature is always kept orthogonal to that of the coil. As regards electromagnetic interaction, such a motor operates in a way similar to that of a permanent-magnet DC motor where the brushes are at an angle of $\pi/2$ with respect to the direction of the excitation flux. In fact, feeding the armature coil makes the rotor rotate, and commutation of brushes from one plate of the commutator to the other allows the rotor to be maintained in rotation. The role played by the brushes and commutator in a permanent-magnet DC motor is analogous to that played by the position sensor and electronic control module in a brushless DC motor.

The main reason for using a brushless DC motor is to eliminate the problems due to mechanical commutation of the brushes in a permanent-magnet DC motor. In fact, the presence of the commutator limits the performance of a permanent-magnet DC motor, since this provokes electric loss due to voltage drops at the contact between the brushes and plates, and mechanical loss due to friction and arcing during commutation from one plate to the next one caused by the inductance of the winding. The elimination of the causes provoking such inconveniences, i.e., the brushes and plates, allows an improvement of motor performance in terms of higher speeds and less material wear.

The inversion between the functions of stator and rotor leads to further advantages. The presence of a winding on the stator instead of the rotor facilitates heat disposal. The absence of a rotor winding, together with the possibility of using rare-earth permanent magnets, allows construction of more compact rotors which are, in turn, characterized by a low moment of inertia. Therefore, the size of a brushless DC motor is smaller than that of a permanent-magnet DC motor of the same power; an improvement of dynamic performance can also be obtained by using a brushless DC motor. For the choice of the most suitable servomotor for a specific application, the cost factor plays a relevant role.

Not uncommon are also stepper motors. These actuators are controlled by suitable excitation sequences and their operating principle does not require measurement of motor shaft angular position. The dynamic behaviour of stepper motors is greatly influenced by payload, though. Also, they induce vibration of the mechanical structure of the manipulator. Such inconveniences confine the use of stepper motors to the field of micromanipulators, for which low-cost implementation prevails over the need for high dynamic performance.

A certain number of applications features the employment of *hydraulic servomotors*, which are based on the simple operating principle of volume variation under the action of compressed fluid. From a construction viewpoint, they are characterized by one or more chambers made by pistons (cylinders reciprocating in tubular housings). Linear servomotors have a limited range and are constituted by a single piston. Rotary servomotors have unlimited range and are constituted by several pistons (usually an odd number) with an axial or radial disposition with respect to the motor axis of rotation. These servomotors offer a static and dynamic performance comparable with that offered by electric servomotors.

The differences between electric and hydraulic servomotors can be fundamentally observed from a plant viewpoint. In this respect, *electric servomotors* present the following *advantages*:

- widespread availability of power supply,
- low cost and wide range of products,
- high power conversion efficiency,
- easy maintenance,
- no pollution of working environment.

Instead, they present the following *limitations*:

- burnout problems at static situations caused by the effect of gravity on the manipulator; emergency brakes are then required,
- need for special protection when operating in flammable environments.

Hydraulic servomotors present the following *drawbacks*:

- need for a hydraulic power station,
- high cost, narrow range of products, and difficulty of miniaturization,
- low power conversion efficiency,
- need for operational maintenance,
- pollution of working environment due to oil leakage.

In their *favour* it is worth pointing out that they:

- do not suffer from burnout in static situations,
- are self-lubricated and the circulating fluid facilitates heat disposal,
- are inherently safe in harmful environments,
- have excellent power-to-weight ratios.

From an operational viewpoint, it can be observed that:

- Both types of servomotors have a good dynamic behaviour, although the electric servomotor has greater control flexibility. The dynamic behaviour of a hydraulic servomotor depends on the temperature of the compressed fluid.
- The electric servomotor is typically characterized by high speeds and low torques, and as such it requires the use of gear transmissions (causing elasticity and backlash). On the other hand, the hydraulic servomotor is capable of generating high torques at low speeds.

In view of the above remarks, hydraulic servomotors are specifically employed for manipulators that have to carry heavy payloads; in this case, not only is the hydraulic servomotor the most suitable actuator, but also the cost of the plant accounts for a reduced percentage on the total cost of the manipulation system.

5.1.3 Power Amplifiers

The *power amplifier* has the task of modulating, under the action of a control signal, the power flow which is provided by the primary supply and has to be delivered to the actuators for the execution of the desired motion. In other words, the amplifier takes a fraction of the power available at the source which is proportional to the control signal; then it transmits this power to the motor in terms of suitable force and flow quantities.

The inputs to the amplifier are the power taken from the primary source P_p and the power associated with the control signal P_c. The total power is partly delivered to the actuator (P_a) and partly lost in dissipation (P_{da}).

Given the typical use of electric and hydraulic servomotors, the operational principles of the respective amplifiers are discussed.

To control an *electric servomotor*, it is necessary to provide it with a voltage or current of suitable form depending on the kind of servomotor employed. Voltage (or current) is direct for permanent-magnet DC servomotors, while it is alternating for brushless DC servomotors. The value of voltage for permanent-magnet DC servomotors or the values of voltage and frequency for brushless DC servomotors are determined by the control signal of the amplifier, so as to make the motor execute the desired motion.

For the power ranges typically required by joint motions (of the order of a few kilowatts), transistor amplifiers are employed which are suitably switched by using pulse-width modulation (PWM) techniques. They allow the achievement of a power conversion efficiency $P_a/(P_p + P_c)$ greater than 0.9 and a power gain P_a/P_c of the order of 10^6. The amplifiers employed to control permanent-magnet DC servomotors are DC-to-DC converters (*choppers*), whereas those employed to control brushless DC servomotors are DC-to-AC converters (*inverters*).

Control of a *hydraulic servomotor* is performed by varying the flow rate of the compressed fluid delivered to the motor. The task of modulating the flow rate is typically entrusted to an interface (electro-hydraulic servovalve). This allows a relationship to be established between the electric control signal and the position of a distributor which is able to vary the flow rate of the fluid transferred from the primary source to the motor. The electric control signal is usually current-amplified and feeds a solenoid which moves (directly or indirectly) the distributor, whose position is measured by a suitable transducer. In this way, a position servo on the valve stem is obtained which reduces occurrence of any stability problem that may arise on motor control. The magnitude of the control signal determines the flow rate of the compressed fluid through the distributor, according to a characteristic which is possibly made linear by means of a keen mechanical design.

5.1.4 Power Supply

The task of the *power supply* is to supply the primary power to the amplifier which is needed for operation of the actuating system.

In the case of *electric servomotors*, the power supply consists of a transformer and a typically uncontrolled bridge rectifier. These allow the alternating voltage available from the distribution to be converted into a direct voltage of suitable magnitude which is required to feed the power amplifier.

In the case of *hydraulic servomotors*, the power supply is obviously more complex. In fact, a gear or piston pump is employed to compress the fluid which is driven by a primary motor operating at constant speed, typically a three-phase nonsynchronous motor. To reduce the unavoidable pressure oscillations provoked by a flow rate demand depending on operational conditions of the motor, a reservoir is interfaced to store hydraulic energy. Such a reservoir, in turn, plays the same role as the filter capacitor used at the output of a bridge rectifier. The hydraulic power station is completed by the use of various components (filters, pressure valves, and check valves) that ensure proper operation of the system. Finally, it can be inferred how the presence of complex hydraulic circuits operating at high pressures (of the order of 100 atm) causes an appreciable pollution of the working environment.

5.2 Drives

This section presents the operation of the *electric drives* and the *hydraulic drives* for the actuation of a manipulator's joints. Starting from the mathematical models describing the dynamic behaviour, the block schemes are derived which allow an emphasis on the control features and the effects of the use of a mechanical transmission.

5.2.1 Electric Drives

From a modelling viewpoint, a permanent-magnet DC motor and a brushless DC motor provided with the commutation module and position sensor can be described by the same differential equations. In the domain of the complex variable s, the electric balance of the armature is described by the equations

$$V_a = (R_a + sL_a)I_a + V_g \tag{5.1}$$

$$V_g = k_v \Omega_m \tag{5.2}$$

where V_a and I_a respectively denote armature voltage and current, R_a and L_a are respectively the armature resistance and inductance, and V_g denotes

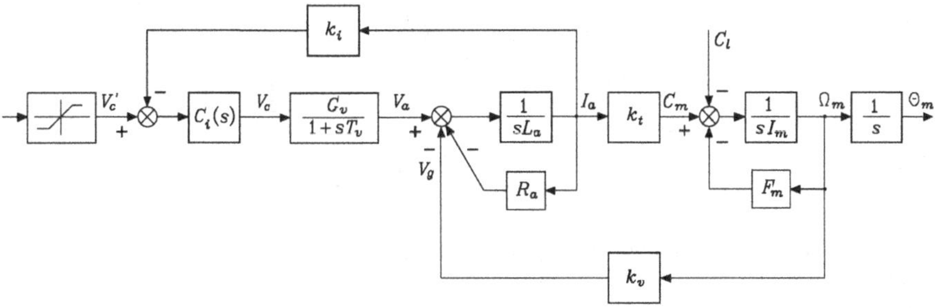

Fig. 5.2. Block scheme of an electric drive

the back electromotive force which is proportional to the angular velocity Ω_m through the voltage constant k_v that depends on the construction details of the motor as well as on the magnetic flux of the coil.

The mechanical balance is described by the equations

$$C_m = (sI_m + F_m)\Omega_m + C_l \tag{5.3}$$

$$C_m = k_t I_a \tag{5.4}$$

where C_m and C_l respectively denote the driving torque and load reaction torque, I_m and F_m are respectively the moment of inertia and viscous friction coefficient at the motor shaft, and the torque constant k_t is numerically equal to k_v in the SI unit system for a compensated motor.

Concerning the power amplifier, the input/output relationship between the control voltage V_c and the armature voltage V_a is given by the transfer function

$$\frac{V_a}{V_c} = \frac{G_v}{1 + sT_v} \tag{5.5}$$

where G_v denotes the voltage gain and T_v is a time constant that can be neglected with respect to the other time constants of the system. In fact, by using a modulation frequency in the range of 10 to 100 kHz, the time constant of the amplifier is in the range of 10^{-5} to 10^{-4}) s.

The block scheme of the servomotor with power amplifier (*electric drive*) is illustrated in Fig. 5.2. In such a scheme, besides the blocks corresponding to the above relations, there is an armature *current feedback* loop where current is thought of as measured by a transducer k_i between the power amplifier and the armature winding of the motor. Further, the scheme features a current regulator $C_i(s)$ as well as an element with a nonlinear saturation characteristic. The aim of such feedback is twofold. On one hand, the voltage V_c' plays the role of a current reference and thus, by means of a suitable choice of the regulator $C_i(s)$, the lag between the current I_a and the voltage V_c' can be reduced with respect to the lag between I_a and V_c. On the other hand, the introduction of a saturation nonlinearity allows the limitation of the magni-

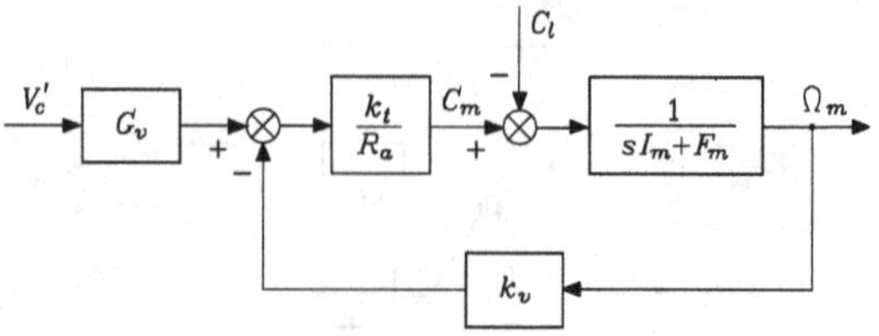

Fig. 5.3. Block scheme of an electric drive as a velocity-controlled generator

tude of V_c', and then it works like a current limit which ensures protection of the power amplifier whenever abnormal operating conditions occur.

The choice of the regulator $C_i(s)$ of the current loop allows a velocity-controlled or torque-controlled behaviour to be obtained from the electric drive, depending on the values attained by the loop gain. In fact, in the case of $k_i = 0$, recalling that the mechanical viscous friction coefficient is negligible with respect to the electrical friction coefficient

$$F_m \ll \frac{k_v k_t}{R_a}, \tag{5.6}$$

assuming a unit gain constant for $C_i(s)$[1] and $C_l = 0$ yields

$$\omega_m \approx \frac{G_v}{k_v} v_c' \tag{5.7}$$

and thus the drive system behaves like a *velocity-controlled generator*.

Instead, when $k_i \neq 0$, choosing a large loop gain for the current loop ($Kk_i \gg R_a$) leads at steady state to

$$c_m \approx \frac{k_t}{k_i}\left(v_c' - \frac{k_v}{G_v}\omega_m\right); \tag{5.8}$$

the drive behaves like a *torque-controlled generator* since, in view of the large value of G_v, the driving torque is practically independent of the angular velocity.

As regards the dynamic behaviour, it is worth considering a *reduced-order model* which can be obtained by neglecting the electric time constant L_a/R_a with respect to the mechanical time constant I_m/F_m, assuming $T_v \approx 0$ and a purely proportional controller. These assumptions, together with $k_i = 0$, lead to the block scheme in Fig. 5.3 for the velocity-controlled generator. On the other hand, if it is assumed $Kk_i \gg R_a$ and $k_v\Omega/Kk_i \approx 0$, the resulting block scheme of the torque-controlled generator is that in Fig. 5.4. From the

[1] It is assumed $C_i(0) = 1$; in the case of presence of an integral action in $C_i(s)$, it should be $\lim_{s\to 0} sC(s) = 1$.

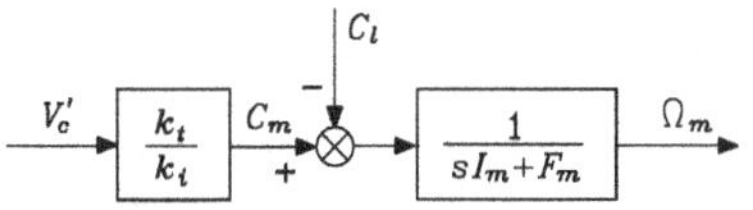

Fig. 5.4. Block scheme of an electric drive as a torque-controlled generator

above schemes, the following input/output relations between control voltage, reaction torque, and angular velocity can be derived:

$$\Omega_m = \frac{\dfrac{1}{k_v}}{1 + s\dfrac{R_a I_m}{k_v k_t}} G_v V_c' - \frac{\dfrac{R_a}{k_v k_t}}{1 + s\dfrac{R_a I_m}{k_v k_t}} C_l \tag{5.9}$$

for the velocity-controlled generator, and

$$\Omega_m = \frac{\dfrac{k_t}{k_i F_m}}{1 + s\dfrac{I_m}{F_m}} V_c' - \frac{\dfrac{1}{F_m}}{1 + s\dfrac{I_m}{F_m}} C_l \tag{5.10}$$

for the torque-controlled generator. These transfer functions show how, without current feedback, the system has a better rejection of disturbance torques in terms of both equivalent gain ($R_a/k_v k_t \ll 1/F_m$) and time response ($R_a I_m/k_v k_t \ll I_m/F_m$).

The relationship between the control input and the actuator position output can be expressed in a unified manner by the transfer function

$$M(s) = \frac{k_m}{s(1 + sT_m)} \tag{5.11}$$

where

$$k_m = \frac{1}{k_v} \qquad T_m = \frac{R_a I_m}{k_v k_t} \tag{5.12}$$

for the velocity-controlled generator, while for the torque-controlled generator it is

$$k_m = \frac{k_t}{k_i F_m} \qquad T_m = \frac{I_m}{F_m}. \tag{5.13}$$

Notice how the power amplifier, in the velocity control case, contributes to the input/output relation with the constant G_v, while in the case of current control the amplifier, being inside a local feedback loop, does not appear as a stand alone but rather in the expression of k_m with a factor $1/k_i$.

These considerations lead to the following conclusions. In all such applications where the drive system has to provide high rejection of disturbance torques (as in the case of independent joint control, see Sect. 8.3) it is not advisable to have a current feedback in the loop, at least when all quantities

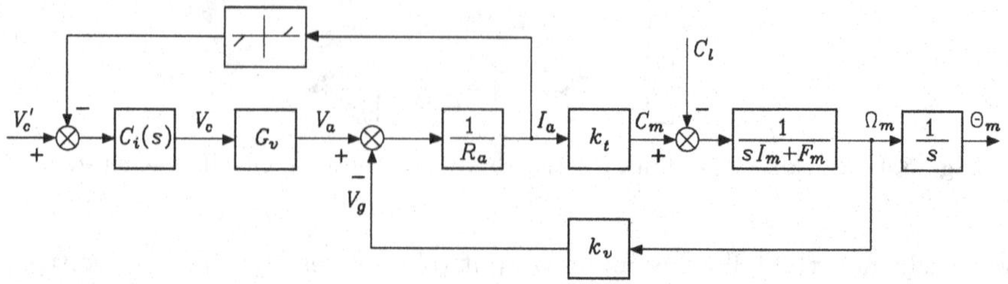

Fig. 5.5. Block scheme of an electric drive with nonlinear current feedback

are within their nominal values. In this case, the problem of setting a protection can be solved by introducing a current limit that is not performed by a saturation on the control signal but it exploits a current feedback with a dead-zone nonlinearity on the feedback path, as shown in Fig. 5.5. Therefore, an actual current limit is obtained whose precision is as high as the slope of the dead zone; it is understood that stability of the current loop is to be addressed when operating in this way.

As will be shown in Sect. 8.5, centralized control schemes, instead, demand the drive system to behave as a torque-controlled generator. It is then clear that a current feedback with a suitable regulator $C_i(s)$ should be used so as to confer a good static and dynamic behaviour to the current loop. In this case, servoing of the driving torque is achieved indirectly, since it is based on a current measurement which is related to the driving torque by means of gain $1/k_t$.

5.2.2 Hydraulic Drives

No matter how a hydraulic servomotor is constructed, the derivation of its input/output mathematical model refers to the basic equations describing the relationship between flow rate and pressure, the relationship between the fluid and the parts in motion, and the mechanical balance of the parts in motion. Let Q represent the volume flow rate supplied by the distributor; the flow rate balance is given by the equation

$$Q = Q_m + Q_l + Q_c \tag{5.14}$$

where Q_m is the flow rate transferred to the motor, Q_l is the flow rate due to leakage, and Q_c is the flow rate related to fluid compressibility. The terms Q_l and Q_c are taken into account in view of the high operating pressures (of the order of 100 atm).

Let P denote the differential pressure of the servomotor due to the load; then it can be assumed that

$$Q_l = k_l P. \tag{5.15}$$

Regarding the loss for compressibility, if V denotes the instantaneous volume of the fluid, one has

$$Q_c = \gamma V s P \tag{5.16}$$

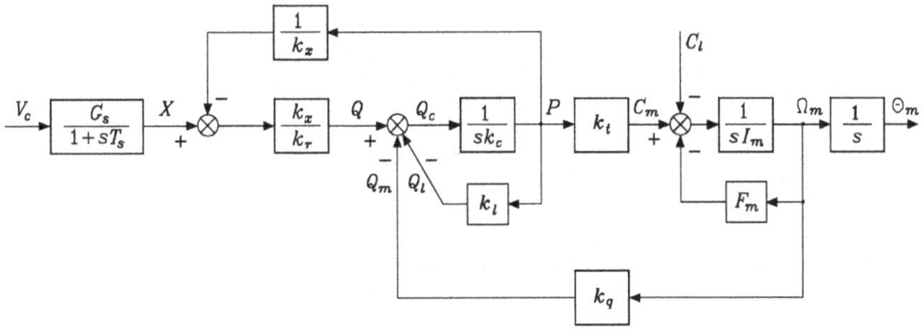

Fig. 5.6. Block scheme of a hydraulic drive

where γ is the uniform compressibility coefficient of the fluid. Notice that the proportional factor $k_c = \gamma V$ between the time derivative of the pressure and the flow rate due to compressibility depends on the volume of the fluid; therefore, in the case of rotary servomotors, k_c is a constant, whereas in the case of a linear servomotor, the volume of fluid varies and thus the characteristic of the response depends on the operating point.

The volume flow rate transferred to the motor is proportional to the volume variation in the chambers per time unit; with reference from now on to a rotary servomotor, such variation is proportional to the angular velocity, and then

$$Q_m = k_q \Omega_m. \tag{5.17}$$

The mechanical balance of the parts in motion is described by

$$C_m = (sI_m + F_m)\Omega_m + C_l \tag{5.18}$$

with obvious meaning of the symbols. Finally, the driving torque is proportional to the differential pressure of the servomotor due to the load, i.e.,

$$C_m = k_t P. \tag{5.19}$$

Concerning the servovalve, the transfer function between the stem position X and the control voltage V_c is expressed by

$$\frac{X}{V_c} = \frac{G_s}{1 + sT_s} \tag{5.20}$$

thanks to the linearizing effect achieved by position feedback; G_s is the equivalent gain of the servovalve, whereas its time constant T_s is of the order of ms and thus it can be neglected with respect to the other time constants of the system.

Finally, regarding the distributor, the relationship between the differential pressure, the flow rate, and the stem displacement is highly nonlinear; linearization about an operating point leads to the equation

$$P = k_x X - k_r Q. \tag{5.21}$$

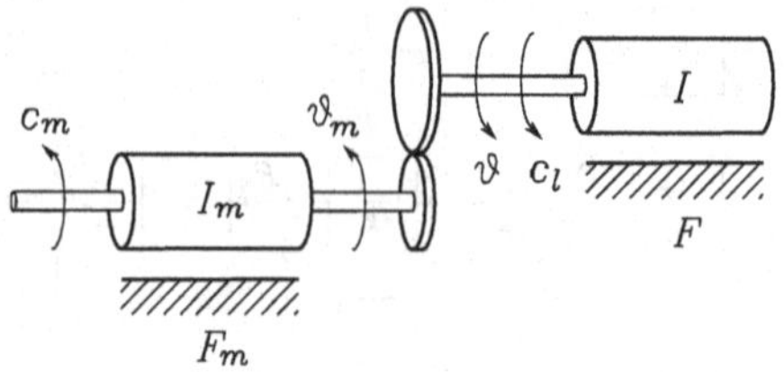

Fig. 5.7. Schematic representation of a mechanical gear

By virtue of (5.14)–(5.21), the servovalve/distributor/motor complex (*hydraulic drive*) is represented by the block scheme of Fig. 5.6. A comparison between the schemes in Figs. 5.2 and 5.6 clearly shows the formal analogy in the dynamic behaviour of an electric and a hydraulic servomotor. Nevertheless, such analogy should not induce one to believe that it is possible to make a hydraulic drive play the role of a velocity- or torque-controlled generator, as for an electric drive. In this case, the pressure feedback loop (formally analogous to the current feedback loop) is indeed a structural characteristic of the system and, as such, it cannot be modified but with the introduction of suitable transducers and the realization of the relative control circuitry.

5.2.3 Transmission Effects

In order to describe quantitatively the effects introduced by the use of a transmission (*mechanical gear*) between the servomotor and the actuated joint, it is worth referring to the mechanical coupling realized by a pair of spur gears of radius r_m and r, which is schematically represented in Fig. 5.7; the kinematic pair is assumed to be ideal (without backlash) and connects the rotation axis of the servomotor with the axis of the corresponding joint.

With reference to an electric servomotor, it is assumed that the rotor of the servomotor is characterized by an inertia moment I_m about its rotation axis and a viscous friction coefficient F_m; likewise, I and F denote the inertia moment and the viscous friction coefficient of the load. The inertia moments and the friction coefficients of the gears are assumed to have been included in the corresponding parameters of the motor (for the gear of radius r_m) and of the load (for the gear of radius r). Let c_m denote the driving torque of the motor and c_l the reaction torque applied to the load axis. Also let ω_m and ϑ_m denote the angular velocity and position of the motor axis, while ω and ϑ denote the corresponding quantities at the load side. Finally, f indicates the force exchanged at the contact between the teeth of the two gears.[2]

[2] In the case considered, it has been assumed that both the motor and the load are characterized by revolute motions; if the load should exhibit a translation motion, the following arguments can be easily extended, with analogous results, by replacing the angular displacements with linear displacements and the inertia moments with masses at the load side.

The gear reduction ratio is defined as

$$k_r = \frac{r}{r_m} = \frac{\vartheta_m}{\vartheta} = \frac{\omega_m}{\omega} \tag{5.22}$$

since, in the absence of slipping in the kinematic coupling, it is $r_m \vartheta_m = r\vartheta$.

The gear reduction ratio, in the case when it is representative of the coupling between a servomotor and the joint of a robot manipulator, attains values much larger than unity ($r_m \ll r$) — typically from a few tens to a few hundreds.

The force f exchanged between the two gears generates a reaction torque $f \cdot r_m$ for the motion at the motor axis and a driving torque $f \cdot r$ for the rotation motion of the load.

The mechanical balances at the motor side and the load side are respectively:

$$c_m = I_m \dot{\omega}_m + F_m \omega_m + f r_m \tag{5.23}$$

$$f r = I\dot{\omega} + F\omega + c_l. \tag{5.24}$$

To describe the motion with reference to the motor angular velocity, in view of (5.22), combining the two equations gives at the motor side

$$c_m = I_{eq} \dot{\omega}_m + F_{eq} \omega_m + \frac{c_l}{k_r} \tag{5.25}$$

where

$$I_{eq} = \left(I_m + \frac{I}{k_r^2} \right) \qquad F_{eq} = \left(F_m + \frac{F}{k_r^2} \right). \tag{5.26}$$

The expressions (5.25), (5.26) show how, in the case of a gear with large reduction ratio, the inertia moment and the viscous friction coefficient of the load are reflected at the motor axis with a reduction of a factor $1/k_r^2$; the reaction torque, instead, is reduced by a factor $1/k_r$. If this torque depends on ϑ in a nonlinear fashion, then the presence of a large reduction ratio tends to linearize the dynamic equation.

Example 5.1

In Fig. 5.8 a rigid pendulum is represented, which is actuated by the torque $f \cdot r$ to the load axis after the gear. In this case, the dynamic equations of the system are

$$c_m = I_m \dot{\omega}_m + F_m \omega_m + f r_m \tag{5.27}$$

$$f r = I\dot{\omega} + F\omega + mg\ell \sin\vartheta \tag{5.28}$$

where I is the inertia moment of the pendulum at the load axis, F is the viscous friction coefficient, m is the pendulum mass, ℓ its length and g the gravity acceleration. Reporting (5.28) to the motor axis gives

$$c_m = I_{eq} \dot{\omega}_m + F_{eq} \omega_m + \left(\frac{mg\ell}{k_r} \right) \sin\left(\frac{\vartheta_m}{k_r} \right) \tag{5.29}$$

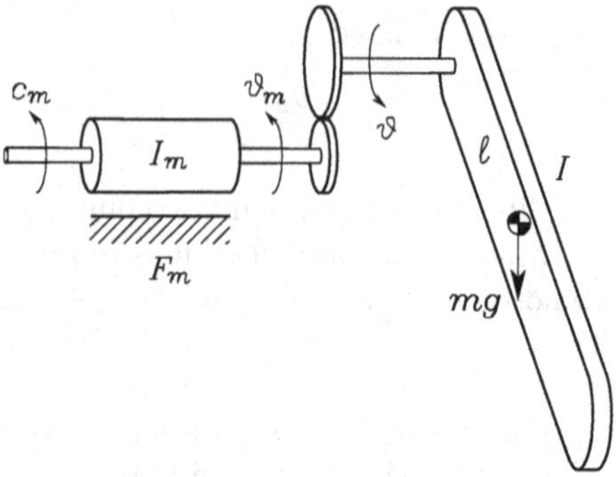

Fig. 5.8. Pendulum actuated via mechanical gear

from which it is clear how the contribution of the nonlinear term is reduced by the factor k_r.

The example of the pendulum has been considered to represent an n-link manipulator with revolute joints, for which each link, considered as isolated from the others, can be considered as a simple rigid pendulum. The connection with other links introduces, in reality, other nonlinear effects which complicate the input/output model; in this regard, it is sufficient to notice that, in the case of a double pendulum, the inertia moment at the motor side of the first link depends also on the angular position of the second link.

In Chap. 7 the effect introduced by the presence of transmissions in a generic n-link manipulator structure will be studied in detail. Nevertheless, it can already be understood how the nonlinear couplings between the motors of the various links will be reduced by the presence of transmissions with large reduction ratios.

5.2.4 Position Control

After having examined the modalities to control the angular velocity of an electric or hydraulic drive, the motion control problem for a link of a generic manipulator is to be solved. A structure is sought which must be capable of determining, in an automatic way, the time evolution of the quantity chosen to control the drive, so that the actuated joint executes the required motion allowing the end-effector to execute a given task.

Once a trajectory has been specified for the end-effector pose, the solution of the inverse kinematics problem allows the computation of the desired trajectories for the various joints, which thus can be considered as available.

Several control techniques can be adopted to control the manipulator motion; the choice of a particular solution depends on the required dynamic performance, the kind of motion to execute, the kinematic structure, and the

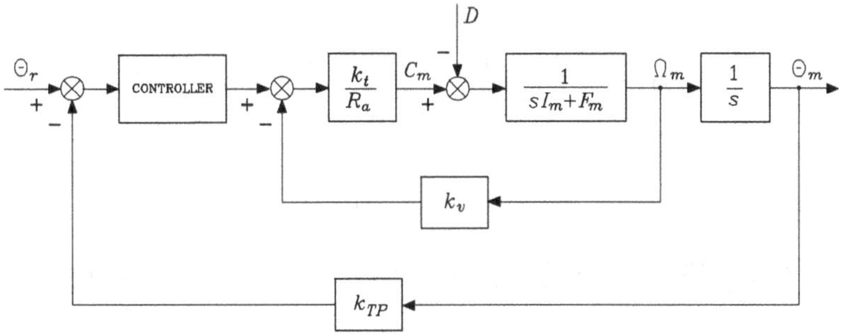

Fig. 5.9. General block scheme of electric drive control

choice to utilize either servomotors with transmissions or torque motors with joint direct drive.

The simplest solution is to consider, at first approximation, the motion of a joint independent of the motion of the other joints, i.e., the interaction can be regarded as a disturbance. Assume the reference trajectory $\vartheta_r(t)$ is available. According to classical automatic control theory, to ensure that the angular motor position ϑ_m, properly measured by means of a transducer with constant k_{TP}, follows ϑ_r, it is worth resorting to a feedback control system providing 'robustness' with respect to both model uncertainty on the motor and the load, and the presence of a disturbance. A more detailed treatment is deferred to Chap. 8, where the most congenial solutions to solve the above problems will be presented.

In the following, the problem of joint *position control* is tackled by assuming an electric DC servomotor; the choice is motivated by the diffusion of this technology, due to the high flexibility of these actuators providing optimal responses in the large majority of motion control applications.

The choice of a feedback control system to realize a position servo at the motor axis requires the adoption of a *controller*; this device generates a signal which, applied to the power amplifier, automatically generates the driving torque producing an axis motion very close to the desired motion ϑ_r. Its structure should be so that the error between the reference input and the measured output is minimized, even in the case of inaccurate knowledge of the dynamics of the motor, the load, and a disturbance. The *rejection* action of the disturbance is the more efficient, the smaller the magnitude of the disturbance.

On the other hand, according to (5.9), the disturbance is minimized, provided the drive is velocity-controlled. In this case, in view of (5.6), the reaction torque influences the motor axis velocity with a coefficient equal to $R_a/k_v k_t$ which is much smaller than $1/F_m$, which represents instead the weight on the reaction torque in the case when the drive is torque-controlled. Therefore, with reference to Fig. 5.3, the general scheme of drive control with *position feedback* is illustrated in Fig. 5.9, where the disturbance d represents the load

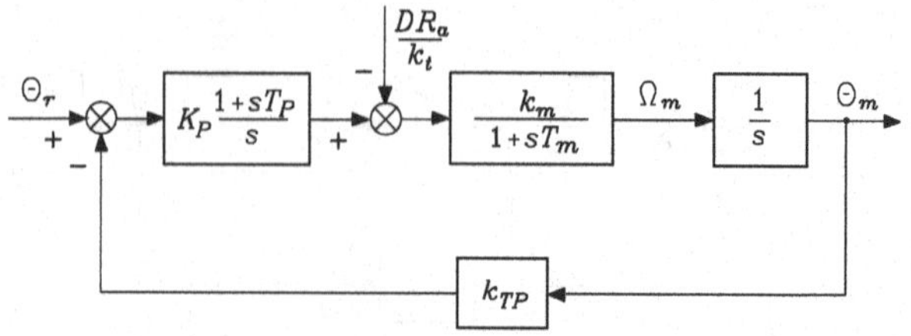

Fig. 5.10. Block scheme of drive control with position feedback

torque and the value of the power amplifier gain has been included in the control action.

Besides reducing the effects of the disturbance on the output, the structure of the controller must ensure an optimal trade-off between the stability of the feedback control system and the capability of the output to dynamically track the reference with a reduced error.

The reduction of the disturbance effects on the output can be achieved by conferring a large value of the gain before the point of intervention of the disturbance, without affecting stability. If, at steady state (ϑ_r = cost, c_l = cost), it is desired to cancel the disturbance effect on the output, the controller must act an *integral action* on the error given by the difference between ϑ_r and $k_{TP}\vartheta_m$.

The above requirements suggest the use of a simple controller with an integral and a proportional action on the error; the *proportional action* is added to realize a stabilizing action, which, however, cannot confer to the closed-loop system a damped transient response with a sufficiently short sampling time. This behaviour is due to the presence of a double pole at the origin of the transfer function of the forward path.

The resulting control scheme is illustrated in Fig. 5.10, where k_m and T_m are respectively the voltage-to-velocity gain constant and the characteristic time constant of the motor in (5.12). The parameters of the controller K_P and T_P should be keenly chosen so as to ensure stability of the feedback control system and obtain a good dynamic behaviour.

To improve the transient response, the industrial drives employed for position servoing may also include a local feedback loop based on the angular velocity measurement (tachometer feedback). The general scheme with *position and velocity feedback* is illustrated in Fig. 5.11; besides the position transducer, a velocity transducer is used with constant k_{TV}, as well as a simple proportional controller with gain K_P. With the adoption of the tachometer feedback, the proportional-integral controller with parameters K_V and T_V is retained in the internal velocity loop so as to cancel the effects of the disturbance on the position ϑ_m at steady state. The presence of two feedback loops, in lieu of one, around the intervention point of the disturbance is expected to lead to a further reduction of the disturbance effects on the output also during the transients.

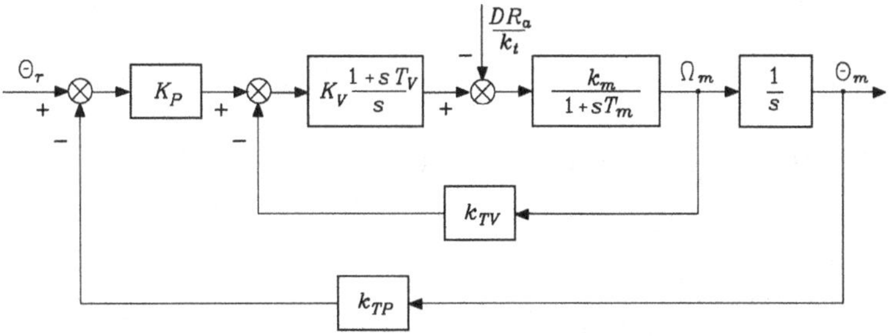

Fig. 5.11. Block scheme of drive control with position and velocity feedback

The adoption of tachometer feedback may also improve the transient response of the whole control system with respect to the previous case. With a keen choice of the controller parameters, indeed, it is possible to achieve a transfer function between ϑ_m and ϑ_r with a larger bandwidth and reduced resonance phenomena. The result is a faster transient response with reduced oscillations, thus improving the capability of $\vartheta_m(t)$ to track more demanding reference trajectories $\vartheta_r(t)$.

The above analysis will be further detailed in Sect. 8.3.

The position servo may also utilize a current-controller motor; the schemes in Figs. 5.9–5.11 can be adopted, provided that the constants in (5.13) are used in the transfer function (5.11) and the disturbance D is weighed with the quantity k_i/k_t in lieu of R_a/k_t. In that case, the voltage gain G_v of the power amplifier will not contribute to the control action.

As a final consideration, the general control structure presented above may be extended to the case when the motor is coupled to a load via a gear reduction. In such a case, it is sufficient to account for (5.25) and (5.26), i.e., replace I_m and F_m with the quantities I_{eq} and F_{eq}, and scale the disturbance by the factor $1/k_r$.

5.3 Proprioceptive Sensors

The adoption of *sensors* is of crucial importance to achieve high-performance robotic systems. It is worth classifying sensors into *proprioceptive* sensors that measure the internal state of the manipulator, and *exteroceptive* sensors that provide the robot with knowledge of the surrounding environment.

In order to guarantee that a coordinated motion of the mechanical structure is obtained in correspondence of the task planning, suitable parameter identification and control algorithms are used which require the on-line measurement, by means of proprioceptive sensors, of the quantities characterizing the internal state of the manipulator, i.e.:

- joint positions,
- joint velocities,
- joint torques.

On the other hand, typical exteroceptive sensors include:

- force sensors,
- tactile sensors,
- proximity sensors,
- range sensors,
- vision sensors.

The goal of such sensors is to extract the features characterizing the interaction of the robot with the objects in the environment, so as to enhance the degree of autonomy of the system. To this class also belong those sensors which are specific for the robotic application, such as sound, humidity, smoke, pressure, and temperature sensors. Fusion of the available sensory data can be used for (high-level) task planning, which in turn characterizes a *robot* as the *intelligent connection of perception to action*.

In the following, the main features of the proprioceptive sensors are illustrated, while those of the exteroceptive sensors will be presented in the next section.

5.3.1 Position Transducers

The aim of *position transducers* is to provide an electric signal proportional to the linear or angular displacement of a mechanical apparatus with respect to a given reference position. They are mostly utilized for control of machine tools, and thus their range is wide. Potentiometers, linear variable-differential transformers (LVDT), and inductosyns may be used to measure linear displacements. Potentiometers, encoders, resolvers and synchros may be used to measure angular displacements.

Angular displacement transducers are typically employed in robotics applications since, also for prismatic joints, the servomotor is of a rotary type. In view of their precision, robustness and reliability, the most common transducers are the *encoders* and *resolvers*, whose operating principles are detailed in what follows.

On the other hand, linear displacement transducers (LVDT's and inductosyns) are mainly employed in measuring robots.

Encoder

There are two types of encoder: absolute and incremental. The *absolute encoder* consists of an optical-glass disk on which concentric circles (tracks) are disposed; each track has an alternating sequence of transparent sectors and matte sectors obtained by deposit of a metallic film. A light beam is emitted in

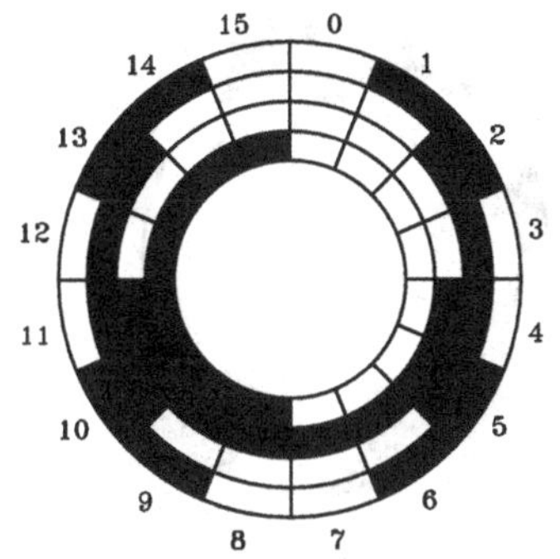

Fig. 5.12. Schematic representation of an absolute encoder

correspondence of each track which is intercepted by a photodiode or a phototransistor located on the opposite side of the disk. By a suitable arrangement of the transparent and matte sectors, it is possible to convert a finite number of angular positions into corresponding digital data. The number of tracks determines the length of the word, and thus the resolution of the encoder.

To avoid problems of incorrect measurement in correspondence of a simultaneous multiple transition between matte and transparent sectors, it is worth utilizing a Gray-code encoder whose schematic representation is given in Fig. 5.12 with reference to the implementation of 4 tracks that allow the discrimination of 16 angular positions. It can be noticed that measurement ambiguity is eliminated, since only one change of contrast occurs at each transition (Table 5.1). For the typical resolution required for joint control, absolute encoders with a minimum number of 12 tracks (bits) are employed (resolution of 1/4096 per circle). Such encoders can provide unambiguous measurements only in a circle. If a gear reduction is present, a circle at the joint side corresponds to several circles at the motor side, and thus a simple electronics is needed to count and store the number of actual circles.

Table 5.1. Coding table with Gray-code

#	Code	#	Code
0	0000	8	1100
1	0001	9	1101
2	0011	10	1111
3	0010	11	1110
4	0110	12	1010
5	0111	13	1011
6	0101	14	1001
7	0100	15	1000

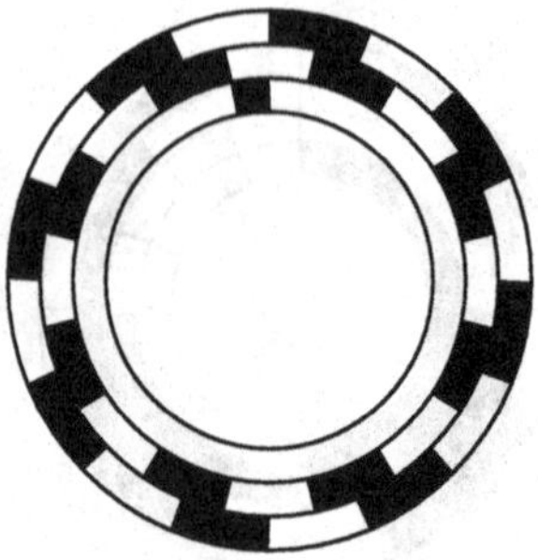

Fig. 5.13. Schematic representation of an incremental encoder

Incremental encoders have a wider use than absolute encoders, since they are simpler from a construction viewpoint and thus cheaper. Like the absolute one, the incremental encoder consists of an optical disk on which two tracks are disposed, whose transparent and matte sectors (in equal number on the two tracks) are mutually in quadrature. The presence of two tracks also allows, besides the number of transitions associated with any angular rotation, the detection of the sign of rotation. Often a third track is present with one single matte sector which allows the definition of an absolute mechanical zero as a reference for angular position. A schematic representation is illustrated in Fig. 5.13.

The use of an incremental encoder for a joint actuating system clearly demands the evaluation of absolute positions. This is performed by means of suitable counting and storing electronic circuits. To this end, it is worth noticing that the position information is available on volatile memories, and thus it can be corrupted due to the effect of disturbances acting on the electronic circuit, or else fluctuations in the supply voltage. Such limitation obviously does not occur for absolute encoders, since the angular position information is coded directly on the optical disk.

The optical encoder has its own signal processing electronics inside the case, which provides direct digital position measurements to be interfaced with the control computer. If an external circuitry is employed, velocity measurements can be reconstructed from position measurements. In fact, if a pulse is generated at each transition, a velocity measurement can be obtained in three possible ways, namely, by using a voltage-to-frequency converter (with analog output), by (digitally) measuring the frequency of the pulse train, or by (digitally) measuring the sampling time of the pulse train. Between these last two techniques, the former is suitable for high-speed measurements while the latter is suitable for low-speed measurements.

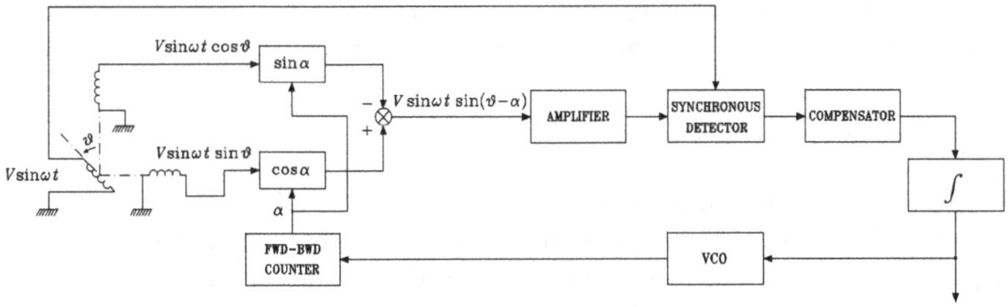

Fig. 5.14. Electric scheme of a resolver with functional diagram of a tracking-type RDC

Resolver

The resolver is an electromechanical position transducer which is compact and robust. Its operating principle is based on the mutual induction between two electric circuits which allow continuous transmission of angular position without mechanical limits. The information on the angular position is associated with the magnitude of two sinusoidal voltages, which are treated by a suitable resolver-to-digital converter (RDC) to obtain the digital data corresponding to the position measurement. The electric scheme of a resolver with the functional diagram of a tracking-type RDC is illustrated in Fig. 5.14.

From a construction viewpoint, the resolver is a small electric machine with a rotor and a stator; the inductance coil is on the rotor while the stator has two windings at 90 electrical degrees one from the other. By feeding the rotor with a sinusoidal voltage $V \sin \omega t$ (with typical frequencies in the range of 0.4 to 10 kHz), a voltage is induced on the stator windings whose magnitude depends on the rotation angle θ. The two voltages are fed to two digital multipliers, whose input is α and whose outputs are algebraically summed to achieve $V \sin \omega t \sin(\theta - \alpha)$; this signal is then amplified and sent to the input of a synchronous detector, whose filtered output is proportional to the quantity $\sin(\theta - \alpha)$. The resulting signal, after a suitable compensating action, is integrated and then sent to the input of a voltage-controlled oscillator (VCO) (a voltage-to-frequency converter) whose output pulses are input to a forward-backward counter. Digital data of the quantity α are available on the output register of the counter, which represent a measurement of the angle θ.

It can be recognized that the converter works according to a feedback principle. The presence of two integrators (one is represented by the forward-backward counter) in the loop ensures that the (digital) position and (analog) velocity measurements are error-free as long as the rotor rotates at constant speed; actually, a round-off error occurs on the word α and thus affects the position measurement. The compensating action is needed to confer suitable stability properties and bandwidth to the system. Whenever digital data are wished also for velocity measurements, it is necessary to use an analog-to-digital converter. Since the resolver is a very precise transducer, a resolution of 1 bit out of 16 can be obtained at the output of the RDC.

5.3.2 Velocity Transducers

Even though velocity measurements can be reconstructed from position transducers, it is often preferred to resort to direct measurements of velocity, by means of suitable transducers. *Velocity transducers* are employed in a wide number of applications and are termed *tachometers*. The most common devices of this kind are based on the operating principles of electric machines. The two basic types of tachometers are the *direct-current (DC) tachometer* and the *alternating-current (AC) tachometer*.

DC tachometer

The direct-current tachometer is the most used transducer in the applications. It is a small DC generator whose magnetic field is provided by a permanent magnet. Special care is paid to its construction, so as to achieve a linear input/output relationship and to reduce the effects of magnetic hysteresis and temperature. Since the field flux is constant, when the rotor is set in rotation, its output voltage is proportional to angular speed according to the constant characteristic of the machine.

Because of the presence of a commutator, the output voltage has a residual ripple which cannot be eliminated by proper filtering, since its frequency depends on angular speed. A linearity range of 0.1 to 1% can be obtained, whereas the residual ripple coefficient is of 2 to 5% of the mean value of the output signal.

AC tachometer

In order to avoid the drawbacks caused by the presence of a residual ripple in the output of a DC tachometer, one may resort to an AC tachometer. While the DC tachometer is a true DC generator, the AC tachometer differs from a generator. In fact, if a synchronous generator would be used, the frequency of the output signal would be proportional to the angular speed.

To obtain an alternating voltage whose magnitude is proportional to speed, one may resort to an electric machine that is structurally different from the synchronous generator. The AC tachometer has two windings on the stator mutually in quadrature and a cup rotor. If one of the windings is fed by a constant-magnitude sinusoidal voltage, a sinusoidal voltage is induced on the other winding which has the same frequency, a magnitude proportional to angular speed, and a phase equal or opposite to that of the input voltage according to the sign of rotation; the exciting frequency is usually set to 400 Hz. The use of a synchronous detector then yields an analog measurement

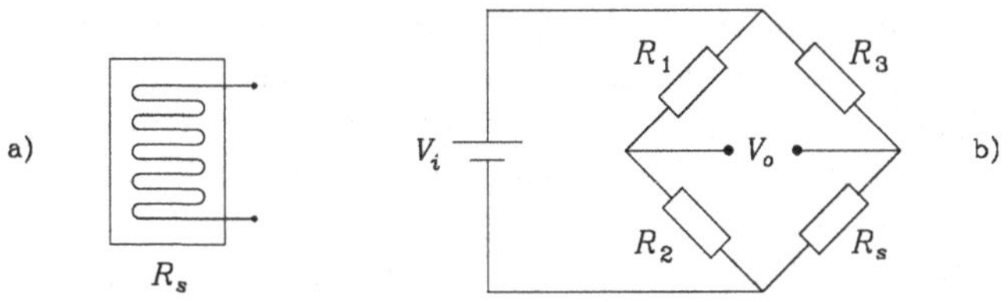

Fig. 5.15. a) Schematic representation of a strain gauge. **b)** Its insertion in a Wheatstone bridge

of angular velocity. In this case, the output ripple can be eliminated by a proper filter, since its fundamental frequency is twice as much as the supply frequency.

The performance of AC tachometers is comparable to that of DC tachometers. Two further advantages of AC tachometers are the lack of wiping contacts and the presence of a low moment of inertia, in view of the use of a lightweight cup rotor. However, a residual voltage occurs, even when the rotor is still, because of the unavoidable parasitic couplings between the stator coil and the measurement circuitry.

5.4 Exteroceptive Sensors

5.4.1 Force Sensors

Measurement of a force or torque is usually reduced to measurement of the strain induced by the force (torque) applied to an extensible element of suitable features. Therefore, an indirect measurement of force is obtained by means of measurements of small displacements. The basic component of a force sensor is the *strain gauge* which uses the change of electric resistance of a wire under strain.

Strain gauge

The strain gauge consists of a wire of low temperature coefficient. The wire is disposed on an insulated support (Fig. 5.15a) which is glued to the element subject to strain under the action of a stress. Dimensions of the wire change and then they cause a change of electric resistance.

The strain gauge is chosen in such a way that the resistance R_s changes linearly in the range of admissible strain for the extensible element. To transform changes of resistance into an electric signal, the strain gauge is inserted in one arm of a Wheatstone bridge which is balanced in the absence of stress on the strain gauge itself. From Fig. 5.15b it can be understood that the voltage balance in the bridge is described by

$$V_o = \left(\frac{R_2}{R_1 + R_2} - \frac{R_s}{R_3 + R_s} \right) V_i. \tag{5.30}$$

If temperature variations occur, the wire changes its dimension without application of any external stress. To reduce the effect of temperature variations on the measurement output, it is worth inserting another strain gauge in an adjacent arm of the bridge, which is glued on a portion of the extensible element not subject to strain.

Finally, to increase bridge sensitivity, two strain gauges may be used which have to be glued on the extensible element in such a way that one strain gauge is subject to traction and the other to compression; the two strain gauges then have to be inserted in two adjacent arms of the bridge.

Shaft torque sensor

In order to employ a servomotor as a torque-controlled generator, an indirect measurement of the driving torque is typically used, e.g., through the measurement of armature current in a permanent-magnet DC servomotor. If it is desired to guarantee insensitivity to change of parameters relating torque to the measured physical quantities, it is necessary to resort to a direct torque measurement.

The torque delivered by the servomotor to the joint can be measured by strain gauges mounted on an extensible apparatus interposed between the motor and the joint, e.g., a hollow shafting. Such apparatus must have low torsional stiffness and high bending stiffness, and it must ensure a proportional relationship between the applied torque and the induced strain.

By connecting the strain gauges mounted on the hollow shafting (in a Wheatstone bridge configuration) to a slip ring by means of graphite brushes, it is possible to feed the bridge and measure the resulting unbalanced signal which is proportional to the applied torque.

The measured torque is that delivered by the servomotor to the joint, and thus it does not coincide with the driving torque C_m in the block schemes of the actuating systems in Fig. 5.2 and in Fig. 5.6. In fact, such measurement does not account for the inertial and friction torque contributions as well as for the transmission located upstream of the measurement point.

Wrist force sensor

When the manipulator's end-effector is in contact with the working environment, the *force sensor* allows the measurement of the three components of a force and the three components of a moment with respect to a frame attached to it.

As illustrated in Fig. 5.16, the sensor is employed as a connecting apparatus at the wrist between the outer link of the manipulator and the end-effector. The connection is made by means of a suitable number of extensible elements subject to strain under the action of a force and a moment. Strain gauges are glued on each element which provide strain measurements. The elements

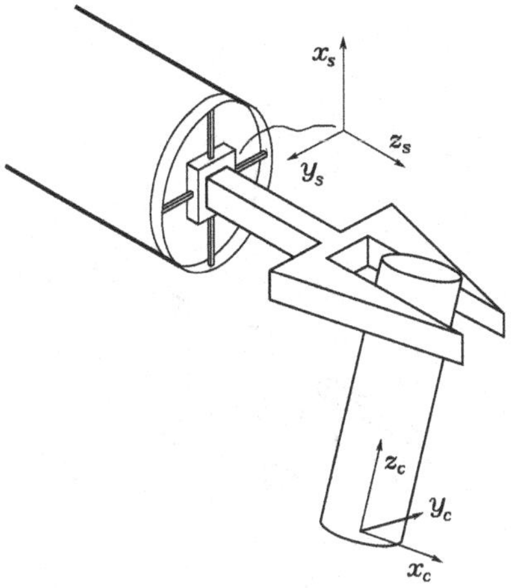

Fig. 5.16. Use of a force sensor on the outer link of a manipulator

have to be disposed in a keen way so that at least one element is appreciably deformed for any possible orientation of forces and moments.

Furthermore, the single force component with respect to the frame attached to the sensor should induce the least possible number of deformations, so as to obtain good structural decoupling of force components. Since a complete decoupling cannot be achieved, the number of significant deformations to reconstruct the six components of the force and moment vector is greater than six.

A typical force sensor is that where the extensible elements are disposed as in a Maltese cross; this is schematically indicated in Fig. 5.17. The elements connecting the outer link with the end-effector are four bars with a rectangular parallelepiped shape. On the opposite sides of each bar, a pair of strain gauges is glued that constitute two arms of a Wheatstone bridge; there is a total of eight bridges and thus the possibility of measuring eight strains.

The matrix relating strain measurements to the force components expressed in a Frame s attached to the sensor is termed sensor *calibration matrix*. Let w_i, for $i = 1, \ldots, 8$, denote the outputs of the eight bridges providing measurement of the strains induced by the applied forces on the bars according

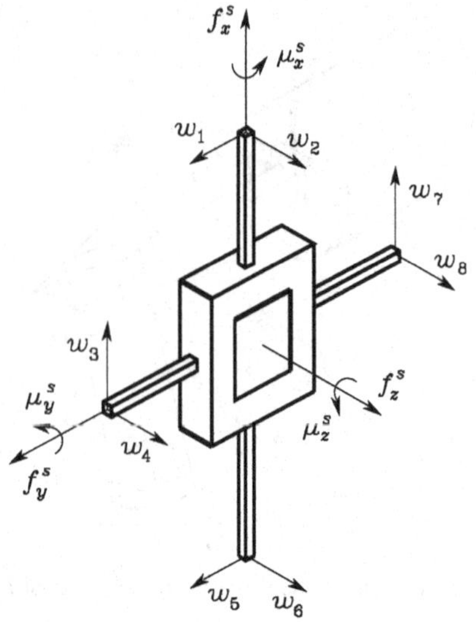

Fig. 5.17. Schematic representation of a Maltese-cross force sensor

to the directions specified in Fig. 5.17. Then, the calibration matrix is given by the transformation

$$\begin{bmatrix} f_x^s \\ f_y^s \\ f_z^s \\ \mu_x^s \\ \mu_y^s \\ \mu_z^s \end{bmatrix} = \begin{bmatrix} 0 & 0 & c_{13} & 0 & 0 & 0 & c_{17} & 0 \\ c_{21} & 0 & 0 & 0 & c_{25} & 0 & 0 & 0 \\ 0 & c_{32} & 0 & c_{34} & 0 & c_{36} & 0 & c_{38} \\ 0 & 0 & 0 & c_{44} & 0 & 0 & 0 & c_{48} \\ 0 & c_{52} & 0 & 0 & 0 & c_{56} & 0 & 0 \\ c_{61} & 0 & c_{63} & 0 & c_{65} & 0 & c_{67} & 0 \end{bmatrix} \begin{bmatrix} w_1 \\ w_2 \\ w_3 \\ w_4 \\ w_5 \\ w_6 \\ w_7 \\ w_8 \end{bmatrix}. \tag{5.31}$$

Reconstruction of force measurements through the calibration matrix is entrusted to suitable signal processing circuitry available in the sensor.

Typical sensors have a diameter of about 10 cm and a height of about 5 cm, with a measurement range of 50 to 500 N for the forces and of 5 to 70 N·m for the torques, and a resolution of the order of 0.1% of the maximum force and of 0.05% of the maximum torque, respectively; the sampling frequency at the output of the processing circuitry is of the order of 1 kHz.

Finally, it is worth noticing that force sensor measurements cannot be directly used by a force/motion control algorithm, since they describe the equivalent forces acting on the sensors which differ from the forces applied to the manipulator's end-effector (Fig. 5.16). It is therefore necessary to transform those forces from the sensor Frame s into the constraint Frame c; in view of the transformation in (3.116), one has

$$\begin{bmatrix} \boldsymbol{f}_c^c \\ \boldsymbol{\mu}_c^c \end{bmatrix} = \begin{bmatrix} \boldsymbol{R}_s^c & \boldsymbol{O} \\ \boldsymbol{S}(\boldsymbol{r}_{cs}^c)\boldsymbol{R}_s^c & \boldsymbol{R}_s^c \end{bmatrix} \begin{bmatrix} \boldsymbol{f}_s^s \\ \boldsymbol{\mu}_s^s \end{bmatrix} \tag{5.32}$$

which requires knowledge of the position $\boldsymbol{r}_{cs}^c$ of the origin of Frame s with respect to Frame c as well as of the orientation $\boldsymbol{R}_s^c$ of Frame s with respect to Frame c. Both such quantities are expressed in Frame c, and thus they are constant only if the end-effector is still, once contact has been achieved.

5.4.2 Range Sensors

The primary function of the exteroceptive sensors is to provide the robot with the information needed to execute 'intelligent' actions in an autonomous way. To this end, it is crucial to detect the presence of an object in the workspace and eventually to measure its range from the robot along a given direction.

The former kind of data is provided by the *proximity sensors*, a simplified type of *range sensors*, capable of detecting only the presence of objects nearby the sensitive part of the sensor, without a physical contact. The distance within which such sensors detect objects is defined *sensitive range*.

In the more general case, range sensors are capable of providing structured data, given by the distance of the measured object and the corresponding measurement direction, i.e., the position in space of the detected object with respect to the sensor.

The data provided by the range sensors are used in robotics to avoid obstacles, build maps of the environment, recognize objects.

The most popular range sensors in robotics applications are those based on sound propagation through an elastic fluid, the so-called *sonars* (SOund NAvigation and Ranging), and those exploiting light propagation features, the so-called *lasers* (Light Amplification by Stimulated Emission of Radiation). In the following, the main features of these two sensors are illustrated.

Sonars

The sonars employ acoustic pulses and their echoes to measure the range to an object. Since the sound speed is usually known for a given media (air, water), the range to an object is proportional to the echo travel time, commonly called *time-of-flight*, i.e., the time which the acoustic wave takes to cover the distance sensor-object-sensor. Sonars are widely utilized in robotics, and especially in mobile and underwater robotics. Their popularity is due to their low cost, light weight, low power consumption, and low computational effort, compared to other ranging sensors. In some applications, such as in underwater and low-visibility environments, the sonar is often the only viable sensing modality.

Despite a few rare examples of sonars operating at audible frequencies for human ears (about 20 Hz to 20 KHz), the ultrasound frequencies (higher

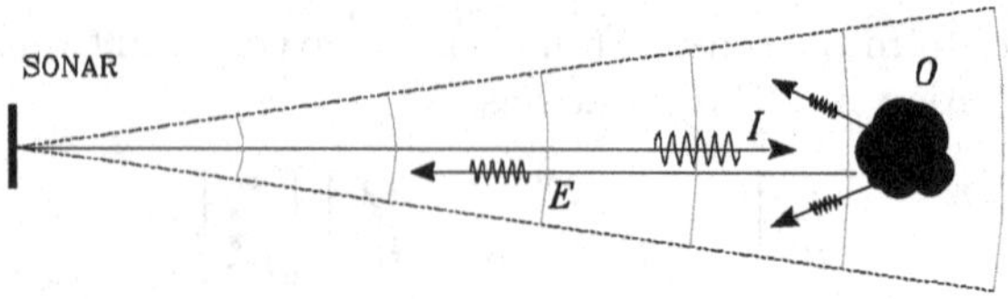

Fig. 5.18. Sonar ranging principle

than 20 KHz) are the most widely used to realize this type of sensor. Typical frequencies in robotics range from 20 KHz to 200 KHz, even though higher values (of the order of MHz) can be achieved utilizing piezoelectric quartz crystals. In this range, the energy of the wave emitted by the sonar can be regarded as concentrated in a conical volume whose beamwidth depends on the frequency as well as on the transducer diameter. Further to measuring range, sonars provide qualitative directional data on the object which has generated the echo. For the most common sensors in robotics, the beamwidth of the energy beam is typically not smaller than 15 deg. Obviously, for smaller beamwidths, higher angular resolutions can be obtained.

The main components of a sonar measurement system are a transducer, which is vibrated and transforms acoustic energy into electric energy and vice versa, and a circuitry for the excitation of the transducer and the detection of the reflected signal. Figure 5.18 schematically illustrates the operating principle: the pulse I emitted by the transducer, after hitting the object O found in the emission cone of the sensor, is partly reflected (echo E) towards the sound source and thus detected. The time-of-flight t_v is the time between the emission of the ultrasound pulse and the reception of the echo. The object range d_O can be computed from t_v using the relation

$$d_O = \frac{c_s t_v}{2} \tag{5.33}$$

where c_s is sound speed, which in low-humidity air depends on the temperature T (measured in centigrade) according to the expression

$$c_s \approx 20.05\sqrt{T + 273.16} \text{ m/s}. \tag{5.34}$$

In the scheme of Fig. 5.18 the use of a sole transducer is represented for the transmission of the pulse and the reception of the echo. This configuration requires that the commutation from transmitter to receiver takes place after a certain latency time which depends not only on the duration of the transmitted pulse but also on the mechanical inertia of the transducer.

Despite the low cost and ease of use, however, these sensors have non-negligible limits with respect to the angular and radial resolution, as well as to the minimum and maximum measurement range that can be achieved. In particular, the width of the radiation cone decreases as frequency increases with improved angular resolution. A higher frequency leads to greater radial

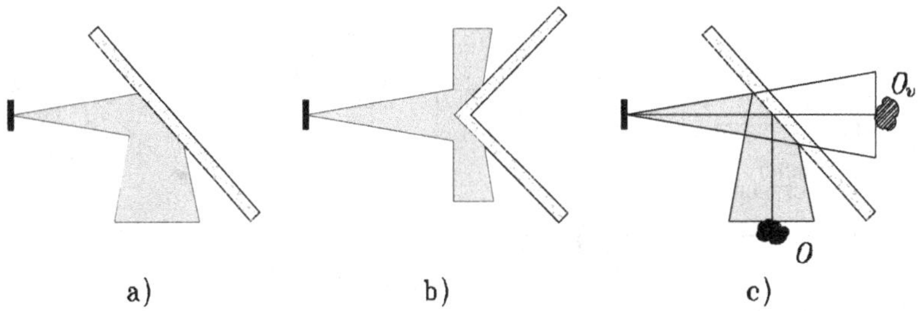

Fig. 5.19. Reflector models on smooth surfaces: **a)** non-detected plane. **b)** non-detected corner. **c)** plane with false detection (O real object, O_v virtual object detected)

resolution and contributes to reducing the minimum range that can be detected by the sonar. Nevertheless, there is a lower limit because of the lapse time when reception is inhibited to avoid interference with the reflected signal — in certain cases better performance can be obtained by employing two distinct transducers for the emission and the detection. On the other hand, too high frequencies may exasperate absorbtion phenomena, depending on the features of the surface generating the echo. Such phenomena further reduce the power of the transmitted signal — decreasing with the square of the range covered by the ultrasound wave — thus reducing the maximum limit of the measurement time.

Piezoelectric and electrostatic transducers are the two major types available that operate in air and can in principle operate both as a transmitter and receiver.

The *piezoelectric transducers* exploit the property of some crystal materials to deform under the action of an electric field and vibrate when a voltage is applied at the resonant frequency of the crystal. The efficiency of the acoustic match of these transducers with compressible fluids such as air is rather low. Often a conical concave horn is mounted on the crystal to match acoustically the crystal acoustic impedance to that of air. Being of resonant type, these transducers are characterized by a rather low bandwidth and show a significant mechanical inertia which severely limits the minimum detectable range, thus justifying the use of two distinct transducers as transmitter and receiver.

The *electrostatic transducers* operate as capacitors whose capacitance varies moving and/or deforming one of its plates. A typical construction consists of a gold-coated plastic foil membrane (*mobile* plate) stretched across a round grooved aluminium back plate (*fixed* plate). When the transducer operates as receiver, the change of capacitance, induced by the deformation of the membrane under the acoustic pressure, produces a proportional change of the voltage across the capacitor, assuming that the foil charge is constant. As a transmitter, the transducer membrane is vibrated by applying a sequence of electric pulses across the capacitor. The electric oscillations generate, as

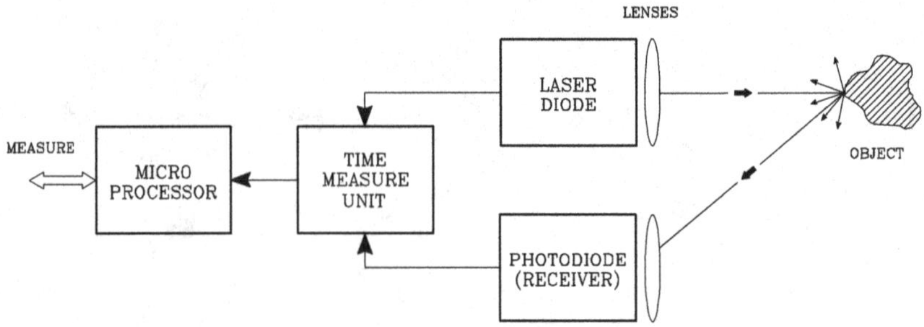

Fig. 5.20. Time-of-flight laser sensor operating principle

a result of the induced electric field, a mechanical force which vibrates the mobile plate.

Since the electrostatic transducers can operate at different frequencies, they are characterized by large bandwidth and high sensitivity, low mechanical inertia and rather efficient acoustic match with air. As compared to the piezoelectric transducers, however, they can operate at lower maximum frequencies (a few hundreds kHz vs a few MHz) and require a bias voltage which complicates the control electronics. Among the ultrasound measurement systems with capacitive transducers, it is worth mentioning the Polaroid sonar, initially developed for autofocus systems and later widely employed as range sensors in several robotic applications. The 600 series sensor utilizes a capacitive transducer of the type described above with a diameter of almost 4 cm, operates at 50 kHz frequency and is characterized by a beamwidth of 15 deg, can detect a maximum range of about 10 m and a mimimum range of about 15 cm with an accuracy of $\pm 1\%$ across the measurement range. The bias voltage is 200 V with current absorbtion peaks of 2 A in transmission.

Accuracy of ultrasound range sensors depends on the features of the transducer and the excitation/detection circuitry, as well as on the reflective properties of the surfaces hit by the acoustic waves.

Smooth surfaces, i.e., those characterized by irregularities of comparable size to that of the wavelength corresponding to the employed frequency, may produce a non-detectable echo at the sensor (Figura 5.19a,b) if the incident angle of the ultrasound beam exceeds a given critical angle which depends on the operational frequency and the reflective material. In the case of the Polaroid sensors, this angle is equal to 65 deg, i.e., 25 deg from the normal to the reflective surface, for a smooth surface in plywood. When operating in complex environments, such mirror reflections may give rise to multiple reflections, thus causing range measurement errors or false detection (Fig. 5.19c).

Lasers

In the construction of optical measurement systems, the laser beam is usually preferred to other light sources for the following reasons:

- They can easily generate bright beams with lightweight sources.
- The infrared beams can be used unobtrusively.
- They focus well to give narrow beams.
- Single-frequency sources allow easier rejection filtering of unwanted frequencies, and do not disperse from refraction as much as full spectrum sources.

There are two types of laser-based range sensors in common use: the time-of-flight sensors and the triangulation sensors.

The *time-of-flight sensors* compute distance by measuring the time that a pulse of light takes to travel from the source to the observed target and then to the detector (usually collocated with the source). The travel time multiplied by the speed of light (properly adjusted for the air temperature) gives the distance measurement. The operating principle of a time-of-flight laser sensor is illustrated in Fig. 5.20.

Limitations on the accuracy of these sensors are based on the minimum observation time — and thus the minimum distance observable, the temporal accuracy (or quantization) of the receiver, and the temporal width of the laser pulse. Such limitations are not only of a technological nature. In many cases, cost is the limiting factor of these measurement devices. For instance, to obtain 1 mm resolution, a time accuracy of about 3 ps, which can be achieved only by using rather expensive technology.

Many time-of-flight sensors used have what is called an *ambiguity interval*. The sensor emits pulses of light periodically, and computes an average target distance from the time of the returning pulses. Typically, to simplify the detection electronics of these sensors, the receiver only accepts signals that arrive within time Δt, but this time window might also observe previous pulses reflected by more distant surfaces. This means that a measurement is ambiguous to the multiple of $\frac{1}{2}c\Delta t$, where c is the speed of light. Typical values of $\frac{1}{2}c\Delta t$ are 20–40 m.

In certain conditions, suitable algorithms can be employed to recover the true depth by assuming that the distances should be changing smoothly.

The time-of-flight sensors transmit only a single beam, thus range measurements are only obtained from a single surface point. In order to obtain more information, the range data is usually supplied as a vector of range to surfaces lying in a plane or as an image. To obtain these denser representations, the laser beam is swept across the scene. Normally the beam is swept by a set of mirrors rather than moving the laser and detector themselves — mirrors are lighter and less prone to motion damage.

Typical time-of-flight sensors suitable for mobile robotics applications have a range of 5–100 m, an accuracy of 5–10 mm, and a frequency of data acquisition per second of 1000–25000 Hz.

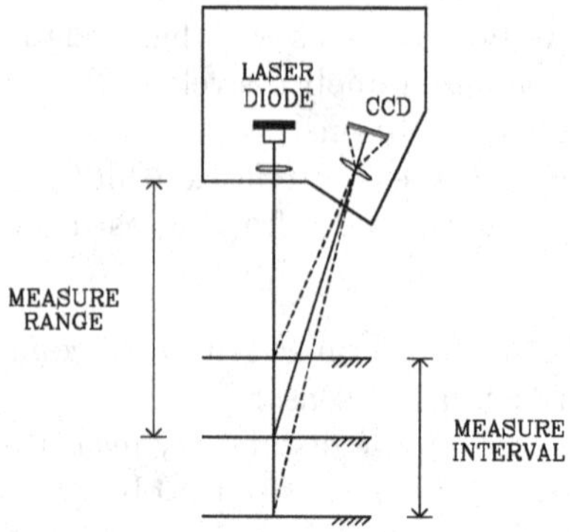

Fig. 5.21. Triangulation laser sensor operating principle

The operating principle of *triangulation laser sensors*[3] is illustrated in Fig. 5.21.

The laser beam emitted by a photodiode is projected onto the observed surface. The reflected beam is focused on a CCD sensor by means of a suitable lens. Obviously, reflection must be diffused. The position of the focused beam reflected to the receiver gives rise to a signal which is proportional to the distance of the transmitter from the object. In fact, from the measurement of the CCD sensor it is possible to resort to the angle at which the reflected energy hits the sensor. Once the relative position and orientation of the CCD sensor with respect to the photodiode are known, as e.g. through a suitable calibration procedure, it is possible to compute the distance from the object with simple geometry.

Accuracy can be influenced by certain object surfaces not favouring reflection, differences or changes of colour. Such occurrences can be mitigated or even eliminated with modern electronic technology and automatic regulation of light intensity.

The possibility of controlling the laser beam light brings the following advantages:

[3] The triangulation method is based on the trigonometric properties of triangles and in particular on the cosine theorem. The method allows the computation of the distance between two non-directly accessible points, i.e., once two angles and one side of a triangle are known, it is possible to determine the other two sides. For the case at issue, one side is given by the distance between the emitter (laser) and the receiver (the CCD sensor), one angle is given by the orientation of the emitter with respect to that side and the other angle can be computed from the position of the laser beam on the image plane. In practice, it is not easy to compute the above quantities, and suitable calibration techniques are to be employed which avoid such computation to determine the distance measurement.

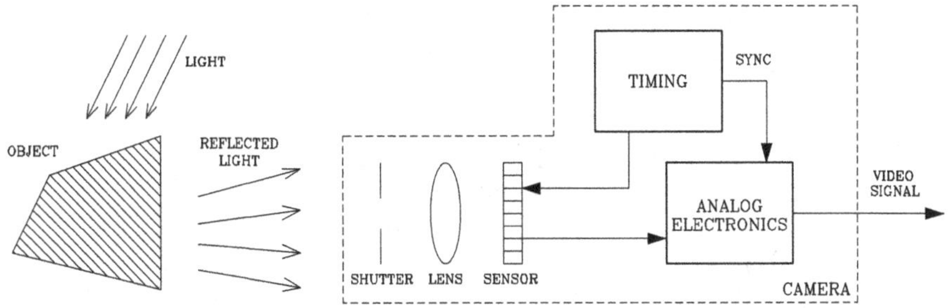

Fig. 5.22. Schematic representation of a vision system

- If the laser beam wavelength is known, e.g. that of the visible red 670 nm, highly selective filters can be used which are set to the same frequency to reduce the effects of other light sources.
- The laser beam may be remodelled through lenses and mirrors so as to create multiple beams or laser strips to measure multiple 3D points simultaneously.
- The direction of the laser beam can be controlled directly by the control system to observe selectively only those portions of the scene of interest.

The main limitations of this type of sensors are the potential eye safety risks from the power of lasers, particularly when invisible laser frequencies are used (commonly infrared), as well as the false specular reflections from metallic and polished objects.

5.4.3 Vision Sensors

The task of a camera as a *vision sensor* is to measure the intensity of the light reflected by an object. To this end, a photosensitive element, termed *pixel* (or *photosite*), is employed, which is capable of transforming light energy into electric energy. Different types of sensors are available depending on the physical principle exploited to realize the energy transformation. The most widely used devices are CCD and CMOS sensors based on the photoelectric effect of semiconductors.

CCD

A CCD (Charge Coupled Device) sensor consists of a rectangular array of photosites. Due to the photoelectric effect, when a photon hits the semiconductor surface, a number of free electrons are created, so that each element accumulates a charge depending on the time integral of the incident illumination over the photosensitive element. This charge is then passed by a transport mechanism (similar to an analog shift register) to the output amplifier, while at

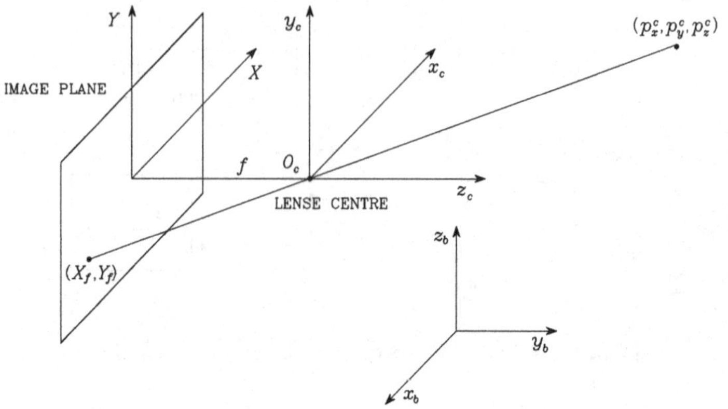

Fig. 5.23. Perspective transformation

the same time the photosite is discharged. The electric signal is to be further processed in order to produce the real *video signal*.

CMOS

A CMOS (Complementary Metal Oxide Semiconductor) sensor consists of a rectangular array of photodiodes. The junction of each photodiode is precharged and it is discharged when hit by photons. An amplifier integrated in each pixel can transform this charge into a voltage or current level. The main difference with the CCD sensor is that the pixels of a CMOS sensor are non-integrating devices; after being activated they measure throughput, not volume. In this manner, a saturated pixel will never overflow and influence a neighboring pixel. This prevents the effect of *blooming*, which indeed affects CCD sensors.

Camera

As sketched in Fig. 5.22, a camera is a complex system comprising several devices other than the photosensitive sensor, i.e., a *shutter*, a *lens* and *analog preprocessing electronics*. The lens is responsible for focusing the light reflected by the object on the plane where the photosensitive sensor lies, called the *image plane*.

With reference to Fig. 5.23, consider a frame O_c–$x_c y_c z_c$ attached to the camera, whose location with respect to the base frame is identified by the homogeneous transformation matrix $\boldsymbol{T}_c^b$. Take a point of the object of coordinates $\boldsymbol{p}^c = [\,p_x^c \quad p_y^c \quad p_z^c\,]^T$; typically, the centroid of the object is chosen. Then, the coordinate transformation from the base frame to the camera frame is described as

$$\widetilde{\boldsymbol{p}}^c = \boldsymbol{T}_b^c \widetilde{\boldsymbol{p}}, \tag{5.35}$$

where $\boldsymbol{p}$ denotes the object position with respect to the base frame and homogeneous representations of vectors have been used.

A reference frame can be introduced on the image plane, whose axes X and Y are parallel to the axes x_c and y_c of the camera frame, and the origin is at the intersection of the optical axis with the image plane, termed principal point. Due to the refraction phenomenon, the point in the camera frame is transformed into a point in the image plane via the *perspective transformation*, i.e.,

$$X_f = -\frac{fp_x^c}{p_z^c}$$
$$Y_f = -\frac{fp_y^c}{p_z^c}$$

where (X_f, Y_f) are the new coordinates in the frame defined on the image plane, and f is the *focal length* of the lens. Notice that these coordinates are expressed in metric units and the above transformation is singular at $p_z^c = 0$.

The presence of the minus sign in the equations of the perspective transformation is consistent with the fact that the image of an object appears upside down on the image plane of the camera. Such an effect can be avoided, for computational ease, by considering a virtual image plane positioned before the lens, in correspondence of the plane $z_c = f$ of the camera frame. In this way, the model represented in Fig. 5.24 is obtained, which is characterized by the *frontal* perspective transformation

$$X_f = \frac{fp_x^c}{p_z^c} \tag{5.36}$$
$$Y_f = \frac{fp_y^c}{p_z^c} \tag{5.37}$$

where, with abuse of notation, the name of the variables on the virtual plane has not been changed.

These relationships hold only in theory, since the real lenses are always affected by imperfections, which cause image quality degradation. Two types of distortions can be recognized, namely, *aberrations* and *geometric distortion*. The former can be reduced by restricting the light rays to a small central region of the lens; the effects of the latter can be compensated on the basis of a suitable model whose parameters are to be identified.

A visual information is typically elaborated by a digital processor, and thus the measurement principle is to transform the light intensity $I(X, Y)$ of each point in the image plane into a number. It is clear that a *spatial sampling* is needed since an infinite number of points in the image plane exist, as well as a *temporal sampling* since the image can change during time. The CCD or CMOS sensors play the role of spatial samplers, while the shutter in front of the lens plays the role of the temporal sampler.

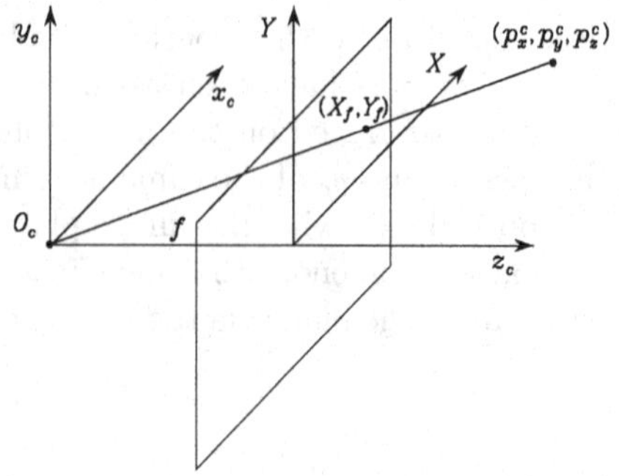

Fig. 5.24. Frontal perspective transformation

The spatial sampling unit is the pixel, and thus the coordinates (X, Y) of a point in the image plane are to be expressed in pixels, i.e., (X_I, Y_I). Due to the photosite finite dimensions, the pixel coordinates of the point are related to the coordinates in metric units through two scale factors α_x and α_y, namely,

$$X_I = \frac{\alpha_x f p_x^c}{p_z^c} + X_0 \tag{5.38}$$

$$Y_I = \frac{\alpha_y f p_y^c}{p_z^c} + Y_0, \tag{5.39}$$

where X_0 and Y_0 are the offsets which take into account the position of the origin of the pixel coordinate system with respect to the optical axis. This nonlinear transformation can be written in a linear form by resorting to the homogeneous representation of the point (x_I, y_I, z_I) via the relationships

$$X_I = \frac{x_I}{\lambda}$$

$$Y_I = \frac{y_I}{\lambda}$$

where $\lambda > 0$. As a consequence, (5.38), (5.39) can be rewritten as

$$\begin{bmatrix} x_I \\ y_I \\ \lambda \end{bmatrix} = \lambda \begin{bmatrix} X_I \\ Y_I \\ 1 \end{bmatrix} = \boldsymbol{\Omega \Pi} \begin{bmatrix} p_x^c \\ p_y^c \\ p_z^c \\ 1 \end{bmatrix} \tag{5.40}$$

where

$$\boldsymbol{\Omega} = \begin{bmatrix} f\alpha_x & 0 & X_0 \\ 0 & f\alpha_y & Y_0 \\ 0 & 0 & 1 \end{bmatrix} \tag{5.41}$$

$$\boldsymbol{\Pi} = \begin{bmatrix} 1 & 0 & 0 & 0 \\ 0 & 1 & 0 & 0 \\ 0 & 0 & 1 & 0 \end{bmatrix}. \tag{5.42}$$

At this point, the overall transformation from the Cartesian space of the observed object to the *image space* of its image in pixels is characterized by composing the transformations in (5.35), (5.40) as

$$\boldsymbol{\Xi} = \boldsymbol{\Omega}\boldsymbol{\Pi}\boldsymbol{T}_b^c \tag{5.43}$$

which represents the so-called *camera calibration* matrix. It is worth pointing out that such a matrix contains *intrinsic parameters* $(\alpha_x, \alpha_y, X_0, Y_0, f)$ in $\boldsymbol{\Omega}$ depending on the sensor and lens characteristics as well as *extrinsic parameters* in $\boldsymbol{T}_c^b$ depending on the relative position and orientation of the camera with respect to the base frame. Several calibration techniques exist to identify these parameters in order to compute the transformation between the Cartesian space and the image space as accurately as possible.

If the intrinsic parameters of a camera are known, from a computationally viewpoint, it is convenient to refer to the *normalized coordinates* (X, Y), defined by the normalized perspective transformation

$$\lambda \begin{bmatrix} X \\ Y \\ 1 \end{bmatrix} = \boldsymbol{\Pi} \begin{bmatrix} p_x^c \\ p_y^c \\ p_z^c \\ 1 \end{bmatrix}. \tag{5.44}$$

These coordinates are defined in metrical units and coincide with the coordinates (5.36), (5.37) in the case when $f = 1$. Comparing (5.40) with (5.44) yields the invertible transformation

$$\begin{bmatrix} X_I \\ Y_I \\ 1 \end{bmatrix} = \boldsymbol{\Omega} \begin{bmatrix} X \\ Y \\ 1 \end{bmatrix} \tag{5.45}$$

relating the normalized coordinates to those expressed in pixels through the matrix of intrinsic parameters.

If a monochrome CCD camera[4] is of concern, the output amplifier of the sensor produces a signal which is processed by a timing analog electronics in order to generate an electric signal according to one of the existing *video standards*, i.e., the CCIR European and Australian standard, or the RS170 American and Japanese standard. In any case, the video signal is a voltage of 1 V peak-to-peak whose amplitude represents sequentially the image intensity.

The entire image is divided into a number of lines (625 for the CCIR standard and 525 for the RS170 standard) to be sequentially scanned. The raster scan proceeds horizontally across each line and each line from top to bottom, but first all the even lines, forming the first *field*, and then all the odd lines, forming the second *field*, so that a *frame* is composed of two successive

[4] Colour cameras are equipped with special CCDs sensitive to three basic colours (RGB); the most sophisticated cameras have three separate sensors, one per each basic colour.

fields. This technique, called *interlacing*, allows the image to be updated either at frame rate or at field rate; in the former case the update frequency is that of the entire frame (25 Hz for the CCIR standard and 30 Hz for the RS170 standard), while in the latter case the update frequency can be doubled as long as half the vertical resolution can be tolerated.

The last step of the measurement process is to digitize the analog video signal. The special analog-to-digital converters adopted for video signal acquisition are called *frame grabbers*. By connecting the output of the camera to the frame grabber, the video waveform is sampled and quantized and the values stored in a two-dimensional memory array representing the spatial sample of the image, known as *framestore*; this array is then updated at field or frame rate.

In the case of CMOS cameras (currently available only for monochrome images), thanks to CMOS technology which allows the integration of the analog-to-digital converter in each pixel, the output of the camera is directly a two-dimensional array, whose elements can be accessed randomly. Such advantage, with respect to CCD cameras, leads to the possibility of higher frame rates if only parts of the entire frame are accessed.

The sequence of steps from image formation to image acquisition described above can be classified as a process of *low-level vision*; this includes the extraction of elementary image features, e.g., centroid and intensity discontinuities. On the other hand, a robotic system can be considered really autonomous only if procedures for emulating cognition are available, e.g., recognizing an observed object among a set of CAD models stored into a data base. In this case, the artificial vision process can be referred to as *high-level vision*.

Bibliography

Scientific literature on actuating systems and sensors is wide and continuously updated. The mechanical aspects on the joint actuating systems can be probed further in e.g. [186]. Details about electric servomotors can be found in [22], while in [156] construction and control problems for hydraulic motors are extensively treated. Control of electric drives is discussed in [128]; for direct drives see [12]. Joint control problems are discussed in [89].

A wide and detailed survey on sensors and in particular on proprioceptive sensors is given in [81]. In [220] force sensors are accurately described, with special attention to wrist force sensors. Further details about range sensors, with reference to mobile robotics applications, are available in [210]. Finally, a general introduction on vision sensors is contained in [48], while in [233] one of the most common calibration techniques for vision systems is described.

Problems

5.1. Prove (5.7)–(5.10).

5.2. Consider the DC servomotor with the data: $I_m = 0.0014\,\text{kg·m}^2$, $F_m = 0.01\,\text{N·m·s/rad}$, $L_a = 2\,\text{mH}$, $R_a = 0.2\,\text{ohm}$, $k_t = 0.2\,\text{N·m/A}$, $k_v = 0.2\,\text{V·s/rad}$, $C_iG_v = 1$, $T_v = 0.1\,\text{ms}$, $k_i = 0$. Perform a computer simulation of the current and velocity response to a unit step voltage input V_c'. Adopt a sampling time of 1 ms.

5.3. For the servomotor of the previous problem, design the controller of the current loop $C_i(s)$ so that the current response to a unit step voltage input V_c' is characterized by a settling time of 2 ms. Compare the velocity response with that obtained in Problem 5.2.

5.4. Find the control voltage/output position and reaction torque/output position transfer functions for the scheme of Fig. 5.6.

5.5. For a Gray-code optical encoder, find the interconversion logic circuit which yields a binary-coded output word.

5.6. With reference to a contact situation of the kind illustrated in Fig. 5.16, let

$$\boldsymbol{r}_{cs}^{c} = [\,-0.3 \quad 0 \quad 0.2\,]^T \text{ m} \qquad \boldsymbol{R}_s^c = \begin{bmatrix} 0 & 0 & 1 \\ 0 & -1 & 0 \\ 1 & 0 & 0 \end{bmatrix}$$

and let the force sensor measurement be

$$\boldsymbol{f}_s^s = [\,20 \quad 0 \quad 0\,]^T \text{ N} \qquad \boldsymbol{\mu}_s^s = [\,0 \quad 6 \quad 0\,]^T \text{ N·m}.$$

Compute the equivalent force and moment in the contact frame.

5.7. Consider the SCARA manipulator in Fig. 2.34 with link lengths $a_1 = a_2 = 0.5\,\text{m}$. Let the base frame be located at the intersection between the first link and the base link with axis z pointing downward and axis x in the direction of the first link when $\vartheta_1 = 0$. Assume that a CCD camera is mounted on the wrist so that the camera frame is aligned with the end-effector frame. The camera parameters are $f = 8\,\text{mm}$, $\alpha_x = 79.2\,\text{pixel/mm}$, $\alpha_y = 120.5\,\text{pixel/mm}$, $X_0 = 250$, $Y_0 = 250$. An object is observed by the camera and is described by the point of coordinates $\boldsymbol{p} = [\,0.8 \quad 0.5 \quad 0.9\,]^T$ m. Compute the pixel coordinates of the point when the manipulator is at the configuration $\boldsymbol{q} = [\,0 \quad \pi/4 \quad 0.1 \quad 0\,]^T$.

6

Control Architecture

This chapter is devoted to presenting a reference model for the *functional architecture* of an industrial robot's *control system*. The *hierarchical structure* and its articulation into *functional modules* allows the determination of the requirements and characteristics of the *programming environment* and the *hardware architecture*. The architecture refers to robot manipulators, yet its articulation in *levels* also holds for mobile robots.

6.1 Functional Architecture

The *control system* to supervise the activities of a robotic system should be endowed with a number of tools providing the following functions:

- capability of moving physical objects in the working environment, i.e., *manipulation* ability;
- capability of obtaining information on the state of the system and working environment, i.e., *sensory* ability;
- capability of exploiting information to modify system behaviour in a pre-programmed manner, i.e., *intelligence* ability;
- capability of storing, elaborating and providing data on system activity, i.e., *data processing* ability.

An effective implementation of these functions can be obtained by means of a *functional architecture* which is thought of as the superposition of several *activity levels* arranged in a *hierarchical structure*. The lower levels of the structure are oriented to physical motion execution, whereas the higher levels are oriented to logical action planning. The levels are connected by data flows; those directed towards the higher levels regard measurements and/or results of actions, while those directed towards the lower levels regard transmission of directions.

With reference to the control system functions implementing management of the above listed system activities, in general it is worth allocating three *functional models* at each level. A first module is devoted to sensory data management (sensory module). A second module is devoted to provide knowledge of the relevant world (modelling module). A third module is devoted to decide the policy of the action (decision module).

More specifically, the *sensory modules* acquire, elaborate, correlate and integrate sensory data in time and space, in order to recognize and measure the system state and environment characteristic; clearly, the functions of each module are oriented to the management of the relevant sensory data for that level.

On the other hand, the *modelling modules* contain models derived on the basis of a priori knowledge of system and environment; these models are updated by the information coming from the sensory modules, while the activation of the required functions is entrusted to the decision modules.

Finally, the *decision modules* perform breakdown of high-level tasks into low-level actions; such task breakdown concerns both breakdown in time of sequential actions and breakdown in space of concurrent actions. Each decision module is entrusted with the functions concerning management of elementary action assignments, task planning and execution.

The functions of a decision module characterize the level of the hierarchy and determine the functions required to the modelling and sensory modules operating at the same level. This implies that the contents of these two modules do not uniquely allow the determination of the hierarchical level, since the same function may be present at more levels depending on the needs of the decision modules at the relative levels.

The functional architecture needs an *operator interface* at each level of the hierarchy, so as to allow an operator to perform supervisory and intervention functions on the robotic system.

The instructions imparted to the decision module at a certain level may be provided either by the decision module at the next higher level or by the operator interface, or else by a combination of the two. Moreover, the operator, by means of suitable communication tools, can be informed on the system state and thus can contribute his/her own knowledge and decisions to the modelling and sensory modules.

In view of the high data flow concerning the exchange of information between the various levels and modules of the functional architecture, it is worth allocating a shared *global memory* which contains the updated estimates on the state of the whole system and environment.

The structure of the reference model for the functional architecture is represented in Fig. 6.1, where the *four hierarchical levels* potentially relevant for robotic systems in industrial applications are illustrated. Such levels regard definition of the *task*, its breakdown into elementary *actions*, assignment of *primitives* to the actions, and implementation of control actions on the *servo*-manipulator. In the following, the general functions of the three modules at each level are described.

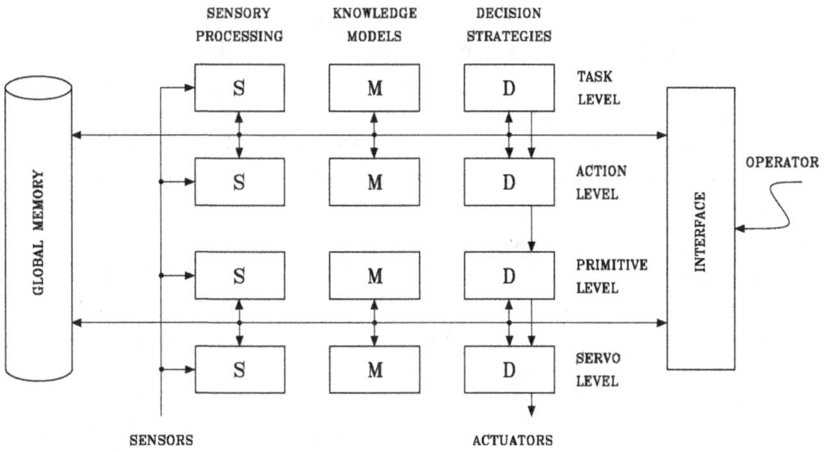

Fig. 6.1. Reference model for a control system functional architecture

At the *task level*, the user specifies the task which the robotic system should execute; this specification is performed at a high level of abstraction. The goal of the desired task is analyzed and broken down into a sequence of actions which are coordinated in space and time and allow implementation of the task. The choice of actions is performed on the basis of knowledge models as well as of the scene of interest for the task. For instance, consider the application of a robot installed in an assembly line which is required to perform a specific assembly task. To define the elementary actions that have to be transmitted to the decision module at the next lower level, the decision module should consult its knowledge base available in the modelling module, e.g., type of assembly, components of the object to assembly, assembly sequence, and choice of tools. This knowledge base should be continuously updated by the information provided by the sensory module concerning location of the parts to assembly; such information is available by means of a high-level vision system operating in a scarcely structured environment, or else by means of simple sensors detecting the presence of an object in a structured environment.

At the *action level*, the symbolic commands coming from the task level are translated into sequences of intermediate configurations which characterize a motion path for each elementary action. The choice of the sequences is performed on the basis of models of the manipulator and environment where the action is to take place. With reference to one of the actions generated by the above assembly task, the decision module chooses the most appropriate coordinate system to compute manipulator's end-effector poses, by separating translation from rotation if needed; it decides whether to operate in the joint or operational space, it computes the path or via points, and for the latter it defines the interpolation functions. By doing so, the decision module should compare the sequence of configurations with a model of the manipulator as well as with a geometric description of the environment, which are both available in the modelling model. In this way, action feasibility is ascertained in terms of obstacle-collision avoidance, motion in the neighbourhood of kinematically singular configurations, occurrence of mechanical joint limits, and eventually utilization of available redundant DOFs. The knowledge base is updated by the information on the portion of scene where the single action takes place which is provided by the sensory module, e.g., by means of a low-level vision system or range sensors.

At the *primitive level*, on the basis of the sequence of configurations received by the action level, admissible motion trajectories are computed and the control strategy is decided. The motion trajectory is interpolated so as to generate the references for the servo level. The choice of motion and control primitives is conditioned by the features of the mechanical structure and its degree of interaction with the environment. Still with reference to the above case study, the decision module computes the geometric path and the relative trajectory on the basis of the knowledge of the manipulator dynamic model available in the modelling module. Moreover, it defines the type of control algorithm, e.g., decentralized control, centralized control, or interaction control; it specifies the relative gains; and it performs proper coordinate transformations, e.g., kinematic inversion if needed. The sensory module provides information on the occurrence of conflicts between motion planning and execution, by means of, e.g., force sensors, low-level vision systems and proximity sensors.

At the *servo level*, on the basis of the motion trajectories and control strategies imparted by the primitive level, control algorithms are implemented which provide the driving signals to the joint servomotors. The control algorithm operates on error signals between the reference and the actual values of the controlled quantities, by utilizing knowledge of manipulator dynamic model, and of kinematics if needed. In particular, the decision module performs a microinterpolation on the reference trajectory to exploit fully the dynamic characteristic of the drives; it computes the control law, and it generates the (voltage or current) signals for controlling the specific drives. The modelling module elaborates the terms of the control law depending on the manipulator current configuration and pass them to the decision module; such terms are computed on the basis of knowledge of manipulator dynamic model. Finally, the sensory module provides measurements of the proprioceptive sensors (position, velocity and contact force if needed); these measurements are used by the decision module to compute the servo errors and, if required, by the modelling module to update the configuration-dependent terms in the model.

The specification of the functions associated with each level points out that the implementation of such functions should be performed at different time rates, in view of their complexity and requirements. On one hand, the functions associated with the higher levels are not subject to demanding real-time constraints, since they regard planning activities. On the other hand, their

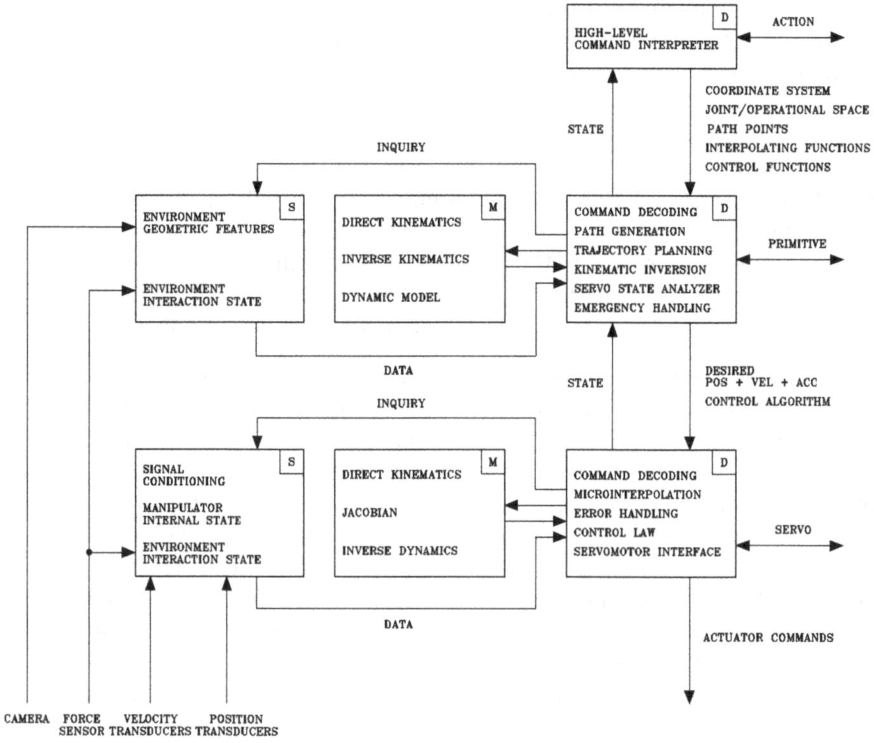

Fig. 6.2. Hierarchical levels of a functional architecture for industrial robots

complexity is notable, since scheduling, optimization, resource management and high-level sensory system data processing are required to update complex models.

At the lowest level, demanding real-time operation prevails in order to obtain high dynamic performance of the mechanical structure. The above remarks lead to the conclusion that, at the servo level, it is necessary to provide the driving commands to the motors and to detect the proprioceptive sensory data at sampling rates of the order of the millisecond, while sampling rates of the order of the minute are admissible at the task level.

With respect to this reference model of functional architecture, current industrial robot's control systems are not endowed with all the functions illustrated, because of both technology and cost limitations. In this regard, the task level is not implemented at all since there do not yet exist effective and reliable application software packages allowing support of the complex functions required at this level.

It is worth characterizing those functional levels of the reference models which are typically implemented in *advanced industrial robot's control systems*. The details of Fig. 6.2 show what follows:

- The modelling and sensory modules are always present at the lowest level, because of demanding requirements at the servo level for high dynamic performance robots to be employed even in relatively simple applications.
- At the primitive level, the modelling module is usually present while the sensory module is present only in a reduced number of applications that require robot interaction with a less structured environment.
- At the action level, the decision module is present only as an interpreter of the high-level commands imparted by the operator. All the task breakdown functions are entrusted to the operator, and thus the modelling and sensory module are absent at this level. Possible checking of action feasibility is moved down to the primitive level where a modelling module exists.

In view of the highly-structured reference model of functional architecture illustrated above, evolution of the control system towards more and more powerful capabilities is possible. In fact, one may foresee that information technology progress may allow the addition of hierarchically higher levels than the task level. These should functionally characterize complex tasks to be broken down into elementary tasks and yet, at an even higher level, missions to be broken down into complex tasks. A six-level hierarchical structure of the above kind has been proposed as the reference model for the functional architecture of the control system of a service robotic system for space applications (NASREM). In this framework, one may allocate the functions required to *advanced robotics* systems devoted to field or service applications, as discussed in Sect. 1.4.

6.2 Programming Environment

Programming a robotic system requires definition of a *programming environment* supported by suitable *languages*, which allows the operator imparting the task directions that the robot should execute. The programming environment is entrusted not only with the function of translating statements by means of a suitable language, but also with the function of checking correct execution of a task being executed by the robot. Therefore, robot programming environments, besides having some features in common with computer programming environments, present a number of issues related to the observation that program execution produces effects on the physical world. In other words, even if a very accurate description of physical reality is available in the programming environment, a number of situations will unavoidably occur which have not been or cannot be predicted.

As a consequence, a robot programming environment should be endowed with the following features:

- real-time operating system,
- world modelling,
- motion control,

- sensory data reading,
- interaction with physical system,
- error detection capability,
- recovery of correct operational functions,
- specific language structure.

Therefore, the requirements on a programming environment may naturally stem from the articulation into models of the preceding reference model of functional architecture. Such an environment will be clearly conditioned by the level of the architecture at which operator access is allowed. In the following, the requirements imposed on the programming environment by the functions respectively characterizing the sensory, modelling and decision modules are presented with reference to the hierarchical levels of the functional architecture.

Sensory data handling is the determining factor which qualifies a programming environment. At the servo level, real-time proprioceptive sensory data conditioning is required. At the primitive level, sensory data have to be expressed in the relevant reference frames. At the action level, geometric features of the objects interested to the action have to be extracted by high-level sensory data. At the task level, tools allowing recognition of the objects present in the scene are required.

The ability of *consulting knowledge models* is a support for a programming environment. At the servo level, on-line numerical computation of the models utilized by control algorithms is to be performed on the basis of sensory data. At the primitive level, coordinate transformations have to be operated. At the action level, it is crucial to have tools allowing system simulation and CAD modelling of elementary objects. At the task level, the programming environment should assume the functions of an expert system.

Decision functions play a fundamental role in a programming environment, since they allow the definition of the flow charts. At the servo level, on-line computation ability is required to generate the driving signals for the mechanical system. At the primitive level, logic conditioning is to be present. At the action level, process synchronization options should be available in order to implement nested loops, parallel computation and interrupt system. At the task level, the programming environment should allow management of concurrent processes, and it should be endowed with tools to test for, locate and remove mistakes from a program (debuggers) at a high-interactive level.

The evolution of programming environments has been conditioned by technology development of computer science. An analysis of this evolution leads to finding three generations of environments with respect to their functional characteristics, namely, *teaching-by-showing*, *robot-oriented programming*, and *object-oriented programming*. In the evolution of the environments, the next generation usually incorporates the functional characteristics of the previous generation.

This classification regards those features of the programming environment relative to the operator interface, and thus it has a direct correspondence with the hierarchical levels of the reference model of functional architecture. The functions associated with the servo level lead to understanding that a programming environment problem does not really exist for the operator. In fact, low-level programming concerns the use of traditional programming languages (Assembly, C) for development of real-time systems. The operator is only left with the possibility of intervening by means of simple command actuation (point-to-point, reset), reading of proprioceptive sensory data, and limited editing capability.

6.2.1 Teaching-by-Showing

The first generation has been characterized by programming techniques of *teaching-by-showing* type. The operator guides the manipulator manually or by means of a teach pendant along the desired motion path. During motion execution, the data read by joint position transducers are stored and thus they can be utilized later as references for the joint drive servos; in this way, the mechanical structure is capable of executing (playing back) the motion taught by a direct acquisition on the spot.

The programming environment does not allow implementation of logic conditioning and queuing, and thus the associated computational hardware plays elementary functions. The operator is not required to have special programming skill, and thus he/she can be a plant technician. The set-up of a working program obviously requires the robot to be available to the operator at the time of teaching, and thus the robot itself has to be taken off production. Typical applications that can be solved by this programming technique include spot welding, spray painting and, in general, simple palletizing.

With regard to the reference model of functional architecture, a programming environment based on the teaching-by-showing technique allows operator access at the primitive level.

The drawbacks of such an environment may be partially overcome by the adoption of simple programming languages which allow:

- the acquisition of a meaningful posture by teaching,
- the computation of the end-effector pose with respect to a reference frame, by means of a direct kinematics transformation,
- the assignment of a motion primitive and the trajectory parameters (usually, velocity as a percentage of the maximum velocity),
- the computation of the servo references, by means of an inverse kinematics transformation,
- the teaching sequences to be conditioned to the use of simple external sensors (presence of an object at the gripper),
- the correction of motion sequences by using simple text editors,
- simple connections to be made between subsets of elementary sequences.

Providing a teaching-by-showing environment with the the above-listed functions can be framed as an attempt to develop a structured programming environment.

6.2.2 Robot-oriented Programming

Following the advent of efficient low-cost computational means, *robot-oriented* programming environments have been developed. The need for interaction of the environment with physical reality has imposed integration of several functions, typical of high-level programming languages (BASIC, PASCAL), with those specifically required by robotic applications. In fact, many robot-oriented languages have retained the teaching-by-showing programming mode, in view of its natural characteristic of accurate interface with the physical world.

Since the general framework is that of a computer programming environment, two alternatives have been considered:

- to develop *ad hoc languages* for robotic applications,
- to develop robot *program libraries* supporting standard programming languages.

The current situation features the existence of numerous new proprietary languages, whereas it would be desirable to develop either robotic libraries to be used in the context of consolidated standards, or new general-purpose languages for industrial automation applications.

Robot-oriented languages are *structured programming* languages which incorporate high-level statements and have the characteristic of an interpreted language, in order to obtain an interactive environment allowing the programmer to check the execution of each source program statement before proceeding to the next one. Common features of such languages are:

- text editor,
- complex data representation structures,
- extensive use of predefined state variable,
- execution of matrix algebra operations,
- extensive use of symbolic representations for coordinate frames,
- possibility to specify the coordinated motion of more frames rigidly attached to objects by means of a single frame,
- inclusion of subroutines with data and parameter exchange,
- use of logic conditioning and queuing by means of flags,
- capability of parallel computing,
- functions of programmable logic controller (PLC).

With respect to the reference model of functional architecture, it can be recognized that a robot-oriented programming environment allows operator access at the action level.

In view of the structured language characteristic, the operator in this case should be an expert language programmer. Editing an application program may be performed off line, i.e., without physical availability of the robot to the operator; off-line programming demands a perfectly structured environment, though. A robotic system endowed with a robot-oriented programming language allows execution of complex applications where the robot is inserted in a work cell and interacts with other machines and devices to perform complex tasks, such as part assembly.

Finally, a programming environment that allows access at the task level of a reference model of functional architecture is characterized by an *object-oriented* language. Such an environment should have the capability of specifying a task by means of high-level statements allowing automatic execution of a number of actions on the objects present in the scene. Robot programming languages belonging to this generation are currently under development and thus they are not yet available on the market. They can be framed in the field of expert systems and artificial intelligence.

6.3 Hardware Architecture

The hierarchical structure of the functional architecture adopted as a reference model for an industrial robot's control system, together with its articulation into different functional modules, suggests hardware implementation which exploits distributed computational resources interconnected by means of suitable communication channels. To this end, it is worth recalling that the functions implemented in current control systems regard the three levels from servo to action, with a typically limited development of the functions implemented at the action level. At the servo and primitive levels, computational capabilities are required with demanding real-time constraints.

A general model of the *hardware architecture* for the control system of an industrial robot is illustrated in Fig. 6.3. In this figure, proper *boards* with autonomous computational capabilities have been associated with the functions indicated in the reference model of functional architecture of Fig. 9.2. The boards are connected to a *bus*, e.g., a VME bus, which allows support of the communication data flow; the bus bandwidth should be wide enough so as to satisfy the requirements imposed by real-time constraints.

The *system* board is typically a CPU endowed with:

- a microprocessor with mathematical coprocessor,
- a bootstrap EPROM memory,
- a local RAM memory,
- a RAM memory shared with the other boards through the bus,
- a number of serial and parallel ports interfacing the bus and the external world,
- counters, registers and timers,

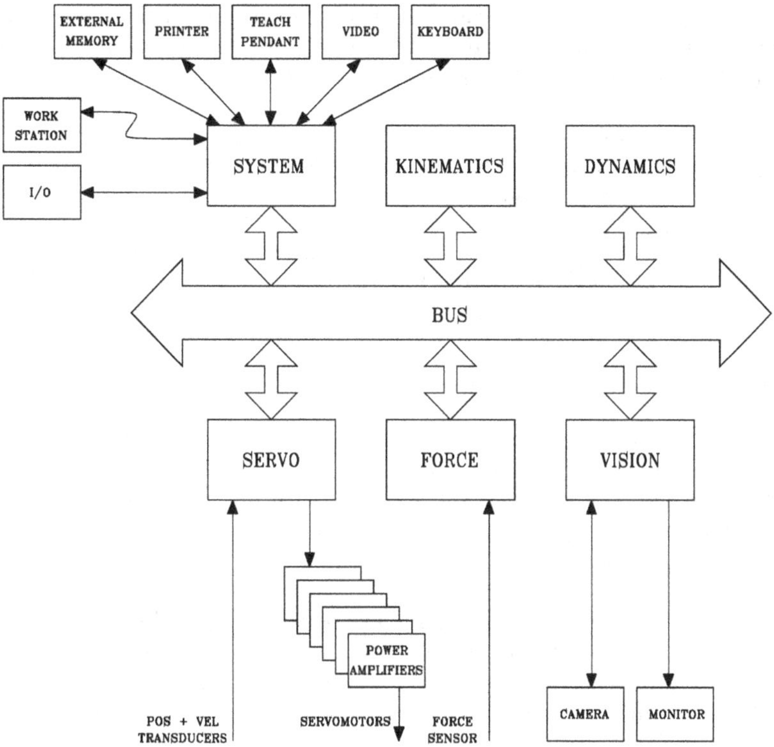

Fig. 6.3. General model of the hardware architecture of an industrial robot's control system

- an interrupt system.

The following functions are to be implemented in the system board:

- operator interface through teach pendant, keyboard, video and printer,
- interface with an external memory (hard disk) used to store data and application programs,
- interface with workstations and other control systems by means of a local communication network, e.g., Ethernet,
- I/O interface with peripheral devices in the working area, e.g., feeders, conveyors and ON/OFF sensors,
- system bootstrap,
- programming language interpreter,
- bus arbiter.

The other boards facing the bus may be endowed, besides the basic components of the system board, with a supplementary or alternative processor

(DSP, Transputer) for implementation of computationally demanding or dedicated functions. With reference to the architecture in Fig. 6.3, the following functions are implemented in the *kinematics* board:

- computation of motion primitives,
- computation of direct kinematics, inverse kinematics and Jacobian,
- test for trajectory feasibility,
- handling of kinematic redundancy.

The *dynamics* board is devoted to

- computation of inverse dynamics.

The *servo* board has the functions of:

- microinterpolation of references,
- computation of control algorithm,
- digital-to-analog conversion and interface with power amplifiers,
- handling of position and velocity transducer data,
- motion interruption in case of malfunction.

The remaining boards in the figure have been considered for the sake of an example to illustrate how the use of sensors may require local processing capabilities to retrieve significant information from the given data which can be effectively used in the sensory system. The *force* board performs the following operations:

- conditioning of data provided by the force sensor,
- representation of forces in a given coordinate frame.

The *vision* board is in charge of:

- processing data provided by the camera,
- extracting geometric features of the scene,
- localizing objects in given coordinate frames.

Although the boards face the same bus, the frequency at which data are exchanged needs not to be the same for each board. Those boards connected to the proprioceptive sensors indeed need to exchange date with the robot at the highest possible frequency (from 100 to 1000 Hz) to ensure high dynamic performance to motion control as well as to reveal end-effector contact in a very short time.

On the other hand, the kinematics and dynamics boards implement modelling functions and, as such, they do not require data update at a rate as high as that required by the servo board. In fact, manipulator postures do not vary appreciably in a very short time, at least with respect to typical operational velocities and/or accelerations of current industrial robots. Common sampling frequencies are in the range of 10 to 100 Hz.

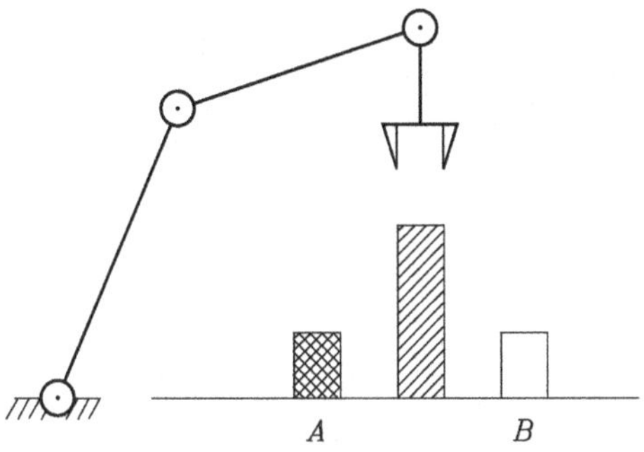

Fig. 6.4. Object pick-and-place task

Also the vision board does not require a high update rate, both because the scene is generally quasi-static, and because processing of interpretive functions are typically complex. Typical frequencies are in the range of 1 to 10 Hz.

In summary, the board access to the communication bus of a hardware control architecture may be performed according to a multirate logic which allows the solution of bus data overflow problems.

Bibliography

The features of robot control architectures are presented in [230, 25]. The NASREM architecture model has been proposed in [3]. For robot programming see [225, 139, 91]. More advanced control architectures based on artificial intelligence concepts are discussed in [8, 158].

Problems

6.1. With reference to the situation illustrated in Fig. 6.4, describe the sequence of actions required from the manipulator to pick up an object at location A and place it at location B.

6.2. For the situation of Problem 6.1, find the motion primitives in the cases of given via points and given path points.

6.3. The planar arm indicated in Fig. 6.5 is endowed with a wrist force sensor which allows the measurement of the relevant force and moment components for the execution of a peg-in-hole task. Draw the flow chart for writing a program to execute the described task.

6.4. A palletizing problem is represented in Fig. 6.6. Sixteen equal objects have to be loaded on the pallet. The manipulator's end-effector has to pick

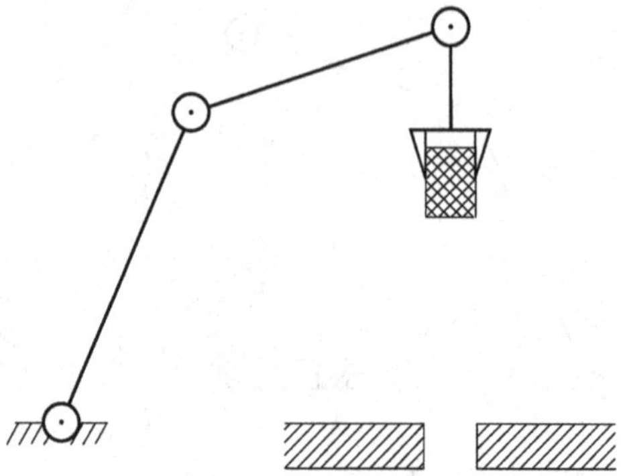

Fig. 6.5. Peg-in-hole task

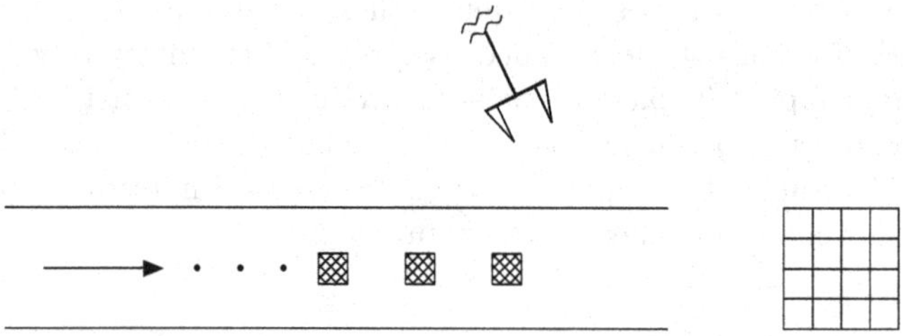

Fig. 6.6. Palletizing task of objects available on a conveyor

up the objects from a conveyor, whose feeding is commanded by the robot in such a way that the objects are always found in the same location to be picked. Write a PASCAL program to execute the task.

7

Dynamics

Derivation of the *dynamic model* of a manipulator plays an important role for simulation of motion, analysis of manipulator structures, and design of control algorithms. Simulating manipulator motion allows control strategies and motion planning techniques to be tested without the need to use a physically available system. The analysis of the dynamic model can be helpful for mechanical design of prototype arms. Computation of the forces and torques required for the execution of typical motions provides useful information for designing joints, transmissions and actuators. The goal of this chapter is to present two methods for derivation of the equations of motion of a manipulator in the *joint space*. The first method is based on the *Lagrange formulation* and is conceptually simple and systematic. The second method is based on the *Newton–Euler formulation* and yields the model in a recursive form; it is computationally more efficient since it exploits the typically open structure of the manipulator kinematic chain. Then, a technique for *dynamic parameter identification* is presented. Further, the problems of *direct dynamics* and *inverse dynamics* are formalized, and a technique for trajectory *dynamic scaling* is introduced, which adapts trajectory planning to the dynamic characteristics of the manipulator. The chapter ends with the derivation of the *dynamic model* of a manipulator in the *operational space* and the definition of the *dynamic manipulability ellipsoid*.

7.1 Lagrange Formulation

The dynamic model of a manipulator provides a description of the relationship between the joint actuator torques and the motion of the structure.

With *Lagrange formulation*, the equations of motion can be derived in a systematic way independently of the reference coordinate frame. Once a set of variables q_i, $i = 1, \ldots, n$, termed *generalized coordinates*, are chosen which effectively describe the link positions of an n-DOF manipulator, the

Lagrangian of the mechanical system can be defined as a function of the generalized coordinates:

$$\mathcal{L} = \mathcal{T} - \mathcal{U} \tag{7.1}$$

where $\mathcal{T}$ and $\mathcal{U}$ respectively denote the total *kinetic energy* and *potential energy* of the system.

The Lagrange equations are expressed by

$$\frac{d}{dt}\frac{\partial \mathcal{L}}{\partial \dot{q}_i} - \frac{\partial \mathcal{L}}{\partial q_i} = \xi_i \qquad i = 1, \ldots, n \tag{7.2}$$

where ξ_i is the *generalized force* associated with the generalized coordinate q_i. Equations (7.2) can be written in compact form as

$$\frac{d}{dt}\left(\frac{\partial \mathcal{L}}{\partial \dot{\boldsymbol{q}}}\right)^T - \left(\frac{\partial \mathcal{L}}{\partial \boldsymbol{q}}\right)^T = \boldsymbol{\xi} \tag{7.3}$$

where, for a manipulator with an open kinematic chain, the generalized coordinates are gathered in the vector of *joint variables* $\boldsymbol{q}$. The contributions to the generalized forces are given by the nonconservative forces, i.e., the joint actuator torques, the joint friction torques, as well as the joint torques induced by end-effector forces at the contact with the environment.[1]

The equations in (7.2) establish the relations existing between the generalized forces applied to the manipulator and the joint positions, velocities and accelerations. Hence, they allow the derivation of the dynamic model of the manipulator starting from the determination of kinetic energy and potential energy of the mechanical system.

Example 7.1

In order to understand the Lagrange formulation technique for deriving the dynamic model, consider again the simple case of the pendulum in Example 5.1. With reference to Fig. 5.8, let ϑ denote the angle with respect to the reference position of the body hanging down ($\vartheta = 0$). By choosing ϑ as the generalized coordinate, the kinetic energy of the system is given by

$$\mathcal{T} = \frac{1}{2}I\dot{\vartheta}^2 + \frac{1}{2}I_m k_r^2 \dot{\vartheta}^2.$$

The system potential energy, defined at less than a constant, is expressed by

$$\mathcal{U} = mg\ell(1 - \cos\vartheta).$$

Therefore, the Lagrangian of the system is

$$\mathcal{L} = \frac{1}{2}I\dot{\vartheta}^2 + \frac{1}{2}I_m k_r^2 \dot{\vartheta}^2 - mg\ell(1 - \cos\vartheta).$$

[1] The term *torque* is used as a synonym of joint *generalized force*.

Substituting this expression in the Lagrange equation in (7.2) yields

$$(I + I_m k_r^2)\ddot{\vartheta} + mg\ell \sin\vartheta = \xi.$$

The generalized force ξ is given by the contributions of the actuation torque τ at the joint and of the viscous friction torques $-F\dot{\vartheta}$ and $-F_m k_r^2 \dot{\vartheta}$, where the latter has been reported to the joint side. Hence, it is

$$\xi = \tau - F\dot{\vartheta} - F_m k_r^2 \dot{\vartheta}$$

leading to the complete dynamic model of the system as the second-order differential equation

$$(I + I_m k_r^2)\ddot{\vartheta} + (F + F_m k_r^2)\dot{\vartheta} + mg\ell \sin\vartheta = \tau.$$

It is easy to verify how this equation is equivalent to (5.25) when reported to the joint side.

7.1.1 Computation of Kinetic Energy

Consider a manipulator with n *rigid links*. The total kinetic energy is given by the sum of the contributions relative to the motion of each link and the contributions relative to the motion of each joint actuator:[2]

$$\mathcal{T} = \sum_{i=1}^{n} (\mathcal{T}_{\ell_i} + \mathcal{T}_{m_i}), \tag{7.4}$$

where $\mathcal{T}_{\ell_i}$ is the kinetic energy of Link i and $\mathcal{T}_{m_i}$ is the kinetic energy of the motor actuating Joint i.

The kinetic energy contribution of Link i is given by

$$\mathcal{T}_{\ell_i} = \frac{1}{2} \int_{V_{\ell_i}} \dot{\boldsymbol{p}}_i^{*T} \dot{\boldsymbol{p}}_i^* \rho dV, \tag{7.5}$$

where $\dot{\boldsymbol{p}}_i^*$ denotes the linear velocity vector and ρ is the density of the elementary particle of volume dV; V_{ℓ_i} is the volume of Link i.

Consider the position vector $\boldsymbol{p}_i^*$ of the elementary particle and the position vector $\boldsymbol{p}_{C_i}$ of the link centre of mass, both expressed in the *base frame*. One has

$$\boldsymbol{r}_i = [\, r_{ix} \quad r_{iy} \quad r_{iz} \,]^T = \boldsymbol{p}_i^* - \boldsymbol{p}_{\ell_i} \tag{7.6}$$

with

$$\boldsymbol{p}_{\ell_i} = \frac{1}{m_{\ell_i}} \int_{V_{\ell_i}} \boldsymbol{p}_i^* \rho dV \tag{7.7}$$

[2] Link 0 is fixed and thus gives no contribution.

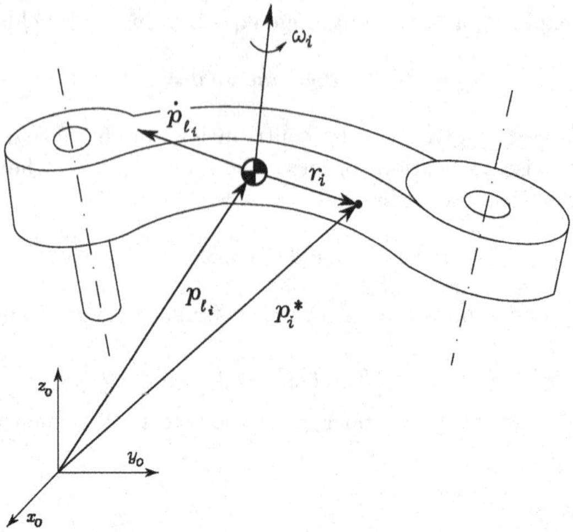

Fig. 7.1. Kinematic description of Link i for Lagrange formulation

where m_{ℓ_i} is the link mass. As a consequence, the link point velocity can be expressed as

$$\begin{aligned}\dot{\boldsymbol{p}}_i^* &= \dot{\boldsymbol{p}}_{\ell_i} + \boldsymbol{\omega}_i \times \boldsymbol{r}_i \\ &= \dot{\boldsymbol{p}}_{\ell_i} + \boldsymbol{S}(\boldsymbol{\omega}_i)\boldsymbol{r}_i,\end{aligned} \tag{7.8}$$

where $\dot{\boldsymbol{p}}_{\ell_i}$ is the linear velocity of the centre of mass and $\boldsymbol{\omega}_i$ is the angular velocity of the link (Fig. 7.1).

By substituting the velocity expression (7.8) into (7.5), it can be recognized that the kinetic energy of each link is formed by the following contributions.

Translational

The contribution is

$$\frac{1}{2}\int_{V_{\ell_i}} \dot{\boldsymbol{p}}_{\ell_i}^T \dot{\boldsymbol{p}}_{\ell_i} \rho dV = \frac{1}{2} m_{\ell_i} \dot{\boldsymbol{p}}_{\ell_i}^T \dot{\boldsymbol{p}}_{\ell_i}. \tag{7.9}$$

Mutual

The contribution is

$$2\left(\frac{1}{2}\int_{V_{\ell_i}} \dot{\boldsymbol{p}}_{\ell_i}^T \boldsymbol{S}(\boldsymbol{\omega}_i)\boldsymbol{r}_i \rho dV\right) = 2\left(\frac{1}{2}\dot{\boldsymbol{p}}_{\ell_i}^T \boldsymbol{S}(\boldsymbol{\omega}_i)\int_{V_{\ell_i}} (\boldsymbol{p}_i^* - \boldsymbol{p}_{\ell_i})\rho dV\right) = 0$$

since, by virtue of (7.7), it is

$$\int_{V_{\ell_i}} \boldsymbol{p}_i^* \rho dV = \boldsymbol{p}_{\ell_i} \int_{V_{\ell_i}} \rho dV.$$

Rotational

The contribution is

$$\frac{1}{2}\int_{V_{\ell_i}} \boldsymbol{r}_i^T \boldsymbol{S}^T(\boldsymbol{\omega}_i)\boldsymbol{S}(\boldsymbol{\omega}_i)\boldsymbol{r}_i \rho dV = \frac{1}{2}\boldsymbol{\omega}_i^T\left(\int_{V_{\ell_i}} \boldsymbol{S}^T(\boldsymbol{r}_i)\boldsymbol{S}(\boldsymbol{r}_i)\rho dV\right)\boldsymbol{\omega}_i$$

where the property $\boldsymbol{S}(\boldsymbol{\omega}_i)\boldsymbol{r}_i = -\boldsymbol{S}(\boldsymbol{r}_i)\boldsymbol{\omega}_i$ has been exploited. In view of the expression of the matrix operator $\boldsymbol{S}(\cdot)$

$$\boldsymbol{S}(\boldsymbol{r}_i) = \begin{bmatrix} 0 & -r_{iz} & r_{iy} \\ r_{iz} & 0 & -r_{ix} \\ -r_{iy} & r_{ix} & 0 \end{bmatrix},$$

it is

$$\frac{1}{2}\int_{V_{\ell_i}} \boldsymbol{r}_i^T \boldsymbol{S}^T(\boldsymbol{\omega}_i)\boldsymbol{S}(\boldsymbol{\omega}_i)\boldsymbol{r}_i \rho dV = \frac{1}{2}\boldsymbol{\omega}_i^T \boldsymbol{I}_{\ell_i}\boldsymbol{\omega}_i. \tag{7.10}$$

The matrix

$$\boldsymbol{I}_{\ell_i} = \begin{bmatrix} \int(r_{iy}^2 + r_{iz}^2)\rho dV & -\int r_{ix}r_{iy}\rho dV & -\int r_{ix}r_{iz}\rho dV \\ * & \int(r_{ix}^2 + r_{iz}^2)\rho dV & -\int r_{iy}r_{iz}\rho dV \\ * & * & \int(r_{ix}^2 + r_{iy}^2)\rho dV \end{bmatrix} \tag{7.11}$$

$$= \begin{bmatrix} I_{\ell_i xx} & -I_{\ell_i xy} & -I_{\ell_i xz} \\ * & I_{\ell_i yy} & -I_{\ell_i yz} \\ * & * & I_{\ell_i zz} \end{bmatrix}.$$

is symmetric[3] and represents the *inertia tensor* relative to the centre of mass of Link i when expressed in the base frame. Notice that the position of Link i depends on the manipulator configuration and thus the inertia tensor, when expressed in the base frame, is configuration-dependent. If the angular velocity of Link i is expressed with reference to a frame attached to the link (as in the Denavit–Hartenberg convention), it is

$$\boldsymbol{\omega}_i^i = \boldsymbol{R}_i^T \boldsymbol{\omega}_i$$

where $\boldsymbol{R}_i$ is the rotation matrix from Link i frame to the base frame. When referred to the link frame, the inertia tensor is constant. Let $\boldsymbol{I}_{\ell_i}^i$ denote such tensor; then it is easy to verify the following relation:

$$\boldsymbol{I}_{\ell_i} = \boldsymbol{R}_i \boldsymbol{I}_{\ell_i}^i \boldsymbol{R}_i^T. \tag{7.12}$$

If the axes of Link i frame coincide with the central axes of inertia, then the inertia products are null and the inertia tensor relative to the centre of mass is a diagonal matrix.

[3] The symbol '$*$' has been used to avoid rewriting the symmetric elements.

By summing the translational and rotational contributions (7.9) and (7.10), the kinetic energy of Link i is

$$\mathcal{T}_{\ell_i} = \frac{1}{2} m_{\ell_i} \dot{\boldsymbol{p}}_{\ell_i}^T \dot{\boldsymbol{p}}_{\ell_i} + \frac{1}{2} \boldsymbol{\omega}_i^T \boldsymbol{R}_i \boldsymbol{I}_{\ell_i}^i \boldsymbol{R}_i^T \boldsymbol{\omega}_i. \tag{7.13}$$

At this point, it is necessary to express the kinetic energy as a function of the generalized coordinates of the system, that are the joint variables. To this end, the geometric method for Jacobian computation can be applied to the intermediate link other than the end-effector, yielding

$$\dot{\boldsymbol{p}}_{\ell_i} = \boldsymbol{j}_{P1}^{(\ell_i)} \dot{q}_1 + \ldots + \boldsymbol{j}_{Pi}^{(\ell_i)} \dot{q}_i = \boldsymbol{J}_P^{(\ell_i)} \dot{\boldsymbol{q}} \tag{7.14}$$

$$\boldsymbol{\omega}_i = \boldsymbol{j}_{O1}^{(\ell_i)} \dot{q}_1 + \ldots + \boldsymbol{j}_{Oi}^{(\ell_i)} \dot{q}_i = \boldsymbol{J}_O^{(\ell_i)} \dot{\boldsymbol{q}}, \tag{7.15}$$

where the contributions of the Jacobian columns relative to the joint velocities have been taken into account up to current Link i. The Jacobians to consider are then:

$$\boldsymbol{J}_P^{(\ell_i)} = \begin{bmatrix} \boldsymbol{j}_{P1}^{(\ell_i)} & \ldots & \boldsymbol{j}_{Pi}^{(\ell_i)} & \boldsymbol{0} & \ldots & \boldsymbol{0} \end{bmatrix} \tag{7.16}$$

$$\boldsymbol{J}_O^{(\ell_i)} = \begin{bmatrix} \boldsymbol{j}_{O1}^{(\ell_i)} & \ldots & \boldsymbol{j}_{Oi}^{(\ell_i)} & \boldsymbol{0} & \ldots & \boldsymbol{0} \end{bmatrix}; \tag{7.17}$$

the columns of the matrices in (7.16) and (7.17) can be computed according to (3.30), giving

$$\boldsymbol{j}_{Pj}^{(\ell_i)} = \begin{cases} \boldsymbol{z}_{j-1} & \text{for a } \textit{prismatic} \text{ joint} \\ \boldsymbol{z}_{j-1} \times (\boldsymbol{p}_{\ell_i} - \boldsymbol{p}_{j-1}) & \text{for a } \textit{revolute} \text{ joint} \end{cases} \tag{7.18}$$

$$\boldsymbol{j}_{Oj}^{(\ell_i)} = \begin{cases} \boldsymbol{0} & \text{for a } \textit{prismatic} \text{ joint} \\ \boldsymbol{z}_{j-1} & \text{for a } \textit{revolute} \text{ joint.} \end{cases} \tag{7.19}$$

where $\boldsymbol{p}_{j-1}$ is the position vector of the origin of Frame $j-1$ and $\boldsymbol{z}_{j-1}$ is the unit vector of axis z of Frame $j-1$. It follows that the kinetic energy of Link i in (7.13) can be written as

$$\mathcal{T}_{\ell_i} = \frac{1}{2} m_{\ell_i} \dot{\boldsymbol{q}}^T \boldsymbol{J}_P^{(\ell_i)T} \boldsymbol{J}_P^{(\ell_i)} \dot{\boldsymbol{q}} + \frac{1}{2} \dot{\boldsymbol{q}}^T \boldsymbol{J}_O^{(\ell_i)T} \boldsymbol{R}_i \boldsymbol{I}_{\ell_i}^i \boldsymbol{R}_i^T \boldsymbol{J}_O^{(\ell_i)} \dot{\boldsymbol{q}}. \tag{7.20}$$

The kinetic energy contribution of the motor of Joint i can be computed in a formally analogous way to that of the link. Consider the typical case of rotary electric motors (that can actuate both revolute and prismatic joints by means of suitable transmissions). It can be assumed that the contribution of the fixed part (stator) is included in that of the link on which such motor is located, and thus the sole contribution of the rotor is to be computed.

With reference to Fig. 7.2, the motor of Joint i is assumed to be located on Link $i-1$. In practice, in the design of the mechanical structure of an open kinematic chain manipulator one attempts to locate the motors as close as possible to the base of the manipulator so as to lighten the dynamic load of

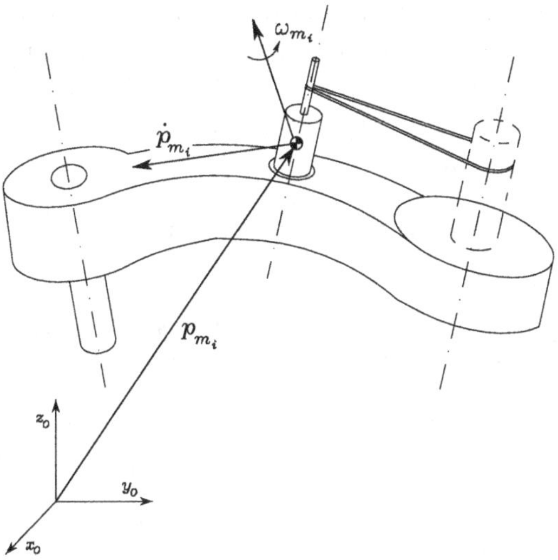

Fig. 7.2. Kinematic description of Motor i

the first joints of the chain. The joint actuator torques are delivered by the motors by means of mechanical transmissions (gears).[4] The contribution of the gears to the kinetic energy can be suitably included in that of the motor. It is assumed that no induced motion occurs, i.e., the motion of Joint i does not actuate the motion of other joints.

The kinetic energy of Rotor i can be written as

$$\mathcal{T}_{m_i} = \frac{1}{2}m_{m_i}\dot{\boldsymbol{p}}_{m_i}^T\dot{\boldsymbol{p}}_{m_i} + \frac{1}{2}\boldsymbol{\omega}_{m_i}^T\boldsymbol{I}_{m_i}\boldsymbol{\omega}_{m_i}, \tag{7.21}$$

where m_{m_i} is the mass of the rotor, $\dot{\boldsymbol{p}}_{m_i}$ denotes the linear velocity of the centre of mass of the rotor, $\boldsymbol{I}_{m_i}$ is the inertia tensor of the rotor relative to its centre of mass, and $\boldsymbol{\omega}_{m_i}$ denotes the angular velocity of the rotor.

Let ϑ_{m_i} denote the angular position of the rotor. On the assumption of a *rigid transmission*, one has

$$k_{ri}\dot{q}_i = \dot{\vartheta}_{m_i} \tag{7.22}$$

where k_{ri} is the gear reduction ratio. Notice that, in the case of actuation of a prismatic joint, the gear reduction ratio is a dimensional quantity.

According to the angular velocity composition rule (3.18) and the relation (7.22), the total angular velocity of the rotor is

$$\boldsymbol{\omega}_{m_i} = \boldsymbol{\omega}_{i-1} + k_{ri}\dot{q}_i\boldsymbol{z}_{m_i} \tag{7.23}$$

where $\boldsymbol{\omega}_{i-1}$ is the angular velocity of Link $i-1$ on which the motor is located, and $\boldsymbol{z}_{m_i}$ denotes the unit vector along the rotor axis.

[4] Alternatively, the joints may be actuated by torque motors directly coupled to the rotation axis without gears.

To express the rotor kinetic energy as a function of the joint variables, it is worth expressing the linear velocity of the rotor centre of mass — similarly to (7.14) — as

$$\dot{\boldsymbol{p}}_{m_i} = \boldsymbol{J}_P^{(m_i)} \dot{\boldsymbol{q}}. \tag{7.24}$$

The Jacobian to compute is then

$$\boldsymbol{J}_P^{(m_i)} = \begin{bmatrix} \boldsymbol{j}_{P1}^{(m_i)} & \dots & \boldsymbol{j}_{P,i-1}^{(m_i)} & \boldsymbol{0} & \dots & \boldsymbol{0} \end{bmatrix} \tag{7.25}$$

whose columns are given by

$$\boldsymbol{j}_{Pj}^{(m_i)} = \begin{cases} \boldsymbol{z}_{j-1} & \text{for a } \textit{prismatic} \text{ joint} \\ \boldsymbol{z}_{j-1} \times (\boldsymbol{p}_{m_i} - \boldsymbol{p}_{j-1}) & \text{for a } \textit{revolute} \text{ joint} \end{cases} \tag{7.26}$$

where $\boldsymbol{p}_{j-1}$ is the position vector of the origin of Frame $j-1$. Notice that $\boldsymbol{j}_{Pi}^{(m_i)} = \boldsymbol{0}$ in (7.25), since the centre of mass of the rotor has been taken along its axis of rotation.

The angular velocity in (7.23) can be expressed as a function of the joint variables, i.e.,

$$\boldsymbol{\omega}_{m_i} = \boldsymbol{J}_O^{(m_i)} \dot{\boldsymbol{q}}. \tag{7.27}$$

The Jacobian to compute is then

$$\boldsymbol{J}_O^{(m_i)} = \begin{bmatrix} \boldsymbol{j}_{O1}^{(m_i)} & \dots & \boldsymbol{j}_{O,i-1}^{(m_i)} & \boldsymbol{j}_{Oi}^{(m_i)} & \boldsymbol{0} & \dots & \boldsymbol{0} \end{bmatrix} \tag{7.28}$$

whose columns, in view of (7.23), (7.15), are respectively given by

$$\boldsymbol{j}_{Oj}^{(m_i)} = \begin{cases} \boldsymbol{j}_{Oj}^{(\ell_i)} & j = 1, \dots, i-1 \\ k_{ri} \boldsymbol{z}_{m_i} & j = i. \end{cases} \tag{7.29}$$

To compute the second relation in (7.29), it is sufficient to know the components of the unit vector of the rotor rotation axis $\boldsymbol{z}_{m_i}$ with respect to the base frame. Hence, the kinetic energy of Rotor i can be written as

$$\mathcal{T}_{m_i} = \frac{1}{2} m_{m_i} \dot{\boldsymbol{q}}^T \boldsymbol{J}_P^{(m_i)T} \boldsymbol{J}_P^{(m_i)} \dot{\boldsymbol{q}} + \frac{1}{2} \dot{\boldsymbol{q}}^T \boldsymbol{J}_O^{(m_i)T} \boldsymbol{R}_{m_i} \boldsymbol{I}_{m_i}^{m_i} \boldsymbol{R}_{m_i}^T \boldsymbol{J}_O^{(m_i)} \dot{\boldsymbol{q}}. \tag{7.30}$$

Finally, by summing the various contributions relative to the single links (7.20) and single rotors (7.30) as in (7.4), the total kinetic energy of the manipulator with actuators is given by the quadratic form

$$\mathcal{T} = \frac{1}{2} \sum_{i=1}^{n} \sum_{j=1}^{n} b_{ij}(\boldsymbol{q}) \dot{q}_i \dot{q}_j = \frac{1}{2} \dot{\boldsymbol{q}}^T \boldsymbol{B}(\boldsymbol{q}) \dot{\boldsymbol{q}} \tag{7.31}$$

where

$$\begin{aligned} \boldsymbol{B}(\boldsymbol{q}) = \sum_{i=1}^{n} \Big(& m_{\ell_i} \boldsymbol{J}_P^{(\ell_i)T} \boldsymbol{J}_P^{(\ell_i)} + \boldsymbol{J}_O^{(\ell_i)T} \boldsymbol{R}_i \boldsymbol{I}_{\ell_i}^i \boldsymbol{R}_i^T \boldsymbol{J}_O^{(\ell_i)} \\ & + m_{m_i} \boldsymbol{J}_P^{(m_i)T} \boldsymbol{J}_P^{(m_i)} + \boldsymbol{J}_O^{(m_i)T} \boldsymbol{R}_{m_i} \boldsymbol{I}_{m_i}^{m_i} \boldsymbol{R}_{m_i}^T \boldsymbol{J}_O^{(m_i)} \Big) \end{aligned} \tag{7.32}$$

is the $(n \times n)$ *inertia matrix* which is:

- *symmetric*,
- *positive definite*,
- (in general) *configuration-dependent*.

7.1.2 Computation of Potential Energy

As done for kinetic energy, the potential energy stored in the manipulator is given by the sum of the contributions relative to each link as well as to each rotor:

$$\mathcal{U} = \sum_{i=1}^{n} (\mathcal{U}_{\ell_i} + \mathcal{U}_{m_i}). \tag{7.33}$$

On the assumption of *rigid links*, the contribution due only to gravitational forces[5] is expressed by

$$\mathcal{U}_{\ell_i} = -\int_{V_{\ell_i}} \boldsymbol{g}_0^T \boldsymbol{p}_i^* \rho dV = -m_{\ell_i} \boldsymbol{g}_0^T \boldsymbol{p}_{\ell_i} \tag{7.34}$$

where $\boldsymbol{g}_0$ is the gravity acceleration vector in the base frame (e.g., $\boldsymbol{g}_0 = [0 \quad 0 \quad -g]^T$ if z is the vertical axis), and (7.7) has been utilized for the coordinates of the centre of mass of Link i. As regards the contribution of Rotor i, similarly to (7.34), one has

$$\mathcal{U}_{m_i} = -m_{m_i} \boldsymbol{g}_0^T \boldsymbol{p}_{m_i}. \tag{7.35}$$

By substituting (7.34), (7.35) into (7.33), the *potential energy* is given by

$$\mathcal{U} = -\sum_{i=1}^{n} (m_{\ell_i} \boldsymbol{g}_0^T \boldsymbol{p}_{\ell_i} + m_{m_i} \boldsymbol{g}_0^T \boldsymbol{p}_{m_i}) \tag{7.36}$$

which reveals that potential energy, through the vectors $\boldsymbol{p}_{\ell_i}$ and $\boldsymbol{p}_{m_i}$ is a function only of the joint variables $\boldsymbol{q}$, and *not* of the joint velocities $\dot{\boldsymbol{q}}$.

7.1.3 Equations of Motion

Having computed the total kinetic and potential energy of the system as in (7.31), (7.36), the Lagrangian (7.1) for the manipulator can be written as

$$\mathcal{L}(\boldsymbol{q}, \dot{\boldsymbol{q}}) = \mathcal{T}(\boldsymbol{q}, \dot{\boldsymbol{q}}) - \mathcal{U}(\boldsymbol{q}). \tag{7.37}$$

Taking the derivatives required by Lagrange equations in (7.3) and recalling that U does not depend on $\dot{\boldsymbol{q}}$ yields

$$\boldsymbol{B}(\boldsymbol{q})\ddot{q} + \boldsymbol{n}(\boldsymbol{q}, \dot{\boldsymbol{q}}) = \boldsymbol{\xi} \tag{7.38}$$

[5] In the case of link flexibility, one would have an additional contribution due to elastic forces.

where

$$n(q, \dot{q}) = \dot{B}(q)\dot{q} - \frac{1}{2}\left(\frac{\partial}{\partial q}\left(\dot{q}^T B(q)\dot{q}\right)\right)^T + \left(\frac{\partial \mathcal{U}(q)}{\partial q}\right)^T.$$

In detail, noticing that $\mathcal{U}$ in (7.36) does not depend on $\dot{q}$ and accounting for (7.31) yields

$$\begin{aligned}\frac{d}{dt}\left(\frac{\partial \mathcal{L}}{\partial \dot{q}_i}\right) = \frac{d}{dt}\left(\frac{\partial \mathcal{T}}{\partial \dot{q}_i}\right) &= \sum_{j=1}^{n} b_{ij}(q)\ddot{q}_j + \sum_{j=1}^{n} \frac{db_{ij}(q)}{dt}\dot{q}_j \\ &= \sum_{j=1}^{n} b_{ij}(q)\ddot{q}_j + \sum_{j=1}^{n}\sum_{k=1}^{n} \frac{\partial b_{ij}(q)}{\partial q_k}\dot{q}_k\dot{q}_j\end{aligned}$$

and

$$\frac{\partial \mathcal{T}}{\partial q_i} = \frac{1}{2}\sum_{j=1}^{n}\sum_{k=1}^{n} \frac{\partial b_{jk}(q)}{\partial q_i}\dot{q}_k\dot{q}_j$$

where the indices of summation have been conveniently switched. Further, in view of (7.14), (7.24), it is

$$\begin{aligned}\frac{\partial \mathcal{U}}{\partial q_i} &= -\sum_{j=1}^{n}\left(m_{\ell_j} g_0^T \frac{\partial p_{\ell_j}}{\partial q_i} + m_{m_j} g_0^T \frac{\partial p_{m_j}}{\partial q_i}\right) \\ &= -\sum_{j=1}^{n}\left(m_{\ell_j} g_0^T j_{Pi}^{(\ell_j)}(q) + m_{m_j} g_0^T j_{Pi}^{(m_j)}(q)\right) = g_i(q)\end{aligned} \tag{7.39}$$

where, again, the index of summation has been changed.

As a result, the equations of motion are

$$\sum_{j=1}^{n} b_{ij}(q)\ddot{q}_j + \sum_{j=1}^{n}\sum_{k=1}^{n} h_{ijk}(q)\dot{q}_k\dot{q}_j + g_i(q) = \xi_i \qquad i = 1, \ldots, n. \tag{7.40}$$

where

$$h_{ijk} = \frac{\partial b_{ij}}{\partial q_k} - \frac{1}{2}\frac{\partial b_{jk}}{\partial q_i}. \tag{7.41}$$

A physical interpretation of (7.40) reveals that:

- For the *acceleration terms*:
  - The coefficient b_{ii} represents the moment of inertia at Joint i axis, in the current manipulator configuration, when the other joints are blocked.
  - The coefficient b_{ij} accounts for the effect of acceleration of Joint j on Joint j.
- For the *quadratic velocity terms*:
  - The term $h_{ijj}\dot{q}_j^2$ represents the *centrifugal* effect induced on Joint i by velocity of Joint j; notice that $h_{iii} = 0$, since $\partial b_{ii}/\partial q_i = 0$.

   - The term $h_{ijk}\dot{q}_j\dot{q}_k$ represents the *Coriolis* effect induced on Joint i by velocities of Joints j and k.
- For the *configuration-dependent terms*:
   - The term g_i represents the moment generated at Joint i axis of the manipulator, in the current configuration, by the presence of gravity.

Some joint dynamic couplings, e.g., coefficients b_{ij} and h_{ijk}, may be reduced or zeroed when designing the structure, so as to simplify the control problem.

Regarding the nonconservative forces doing work at the manipulator joints, these are given by the *actuation torques* $\boldsymbol{\tau}$ minus the *viscous friction* torques $\boldsymbol{F}_v\dot{\boldsymbol{q}}$ and the static friction torques $\boldsymbol{f}_s(\boldsymbol{q},\dot{\boldsymbol{q}})$: $\boldsymbol{F}_v$ denotes the $(n \times n)$ diagonal matrix of viscous friction coefficients. As a simplified model of static friction torques, one may consider the *Coulomb friction* torques $\boldsymbol{F}_s\,\mathbf{sgn}\,(\dot{\boldsymbol{q}})$, where $\boldsymbol{F}_s$ is an $(n \times n)$ diagonal matrix and $\mathbf{sgn}\,(\dot{\boldsymbol{q}})$ denotes the $(n \times 1)$ vector whose components are given by the sign functions of the single joint velocities.

If the manipulator's end-effector is in contact with an environment, a portion of the actuation torques is used to balance the torques induced at the joints by the contact forces. According to a relation formally analogous to (3.111), such torques are given by $\boldsymbol{J}^T(\boldsymbol{q})\boldsymbol{h}_e$ where $\boldsymbol{h}_e$ denotes the vector of force and moment exerted by the end-effector on the environment.

In summary, the equations of motion (7.38) can be rewritten in the compact matrix form which represents the *joint space dynamic model*:

$$\boldsymbol{B}(\boldsymbol{q})\ddot{\boldsymbol{q}} + \boldsymbol{C}(\boldsymbol{q},\dot{\boldsymbol{q}})\dot{\boldsymbol{q}} + \boldsymbol{F}_v\dot{\boldsymbol{q}} + \boldsymbol{F}_s\,\mathbf{sgn}\,(\dot{\boldsymbol{q}}) + \boldsymbol{g}(\boldsymbol{q}) = \boldsymbol{\tau} - \boldsymbol{J}^T(\boldsymbol{q})\boldsymbol{h}_e \tag{7.42}$$

where $\boldsymbol{C}$ is a suitable $(n \times n)$ matrix such that its elements c_{ij} satisfy the equation

$$\sum_{j=1}^{n} c_{ij}\dot{q}_j = \sum_{j=1}^{n}\sum_{k=1}^{n} h_{ijk}\dot{q}_k\dot{q}_j. \tag{7.43}$$

7.2 Notable Properties of Dynamic Model

In the following, two *notable properties* of the dynamic model are presented which will be useful for dynamic parameter identification as well as for deriving control algorithms.

7.2.1 Skew-symmetry of Matrix $\dot{\boldsymbol{B}} - 2\boldsymbol{C}$

The choice of the matrix $\boldsymbol{C}$ is not unique, since there exist several matrices $\boldsymbol{C}$ whose elements satisfy (7.43). A particular choice can be obtained by elaborating the term on the right-hand side of (7.43) and accounting for the expressions of the coefficients h_{ijk} in (7.41). To this end, one has

$$\begin{aligned}\sum_{j=1}^{n} c_{ij}\dot{q}_j &= \sum_{j=1}^{n}\sum_{k=1}^{n} h_{ijk}\dot{q}_k\dot{q}_j \\ &= \sum_{j=1}^{n}\sum_{k=1}^{n}\left(\frac{\partial b_{ij}}{\partial q_k} - \frac{1}{2}\frac{\partial b_{jk}}{\partial q_i}\right)\dot{q}_k\dot{q}_j.\end{aligned}$$

Splitting the first term on the right-hand side by an opportune switch of summation between j and k yields

$$\sum_{j=1}^{n} c_{ij}\dot{q}_j = \frac{1}{2}\sum_{j=1}^{n}\sum_{k=1}^{n}\frac{\partial b_{ij}}{\partial q_k}\dot{q}_k\dot{q}_j + \frac{1}{2}\sum_{j=1}^{n}\sum_{k=1}^{n}\left(\frac{\partial b_{ik}}{\partial q_j} - \frac{\partial b_{jk}}{\partial q_i}\right)\dot{q}_k\dot{q}_j.$$

As a consequence, the generic element of $\boldsymbol{C}$ is

$$c_{ij} = \sum_{k=1}^{n} c_{ijk}\dot{q}_k \tag{7.44}$$

where the coefficients

$$c_{ijk} = \frac{1}{2}\left(\frac{\partial b_{ij}}{\partial q_k} + \frac{\partial b_{ik}}{\partial q_j} - \frac{\partial b_{jk}}{\partial q_i}\right) \tag{7.45}$$

are termed *Christoffel symbols of the first type*. Notice that, in view of the symmetry of $\boldsymbol{B}$, it is

$$c_{ijk} = c_{ikj}. \tag{7.46}$$

This choice for the matrix $\boldsymbol{C}$ leads to deriving the following notable property of the equations of motion (7.42). The matrix

$$\boldsymbol{N}(\boldsymbol{q},\dot{\boldsymbol{q}}) = \dot{\boldsymbol{B}}(\boldsymbol{q}) - 2\boldsymbol{C}(\boldsymbol{q},\dot{\boldsymbol{q}}) \tag{7.47}$$

is *skew-symmetric*; that is, given any $(n \times 1)$ vector $\boldsymbol{w}$, the following relation holds:

$$\boldsymbol{w}^T\boldsymbol{N}(\boldsymbol{q},\dot{\boldsymbol{q}})\boldsymbol{w} = 0. \tag{7.48}$$

In fact, substituting the coefficient (7.45) into (7.44) gives

$$\begin{aligned}c_{ij} &= \frac{1}{2}\sum_{k=1}^{n}\frac{\partial b_{ij}}{\partial q_k}\dot{q}_k + \frac{1}{2}\sum_{k=1}^{n}\left(\frac{\partial b_{ik}}{\partial q_j} - \frac{\partial b_{jk}}{\partial q_i}\right)\dot{q}_k \\ &= \frac{1}{2}\dot{b}_{ij} + \frac{1}{2}\sum_{k=1}^{n}\left(\frac{\partial b_{ik}}{\partial q_j} - \frac{\partial b_{jk}}{\partial q_i}\right)\dot{q}_k\end{aligned}$$

and then the expression of the generic element of the matrix $\boldsymbol{N}$ in (7.47) is

$$n_{ij} = \dot{b}_{ij} - 2c_{ij} = \sum_{k=1}^{n}\left(\frac{\partial b_{jk}}{\partial q_i} - \frac{\partial b_{ik}}{\partial q_j}\right)\dot{q}_k.$$

The result follows by observing that

$$n_{ij} = -n_{ji}.$$

An interesting property which is a direct implication of the skew-symmetry of $\boldsymbol{N}(\boldsymbol{q}, \dot{\boldsymbol{q}})$ is that, by setting $\boldsymbol{w} = \dot{\boldsymbol{q}}$,

$$\dot{\boldsymbol{q}}^T \boldsymbol{N}(\boldsymbol{q}, \dot{\boldsymbol{q}})\dot{\boldsymbol{q}} = 0; \tag{7.49}$$

notice that (7.49) does not imply (7.48), since $\boldsymbol{N}$ is a function of $\dot{\boldsymbol{q}}$, too.

It can be shown that (7.49) holds for any choice of the matrix $\boldsymbol{C}$, since it is a result of the principle of conservation of energy (*Hamilton*). By virtue of this principle, the total time derivative of kinetic energy is balanced by the power generated by all the forces acting on the manipulator joints. For the mechanical system at issue, one may write

$$\frac{1}{2}\frac{d}{dt}\left(\dot{\boldsymbol{q}}^T \boldsymbol{B}(\boldsymbol{q})\dot{\boldsymbol{q}}\right) = \dot{\boldsymbol{q}}^T\left(\boldsymbol{\tau} - \boldsymbol{F}_v\dot{\boldsymbol{q}} - \boldsymbol{F}_s\,\mathbf{sgn}\,(\dot{\boldsymbol{q}}) - \boldsymbol{g}(\boldsymbol{q}) - \boldsymbol{J}^T(\boldsymbol{q})\boldsymbol{h}_e\right). \tag{7.50}$$

Taking the derivative on the left-hand side of (7.50) gives

$$\frac{1}{2}\dot{\boldsymbol{q}}^T \dot{\boldsymbol{B}}(\boldsymbol{q})\dot{\boldsymbol{q}} + \dot{\boldsymbol{q}}^T \boldsymbol{B}(\boldsymbol{q})\ddot{\boldsymbol{q}}$$

and substituting the expression of $\boldsymbol{B}(\boldsymbol{q})\ddot{\boldsymbol{q}}$ in (7.42) yields

$$\begin{aligned}\frac{1}{2}\frac{d}{dt}\left(\dot{\boldsymbol{q}}^T \boldsymbol{B}(\boldsymbol{q})\dot{\boldsymbol{q}}\right) &= \frac{1}{2}\dot{\boldsymbol{q}}^T\left(\dot{\boldsymbol{B}}(\boldsymbol{q}) - 2\boldsymbol{C}(\boldsymbol{q}, \dot{\boldsymbol{q}})\right)\dot{\boldsymbol{q}} \\ &\quad + \dot{\boldsymbol{q}}^T\left(\boldsymbol{\tau} - \boldsymbol{F}_v\dot{\boldsymbol{q}} - \boldsymbol{F}_s\,\mathbf{sgn}\,(\dot{\boldsymbol{q}}) - \boldsymbol{g}(\boldsymbol{q}) - \boldsymbol{J}^T(\boldsymbol{q})\boldsymbol{h}_e\right).\end{aligned} \tag{7.51}$$

A direct comparison of the right-hand sides of (7.50) and (7.51) leads to the result established by (7.49).

To summarize, the relation (7.49) holds for any choice of the matrix $\boldsymbol{C}$, since it is a direct consequence of the physical properties of the system, whereas the relation (7.48) holds only for the particular choice of the elements of $\boldsymbol{C}$ as in (7.44), (7.45).

7.2.2 Linearity in the Dynamic Parameters

An important property of the dynamic model is the *linearity* with respect to the *dynamic parameters* characterizing the manipulator links and rotors.

In order to determine such parameters, it is worth associating the kinetic and potential energy contributions of each rotor with those of the link on which it is located. Hence, by considering the union of Link i and Rotor $i+1$ (*augmented Link* i), the kinetic energy contribution is given by

$$\mathcal{T}_i = \mathcal{T}_{\ell_i} + \mathcal{T}_{m_{i+1}} \tag{7.52}$$

where

$$\mathcal{T}_{\ell_i} = \frac{1}{2} m_{\ell i} \dot{\boldsymbol{p}}_{\ell_i}^T \dot{\boldsymbol{p}}_{\ell_i} + \frac{1}{2} \boldsymbol{\omega}_i^T \boldsymbol{I}_{\ell_i} \boldsymbol{\omega}_i \tag{7.53}$$

and

$$\mathcal{T}_{m_{i+1}} = \frac{1}{2} m_{m_{i+1}} \dot{\boldsymbol{p}}_{m_{i+1}}^T \dot{\boldsymbol{p}}_{m_{i+1}} + \frac{1}{2} \boldsymbol{\omega}_{m_{i+1}}^T \boldsymbol{I}_{m_{i+1}} \boldsymbol{\omega}_{m_{i+1}}. \tag{7.54}$$

With reference to the centre of mass of the augmented link, the linear velocities of the link and rotor can be expressed according to (3.26) as

$$\dot{\boldsymbol{p}}_{\ell_i} = \dot{\boldsymbol{p}}_{C_i} + \boldsymbol{\omega}_i \times \boldsymbol{r}_{C_i,\ell_i} \tag{7.55}$$

$$\dot{\boldsymbol{p}}_{m_{i+1}} = \dot{\boldsymbol{p}}_{C_i} + \boldsymbol{\omega}_i \times \boldsymbol{r}_{C_i,m_{i+1}} \tag{7.56}$$

with

$$\boldsymbol{r}_{C_i,\ell_i} = \boldsymbol{p}_{\ell_i} - \boldsymbol{p}_{C_i} \tag{7.57}$$

$$\boldsymbol{r}_{C_i,m_{i+1}} = \boldsymbol{p}_{m_{i+1}} - \boldsymbol{p}_{C_i}, \tag{7.58}$$

where $\boldsymbol{p}_{C_i}$ denotes the position vector of the centre of mass of augmented Link i.

Substituting (7.55) into (7.53) gives

$$\begin{aligned} \mathcal{T}_{\ell_i} &= \frac{1}{2} m_{\ell i} \dot{\boldsymbol{p}}_{C_i}^T \dot{\boldsymbol{p}}_{C_i} + \dot{\boldsymbol{p}}_{C_i}^T \boldsymbol{S}(\boldsymbol{\omega}_i) m_{\ell_i} \boldsymbol{r}_{C_i,\ell_i} \\ &\quad + \frac{1}{2} m_{\ell_i} \boldsymbol{\omega}_i^T \boldsymbol{S}^T(\boldsymbol{r}_{C_i,\ell_i}) \boldsymbol{S}(\boldsymbol{r}_{C_i,\ell_i}) \boldsymbol{\omega}_i + \frac{1}{2} \boldsymbol{\omega}_i^T \boldsymbol{I}_{\ell_i} \boldsymbol{\omega}_i. \end{aligned} \tag{7.59}$$

By virtue of *Steiner theorem*, the matrix

$$\bar{\boldsymbol{I}}_{\ell_i} = \boldsymbol{I}_{\ell_i} + m_{\ell_i} \boldsymbol{S}^T(\boldsymbol{r}_{C_i,\ell_i}) \boldsymbol{S}(\boldsymbol{r}_{C_i,\ell_i}) \tag{7.60}$$

represents the inertia tensor relative to the overall centre of mass $\boldsymbol{p}_{C_i}$, which contains an additional contribution due to the translation of the pole with respect to which the tensor is evaluated, as in (7.57). Therefore, (7.59) can be written as

$$\mathcal{T}_{\ell_i} = \frac{1}{2} m_{\ell i} \dot{\boldsymbol{p}}_{C_i}^T \dot{\boldsymbol{p}}_{C_i} + \dot{\boldsymbol{p}}_{C_i}^T \boldsymbol{S}(\boldsymbol{\omega}_i) m_{\ell_i} \boldsymbol{r}_{C_i,\ell_i} + \frac{1}{2} \boldsymbol{\omega}_i^T \bar{\boldsymbol{I}}_{\ell_i} \boldsymbol{\omega}_i. \tag{7.61}$$

In a similar fashion, substituting (7.56) into (7.54) and exploiting (7.23) yields

$$\begin{aligned} \mathcal{T}_{m_{i+1}} &= \frac{1}{2} m_{m_{i+1}} \dot{\boldsymbol{p}}_{C_i}^T \dot{\boldsymbol{p}}_{C_i} + \dot{\boldsymbol{p}}_{C_i}^T \boldsymbol{S}(\boldsymbol{\omega}_i) m_{m_{i+1}} \boldsymbol{r}_{C_i,m_{i+1}} + \frac{1}{2} \boldsymbol{\omega}_i^T \bar{\boldsymbol{I}}_{m_{i+1}} \boldsymbol{\omega}_i \\ &\quad + k_{r,i+1} \dot{q}_{i+1} \boldsymbol{z}_{m_{i+1}}^T \boldsymbol{I}_{m_{i+1}} \boldsymbol{\omega}_i + \frac{1}{2} k_{r,i+1}^2 \dot{q}_{i+1}^2 \boldsymbol{z}_{m_{i+1}}^T \boldsymbol{I}_{m_{i+1}} \boldsymbol{z}_{m_{i+1}}, \end{aligned} \tag{7.62}$$

where

$$\bar{\boldsymbol{I}}_{m_{i+1}} = \boldsymbol{I}_{m_{i+1}} + m_{m_{i+1}} \boldsymbol{S}^T(\boldsymbol{r}_{C_i,m_{i+1}}) \boldsymbol{S}(\boldsymbol{r}_{C_i,m_{i+1}}). \tag{7.63}$$

Summing the contributions in (7.61), (7.62) as in (7.52) gives the expression of the kinetic energy of augmented Link i in the form

$$\mathcal{T}_i = \frac{1}{2}m_i\dot{\boldsymbol{p}}_{C_i}^T\dot{\boldsymbol{p}}_{C_i} + \frac{1}{2}\boldsymbol{\omega}_i^T\bar{\boldsymbol{I}}_i\boldsymbol{\omega}_i + k_{r,i+1}\dot{q}_{i+1}\boldsymbol{z}_{m_{i+1}}^T\boldsymbol{I}_{m_{i+1}}\boldsymbol{\omega}_i + \frac{1}{2}k_{r,i+1}^2\dot{q}_{i+1}^2\boldsymbol{z}_{m_{i+1}}^T\boldsymbol{I}_{m_{i+1}}\boldsymbol{z}_{m_{i+1}}, \tag{7.64}$$

where $m_i = m_{\ell_i} + m_{m_{i+1}}$ and $\bar{\boldsymbol{I}}_i = \bar{\boldsymbol{I}}_{\ell_i} + \bar{\boldsymbol{I}}_{m_{i+1}}$ are respectively the overall mass and inertia tensor. In deriving (7.64), the relations in (7.57), (7.58) have been utilized as well as the following relation between the positions of the centres of mass:

$$m_{\ell_i}\boldsymbol{p}_{\ell_i} + m_{m_{i+1}}\boldsymbol{p}_{m_{i+1}} = m_i\boldsymbol{p}_{C_i}. \tag{7.65}$$

Notice that the first two terms on the right-hand side of (7.64) represent the kinetic energy contribution of the rotor when this is still, whereas the remaining two terms account for the rotor's own motion.

On the assumption that the rotor has a symmetric mass distribution about its axis of rotation, its inertia tensor expressed in a frame $\boldsymbol{R}_{m_i}$ with origin at the centre of mass and axis z_{m_i} aligned with the rotation axis can be written as

$$\boldsymbol{I}_{m_i}^{m_i} = \begin{bmatrix} I_{m_ixx} & 0 & 0 \\ 0 & I_{m_iyy} & 0 \\ 0 & 0 & I_{m_izz} \end{bmatrix} \tag{7.66}$$

where $I_{m_iyy} = I_{m_ixx}$. As a consequence, the inertia tensor is invariant with respect to any rotation about axis z_{m_i} and is, anyhow, constant when referred to any frame attached to Link $i-1$.

Since the aim is to determine a set of dynamic parameters independent of the manipulator joint configuration, it is worth referring the inertia tensor of the link $\bar{\boldsymbol{I}}_i$ to frame $\boldsymbol{R}_i$ attached to the link and the inertia tensor $\boldsymbol{I}_{m_{i+1}}$ to frame $\boldsymbol{R}_{m_{i+1}}$ so that it is diagonal. In view of (7.66) one has

$$\boldsymbol{I}_{m_{i+1}}\boldsymbol{z}_{m_{i+1}} = \boldsymbol{R}_{m_{i+1}}\boldsymbol{I}_{m_{i+1}}^{m_{i+1}}\boldsymbol{R}_{m_{i+1}}^T\boldsymbol{z}_{m_{i+1}} = I_{m_{i+1}}\boldsymbol{z}_{m_{i+1}} \tag{7.67}$$

where $I_{m_{i+1}} = I_{m_{i+1}zz}$ denotes the constant scalar moment of inertia of the rotor about its rotation axis.

Therefore, the kinetic energy (7.64) becomes

$$\mathcal{T}_i = \frac{1}{2}m_i\dot{\boldsymbol{p}}_{C_i}^{iT}\dot{\boldsymbol{p}}_{C_i}^i + \frac{1}{2}\boldsymbol{\omega}_i^{iT}\bar{\boldsymbol{I}}_i^i\boldsymbol{\omega}_i^i + k_{r,i+1}\dot{q}_{i+1}I_{m_{i+1}}\boldsymbol{z}_{m_{i+1}}^{iT}\boldsymbol{\omega}_i^i + \frac{1}{2}k_{r,i+1}^2\dot{q}_{i+1}^2I_{m_{i+1}}. \tag{7.68}$$

According to the linear velocity composition rule for Link i in (3.15), one may write

$$\dot{\boldsymbol{p}}_{C_i}^i = \dot{\boldsymbol{p}}_i^i + \boldsymbol{\omega}_i^i \times \boldsymbol{r}_{i,C_i}^i, \tag{7.69}$$

where all the vectors have been referred to Frame i; note that $\boldsymbol{r}_{i,C_i}^i$ is fixed in such a frame. Substituting (7.69) into (7.68) gives

$$\begin{aligned}\mathcal{T}_i = & \frac{1}{2} m_i \dot{\boldsymbol{p}}_i^{iT} \dot{\boldsymbol{p}}_i^i + \dot{\boldsymbol{p}}_i^{iT} \boldsymbol{S}(\boldsymbol{\omega}_i^i) m_i \boldsymbol{r}_{i,C_i}^i + \frac{1}{2} \boldsymbol{\omega}_i^{iT} \widehat{\boldsymbol{I}}_i^i \boldsymbol{\omega}_i^i \\ & + k_{r,i+1} \dot{q}_{i+1} I_{m_{i+1}} \boldsymbol{z}_{m_{i+1}}^{iT} \boldsymbol{\omega}_i^i + \frac{1}{2} k_{r,i+1}^2 \dot{q}_{i+1}^2 I_{m_{i+1}},\end{aligned} \tag{7.70}$$

where

$$\widehat{\boldsymbol{I}}_i^i = \bar{\boldsymbol{I}}_i^i + m_i \boldsymbol{S}^T(\boldsymbol{r}_{i,C_i}^i) \boldsymbol{S}(\boldsymbol{r}_{i,C_i}^i) \tag{7.71}$$

represents the inertia tensor with respect to the origin of Frame i according to Steiner theorem.

Let $\boldsymbol{r}_{i,C_i}^i = [\,\ell_{C_ix} \quad \ell_{C_iy} \quad \ell_{C_iz}\,]^T$. The *first moment of inertia* is

$$m_i \boldsymbol{r}_{i,C_i}^i = \begin{bmatrix} m_i \ell_{C_ix} \\ m_i \ell_{C_iy} \\ m_i \ell_{C_iz} \end{bmatrix}. \tag{7.72}$$

From (7.71) the inertia tensor of augmented Link i is

$$\begin{aligned}\widehat{\boldsymbol{I}}_i^i = & \begin{bmatrix} \bar{I}_{ixx} + m_i(\ell_{C_iy}^2 + \ell_{C_iz}^2) & -\bar{I}_{ixy} - m_i \ell_{C_ix} \ell_{C_iy} & -\bar{I}_{ixz} - m_i \ell_{C_ix} \ell_{C_iz} \\ * & \bar{I}_{iyy} + m_i(\ell_{C_ix}^2 + \ell_{C_iz}^2) & -\bar{I}_{iyz} - m_i \ell_{C_iy} \ell_{C_iz} \\ * & * & \bar{I}_{izz} + m_i(\ell_{C_ix}^2 + \ell_{C_iy}^2) \end{bmatrix} \\ = & \begin{bmatrix} \widehat{I}_{ixx} & -\widehat{I}_{ixy} & -\widehat{I}_{ixz} \\ * & \widehat{I}_{iyy} & -\widehat{I}_{iyz} \\ * & * & \widehat{I}_{izz} \end{bmatrix}.\end{aligned} \tag{7.73}$$

Therefore, the kinetic energy of the augmented link is linear with respect to the dynamic parameters, namely, the *mass*, the *three components of the first moment of inertia* in (7.72), the *six components of the inertia tensor* in (7.73), and the *moment of inertia of the rotor*.

As regards potential energy, it is worth referring to the centre of mass of augmented Link i defined as in (7.65), and thus the single contribution of potential energy can be written as

$$\mathcal{U}_i = -m_i \boldsymbol{g}_0^{iT} \boldsymbol{p}_{C_i}^i \tag{7.74}$$

where the vectors have been referred to Frame i. According to the relation

$$\boldsymbol{p}_{C_i}^i = \boldsymbol{p}_i^i + \boldsymbol{r}_{i,C_i}^i.$$

The expression in (7.74) can be rewritten as

$$\mathcal{U}_i = -\boldsymbol{g}_0^{iT} (m_i \boldsymbol{p}_i^i + m_i \boldsymbol{r}_{i,C_i}^i) \tag{7.75}$$

that is, the potential energy of the augmented link is linear with respect to the mass and the three components of the first moment of inertia in (7.72).

By summing the contributions of kinetic energy and potential energy for all augmented links, the Lagrangian of the system (7.1) can be expressed in the form

$$\mathcal{L} = \sum_{i=1}^{n} (\boldsymbol{\beta}_{\mathcal{T}i}^T - \boldsymbol{\beta}_{\mathcal{U}i}^T)\boldsymbol{\pi}_i \tag{7.76}$$

where $\boldsymbol{\pi}_i$ is the (11×1) vector of dynamic parameters

$$\boldsymbol{\pi}_i = [\, m_i \;\; m_i\ell_{C_ix} \;\; m_i\ell_{C_iy} \;\; m_i\ell_{C_iz} \;\; \widehat{I}_{ixx} \;\; \widehat{I}_{ixy} \;\; \widehat{I}_{ixz} \;\; \widehat{I}_{iyy} \;\; \widehat{I}_{iyz} \;\; \widehat{I}_{izz} \;\; I_{m_i} \,]^T, \tag{7.77}$$

in which the moment of inertia of Rotor i has been associated with the parameters of Link i so as to simplify the notation.

In (7.76), $\boldsymbol{\beta}_{\mathcal{T}i}$ and $\boldsymbol{\beta}_{\mathcal{U}i}$ are two (11×1) vectors that allow the Lagrangian to be written as a function of $\boldsymbol{\pi}_i$. Such vectors are a function of the generalized coordinates of the mechanical system (and also of their derivatives as regards $\boldsymbol{\beta}_{\mathcal{T}i}$). In particular, it can be shown that $\boldsymbol{\beta}_{\mathcal{T}i} = \boldsymbol{\beta}_{\mathcal{T}i}(q_1, q_2, \ldots, q_i, \dot{q}_1, \dot{q}_2, \ldots, \dot{q}_i)$ and $\boldsymbol{\beta}_{\mathcal{U}i} = \boldsymbol{\beta}_{\mathcal{U}i}(q_1, q_2, \ldots, q_i)$, i.e., they do not depend on the variables of the joints subsequent to Link i.

At this point, it should be observed how the derivations required by the Lagrange equations in (7.2) do not alter the property of linearity in the parameters, and then the generalized force at Joint i can be written as

$$\xi_i = \sum_{j=1}^{n} \boldsymbol{y}_{ij}^T \boldsymbol{\pi}_j \tag{7.78}$$

where

$$\boldsymbol{y}_{ij} = \frac{d}{dt}\frac{\partial \boldsymbol{\beta}_{\mathcal{T}j}}{\partial \dot{q}_i} - \frac{\partial \boldsymbol{\beta}_{\mathcal{T}j}}{\partial q_i} + \frac{\partial \boldsymbol{\beta}_{\mathcal{U}j}}{\partial q_i}. \tag{7.79}$$

Since the partial derivatives of $\boldsymbol{\beta}_{\mathcal{T}j}$ and $\boldsymbol{\beta}_{\mathcal{U}j}$ appearing in (7.79) vanish for $j < i$, the following notable result is obtained:

$$\begin{bmatrix} \xi_1 \\ \xi_2 \\ \vdots \\ \xi_n \end{bmatrix} = \begin{bmatrix} \boldsymbol{y}_{11}^T & \boldsymbol{y}_{12}^T & \ldots & \boldsymbol{y}_{1n}^T \\ \boldsymbol{0}^T & \boldsymbol{y}_{22}^T & \ldots & \boldsymbol{y}_{2n}^T \\ \vdots & \vdots & \ddots & \vdots \\ \boldsymbol{0}^T & \boldsymbol{0}^T & \ldots & \boldsymbol{y}_{nn}^T \end{bmatrix} \begin{bmatrix} \boldsymbol{\pi}_1 \\ \boldsymbol{\pi}_2 \\ \vdots \\ \boldsymbol{\pi}_n \end{bmatrix} \tag{7.80}$$

which yields the property of *linearity of the model* of a manipulator with respect to a suitable set of *dynamic parameters*.

In the simple case of no contact forces ($\boldsymbol{h}_e = \boldsymbol{0}$), it may be worth including the viscous friction coefficient F_{vi} and Coulomb friction coefficient F_{si} in the parameters of the vector $\boldsymbol{\pi}_i$, thus leading to a total number of 13 parameters per joint. In summary, (7.80) can be compactly written as

$$\boldsymbol{\tau} = \boldsymbol{Y}(\boldsymbol{q}, \dot{\boldsymbol{q}}, \ddot{\boldsymbol{q}})\boldsymbol{\pi} \tag{7.81}$$

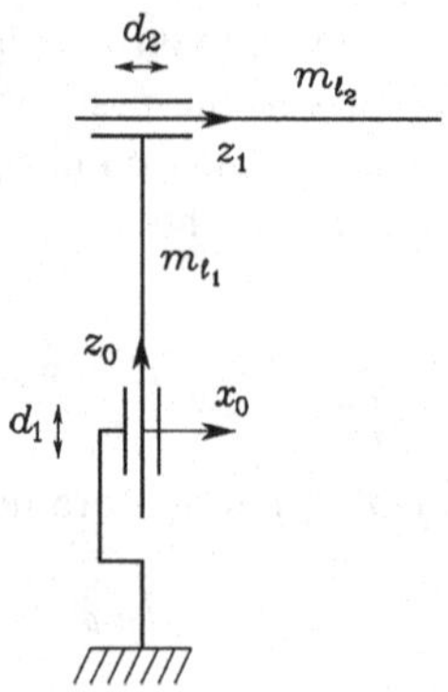

Fig. 7.3. Two-link Cartesian arm

where $\boldsymbol{\pi}$ is a $(p \times 1)$ vector of *constant* parameters and $\boldsymbol{Y}$ is an $(n \times p)$ matrix which is a *function of joint positions, velocities and accelerations*; this matrix is usually called *regressor*. Regarding the dimension of the parameter vector, notice that $p \leq 13n$ since not all the thirteen parameters for each joint may explicitly appear in (7.81).

7.3 Dynamic Model of Simple Manipulator Structures

In the following, three examples of dynamic model computation are illustrated for simple two-DOF manipulator structures. Two DOFs, in fact, are enough to understand the physical meaning of all dynamic terms, especially the joint coupling terms. On the other hand, dynamic model computation for manipulators with more DOFs would be quite tedious and prone to errors, when carried out by paper and pencil. In those cases, it is advisable to perform it with the aid of a symbolic programming software package.

7.3.1 Two-link Cartesian Arm

Consider the two-link Cartesian arm in Fig. 7.3, for which the vector of generalized coordinates is $\boldsymbol{q} = [\,d_1 \quad d_2\,]^T$. Let m_{ℓ_1}, m_{ℓ_2} be the masses of the two links, and m_{m_1}, m_{m_2} the masses of the rotors of the two joint motors. Also let I_{m_1}, I_{m_2} be the moments of inertia with respect to the axes of the two rotors. It is assumed that $\boldsymbol{p}_{m_i} = \boldsymbol{p}_{i-1}$ and $\boldsymbol{z}_{m_i} = \boldsymbol{z}_{i-1}$, for $i = 1, 2$, i.e., the motors are located on the joint axes with centres of mass located at the origins of the respective frames.

With the chosen coordinate frames, computation of the Jacobians in (7.16), (7.18) yields

$$\boldsymbol{J}_P^{(\ell_1)} = \begin{bmatrix} 0 & 0 \\ 0 & 0 \\ 1 & 0 \end{bmatrix} \qquad \boldsymbol{J}_P^{(\ell_2)} = \begin{bmatrix} 0 & 1 \\ 0 & 0 \\ 1 & 0 \end{bmatrix}.$$

Obviously, in this case there are no angular velocity contributions for both links.

Computation of the Jacobians in (7.25), (7.26) e (7.28), (7.29) yields

$$\boldsymbol{J}_P^{(m_1)} = \begin{bmatrix} 0 & 0 \\ 0 & 0 \\ 0 & 0 \end{bmatrix} \qquad \boldsymbol{J}_P^{(m_2)} = \begin{bmatrix} 0 & 0 \\ 0 & 0 \\ 1 & 0 \end{bmatrix}$$

$$\boldsymbol{J}_O^{(m_1)} = \begin{bmatrix} 0 & 0 \\ 0 & 0 \\ k_{r1} & 0 \end{bmatrix} \qquad \boldsymbol{J}_O^{(m_2)} = \begin{bmatrix} 0 & k_{r2} \\ 0 & 0 \\ 0 & 0 \end{bmatrix}$$

where k_{ri} is the gear reduction ratio of Motor i. It is obvious to see that $\boldsymbol{z}_1 = [1 \quad 0 \quad 0]^T$, which greatly simplifies computation of the second term in (4.34).

From (7.32), the inertia matrix is

$$\boldsymbol{B} = \begin{bmatrix} m_{\ell_1} + m_{m_2} + k_{r1}^2 I_{m_1} + m_{\ell_2} & 0 \\ 0 & m_{\ell_2} + k_{r2}^2 I_{m_2} \end{bmatrix}.$$

It is worth observing that $\boldsymbol{B}$ is *constant*, i.e., it does not depend on the arm configuration. This implies also that $\boldsymbol{C} = \boldsymbol{O}$, i.e., there are no contributions of centrifugal and Coriolis forces. As for the gravitational terms, since $\boldsymbol{g}_0 = [0 \quad 0 \quad -g]^T$ (g is gravity acceleration), (7.39) with the above Jacobians gives

$$g_1 = (m_{\ell_1} + m_{m_2} + m_{\ell_2})g \qquad g_2 = 0.$$

In the absence of friction and tip contact forces, the resulting equations of motion are

$$\begin{aligned} (m_{\ell_1} + m_{m_2} + k_{r1}^2 I_{m_1} + m_{\ell_2})\ddot{d}_1 + (m_{\ell_1} + m_{m_2} + m_{\ell_2})g &= \tau_1 \\ (m_{\ell_2} + k_{r2}^2 I_{m_2})\ddot{d}_2 &= \tau_2 \end{aligned}$$

where τ_1 and τ_2 denote the forces applied to the two joints. Notice that a completely decoupled dynamics has been obtained. This is a consequence not only of the Cartesian structures but also of the particular geometry; in other words, if the second joint axis were not at a right angle with the first joint axis, the resulting inertia matrix would not be diagonal (see Problem 7.1).

7.3.2 Two-link Planar Arm

Consider the two-link planar arm in Fig. 7.4, for which the vector of generalized coordinates is $\boldsymbol{q} = [\vartheta_1 \quad \vartheta_2]^T$. Let ℓ_1, ℓ_2 be the distances of the centres of mass of the two links from the respective joint axes. Also let m_{ℓ_1}, m_{ℓ_2} be the masses of the two links, and m_{m_1}, m_{m_2} the masses of the rotors of the two joint motors. Finally, let I_{m_1}, I_{m_2} be the moments of inertia with respect to the axes of the two rotors, and I_{ℓ_1}, I_{ℓ_2} the moments of inertia relative to the

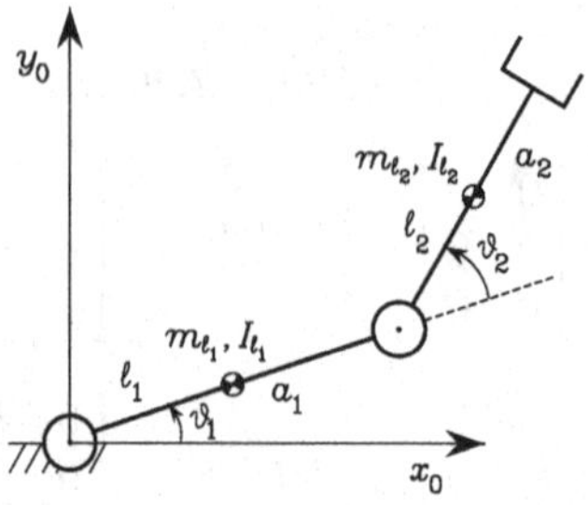

Fig. 7.4. Two-link planar arm

centres of mass of the two links, respectively. It is assumed that $\boldsymbol{p}_{m_i} = \boldsymbol{p}_{i-1}$ and $\boldsymbol{z}_{m_i} = \boldsymbol{z}_{i-1}$, for $i = 1, 2$, i.e., the motors are located on the joint axes with centres of mass located at the origins of the respective frames.

With the chosen coordinate frames, computation of the Jacobians in (7.16), (7.18) yields

$$\boldsymbol{J}_P^{(\ell_1)} = \begin{bmatrix} -\ell_1 s_1 & 0 \\ \ell_1 c_1 & 0 \\ 0 & 0 \end{bmatrix} \qquad \boldsymbol{J}_P^{(\ell_2)} = \begin{bmatrix} -a_1 s_1 - \ell_2 s_{12} & -\ell_2 s_{12} \\ a_1 c_1 + \ell_2 c_{12} & \ell_2 c_{12} \\ 0 & 0 \end{bmatrix},$$

whereas computation of the Jacobians in (7.17), (7.19) yields

$$\boldsymbol{J}_O^{(\ell_1)} = \begin{bmatrix} 0 & 0 \\ 0 & 0 \\ 1 & 0 \end{bmatrix} \qquad \boldsymbol{J}_O^{(\ell_2)} = \begin{bmatrix} 0 & 0 \\ 0 & 0 \\ 1 & 1 \end{bmatrix}.$$

Notice that $\boldsymbol{\omega}_i$, for $i = 1, 2$, is aligned with z_0, and thus $\boldsymbol{R}_i$ has *no* effect. It is then possible to refer to the scalar moments of inertia I_{ℓ_i}.

Computation of the Jacobians in (7.25), (7.26) yields

$$\boldsymbol{J}_P^{(m_1)} = \begin{bmatrix} 0 & 0 \\ 0 & 0 \\ 0 & 0 \end{bmatrix} \qquad \boldsymbol{J}_P^{(m_2)} = \begin{bmatrix} -a_1 s_1 & 0 \\ a_1 c_1 & 0 \\ 0 & 0 \end{bmatrix},$$

whereas computation of the Jacobians in (7.28), (7.29) yields

$$\boldsymbol{J}_O^{(m_1)} = \begin{bmatrix} 0 & 0 \\ 0 & 0 \\ k_{r1} & 0 \end{bmatrix} \qquad \boldsymbol{J}_O^{(m_2)} = \begin{bmatrix} 0 & 0 \\ 0 & 0 \\ 1 & k_{r2} \end{bmatrix}$$

where k_{ri} is the gear reduction ratio of Motor i.

From (7.32), the inertia matrix is

$$\boldsymbol{B}(\boldsymbol{q}) = \begin{bmatrix} b_{11}(\vartheta_2) & b_{12}(\vartheta_2) \\ b_{21}(\vartheta_2) & b_{22} \end{bmatrix}$$

$$\begin{aligned}
b_{11} &= I_{\ell_1} + m_{\ell_1}\ell_1^2 + k_{r1}^2 I_{m_1} + I_{\ell_2} + m_{\ell_2}(a_1^2 + \ell_2^2 + 2a_1\ell_2 c_2) \\
&\quad + I_{m_2} + m_{m_2} a_1^2 \\
b_{12} &= b_{21} = I_{\ell_2} + m_{\ell_2}(\ell_2^2 + a_1\ell_2 c_2) + k_{r2} I_{m_2} \\
b_{22} &= I_{\ell_2} + m_{\ell_2}\ell_2^2 + k_{r2}^2 I_{m_2}.
\end{aligned}$$

Compared to the previous example, the inertia matrix is now configuration-dependent. Notice that the term $k_{r2}I_{m_2}$ in the off-diagonal term of the inertia matrix derives from having considered the rotational part of the motor kinetic energy as due to the total angular velocity, i.e., its own angular velocity and that of the preceding link in the kinematic chain. At first approximation, especially in the case of high values of the gear reduction ratio, this contribution could be neglected; in the resulting reduced model, motor inertias would appear uniquely in the elements on the diagonal of the inertia matrix with terms of the type $k_{ri}^2 I_{m_i}$.

The computation of Christoffel symbols as in (7.45) gives

$$\begin{aligned}
c_{111} &= \frac{1}{2}\frac{\partial b_{11}}{\partial q_1} = 0 \\
c_{112} &= c_{121} = \frac{1}{2}\frac{\partial b_{11}}{\partial q_2} = -m_{\ell_2} a_1 \ell_2 s_2 = h \\
c_{122} &= \frac{\partial b_{12}}{\partial q_2} - \frac{1}{2}\frac{\partial b_{22}}{\partial q_1} = h \\
c_{211} &= \frac{\partial b_{21}}{\partial q_1} - \frac{1}{2}\frac{\partial b_{11}}{\partial q_2} = -h \\
c_{212} &= c_{221} = \frac{1}{2}\frac{\partial b_{22}}{\partial q_1} = 0 \\
c_{222} &= \frac{1}{2}\frac{\partial b_{22}}{\partial q_2} = 0,
\end{aligned}$$

leading to the matrix

$$\boldsymbol{C}(\boldsymbol{q}, \dot{\boldsymbol{q}}) = \begin{bmatrix} h\dot{\vartheta}_2 & h(\dot{\vartheta}_1 + \dot{\vartheta}_2) \\ -h\dot{\vartheta}_1 & 0 \end{bmatrix}.$$

Computing the matrix $\boldsymbol{N}$ in (7.47) gives

$$\begin{aligned}
\boldsymbol{N}(\boldsymbol{q}, \dot{\boldsymbol{q}}) &= \dot{\boldsymbol{B}}(\boldsymbol{q}) - 2\boldsymbol{C}(\boldsymbol{q}, \dot{\boldsymbol{q}}) \\
&= \begin{bmatrix} 2h\dot{\vartheta}_2 & h\dot{\vartheta}_2 \\ h\dot{\vartheta}_2 & 0 \end{bmatrix} - 2\begin{bmatrix} h\dot{\vartheta}_2 & h(\dot{\vartheta}_1 + \dot{\vartheta}_2) \\ -h\dot{\vartheta}_1 & 0 \end{bmatrix} \\
&= \begin{bmatrix} 0 & -2h\dot{\vartheta}_1 - h\dot{\vartheta}_2 \\ 2h\dot{\vartheta}_1 + h\dot{\vartheta}_2 & 0 \end{bmatrix}
\end{aligned}$$

that allows the verification of the skew-symmetry property expressed by (7.48). See also Problem 7.2.

As for the gravitational terms, since $\boldsymbol{g}_0 = [0 \quad -g \quad 0]^T$, (7.39) with the above Jacobians gives

$$
\begin{aligned}
g_1 &= (m_{\ell_1}\ell_1 + m_{m_2}a_1 + m_{\ell_2}a_1)gc_1 + m_{\ell_2}\ell_2 gc_{12} \\
g_2 &= m_{\ell_2}\ell_2 gc_{12}.
\end{aligned}
$$

In the absence of friction and tip contact forces, the resulting equations of motion are

$$
\begin{aligned}
&\big(I_{\ell_1} + m_{\ell_1}\ell_1^2 + k_{r1}^2 I_{m_1} + I_{\ell_2} + m_{\ell_2}(a_1^2 + \ell_2^2 + 2a_1\ell_2 c_2) + I_{m_2} + m_{m_2}a_1^2\big)\ddot{\vartheta}_1 \\
&\quad + \big(I_{\ell_2} + m_{\ell_2}(\ell_2^2 + a_1\ell_2 c_2) + k_{r2}I_{m_2}\big)\ddot{\vartheta}_2 \\
&\quad -2m_{\ell_2}a_1\ell_2 s_2\dot{\vartheta}_1\dot{\vartheta}_2 - m_{\ell_2}a_1\ell_2 s_2\dot{\vartheta}_2^2 \\
&\quad +(m_{\ell_1}\ell_1 + m_{m_2}a_1 + m_{\ell_2}a_1)gc_1 + m_{\ell_2}\ell_2 gc_{12} = \tau_1 \\
&\big(I_{\ell_2} + m_{\ell_2}(\ell_2^2 + a_1\ell_2 c_2) + k_{r2}I_{m_2}\big)\ddot{\vartheta}_1 + \big(I_{\ell_2} + m_{\ell_2}\ell_2^2 + k_{r2}^2 I_{m_2}\big)\ddot{\vartheta}_2 \\
&\quad +m_{\ell_2}a_1\ell_2 s_2\dot{\vartheta}_1^2 + m_{\ell_2}\ell_2 gc_{12} = \tau_2
\end{aligned}
\tag{7.82}
$$

where τ_1 and τ_2 denote the torques applied to the joints.

Finally, it is wished to derive a parameterization of the dynamic model (7.82) according to the relation (7.81). By direct inspection of the expressions of the joint torques, it is possible to find the following parameter vector:

$$
\boldsymbol{\pi} = [\pi_1 \quad \pi_2 \quad \pi_3 \quad \pi_4 \quad \pi_5 \quad \pi_6 \quad \pi_7 \quad \pi_8]^T \tag{7.83}
$$

$$
\begin{aligned}
\pi_1 &= m_1 = m_{\ell_1} + m_{m_2} \\
\pi_2 &= m_1\ell_{C_1} = m_{\ell_1}(\ell_1 - a_1) \\
\pi_3 &= \widehat{I}_1 = I_{\ell_1} + m_{\ell_1}(\ell_1 - a_1)^2 + I_{m_2} \\
\pi_4 &= I_{m_1} \\
\pi_5 &= m_2 = m_{\ell_2} \\
\pi_6 &= m_2\ell_{C_2} = m_{\ell_2}(\ell_2 - a_2) \\
\pi_7 &= \widehat{I}_2 = I_{\ell_2} + m_{\ell_2}(\ell_2 - a_2)^2 \\
\pi_8 &= I_{m_2},
\end{aligned}
$$

where the parameters for the augmented links have been found according to (7.77). It can be recognized that the number of non-null parameters is less than the maximum number of twenty-two parameters allowed in this case.[6] The regressor in (7.81) is

$$
\boldsymbol{Y} = \begin{bmatrix} y_{11} & y_{12} & y_{13} & y_{14} & y_{15} & y_{16} & y_{17} & y_{18} \\ y_{21} & y_{22} & y_{23} & y_{24} & y_{25} & y_{26} & y_{27} & y_{28} \end{bmatrix} \tag{7.84}
$$

[6] The number of parameters can be further reduced by resorting to a more accurate inspection, which leads to finding a minimum number of five parameters; those turn out to be a linear combination of the parameters in (7.83) (see Problem 7.4).

$$
\begin{aligned}
y_{11} &= a_1^2\ddot{\vartheta}_1 + a_1 g c_1 \\
y_{12} &= 2a_1\ddot{\vartheta}_1 + g c_1 \\
y_{13} &= \ddot{\vartheta}_1 \\
y_{14} &= k_{r1}^2\ddot{\vartheta}_1 \\
y_{15} &= (a_1^2 + 2a_1 a_2 c_2 + a_2^2)\ddot{\vartheta}_1 + (a_1 a_2 c_2 + a_2^2)\ddot{\vartheta}_2 - 2a_1 a_2 s_2 \dot{\vartheta}_1\dot{\vartheta}_2 \\
&\quad - a_1 a_2 s_2 \dot{\vartheta}_2^2 + a_1 g c_1 + a_2 g c_{12} \\
y_{16} &= (2a_1 c_2 + 2a_2)\ddot{\vartheta}_1 + (a_1 c_2 + 2a_2)\ddot{\vartheta}_2 - 2a_1 s_2 \dot{\vartheta}_1\dot{\vartheta}_2 - a_1 s_2 \dot{\vartheta}_2^2 \\
&\quad + g c_{12} \\
y_{17} &= \ddot{\vartheta}_1 + \ddot{\vartheta}_2 \\
y_{18} &= k_{r2}\ddot{\vartheta}_2
\end{aligned}
$$

$$
\begin{aligned}
y_{21} &= 0 \\
y_{22} &= 0 \\
y_{23} &= 0 \\
y_{24} &= 0 \\
y_{25} &= (a_1 a_2 c_2 + a_2^2)\ddot{\vartheta}_1 + a_2^2\ddot{\vartheta}_2 + a_1 a_2 s_2 \dot{\vartheta}_1^2 + a_2 g c_{12} \\
y_{26} &= (a_1 c_2 + 2a_2)\ddot{\vartheta}_1 + 2a_2\ddot{\vartheta}_2 + a_1 s_2 \dot{\vartheta}_1^2 + g c_{12} \\
y_{27} &= \ddot{\vartheta}_1 + \ddot{\vartheta}_2 \\
y_{28} &= k_{r2}\ddot{\vartheta}_1 + k_{r2}^2\ddot{\vartheta}_2.
\end{aligned}
$$

Example 7.2

In order to understand the relative weight of the various torque contributions in the dynamic model (7.82), consider a two-link planar arm with the following data:

$$
a_1 = a_2 = 1\,\text{m} \quad \ell_1 = \ell_2 = 0.5\,\text{m} \quad m_{\ell_1} = m_{\ell_2} = 50\,\text{kg} \quad I_{\ell_1} = I_{\ell_2} = 10\,\text{kg}\cdot\text{m}^2
$$

$$
k_{r1} = k_{r2} = 100 \quad m_{m_1} = m_{m_2} = 5\,\text{kg} \quad I_{m_1} = I_{m_2} = 0.01\,\text{kg}\cdot\text{m}^2.
$$

The two links have been chosen equal to illustrate better the dynamic interaction between the two joints.

Figure 7.5 shows the time history of positions, velocities, accelerations and torques resulting from joint trajectories with typical triangular velocity profile and equal time duration. The initial arm configuration is so that the tip is located at the point $(0.2, 0)$ m with a lower elbow posture. Both joints make a rotation of $\pi/2$ rad in a time of 0.5 s.

From the time history of the single torque contributions in Fig. 7.6 it can be recognized that:

- The inertia torque at Joint 1 due to Joint 1 acceleration follows the time history of the acceleration.
- The inertia torque at Joint 2 due to Joint 2 acceleration is piecewise constant, since the inertia moment at Joint 2 axis is constant.

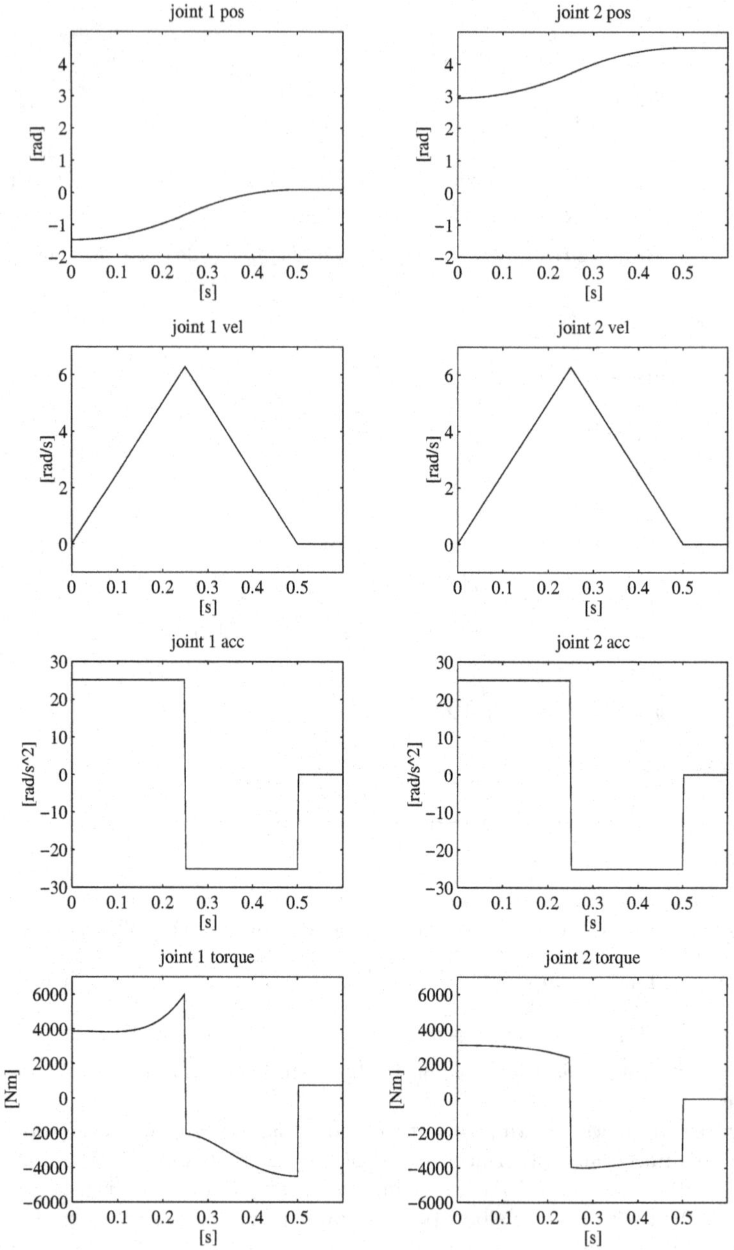

Fig. 7.5. Time history of positions, velocities, accelerations and torques with joint trajectories of equal duration

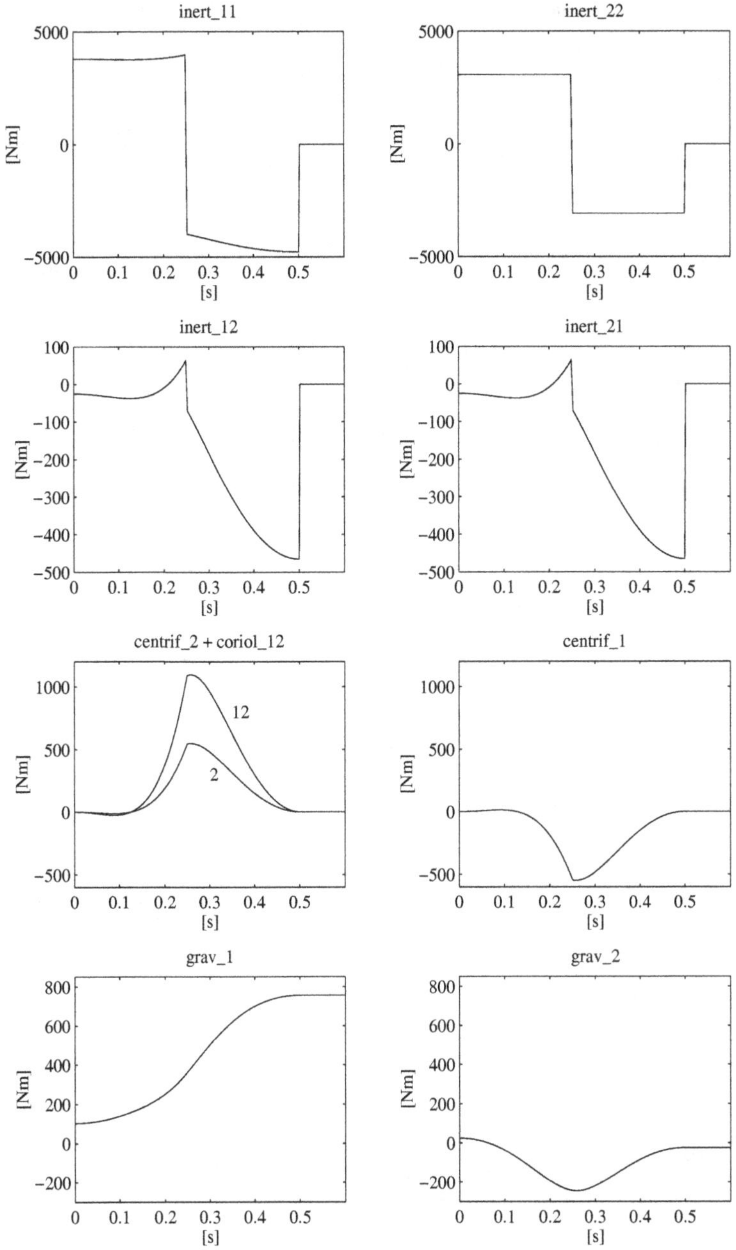

Fig. 7.6. Time history of torque contributions with joint trajectories of equal duration

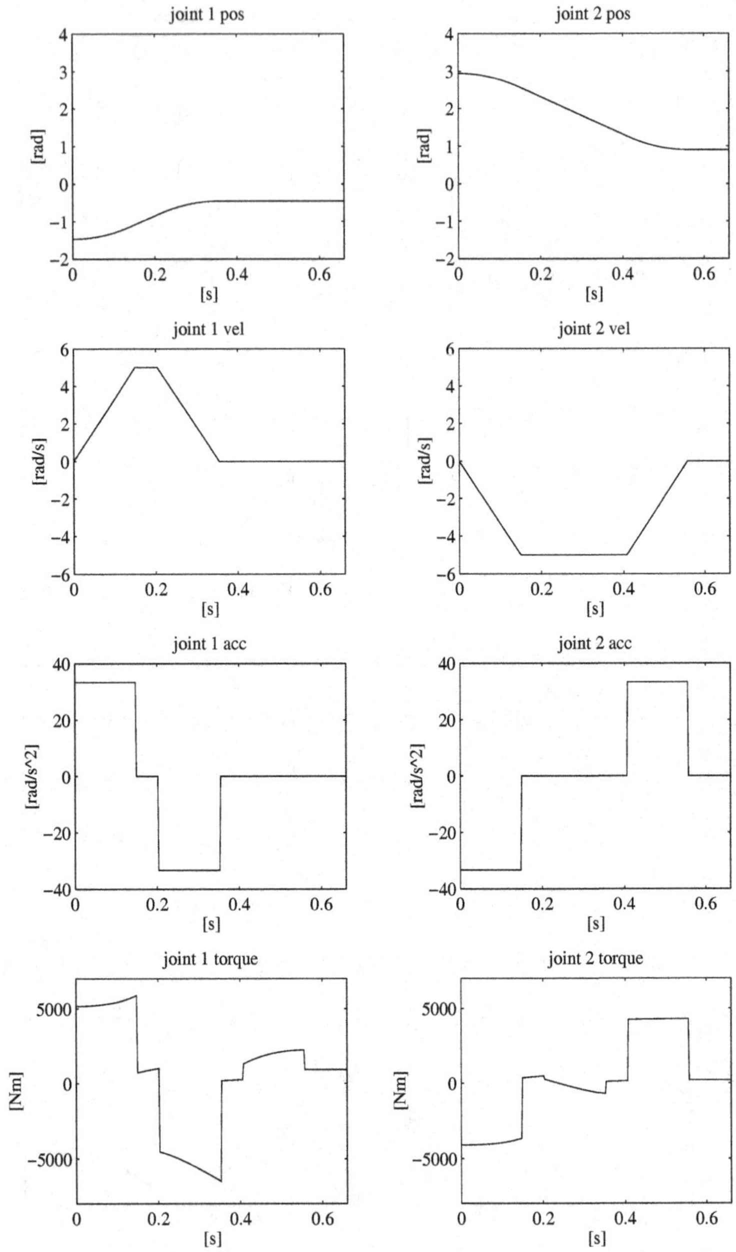

Fig. 7.7. Time history of positions, velocities, accelerations and torques with joint trajectories of different duration

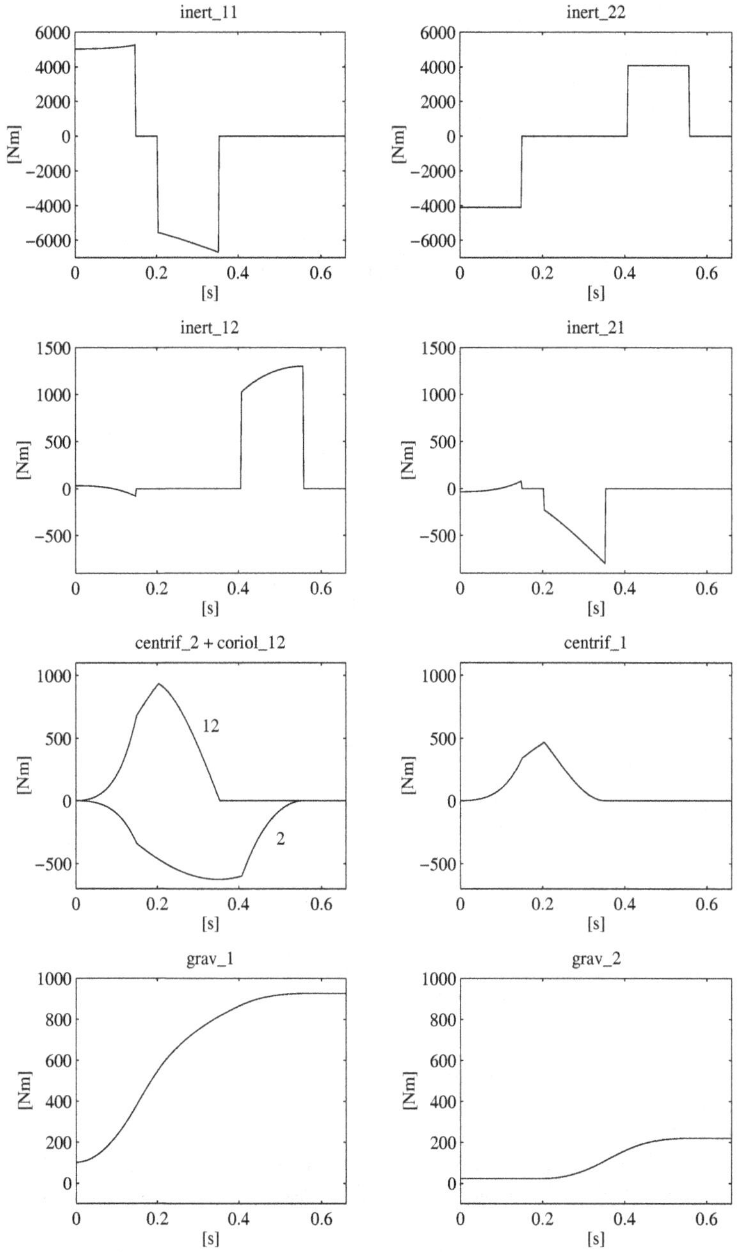

Fig. 7.8. Time history of torque contributions with joint trajectories of different duration

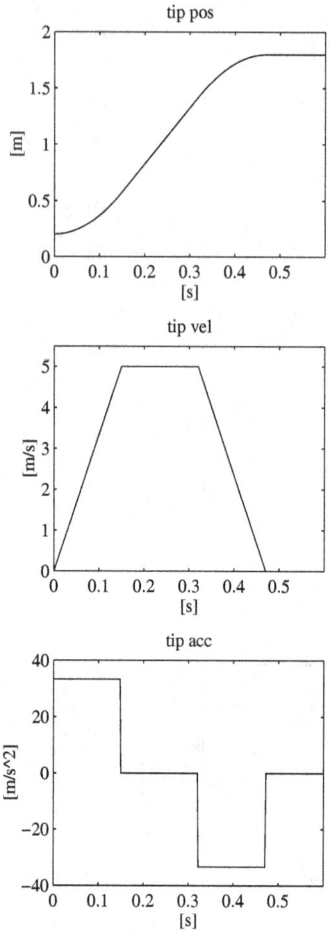

Fig. 7.9. Time history of tip position, velocity and acceleration with a straight line tip trajectory along the horizontal axis

- The inertia torques at each joint due to acceleration of the other joint confirm the symmetry of the inertia matrix, since the acceleration profiles are the same for both joints.
- The Coriolis effect is present only at Joint 1, since the arm tip moves with respect to the mobile frame attached to Link 1 but is fixed with respect to the frame attached to Link 2.
- The centrifugal and Coriolis torques reflect the above symmetry.

Figure 7.7 shows the time history of positions, velocities, accelerations and torques resulting from joint trajectories with typical trapezoidal velocity profile and different time duration. The initial configuration is the same as in the previous case. The two joints make a rotation so as to take the tip to the point $(1.8, 0)$ m. The acceleration time is 0.15 s and the maximum velocity is 5 rad/s for both joints.

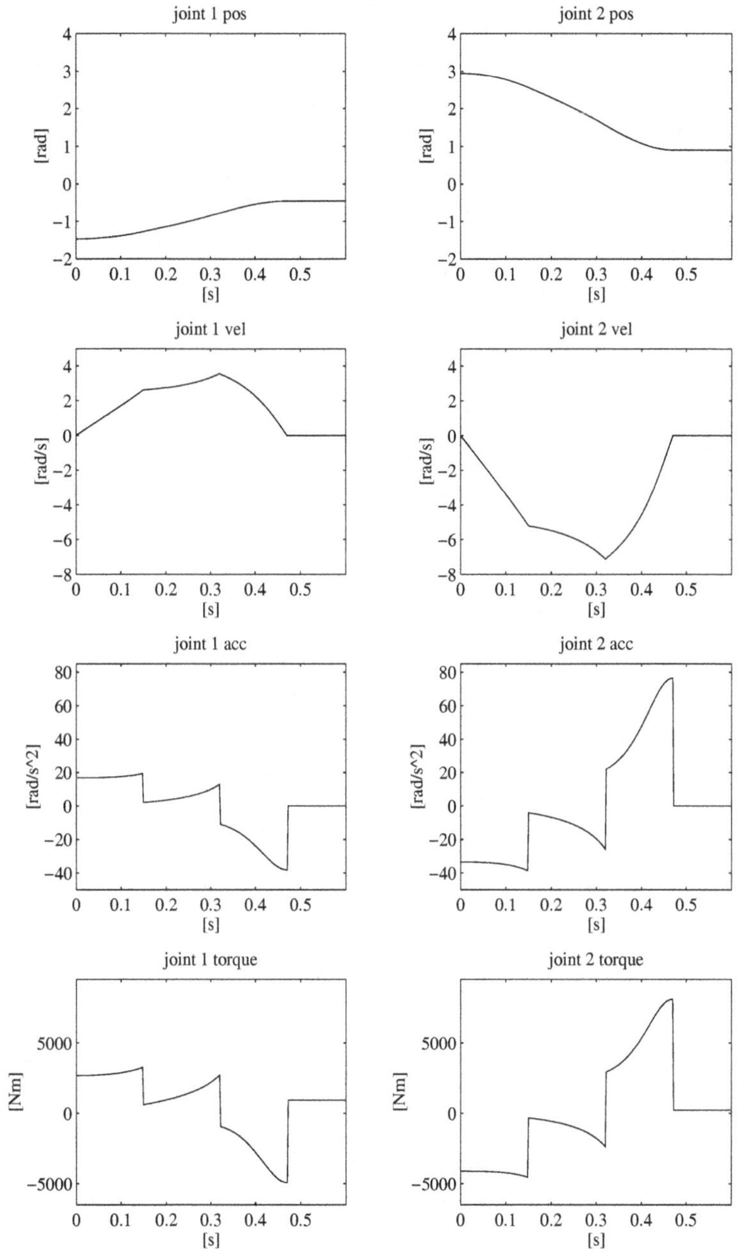

Fig. 7.10. Time history of joint positions, velocities, accelerations, and torques with a straight line tip trajectory along the horizontal axis

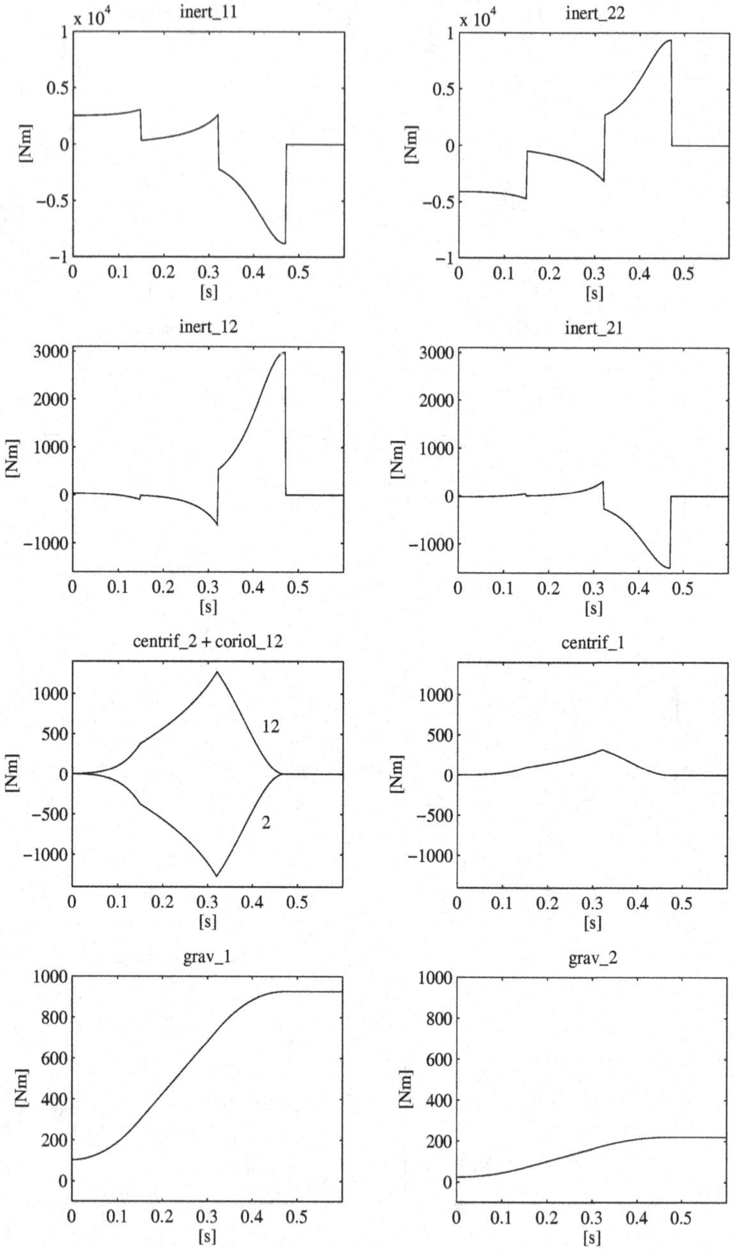

Fig. 7.11. Time history of joint torque contributions with a straight line tip trajectory along the horizontal axis

From the time history of the single torque contributions in Fig. 7.8 it can be recognized that:

- The inertia torque at Joint 1 due to Joint 2 acceleration is opposite to that at Joint 2 due to Joint 1 acceleration in that portion of trajectory when the two accelerations have the same magnitude but opposite sign.
- The different velocity profiles imply that the centrifugal effect induced at Joint 1 by Joint 2 velocity dies out later than the centrifugal effect induced at Joint 2 by Joint 1 velocity.
- The gravitational torque at Joint 2 is practically constant in the first portion of the trajectory, since Link 2 is almost kept in the same posture. As for the gravitational torque at Joint 1, instead, the centre of mass of the articulated system moves away from the origin of the axes.

Finally, Fig. 7.9 shows the time history of tip position, velocity and acceleration for a trajectory with a trapezoidal velocity profile. Starting from the same initial posture as above, the arm tip makes a translation of 1.6 m along the horizontal axis; the acceleration time is 0.15 s and the maximum velocity is 5 m/s.

As a result of an inverse kinematics procedure, the time history of joint positions, velocities and accelerations have been computed which are illustrated in Fig. 7.10, together with the joint torques that are needed to execute the assigned trajectory. It can be noticed that the time history of the represented quantities differs from the corresponding ones in the operational space, in view of the nonlinear effects introduced by kinematic relations.

For what concerns the time history of the individual torque contributions in Fig. 7.11, it is possible to make a number of remarks similar to those made above for trajectories assigned directly in the joint space.

7.3.3 Parallelogram Arm

Consider the parallelogram arm in Fig. 7.12. Because of the presence of the closed chain, the equivalent tree-structured open-chain arm is initially taken into account. Let $\ell_{1'}$, $\ell_{2'}$, $\ell_{3'}$ and $\ell_{1''}$ be the distances of the centres of mass of the three links along one branch of the tree, and of the single link along the other branch, from the respective joint axes. Also let $m_{\ell_{1'}}$, $m_{\ell_{2'}}$, $m_{\ell_{3'}}$ and $m_{\ell_{1''}}$ be the masses of the respective links, and $I_{\ell_{1'}}$, $I_{\ell_{2'}}$, $I_{\ell_{3'}}$ and $I_{\ell_{1''}}$ the moments of inertia relative to the centres of mass of the respective links. For the sake of simplicity, the contributions of the motors are neglected.

With the chosen coordinate frames, computation of the Jacobians in (7.16) (7.18) yields

$$\boldsymbol{J}_P^{(\ell_{1'})} = \begin{bmatrix} -\ell_{1'}s_{1'} & 0 & 0 \\ \ell_{1'}c_{1'} & 0 & 0 \\ 0 & 0 & 0 \end{bmatrix} \qquad \boldsymbol{J}_P^{(\ell_{2'})} = \begin{bmatrix} -a_{1'}s_{1'} - \ell_{2}s_{1'2'} & -\ell_{2'}s_{1'2'} & 0 \\ a_{1'}c_{1'} + \ell_{2'}c_{1'2'} & \ell_{2}c_{1'2'} & 0 \\ 0 & 0 & 0 \end{bmatrix}$$

$$\boldsymbol{J}_P^{(\ell_{3'})} = \begin{bmatrix} -a_{1'}s_{1'} - a_{2'}s_{1'2'} - \ell_{3'}s_{1'2'3'} & -a_{2'}s_{1'2'} - \ell_{3'}s_{1'2'3'} & -\ell_{3'}s_{1'2'3'} \\ a_{1'}c_{1'} + a_{2'}c_{1'2'} + \ell_{3'}c_{1'2'3'} & a_{2'}c_{1'2'} + \ell_{3'}c_{1'2'3'} & \ell_{3'}c_{1'2'3'} \\ 0 & 0 & 0 \end{bmatrix}$$

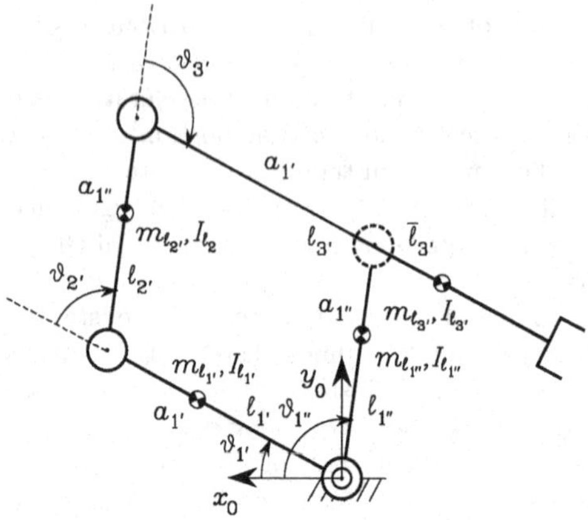

Fig. 7.12. Parallelogram arm

and

$$\boldsymbol{J}_P^{(\ell_{1''})} = \begin{bmatrix} -\ell_{1''} s_{1''} \\ \ell_{1''} c_{1''} \\ 0 \end{bmatrix},$$

whereas computation of the Jacobians in (7.17), (7.19) yields

$$\boldsymbol{J}_O^{(\ell_{1'})} = \begin{bmatrix} 0 & 0 & 0 \\ 0 & 0 & 0 \\ 1 & 0 & 0 \end{bmatrix} \qquad \boldsymbol{J}_O^{(\ell_{2'})} = \begin{bmatrix} 0 & 0 & 0 \\ 0 & 0 & 0 \\ 1 & 1 & 0 \end{bmatrix} \qquad \boldsymbol{J}_O^{(\ell_{3'})} = \begin{bmatrix} 0 & 0 & 0 \\ 0 & 0 & 0 \\ 1 & 1 & 1 \end{bmatrix}$$

and

$$\boldsymbol{J}_O^{(\ell_{1''})} = \begin{bmatrix} 0 \\ 0 \\ 1 \end{bmatrix}.$$

From (7.32), the inertia matrix of the virtual arm composed of joints $\vartheta_{1'}$, $\vartheta_{2'}$, $\vartheta_{3'}$ is

$$\boldsymbol{B}'(\boldsymbol{q}') = \begin{bmatrix} b_{1'1'}(\vartheta_{2'},\vartheta_{3'}) & b_{1'2'}(\vartheta_{2'},\vartheta_{3'}) & b_{1'3'}(\vartheta_{2'},\vartheta_{3'}) \\ b_{2'1'}(\vartheta_{2'},\vartheta_{3'}) & b_{2'2'}(\vartheta_{3'}) & b_{2'3'}(\vartheta_{3'}) \\ b_{3'1'}(\vartheta_{2'},\vartheta_{3'}) & b_{3'2'}(\vartheta_{3'}) & b_{3'3'} \end{bmatrix}$$

$$\begin{aligned}
b_{1'1'} &= I_{\ell_{1'}} + m_{\ell_{1'}}\ell_{1'}^2 + I_{\ell_{2'}} + m_{\ell_{2'}}(a_{1'}^2 + \ell_{2'}^2 + 2a_{1'}\ell_{2'}c_{2'}) + I_{\ell_{3'}} \\
&\quad + m_{\ell_{3'}}(a_{1'}^2 + a_{2'}^2 + \ell_{3'}^2 + 2a_{1'}a_{2'}c_{2'} + 2a_{1'}\ell_{3'}c_{2'3'} + 2a_{2'}\ell_{3'}c_{3'}) \\
b_{1'2'} &= b_{2'1'} = I_{\ell_{2'}} + m_{\ell_{2'}}(\ell_{2'}^2 + a_{1'}\ell_{2'}c_{2'}) + I_{\ell_{3'}} \\
&\quad + m_{\ell_{3'}}(a_{2'}^2 + \ell_{3'}^2 + a_{1'}a_{2'}c_{2'} + a_{1'}\ell_{3'}c_{2'3'} + 2a_{2'}\ell_{3'}c_{3'}) \\
b_{1'3'} &= b_{31} = I_{\ell_{3'}} + m_{\ell_{3'}}(\ell_{3'}^2 + a_{1'}\ell_{3'}c_{2'3'} + a_{2'}\ell_{3'}c_{3'}) \\
b_{2'2'} &= I_{\ell_{2'}} + m_{\ell_{2'}}\ell_{2'}^2 + I_{\ell_{3'}} + m_{\ell_{3'}}(a_{2'}^2 + \ell_{3'}^2 + 2a_{2'}\ell_{3'}c_{3'}) \\
b_{2'3'} &= I_{\ell_{3'}} + m_{\ell_{3'}}(\ell_{3'}^2 + a_{2'}\ell_{3'}c_{3'}) \\
b_{3'3'} &= I_{\ell_{3'}} + m_{\ell_{3'}}\ell_{3'}^2
\end{aligned}$$

while the moment of inertia of the virtual arm composed of just joint $\vartheta_{1''}$ is

$$b_{1''1''} = I_{\ell_{1''}} + m_{\ell_{1''}}\ell_{1''}^2.$$

Therefore, the inertial torque contributions of the two virtual arms are respectively:

$$\tau_{i'} = \sum_{j'=1'}^{3'} b_{i'j'}\ddot{\vartheta}_{j'} \qquad \tau_{1''} = b_{1''1''}\ddot{\vartheta}_{1''}.$$

At this point, in view of (2.64) and (3.121), the inertial torque contributions at the actuated joints for the closed-chain arm turn out to be

$$\boldsymbol{\tau}_a = \boldsymbol{B}_a\ddot{\boldsymbol{q}}_a$$

where $\boldsymbol{q}_a = \begin{bmatrix}\vartheta_{1'} & \vartheta_{1''}\end{bmatrix}^T$, $\boldsymbol{\tau}_a = \begin{bmatrix}\tau_{a1} & \tau_{a2}\end{bmatrix}^T$ and

$$\boldsymbol{B}_a = \begin{bmatrix} b_{a11} & b_{a12} \\ b_{a21} & b_{a22} \end{bmatrix}$$

$$\begin{aligned}
b_{a11} &= I_{\ell_{1'}} + m_{\ell_{1'}}\ell_{1'}^2 + m_{\ell_{2'}}a_{1'}^2 + I_{\ell_{3'}} + m_{\ell_{3'}}\ell_{3'}^2 + m_{\ell_{3'}}a_{1'}^2 \\
&\quad -2a_{1'}m_{\ell_{3'}}\ell_{3'} \\
b_{a12} &= b_{a21} = \big(a_{1'}m_{\ell_{2'}}\ell_{2'} + a_{1''}m_{\ell_{3'}}(a_{1'} - \ell_{3'})\big)\cos(\vartheta_{1''} - \vartheta_{1'}) \\
b_{a22} &= I_{\ell_{1'}} + m_{\ell_{1'}}\ell_{1'}^2 + I_{\ell_{2'}} + m_{\ell_{2'}}\ell_{2'}^2 + m_{\ell_{3'}}a_{1''}^2.
\end{aligned}$$

This expression reveals the possibility of obtaining a *configuration-independent* and *decoupled* inertia matrix; to this end it is sufficient to design the four links of the parallelogram so that

$$\frac{m_{\ell_{3'}}\bar{\ell}_{3'}}{m_{\ell_{2'}}\ell_{2'}} = \frac{a_{1'}}{a_{1''}}$$

where $\bar{\ell}_{3'} = \ell_{3'} - a_{1'}$ is the distance of the centre of mass of Link $3'$ from the axis of Joint 4. If this condition is satisfied, then the inertia matrix is diagonal $(b_{a12} = b_{a21} = 0)$ with

$$\begin{aligned}
b_{a11} &= I_{\ell_{1'}} + m_{\ell_{1'}}\ell_{1'}^2 + m_{\ell_{2'}}a_{1'}^2\left(1 + \frac{\ell_{2'}\bar{\ell}_{3'}}{a_{1'}a_{1''}}\right) + I_{\ell_{3'}} \\
b_{a22} &= I_{\ell_{1'}} + m_{\ell_{1'}}\ell_{1'}^2 + I_{\ell_{2'}} + m_{\ell_{2'}}\ell_{2'}^2\left(1 + \frac{a_{1'}a_{1''}}{\ell_{2'}\bar{\ell}_{3'}}\right).
\end{aligned}$$

As a consequence, no contributions of Coriolis and centrifugal torques are obtained. Such a result could not be achieved with the previous two-link planar arm, no matter how the design parameters were chosen.

As for the gravitational terms, since $\boldsymbol{g}_0 = [0 \quad -g \quad 0]^T$, (7.39) with the above Jacobians gives

$$\begin{aligned} g_{1'} \quad & (m_{\ell_{1'}}\ell_{1'} + m_{\ell_{2'}}a_{1'} + m_{\ell_{3'}}a_{1'})gc_{1'} + (m_{\ell_{2'}}\ell_{2'} + m_{\ell_{3'}}a_{2'})gc_{1'2'} \\ & + m_{\ell_{3'}}\ell_{3'}gc_{1'2'3} \\ g_{2'} \quad & (m_{\ell_{2'}}\ell_{2'} + m_{\ell_{3'}}a_{2'})gc_{1'2'} + m_{\ell_{3'}}\ell_{3'}gc_{1'2'3} \\ g_{3'} \quad & m_{\ell_{3'}}\ell_{3'}gc_{1'2'3} \end{aligned}$$

and

$$g_{1''} = m_{\ell_{1''}}\ell_{1''}gc_{1''}.$$

Composing the various contributions as done above yields

$$\boldsymbol{g}_a = \begin{bmatrix} (m_{\ell_{1'}}\ell_{1'} + m_{\ell_{2'}}a_{1'} - m_{\ell_{3'}}\bar{\ell}_{3'})gc_{1'} \\ (m_{\ell_{1''}}\ell_{1''} + m_{\ell_{2'}}\ell_{2'} + m_{\ell_{3'}}a_{1''})gc_{1''} \end{bmatrix}$$

which, together with the inertial torques, completes the derivation of the sought dynamic model.

A final comment is in order. In spite of its kinematic equivalence with the two-link planar arm, the dynamic model of the parallelogram is remarkably lighter. This property is quite advantageous for trajectory planning and control purposes. For this reason, apart from obvious considerations related to manipulation of heavy payloads, the adoption of closed kinematic chains in the design of industrial robots has received a great deal of attention.

7.4 Dynamic Parameter Identification

The use of the dynamic model for solving simulation and control problems demands the knowledge of the values of dynamic parameters of the manipulator model.

Computing such parameters from the design data of the mechanical structure is not simple. CAD modelling techniques can be adopted which allow the computation of the values of the inertial parameters of the various components (links, actuators and transmissions) on the basis of their geometry and type of materials employed. Nevertheless, the estimates obtained by such techniques are inaccurate because of the simplification typically introduced by geometric modelling; moreover, complex dynamic effects, such as joint friction, cannot be taken into account.

A heuristic approach could be to dismantle the various components of the manipulator and perform a series of measurements to evaluate the inertial parameters. Such technique is not easy to implement and may be troublesome to measure the relevant quantities.

In order to find accurate estimates of dynamic parameters, it is worth resorting to *identification* techniques which conveniently exploit the *property of linearity* (7.81) of the manipulator model with respect to a suitable set of *dynamic parameters*. Such techniques allow the computation of the parameter vector $\boldsymbol{\pi}$ from the measurements of joint torques $\boldsymbol{\tau}$ and of relevant quantities for the evaluation of the matrix $\boldsymbol{Y}$, when suitable motion trajectories are imposed to the manipulator.

On the assumption that the kinematic parameters in the matrix $\boldsymbol{Y}$ are known with good accuracy, e.g., as a result of a kinematic calibration, measurements of joint positions $\boldsymbol{q}$, velocities $\dot{\boldsymbol{q}}$ and accelerations $\ddot{\boldsymbol{q}}$ are required. Joint positions and velocities can be actually measured while numerical reconstruction of accelerations is needed; this can be performed on the basis of the position and velocity values recorded during the execution of the trajectories. The reconstructing filter does not work in real time and thus it can also be anti-causal, allowing an accurate reconstruction of the accelerations.

As regards joint torques, in the unusual case of torque sensors at the joint, these can be measured directly. Otherwise, they can be evaluated from either wrist force measurements or current measurements in the case of electric actuators.

If measurements of joint torques, positions, velocities and accelerations have been obtained at given time instants $t_1, \ldots, t_N$ along a given trajectory, one may write

$$\bar{\boldsymbol{\tau}} = \begin{bmatrix} \boldsymbol{\tau}(t_1) \\ \vdots \\ \boldsymbol{\tau}(t_N) \end{bmatrix} = \begin{bmatrix} \boldsymbol{Y}(t_1) \\ \vdots \\ \boldsymbol{Y}(t_N) \end{bmatrix} \boldsymbol{\pi} = \bar{\boldsymbol{Y}} \boldsymbol{\pi}. \tag{7.85}$$

The number of time instants sets the number of measurements to perform and should be large enough (typically $Nn \gg p$) so as to avoid ill-conditioning of matrix $\bar{\boldsymbol{Y}}$. Solving (7.85) by a least-squares technique leads to the solution in the form

$$\boldsymbol{\pi} = (\bar{\boldsymbol{Y}}^T \bar{\boldsymbol{Y}})^{-1} \bar{\boldsymbol{Y}}^T \bar{\boldsymbol{\tau}} \tag{7.86}$$

where $(\bar{\boldsymbol{Y}}^T \bar{\boldsymbol{Y}})^{-1} \bar{\boldsymbol{Y}}^T$ is the *left pseudo-inverse* matrix of $\bar{\boldsymbol{Y}}$.

It should be noticed that, in view of the block triangular structure of matrix $\boldsymbol{Y}$ in (7.80), computation of parameter estimates could be simplified by resorting to a sequential procedure. Take the equation $\tau_n = \boldsymbol{y}_{nn}^T \boldsymbol{\pi}_n$ and solve it for $\boldsymbol{\pi}_n$ by specifying τ_n and $\boldsymbol{y}_{nn}^T$ for a given trajectory on Joint n. By iterating the procedure, the manipulator parameters can be identified on the basis of measurements performed joint by joint from the outer link to the base. Such procedure, however, may have the inconvenience to accumulate any error due to ill-conditioning of the matrices involved step by step. It may then be worth operating with a global procedure by imposing motions on all manipulator joints at the same time.

Regarding the rank of matrix $\bar{\boldsymbol{Y}}$, it is possible to identify only the dynamic parameters of the manipulator that contribute to the dynamic model. Example 7.2 has indeed shown that for the two-link planar arm considered, only 8 out of the 22 possible dynamic parameters appear in the dynamic model. Hence, there exist some dynamic parameters which, in view of the disposition of manipulator links and joints, are *non-identifiable*, since for any trajectory assigned to the structure they do not contribute to the equations of motion. A direct consequence is that the columns of the matrix $\boldsymbol{Y}$ in (7.80) corresponding to such parameters are null and thus they have to be removed from the matrix itself; e.g., the resulting (2×8) matrix in (7.84).

Another issue to consider about determination of the effective number of parameters that can be identified by (7.86) is that some parameters can be *identified in linear combinations* whenever they do not appear isolated in the equations. In such a case, it is necessary, for each linear combination, to remove as many columns of the matrix $\boldsymbol{Y}$ as the number of parameters in the linear combination minus one.

For the determination of the minimum number of identifiable parameters that allow direct application of the least-squares technique based on (7.86), it is possible to inspect directly the equations of the dynamic model, as long as the manipulator has few joints. Otherwise, numerical techniques based on singular value decomposition of matrix $\bar{\boldsymbol{Y}}$ have to be used. If the matrix $\bar{\boldsymbol{Y}}$ resulting from a series of measurements is not full-rank, one has to resort to a *damped least-squares inverse* of $\bar{\boldsymbol{Y}}$ where solution accuracy depends on the weight of the damping factor.

In the above discussion, the type of trajectory imposed to the manipulator joints has not been explicitly addressed. It can be generally ascertained that the choice should be oriented in favor of polynomial type trajectories which are sufficiently *rich* to allow an accurate evaluation of the identifiable parameters. This corresponds to achieving a low condition number of the matrix $\bar{\boldsymbol{Y}}^T\bar{\boldsymbol{Y}}$ along the trajectory. On the other hand, such trajectories should not excite any unmodelled dynamic effects such as joint elasticity or link flexibility that would naturally lead to unreliable estimates of the dynamic parameters to identify.

Finally, it is worth observing that the technique presented above can also be extended to the identification of the dynamic parameters of an unknown payload at the manipulator's end-effector. In such a case, the payload can be regarded as a structural modification of the last link and one may proceed to identify the dynamic parameters of the modified link. To this end, if a force sensor is available at the manipulator's wrist, it is possible to characterize directly the dynamic parameters of the payload starting from force sensor measurements.

7.5 Newton–Euler Formulation

In the Lagrange formulation, the manipulator dynamic model is derived starting from the total Lagrangian of the system. On the other hand, the *Newton–Euler* formulation is based on a balance of all the forces acting on the generic link of the manipulator. This leads to a set of equations whose structure allows a recursive type of solution; a forward recursion is performed for propagating

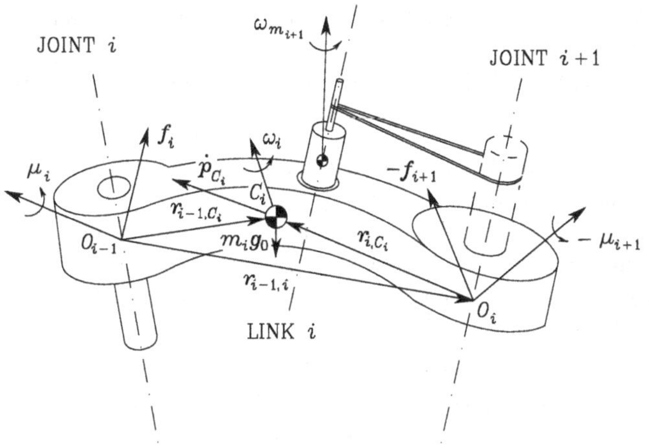

Fig. 7.13. Characterization of Link i for Newton–Euler formulation

link velocities and accelerations, followed by a backward recursion for propagating forces.

Consider the generic *augmented Link i* (Link i plus motor of Joint $i+1$) of the manipulator kinematic chain (Fig. 7.13). According to what was presented in Sect. 7.2.2, one can refer to the centre of mass C_i of the augmented link to characterize the following parameters:

- m_i mass of augmented link,
- $\bar{\boldsymbol{I}}_i$ inertia tensor of augmented link,
- I_{m_i} moment of inertia of rotor,
- $\boldsymbol{r}_{i-1,C_i}$ vector from origin of Frame $(i-1)$ to centre of mass C_i,
- $\boldsymbol{r}_{i,C_i}$ vector from origin of Frame i to centre of mass C_i,
- $\boldsymbol{r}_{i-1,i}$ vector from origin of Frame $(i-1)$ to origin of Frame i.

The velocities and accelerations to be considered are:

- $\dot{\boldsymbol{p}}_{C_i}$ linear velocity of centre of mass C_i,
- $\dot{\boldsymbol{p}}_i$ linear velocity of origin of Frame i,
- $\boldsymbol{\omega}_i$ angular velocity of link,
- $\boldsymbol{\omega}_{m_i}$ angular velocity of rotor,
- $\ddot{\boldsymbol{p}}_{C_i}$ linear acceleration of centre of mass C_i,
- $\ddot{\boldsymbol{p}}_i$ linear acceleration of origin of Frame i,
- $\dot{\boldsymbol{\omega}}_i$ angular acceleration of link,
- $\dot{\boldsymbol{\omega}}_{m_i}$ angular acceleration of rotor,
- $\boldsymbol{g}_0$ gravity acceleration.

The forces and moments to be considered are:

- $\boldsymbol{f}_i$ force exerted by Link $i-1$ on Link i,
- $-\boldsymbol{f}_{i+1}$ force exerted by Link $i+1$ on Link i,

- $\boldsymbol{\mu}_i$ moment exerted by Link $i-1$ on Link i with respect to origin of Frame $i-1$,
- $-\boldsymbol{\mu}_{i+1}$ moment exerted by Link $i+1$ on Link i with respect to origin of Frame i.

Initially, all the vectors and matrices are assumed to be expressed with reference to the *base frame*.

As already anticipated, the Newton–Euler formulation describes the motion of the link in terms of a balance of forces and moments acting on it.

The *Newton* equation for the *translational* motion of the centre of mass can be written as

$$\boldsymbol{f}_i - \boldsymbol{f}_{i+1} + m_i\boldsymbol{g}_0 = m_i\ddot{\boldsymbol{p}}_{C_i}. \tag{7.87}$$

The *Euler* equation for the *rotational* motion of the link (referring moments to the centre of mass) can be written as

$$\boldsymbol{\mu}_i + \boldsymbol{f}_i \times \boldsymbol{r}_{i-1,C_i} - \boldsymbol{\mu}_{i+1} - \boldsymbol{f}_{i+1} \times \boldsymbol{r}_{i,C_i} = \frac{d}{dt}(\bar{\boldsymbol{I}}_i\boldsymbol{\omega}_i + k_{r,i+1}\dot{q}_{i+1}I_{m_{i+1}}\boldsymbol{z}_{m_{i+1}}), \tag{7.88}$$

where (7.67) has been used for the angular momentum of the rotor. Notice that the gravitational force $m_i\boldsymbol{g}_0$ does not generate any moment, since it is concentrated at the centre of mass.

As pointed out in the above Lagrange formulation, it is convenient to express the inertia tensor in the current frame (constant tensor). Hence, according to (7.12), one has $\bar{\boldsymbol{I}}_i = \boldsymbol{R}_i\bar{\boldsymbol{I}}_i^i\boldsymbol{R}_i^T$, where $\boldsymbol{R}_i$ is the rotation matrix from Frame i to the base frame. Substituting this relation in the first term on the right-hand side of (7.88) yields

$$\begin{aligned}\frac{d}{dt}(\bar{\boldsymbol{I}}_i\boldsymbol{\omega}_i) &= \dot{\boldsymbol{R}}_i\bar{\boldsymbol{I}}_i^i\boldsymbol{R}_i^T\boldsymbol{\omega}_i + \boldsymbol{R}_i\bar{\boldsymbol{I}}_i^i\dot{\boldsymbol{R}}_i^T\boldsymbol{\omega}_i + \boldsymbol{R}_i\bar{\boldsymbol{I}}_i^i\boldsymbol{R}_i^T\dot{\boldsymbol{\omega}}_i \\ &= \boldsymbol{S}(\boldsymbol{\omega}_i)\boldsymbol{R}_i\bar{\boldsymbol{I}}_i^i\boldsymbol{R}_i^T\boldsymbol{\omega}_i + \boldsymbol{R}_i\bar{\boldsymbol{I}}_i^i\boldsymbol{R}_i^T\boldsymbol{S}^T(\boldsymbol{\omega}_i)\boldsymbol{\omega}_i + \boldsymbol{R}_i\bar{\boldsymbol{I}}_i^i\boldsymbol{R}_i^T\dot{\boldsymbol{\omega}}_i \\ &= \bar{\boldsymbol{I}}_i\dot{\boldsymbol{\omega}}_i + \boldsymbol{\omega}_i \times (\bar{\boldsymbol{I}}_i\boldsymbol{\omega}_i)\end{aligned} \tag{7.89}$$

where the second term represents the *gyroscopic* torque induced by the dependence of $\bar{\boldsymbol{I}}_i$ on link orientation.[7] Moreover, by observing that the unit vector $\boldsymbol{z}_{m_{i+1}}$ rotates accordingly to Link i, the derivative needed in the second term on the right-hand side of (7.88) is

$$\frac{d}{dt}(\dot{q}_{i+1}I_{m_{i+1}}\boldsymbol{z}_{m_{i+1}}) = \ddot{q}_{i+1}I_{m_{i+1}}\boldsymbol{z}_{m_{i+1}} + \dot{q}_{i+1}I_{m_{i+1}}\boldsymbol{\omega}_i \times \boldsymbol{z}_{m_{i+1}} \tag{7.90}$$

By substituting (7.89), (7.90) in (7.88), the resulting Euler equation is

$$\begin{aligned}\boldsymbol{\mu}_i + \boldsymbol{f}_i \times \boldsymbol{r}_{i-1,C_i} - \boldsymbol{\mu}_{i+1} - \boldsymbol{f}_{i+1} \times \boldsymbol{r}_{i,C_i} &= \bar{\boldsymbol{I}}_i\dot{\boldsymbol{\omega}}_i + \boldsymbol{\omega}_i \times (\bar{\boldsymbol{I}}_i\boldsymbol{\omega}_i) \\ &\quad + k_{r,i+1}\ddot{q}_{i+1}I_{m_{i+1}}\boldsymbol{z}_{m_{i+1}} + k_{r,i+1}\dot{q}_{i+1}I_{m_{i+1}}\boldsymbol{\omega}_i \times \boldsymbol{z}_{m_{i+1}}.\end{aligned} \tag{7.91}$$

---

[7] In deriving (7.89), the operator $\boldsymbol{S}$ has been introduced to compute the derivative of $\boldsymbol{R}_i$, as in (3.8); also, the property $\boldsymbol{S}^T(\boldsymbol{\omega}_i)\boldsymbol{\omega}_i = \boldsymbol{0}$ has been utilized.

The generalized force at Joint i can be computed by projecting the force $\boldsymbol{f}_i$ for a prismatic joint, or the moment $\boldsymbol{\mu}_i$ for a revolute joint, along the joint axis. In addition, there is the contribution of the rotor inertia torque $k_{ri}I_{m_i}\dot{\boldsymbol{\omega}}_{m_i}^T\boldsymbol{z}_{m_i}$. Hence, the generalized force at Joint i is expressed by

$$\tau_i = \begin{cases} \boldsymbol{f}_i^T\boldsymbol{z}_{i-1} + k_{ri}I_{m_i}\dot{\boldsymbol{\omega}}_{m_i}^T\boldsymbol{z}_{m_i} & \text{for a } prismatic \text{ joint} \\ \boldsymbol{\mu}_i^T\boldsymbol{z}_{i-1} + k_{ri}I_{m_i}\dot{\boldsymbol{\omega}}_{m_i}^T\boldsymbol{z}_{m_i} & \text{for a } revolute \text{ joint.} \end{cases} \tag{7.92}$$

7.5.1 Link Accelerations

The Newton–Euler equations in (7.87), (7.91) and the equation in (7.92) require the computation of linear and angular acceleration of Link i and Rotor i. This computation can be carried out on the basis of the relations expressing the linear and angular velocities previously derived. The equations in (3.21), (3.22), (3.25), (3.26) can be briefly rewritten as

$$\boldsymbol{\omega}_i = \begin{cases} \boldsymbol{\omega}_{i-1} & \text{for a } prismatic \text{ joint} \\ \boldsymbol{\omega}_{i-1} + \dot{\vartheta}_i\boldsymbol{z}_{i-1} & \text{for a } revolute \text{ joint} \end{cases} \tag{7.93}$$

and

$$\dot{\boldsymbol{p}}_i = \begin{cases} \dot{\boldsymbol{p}}_{i-1} + \dot{d}_i\boldsymbol{z}_{i-1} + \boldsymbol{\omega}_i \times \boldsymbol{r}_{i-1,i} & \text{for a } prismatic \text{ joint} \\ \dot{\boldsymbol{p}}_{i-1} + \boldsymbol{\omega}_i \times \boldsymbol{r}_{i-1,i} & \text{for a } revolute \text{ joint.} \end{cases} \tag{7.94}$$

As for the angular acceleration of the link, it can be seen that, for a prismatic joint, differentiating (3.21) with respect to time gives

$$\dot{\boldsymbol{\omega}}_i = \dot{\boldsymbol{\omega}}_{i-1}, \tag{7.95}$$

whereas, for a revolute joint, differentiating (3.25) with respect to time gives

$$\dot{\boldsymbol{\omega}}_i = \dot{\boldsymbol{\omega}}_{i-1} + \ddot{\vartheta}_i\boldsymbol{z}_{i-1} + \dot{\vartheta}_i\boldsymbol{\omega}_{i-1} \times \boldsymbol{z}_{i-1}. \tag{7.96}$$

As for the linear acceleration of the link, for a prismatic joint, differentiating (3.22) with respect to time gives

$$\begin{aligned} \ddot{\boldsymbol{p}}_i = \ddot{\boldsymbol{p}}_{i-1} &+ \ddot{d}_i\boldsymbol{z}_{i-1} + \dot{d}_i\boldsymbol{\omega}_{i-1} \times \boldsymbol{z}_{i-1} + \dot{\boldsymbol{\omega}}_i \times \boldsymbol{r}_{i-1,i} \\ &+ \boldsymbol{\omega}_i \times \dot{d}_i\boldsymbol{z}_{i-1} + \boldsymbol{\omega}_i \times (\boldsymbol{\omega}_{i-1} \times \boldsymbol{r}_{i-1,i}) \end{aligned} \tag{7.97}$$

where the relation $\dot{\boldsymbol{r}}_{i-1,i} = \dot{d}_i\boldsymbol{z}_{i-1} + \boldsymbol{\omega}_{i-1} \times \boldsymbol{r}_{i-1,i}$ has been used. Hence, in view of (3.21), the equation in (7.97) can be rewritten as

$$\ddot{\boldsymbol{p}}_i = \ddot{\boldsymbol{p}}_{i-1} + \ddot{d}_i\boldsymbol{z}_{i-1} + 2\dot{d}_i\boldsymbol{\omega}_i \times \boldsymbol{z}_{i-1} + \dot{\boldsymbol{\omega}}_i \times \boldsymbol{r}_{i-1,i} + \boldsymbol{\omega}_i \times (\boldsymbol{\omega}_i \times \boldsymbol{r}_{i-1,i}). \tag{7.98}$$

Also, for a revolute joint, differentiating (3.26) with respect to time gives

$$\ddot{\boldsymbol{p}}_i = \ddot{\boldsymbol{p}}_{i-1} + \dot{\boldsymbol{\omega}}_i \times \boldsymbol{r}_{i-1,i} + \boldsymbol{\omega}_i \times (\boldsymbol{\omega}_i \times \boldsymbol{r}_{i-1,i}). \tag{7.99}$$

In summary, the equations in (7.95), (7.96), (7.98), (7.99) can be compactly rewritten as

$$\dot{\boldsymbol{\omega}}_i = \begin{cases} \dot{\boldsymbol{\omega}}_{i-1} & \text{for a } prismatic \text{ joint} \\ \dot{\boldsymbol{\omega}}_{i-1} + \ddot{\vartheta}_i \boldsymbol{z}_{i-1} + \dot{\vartheta}_i \boldsymbol{\omega}_{i-1} \times \boldsymbol{z}_{i-1} & \text{for a } revolute \text{ joint} \end{cases} \tag{7.100}$$

and

$$\ddot{\boldsymbol{p}}_i = \begin{cases} \ddot{\boldsymbol{p}}_{i-1} + \ddot{d}_i \boldsymbol{z}_{i-1} + 2\dot{d}_i \boldsymbol{\omega}_i \times \boldsymbol{z}_{i-1} & \text{for a } prismatic \text{ joint} \\ \quad + \dot{\boldsymbol{\omega}}_i \times \boldsymbol{r}_{i-1,i} + \boldsymbol{\omega}_i \times (\boldsymbol{\omega}_i \times \boldsymbol{r}_{i-1,i}) & \\ \ddot{\boldsymbol{p}}_{i-1} + \dot{\boldsymbol{\omega}}_i \times \boldsymbol{r}_{i-1,i} & \text{for a } revolute \text{ joint.} \\ \quad + \boldsymbol{\omega}_i \times (\boldsymbol{\omega}_i \times \boldsymbol{r}_{i-1,i}) & \end{cases} \tag{7.101}$$

The acceleration of the centre of mass of Link i required by the Newton equation in (7.87) can be derived from (3.15), since $\dot{\boldsymbol{r}}^i_{i,C_i} = \boldsymbol{0}$; by differentiating (3.15) with respect to time, the acceleration of the centre of mass C_i can be expressed as a function of the velocity and acceleration of the origin of Frame i, i.e.,

$$\ddot{\boldsymbol{p}}_{C_i} = \ddot{\boldsymbol{p}}_i + \dot{\boldsymbol{\omega}}_i \times \boldsymbol{r}_{i,C_i} + \boldsymbol{\omega}_i \times (\boldsymbol{\omega}_i \times \boldsymbol{r}_{i,C_i}). \tag{7.102}$$

Finally, the angular acceleration of the rotor can be obtained by time differentiation of (7.23), i.e.,

$$\dot{\boldsymbol{\omega}}_{m_i} = \dot{\boldsymbol{\omega}}_{i-1} + k_{ri}\ddot{q}_i \boldsymbol{z}_{m_i} + k_{ri}\dot{q}_i \boldsymbol{\omega}_{i-1} \times \boldsymbol{z}_{m_i}. \tag{7.103}$$

7.5.2 Recursive Algorithm

It is worth remarking that the resulting Newton–Euler equations of motion are *not* in *closed form*, since the motion of a single link is coupled to the motion of the other links through the kinematic relationship for velocities and accelerations.

Once the joint positions, velocities and accelerations are known, one can compute the link velocities and accelerations, and the Newton–Euler equations can be utilized to find the forces and moments acting on each link in a recursive fashion, starting from the force and moment applied to the end-effector. On the other hand, also link and rotor velocities and accelerations can be computed recursively starting from the velocity and acceleration of the base link. In summary, a computationally *recursive algorithm* can be constructed that features a *forward recursion* relative to the propagation of *velocities and accelerations* and a *backward recursion* for the propagation of *forces and moments* along the structure.

For the forward recursion, once $\boldsymbol{q}$, $\dot{\boldsymbol{q}}$, $\ddot{\boldsymbol{q}}$, and the velocity and acceleration of the base link $\boldsymbol{\omega}_0$, $\ddot{\boldsymbol{p}}_0 - \boldsymbol{g}_0$, $\dot{\boldsymbol{\omega}}_0$ are specified, $\boldsymbol{\omega}_i$, $\dot{\boldsymbol{\omega}}_i$, $\ddot{\boldsymbol{p}}_i$, $\ddot{\boldsymbol{p}}_{C_i}$, $\dot{\boldsymbol{\omega}}_{m_i}$ can be computed using (7.93), (7.100), (7.101), (7.102), (7.103), respectively. Notice that the linear acceleration has been taken as $\ddot{\boldsymbol{p}}_0 - \boldsymbol{g}_0$ so as to incorporate the term $-\boldsymbol{g}_0$ in the computation of the acceleration of the centre of mass $\ddot{\boldsymbol{p}}_{C_i}$ via (7.101), (7.102).

Having computed the velocities and accelerations with the forward recursion from the base link to the end-effector, a backward recursion can be carried out for the forces. In detail, once $\boldsymbol{h}_e = [\,\boldsymbol{f}_{n+1}^T \quad \boldsymbol{\mu}_{n+1}^T\,]^T$ is given (eventually $\boldsymbol{h}_e = \boldsymbol{0}$), the Newton equation in (7.87) to be used for the recursion can be rewritten as

$$\boldsymbol{f}_i = \boldsymbol{f}_{i+1} + m_i \ddot{\boldsymbol{p}}_{C_i} \tag{7.104}$$

since the contribution of gravity acceleration has already been included in $\ddot{\boldsymbol{p}}_{C_i}$. Further, the Euler equation gives

$$\begin{aligned}\boldsymbol{\mu}_i = &-\boldsymbol{f}_i \times (\boldsymbol{r}_{i-1,i} + \boldsymbol{r}_{i,C_i}) + \boldsymbol{\mu}_{i+1} + \boldsymbol{f}_{i+1} \times \boldsymbol{r}_{i,C_i} + \bar{\boldsymbol{I}}_i \dot{\boldsymbol{\omega}}_i + \boldsymbol{\omega}_i \times (\bar{\boldsymbol{I}}_i \boldsymbol{\omega}_i)\\ &+ k_{r,i+1} \ddot{q}_{i+1} I_{m_{i+1}} \boldsymbol{z}_{m_{i+1}} + k_{r,i+1} \dot{q}_{i+1} I_{m_{i+1}} \boldsymbol{\omega}_i \times \boldsymbol{z}_{m_{i+1}}\end{aligned} \tag{7.105}$$

which derives from (7.91), where $\boldsymbol{r}_{i-1,C_i}$ has been expressed as the sum of the two vectors appearing already in the forward recursion. Finally, the generalized forces resulting at the joints can be computed from (7.92) as

$$\tau_i = \begin{cases} \boldsymbol{f}_i^T \boldsymbol{z}_{i-1} + k_{ri} I_{m_i} \dot{\boldsymbol{\omega}}_{m_i}^T \boldsymbol{z}_{m_i} \\ \quad + F_{vi} \dot{d}_i + F_{si} \operatorname{sgn}(\dot{d}_i) & \text{for a } \textit{prismatic} \text{ joint} \\ \boldsymbol{\mu}_i^T \boldsymbol{z}_{i-1} + k_{ri} I_{m_i} \dot{\boldsymbol{\omega}}_{m_i}^T \boldsymbol{z}_{m_i} \\ \quad + F_{vi} \dot{\vartheta}_i + F_{si} \operatorname{sgn}(\dot{\vartheta}_i) & \text{for a } \textit{revolute} \text{ joint,} \end{cases} \tag{7.106}$$

where joint viscous and Coulomb friction torques have been included.

In the above derivation, it has been assumed that all vectors were referred to the base frame. To simplify greatly computation, however, the recursion is computationally more efficient if all vectors are referred to the current frame on Link i. This implies that all vectors that need to be transformed from Frame $i+1$ into Frame i have to be multiplied by the rotation matrix $\boldsymbol{R}_{i+1}^i$, whereas all vectors that need to be transformed from Frame $i-1$ into Frame i have to be multiplied by the rotation matrix $\boldsymbol{R}_i^{i-1T}$. Therefore, the equations in (7.93), (7.100), (7.101), (7.102), (7.103), (7.104), (7.105), (7.106) can be rewritten as:

$$\boldsymbol{\omega}_i^i = \begin{cases} \boldsymbol{R}_i^{i-1T} \boldsymbol{\omega}_{i-1}^{i-1} & \text{for a } \textit{prismatic} \text{ joint} \\ \boldsymbol{R}_i^{i-1T} (\boldsymbol{\omega}_{i-1}^{i-1} + \dot{\vartheta}_i \boldsymbol{z}_0) & \text{for a } \textit{revolute} \text{ joint} \end{cases} \tag{7.107}$$

$$\dot{\boldsymbol{\omega}}_i^i = \begin{cases} \boldsymbol{R}_i^{i-1T} \dot{\boldsymbol{\omega}}_{i-1}^{i-1} & \text{for a } \textit{prismatic} \text{ joint} \\ \boldsymbol{R}_i^{i-1T} (\dot{\boldsymbol{\omega}}_{i-1}^{i-1} + \ddot{\vartheta}_i \boldsymbol{z}_0 + \dot{\vartheta}_i \boldsymbol{\omega}_{i-1}^{i-1} \times \boldsymbol{z}_0) & \text{for a } \textit{revolute} \text{ joint} \end{cases} \tag{7.108}$$

$$\ddot{\boldsymbol{p}}_i^i = \begin{cases} \boldsymbol{R}_i^{i-1T} (\ddot{\boldsymbol{p}}_{i-1}^{i-1} + \ddot{d}_i \boldsymbol{z}_0) + 2\dot{d}_i \boldsymbol{\omega}_i^i \times \boldsymbol{R}_i^{i-1T} \boldsymbol{z}_0 \\ \quad + \dot{\boldsymbol{\omega}}_i^i \times \boldsymbol{r}_{i-1,i}^i + \boldsymbol{\omega}_i^i \times (\boldsymbol{\omega}_i^i \times \boldsymbol{r}_{i-1,i}^i) & \text{for a } \textit{prismatic} \text{ joint} \\ \boldsymbol{R}_i^{i-1T} \ddot{\boldsymbol{p}}_{i-1}^{i-1} + \dot{\boldsymbol{\omega}}_i^i \times \boldsymbol{r}_{i-1,i}^i \\ \quad + \boldsymbol{\omega}_i^i \times (\boldsymbol{\omega}_i^i \times \boldsymbol{r}_{i-1,i}^i) & \text{for a } \textit{revolute} \text{ joint} \end{cases} \tag{7.109}$$

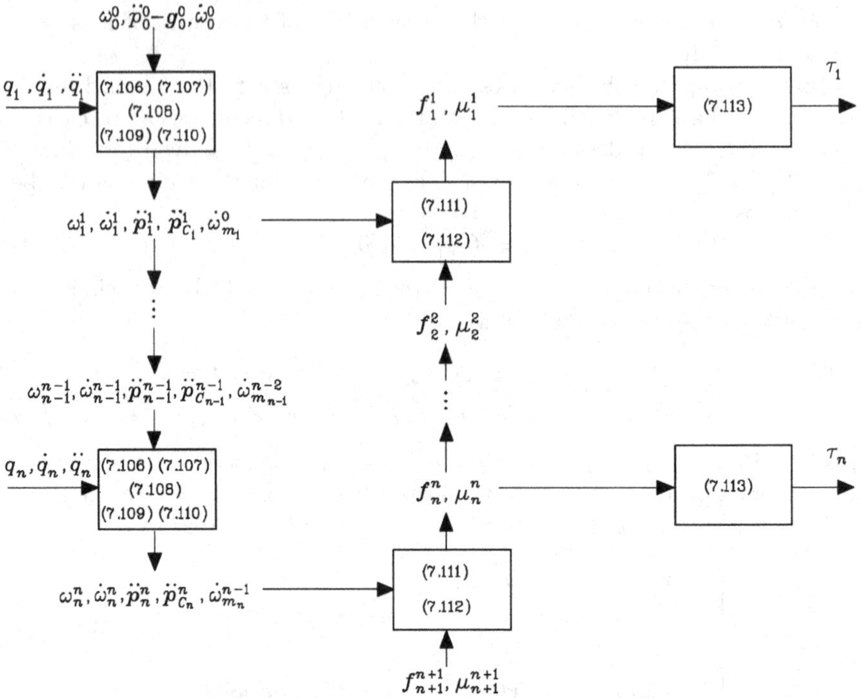

Fig. 7.14. Computational structure of the Newton–Euler recursive algorithm

$$\ddot{\boldsymbol{p}}_{C_i}^i = \ddot{\boldsymbol{p}}_i^i + \dot{\boldsymbol{\omega}}_i^i \times \boldsymbol{r}_{i,C_i}^i + \boldsymbol{\omega}_i^i \times (\boldsymbol{\omega}_i^i \times \boldsymbol{r}_{i,C_i}^i) \tag{7.110}$$

$$\dot{\boldsymbol{\omega}}_{m_i}^{i-1} = \dot{\boldsymbol{\omega}}_{i-1}^{i-1} + k_{ri}\ddot{q}_i\boldsymbol{z}_{m_i}^{i-1} + k_{ri}\dot{q}_i\boldsymbol{\omega}_{i-1}^{i-1} \times \boldsymbol{z}_{m_i}^{i-1} \tag{7.111}$$

$$\boldsymbol{f}_i^i = \boldsymbol{R}_{i+1}^i \boldsymbol{f}_{i+1}^{i+1} + m_i \ddot{\boldsymbol{p}}_{C_i}^i \tag{7.112}$$

$$\begin{aligned}\boldsymbol{\mu}_i^i = & -\boldsymbol{f}_i^i \times (\boldsymbol{r}_{i-1,i}^i + \boldsymbol{r}_{i,C_i}^i) + \boldsymbol{R}_{i+1}^i \boldsymbol{\mu}_{i+1}^{i+1} + \boldsymbol{R}_{i+1}^i \boldsymbol{f}_{i+1}^{i+1} \times \boldsymbol{r}_{i,C_i}^i \\ & + \bar{\boldsymbol{I}}_i^i \dot{\boldsymbol{\omega}}_i^i + \boldsymbol{\omega}_i^i \times (\bar{\boldsymbol{I}}_i^i \boldsymbol{\omega}_i^i) \\ & + \boldsymbol{\omega}_i^i \times (\bar{\boldsymbol{I}}_i^i \boldsymbol{\omega}_i^i) + k_{r,i+1}\ddot{q}_{i+1} I_{m_{i+1}} \boldsymbol{z}_{m_{i+1}}^i + k_{r,i+1}\dot{q}_{i+1} I_{m_{i+1}} \boldsymbol{\omega}_i^i \times \boldsymbol{z}_{m_{i+1}}^i\end{aligned} \tag{7.113}$$

$$\tau_i = \begin{cases} \boldsymbol{f}_i^{iT} \boldsymbol{R}_i^{i-1T} \boldsymbol{z}_0 + k_{ri} I_{m_i} \dot{\boldsymbol{\omega}}_{m_i}^{i-1T} \boldsymbol{z}_{m_i}^{i-1} & \\ \quad + F_{vi}\dot{d}_i + F_{si}\,\mathrm{sgn}\,(\dot{d}_i) & \text{for a } \textit{prismatic} \text{ joint} \\ \boldsymbol{\mu}_i^{iT} \boldsymbol{R}_i^{i-1T} \boldsymbol{z}_0 + k_{ri} I_{m_i} \dot{\boldsymbol{\omega}}_{m_i}^{i-1T} \boldsymbol{z}_{m_i}^{i-1} & \\ \quad + F_{vi}\dot{\vartheta}_i + F_{si}\,\mathrm{sgn}\,(\dot{\vartheta}_i) & \text{for a } \textit{revolute} \text{ joint.} \end{cases} \tag{7.114}$$

The above equations have the advantage that the quantities $\bar{\boldsymbol{I}}_i^i$, $\boldsymbol{r}_{i,C_i}^i$, $\boldsymbol{z}_{m_i}^{i-1}$ are *constant*; further, it is $\boldsymbol{z}_0 = [\,0 \quad 0 \quad 1\,]^T$.

To summarize, for given joint positions, velocities and accelerations, the recursive algorithm is carried out in the following two phases:

- With known initial conditions $\boldsymbol{\omega}_0^0$, $\ddot{\boldsymbol{p}}_0^0 - \boldsymbol{g}_0^0$, and $\dot{\boldsymbol{\omega}}_0^0$, use (7.107), (7.108), (7.109), (7.110), (7.111), for $i = 1, \ldots, n$, to compute $\boldsymbol{\omega}_i^i$, $\dot{\boldsymbol{\omega}}_i^i$, $\ddot{\boldsymbol{p}}_i^i$, $\ddot{\boldsymbol{p}}_{C_i}^i$, $\dot{\boldsymbol{\omega}}_{m_i}^{i-1}$.
- With known terminal conditions $\boldsymbol{f}_{n+1}^{n+1}$ and $\boldsymbol{\mu}_{n+1}^{n+1}$, use (7.112), (7.113), for $i = n, \ldots, 1$, to compute $\boldsymbol{f}_i^i$, $\boldsymbol{\mu}_i^i$, and then (7.114) to compute τ_i.

The computational structure of the algorithm is schematically illustrated in Fig. 7.14.

7.5.3 Example

In the following, an example to illustrate the single steps of the Newton–Euler algorithm is developed. Consider the two-link planar arm whose dynamic model has already been derived in Example 7.2.

Start by imposing the initial conditions for the velocities and accelerations:

$$\ddot{\boldsymbol{p}}_0^0 - \boldsymbol{g}_0^0 = [0 \quad g \quad 0]^T \qquad \boldsymbol{\omega}_0^0 = \dot{\boldsymbol{\omega}}_0^0 = \boldsymbol{0},$$

and the terminal conditions for the forces:

$$\boldsymbol{f}_3^3 = \boldsymbol{0} \qquad \boldsymbol{\mu}_3^3 = \boldsymbol{0}.$$

All quantities are referred to the current link frame. As a consequence, the following constant vectors are obtained:

$$\boldsymbol{r}_{1,C_1}^1 = \begin{bmatrix} \ell_{C_1} \\ 0 \\ 0 \end{bmatrix} \quad \boldsymbol{r}_{0,1}^1 = \begin{bmatrix} a_1 \\ 0 \\ 0 \end{bmatrix} \quad \boldsymbol{r}_{2,C_2}^2 = \begin{bmatrix} \ell_{C_2} \\ 0 \\ 0 \end{bmatrix} \quad \boldsymbol{r}_{1,2}^2 = \begin{bmatrix} a_2 \\ 0 \\ 0 \end{bmatrix}$$

where ℓ_{C_1} and ℓ_{C_2} are both negative quantities. The rotation matrices needed for vector transformation from one frame to another are

$$\boldsymbol{R}_i^{i-1} = \begin{bmatrix} c_i & -s_i & 0 \\ s_i & c_i & 0 \\ 0 & 0 & 1 \end{bmatrix} \qquad i = 1, 2 \qquad\qquad \boldsymbol{R}_3^2 = \boldsymbol{I}.$$

Further, it is assumed that the axes of rotation of the two rotors coincide with the respective joint axes, i.e., $\boldsymbol{z}_{m_i}^{i-1} = \boldsymbol{z}_0 = [0 \quad 0 \quad 1]^T$ for $i = 1, 2$.

According to (7.107)–(7.114), the Newton–Euler algorithm requires the execution of the following steps:

- Forward recursion: Link 1

$$\boldsymbol{\omega}_1^1 = \begin{bmatrix} 0 \\ 0 \\ \dot{\vartheta}_1 \end{bmatrix}$$

$$\dot{\boldsymbol{\omega}}_1^1 = \begin{bmatrix} 0 \\ 0 \\ \ddot{\vartheta}_1 \end{bmatrix}$$

$$\ddot{\boldsymbol{p}}_1^1 = \begin{bmatrix} -a_1\dot{\vartheta}_1^2 + gs_1 \\ a_1\ddot{\vartheta}_1 + gc_1 \\ 0 \end{bmatrix}$$

$$\ddot{\boldsymbol{p}}_{C_1}^1 = \begin{bmatrix} -(\ell_{C_1} + a_1)\dot{\vartheta}_1^2 + gs_1 \\ (\ell_{C_1} + a_1)\ddot{\vartheta}_1 + gc_1 \\ 0 \end{bmatrix}$$

$$\dot{\boldsymbol{\omega}}_{m_1}^0 = \begin{bmatrix} 0 \\ 0 \\ k_{r1}\ddot{\vartheta}_1 \end{bmatrix}.$$

- Forward recursion: Link 2

$$\boldsymbol{\omega}_2^2 = \begin{bmatrix} 0 \\ 0 \\ \dot{\vartheta}_1 + \dot{\vartheta}_2 \end{bmatrix}$$

$$\dot{\boldsymbol{\omega}}_2^2 = \begin{bmatrix} 0 \\ 0 \\ \ddot{\vartheta}_1 + \ddot{\vartheta}_2 \end{bmatrix}$$

$$\ddot{\boldsymbol{p}}_2^2 = \begin{bmatrix} a_1s_2\ddot{\vartheta}_1 - a_1c_2\dot{\vartheta}_1^2 - a_2(\dot{\vartheta}_1 + \dot{\vartheta}_2)^2 + gs_{12} \\ a_1c_2\ddot{\vartheta}_1 + a_2(\ddot{\vartheta}_1 + \ddot{\vartheta}_2) + a_1s_2\dot{\vartheta}_1^2 + gc_{12} \\ 0 \end{bmatrix}$$

$$\ddot{\boldsymbol{p}}_{C_2}^2 = \begin{bmatrix} a_1s_2\ddot{\vartheta}_1 - a_1c_2\dot{\vartheta}_1^2 - (\ell_{C_2} + a_2)(\dot{\vartheta}_1 + \dot{\vartheta}_2)^2 + gs_{12} \\ a_1c_2\ddot{\vartheta}_1 + (\ell_{C_2} + a_2)(\ddot{\vartheta}_1 + \ddot{\vartheta}_2) + a_1s_2\dot{\vartheta}_1^2 + gc_{12} \\ 0 \end{bmatrix}$$

$$\dot{\boldsymbol{\omega}}_{m_2}^1 = \begin{bmatrix} 0 \\ 0 \\ \ddot{\vartheta}_1 + k_{r2}\ddot{\vartheta}_2 \end{bmatrix}.$$

- Backward recursion: Link 2

$$
\boldsymbol{f}_2^2 = \begin{bmatrix} m_2\big(a_1 s_2 \ddot{\vartheta}_1 - a_1 c_2 \dot{\vartheta}_1^2 - (\ell_{C_2} + a_2)(\dot{\vartheta}_1 + \dot{\vartheta}_2)^2 + g s_{12}\big) \\ m_2\big(a_1 c_2 \ddot{\vartheta}_1 + (\ell_{C_2} + a_2)(\ddot{\vartheta}_1 + \ddot{\vartheta}_2) + a_1 s_2 \dot{\vartheta}_1^2 + g c_{12}\big) \\ 0 \end{bmatrix}
$$

$$
\boldsymbol{\mu}_2^2 = \begin{bmatrix} * \\ * \\ \bar{I}_{2zz}(\ddot{\vartheta}_1 + \ddot{\vartheta}_2) + m_2(\ell_{C_2} + a_2)^2(\ddot{\vartheta}_1 + \ddot{\vartheta}_2) + m_2 a_1(\ell_{C_2} + a_2) c_2 \ddot{\vartheta}_1 \\ + m_2 a_1 (\ell_{C_2} + a_2) s_2 \dot{\vartheta}_1^2 + m_2(\ell_{C_2} + a_2) g c_{12} \end{bmatrix}
$$

$$
\begin{aligned}
\tau_2 = {} & \big(\bar{I}_{2zz} + m_2\big((\ell_{C_2} + a_2)^2 + a_1(\ell_{C_2} + a_2)c_2\big) + k_{r2} I_{m_2}\big)\ddot{\vartheta}_1 \\
& + \big(\bar{I}_{2zz} + m_2(\ell_{C_2} + a_2)^2 + k_{r2}^2 I_{m_2}\big)\ddot{\vartheta}_2 \\
& + m_2 a_1(\ell_{C_2} + a_2) s_2 \dot{\vartheta}_1^2 + m_2(\ell_{C_2} + a_2) g c_{12}.
\end{aligned}
$$

- Backward recursion: Link 1

$$
\boldsymbol{f}_1^1 = \begin{bmatrix} -m_2(\ell_{C_2} + a_2) s_2 (\ddot{\vartheta}_1 + \ddot{\vartheta}_2) - m_1(\ell_{C_1} + a_1)\dot{\vartheta}_1^2 - m_2 a_1 \dot{\vartheta}_1^2 \\ - m_2(\ell_{C_2} + a_2) c_2 (\dot{\vartheta}_1 + \dot{\vartheta}_2)^2 + (m_1 + m_2) g s_1 \\ m_1(\ell_{C_1} + a_1)\ddot{\vartheta}_1 + m_2 a_1 \ddot{\vartheta}_1 + m_2(\ell_{C_2} + a_2) c_2 (\ddot{\vartheta}_1 + \ddot{\vartheta}_2) \\ - m_2(\ell_{C_2} + a_2) s_2 (\dot{\vartheta}_1 + \dot{\vartheta}_2)^2 + (m_1 + m_2) g c_1 \\ 0 \end{bmatrix}
$$

$$
\boldsymbol{\mu}_1^1 = \begin{bmatrix} * \\ * \\ \bar{I}_{1zz}\ddot{\vartheta}_1 + m_2 a_1^2 \ddot{\vartheta}_1 + m_1(\ell_{C_1} + a_1)^2 \ddot{\vartheta}_1 + m_2 a_1(\ell_{C_2} + a_2) c_2 \ddot{\vartheta}_1 \\ + \bar{I}_{2zz}(\ddot{\vartheta}_1 + \ddot{\vartheta}_2) + m_2 a_1 (\ell_{C_2} + a_2) c_2 (\ddot{\vartheta}_1 + \ddot{\vartheta}_2) \\ + m_2(\ell_{C_2} + a_2)^2(\ddot{\vartheta}_1 + \ddot{\vartheta}_2) + k_{r2} I_{m_2} \ddot{\vartheta}_2 \\ + m_2 a_1(\ell_{C_2} + a_2) s_2 \dot{\vartheta}_1^2 - m_2 a_1 (\ell_{C_2} + a_2) s_2 (\dot{\vartheta}_1 + \dot{\vartheta}_2)^2 \\ + m_1(\ell_{C_1} + a_1) g c_1 + m_2 a_1 g c_1 + m_2(\ell_{C_2} + a_2) g c_{12} \end{bmatrix}
$$

$$
\begin{aligned}
\tau_1 = {} & \Big(\bar{I}_{1zz} + m_1(\ell_{C_1} + a_1)^2 + k_{r1}^2 I_{m_1} + \bar{I}_{2zz} \\
& \quad + m_2\big(a_1^2 + (\ell_{C_2} + a_2)^2 + 2a_1(\ell_{C_2} + a_2)c_2\big)\Big)\ddot{\vartheta}_1 \\
& + \Big(\bar{I}_{2zz} + m_2\big((\ell_{C_2} + a_2)^2 + a_1(\ell_{C_2} + a_2)c_2\big) + k_{r2} I_{m_2}\Big)\ddot{\vartheta}_2 \\
& - 2 m_2 a_1 (\ell_{C_2} + a_2) s_2 \dot{\vartheta}_1 \dot{\vartheta}_2 - m_2 a_1 (\ell_{C_2} + a_2) s_2 \dot{\vartheta}_2^2 \\
& + \big(m_1(\ell_{C_1} + a_1) + m_2 a_1\big) g c_1 + m_2(\ell_{C_2} + a_2) g c_{12}.
\end{aligned}
$$

As for the moment components, those marked by the symbol '*' have not been computed, since they are not related to the joint torques τ_2 and τ_1.

Expressing the dynamic parameters in the above torques as a function of the link and rotor parameters as in (7.83) yields

$$\begin{aligned}
m_1 &= m_{\ell_1} + m_{m_2} \\
m_1 \ell_{C_1} &= m_{\ell_1}(\ell_1 - a_1) \\
\bar{I}_{1zz} + m_1 \ell_{C_1}^2 = \widehat{I}_1 &= I_{\ell_1} + m_{\ell_1}(\ell_1 - a_1)^2 + I_{m_2} \\
m_2 &= m_{\ell_2} \\
m_2 \ell_{C_2} &= m_{\ell_2}(\ell_2 - a_2) \\
\bar{I}_{2zz} + m_2 \ell_{C_2}^2 = \widehat{I}_2 &= I_{\ell_2} + m_{\ell_2}(\ell_2 - a_2)^2.
\end{aligned}$$

On the basis of these relations, it can be verified that the resulting dynamic model coincides with the model derived in (7.82) with Lagrange formulation.

7.6 Direct Dynamics and Inverse Dynamics

Both Lagrange formulation and Newton–Euler formulation allow the computation of the relationship between the joint torques — and, if present, the end-effector forces — and the motion of the structure. A comparison between the two approaches reveals what follows. The *Lagrange* formulation has the following advantages:

- It is *systematic* and of immediate comprehension.
- It provides the equations of motion in a compact *analytical form* containing the inertia matrix, the matrix in the centrifugal and Coriolis forces, and the vector of gravitational forces. Such a form is advantageous for *control design*.
- It is effective if it is wished to include more complex mechanical effects such as flexible link deformation.

The *Newton–Euler* formulation has the following fundamental advantage:

- It is an inherently *recursive* method that is computationally efficient.

In the study of dynamics, it is relevant to find a solution to two kinds of problems concerning computation of direct dynamics and inverse dynamics.

The *direct dynamics* problem consists of determining, for $t > t_0$, the joint accelerations $\ddot{\boldsymbol{q}}(t)$ (and thus $\dot{\boldsymbol{q}}(t)$, $\boldsymbol{q}(t)$) resulting from the given joint torques $\boldsymbol{\tau}(t)$ — and the possible end-effector forces $\boldsymbol{h}_e(t)$ — once the initial positions $\boldsymbol{q}(t_0)$ and velocities $\dot{\boldsymbol{q}}(t_0)$ are known (initial state of the system).

The *inverse dynamics* problem consists of determining the joint torques $\boldsymbol{\tau}(t)$ which are needed to generate the motion specified by the joint accelerations $\ddot{\boldsymbol{q}}(t)$, velocities $\dot{\boldsymbol{q}}(t)$, and positions $\boldsymbol{q}(t)$ — once the possible end-effector forces $\boldsymbol{h}_e(t)$ are known.

Solving the direct dynamics problem is useful for manipulator *simulation*. Direct dynamics allows the motion of the real physical system to be described in terms of the joint accelerations, when a set of assigned joint torques is applied to the manipulator; joint velocities and positions can be obtained by integrating the system of nonlinear differential equations.

Since the equations of motion obtained with Lagrange formulation give the analytical relationship between the joint torques (and the end-effector forces) and the joint positions, velocities and accelerations, these can be computed from (7.42) as

$$\ddot{\boldsymbol{q}} = \boldsymbol{B}^{-1}(\boldsymbol{q})(\boldsymbol{\tau} - \boldsymbol{\tau}') \tag{7.115}$$

where

$$\boldsymbol{\tau}'(\boldsymbol{q}, \dot{\boldsymbol{q}}) = \boldsymbol{C}(\boldsymbol{q}, \dot{\boldsymbol{q}})\dot{\boldsymbol{q}} + \boldsymbol{F}_v\dot{\boldsymbol{q}} + \boldsymbol{F}_s\,\text{sgn}\,(\dot{\boldsymbol{q}}) + \boldsymbol{g}(\boldsymbol{q}) + \boldsymbol{J}^T(\boldsymbol{q})\boldsymbol{h}_e \tag{7.116}$$

denotes the torque contributions depending on joint positions and velocities. Therefore, for simulation of manipulator motion, once the state at the time instant t_k is known in terms of the position $\boldsymbol{q}(t_k)$ and velocity $\dot{\boldsymbol{q}}(t_k)$, the acceleration $\ddot{\boldsymbol{q}}(t_k)$ can be computed by (7.115). Then using a numerical integration method, e.g., Runge–Kutta, with integration step Δt, the velocity $\dot{\boldsymbol{q}}(t_{k+1})$ and position $\boldsymbol{q}(t_{k+1})$ at the instant $t_{k+1} = t_k + \Delta t$ can be computed.

If the equations of motion are obtained with Newton–Euler formulation, it is possible to compute direct dynamics by using a computationally more efficient method. In fact, for given $\boldsymbol{q}$ and $\dot{\boldsymbol{q}}$, the torques $\boldsymbol{\tau}'(\boldsymbol{q}, \dot{\boldsymbol{q}})$ in (7.116) can be computed as the torques given by the algorithm of Fig. 7.14 with $\ddot{\boldsymbol{q}} = \boldsymbol{0}$. Further, column $\boldsymbol{b}_i$ of matrix $\boldsymbol{B}(\boldsymbol{q})$ can be computed as the torque vector given by the algorithm of Fig. 7.14 with $\boldsymbol{g}_0 = \boldsymbol{0}$, $\dot{\boldsymbol{q}} = \boldsymbol{0}$, $\ddot{q}_i = 1$ and $\ddot{q}_j = 0$ for $j \neq i$; iterating this procedure for $i = 1, \ldots, n$ leads to constructing the matrix $\boldsymbol{B}(\boldsymbol{q})$. Hence, from the current values of $\boldsymbol{B}(\boldsymbol{q})$ and $\boldsymbol{\tau}'(\boldsymbol{q}, \dot{\boldsymbol{q}})$, and the given $\boldsymbol{\tau}$, the equations in (7.115) can be integrated as illustrated above.

Solving the inverse dynamics problem is useful for manipulator trajectory planning and control algorithm implementation. Once a joint trajectory is specified in terms of positions, velocities and accelerations (typically as a result of an inverse kinematics procedure), and if the end-effector forces are known, inverse dynamics allows computation of the torques to be applied to the joints to obtain the desired motion. This computation turns out to be useful both for verifying feasibility of the imposed trajectory and for compensating nonlinear terms in the dynamic model of a manipulator. To this end, Newton–Euler formulation provides a computationally efficient recursive method for on-line computation of inverse dynamics. Nevertheless, it can be shown that also Lagrange formulation is liable to a computationally efficient recursive implementation, though with a nonnegligible reformulation effort.

For an n-joint manipulator the *number of operations* required is:[8]

[8] See Sect. E.1 for the definition of computational complexity of an algorithm.

- $O(n^2)$ for computing *direct dynamics*,
- $O(n)$ for computing *inverse dynamics*.

7.7 Dynamic Scaling of Trajectories

The existence of *dynamic constraints* to be taken into account for trajectory generation has been mentioned in Sect. 4.1. In practice, with reference to the given trajectory time or path shape (segments with high curvature), the trajectories that can be obtained with any of the previously illustrated methods may impose too severe dynamic performance for the manipulator. A typical case is that when the required torques to generate the motion are larger than the maximum torques the actuators can supply. In this case, an infeasible trajectory has to be suitably time-scaled.

Suppose a trajectory has been generated for all the manipulator joints as $\boldsymbol{q}(t)$, for $t \in [0, t_f]$. Computing inverse dynamics allows the evaluation of the time history of the torques $\boldsymbol{\tau}(t)$ required for the execution of the given motion. By comparing the obtained torques with the *torque limits* available at the actuators, it is easy to check whether or not the trajectory is actually executable. The problem is then to seek an automatic trajectory *dynamic scaling* technique — avoiding inverse dynamics recomputation — so that the manipulator can execute the motion on the specified path with a proper timing law without exceeding the torque limits.

Consider the manipulator dynamic model as given in (7.42) with $\boldsymbol{F}_v = \boldsymbol{O}$, $\boldsymbol{F}_s = \boldsymbol{O}$ and $\boldsymbol{h}_e = \boldsymbol{0}$, for simplicity. The term $\boldsymbol{C}(\boldsymbol{q}, \dot{\boldsymbol{q}})$ accounting for centrifugal and Coriolis forces has a quadratic dependence on joint velocities, and thus it can be formally rewritten as

$$\boldsymbol{C}(\boldsymbol{q}, \dot{\boldsymbol{q}})\dot{\boldsymbol{q}} = \boldsymbol{\Gamma}(\boldsymbol{q})[\dot{\boldsymbol{q}}\dot{\boldsymbol{q}}], \tag{7.117}$$

where $[\dot{\boldsymbol{q}}\dot{\boldsymbol{q}}]$ indicates the symbolic notation of the $(n(n+1)/2 \times 1)$ vector

$$[\dot{\boldsymbol{q}}\dot{\boldsymbol{q}}] = [\,\dot{q}_1^2 \quad \dot{q}_1\dot{q}_2 \quad \dots \quad \dot{q}_{n-1}\dot{q}_n \quad \dot{q}_n^2\,]^T;$$

$\boldsymbol{\Gamma}(\boldsymbol{q})$ is a proper $(n \times n(n+1)/2)$ matrix that satisfies (7.117). In view of such position, the manipulator dynamic model can be expressed as

$$\boldsymbol{B}(\boldsymbol{q}(t))\ddot{\boldsymbol{q}}(t) + \boldsymbol{\Gamma}(\boldsymbol{q}(t))[\dot{\boldsymbol{q}}(t)\dot{\boldsymbol{q}}(t)] + \boldsymbol{g}(\boldsymbol{q}(t)) = \boldsymbol{\tau}(t), \tag{7.118}$$

where the explicit dependence on time t has been shown.

Consider the new variable $\bar{\boldsymbol{q}}(r(t))$ satisfying the equation

$$\boldsymbol{q}(t) = \bar{\boldsymbol{q}}(r(t)), \tag{7.119}$$

where $r(t)$ is a strictly increasing scalar function of time with $r(0) = 0$ and $r(t_f) = \bar{t}_f$.

Differentiating (7.119) twice with respect to time provides the following relations:

$$\dot{q} = \dot{r}\bar{\boldsymbol{q}}'(r) \tag{7.120}$$

$$\ddot{q} = \dot{r}^2\bar{\boldsymbol{q}}''(r) + \ddot{r}\bar{\boldsymbol{q}}'(r) \tag{7.121}$$

where the prime denotes the derivative with respect to r. Substituting (7.120), (7.121) into (7.118) yields

$$\dot{r}^2\Big(\boldsymbol{B}(\bar{\boldsymbol{q}}(r))\bar{\boldsymbol{q}}''(r) + \boldsymbol{\Gamma}(\bar{\boldsymbol{q}}(r))[\bar{\boldsymbol{q}}'(r)\bar{\boldsymbol{q}}'(r)]\Big) + \ddot{r}\boldsymbol{B}(\bar{\boldsymbol{q}}(r))\bar{\boldsymbol{q}}'(r) + \boldsymbol{g}(\bar{\boldsymbol{q}}(r)) = \boldsymbol{\tau}. \tag{7.122}$$

In (7.118) it is possible to identify the term

$$\boldsymbol{\tau}_s(t) = \boldsymbol{B}(\boldsymbol{q}(t))\ddot{\boldsymbol{q}}(t) + \boldsymbol{\Gamma}(\boldsymbol{q}(t))[\dot{\boldsymbol{q}}(t)\dot{\boldsymbol{q}}(t)], \tag{7.123}$$

representing the torque contribution that depends on velocities and accelerations. Correspondingly, in (7.122) one can set

$$\boldsymbol{\tau}_s(t) = \dot{r}^2\Big(\boldsymbol{B}(\bar{\boldsymbol{q}}(r))\bar{\boldsymbol{q}}''(r) + \boldsymbol{\Gamma}(\bar{\boldsymbol{q}}(r))[\bar{\boldsymbol{q}}'(r)\bar{\boldsymbol{q}}'(r)]\Big) + \ddot{r}\boldsymbol{B}(\bar{\boldsymbol{q}}(r))\bar{\boldsymbol{q}}'(r). \tag{7.124}$$

By analogy with (7.123), it can be written

$$\bar{\boldsymbol{\tau}}_s(r) = \boldsymbol{B}(\bar{\boldsymbol{q}}(r))\bar{\boldsymbol{q}}''(r) + \boldsymbol{\Gamma}(\bar{\boldsymbol{q}}(r))[\bar{\boldsymbol{q}}'(r)\bar{\boldsymbol{q}}'(r)] \tag{7.125}$$

and then (7.124) becomes

$$\boldsymbol{\tau}_s(t) = \dot{r}^2\bar{\boldsymbol{\tau}}_s(r) + \ddot{r}\boldsymbol{B}(\bar{\boldsymbol{q}}(r))\bar{\boldsymbol{q}}'(r). \tag{7.126}$$

The expression in (7.126) gives the relationship between the torque contributions depending on velocities and accelerations required by the manipulator when this is subject to motions having the same path but different timing laws, obtained through a time scaling of joint variables as in (7.119).

Gravitational torques have not been considered, since they are a function of the joint positions only, and thus their contribution is not influenced by time scaling.

The simplest choice for the scaling function $r(t)$ is certainly the *linear* function

$$r(t) = ct$$

with c a positive constant. In this case, (7.126) becomes

$$\boldsymbol{\tau}_s(t) = c^2\bar{\boldsymbol{\tau}}_s(ct),$$

which reveals that a linear time scaling by c causes a scaling of the magnitude of the torques by the coefficient c^2. Let $c > 1$: (7.119) shows that the trajectory described by $\bar{\boldsymbol{q}}(r(t))$, assuming $r = ct$ as the independent variable, has a duration $\bar{t}_f > t_f$ to cover the entire path specified by $\boldsymbol{q}$. Correspondingly, the torque contributions $\bar{\boldsymbol{\tau}}_s(ct)$ computed as in (7.125) are scaled by the factor c^2 with respect to the torque contributions $\boldsymbol{\tau}_s(t)$ required to execute the original trajectory $\boldsymbol{q}(t)$.

With the use of a recursive algorithm for inverse dynamics computation, it is possible to check whether the torques exceed the allowed limits during trajectory execution; obviously, limit violation should not be caused by the sole gravity torques. It is necessary to find the joint for which the torque has exceeded the limit more than the others, and to compute the torque contribution subject to scaling, which in turn determines the factor c^2. It is then possible to compute the time-scaled trajectory as a function of the new time variable $r = ct$ which no longer exceeds torque limits. It should be pointed out, however, that with this kind of linear scaling the entire trajectory may be penalized, even when a torque limit on a single joint is exceeded only for a short interval of time.

7.8 Operational Space Dynamic Model

As an alternative to the joint space dynamic model, the equations of motion of the system can be expressed directly in the operational space; to this end it is necessary to find a *dynamic model* which describes the relationship between the generalized forces acting on the manipulator and the number of minimal variables chosen to describe the end-effector position and orientation in the *operational space*.

Similar to kinematic description of a manipulator in the operational space, the presence of redundant DOFs and/or kinematic and representation singularities deserves careful attention in the derivation of an operational space dynamic model.

The determination of the dynamic model with Lagrange formulation using operational space variables allows a complete description of the system motion only in the case of a *nonredundant* manipulator, when the above variables constitute a set of *generalized coordinates* in terms of which the kinetic energy, the potential energy, and the nonconservative forces doing work on them can be expressed.

This way of proceeding does not provide a complete description of dynamics for a *redundant* manipulator; in this case, in fact, it is reasonable to expect the occurrence of *internal motions* of the structure caused by those joint generalized forces which do not affect the end-effector motion.

To develop an operational space model which can be adopted for both redundant and nonredundant manipulators, it is then convenient to start from the joint space model which is in all general. In fact, solving (7.42) for the joint accelerations, and neglecting the joint friction torques for simplicity, yields

$$\ddot{\boldsymbol{q}} = -\boldsymbol{B}^{-1}(\boldsymbol{q})\boldsymbol{C}(\boldsymbol{q},\dot{\boldsymbol{q}})\dot{\boldsymbol{q}} - \boldsymbol{B}^{-1}(\boldsymbol{q})\boldsymbol{g}(\boldsymbol{q}) + \boldsymbol{B}^{-1}(\boldsymbol{q})\boldsymbol{J}^T(\boldsymbol{q})(\boldsymbol{\gamma}_e - \boldsymbol{h}_e), \quad (7.127)$$

where the joint torques $\boldsymbol{\tau}$ have been expressed in terms of the equivalent end-effector forces $\boldsymbol{\gamma}$ according to (3.111). It is worth noting that $\boldsymbol{h}$ represents the contribution of the end-effector forces due to contact with the environment, whereas $\boldsymbol{\gamma}$ expresses the contribution of the end-effector forces due to joint actuation.

On the other hand, the second-order differential kinematics equation in (3.98) describes the relationship between joint space and operational space accelerations, i.e.,

$$\ddot{\boldsymbol{x}}_e = \boldsymbol{J}_A(\boldsymbol{q})\ddot{\boldsymbol{q}} + \dot{\boldsymbol{J}}_A(\boldsymbol{q}, \dot{\boldsymbol{q}})\dot{\boldsymbol{q}}.$$

The solution in (7.127) features the geometric Jacobian $\boldsymbol{J}$, whereas the analytical Jacobian $\boldsymbol{J}_A$ appears in (3.98). For notation uniformity, in view of (3.66), one can set

$$\boldsymbol{T}_A^T(\boldsymbol{x}_e)\boldsymbol{\gamma}_e = \boldsymbol{\gamma}_A \qquad \boldsymbol{T}_A^T(\boldsymbol{x}_e)\boldsymbol{h}_e = \boldsymbol{h}_A \tag{7.128}$$

where $\boldsymbol{T}_A$ is the transformation matrix between the two Jacobians. Substituting (7.127) into (3.98) and accounting for (7.128) gives

$$\ddot{\boldsymbol{x}}_e = -\boldsymbol{J}_A\boldsymbol{B}^{-1}\boldsymbol{C}\dot{\boldsymbol{q}} - \boldsymbol{J}_A\boldsymbol{B}^{-1}\boldsymbol{g} + \dot{\boldsymbol{J}}_A\dot{\boldsymbol{q}} + \boldsymbol{J}_A\boldsymbol{B}^{-1}\boldsymbol{J}_A^T(\boldsymbol{\gamma}_A - \boldsymbol{h}_A). \tag{7.129}$$

where the dependence on $\boldsymbol{q}$ and $\dot{\boldsymbol{q}}$ has been omitted. With the positions

$$\boldsymbol{B}_A = (\boldsymbol{J}_A\boldsymbol{B}^{-1}\boldsymbol{J}_A^T)^{-1} \tag{7.130}$$

$$\boldsymbol{C}_A\dot{\boldsymbol{x}}_e = \boldsymbol{B}_A\boldsymbol{J}_A\boldsymbol{B}^{-1}\boldsymbol{C}\dot{\boldsymbol{q}} - \boldsymbol{B}_A\dot{\boldsymbol{J}}_A\dot{\boldsymbol{q}} \tag{7.131}$$

$$\boldsymbol{g}_A = \boldsymbol{B}_A\boldsymbol{J}_A\boldsymbol{B}^{-1}\boldsymbol{g}, \tag{7.132}$$

the expression in (7.129) can be rewritten as

$$\boldsymbol{B}_A(\boldsymbol{x}_e)\ddot{\boldsymbol{x}}_e + \boldsymbol{C}_A(\boldsymbol{x}_e, \dot{\boldsymbol{x}}_e)\dot{\boldsymbol{x}}_e + \boldsymbol{g}_A(\boldsymbol{x}_e) = \boldsymbol{\gamma}_A - \boldsymbol{h}_A, \tag{7.133}$$

which is formally analogous to the joint space dynamic model (7.42). Notice that the matrix $\boldsymbol{J}_A\boldsymbol{B}^{-1}\boldsymbol{J}_A^T$ is invertible if and only if $\boldsymbol{J}_A$ is full-rank, that is, in the absence of both kinematic and representation singularities.

For a nonredundant manipulator in a nonsingular configuration, the expressions in (7.130)–(7.132) become:

$$\boldsymbol{B}_A = \boldsymbol{J}_A^{-T}\boldsymbol{B}\boldsymbol{J}_A^{-1} \tag{7.134}$$

$$\boldsymbol{C}_A\dot{\boldsymbol{x}}_e = \boldsymbol{J}_A^{-T}\boldsymbol{C}\dot{\boldsymbol{q}} - \boldsymbol{B}_A\dot{\boldsymbol{J}}_A\dot{\boldsymbol{q}} \tag{7.135}$$

$$\boldsymbol{g}_A = \boldsymbol{J}_A^{-T}\boldsymbol{g}. \tag{7.136}$$

As anticipated above, the main feature of the obtained model is its formal validity also for a redundant manipulator, even though the variables $\boldsymbol{x}_e$ do not constitute a set of generalized coordinates for the system; in this case, the matrix $\boldsymbol{B}_A$ is representative of a *kinetic pseudo-energy*.

In the following, the utility of the operational space dynamic model in (7.133) for solving direct and inverse dynamics problems is investigated. The following derivation is meaningful for redundant manipulators; for a nonredundant manipulator, in fact, using (7.133) does not pose specific problems as long as $\boldsymbol{J}_A$ is nonsingular ((7.134)–(7.136)).

With reference to operational space, the *direct dynamics* problem consists of determining the resulting end-effector accelerations $\ddot{\boldsymbol{x}}_e(t)$ (and thus $\dot{\boldsymbol{x}}_e(t)$, $\boldsymbol{x}_e(t)$) from the given joint torques $\boldsymbol{\tau}(t)$ and end-effector forces $\boldsymbol{h}_e(t)$. For a redundant manipulator, (7.133) cannot be directly used, since (3.111) has a solution in $\boldsymbol{\gamma}_e$ only if $\boldsymbol{\tau} \in \mathcal{R}(\boldsymbol{J}^T)$. It follows that for simulation purposes, the solution to the problem is naturally obtained in the joint space; in fact, the expression in (7.42) allows the computation of $\boldsymbol{q}$, $\dot{\boldsymbol{q}}$, $\ddot{\boldsymbol{q}}$ which, substituted into the direct kinematics equations in ((2.82), (3.62), (3.98), give $\boldsymbol{x}_e$, $\dot{\boldsymbol{x}}_e$, $\ddot{\boldsymbol{x}}_e$, respectively.

Formulation of an *inverse dynamics* problem in the operational space requires the determination of the joint torques $\boldsymbol{\tau}(t)$ that are needed to generate a specific motion assigned in terms of $\ddot{\boldsymbol{x}}_e(t)$, $\dot{\boldsymbol{x}}_e(t)$, $\boldsymbol{x}_e(t)$, for given end-effector forces $\boldsymbol{h}_e(t)$. A possible way of solution is to solve a complete inverse kinematics problem for (2.82), (3.62), (3.98), and then compute the required torques with the joint space inverse dynamics as in (7.42). Hence, for redundant manipulators, redundancy resolution is performed at kinematic level.

An alternative solution to the inverse dynamics problem consists of computing $\boldsymbol{\gamma}_A$ as in (7.133) and the joint torques $\boldsymbol{\tau}$ as in (3.111). In this way, however, the presence of redundant DOFs is not exploited at all, since the computed torques do not generate internal motions of the structure.

If it is desired to find a formal solution that allows redundancy resolution at dynamic level, it is necessary to determine those torques corresponding to the equivalent end-effector forces computed as in (7.133). By analogy with the differential kinematics solution (3.54), the expression of the torques to be determined will feature the presence of a minimum-norm term and a homogeneous term. Since the joint torques have to be computed, it is convenient to express the model (7.133) in terms of $\boldsymbol{q}$, $\dot{\boldsymbol{q}}$, $\ddot{\boldsymbol{q}}$. By recalling the positions (7.131), (7.132), the expression in (7.133) becomes

$$\boldsymbol{B}_A(\ddot{\boldsymbol{x}}_e - \dot{\boldsymbol{J}}_A\dot{\boldsymbol{q}}) + \boldsymbol{B}_A\boldsymbol{J}_A\boldsymbol{B}^{-1}\boldsymbol{C}\dot{\boldsymbol{q}} + \boldsymbol{B}_A\boldsymbol{J}_A\boldsymbol{B}^{-1}\boldsymbol{g} = \boldsymbol{\gamma}_A - \boldsymbol{h}_A$$

and, in view of (3.98),

$$\boldsymbol{B}_A\boldsymbol{J}_A\ddot{\boldsymbol{q}} + \boldsymbol{B}_A\boldsymbol{J}_A\boldsymbol{B}^{-1}\boldsymbol{C}\dot{\boldsymbol{q}} + \boldsymbol{B}_A\boldsymbol{J}_A\boldsymbol{B}^{-1}\boldsymbol{g} = \boldsymbol{\gamma}_A - \boldsymbol{h}_A. \tag{7.137}$$

By setting

$$\bar{\boldsymbol{J}}_A(\boldsymbol{q}) = \boldsymbol{B}^{-1}(\boldsymbol{q})\boldsymbol{J}_A^T(\boldsymbol{q})\boldsymbol{B}_A(\boldsymbol{q}), \tag{7.138}$$

the expression in (7.137) becomes

$$\bar{\boldsymbol{J}}_A^T(\boldsymbol{B}\ddot{\boldsymbol{q}} + \boldsymbol{C}\dot{\boldsymbol{q}} + \boldsymbol{g}) = \boldsymbol{\gamma}_A - \boldsymbol{h}_A. \tag{7.139}$$

At this point, from the joint space dynamic model in (7.42), it is easy to recognize that (7.139) can be written as

$$\bar{\boldsymbol{J}}_A^T(\boldsymbol{\tau} - \boldsymbol{J}_A^T\boldsymbol{h}_A) = \boldsymbol{\gamma}_A - \boldsymbol{h}_A$$

from which

$$\bar{\boldsymbol{J}}_A^T \boldsymbol{\tau} = \boldsymbol{\gamma}_A. \tag{7.140}$$

The general solution to (7.140) is of the form (see Problem 7.10)

$$\boldsymbol{\tau} = \boldsymbol{J}_A^T(\boldsymbol{q})\boldsymbol{\gamma}_A + \big(\boldsymbol{I}_n - \boldsymbol{J}_A^T(\boldsymbol{q})\bar{\boldsymbol{J}}_A^T(\boldsymbol{q})\big)\boldsymbol{\tau}_0, \tag{7.141}$$

that can be derived by observing that $\boldsymbol{J}_A^T$ in (7.138) is a *right pseudo-inverse* of $\bar{\boldsymbol{J}}_A^T$ weighted by the inverse of the inertia matrix $\boldsymbol{B}^{-1}$. The $(n \times 1)$ vector of arbitrary torques $\boldsymbol{\tau}_0$ in (7.141) does not contribute to the end-effector forces, since it is projected in the null space of $\bar{\boldsymbol{J}}_A^T$.

To summarize, for given $\boldsymbol{x}_e$, $\dot{\boldsymbol{x}}_e$, $\ddot{\boldsymbol{x}}_e$ and $\boldsymbol{h}_A$, the expression in (7.133) allows the computation of $\boldsymbol{\gamma}_A$. Then, (7.141) gives the torques $\boldsymbol{\tau}$ which, besides executing the assigned end-effector motion, generate internal motions of the structure to be employed for handling redundancy at dynamic level through a suitable choice of $\boldsymbol{\tau}_0$.

7.9 Dynamic Manipulability Ellipsoid

The availability of the dynamic model allows formulation of the *dynamic manipulability ellipsoid* which provides a useful tool for manipulator dynamic performance analysis. This can be used for mechanical structure design as well as for seeking optimal manipulator configurations.

Consider the set of joint torques of constant (unit) norm

$$\boldsymbol{\tau}^T\boldsymbol{\tau} = 1 \tag{7.142}$$

describing the points on the surface of a sphere. It is desired to describe the operational space accelerations that can be generated by the given set of joint torques.

For studying dynamic manipulability, suppose to consider the case of a manipulator standing still ($\dot{\boldsymbol{q}} = \boldsymbol{0}$), not in contact with the environment ($\boldsymbol{h}_e = \boldsymbol{0}$). The simplified model is

$$\boldsymbol{B}(\boldsymbol{q})\ddot{\boldsymbol{q}} + \boldsymbol{g}(\boldsymbol{q}) = \boldsymbol{\tau}. \tag{7.143}$$

The joint accelerations $\ddot{\boldsymbol{q}}$ can be computed from the second-order differential kinematics that can be obtained by differentiating (3.39), and imposing successively $\dot{\boldsymbol{q}} = \boldsymbol{0}$, leading to

$$\dot{\boldsymbol{v}}_e = \boldsymbol{J}(\boldsymbol{q})\ddot{\boldsymbol{q}}. \tag{7.144}$$

Solving for minimum-norm accelerations only, for a *nonsingular Jacobian*, and substituting in (7.143) yields the expression of the torques

$$\boldsymbol{\tau} = \boldsymbol{B}(\boldsymbol{q})\boldsymbol{J}^{\dagger}(\boldsymbol{q})\dot{\boldsymbol{v}}_e + \boldsymbol{g}(\boldsymbol{q}) \tag{7.145}$$

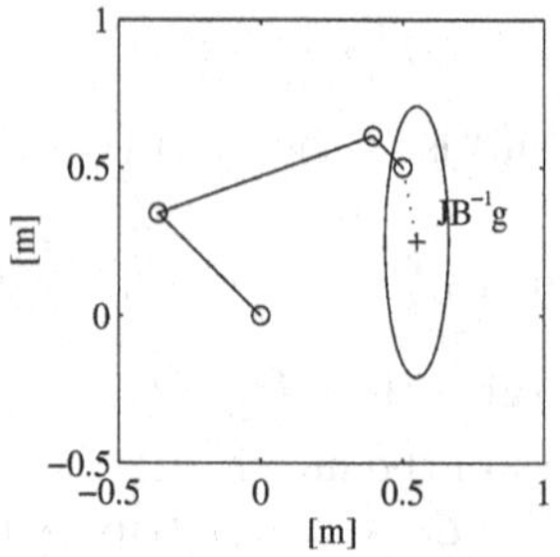

Fig. 7.15. Effect of gravity on the dynamic manipulability ellipsoid for a three-link planar arm

needed to derive the ellipsoid. In fact, substituting (7.145) into (7.142) gives

$$\left(\boldsymbol{B}(\boldsymbol{q})\boldsymbol{J}^{\dagger}(\boldsymbol{q})\dot{\boldsymbol{v}}_e + \boldsymbol{g}(\boldsymbol{q})\right)^T\left(\boldsymbol{B}(\boldsymbol{q})\boldsymbol{J}^{\dagger}(\boldsymbol{q})\dot{\boldsymbol{v}}_e + \boldsymbol{g}(\boldsymbol{q})\right) = 1.$$

The vector on the right-hand side of (7.145) can be rewritten as

$$\begin{aligned}\boldsymbol{B}\boldsymbol{J}^{\dagger}\dot{\boldsymbol{v}}_e + \boldsymbol{g} &= \boldsymbol{B}(\boldsymbol{J}^{\dagger}\dot{\boldsymbol{v}}_e + \boldsymbol{B}^{-1}\boldsymbol{g}) \\ &= \boldsymbol{B}(\boldsymbol{J}^{\dagger}\dot{\boldsymbol{v}}_e + \boldsymbol{B}^{-1}\boldsymbol{g} + \boldsymbol{J}^{\dagger}\boldsymbol{J}\boldsymbol{B}^{-1}\boldsymbol{g} - \boldsymbol{J}^{\dagger}\boldsymbol{J}\boldsymbol{B}^{-1}\boldsymbol{g}) \\ &= \boldsymbol{B}\left(\boldsymbol{J}^{\dagger}\dot{\boldsymbol{v}}_e + \boldsymbol{J}^{\dagger}\boldsymbol{J}\boldsymbol{B}^{-1}\boldsymbol{g} + (\boldsymbol{I}_n - \boldsymbol{J}^{\dagger}\boldsymbol{J})\boldsymbol{B}^{-1}\boldsymbol{g}\right),\end{aligned} \tag{7.146}$$

where the dependence on $\boldsymbol{q}$ has been omitted. According to what was done for solving (7.144), one can neglect the contribution of the accelerations given by $\boldsymbol{B}^{-1}\boldsymbol{g}$ which are in the null space of $\boldsymbol{J}$ and then produce no end-effector acceleration. Hence, (7.146) becomes

$$\boldsymbol{B}\boldsymbol{J}^{\dagger}\dot{\boldsymbol{v}}_e + \boldsymbol{g} = \boldsymbol{B}\boldsymbol{J}^{\dagger}(\dot{\boldsymbol{v}}_e + \boldsymbol{J}\boldsymbol{B}^{-1}\boldsymbol{g}) \tag{7.147}$$

and the dynamic manipulability ellipsoid can be expressed in the form

$$(\dot{\boldsymbol{v}}_e + \boldsymbol{J}\boldsymbol{B}^{-1}\boldsymbol{g})^T\boldsymbol{J}^{\dagger T}\boldsymbol{B}^T\boldsymbol{B}\boldsymbol{J}^{\dagger}(\dot{\boldsymbol{v}}_e + \boldsymbol{J}\boldsymbol{B}^{-1}\boldsymbol{g}) = 1. \tag{7.148}$$

The core of the quadratic form $\boldsymbol{J}^{\dagger T}\boldsymbol{B}^T\boldsymbol{B}\boldsymbol{J}^{\dagger}$ depends on the geometrical and inertial characteristics of the manipulator and determines the volume and principal axes of the ellipsoid. The vector $-\boldsymbol{J}\boldsymbol{B}^{-1}\boldsymbol{g}$, describing the contribution of gravity, produces a constant translation of the centre of the ellipsoid (for each manipulator configuration) with respect to the origin of the reference frame; see the example in Fig. 7.15 for a three-link planar arm.

The meaning of the dynamic manipulability ellipsoid is conceptually similar to that of the ellipsoids considered with reference to kineto-statics duality. In fact, the distance of a point on the surface of the ellipsoid from the end-effector gives a measure of the accelerations which can be imposed to the end-effector along the given direction, with respect to the constraint (7.142). With reference to Fig. 7.15, it is worth noticing how the presence of gravity acceleration allows the execution of larger accelerations downward, as natural to predict.

In the case of a nonredundant manipulator, the ellipsoid reduces to

$$(\dot{\boldsymbol{v}}_e + \boldsymbol{J}\boldsymbol{B}^{-1}\boldsymbol{g})^T \boldsymbol{J}^{-T}\boldsymbol{B}^T\boldsymbol{B}\boldsymbol{J}^{-1}(\dot{\boldsymbol{v}}_e + \boldsymbol{J}\boldsymbol{B}^{-1}\boldsymbol{g}) = 1. \tag{7.149}$$

Bibliography

The derivation of the dynamic model for rigid manipulators can be found in several classical robotics texts, such as [180, 10, 248, 53, 217, 111].

The first works on the computation of the dynamic model of open-chain manipulators based on the Lagrange formulation are [234, 19, 221, 236]. A computationally efficient formulation is presented in [96].

Dynamic model computation for robotic systems having a closed-chain or a tree kinematic structure can be found in [11, 144] and [112], respectively. Joint friction models are analyzed in [9].

The notable properties of the dynamic model deriving from the principle of energy conservation are underlined in [213], on the basis of the work in [119]. Algorithms to find the parameterization of the dynamic model in terms of a minimum number of parameters are considered in [115], which utilizes the results in [166]. Methods for symbolic computation of those parameters are presented in [85] for open kinematic chains and [110] for closed kinematic chains. Parameter identification methods based on least-squares techniques are given in [13].

The Newton–Euler formulation is proposed in [172], and a computationally efficient version for inverse dynamics can be found in [142]; an analogous formulation is employed for direct dynamics computation in [237]. The Lagrange and Newton–Euler formulations are compared by a computational viewpoint in [211], while they are utilized in [201] for dynamic model computation with inclusion of inertial and gyroscopic effects of actuators. Efficient algorithms for direct dynamics computation are given in [76, 77].

The trajectory dynamic scaling technique is presented in [97]. The operational space dynamic model is illustrated in [114] and the concept of weighted pseudo-inverse of the inertia matrix is introduced in [78]. The manipulability ellipsoids are analyzed in [246, 38].

Problems

7.1. Find the dynamic model of a two-link Cartesian arm in the case when the second joint axis forms an angle of $\pi/4$ with the first joint axis; compare the result with the model of the manipulator in Fig. 7.3.

7.2. For the two-link planar arm of Sect. 7.3.2, prove that with a different choice of the matrix $\boldsymbol{C}$, (7.49) holds true while (7.48) does not.

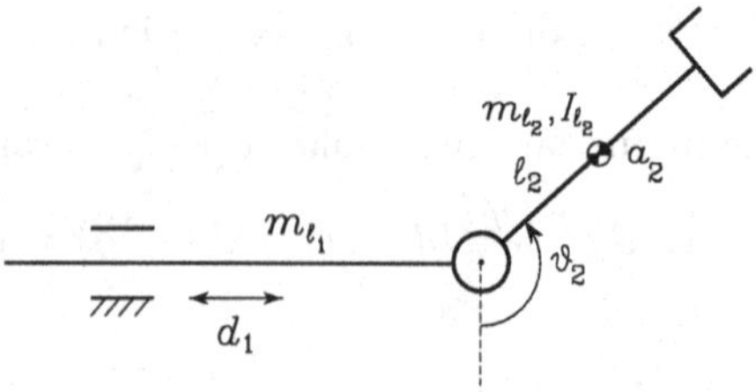

Fig. 7.16. Two-link planar arm with a prismatic joint and a revolute joint

7.3. Find the dynamic model of the SCARA manipulator in Fig. 2.36.

7.4. For the planar arm of Sect. 7.3.2, find a minimal parameterization of the dynamic model in (7.82).

7.5. Find the dynamic model of the two-link planar arm with a prismatic joint and a revolute joint in Fig. 7.16 with the Lagrange formulation. Then, consider the addition of a concentrated tip payload of mass m_L, and express the resulting model in a linear form with respect to a suitable set of dynamic parameters as in (7.81).

7.6. For the two-link planar arm of Fig. 7.4, find the dynamic model with the Lagrange formulation when the absolute angles with respect to the base frame are chosen as generalized coordinates. Discuss the result in view of a comparison with the model derived in (7.82).

7.7. Compute the joint torques for the two-link planar arm of Fig. 7.4 with the data and along the trajectories of Example 7.2, in the case of tip forces $\boldsymbol{f} = [\,500 \quad 500\,]^T$ N.

7.8. Find the dynamic model of the two-link planar arm with a prismatic joint and a revolute joint in Fig. 7.16 by using the recursive Newton–Euler algorithm.

7.9. Show that for the operational space dynamic model (7.133) a skew-symmetry property holds which is analogous to (7.48).

7.10. Show how to obtain the general solution to (7.140) in the form (7.141).

7.11. For a nonredundant manipulator, compute the relationship between the dynamic manipulability measure that can be defined for the dynamic manipulability ellipsoid and the manipulability measure defined in (3.56).

8

Motion Control

In Chap. 4, trajectory planning techniques have been presented which allow the generation of the reference inputs to the motion control system. The problem of controlling a manipulator can be formulated as that to determine the time history of the generalized forces (forces or torques) to be developed by the joint actuators, so as to guarantee execution of the commanded task while satisfying given transient and steady-state requirements. The task may regard either the execution of specified motions for a manipulator operating in free space, or the execution of specified motions and contact forces for a manipulator whose end-effector is constrained by the environment. In view of problem complexity, the two aspects will be treated separately; first, motion control in free space, and then control of the interaction with the environment. The problem of *motion control* of a manipulator is the topic of this chapter. A number of *joint space* control techniques are presented. These can be distinguished between *decentralized control* schemes, i.e., when the single manipulator joint is controlled independently of the others, and *centralized control* schemes, i.e., when the dynamic interaction effects between the joints are taken into account. Finally, as a premise to the interaction control problem, the basic features of *operational space* control schemes are illustrated.

8.1 The Control Problem

Several techniques can be employed for controlling a manipulator. The technique followed, as well as the way it is implemented, may have a significant influence on the manipulator performance and then on the possible range of applications. For instance, the need for trajectory tracking control in the operational space may lead to hardware/software implementations, which differ from those allowing point-to-point control, where only reaching of the final position is of concern.

On the other hand, the manipulator mechanical design has an influence on the kind of control scheme utilized. For instance, the control problem of

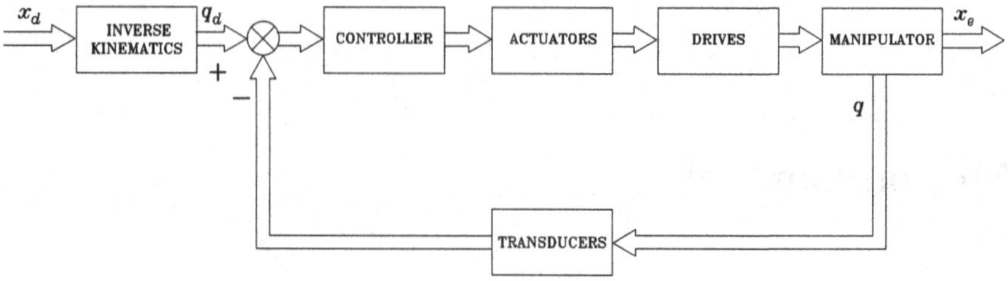

Fig. 8.1. General scheme of joint space control

a Cartesian manipulator is substantially different from that of an anthropomorphic manipulator.

The driving system of the joints also has an effect on the type of control strategy used. If a manipulator is actuated by electric motors with reduction gears of high ratios, the presence of gears tends to linearize system dynamics, and thus to decouple the joints in view of the reduction of nonlinearity effects. The price to pay, however, is the occurrence of joint friction, elasticity and backlash that may limit system performance more than it is due to configuration-dependent inertia, Coriolis and centrifugal forces, and so forth. On the other hand, a robot actuated with direct drives eliminates the drawbacks due to friction, elasticity and backlash, but the weight of nonlinearities and couplings between the joints becomes relevant. As a consequence, different control strategies have to be thought of to obtain high performance.

Without any concern to the specific type of mechanical manipulator, it is worth remarking that task specification (end-effector motion and forces) is usually carried out in the operational space, whereas control actions (joint actuator generalized forces) are performed in the joint space. This fact naturally leads to considering two kinds of general control schemes, namely, a *joint space control* scheme (Fig. 8.1) and an *operational space control* scheme (Fig. 8.2). In both schemes, the control structure has closed loops to exploit the good features provided by feedback, i.e., robustness to modelling uncertainties and reduction of disturbance effects. In general terms, the following considerations should be made.

The *joint space control* problem is actually articulated in two subproblems. First, manipulator inverse kinematics is solved to transform the motion requirements $\boldsymbol{x}_d$ from the operational space into the corresponding motion $\boldsymbol{q}_d$ in the joint space. Then, a joint space control scheme is designed that allows the actual motion $\boldsymbol{q}$ to track the reference inputs. However, this solution has the drawback that a joint space control scheme does not influence the operational space variables $\boldsymbol{x}_e$ which are controlled in an open-loop fashion through the manipulator mechanical structure. It is then clear that any uncertainty of the structure (construction tolerance, lack of calibration, gear backlash, elasticity) or any imprecision in the knowledge of the end-effector pose relative

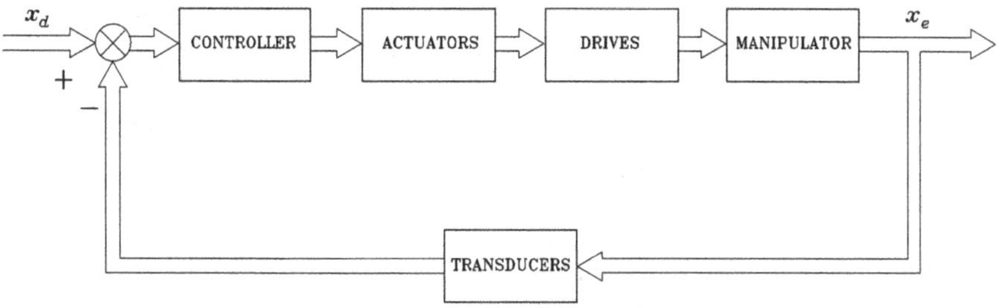

Fig. 8.2. General scheme of operational space control

to an object to manipulate causes a loss of accuracy on the operational space variables.

The *operational space control* problem follows a global approach that requires a greater algorithmic complexity; notice that inverse kinematics is now embedded into the feedback control loop. Its conceptual advantage regards the possibility of acting directly on operational space variables; this is somewhat only a potential advantage, since measurement of operational space variables is often performed not directly, but through the evaluation of direct kinematics functions starting from measured joint space variables.

On the above premises, in the following, joint space control schemes for manipulator motion in the free space are presented first. In the sequel, operational space control schemes will be illustrated which are logically at the basis of control of the interaction with the environment.

8.2 Joint Space Control

In Chap. 7, it was shown that the equations of motion of a manipulator in the absence of external end-effector forces and, for simplicity, of static friction (difficult to model accurately) are described by

$$\boldsymbol{B}(\boldsymbol{q})\ddot{\boldsymbol{q}} + \boldsymbol{C}(\boldsymbol{q}, \dot{\boldsymbol{q}})\dot{\boldsymbol{q}} + \boldsymbol{F}_v\dot{\boldsymbol{q}} + \boldsymbol{g}(\boldsymbol{q}) = \boldsymbol{\tau} \tag{8.1}$$

with obvious meaning of the symbols. To control the motion of the manipulator in free space means to determine the n components of generalized forces — torques for revolute joints, forces for prismatic joints — that allow execution of a motion $\boldsymbol{q}(t)$ so that

$$\boldsymbol{q}(t) = \boldsymbol{q}_d(t),$$

as closely as possible, where $\boldsymbol{q}_d(t)$ denotes the vector of desired joint trajectory variables.

The generalized forces are supplied by the actuators through proper transmissions to transform the motion characteristics. Let $\boldsymbol{q}_m$ denote the vector of joint actuator displacements; the transmissions — assumed to be rigid and with no backlash — establish the following relationship:

$$\boldsymbol{K}_r\boldsymbol{q} = \boldsymbol{q}_m, \tag{8.2}$$

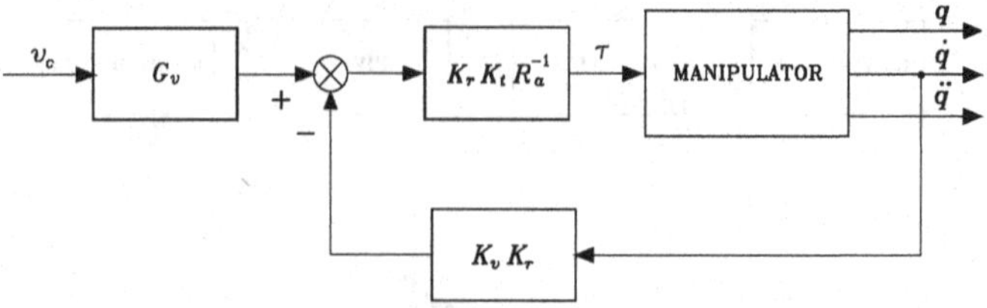

Fig. 8.3. Block scheme of the manipulator and drives system as a voltage-controlled system

where $\boldsymbol{K}_r$ is an $(n \times n)$ diagonal matrix, whose elements are defined in (7.22) and are much greater than unity.[1]

In view of (8.2), if $\boldsymbol{\tau}_m$ denotes the vector of actuator driving torques, one can write

$$\boldsymbol{\tau}_m = \boldsymbol{K}_r^{-1}\boldsymbol{\tau}. \tag{8.3}$$

With reference to (5.1)–(5.4), the n driving systems can be described in compact matrix form by the equations:

$$\boldsymbol{K}_r^{-1}\boldsymbol{\tau} = \boldsymbol{K}_t \boldsymbol{i}_a \tag{8.4}$$
$$\boldsymbol{v}_a = \boldsymbol{R}_a \boldsymbol{i}_a + \boldsymbol{K}_v \dot{\boldsymbol{q}}_m \tag{8.5}$$
$$\boldsymbol{v}_a = \boldsymbol{G}_v \boldsymbol{v}_c. \tag{8.6}$$

In (8.4), $\boldsymbol{K}_t$ is the diagonal matrix of torque constants and $\boldsymbol{i}_a$ is the vector of armature currents of the n motors; in (8.5), $\boldsymbol{v}_a$ is the vector of armature voltages, $\boldsymbol{R}_a$ is the diagonal matrix of armature resistances,[2] and $\boldsymbol{K}_v$ is the diagonal matrix of voltage constants of the n motors; in (8.6), $\boldsymbol{G}_v$ is the diagonal matrix of gains of the n amplifiers and $\boldsymbol{v}_c$ is the vector of control voltages of the n servomotors.

On reduction of (8.1), (8.2), (8.4), (8.5), (8.6), the dynamic model of the system given by the manipulator and drives is described by

$$\boldsymbol{B}(\boldsymbol{q})\ddot{\boldsymbol{q}} + \boldsymbol{C}(\boldsymbol{q}, \dot{\boldsymbol{q}})\dot{\boldsymbol{q}} + \boldsymbol{F}\dot{\boldsymbol{q}} + \boldsymbol{g}(\boldsymbol{q}) = \boldsymbol{u} \tag{8.7}$$

where the following positions have been made:

$$\boldsymbol{F} = \boldsymbol{F}_v + \boldsymbol{K}_r \boldsymbol{K}_t \boldsymbol{R}_a^{-1} \boldsymbol{K}_v \boldsymbol{K}_r \tag{8.8}$$
$$\boldsymbol{u} = \boldsymbol{K}_r \boldsymbol{K}_t \boldsymbol{R}_a^{-1} \boldsymbol{G}_v \boldsymbol{v}_c. \tag{8.9}$$

From (8.1), (8.7), (8.8), (8.9) it is

$$\boldsymbol{K}_r \boldsymbol{K}_t \boldsymbol{R}_a^{-1} \boldsymbol{G}_v \boldsymbol{v}_c = \boldsymbol{\tau} + \boldsymbol{K}_r \boldsymbol{K}_t \boldsymbol{R}_a^{-1} \boldsymbol{K}_v \boldsymbol{K}_r \dot{\boldsymbol{q}} \tag{8.10}$$

[1] Assuming a diagonal $\boldsymbol{K}_r$ leads to excluding the presence of kinematic couplings in the transmission, that is the motion of each actuator does not induce motion on a joint other than that actuated.

[2] The contribution of the inductance has been neglected.

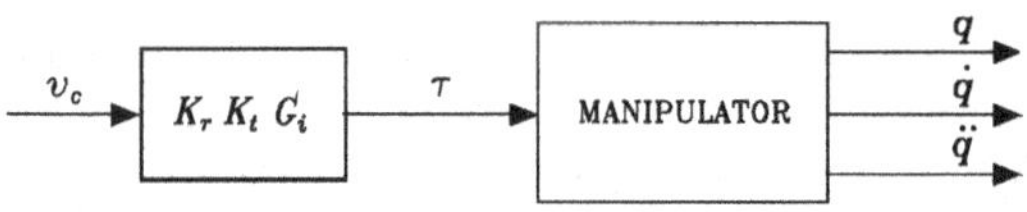

Fig. 8.4. Block scheme of the manipulator and drives system as a torque-controlled system

and thus

$$\boldsymbol{\tau} = \boldsymbol{K}_r \boldsymbol{K}_t \boldsymbol{R}_a^{-1} (\boldsymbol{G}_v \boldsymbol{v}_c - \boldsymbol{K}_v \boldsymbol{K}_r \dot{\boldsymbol{q}}). \tag{8.11}$$

The overall system is then *voltage-controlled* and the corresponding block scheme is illustrated in Fig. 8.3. If the following assumptions hold:

- the elements of matrix $\boldsymbol{K}_r$, characterizing the transmissions, are much greater than unity;
- the elements of matrix $\boldsymbol{R}_a$ are very small, which is typical in the case of high-efficiency servomotors;
- the values of the torques $\boldsymbol{\tau}$ required for the execution of the desired motions are not too large;

then it can be assumed that

$$\boldsymbol{G}_v \boldsymbol{v}_c \approx \boldsymbol{K}_v \boldsymbol{K}_r \dot{\boldsymbol{q}}. \tag{8.12}$$

The proportionality relationship obtained between $\dot{\boldsymbol{q}}$ and $\boldsymbol{v}_c$ is independent of the values attained by the manipulator parameters; the smaller the joint velocities and accelerations, the more valid this assumption. Hence, velocity (or voltage) control shows an inherent robustness with respect to parameter variations of the manipulator model, which is enhanced by the values of the gear reduction ratios.

In this case, the scheme illustrated in Fig. 8.3 can be taken as the reference structure for the design of the control system. Having assumed that

$$\boldsymbol{v}_c \approx \boldsymbol{G}_v^{-1} \boldsymbol{K}_v \boldsymbol{K}_r \dot{\boldsymbol{q}} \tag{8.13}$$

implies that the velocity of the i-th joint depends only on the i-th control voltage, since the matrix $\boldsymbol{G}_v^{-1} \boldsymbol{K}_v \boldsymbol{K}_r$ is diagonal. Therefore, the joint position control system can be designed according to a *decentralized control structure*, since each joint can be controlled independently of the others. The results, evaluated in the terms of the tracking accuracy of the joint variables with respect to the desired trajectories, are improved in the case of higher gear reduction ratios and less demanding values of required speeds and accelerations.

On the other hand, if the desired manipulator motion requires large joint speeds and/or accelerations, the approximation (8.12) no longer holds, in view of the magnitude of the required driving torques; this occurrence is even more evident for direct-drive actuation ($\boldsymbol{K}_r = \boldsymbol{I}$).

In this case, by resorting to an inverse dynamics technique, it is possible to find the joint torques $\boldsymbol{\tau}(t)$ needed to track any specified motion in terms of the joint accelerations $\ddot{\boldsymbol{q}}(t)$, velocities $\dot{\boldsymbol{q}}(t)$ and positions $\boldsymbol{q}(t)$. Obviously, this solution requires the accurate knowledge of the manipulator dynamic model. The determination of the torques to be generated by the drive system can thus refer to a *centralized control structure*, since to compute the torque history at the i-th joint it is necessary to know the time evolution of the motion of all the joints. By recalling that

$$\boldsymbol{\tau} = \boldsymbol{K}_r \boldsymbol{K}_t \boldsymbol{i}_a, \tag{8.14}$$

to find a relationship between the torques $\boldsymbol{\tau}$ and the control voltages $\boldsymbol{v}_c$, using (8.5), (8.6) leads to

$$\boldsymbol{\tau} = \boldsymbol{K}_r \boldsymbol{K}_t \boldsymbol{R}_a^{-1} \boldsymbol{G}_v \boldsymbol{v}_c - \boldsymbol{K}_r \boldsymbol{K}_t \boldsymbol{R}_a^{-1} \boldsymbol{K}_v \boldsymbol{K}_r \dot{\boldsymbol{q}}. \tag{8.15}$$

If the actuators have to provide torque contributions computed on the basis of the manipulator dynamic model, the control voltages — to be determined according to (8.15) — depend on the torque values and also on the joint velocities; this relationship depends on the matrices $\boldsymbol{K}_t$, $\boldsymbol{K}_v$ and $\boldsymbol{R}_a^{-1}$, whose elements are influenced by the operating conditions of the motors. To reduce sensitivity to parameter variations, it is worth considering driving systems characterized by a current control rather than by a voltage control. In this case the actuators behave as torque-controlled generators; the equation in (8.5) becomes meaningless and is replaced by

$$\boldsymbol{i}_a = \boldsymbol{G}_i \boldsymbol{v}_c, \tag{8.16}$$

which gives a proportional relation between the armature currents $\boldsymbol{i}_a$ (and thus the torques $\boldsymbol{\tau}$) and the control voltages $\boldsymbol{v}_c$ established by the constant matrix $\boldsymbol{G}_i$. As a consequence, (8.9) becomes

$$\boldsymbol{\tau} = \boldsymbol{u} = \boldsymbol{K}_r \boldsymbol{K}_t \boldsymbol{G}_i \boldsymbol{v}_c \tag{8.17}$$

which shows a reduced dependence of $\boldsymbol{u}$ on the motor parameters. The overall system is now *torque-controlled* and the resulting block scheme is illustrated in Fig. 8.4.

The above presentation suggests resorting for the decentralized structure — where the need for robustness prevails — to feedback control systems, while for the centralized structure — where the computation of inverse dynamics is needed — it is necessary to refer to control systems with feedforward actions. Nevertheless, it should be pointed out that centralized control still requires the use of error contributions between the desired and the actual trajectory, no matter whether they are implemented in a feedback or in a feedforward fashion. This is a consequence of the fact that the considered dynamic model, even though a quite complex one, is anyhow an idealization of reality which does not include effects, such as joint Coulomb friction, gear backlash, dimension tolerance, and the simplifying assumptions in the model, e.g., link rigidity, and so on.

As already pointed out, the drive systems is anyhow inserted into a feedback control system. In the case of decentralized control, the drive will be characterized by the model describing its behaviour as a velocity-controlled generator. Instead, in the case of centralized control, since the driving torque is to be computed on a complete or reduced manipulator dynamic model, the drive will be characterized as a torque-controlled generator.

8.3 Decentralized Control

The simplest control strategy that can be thought of is one that regards the manipulator as formed by n independent systems (the n joints) and controls each joint axis as a *single-input/single-output system*. Coupling effects between joints due to varying configurations during motion are treated as *disturbance* inputs.

In order to analyze various control schemes and their performance, it is worth considering the model of the system manipulator with drives in terms of mechanical quantities at the motor side; in view of (8.2), (8.3), it is

$$\boldsymbol{K}_r^{-1}\boldsymbol{B}(\boldsymbol{q})\boldsymbol{K}_r^{-1}\ddot{\boldsymbol{q}}_m + \boldsymbol{K}_r^{-1}\boldsymbol{C}(\boldsymbol{q},\dot{\boldsymbol{q}})\boldsymbol{K}_r^{-1}\dot{\boldsymbol{q}}_m + \boldsymbol{K}_r^{-1}\boldsymbol{F}_v\boldsymbol{K}_r^{-1} + \boldsymbol{K}_r^{-1}\boldsymbol{g}(\boldsymbol{q}) = \boldsymbol{\tau}_m. \tag{8.18}$$

By observing that the diagonal elements of $\boldsymbol{B}(\boldsymbol{q})$ are formed by constant terms and configuration-dependent terms (functions of sine and cosine for revolute joints), one can set

$$\boldsymbol{B}(\boldsymbol{q}) = \bar{\boldsymbol{B}} + \Delta\boldsymbol{B}(\boldsymbol{q}) \tag{8.19}$$

where $\bar{\boldsymbol{B}}$ is the *diagonal* matrix whose constant elements represent the resulting average inertia at each joint. Substituting (8.19) into (8.1) yields

$$\boldsymbol{K}_r^{-1}\bar{\boldsymbol{B}}\boldsymbol{K}_r^{-1}\ddot{\boldsymbol{q}}_m + \boldsymbol{F}_m\dot{\boldsymbol{q}}_m + \boldsymbol{d} = \boldsymbol{\tau}_m \tag{8.20}$$

where

$$\boldsymbol{F}_m = \boldsymbol{K}_r^{-1}\boldsymbol{F}_v\boldsymbol{K}_r^{-1} \tag{8.21}$$

represents the matrix of viscous friction coefficients about the motor axes, and

$$\boldsymbol{d} = \boldsymbol{K}_r^{-1}\Delta\boldsymbol{B}(\boldsymbol{q})\boldsymbol{K}_r^{-1}\ddot{\boldsymbol{q}}_m + \boldsymbol{K}_r^{-1}\boldsymbol{C}(\boldsymbol{q},\dot{\boldsymbol{q}})\boldsymbol{K}_r^{-1}\dot{\boldsymbol{q}}_m + \boldsymbol{K}_r^{-1}\boldsymbol{g}(\boldsymbol{q}) \tag{8.22}$$

represents the contribution depending on the configuration.

As illustrated by the block scheme of Fig. 8.5, the system of manipulator with drives is actually constituted by two subsystems; one has $\boldsymbol{\tau}_m$ as input and $\boldsymbol{q}_m$ as output, the other has $\boldsymbol{q}_m$, $\dot{\boldsymbol{q}}_m$, $\ddot{\boldsymbol{q}}_m$ as inputs, and $\boldsymbol{d}$ as output. The former is *linear* and *decoupled*, since each component of $\boldsymbol{\tau}_m$ influences only the corresponding component of $\boldsymbol{q}_m$. The latter is *nonlinear* and *coupled*, since

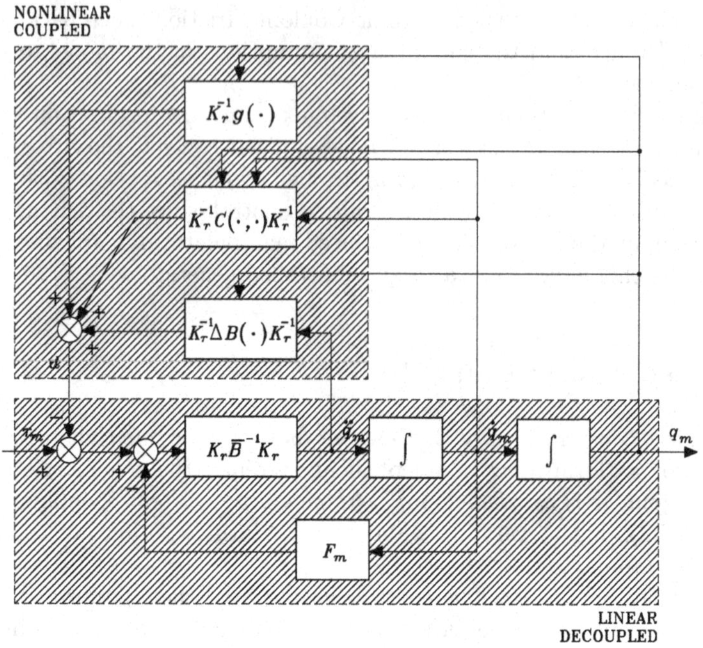

Fig. 8.5. Block scheme of the system of manipulator with drives

it accounts for all those nonlinear and coupling terms of manipulator joint dynamics.

On the basis of the above scheme, several control algorithms can be derived with reference to the detail of knowledge of the dynamic model. The simplest approach that can be followed, in case of high-gear reduction ratios and/or limited performance in terms of required velocities and accelerations, is to consider the component of the nonlinear interacting term $\boldsymbol{d}$ as a *disturbance* for the single joint servo.

The design of the control algorithm leads to a *decentralized control structure*, since each joint is considered independently of the others. The joint controller must guarantee good performance in terms of high disturbance rejection and enhanced trajectory tracking capabilities. The resulting control structure is substantially based on the error between the desired and actual output, while the input control torque at actuator i depends only on the error of output i.

Therefore, the system to control is Joint i drive corresponding to the single-input/single-output system of the decoupled and linear part of the scheme in Fig. 8.5. The interaction with the other joints is described by component i of the vector $\boldsymbol{d}$ in (8.22).

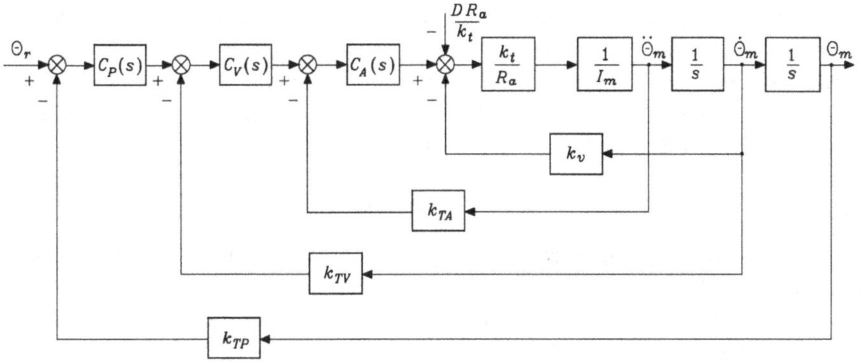

Fig. 8.6. Block scheme of general independent joint control

Assumed that the actuator is a rotary electric DC motor, the general scheme of drive control is that in Fig. 5.9 where I_m is the average inertia reported to the motor axis ($I_{mi} = \bar{b}_{ii}/k_{ri}^2$).[3]

8.3.1 Independent Joint Control

To guide selection of the controller structure, start noticing that an effective rejection of the disturbance d on the output ϑ_m is ensured by:

- a large value of the amplifier gain before the point of intervention of the disturbance,
- the presence of an integral action in the controller so as to cancel the effect of the gravitational component on the output at steady state (constant ϑ_m).

These requisites clearly suggest the use of a *proportional-integral* (PI) control action in the forward path whose transfer function is

$$C(s) = K_c \frac{1 + sT_c}{s}; \tag{8.23}$$

this yields zero error at steady state for a constant disturbance, and the presence of the real zero at $s = -1/T_c$ offers a stabilizing action. To improve dynamic performance, it is worth choosing the controller as a cascade of elementary actions with local feedback loops closed around the disturbance.

Besides closure of a position feedback loop, the most general solution is obtained by closing inner loops on velocity and acceleration. This leads to the scheme in Fig. 8.6, where $C_P(s)$, $C_V(s)$, $C_A(s)$ respectively represent *position*, *velocity*, *acceleration* controllers, and the inmost controller should

[3] Subscript i is to be dropped for notation compactness.

be of PI type as in (8.23) so as to obtain zero error at steady state for a constant disturbance. Further, k_{TP}, k_{TV}, k_{TA} are the respective transducer constants, and the amplifier gain G_v has been embedded in the gain of the inmost controller. In the scheme of Fig. 8.6, notice that ϑ_r is the reference input, which is related to the desired output ϑ_{md} as

$$\vartheta_r = k_{TP}\vartheta_{md}.$$

Further, the disturbance torque D has been suitably transformed into a voltage by the factor R_a/k_t.

In the following, a number of possible solutions that can be derived from the general scheme of Fig. 8.6 are presented; at this stage, the issue arising from possible lack of measurement of physical variables is not considered yet. Three case studies are considered which differ in the number of active feedback loops.[4]

Position feedback

In this case, the control action is characterized by

$$C_P(s) = K_P\frac{1+sT_P}{s} \qquad C_V(s) = 1 \qquad C_A(s) = 1$$

$$k_{TV} = k_{TA} = 0.$$

With these positions, the structure of the control scheme in Fig. 8.6 leads to the scheme illustrated in Fig. 5.10. From this scheme the transfer function of the forward path is

$$P(s) = \frac{k_m K_P(1+sT_P)}{s^2(1+sT_m)},$$

while that of the return path is

$$H(s) = k_{TP}.$$

A root locus analysis can be performed as a function of the gain of the position loop $k_m K_P k_{TP} T_P/T_m$. Three situations are illustrated for the poles of the closed-loop system with reference to the relation between T_P and T_m (Fig. 8.7). Stability of the closed-loop feedback system imposes some constraints on the choice of the parameters of the PI controller. If $T_P < T_m$, the system is inherently unstable (Fig. 8.7a). Then, it must be $T_P > T_m$ (Fig. 8.7b). As T_P increases, the absolute value of the real part of the two roots of the locus tending towards the asymptotes increases too, and the system has faster time response. Hence, it is convenient to render $T_P \gg T_m$ (Fig. 8.7c). In any case, the real part of the dominant poles cannot be less than $-1/2T_m$.

---

[4] See Appendix C for a brief brush-up on control of linear single-input/single-output systems.

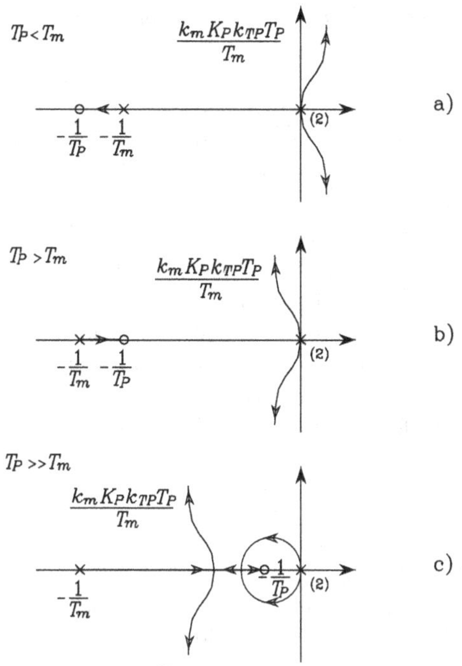

Fig. 8.7. Root loci for the position feedback control scheme

The closed-loop input/output transfer function is

$$\frac{\Theta_m(s)}{\Theta_r(s)} = \frac{\dfrac{1}{k_{TP}}}{1 + \dfrac{s^2(1 + sT_m)}{k_m K_P k_{TP}(1 + sT_P)}}, \tag{8.24}$$

which can be expressed in the form

$$W(s) = \frac{\dfrac{1}{k_{TP}}(1 + sT_P)}{\left(1 + \dfrac{2\zeta s}{\omega_n} + \dfrac{s^2}{\omega_n^2}\right)(1 + s\tau)},$$

where ω_n and ζ are respectively the natural frequency and damping ratio of the pair of complex poles and $-1/\tau$ locates the real pole. These values are assigned to define the joint drive dynamics as a function of the constant T_P; if $T_P > T_m$, then $1/\zeta\omega_n > T_P > \tau$ (Fig. 8.7b); if $T_P \gg T_m$ (Fig. 8.7c), for large values of the loop gain, then $\zeta\omega_n > 1/\tau \approx 1/T_P$ and the zero at $-1/T_P$ in the transfer function $W(s)$ tends to cancel the effect of the real pole.

The closed-loop disturbance/output transfer function is

$$\frac{\Theta_m(s)}{D(s)} = -\frac{\dfrac{sR_a}{k_t K_P k_{TP}(1+sT_P)}}{1+\dfrac{s^2(1+sT_m)}{k_m K_P k_{TP}(1+sT_P)}}, \tag{8.25}$$

which shows that it is worth increasing K_P to reduce the effect of disturbance on the output during the transient. The function in (8.25) has two complex poles $(-\zeta\omega_n, \pm j\sqrt{1-\zeta^2}\omega_n)$, a real pole $(-1/\tau)$, and a zero at the origin. The zero is due to the PI controller and allows the cancellation of the effects of gravity on the angular position when ϑ_m is a constant.

In (8.25), it can be recognized that the term $K_P k_{TP}$ is the reduction factor imposed by the feedback gain on the amplitude of the output due to disturbance; hence, the quantity

$$X_R = K_P k_{TP} \tag{8.26}$$

can be interpreted as the *disturbance rejection factor*, which in turn is determined by the gain K_P. However, it is not advisable to increase K_P too much, because small damping ratios would result leading to unacceptable oscillations of the output. An estimate T_R of the *output recovery time* needed by the control system to recover the effects of the disturbance on the angular position can be evaluated by analyzing the modes of evolution of (8.25). Since $\tau \approx T_P$, such estimate is expressed by

$$T_R = \max\left\{T_P, \frac{1}{\zeta\omega_n}\right\}. \tag{8.27}$$

Position and velocity feedback

In this case, the control action is characterized by

$$C_P(s) = K_P \qquad C_V(s) = K_V\frac{1+sT_V}{s} \qquad C_A(s) = 1$$

$$k_{TA} = 0;$$

with these positions, the structure of the control scheme in Fig. 8.6 leads to scheme illustrated in Fig. 5.11. To carry out a root locus analysis as a function of the velocity feedback loop gain, it is worth reducing the velocity loop in parallel to the position loop by following the usual rules for moving blocks. From the scheme in Fig. 5.11 the transfer function of the forward path is

$$P(s) = \frac{k_m K_P K_V(1+sT_V)}{s^2(1+sT_m)},$$

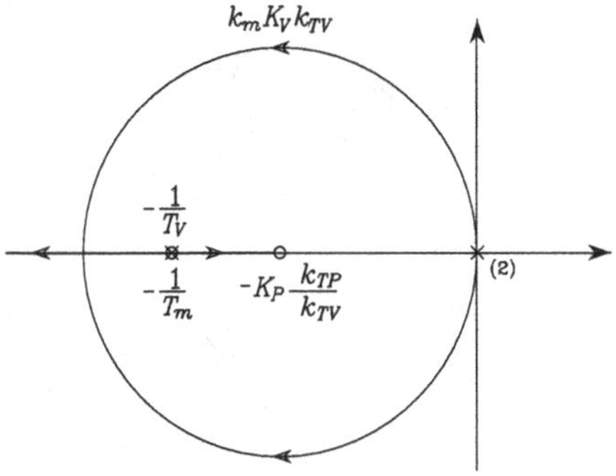

Fig. 8.8. Root locus for the position and velocity feedback control scheme

while that of the return path is

$$H(s) = k_{TP}\left(1 + s\frac{k_{TV}}{K_P k_{TP}}\right).$$

The zero of the controller at $s = -1/T_V$ can be chosen so as to cancel the effects of the real pole of the motor at $s = -1/T_m$. Then, by setting

$$T_V = T_m,$$

the poles of the closed-loop system move on the root locus as a function of the loop gain $k_m K_V k_{TV}$, as shown in Fig. 8.8. By increasing the position feedback gain K_P, it is possible to confine the closed-loop poles into a region of the complex plane with large absolute values of the real part. Then, the actual location can be established by a suitable choice of K_V.

The closed-loop input/output transfer function is

$$\frac{\Theta_m(s)}{\Theta_r(s)} = \frac{\dfrac{1}{k_{TP}}}{1 + \dfrac{sk_{TV}}{K_P k_{TP}} + \dfrac{s^2}{k_m K_P k_{TP} K_V}}, \tag{8.28}$$

which can be compared with the typical transfer function of a second-order system

$$W(s) = \frac{\dfrac{1}{k_{TP}}}{1 + \dfrac{2\zeta s}{\omega_n} + \dfrac{s^2}{\omega_n^2}}. \tag{8.29}$$

It can be recognized that, with a suitable choice of the gains, it is possible to obtain any value of natural frequency ω_n and damping ratio ζ. Hence, if ω_n and ζ are given as design requirements, the following relations can be found:

$$K_V k_{TV} = \frac{2\zeta\omega_n}{k_m} \tag{8.30}$$

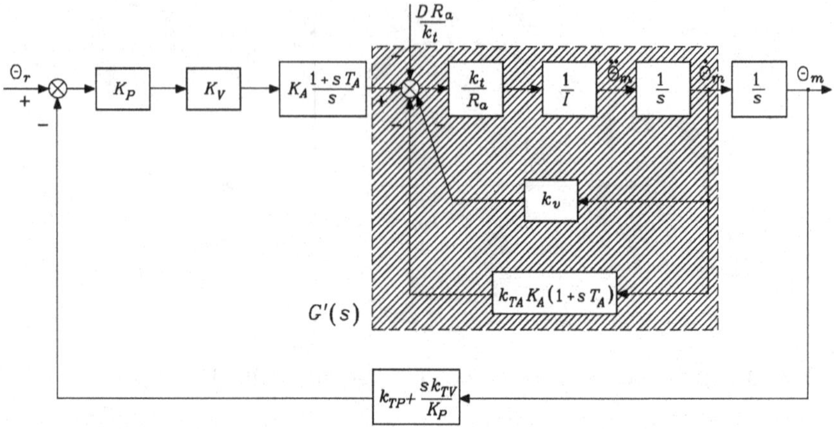

Fig. 8.9. Block scheme of position, velocity and acceleration feedback control

$$K_P k_{TP} K_V = \frac{\omega_n^2}{k_m}. \tag{8.31}$$

For given transducer constants k_{TP} and k_{TV}, once K_V has been chosen to satisfy (8.30), the value of K_P is obtained from (8.31).

The closed-loop disturbance/output transfer function is

$$\frac{\Theta_m(s)}{D(s)} = -\frac{\dfrac{sR_a}{k_t K_P k_{TP} K_V (1 + sT_m)}}{1 + \dfrac{sk_{TV}}{K_P k_{TP}} + \dfrac{s^2}{k_m K_P k_{TP} K_V}}, \tag{8.32}$$

which shows that the *disturbance rejection factor* is

$$X_R = K_P k_{TP} K_V \tag{8.33}$$

and is fixed, once K_P and K_V have been chosen via (8.30), (8.31). Concerning disturbance dynamics, the presence of a zero at the origin introduced by the PI, of a real pole at $s = -1/T_m$, and of a pair of complex poles having real part $-\zeta\omega_n$ should be noticed. Hence, in this case, an estimate of the *output recovery time* is given by the time constant

$$T_R = \max\left\{T_m, \frac{1}{\zeta\omega_n}\right\}; \tag{8.34}$$

which reveals an improvement with respect to the previous case in (8.27), since $T_m \ll T_P$ and the real part of the dominant poles is not constrained by the inequality $\zeta\omega_n < 1/2T_m$.

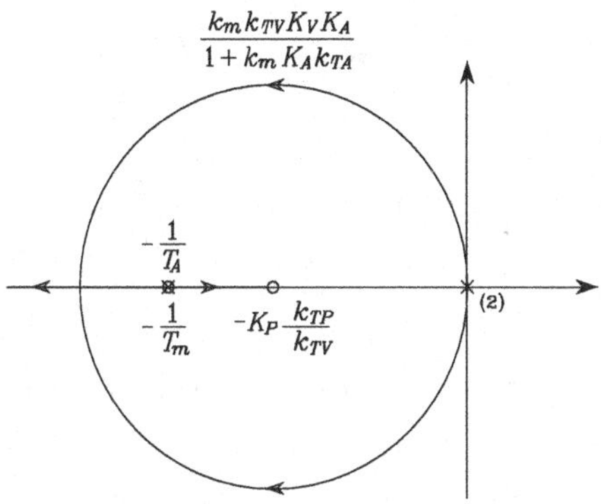

Fig. 8.10. Root locus for the position, velocity and acceleration feedback control scheme

Position, velocity and acceleration feedback

In this case, the control action is characterized by

$$C_P(s) = K_P \qquad C_V(s) = K_V \qquad C_A(s) = K_A \frac{1+sT_A}{s}.$$

After some manipulation, the block scheme of Fig. 8.6 can be reduced to that of Fig. 8.9 where $G'(s)$ indicates the following transfer function:

$$G'(s) = \frac{k_m}{(1+k_m K_A k_{TA})\left(1 + \frac{sT_m\left(1 + k_m K_A k_{TA}\frac{T_A}{T_m}\right)}{(1+k_m K_A k_{TA})}\right)}.$$

The transfer function of the forward path is

$$P(s) = \frac{K_P K_V K_A (1+sT_A)}{s^2} G'(s),$$

while that of the return path is

$$H(s) = k_{TP}\left(1 + \frac{sk_{TV}}{K_P k_{TP}}\right).$$

Also in this case, a suitable pole cancellation is worthy which can be achieved either by setting

$$T_A = T_m,$$

or by making

$$k_m K_A k_{TA} T_A \gg T_m \qquad k_m K_A k_{TA} \gg 1.$$

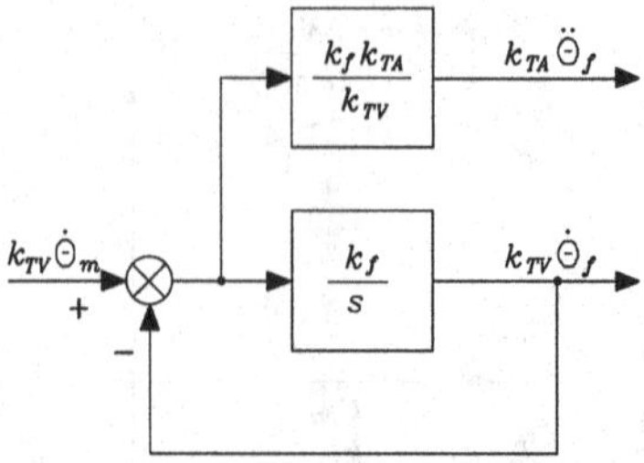

Fig. 8.11. Block scheme of a first-order filter

The two solutions are equivalent as regards dynamic performance of the control system. In both cases, the poles of the closed-loop system are constrained to move on the root locus as a function of the loop gain $k_m K_P K_V K_A/(1 + k_m K_A k_{TA})$ (Fig. 8.10). A close analogy with the previous scheme can be recognized, in that the resulting closed-loop system is again of second-order type.

The closed-loop input/output transfer function is

$$\frac{\Theta_m(s)}{\Theta_r(s)} = \frac{\dfrac{1}{k_{TP}}}{1 + \dfrac{s k_{TV}}{K_P k_{TP}} + \dfrac{s^2(1 + k_m K_A k_{TA})}{k_m K_P k_{TP} K_V K_A}}, \tag{8.35}$$

while the closed-loop disturbance/output transfer function is

$$\frac{\Theta_m(s)}{D(s)} = -\frac{\dfrac{s R_a}{k_t K_P k_{TP} K_V K_A (1 + s T_A)}}{1 + \dfrac{s k_{TV}}{K_P k_{TP}} + \dfrac{s^2(1 + k_m K_A k_{TA})}{k_m K_P k_{TP} K_V K_A}}. \tag{8.36}$$

The resulting *disturbance rejection factor* is given by

$$X_R = K_P k_{TP} K_V K_A, \tag{8.37}$$

while the *output recovery time* is given by the time constant

$$T_R = \max\left\{T_A, \frac{1}{\zeta\omega_n}\right\} \tag{8.38}$$

where T_A can be made less than T_m, as pointed out above.

With reference to the transfer function in (8.29), the following relations can be established for design purposes, once ζ, ω_n, X_R have been specified:

$$\frac{2K_P k_{TP}}{k_{TV}} = \frac{\omega_n}{\zeta} \tag{8.39}$$

$$k_m K_A k_{TA} = \frac{k_m X_R}{\omega_n^2} - 1 \tag{8.40}$$

$$K_P k_{TP} K_V K_A = X_R. \tag{8.41}$$

For given k_{TP}, k_{TV}, k_{TA}, K_P is chosen to satisfy (8.39), K_A is chosen to satisfy (8.40), and then K_V is obtained from (8.41). Notice how admissible solutions for the controller typically require large values for the rejection factor X_R. Hence, in principle, not only does the acceleration feedback allow the achievement of any desired dynamic behaviour but, with respect to the previous case, it also allows the prescription of the disturbance rejection factor as long as $k_m X_R/\omega_n^2 > 1$.

In deriving the above control schemes, the issue of measurement of feedback variables was not considered explicitly. With reference to the typical position control servos that are implemented in industrial practice, there is no problem of measuring position and velocity, while a direct measurement of acceleration, in general, either is not available or is too expensive to obtain. Therefore, for the scheme of Fig. 8.9, an indirect measurement can be obtained by reconstructing acceleration from direct velocity measurement through a first-order *filter* (Fig. 8.11). The filter is characterized by a bandwidth $\omega_{3f} = k_f$. By choosing this bandwidth wide enough, the effects due to measurement lags are not appreciable, and then it is feasible to take the acceleration filter output as the quantity to feed back. Some problem may occur concerning the noise superimposed on the filtered acceleration signal, though.

Resorting to a filtering technique may be useful when only the direct position measurement is available. In this case, by means of a second-order state variable filter, it is possible to reconstruct velocity and acceleration. However, the greater lags induced by the use of a second-order filter typically degrade the performance with respect to the use of a first-order filter, because of limitations imposed on the filter bandwidth by numerical implementation of the controller and filter.

Notice that the above derivation is based on an ideal dynamic model, i.e., when the effects of transmission elasticity as well as those of amplifier and motor electrical time constants are neglected. This implies that satisfaction of design requirements imposing large values of feedback gains may not be verified in practice, since the existence of unmodelled dynamics — such as electric dynamics, elastic dynamics due to non-perfectly rigid transmissions, filter dynamics for the third scheme — might lead to degrading the system and eventually driving it to instability. In summary, the above solutions constitute design guidelines whose limits should be emphasized with regard to the specific application.

8.3.2 Decentralized Feedforward Compensation

When the joint control servos are required to track reference trajectories with high values of speed and acceleration, the tracking capabilities of the scheme in Fig. 8.6 are unavoidably degraded. The adoption of a *decentralized feedforward compensation* allows a reduction of the tracking error. Therefore, in view of the closed-loop input/output transfer functions in (8.24), (8.28), (8.35),

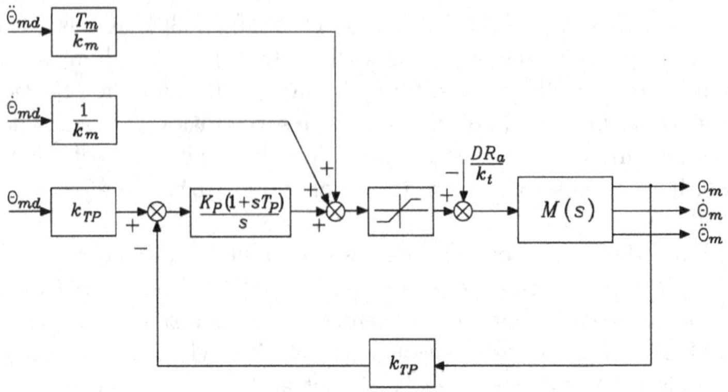

Fig. 8.12. Block scheme of position feedback control with decentralized feedforward compensation

the reference inputs to the three control structures analyzed in the previous section can be respectively modified into

$$\Theta'_r(s) = \left(k_{TP} + \frac{s^2(1+sT_m)}{k_m K_P(1+sT_P)}\right)\Theta_{md}(s) \tag{8.42}$$

$$\Theta'_r(s) = \left(k_{TP} + \frac{sk_{TV}}{K_P} + \frac{s^2}{k_m K_P K_V}\right)\Theta_{md}(s) \tag{8.43}$$

$$\Theta'_r(s) = \left(k_{TP} + \frac{sk_{TV}}{K_P} + \frac{s^2(1+k_m K_A k_{TA})}{k_m K_P K_V K_A}\right)\Theta_{md}(s); \tag{8.44}$$

in this way, tracking of the desired joint position $\Theta_{md}(s)$ is achieved, if not for the effect of disturbances. Notice that computing time derivatives of the desired trajectory is not a problem, once $\vartheta_{md}(t)$ is known analytically. The tracking control schemes, resulting from simple manipulation of (8.42), (8.43), (8.44) are reported respectively in Figs. 8.12, 8.13, 8.14, where $M(s)$ indicates the motor transfer function in (5.11), with k_m and T_m as in (5.12).

All the solutions allow the input trajectory to be tracked within the range of validity and linearity of the models employed. It is worth noticing that, as the number of nested feedback loops increases, a less accurate knowledge of the system model is required to perform feedforward compensation. In fact, T_m and k_m are required for the scheme of Fig. 8.12, only k_m is required for the scheme of Fig. 8.13, and k_m again — but with reduced weight — for the scheme of Fig. 8.14.

It is worth recalling that *perfect* tracking can be obtained only under the assumption of exact matching of the controller and feedforward compensation parameters with the process parameters, as well as of exact modelling and linearity of the physical system. Deviations from the ideal values cause a performance degradation that should be analyzed case by case.

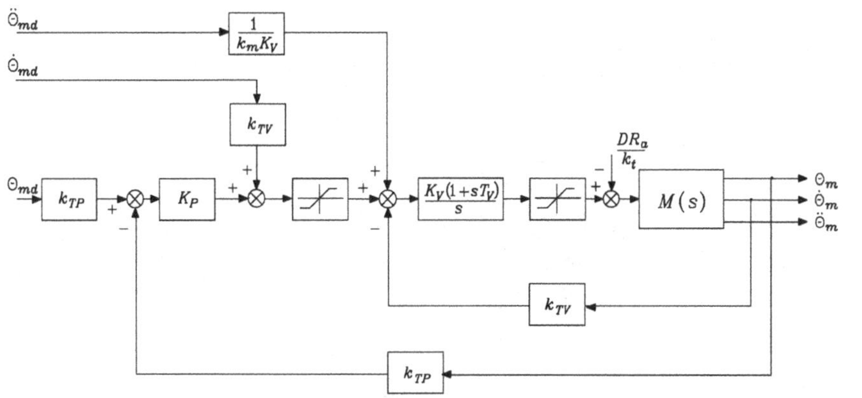

Fig. 8.13. Block scheme of position and velocity feedback control with decentralized feedforward compensation

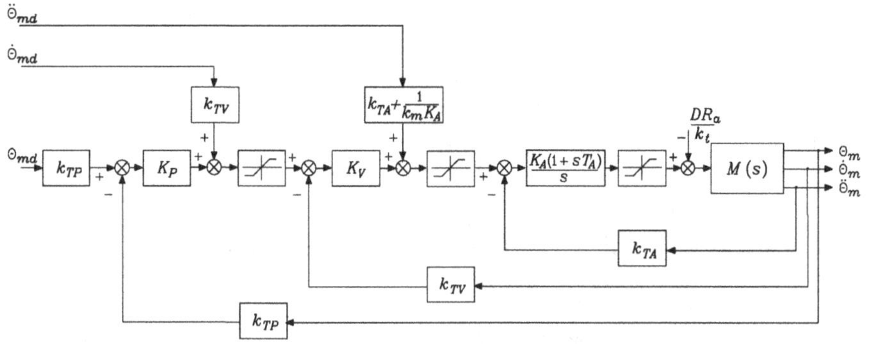

Fig. 8.14. Block scheme of position, velocity and acceleration feedback control with decentralized feedforward compensation

The presence of saturation blocks in the schemes of Figs. 8.12, 8.13, 8.14 is to be intended as intentional nonlinearities whose function is to limit relevant physical quantities during transients; the greater the number of feedback loops, the greater the number of quantities that can be limited (velocity, acceleration, and motor voltage). To this end, notice that trajectory tracking is obviously lost whenever any of the above quantities saturates. This situation often occurs for industrial manipulators required to execute point-to-point motions; in this case, there is less concern about the actual trajectories followed, and the actuators are intentionally taken to operate at the current limits so as to realize the fastest possible motions.

After simple block reduction on the above schemes, it is possible to determine equivalent control structures that utilize position feedback only and *regulators with standard actions*. It should be emphasized that the two solutions are equivalent in terms of disturbance rejection and trajectory tracking.

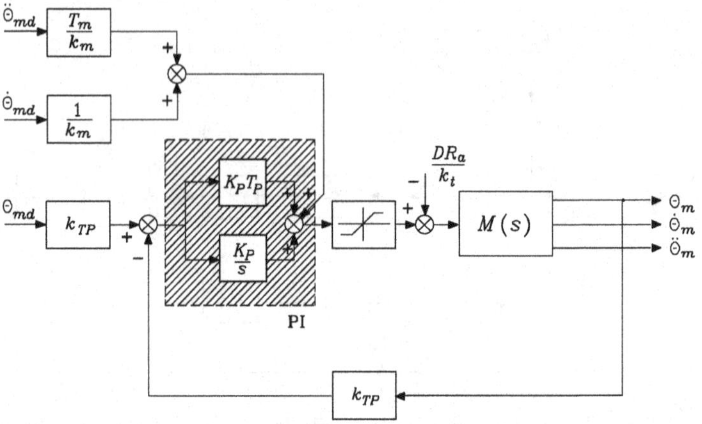

Fig. 8.15. Equivalent control scheme of PI type

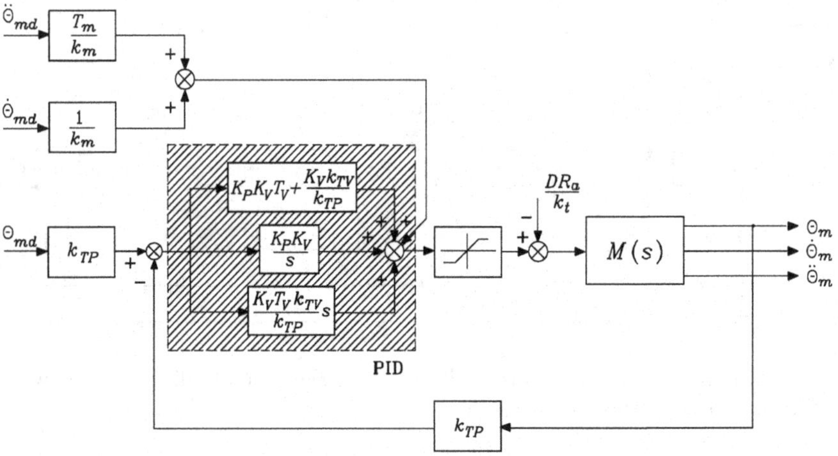

Fig. 8.16. Equivalent control scheme of PID type

However, tuning of regulator parameters is less straightforward, and the elimination of inner feedback loops prevents the possibility of setting saturations on velocity and/or acceleration. The control structures equivalent to those of Figs. 8.12, 8.13, 8.14 are illustrated in Figs. 8.15, 8.16, 8.17, respectively; control actions of PI, PID, PIDD^2 type are illustrated which are respectively equivalent to the cases of: position feedback; position and velocity feedback; position, velocity and acceleration feedback.

It is worth noticing that the equivalent control structures in Figs. 8.15–8.17 are characterized by the presence of the feedforward action $(T_m/k_m)\ddot{\vartheta}_{md} + (1/k_m)\dot{\vartheta}_{md}$. If the motor is current-controlled and not voltage-controlled, by recalling (5.13), the feedforward action is equal to $(k_i/k_t)(I_m\ddot{\vartheta}_{md} + F_m\dot{\vartheta}_{md})$. If $\dot{\vartheta}_m \approx \dot{\vartheta}_{md}$, $\ddot{\vartheta}_m \approx \ddot{\vartheta}_{md}$ and the disturbance is negligible, the term $I_m\ddot{\vartheta}_d +$

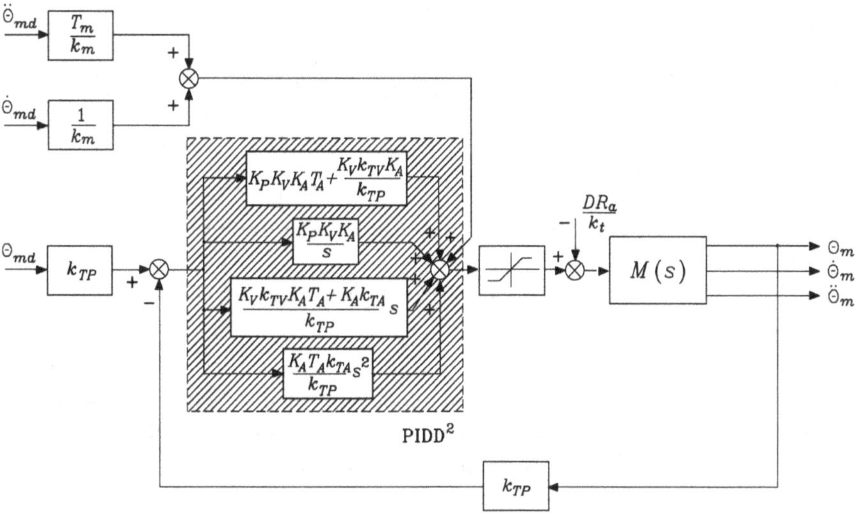

Fig. 8.17. Equivalent control scheme of PIDD^2 type

$F_m \dot{\vartheta}_d$ represents the driving torque providing the desired velocity and acceleration, as indicated by (5.3). By setting

$$i_{ad} = \frac{1}{k_t}(I_m \ddot{\vartheta}_{md} + F_m \dot{\vartheta}_{md}),$$

the feedforward action can be rewritten in the form $k_i i_{ad}$. This shows that, in the case the drive is current-controlled, it is possible to replace the acceleration and velocity feedforward actions with a current and thus a torque feedforward action, which is to be properly computed with reference to the desired motion.

This equivalence is illustrated in Fig. 8.18, where $M(s)$ has been replaced by the block scheme of an electric drive of Fig. 5.2, where the parameters of the current loop are chosen so as to realize a torque-controlled generator. The feedforward action represents a reference for the motor current, which imposes the generation of the nominal torque to execute the desired motion; the presence of the position reference allows the closure of a feedback loop which, in view of the adoption of a standard regulator with transfer function $C_R(s)$, confers robustness to the presented control structure. In summary, the performance that can be achieved with velocity and acceleration feedforward actions and voltage-controlled actuator can be achieved with a current-controlled actuator and a desired torque feedforward action.

The above schemes can incorporate the typical structure of the controllers actually implemented in the control architectures of industrial robots. In these systems it is important to choose the largest possible gains so that model inaccuracy and coupling terms do not appreciably affect positions of the single joints. As pointed out above, the upper limit on the gains is imposed by

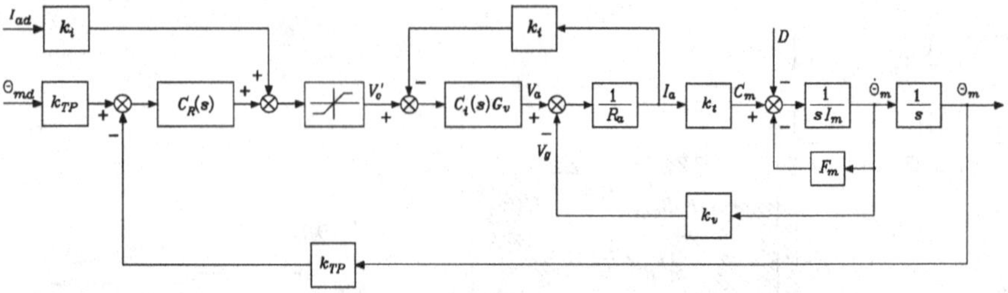

Fig. 8.18. Control scheme with current-controlled drive and current feedforward action

all those factors that have not been modelled, such as implementation of discrete-time controllers in lieu of the continuous-time controllers analyzed in theory, presence of finite sampling time, neglected dynamic effects (e.g., joint elasticity, structural resonance, finite transducer bandwidth), and sensor noise. In fact, the influence of such factors in the implementation of the above controllers may cause a severe system performance degradation for much too large values of feedback gains.

8.4 Computed Torque Feedforward Control

Define the tracking error $e(t) = \vartheta_{md}(t) - \vartheta_m(t)$. With reference to the most general scheme (Fig. 8.17), the output of the PIDD^2 regulator can be written as

$$a_2\ddot{e} + a_1\dot{e} + a_0 e + a_{-1}\int^t e(\varsigma)d\varsigma$$

which describes the time evolution of the error. The constant coefficients a_2, a_1, a_0, a_{-1} are determined by the particular solution adopted. Summing the contribution of the feedforward actions and of the disturbance to this expression yields

$$\frac{T_m}{k_m}\ddot{\vartheta}_{md} + \frac{1}{k_m}\dot{\vartheta}_{md} - \frac{R_a}{k_t}d,$$

where

$$\frac{T_m}{k_m} = \frac{I_m R_a}{k_t} \qquad k_m = \frac{1}{k_v}.$$

The input to the motor (Fig. 8.6) has then to satisfy the following equation:

$$a_2\ddot{e} + a_1\dot{e} + a_0 e + a_{-1}\int^t e(\varsigma)d\varsigma + \frac{T_m}{k_m}\ddot{\vartheta}_{md} + \frac{1}{k_m}\dot{\vartheta}_{md} - \frac{R_a}{k_t}d = \frac{T_m}{k_m}\ddot{\vartheta}_m + \frac{1}{k_m}\dot{\vartheta}_m.$$

With a suitable change of coefficients, this can be rewritten as

$$a_2'\ddot{e} + a_1'\dot{e} + a_0' e + a_{-1}'\int^t e(\varsigma)d\varsigma = \frac{R_a}{k_t}d.$$

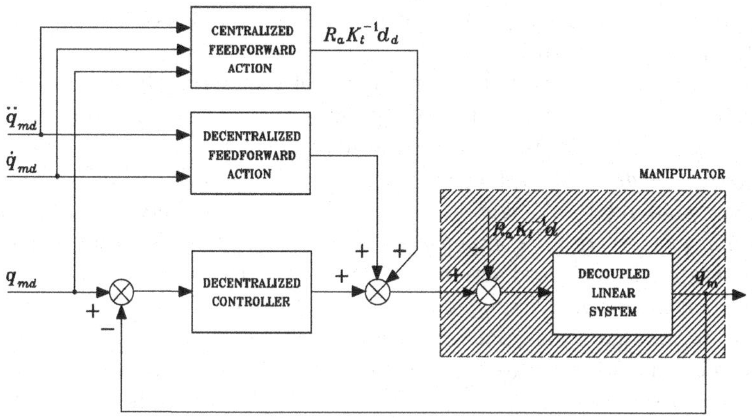

Fig. 8.19. Block scheme of computed torque feedforward control

This equation describes the *error dynamics* and shows that any physically executable trajectory is asymptotically tracked only if the disturbance term $d(t) = 0$. With the term *physically executable* it is meant that the saturation limits on the physical quantities, e.g., current and voltage in electric motors, are not violated in the execution of the desired trajectory.

The presence of the term $d(t)$ causes a tracking error whose magnitude is reduced as much as the disturbance frequency content is located off to the left of the lower limit of the bandwidth of the error system. The disturbance/error transfer function is given by

$$\frac{E(s)}{D(s)} = \frac{\dfrac{R_a}{k_t}s}{a'_2s^3 + a'_1s^2 + a'_0s + a'_{-1}},$$

and thus the adoption of loop gains which are not realizable for the above discussed reasons is often required.

Nevertheless, even if the term $d(t)$ has been introduced as a disturbance, its expression is given by (8.22). It is then possible to add a further term to the previous feedforward actions which is able to compensate the disturbance itself rather than its effects. In other words, by taking advantage of model knowledge, the rejection effort of an independent joint control scheme can be lightened with notable simplification from the implementation viewpoint.

Let $\boldsymbol{q}_d(t)$ be the desired joint trajectory and $\boldsymbol{q}_{md}(t)$ the corresponding actuator trajectory as in (8.2). By adopting an *inverse model* strategy, the *feedforward* action $\boldsymbol{R}_a\boldsymbol{K}_t^{-1}\boldsymbol{d}_d$ can be introduced with

$$\boldsymbol{d}_d = \boldsymbol{K}_r^{-1}\Delta\boldsymbol{B}(\boldsymbol{q}_d)\boldsymbol{K}_r^{-1}\ddot{\boldsymbol{q}}_{md} + \boldsymbol{K}_r^{-1}\boldsymbol{C}(\boldsymbol{q}_d,\dot{\boldsymbol{q}}_d)\boldsymbol{K}_r^{-1}\dot{\boldsymbol{q}}_{md} + \boldsymbol{K}_r^{-1}\boldsymbol{g}(\boldsymbol{q}_d), \quad (8.45)$$

where $\boldsymbol{R}_a$ and $\boldsymbol{K}_t$ denote the diagonal matrices of armature resistances and torque constants of the actuators. This action tends to compensate the actual disturbance expressed by (8.22) and in turn allows the control system to operate in a better condition.

This solution is illustrated in the scheme of Fig. 8.19, which conceptually describes the control system of a manipulator with *computed torque* control. The feedback control system is representative of the n independent joint control servos; it is *decentralized*, since controller i elaborates references and measurements that refer to single Joint i. The interactions between the various joints, expressed by $\boldsymbol{d}$, are compensated by a *centralized* action whose function is to generate a feedforward action that depends on the joint references as well as on the manipulator dynamic model. This action compensates the nonlinear coupling terms due to inertial, Coriolis, centrifugal, and gravitational forces that depend on the structure and, as such, vary during manipulator motion.

Although the residual disturbance term $\widetilde{\boldsymbol{d}} = \boldsymbol{d}_d - \boldsymbol{d}$ vanishes only in the ideal case of perfect tracking ($\boldsymbol{q} = \boldsymbol{q}_d$) and exact dynamic modelling, $\widetilde{\boldsymbol{d}}$ is representative of interaction disturbances of considerably reduced magnitude with respect to $\boldsymbol{d}$. Hence, the computed torque technique has the advantage to alleviate the disturbance rejection task for the feedback control structure and in turn allows limited gains. Notice that expression (8.45) in general imposes a computationally demanding burden on the centralized part of the controller. Therefore, in those applications where the desired trajectory is generated in real time with regard to exteroceptive sensory data and commands from higher hierarchical levels of the robot control architecture,[5] on-line computation of the centralized feedforward action may require too much time.[6]

Since the actual controller is to be implemented on a computer with a finite sampling time, torque computation has to be carried out during this interval of time; in order not to degrade dynamic system performance, typical sampling times are of the order of the millisecond.

Therefore, it may be worth performing only a *partial* feedforward action so as to compensate those terms of (8.45) that give the most relevant contributions during manipulator motion. Since inertial and gravitational terms dominate velocity-dependent terms (at operational joint speeds not greater than a few radians per second), a partial compensation can be achieved by computing only the gravitational torques and the inertial torques due to the diagonal elements of the inertia matrix. In this way, only the terms depending on the global manipulator configuration are compensated while those deriving from motion interaction with the other joints are not.

Finally, it should be pointed out that, for repetitive trajectories, the above compensating contributions can be computed off-line and properly stored on the basis of a trade-off solution between memory capacity and computational requirements of the control architecture.

---

[5] See also Chap. 6.

[6] In this regard, the problem of real-time computation of compensating torques can be solved by resorting to efficient recursive formulations of manipulator inverse dynamics, such as the Newton–Euler algorithm presented in Chap. 7.

8.5 Centralized Control

In the previous sections several techniques have been discussed that allow the design of independent joint controllers. These are based on a single-input/single-output approach, since interaction and coupling effects between the joints have been considered as disturbances acting on each single joint drive system.

On the other hand, when large operational speeds are required or direct-drive actuation is employed ($\boldsymbol{K}_r = \boldsymbol{I}$), the nonlinear coupling terms strongly influence system performance. Therefore, considering the effects of the components of $\boldsymbol{d}$ as a disturbance may generate large tracking errors. In this case, it is advisable to design control algorithms that take advantage of a detailed knowledge of manipulator dynamics so as to compensate for the nonlinear coupling terms of the model. In other words, it is necessary to eliminate the causes rather than to reduce the effects induced by them; that is, to generate compensating torques for the nonlinear terms in (8.22). This leads to *centralized control* algorithms that are based on the (partial or complete) knowledge of the manipulator dynamic model.

Whenever the robot is endowed with the torque sensors at the joint motors presented in Sect. 5.4.1, those measurements can be conveniently utilized to generate the compensation action, thus avoiding the on-line computation of the terms of the dynamic model.

As shown by the dynamic model (8.1), the manipulator is not a set of n decoupled system but it is a multivariable system with n inputs (joint torques) and n outputs (joint positions) interacting between them by means of nonlinear relations.[7]

In order to follow a methodological approach which is consistent with control design, it is necessary to treat the control problem in the context of nonlinear multivariable systems. This approach will obviously account for the manipulator *dynamic model* and lead to finding *nonlinear centralized control* laws, whose implementation is needed for high manipulator dynamic performance. On the other hand, the above computed torque control can be interpreted in this framework, since it provides a model-based nonlinear control term to enhance trajectory tracking performance. Notice, however, that this action is inherently performed off line, as it is computed on the time history of the desired trajectory and not of the actual one.

In the following, the problem of the determination of the control law $\boldsymbol{u}$ ensuring a given performance to the system of manipulator with drives is tackles. Since (8.17) can be considered as a proportional relationship between $\boldsymbol{v}_c$ and $\boldsymbol{u}$, the centralized control schemes below refer directly to the generation of control toques $\boldsymbol{u}$.

[7] See Appendix C for the basic concepts on control of nonlinear mechanical systems.

8.5.1 PD Control with Gravity Compensation

Let a *constant* equilibrium posture be assigned for the system as the vector of desired joint variables $\boldsymbol{q}_d$. It is desired to find the structure of the controller which ensures global asymptotic stability of the above posture.

The determination of the control input which stabilizes the system around the equilibrium posture is based on the Lyapunov direct method.

Take the vector $[\,\widetilde{\boldsymbol{q}}^T \quad \dot{\boldsymbol{q}}^T\,]^T$ as the system state, where

$$\widetilde{\boldsymbol{q}} = \boldsymbol{q}_d - \boldsymbol{q} \tag{8.46}$$

represents the error between the desired and the actual posture. Choose the following positive definite quadratic form as Lyapunov function candidate:

$$V(\dot{\boldsymbol{q}}, \widetilde{\boldsymbol{q}}) = \frac{1}{2}\dot{\boldsymbol{q}}^T \boldsymbol{B}(\boldsymbol{q})\dot{\boldsymbol{q}} + \frac{1}{2}\widetilde{\boldsymbol{q}}^T \boldsymbol{K}_P \widetilde{\boldsymbol{q}} > 0 \qquad \forall \dot{\boldsymbol{q}}, \widetilde{\boldsymbol{q}} \neq \boldsymbol{0} \tag{8.47}$$

where $\boldsymbol{K}_P$ is an $(n \times n)$ symmetric positive definite matrix. An energy-based interpretation of (8.47) reveals a first term expressing the system kinetic energy and a second term expressing the potential energy stored in the system of equivalent stiffness $\boldsymbol{K}_P$ provided by the n position feedback loops.

Differentiating (8.47) with respect to time, and recalling that $\boldsymbol{q}_d$ is constant, yields

$$\dot{V} = \dot{\boldsymbol{q}}^T \boldsymbol{B}(\boldsymbol{q})\ddot{\boldsymbol{q}} + \frac{1}{2}\dot{\boldsymbol{q}}^T \dot{\boldsymbol{B}}(\boldsymbol{q})\dot{\boldsymbol{q}} - \dot{\boldsymbol{q}}^T \boldsymbol{K}_P \widetilde{\boldsymbol{q}}. \tag{8.48}$$

Solving (8.7) for $\boldsymbol{B}\ddot{\boldsymbol{q}}$ and substituting it in (8.48) gives

$$\dot{V} = \frac{1}{2}\dot{\boldsymbol{q}}^T \big(\dot{\boldsymbol{B}}(\boldsymbol{q}) - 2\boldsymbol{C}(\boldsymbol{q}, \dot{\boldsymbol{q}})\big)\dot{\boldsymbol{q}} - \dot{\boldsymbol{q}}^T \boldsymbol{F}\dot{\boldsymbol{q}} + \dot{\boldsymbol{q}}^T \big(\boldsymbol{u} - \boldsymbol{g}(\boldsymbol{q}) - \boldsymbol{K}_P \widetilde{\boldsymbol{q}}\big). \tag{8.49}$$

The first term on the right-hand side is null since the matrix $\boldsymbol{N} = \dot{\boldsymbol{B}} - 2\boldsymbol{C}$ satisfies (7.49). The second term is negative definite. Then, the choice

$$\boldsymbol{u} = \boldsymbol{g}(\boldsymbol{q}) + \boldsymbol{K}_P \widetilde{\boldsymbol{q}}, \tag{8.50}$$

describing a controller with compensation of gravitational terms and a proportional action, leads to a negative semi-definite $\dot{V}$ since

$$\dot{V} = 0 \qquad \dot{\boldsymbol{q}} = \boldsymbol{0}, \forall \widetilde{\boldsymbol{q}}.$$

This result can be obtained also by taking the control law

$$\boldsymbol{u} = \boldsymbol{g}(\boldsymbol{q}) + \boldsymbol{K}_P \widetilde{\boldsymbol{q}} - \boldsymbol{K}_D \dot{\boldsymbol{q}}, \tag{8.51}$$

with $\boldsymbol{K}_D$ positive definite, corresponding to a *nonlinear compensation action of gravitational terms* with a *linear proportional-derivative* (PD) *action*. In fact, substituting (8.51) into (8.49) gives

$$\dot{V} = -\dot{\boldsymbol{q}}^T (\boldsymbol{F} + \boldsymbol{K}_D)\dot{\boldsymbol{q}}, \tag{8.52}$$

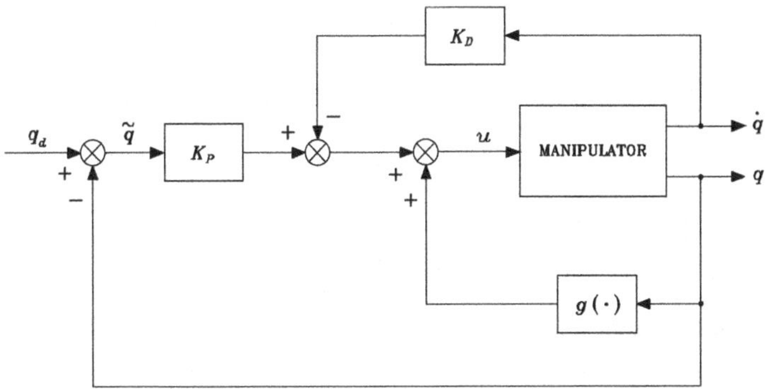

Fig. 8.20. Block scheme of joint space PD control with gravity compensation

which reveals that the introduction of the derivative term causes an increase of the absolute values of $\dot{V}$ along the system trajectories, and then it gives an improvement of system time response. Notice that the inclusion of a derivative action in the controller, as in (8.51), is crucial when direct-drive manipulators are considered. In that case, in fact, mechanical viscous damping is practically null, and current control does not allow the exploitation of the electrical viscous damping provided by voltage-controlled actuators.

According to the above, the function candidate V decreases as long as $\dot{\boldsymbol{q}} \neq \boldsymbol{0}$ for all system trajectories. It can be shown that the system reaches an *equilibrium posture*. To find such posture, notice that $\dot{V} \equiv 0$ only if $\dot{\boldsymbol{q}} \equiv \boldsymbol{0}$. System dynamics under control (8.51) is given by

$$\boldsymbol{B}(\boldsymbol{q})\ddot{\boldsymbol{q}} + \boldsymbol{C}(\boldsymbol{q}, \dot{\boldsymbol{q}})\dot{\boldsymbol{q}} + \boldsymbol{F}\dot{\boldsymbol{q}} + \boldsymbol{g}(\boldsymbol{q}) = \boldsymbol{g}(\boldsymbol{q}) + \boldsymbol{K}_P\widetilde{\boldsymbol{q}} - \boldsymbol{K}_D\dot{\boldsymbol{q}}. \tag{8.53}$$

At the equilibrium ($\dot{\boldsymbol{q}} \equiv \boldsymbol{0}$, $\ddot{\boldsymbol{q}} \equiv \boldsymbol{0}$) it is

$$\boldsymbol{K}_P\widetilde{\boldsymbol{q}} = \boldsymbol{0} \tag{8.54}$$

and then

$$\widetilde{\boldsymbol{q}} = \boldsymbol{q}_d - \boldsymbol{q} \equiv \boldsymbol{0}$$

is the sought equilibrium posture. The above derivation rigorously shows that any manipulator equilibrium posture is *globally asymptotically stable* under a controller with a PD linear action and a nonlinear gravity compensating action. Stability is ensured for any choice of $\boldsymbol{K}_P$ and $\boldsymbol{K}_D$, as long as these are positive definite matrices. The resulting block scheme is shown in Fig. 8.20.

The control law requires the on-line computation of the term $\boldsymbol{g}(\boldsymbol{q})$. If compensation is imperfect, the above discussion does not lead to the same result; this aspect will be revisited later with reference to robustness of controllers performing nonlinear compensation.

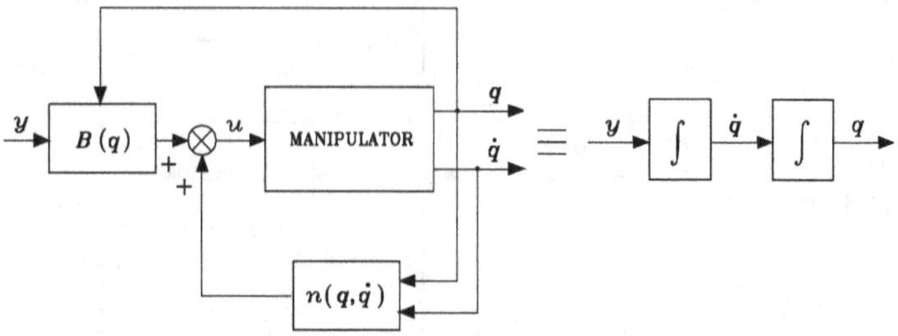

Fig. 8.21. Exact linearization performed by inverse dynamics control

8.5.2 Inverse Dynamics Control

Consider now the problem of tracking a joint space trajectory. The reference framework is that of control of nonlinear multivariable systems. The dynamic model of an n-joint manipulator is expressed by (8.7) which can be rewritten as

$$\boldsymbol{B}(\boldsymbol{q})\ddot{\boldsymbol{q}} + \boldsymbol{n}(\boldsymbol{q}, \dot{\boldsymbol{q}}) = \boldsymbol{u}, \tag{8.55}$$

where for simplicity it has been set

$$\boldsymbol{n}(\boldsymbol{q}, \dot{\boldsymbol{q}}) = \boldsymbol{C}(\boldsymbol{q}, \dot{\boldsymbol{q}})\dot{\boldsymbol{q}} + \boldsymbol{F}\dot{\boldsymbol{q}} + \boldsymbol{g}(\boldsymbol{q}). \tag{8.56}$$

The approach that follows is founded on the idea to find a control vector $\boldsymbol{u}$, as a function of the system state, which is capable of realizing an input/output relationship of linear type; in other words, it is desired to perform not an approximate linearization but an *exact linearization* of system dynamics obtained by means of a *nonlinear state feedback*. The possibility of finding such a linearizing controller is guaranteed by the particular form of system dynamics. In fact, the equation in (8.55) is linear in the control $\boldsymbol{u}$ and has a full-rank matrix $\boldsymbol{B}(\boldsymbol{q})$ which can be inverted for any manipulator configuration.

Taking the control $\boldsymbol{u}$ as a function of the manipulator state in the form

$$\boldsymbol{u} = \boldsymbol{B}(\boldsymbol{q})\boldsymbol{y} + \boldsymbol{n}(\boldsymbol{q}, \dot{\boldsymbol{q}}), \tag{8.57}$$

leads to the system described by

$$\ddot{\boldsymbol{q}} = \boldsymbol{y}$$

where $\boldsymbol{y}$ represents a new input vector whose expression is to be determined yet; the resulting block scheme is shown in Fig. 8.21. The nonlinear control law in (8.57) is termed *inverse dynamics control* since it is based on the computation of manipulator inverse dynamics. The system under control (8.57) is *linear* and *decoupled* with respect to the new input $\boldsymbol{y}$. In other words, the component y_i influences, with a double integrator relationship, only the joint variable q_i, independently of the motion of the other joints.

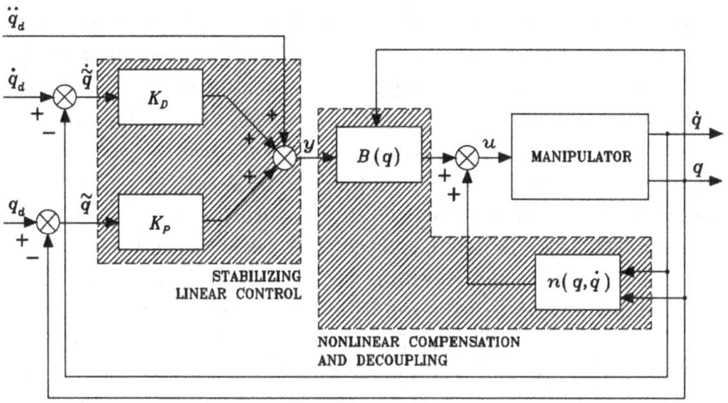

Fig. 8.22. Block scheme of joint space inverse dynamics control

In view of the choice (8.57), the manipulator control problem is reduced to that of finding a stabilizing control law $\boldsymbol{y}$. To this end, the choice

$$\boldsymbol{y} = -\boldsymbol{K}_P\boldsymbol{q} - \boldsymbol{K}_D\dot{\boldsymbol{q}} + \boldsymbol{r} \tag{8.58}$$

leads to the system of second-order equations

$$\ddot{\boldsymbol{q}} + \boldsymbol{K}_D\dot{\boldsymbol{q}} + \boldsymbol{K}_P\boldsymbol{q} = \boldsymbol{r} \tag{8.59}$$

which, under the assumption of positive definite matrices $\boldsymbol{K}_P$ and $\boldsymbol{K}_D$, is asymptotically stable. Choosing $\boldsymbol{K}_P$ and $\boldsymbol{K}_D$ as *diagonal* matrices of the type

$$\boldsymbol{K}_P = \text{diag}\{\omega_{n1}^2, \ldots, \omega_{nn}^2\} \qquad \boldsymbol{K}_D = \text{diag}\{2\zeta_1\omega_{n1}, \ldots, 2\zeta_n\omega_{nn}\},$$

gives a decoupled system. The reference component r_i influences only the joint variable q_i with a second-order input/output relationship characterized by a natural frequency ω_{ni} and a damping ratio ζ_i.

Given any desired trajectory $\boldsymbol{q}_d(t)$, tracking of this trajectory for the output $\boldsymbol{q}(t)$ is ensured by choosing

$$\boldsymbol{r} = \ddot{\boldsymbol{q}}_d + \boldsymbol{K}_D\dot{\boldsymbol{q}}_d + \boldsymbol{K}_P\boldsymbol{q}_d. \tag{8.60}$$

In fact, substituting (8.60) into (8.59) gives the homogeneous second-order differential equation

$$\ddot{\tilde{\boldsymbol{q}}} + \boldsymbol{K}_D\dot{\tilde{\boldsymbol{q}}} + \boldsymbol{K}_P\tilde{\boldsymbol{q}} = \boldsymbol{0} \tag{8.61}$$

expressing the dynamics of position error (8.46) while tracking the given trajectory. Such error occurs only if $\tilde{\boldsymbol{q}}(0)$ and/or $\dot{\tilde{\boldsymbol{q}}}(0)$ are different from zero and converges to zero with a speed depending on the matrices $\boldsymbol{K}_P$ and $\boldsymbol{K}_D$ chosen.

The resulting block scheme is illustrated in Fig. 8.22, in which two feedback loops are represented; an inner loop based on the manipulator dynamic model, and an outer loop operating on the tracking error. The function of the *inner loop* is to obtain a *linear and decoupled input/output relationship*, whereas the *outer loop* is required to *stabilize the overall system*. The controller design for the outer loop is simplified since it operates on a linear and time-invariant system. Notice that the implementation of this control scheme requires computation of the inertia matrix $\boldsymbol{B}(\boldsymbol{q})$ and of the vector of Coriolis, centrifugal, gravitational, and damping terms $\boldsymbol{n}(\boldsymbol{q}, \dot{\boldsymbol{q}})$ in (8.56). Unlike computed torque control, these terms must be computed *on-line* since control is now based on nonlinear feedback of the current system state, and thus it is not possible to precompute the terms off line as for the previous technique.

The above technique of nonlinear compensation and decoupling is very attractive from a control viewpoint since the nonlinear and coupled manipulator dynamics is replaced with n linear and decoupled second-order subsystems. Nonetheless, this technique is based on the assumption of perfect cancellation of dynamic terms, and then it is quite natural to raise questions about sensitivity and robustness problems due to unavoidably imperfect compensation.

Implementation of inverse dynamics control laws indeed requires that parameters of the system dynamic model are accurately known and the complete equations of motion are computed in real time. These conditions are difficult to verify in practice. On one hand, the model is usually known with a certain degree of uncertainty due to imperfect knowledge of manipulator mechanical parameters, existence of unmodelled dynamics, and model dependence on end-effector payloads not exactly known and thus not perfectly compensated. On the other hand, inverse dynamics computation is to be performed at sampling times of the order of a millisecond so as to ensure that the assumption of operating in the continuous time domain is realistic. This may pose severe constraints on the hardware/software architecture of the control system. In such cases, it may be advisable to lighten the computation of inverse dynamics and compute only the dominant terms.

On the basis of the above remarks, from an implementation viewpoint, *compensation* may be *imperfect* both for model uncertainty and for the approximations made in on-line computation of inverse dynamics. In the following, two control techniques are presented which are aimed at counteracting the effects of imperfect compensation. The first consists of the introduction of an additional term to an inverse dynamics controller which provides *robustness* to the control system by counteracting the effects of the approximations made in on-line computation of inverse dynamics. The second *adapts* the parameters of the model used for inverse dynamics computation to those of the true manipulator dynamic model.

8.5.3 Robust Control

In the case of *imperfect compensation*, it is reasonable to assume in (8.55) a control vector expressed by

$$\boldsymbol{u} = \widehat{\boldsymbol{B}}(\boldsymbol{q})\boldsymbol{y} + \widehat{\boldsymbol{n}}(\boldsymbol{q}, \dot{\boldsymbol{q}}) \tag{8.62}$$

where $\widehat{\boldsymbol{B}}$ and $\widehat{\boldsymbol{n}}$ represent the adopted computational model in terms of estimates of the terms in the dynamic model. The error on the estimates, i.e., the *uncertainty*, is represented by

$$\widetilde{\boldsymbol{B}} = \widehat{\boldsymbol{B}} - \boldsymbol{B} \qquad\qquad \widetilde{\boldsymbol{n}} = \widehat{\boldsymbol{n}} - \boldsymbol{n} \tag{8.63}$$

and is due to imperfect model compensation as well as to intentional simplification in inverse dynamics computation. Notice that by setting $\widehat{\boldsymbol{B}} = \bar{\boldsymbol{B}}$ (where $\bar{\boldsymbol{B}}$ is the diagonal matrix of average inertia at the joint axes) and $\widehat{\boldsymbol{n}} = \boldsymbol{0}$, the above decentralized control scheme is recovered where the control action $\boldsymbol{y}$ can be of the general PID type computed on the error.

Using (8.62) as a nonlinear control law gives

$$\boldsymbol{B}\ddot{\boldsymbol{q}} + \boldsymbol{n} = \widehat{\boldsymbol{B}}\boldsymbol{y} + \widehat{\boldsymbol{n}} \tag{8.64}$$

where functional dependence has been omitted. Since the inertia matrix $\boldsymbol{B}$ is invertible, it is

$$\ddot{\boldsymbol{q}} = \boldsymbol{y} + (\boldsymbol{B}^{-1}\widehat{\boldsymbol{B}} - \boldsymbol{I})\boldsymbol{y} + \boldsymbol{B}^{-1}\widetilde{\boldsymbol{n}} = \boldsymbol{y} - \boldsymbol{\eta} \tag{8.65}$$

where

$$\boldsymbol{\eta} = (\boldsymbol{I} - \boldsymbol{B}^{-1}\widehat{\boldsymbol{B}})\boldsymbol{y} - \boldsymbol{B}^{-1}\widetilde{\boldsymbol{n}}. \tag{8.66}$$

Taking as above

$$\boldsymbol{y} = \ddot{\boldsymbol{q}}_d + \boldsymbol{K}_D(\dot{\boldsymbol{q}}_d - \dot{\boldsymbol{q}}) + \boldsymbol{K}_P(\boldsymbol{q}_d - \boldsymbol{q}),$$

leads to

$$\ddot{\widetilde{\boldsymbol{q}}} + \boldsymbol{K}_D\dot{\widetilde{\boldsymbol{q}}} + \boldsymbol{K}_P\widetilde{\boldsymbol{q}} = \boldsymbol{\eta}. \tag{8.67}$$

The system described by (8.67) is still nonlinear and coupled, since $\boldsymbol{\eta}$ is a nonlinear function of $\widetilde{\boldsymbol{q}}$ and $\dot{\widetilde{\boldsymbol{q}}}$; error convergence to zero is not ensured by the term on the left-hand side only.

To find control laws ensuring error convergence to zero while tracking a trajectory even in the face of uncertainties, a linear PD control is no longer sufficient. To this end, the Lyapunov direct method can be utilized again for the design of an outer feedback loop on the error which should be *robust* to the uncertainty $\boldsymbol{\eta}$.

Let the desired trajectory $\boldsymbol{q}_d(t)$ be assigned in the joint space and let $\widetilde{\boldsymbol{q}} = \boldsymbol{q}_d - \boldsymbol{q}$ be the position error. Its first time-derivative is $\dot{\widetilde{\boldsymbol{q}}} = \dot{\boldsymbol{q}}_d - \dot{\boldsymbol{q}}$, while its second time-derivative in view of (8.65) is

$$\ddot{\widetilde{\boldsymbol{q}}} = \ddot{\boldsymbol{q}}_d - \boldsymbol{y} + \boldsymbol{\eta}. \tag{8.68}$$

By taking

$$\boldsymbol{\xi} = \begin{bmatrix} \widetilde{\boldsymbol{q}} \\ \dot{\widetilde{\boldsymbol{q}}} \end{bmatrix}, \tag{8.69}$$

as the system state, the following first-order differential matrix equation is obtained:

$$\dot{\boldsymbol{\xi}} = \boldsymbol{H}\boldsymbol{\xi} + \boldsymbol{D}(\ddot{\boldsymbol{q}}_d - \boldsymbol{y} + \boldsymbol{\eta}), \tag{8.70}$$

where $\boldsymbol{H}$ and $\boldsymbol{D}$ are block matrices of dimensions $(2n \times 2n)$ and $(2n \times n)$, respectively:

$$\boldsymbol{H} = \begin{bmatrix} \boldsymbol{O} & \boldsymbol{I} \\ \boldsymbol{O} & \boldsymbol{O} \end{bmatrix} \qquad \boldsymbol{D} = \begin{bmatrix} \boldsymbol{O} \\ \boldsymbol{I} \end{bmatrix}. \tag{8.71}$$

Then, the problem of tracking a given trajectory can be regarded as the problem of finding a control law $\boldsymbol{y}$ which stabilizes the nonlinear time-varying error system (8.70).

Control design is based on the assumption that, even though the uncertainty $\boldsymbol{\eta}$ is unknown, an estimate on its range of variation is available. The sought control law $\boldsymbol{y}$ should guarantee asymptotic stability of (8.70) for any $\boldsymbol{\eta}$ varying in the above range. By recalling that $\boldsymbol{\eta}$ in (8.66) is a function of $\boldsymbol{q}$, $\dot{\boldsymbol{q}}$, $\ddot{\boldsymbol{q}}_d$, the following assumptions are made:

$$\sup_{t\ge 0} \|\ddot{\boldsymbol{q}}_d\| < Q_M < \infty \quad \forall \ddot{\boldsymbol{q}}_d \tag{8.72}$$

$$\|\boldsymbol{I} - \boldsymbol{B}^{-1}(\boldsymbol{q})\widehat{\boldsymbol{B}}(\boldsymbol{q})\| \le \alpha \le 1 \; \forall \boldsymbol{q} \tag{8.73}$$

$$\|\widetilde{\boldsymbol{n}}\| \le \Phi < \infty \qquad \forall \boldsymbol{q}, \dot{\boldsymbol{q}}. \tag{8.74}$$

Assumption (8.72) is practically satisfied since any planned trajectory cannot require infinite accelerations.

Regarding assumption (8.73), since $\boldsymbol{B}$ is a positive definite matrix with upper and lower limited norms, the following inequality holds:

$$0 < B_m \le \|\boldsymbol{B}^{-1}(\boldsymbol{q})\| \le B_M < \infty \qquad \forall \boldsymbol{q}, \tag{8.75}$$

and then a choice for $\widehat{\boldsymbol{B}}$ always exists which satisfies (8.73). In fact, by setting

$$\widehat{\boldsymbol{B}} = \frac{2}{B_M + B_m}\boldsymbol{I},$$

from (8.73) it is

$$\|\boldsymbol{B}^{-1}\widehat{\boldsymbol{B}} - \boldsymbol{I}\| \le \frac{B_M - B_m}{B_M + B_m} = \alpha < 1. \tag{8.76}$$

If $\widehat{\boldsymbol{B}}$ is a more accurate estimate of the inertia matrix, the inequality is satisfied with values of α that can be made arbitrarily small (in the limit, it is $\widehat{\boldsymbol{B}} = \boldsymbol{B}$ and $\alpha = 0$).

Finally, concerning assumption (8.74), observe that $\widetilde{\boldsymbol{n}}$ is a function of $\boldsymbol{q}$ and $\dot{\boldsymbol{q}}$. For revolute joints a periodical dependence on $\boldsymbol{q}$ is obtained, while for prismatic joints a linear dependence is obtained, but the joint ranges are limited and then the above contribution is also limited. On the other hand, regarding the dependence on $\dot{\boldsymbol{q}}$, unbounded velocities for an unstable system may arise in the limit, but in reality saturations exist on the maximum velocities of the motors. In summary, assumption (8.74) can be realistically satisfied, too.

With reference to (8.65), choose now

$$\boldsymbol{y} = \ddot{\boldsymbol{q}}_d + \boldsymbol{K}_D\dot{\widetilde{\boldsymbol{q}}} + \boldsymbol{K}_P\widetilde{\boldsymbol{q}} + \boldsymbol{w} \tag{8.77}$$

where the PD term ensures stabilization of the error dynamic system matrix, $\ddot{\boldsymbol{q}}_d$ provides a feedforward term, and the term $\boldsymbol{w}$ is to be chosen to guarantee robustness to the effects of uncertainty described by $\boldsymbol{\eta}$ in (8.66).

Using (8.77) and setting $\boldsymbol{K} = [\,\boldsymbol{K}_P \quad \boldsymbol{K}_D\,]$ yields

$$\dot{\boldsymbol{\xi}} = \widetilde{\boldsymbol{H}}\boldsymbol{\xi} + \boldsymbol{D}(\boldsymbol{\eta} - \boldsymbol{w}), \tag{8.78}$$

where

$$\widetilde{\boldsymbol{H}} = (\boldsymbol{H} - \boldsymbol{D}\boldsymbol{K}) = \begin{bmatrix} \boldsymbol{O} & \boldsymbol{I} \\ -\boldsymbol{K}_P & -\boldsymbol{K}_D \end{bmatrix}$$

is a matrix whose eigenvalues all have negative real parts — $\boldsymbol{K}_P$ and $\boldsymbol{K}_D$ being positive definite — which allows the desired error system dynamics to be prescribed. In fact, by choosing $\boldsymbol{K}_P = \mathrm{diag}\{\omega_{n1}^2, \ldots, \omega_{nn}^2\}$ and $\boldsymbol{K}_D = \mathrm{diag}\{2\zeta_1\omega_{n1}, \ldots, 2\zeta_n\omega_{nn}\}$, n decoupled equations are obtained as regards the linear part. If the uncertainty term vanishes, it is obviously $\boldsymbol{w} = \boldsymbol{0}$ and the above result with an exact inverse dynamics controller is recovered ($\widehat{\boldsymbol{B}} = \boldsymbol{B}$ and $\widehat{\boldsymbol{n}} = \boldsymbol{n}$).

To determine $\boldsymbol{w}$, consider the following positive definite quadratic form as Lyapunov function candidate:

$$V(\boldsymbol{\xi}) = \boldsymbol{\xi}^T\boldsymbol{Q}\boldsymbol{\xi} > 0 \qquad \forall\boldsymbol{\xi} \neq \boldsymbol{0}, \tag{8.79}$$

where $\boldsymbol{Q}$ is a $(2n \times 2n)$ positive definite matrix. The derivative of V along the trajectories of the error system (8.78) is

$$\begin{aligned} \dot{V} &= \dot{\boldsymbol{\xi}}^T\boldsymbol{Q}\boldsymbol{\xi} + \boldsymbol{\xi}^T\boldsymbol{Q}\dot{\boldsymbol{\xi}} \\ &= \boldsymbol{\xi}^T(\widetilde{\boldsymbol{H}}^T\boldsymbol{Q} + \boldsymbol{Q}\widetilde{\boldsymbol{H}})\boldsymbol{\xi} + 2\boldsymbol{\xi}^T\boldsymbol{Q}\boldsymbol{D}(\boldsymbol{\eta} - \boldsymbol{w}). \end{aligned} \tag{8.80}$$

Since $\widetilde{\boldsymbol{H}}$ has eigenvalues with all negative real parts, it is well-known that for any symmetric positive definite matrix $\boldsymbol{P}$, the equation

$$\widetilde{\boldsymbol{H}}^T\boldsymbol{Q} + \boldsymbol{Q}\widetilde{\boldsymbol{H}} = -\boldsymbol{P} \tag{8.81}$$

gives a unique solution $\boldsymbol{Q}$ which is symmetric positive definite as well. In view of this, (8.80) becomes

$$\dot{V} = -\boldsymbol{\xi}^T\boldsymbol{P}\boldsymbol{\xi} + 2\boldsymbol{\xi}^T\boldsymbol{Q}\boldsymbol{D}(\boldsymbol{\eta} - \boldsymbol{w}). \tag{8.82}$$

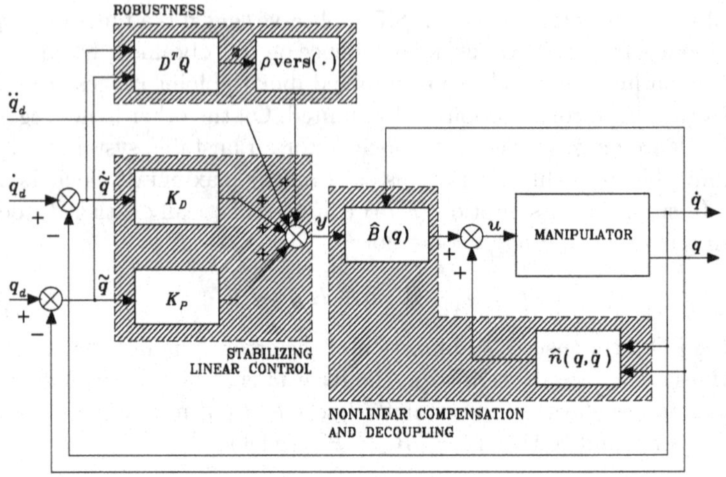

Fig. 8.23. Block scheme of joint space robust control

The first term on the right-hand side of (8.82) is negative definite and then the solutions converge if $\boldsymbol{\xi} \in \mathcal{N}(\boldsymbol{D}^T\boldsymbol{Q})$. If instead $\boldsymbol{\xi} \notin \mathcal{N}(\boldsymbol{D}^T\boldsymbol{Q})$, the control $\boldsymbol{w}$ must be chosen so as to render the second term in (8.82) less than or equal to zero. By setting $\boldsymbol{z} = \boldsymbol{D}^T\boldsymbol{Q}\boldsymbol{\xi}$, the second term in (8.82) can be rewritten as $\boldsymbol{z}^T(\boldsymbol{\eta} - \boldsymbol{w})$. Adopting the control law

$$\boldsymbol{w} = \frac{\rho}{\|\boldsymbol{z}\|}\boldsymbol{z} \qquad \rho > 0 \tag{8.83}$$

gives[8]

$$\begin{aligned} \boldsymbol{z}^T(\boldsymbol{\eta} - \boldsymbol{w}) &= \boldsymbol{z}^T\boldsymbol{\eta} - \frac{\rho}{\|\boldsymbol{z}\|}\boldsymbol{z}^T\boldsymbol{z} \\ &\le \|\boldsymbol{z}\|\|\boldsymbol{\eta}\| - \rho\|\boldsymbol{z}\| \\ &= \|\boldsymbol{z}\|(\|\boldsymbol{\eta}\| - \rho). \end{aligned} \tag{8.84}$$

Then, if ρ is chosen so that

$$\rho \ge \|\boldsymbol{\eta}\| \qquad \forall \boldsymbol{q}, \dot{\boldsymbol{q}}, \ddot{\boldsymbol{q}}_d, \tag{8.85}$$

the control (8.83) ensures that $\dot{V}$ is less than zero along all error system trajectories.

In order to satisfy (8.85), notice that, in view of the definition of $\boldsymbol{\eta}$ in (8.66) and of assumptions (8.72)–(8.74), and being $\|\boldsymbol{w}\| = \rho$, it is

$$\|\boldsymbol{\eta}\| \le \|\boldsymbol{I} - \boldsymbol{B}^{-1}\widehat{\boldsymbol{B}}\|\big(\|\ddot{\boldsymbol{q}}_d\| + \|\boldsymbol{K}\|\,\|\boldsymbol{\xi}\| + \|\boldsymbol{w}\|\big) + \|\boldsymbol{B}^{-1}\|\,\|\widetilde{\boldsymbol{n}}\|$$

[8] Notice that it is necessary to divide $\boldsymbol{z}$ by the norm of $\boldsymbol{z}$ so as to obtain a linear dependence on $\boldsymbol{z}$ of the term containing the control $\boldsymbol{z}^T\boldsymbol{w}$, and thus to effectively counteract, for $\boldsymbol{z} \to \boldsymbol{0}$, the term containing the uncertainty $\boldsymbol{z}^T\boldsymbol{\eta}$ which is linear in $\boldsymbol{z}$.

$$\leq \alpha Q_M + \alpha \|\boldsymbol{K}\| \, \|\boldsymbol{\xi}\| + \alpha \rho + B_M \Phi. \tag{8.86}$$

Therefore, setting

$$\rho \geq \frac{1}{1-\alpha} (\alpha Q_M + \alpha \|\boldsymbol{K}\| \|\boldsymbol{\xi}\| + B_M \Phi) \tag{8.87}$$

gives

$$\dot{V} = -\boldsymbol{\xi}^T \boldsymbol{P} \boldsymbol{\xi} + 2\boldsymbol{z}^T \left(\boldsymbol{\eta} - \frac{\rho}{\|\boldsymbol{z}\|} \boldsymbol{z} \right) < 0 \qquad \forall \boldsymbol{\xi} \neq \boldsymbol{0}. \tag{8.88}$$

The resulting block scheme is illustrated in Fig. 8.23.

To summarize, the presented approach has lead to finding a *control* law which is formed by three different contributions:

- The term $\widehat{\boldsymbol{B}}\boldsymbol{y}+\widehat{\boldsymbol{n}}$ ensures an *approximate compensation of nonlinear effects and joint decoupling.*
- The term $\ddot{\boldsymbol{q}}_d + \boldsymbol{K}_D \dot{\widetilde{\boldsymbol{q}}} + \boldsymbol{K}_P \widetilde{\boldsymbol{q}}$ introduces a *linear feedforward action* ($\ddot{\boldsymbol{q}}_d + \boldsymbol{K}_D \dot{\boldsymbol{q}}_d + \boldsymbol{K}_P \boldsymbol{q}_d$) and *linear feedback action* ($-\boldsymbol{K}_D \dot{\boldsymbol{q}} - \boldsymbol{K}_P \boldsymbol{q}$) *which stabilizes the error system dynamics.*
- The term $\boldsymbol{w} = (\rho / \|\boldsymbol{z}\|)\boldsymbol{z}$ represents the *robust contribution that counteracts the indeterminacy* $\widetilde{\boldsymbol{B}}$ and $\widetilde{\boldsymbol{n}}$ in computing the nonlinear terms that depend on the manipulator state; the greater the uncertainty, the greater the positive scalar ρ. The resulting control law is of the *unit vector* type, since it is described by a vector of magnitude ρ aligned with the unit vector of $\boldsymbol{z} = \boldsymbol{D}^T \boldsymbol{Q} \boldsymbol{\xi}$, $\forall \boldsymbol{\xi}$.

All the resulting trajectories under the above robust control reach the subspace $\boldsymbol{z} = \boldsymbol{D}^T \boldsymbol{Q} \boldsymbol{\xi} = \boldsymbol{0}$ that depends on the matrix $\boldsymbol{Q}$ in the Lyapunov function V. On this *attractive* subspace, termed *sliding subspace*, the control $\boldsymbol{w}$ is ideally commuted at an infinite frequency and all error components tend to zero with a transient depending on the matrices $\boldsymbol{Q}$, $\boldsymbol{K}_P$, $\boldsymbol{K}_D$. A characterization of an error trajectory in the two-dimensional case is given in Fig. 8.24. Notice that in the case $\boldsymbol{\xi}(0) \neq \boldsymbol{0}$, with $\boldsymbol{\xi}(0) \notin \mathcal{N}(\boldsymbol{D}^T \boldsymbol{Q})$, the trajectory is attracted on the sliding hyperplane (a line) $\boldsymbol{z} = \boldsymbol{0}$ and tends towards the origin of the error state space with a time evolution governed by ρ.

In reality, the physical limits on the elements employed in the controller impose a control signal that commutes at a finite frequency, and the trajectories oscillate around the sliding subspace with a magnitude as low as the frequency is high.

Elimination of these high-frequency components (*chattering*) can be achieved by adopting a robust control law which, even if it does not guarantee error convergence to zero, ensures *bounded-norm* errors. A control law of this type is

$$\boldsymbol{w} = \begin{cases} \dfrac{\rho}{\|\boldsymbol{z}\|} \boldsymbol{z} & \text{per } \|\boldsymbol{z}\| \geq \epsilon \\[2ex] \dfrac{\rho}{\epsilon} \boldsymbol{z} & \text{per } \|\boldsymbol{z}\| < \epsilon. \end{cases} \tag{8.89}$$

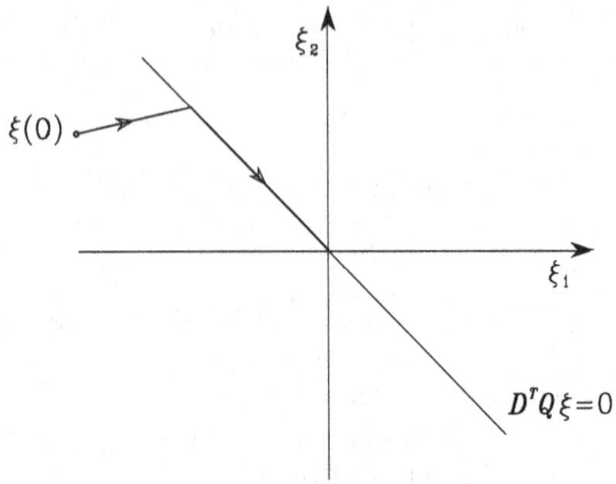

Fig. 8.24. Error trajectory with robust control

In order to provide an intuitive interpretation of this law, notice that (8.89) gives a null control input when the error is in the null space of matrix $\boldsymbol{D}^T\boldsymbol{Q}$. On the other hand, (8.83) has an equivalent gain tending to infinity when $\boldsymbol{z}$ tends to the null vector, thus generating a control input of limited magnitude. Since these inputs commute at an infinite frequency, they force the error system dynamics to stay on the sliding subspace. With reference to the above example, control law (8.89) gives rise to a hyperplane $\boldsymbol{z} = \boldsymbol{0}$ which is no longer attractive, and the error is allowed to vary within a boundary layer whose thickness depends on ϵ (Fig. 8.25).

The introduction of a contribution based on the computation of a suitable linear combination of the generalized error confers robustness to a control scheme based on nonlinear compensation. Even if the manipulator is accurately modeled, indeed, an exact nonlinear compensation may be computationally demanding, and thus it may require either a sophisticated hardware architecture or an increase of the sampling time needed to compute the control law. The solution then becomes weak from an engineering viewpoint, due either to infeasible costs of the control architecture, or to poor performance at decreased sampling rates. Therefore, considering a partial knowledge of the manipulator dynamic model with an accurate, pondered estimate of uncertainty may suggest robust control solutions of the kind presented above. It is understood that an estimate of the uncertainty should be found so as to impose control inputs which the mechanical structure can bear.

8.5.4 Adaptive Control

The computational model employed for computing inverse dynamics typically has the same structure as that of the true manipulator dynamic model, but parameter estimate uncertainty does exist. In this case, it is possible to devise solutions that allow an *on-line adaptation of the computational model to the*

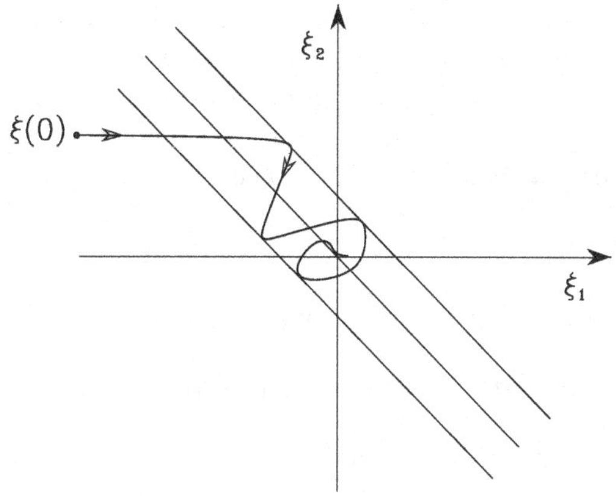

Fig. 8.25. Error trajectory with robust control and chattering elimination

dynamic model, thus performing a control scheme of the inverse dynamics type.

The possibility of finding adaptive control laws is ensured by the property of *linearity in the parameters* of the dynamic model of a manipulator. In fact, it is always possible to express the nonlinear equations of motion in a linear form with respect to a suitable set of constant dynamic parameters as in (7.81). The equation in (8.7) can then be written as

$$\boldsymbol{B}(\boldsymbol{q})\ddot{\boldsymbol{q}} + \boldsymbol{C}(\boldsymbol{q}, \dot{\boldsymbol{q}})\dot{\boldsymbol{q}} + \boldsymbol{F}\dot{\boldsymbol{q}} + \boldsymbol{g}(\boldsymbol{q}) = \boldsymbol{Y}(\boldsymbol{q}, \dot{\boldsymbol{q}}, \ddot{\boldsymbol{q}})\boldsymbol{\pi} = \boldsymbol{u}, \tag{8.90}$$

where $\boldsymbol{\pi}$ is a $(p \times 1)$ vector of constant parameters and $\boldsymbol{Y}$ is an $(n \times p)$ matrix which is a function of joint positions, velocities and accelerations. This property of linearity in the dynamic parameters is fundamental for deriving adaptive control laws, among which the technique illustrated below is one of the simplest.

At first, a control scheme which can be derived through a combined computed torque/inverse dynamics approach is illustrated. The computational model is assumed to coincide with the dynamic model.

Consider the control law

$$\boldsymbol{u} = \boldsymbol{B}(\boldsymbol{q})\ddot{\boldsymbol{q}}_r + \boldsymbol{C}(\boldsymbol{q}, \dot{\boldsymbol{q}})\dot{\boldsymbol{q}}_r + \boldsymbol{F}\dot{\boldsymbol{q}}_r + \boldsymbol{g}(\boldsymbol{q}) + \boldsymbol{K}_D\boldsymbol{\sigma}, \tag{8.91}$$

with $\boldsymbol{K}_D$ a positive definite matrix. The choice

$$\dot{\boldsymbol{q}}_r = \dot{\boldsymbol{q}}_d + \boldsymbol{\Lambda}\widetilde{\boldsymbol{q}} \qquad \ddot{\boldsymbol{q}}_r = \ddot{\boldsymbol{q}}_d + \boldsymbol{\Lambda}\dot{\widetilde{\boldsymbol{q}}}, \tag{8.92}$$

with $\boldsymbol{\Lambda}$ a positive definite (usually diagonal) matrix, allows the nonlinear compensation and decoupling terms to be expressed as a function of the desired velocity and acceleration, corrected by the current state ($\boldsymbol{q}$ and $\dot{\boldsymbol{q}}$) of the manipulator. In fact, notice that the term $\dot{\boldsymbol{q}}_r = \dot{\boldsymbol{q}}_d + \boldsymbol{\Lambda}\widetilde{\boldsymbol{q}}$ weighs the contribution that depends on velocity, not only on the basis of the desired velocity but also on the basis of the position tracking error. A similar argument also holds for the acceleration contribution, where a term depending on the velocity tracking error is considered besides the desired acceleration.

The term $\boldsymbol{K}_D\boldsymbol{\sigma}$ is equivalent to a PD action on the error if $\boldsymbol{\sigma}$ is taken as

$$\boldsymbol{\sigma} = \dot{\boldsymbol{q}}_r - \dot{\boldsymbol{q}} = \dot{\widetilde{\boldsymbol{q}}} + \boldsymbol{\Lambda}\widetilde{\boldsymbol{q}}. \tag{8.93}$$

Substituting (8.91) into (8.90) and accounting for (8.93) yields

$$\boldsymbol{B}(\boldsymbol{q})\dot{\boldsymbol{\sigma}} + \boldsymbol{C}(\boldsymbol{q}, \dot{\boldsymbol{q}})\boldsymbol{\sigma} + \boldsymbol{F}\boldsymbol{\sigma} + \boldsymbol{K}_D\boldsymbol{\sigma} = \boldsymbol{0}. \tag{8.94}$$

Consider the Lyapunov function candidate

$$V(\boldsymbol{\sigma}, \widetilde{\boldsymbol{q}}) = \frac{1}{2}\boldsymbol{\sigma}^T\boldsymbol{B}(\boldsymbol{q})\boldsymbol{\sigma} + \frac{1}{2}\widetilde{\boldsymbol{q}}^T\boldsymbol{M}\widetilde{\boldsymbol{q}} > 0 \qquad \forall \boldsymbol{\sigma}, \widetilde{\boldsymbol{q}} \neq \boldsymbol{0}, \tag{8.95}$$

where $\boldsymbol{M}$ is an $(n \times n)$ symmetric positive definite matrix; the introduction of the second term in (8.95) is necessary to obtain a Lyapunov function of the entire system state which vanishes for $\widetilde{\boldsymbol{q}} = \boldsymbol{0}$ and $\dot{\widetilde{\boldsymbol{q}}} = \boldsymbol{0}$. The time derivative of V along the trajectories of system (8.94) is

$$\begin{aligned} \dot{V} &= \boldsymbol{\sigma}^T\boldsymbol{B}(\boldsymbol{q})\dot{\boldsymbol{\sigma}} + \frac{1}{2}\boldsymbol{\sigma}^T\dot{\boldsymbol{B}}(\boldsymbol{q})\boldsymbol{\sigma} + \widetilde{\boldsymbol{q}}^T\boldsymbol{M}\dot{\widetilde{\boldsymbol{q}}} \\ &= -\boldsymbol{\sigma}^T(\boldsymbol{F} + \boldsymbol{K}_D)\boldsymbol{\sigma} + \widetilde{\boldsymbol{q}}^T\boldsymbol{M}\dot{\widetilde{\boldsymbol{q}}}, \end{aligned} \tag{8.96}$$

where the skew-symmetry propertyv of the matrix $\boldsymbol{N} = \dot{\boldsymbol{B}} - 2\boldsymbol{C}$ has been exploited. In view of the expression of $\boldsymbol{\sigma}$ in (8.93), with diagonal $\boldsymbol{\Lambda}$ and $\boldsymbol{K}_D$, it is convenient to choose $\boldsymbol{M} = 2\boldsymbol{\Lambda}\boldsymbol{K}_D$; this leads to

$$\dot{V} = -\boldsymbol{\sigma}^T\boldsymbol{F}\boldsymbol{\sigma} - \dot{\widetilde{\boldsymbol{q}}}^T\boldsymbol{K}_D\dot{\widetilde{\boldsymbol{q}}} - \widetilde{\boldsymbol{q}}^T\boldsymbol{\Lambda}\boldsymbol{K}_D\boldsymbol{\Lambda}\widetilde{\boldsymbol{q}}. \tag{8.97}$$

This expression shows that the time derivative is negative definite since it vanishes only if $\widetilde{\boldsymbol{q}} \equiv \boldsymbol{0}$ and $\dot{\widetilde{\boldsymbol{q}}} \equiv \boldsymbol{0}$; thus, it follows that the origin of the state space $[\,\widetilde{\boldsymbol{q}}^T \quad \boldsymbol{\sigma}^T\,]^T = \boldsymbol{0}$ is *globally asymptotically stable*. It is worth noticing that, unlike the robust control case, the error trajectory tends to the subspace $\boldsymbol{\sigma} = \boldsymbol{0}$ without the need of a high-frequency control.

On the basis of this notable result, the *control* law can be made *adaptive* with respect to the vector of parameters $\boldsymbol{\pi}$.

Suppose that the computational model has the same structure as that of the manipulator dynamic model, but its parameters are not known exactly. The control law (8.91) is then modified into

$$\begin{aligned} \boldsymbol{u} &= \widehat{\boldsymbol{B}}(\boldsymbol{q})\ddot{\boldsymbol{q}}_r + \widehat{\boldsymbol{C}}(\boldsymbol{q}, \dot{\boldsymbol{q}})\dot{\boldsymbol{q}}_r + \widehat{\boldsymbol{F}}\dot{\boldsymbol{q}}_r + \widehat{\boldsymbol{g}} + \boldsymbol{K}_D\boldsymbol{\sigma} \\ &= \boldsymbol{Y}(\boldsymbol{q}, \dot{\boldsymbol{q}}, \dot{\boldsymbol{q}}_r, \ddot{\boldsymbol{q}}_r)\widehat{\boldsymbol{\pi}} + \boldsymbol{K}_D\boldsymbol{\sigma}, \end{aligned} \tag{8.98}$$

where $\widehat{\boldsymbol{\pi}}$ represents the available estimate on the parameters and, accordingly, $\widehat{\boldsymbol{B}}$, $\widehat{\boldsymbol{C}}$, $\widehat{\boldsymbol{F}}$, $\widehat{\boldsymbol{g}}$ denote the estimated terms in the dynamic model. Substituting control (8.98) into (8.90) gives

$$\begin{aligned}\boldsymbol{B}(\boldsymbol{q})\dot{\boldsymbol{\sigma}}+\boldsymbol{C}(\boldsymbol{q},\dot{\boldsymbol{q}})\boldsymbol{\sigma}+\boldsymbol{F}\boldsymbol{\sigma}+\boldsymbol{K}_D\boldsymbol{\sigma} &= -\widetilde{\boldsymbol{B}}(\boldsymbol{q})\ddot{\boldsymbol{q}}_r-\widetilde{\boldsymbol{C}}(\boldsymbol{q},\dot{\boldsymbol{q}})\dot{\boldsymbol{q}}_r-\widetilde{\boldsymbol{F}}\dot{\boldsymbol{q}}_r-\widetilde{\boldsymbol{g}}(\boldsymbol{q})\\ &= -\boldsymbol{Y}(\boldsymbol{q},\dot{\boldsymbol{q}},\dot{\boldsymbol{q}}_r,\ddot{\boldsymbol{q}}_r)\widetilde{\boldsymbol{\pi}}, \qquad (8.99)\end{aligned}$$

where the property of linearity in the error parameter vector

$$\widetilde{\boldsymbol{\pi}}=\widehat{\boldsymbol{\pi}}-\boldsymbol{\pi} \qquad (8.100)$$

has been conveniently exploited. In view of (8.63), the modelling error is characterized by

$$\widetilde{\boldsymbol{B}}=\widehat{\boldsymbol{B}}-\boldsymbol{B} \qquad \widetilde{\boldsymbol{C}}=\widehat{\boldsymbol{C}}-\boldsymbol{C} \qquad \widetilde{\boldsymbol{F}}=\widehat{\boldsymbol{F}}-\boldsymbol{F} \qquad \widetilde{\boldsymbol{g}}=\widehat{\boldsymbol{g}}-\boldsymbol{g}. \qquad (8.101)$$

It is worth remarking that, in view of position (8.92), the matrix $\boldsymbol{Y}$ does not depend on the actual joint accelerations but only on their desired values; this avoids problems due to direct measurement of acceleration.

At this point, modify the Lyapunov function candidate in (8.95) into the form

$$V(\boldsymbol{\sigma},\widetilde{\boldsymbol{q}},\widetilde{\boldsymbol{\pi}})=\frac{1}{2}\boldsymbol{\sigma}^T\boldsymbol{B}(\boldsymbol{q})\boldsymbol{\sigma}+\widetilde{\boldsymbol{q}}^T\boldsymbol{\Lambda}\boldsymbol{K}_D\widetilde{\boldsymbol{q}}+\frac{1}{2}\widetilde{\boldsymbol{\pi}}^T\boldsymbol{K}_\pi\widetilde{\boldsymbol{\pi}}>0 \qquad \forall\boldsymbol{\sigma},\widetilde{\boldsymbol{q}},\widetilde{\boldsymbol{\pi}}\neq\boldsymbol{0}, \qquad (8.102)$$

which features an additional term accounting for the parameter error (8.100), with $\boldsymbol{K}_\pi$ symmetric positive definite. The time derivative of V along the trajectories of system (8.99) is

$$\dot{V}=-\boldsymbol{\sigma}^T\boldsymbol{F}\boldsymbol{\sigma}-\dot{\widetilde{\boldsymbol{q}}}^T\boldsymbol{K}_D\dot{\widetilde{\boldsymbol{q}}}-\widetilde{\boldsymbol{q}}^T\boldsymbol{\Lambda}\boldsymbol{K}_D\boldsymbol{\Lambda}\widetilde{\boldsymbol{q}}+\widetilde{\boldsymbol{\pi}}^T\big(\boldsymbol{K}_\pi\dot{\widetilde{\boldsymbol{\pi}}}-\boldsymbol{Y}^T(\boldsymbol{q},\dot{\boldsymbol{q}},\dot{\boldsymbol{q}}_r,\ddot{\boldsymbol{q}}_r)\boldsymbol{\sigma}\big). \qquad (8.103)$$

If the estimate of the parameter vector is updated as in the adaptive law

$$\dot{\widehat{\boldsymbol{\pi}}}=\boldsymbol{K}_\pi^{-1}\boldsymbol{Y}^T(\boldsymbol{q},\dot{\boldsymbol{q}},\dot{\boldsymbol{q}}_r,\ddot{\boldsymbol{q}}_r)\boldsymbol{\sigma}, \qquad (8.104)$$

the expression in (8.103) becomes

$$\dot{V}=-\boldsymbol{\sigma}^T\boldsymbol{F}\boldsymbol{\sigma}-\dot{\widetilde{\boldsymbol{q}}}^T\boldsymbol{K}_D\dot{\widetilde{\boldsymbol{q}}}-\widetilde{\boldsymbol{q}}^T\boldsymbol{\Lambda}\boldsymbol{K}_D\boldsymbol{\Lambda}\widetilde{\boldsymbol{q}}$$

since $\dot{\widehat{\boldsymbol{\pi}}}=\dot{\widetilde{\boldsymbol{\pi}}}$ — $\boldsymbol{\pi}$ is constant.

By an argument similar to above, it is not difficult to show that the trajectories of the manipulator described by the model

$$\boldsymbol{B}(\boldsymbol{q})\ddot{\boldsymbol{q}}+\boldsymbol{C}(\boldsymbol{q},\dot{\boldsymbol{q}})\dot{\boldsymbol{q}}+\boldsymbol{F}\dot{\boldsymbol{q}}+\boldsymbol{g}(\boldsymbol{q})=\boldsymbol{u},$$

under the control law

$$\boldsymbol{u}=\boldsymbol{Y}(\boldsymbol{q},\dot{\boldsymbol{q}},\dot{\boldsymbol{q}}_r,\ddot{\boldsymbol{q}}_r)\widehat{\boldsymbol{\pi}}+\boldsymbol{K}_D(\dot{\widetilde{\boldsymbol{q}}}+\boldsymbol{\Lambda}\widetilde{\boldsymbol{q}})$$

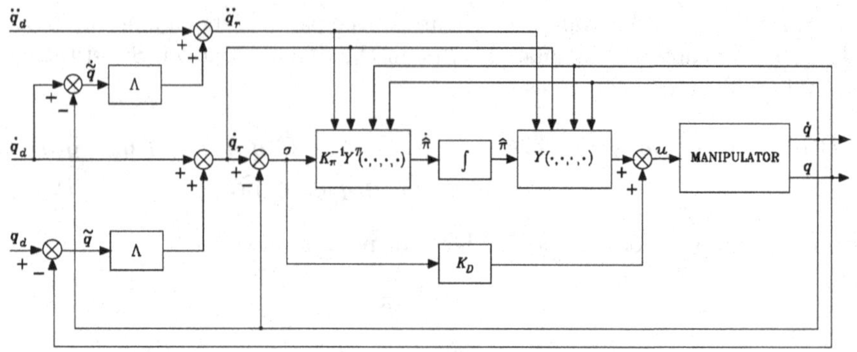

Fig. 8.26. Block scheme of joint space adaptive control

and the parameter adaptive law

$$\dot{\widehat{\boldsymbol{\pi}}} = \boldsymbol{K}_\pi^{-1}\boldsymbol{Y}^T(\boldsymbol{q},\dot{\boldsymbol{q}},\dot{\boldsymbol{q}}_r,\ddot{\boldsymbol{q}}_r)(\dot{\widetilde{\boldsymbol{q}}} + \boldsymbol{\Lambda}\widetilde{\boldsymbol{q}}),$$

globally asymptotically converge to $\boldsymbol{\sigma} = \mathbf{0}$ and $\widetilde{\boldsymbol{q}} = \mathbf{0}$, which implies *convergence to zero of* $\widetilde{\boldsymbol{q}}$, $\dot{\widetilde{\boldsymbol{q}}}$, and *boundedness of* $\widehat{\boldsymbol{\pi}}$. The equation in (8.99) shows that asymptotically it is

$$\boldsymbol{Y}(\boldsymbol{q},\dot{\boldsymbol{q}},\dot{\boldsymbol{q}}_r,\ddot{\boldsymbol{q}}_r)(\widehat{\boldsymbol{\pi}} - \boldsymbol{\pi}) = \mathbf{0}. \tag{8.105}$$

This equation does not imply that $\widehat{\boldsymbol{\pi}}$ tends to $\boldsymbol{\pi}$; indeed, convergence of parameters to their true values depends on the structure of the matrix $\boldsymbol{Y}(\boldsymbol{q},\dot{\boldsymbol{q}},\dot{\boldsymbol{q}}_r,\ddot{\boldsymbol{q}}_r)$ and then on the desired and actual trajectories. Nonetheless, the followed approach is aimed at solving a *direct* adaptive control problem, i.e., finding a control law that ensures limited tracking errors, and not at determining the actual parameters of the system (as in an indirect adaptive control problem). The resulting block scheme is illustrated in Fig. 8.26. To summarize, the above control law is formed by three different contributions:

- The term $\boldsymbol{Y}\widehat{\boldsymbol{\pi}}$ describes a control action of inverse dynamics type which ensures an *approximate compensation of nonlinear effects and joint decoupling*.
- The term $\boldsymbol{K}_D\boldsymbol{\sigma}$ introduces a *stabilizing linear control action of PD type on the tracking error*.
- The vector of parameter estimates $\widehat{\boldsymbol{\pi}}$ is updated by an *adaptive law of gradient type* so as to ensure asymptotic compensation of the terms in the manipulator dynamic model; the matrix $\boldsymbol{K}_\pi$ determines the convergence rate of parameters to their asymptotic values.

Notice that, with $\boldsymbol{\sigma} \approx \mathbf{0}$, the control law (8.98) is equivalent to a pure inverse dynamics compensation of the computed torque type on the basis of desired velocities and accelerations; this is made possible by the fact that $\boldsymbol{Y}\widehat{\boldsymbol{\pi}} \approx \boldsymbol{Y}\boldsymbol{\pi}$.

The control law with parameter adaptation requires the availability of a complete computational model and it does not feature any action aimed at reducing the effects of external disturbances. Therefore, a performance degradation is expected whenever unmodelled dynamic effects, e.g., when a reduced computational model is used, or external disturbances occur. In both cases, the effects induced on the output variables are attributed by the controller to parameter estimate mismatching; as a consequence, the control law attempts to counteract those effects by acting on quantities that did not provoke them originally.

On the other hand, robust control techniques provide a natural rejection to external disturbances, although they are sensitive to unmodelled dynamics; this rejection is provided by a high-frequency commuted control action that constrains the error trajectories to stay on the sliding subspace. The resulting inputs to the mechanical structure may be unacceptable. This inconvenience is in general not observed with the adoption of adaptive control techniques whose action has a naturally smooth time behaviour.

8.6 Operational Space Control

In all the above control schemes, it was always assumed that the desired trajectory is available in terms of the time sequence of the values of joint position, velocity and acceleration. Accordingly, the error for the control schemes was expressed in the joint space.

As often pointed out, motion specifications are usually assigned in the operational space, and then an inverse kinematics algorithm has to be utilized to transform operational space references into the corresponding joint space references. The process of kinematic inversion has an increasing computational load when, besides inversion of direct kinematics, inversion of first-order and second-order differential kinematics is also required to transform the desired time history of end-effector position, velocity and acceleration into the corresponding quantities at the joint level. It is for this reason that current industrial robot control systems compute the joint positions through kinematics inversion, and then perform a numerical differentiation to compute velocities and accelerations.

A different approach consists of considering control schemes developed directly in the operational space. If the motion is specified in terms of operational space variables, the measured joint space variables can be transformed into the corresponding operational space variables through direct kinematics relations. Comparing the desired input with the reconstructed variables allows the design of feedback control loops where trajectory inversion is replaced with a suitable coordinate transformation embedded in the feedback loop.

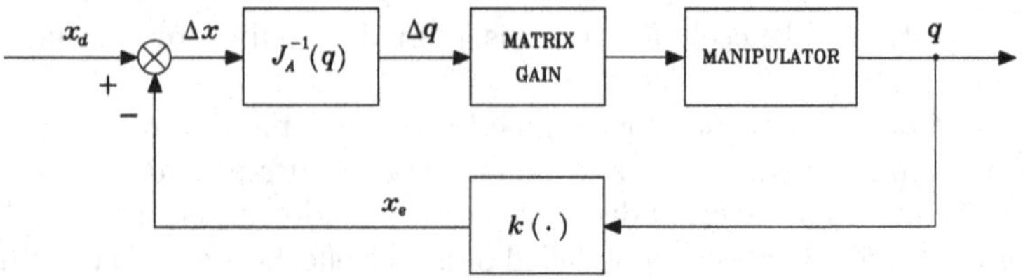

Fig. 8.27. Block scheme of Jacobian inverse control

All operational space control schemes present considerable computational requirements, in view of the necessity to perform a number of computations in the feedback loop which are somewhat representative of inverse kinematics functions. With reference to a numerical implementation, the presence of a computationally demanding load requires sampling times that may lead to degrading the performance of the overall control system.

In the face of the above limitations, it is worth presenting *operational space control* schemes, whose utilization becomes necessary when the problem of controlling interaction between the manipulator and the environment is of concern. In fact, joint space control schemes suffice only for motion control in the free space. When the manipulator's end-effector is constrained by the environment, e.g., in the case of end-effector in contact with an elastic environment, it is necessary to control both positions and contact forces and it is convenient to refer to operational space control schemes. Hence, below some solutions are presented; these are worked out for motion control, but they constitute the premise for the force/position control strategies that will be illustrated in the next chapter.

8.6.1 General Schemes

As pointed out above, operational space control schemes are based on a direct comparison of the inputs, specifying operational space trajectories, with the measurements of the corresponding manipulator outputs. It follows that the control system should incorporate some actions that allow the transformation from the operational space, in which the error is specified, to the joint space, in which control generalized forces are developed.

A possible control scheme that can be devised is the so-called *Jacobian inverse control* (Fig. 8.27). In this scheme, the end-effector pose in the operational space $\boldsymbol{x}_e$ is compared with the corresponding desired quantity $\boldsymbol{x}_d$, and then an operational space deviation $\Delta\boldsymbol{x}$ can be computed. Assumed that this deviation is sufficiently small for a good control system, $\Delta\boldsymbol{x}$ can be transformed into a corresponding joint space deviation $\Delta\boldsymbol{q}$ through the inverse of the manipulator Jacobian. Then, the control input generalized forces can be computed on the basis of this deviation through a suitable feedback matrix gain. The result is a presumable reduction of $\Delta\boldsymbol{q}$ and in turn of $\Delta\boldsymbol{x}$. In other words, the Jacobian inverse control leads to an overall system that intuitively

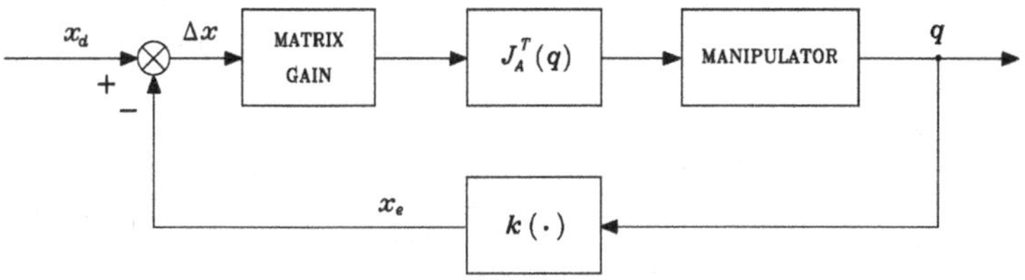

Fig. 8.28. Block scheme of Jacobian transpose control

behaves like a mechanical system with a generalized n-dimensional spring in the joint space, whose constant stiffness is determined by the feedback matrix gain. The role of such system is to take the deviation $\Delta\boldsymbol{q}$ to zero. If the matrix gain is diagonal, the generalized spring corresponds to n independent elastic elements, one for each joint.

A conceptually analogous scheme is the so-called *Jacobian transpose control* (Fig. 8.28). In this case, the operational space error is treated first through a matrix gain. The output of this block can then be considered as the elastic force generated by a generalized spring whose function in the operational space is that to reduce or to cancel the position deviation $\Delta\boldsymbol{x}$. In other words, the resulting force drives the end-effector along a direction so as to reduce $\Delta\boldsymbol{x}$. This operational space force has then to be transformed into the joint space generalized forces, through the transpose of the Jacobian, so as to realize the described behaviour.

Both Jacobian inverse and transpose control schemes have been derived in an intuitive fashion. Hence, there is no guarantee that such schemes are effective in terms of stability and trajectory tracking accuracy. These problems can be faced by presenting two mathematical solutions below, which will be shown to be substantially equivalent to the above schemes.

8.6.2 PD Control with Gravity Compensation

By analogy with joint space stability analysis, given a *constant* end-effector pose $\boldsymbol{x}_d$, it is desired to find the control structure so that the operational space error

$$\widetilde{\boldsymbol{x}} = \boldsymbol{x}_d - \boldsymbol{x}_e \tag{8.106}$$

tends asymptotically to zero. Choose the following positive definite quadratic form as a Lyapunov function candidate:

$$V(\dot{\boldsymbol{q}}, \widetilde{\boldsymbol{x}}) = \frac{1}{2}\dot{\boldsymbol{q}}^T \boldsymbol{B}(\boldsymbol{q})\dot{\boldsymbol{q}} + \frac{1}{2}\widetilde{\boldsymbol{x}}^T \boldsymbol{K}_P \widetilde{\boldsymbol{x}} > 0 \qquad \forall \dot{\boldsymbol{q}}, \widetilde{\boldsymbol{x}} \neq \boldsymbol{0}, \tag{8.107}$$

with $\boldsymbol{K}_P$ a symmetric positive definite matrix. Differentiating (8.107) with respect to time gives

$$\dot{V} = \dot{\boldsymbol{q}}^T \boldsymbol{B}(\boldsymbol{q})\ddot{\boldsymbol{q}} + \frac{1}{2}\dot{\boldsymbol{q}}^T \dot{\boldsymbol{B}}(\boldsymbol{q})\dot{\boldsymbol{q}} + \dot{\widetilde{\boldsymbol{x}}}^T \boldsymbol{K}_P \widetilde{\boldsymbol{x}}.$$

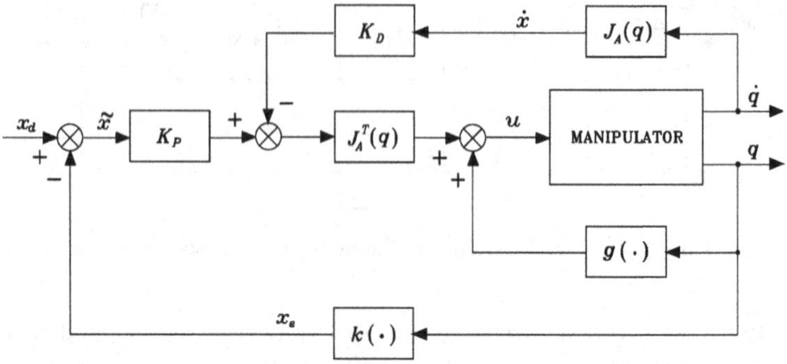

Fig. 8.29. Block scheme of operational space PD control with gravity compensation

Since $\dot{\boldsymbol{x}}_d = \mathbf{0}$, in view of (3.62) it is

$$\dot{\tilde{\boldsymbol{x}}} = -\boldsymbol{J}_A(\boldsymbol{q})\dot{\boldsymbol{q}}$$

and then

$$\dot{V} = \dot{\boldsymbol{q}}^T \boldsymbol{B}(\boldsymbol{q})\ddot{\boldsymbol{q}} + \frac{1}{2}\dot{\boldsymbol{q}}^T \dot{\boldsymbol{B}}(\boldsymbol{q})\dot{\boldsymbol{q}} - \dot{\boldsymbol{q}}^T \boldsymbol{J}_A^T(\boldsymbol{q})\boldsymbol{K}_P\tilde{\boldsymbol{x}}. \tag{8.108}$$

By recalling the expression of the joint space manipulator dynamic model in (8.7) and the property in (7.49), the expression in (8.108) becomes

$$\dot{V} = -\dot{\boldsymbol{q}}^T \boldsymbol{F}\dot{\boldsymbol{q}} + \dot{\boldsymbol{q}}^T\left(\boldsymbol{u} - \boldsymbol{g}(\boldsymbol{q}) - \boldsymbol{J}_A^T(\boldsymbol{q})\boldsymbol{K}_P\tilde{\boldsymbol{x}}\right). \tag{8.109}$$

This equation suggests the structure of the controller; in fact, by choosing the control law

$$\boldsymbol{u} = \boldsymbol{g}(\boldsymbol{q}) + \boldsymbol{J}_A^T(\boldsymbol{q})\boldsymbol{K}_P\tilde{\boldsymbol{x}} - \boldsymbol{J}_A^T(\boldsymbol{q})\boldsymbol{K}_D\boldsymbol{J}_A(\boldsymbol{q})\dot{\boldsymbol{q}} \tag{8.110}$$

with $\boldsymbol{K}_D$ positive definite, (8.109) becomes

$$\dot{V} = -\dot{\boldsymbol{q}}^T \boldsymbol{F}\dot{\boldsymbol{q}} - \dot{\boldsymbol{q}}^T \boldsymbol{J}_A^T(\boldsymbol{q})\boldsymbol{K}_D\boldsymbol{J}_A(\boldsymbol{q})\dot{\boldsymbol{q}}. \tag{8.111}$$

As can be seen from Fig. 8.29, the resulting block scheme reveals an analogy with the scheme of Fig. 8.28. Control law (8.110) performs a *nonlinear compensating action of joint space gravitational forces* and an *operational space linear PD control action*. The last term has been introduced to enhance system damping; in particular, if measurement of $\dot{\boldsymbol{x}}$ is deduced from that of $\dot{\boldsymbol{q}}$, one can simply choose the derivative term as $-\boldsymbol{K}_D\dot{\boldsymbol{q}}$.

The expression in (8.111) shows that, for any system trajectory, the Lyapunov function decreases as long as $\dot{\boldsymbol{q}} \neq \mathbf{0}$. The system then reaches an *equilibrium posture*. By a stability argument similar to that in the joint space (see (8.52)–(8.54)) this posture is determined by

$$\boldsymbol{J}_A^T(\boldsymbol{q})\boldsymbol{K}_P\tilde{\boldsymbol{x}} = \mathbf{0}. \tag{8.112}$$

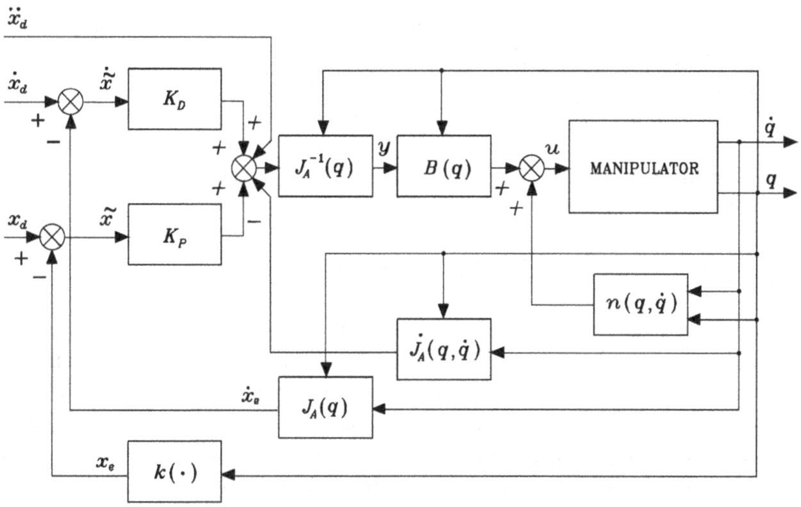

Fig. 8.30. Block scheme of operational space inverse dynamics control

From (8.112) it can be recognized that, under the assumption of *full-rank* Jacobian, it is

$$\tilde{\boldsymbol{x}} = \boldsymbol{x}_d - \boldsymbol{x}_e = \mathbf{0},$$

i.e., the sought result.

If measurements of $\boldsymbol{x}_e$ and $\dot{\boldsymbol{x}}_e$ are made directly in the operational space, $\boldsymbol{k}(\boldsymbol{q})$ and $\boldsymbol{J}_A(\boldsymbol{q})$ in the scheme of Fig. 8.45 are just indicative of direct kinematics functions; it is, however, necessary to measure $\boldsymbol{q}$ to update both $\boldsymbol{J}_A^T(\boldsymbol{q})$ and $\boldsymbol{g}(\boldsymbol{q})$ on-line. If measurements of operational space quantities are indirect, the controller has to compute the direct kinematics functions, too.

8.6.3 Inverse Dynamics Control

Consider now the problem of tracking an operational space trajectory. Recall the manipulator dynamic model in the form (8.55)

$$\boldsymbol{B}(\boldsymbol{q})\ddot{\boldsymbol{q}} + \boldsymbol{n}(\boldsymbol{q}, \dot{\boldsymbol{q}}) = \boldsymbol{u},$$

where $\boldsymbol{n}$ is given by (8.56). As in (8.57), the choice of the *inverse dynamics linearizing control*

$$\boldsymbol{u} = \boldsymbol{B}(\boldsymbol{q})\boldsymbol{y} + \boldsymbol{n}(\boldsymbol{q}, \dot{\boldsymbol{q}})$$

leads to the system of double integrators

$$\ddot{\boldsymbol{q}} = \boldsymbol{y}. \tag{8.113}$$

The new control input $\boldsymbol{y}$ is to be designed so as to yield tracking of a trajectory specified by $\boldsymbol{x}_d(t)$. To this end, the second-order differential equation in the form (3.98)

$$\ddot{\boldsymbol{x}}_e = \boldsymbol{J}_A(\boldsymbol{q})\ddot{\boldsymbol{q}} + \dot{\boldsymbol{J}}_A(\boldsymbol{q}, \dot{\boldsymbol{q}})\dot{\boldsymbol{q}}$$

suggests, for a nonredundant manipulator, the choice of the control law — formally analogous to (3.102) —

$$\boldsymbol{y} = \boldsymbol{J}_A^{-1}(\boldsymbol{q})\big(\ddot{\boldsymbol{x}}_d + \boldsymbol{K}_D\dot{\widetilde{\boldsymbol{x}}} + \boldsymbol{K}_P\widetilde{\boldsymbol{x}} - \dot{\boldsymbol{J}}_A(\boldsymbol{q},\dot{\boldsymbol{q}})\dot{\boldsymbol{q}}\big) \tag{8.114}$$

with $\boldsymbol{K}_P$ and $\boldsymbol{K}_D$ positive definite (diagonal) matrices. In fact, substituting (8.114) into (8.113) gives

$$\ddot{\widetilde{\boldsymbol{x}}} + \boldsymbol{K}_D\dot{\widetilde{\boldsymbol{x}}} + \boldsymbol{K}_P\widetilde{\boldsymbol{x}} = \boldsymbol{0} \tag{8.115}$$

which describes the operational space error dynamics, with $\boldsymbol{K}_P$ and $\boldsymbol{K}_D$ determining the error convergence rate to zero. The resulting inverse dynamics control scheme is reported in Fig. 8.30, which confirms the anticipated analogy with the scheme of Fig. 8.27. Again in this case, besides $\boldsymbol{x}_e$ and $\dot{\boldsymbol{x}}_e$, $\boldsymbol{q}$ and $\dot{\boldsymbol{q}}$ are also to be measured. If measurements of $\boldsymbol{x}_e$ and $\dot{\boldsymbol{x}}_e$ are indirect, the controller must compute the direct kinematics functions $\boldsymbol{k}(\boldsymbol{q})$ and $\boldsymbol{J}_A(\boldsymbol{q})$ on-line.

A critical analysis of the schemes in Figs. 8.29, 8.30 reveals that the design of an operational space controller always requires computation of manipulator Jacobian. As a consequence, controlling a manipulator in the operational space is in general more complex than controlling it in the joint space. In fact, the presence of *singularities* and/or *redundancy* influences the Jacobian, and the induced effects are somewhat difficult to handle with an operational space controller. For instance, if a singularity occurs for the scheme of Fig. 8.29 and the error enters the null space of the Jacobian, the manipulator gets stuck at a different configuration from the desired one. This problem is even more critical for the scheme of Fig. 8.30 which would require the computation of a DLS inverse of the Jacobian. Yet, for a redundant manipulator, a joint space control scheme is naturally transparent to this situation, since redundancy has already been solved by inverse kinematics, whereas an operational space control scheme should incorporate a redundancy handling technique inside the feedback loop.

As a final remark, the above operational space control schemes have been derived with reference to a minimal description of orientation in terms of Euler angles. It is understood that, similar to what is presented in Sect. 3.7.3 for inverse kinematics algorithms, it is possible to adopt different definitions of orientation error, e.g., based on the angle and axis or the unit quaternion. The advantage is the use of the geometric Jacobian in lieu of the analytical Jacobian. The price to pay, however, is a more complex analysis of the stability and convergence characteristics of the closed-loop system. Even the inverse dynamics control scheme will not lead to a homogeneous error equation, and a Lyapunov argument should be invoked to ascertain its stability.

8.7 Comparison Among Various Control Schemes

In order to make a comparison between the various control schemes presented, consider the two-link planar arm with the same data of Example 7.2:

$$a_1 = a_2 = 1\,\text{m} \quad \ell_1 = \ell_2 = 0.5\,\text{m} \quad m_{\ell_1} = m_{\ell_2} = 50\,\text{kg} \quad I_{\ell_1} = I_{\ell_2} = 10\,\text{kg}\cdot\text{m}^2$$

$$k_{r1} = k_{r2} = 100 \qquad m_{m1} = m_{m2} = 5\,\text{kg} \qquad I_{m1} = I_{m2} = 0.01\,\text{kg}\cdot\text{m}^2.$$

The arm is assumed to be driven by two equal actuators with the following data:

$$F_{m1} = F_{m2} = 0.01\,\text{N}\cdot\text{m}\cdot\text{s/rad} \qquad R_{a1} = R_{a2} = 10\,\text{ohm}$$

$$k_{t1} = k_{t2} = 2\,\text{N}\cdot\text{m/A} \qquad k_{v1} = k_{v2} = 2\,\text{V}\cdot\text{s/rad};$$

it can be verified that $F_{m_i} \ll k_{vi}k_{ti}/R_{ai}$ for $i = 1, 2$.

The desired tip trajectories have a typical trapezoidal velocity profile, and thus it is anticipated that sharp torque variations will be induced. The tip path is a motion of 1.6 m along the horizontal axis, as in the path of Example 7.2. In the first case (*fast* trajectory), the acceleration time is 0.6 s and the maximum velocity is 1 m/s. In the second case (*slow* trajectory), the acceleration time is 0.6 s and the maximum velocity is 0.25 m/s. The motion of the controlled arm was simulated on a computer, by adopting a discrete-time implementation of the controller with a sampling time of 1 ms.

The following control schemes in the joint space and in the operational space have been utilized; an (analytic) inverse kinematics solution has been implemented to generate the reference inputs to the joint space control schemes:

A. Independent joint control with position and velocity feedback (Fig. 5.11) with the following data for each joint servo:

$$K_P = 5 \qquad K_V = 10 \qquad k_{TP} = k_{TV} = 1,$$

corresponding to $\omega_n = 5\,\text{rad/s}$ and $\zeta = 0.5$.

B. Independent joint control with position, velocity and acceleration feedback (Fig. 8.9) with the following data for each joint servo:

$$K_P = 5 \qquad K_V = 10 \qquad K_A = 2 \qquad k_{TP} = k_{TV} = k_{TA} = 1,$$

corresponding to $\omega_n = 5\,\text{rad/s}$, $\zeta = 0.5$, $X_R = 100$. To reconstruct acceleration, a first-order filter has been utilized (Fig. 8.11) characterized by $\omega_{3f} = 100\,\text{rad/s}$.

C. As in scheme **A** with the addition of a decentralized feedforward action (Fig. 8.13).

D. As in scheme **B** with the addition of a decentralized feedforward action (Fig. 8.14).

E. Joint space computed torque control (Fig. 8.19) with feedforward compensation of the diagonal terms of the inertia matrix and of gravitational terms, and decentralized feedback controllers as in scheme **A**.

F. Joint space PD control with gravity compensation (Fig. 8.20), modified by the addition of a feedforward velocity term $\boldsymbol{K}_D\dot{\boldsymbol{q}}_d$, with the following data:

$$\boldsymbol{K}_P = 3750\boldsymbol{I}_2 \qquad \boldsymbol{K}_D = 750\boldsymbol{I}_2.$$

G. Joint space inverse dynamics control (Fig. 8.22) with the following data:

$$\boldsymbol{K}_P = 25\boldsymbol{I}_2 \qquad \boldsymbol{K}_D = 5\boldsymbol{I}_2.$$

H. Joint space robust control (Fig. 8.23), under the assumption of constant inertia ($\widehat{\boldsymbol{B}} = \bar{\boldsymbol{B}}$) and compensation of friction and gravity ($\widehat{\boldsymbol{n}} = \boldsymbol{F}_v\dot{\boldsymbol{q}} + \boldsymbol{g}$), with the following data:

$$\boldsymbol{K}_P = 25\boldsymbol{I}_2 \qquad \boldsymbol{K}_D = 5\boldsymbol{I}_2 \qquad \boldsymbol{P} = \boldsymbol{I}_2 \qquad \rho = 70 \qquad \epsilon = 0.004.$$

I. As in case **H** with $\epsilon = 0.01$.

J. Joint space adaptive control (Fig. 8.26) with a parameterization of the arm dynamic model (7.82) as in (7.83), (7.84). The initial estimate of the vector $\widehat{\boldsymbol{\pi}}$ is computed on the basis of the nominal parameters. The arm is supposed to carry a load which causes the following variations on the second link parameters:

$$\Delta m_2 = 10\,\text{kg} \qquad \Delta m_2\ell_{C2} = 11\,\text{kg·m} \qquad \Delta\widehat{I}_2 = 12.12\,\text{kg·m}^2.$$

This information is obviously utilized only to update the simulated arm model. Further, the following data are set:

$$\boldsymbol{\Lambda} = 5\boldsymbol{I}_2 \qquad \boldsymbol{K}_D = 750\boldsymbol{I}_2 \qquad \boldsymbol{K}_\pi = 0.01\boldsymbol{I}_8.$$

K. Operational space PD control with gravity compensation (Fig. 8.29), modified by the addition of a feedforward velocity term $\boldsymbol{K}_D\dot{\boldsymbol{x}}_d$, with the following data:

$$\boldsymbol{K}_P = 16250\boldsymbol{I}_2 \qquad \boldsymbol{K}_D = 3250\boldsymbol{I}_2.$$

L. Operational space inverse dynamics control (Fig. 8.30) with the following data:

$$\boldsymbol{K}_P = 25\boldsymbol{I}_2 \qquad \boldsymbol{K}_D = 5\boldsymbol{I}_2.$$

It is worth remarking that the adopted model of the dynamic system of arm with drives is that described by (8.7). In the decentralized control schemes **A**–**E**, the joints have been voltage-controlled as in the block scheme of Fig. 8.3, with unit amplifier gains ($\boldsymbol{G}_v = \boldsymbol{I}$). On the other hand, in the centralized control schemes **F**–**L**, the joints have been current-controlled as in the block scheme of Fig. 8.4, with unit amplifier gains ($\boldsymbol{G}_i = \boldsymbol{I}$).

Regarding the parameters of the various controllers, these have been chosen in such a way as to allow a significant comparison of the performance of each scheme in response to congruent control actions. In particular, it can be observed that:

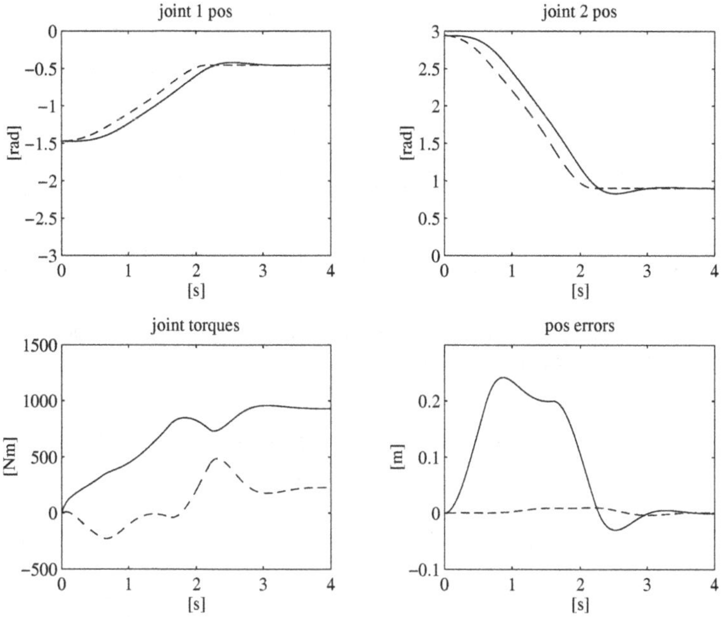

Fig. 8.31. Time history of the joint positions and torques and of the tip position errors for the *fast* trajectory with control scheme **A**

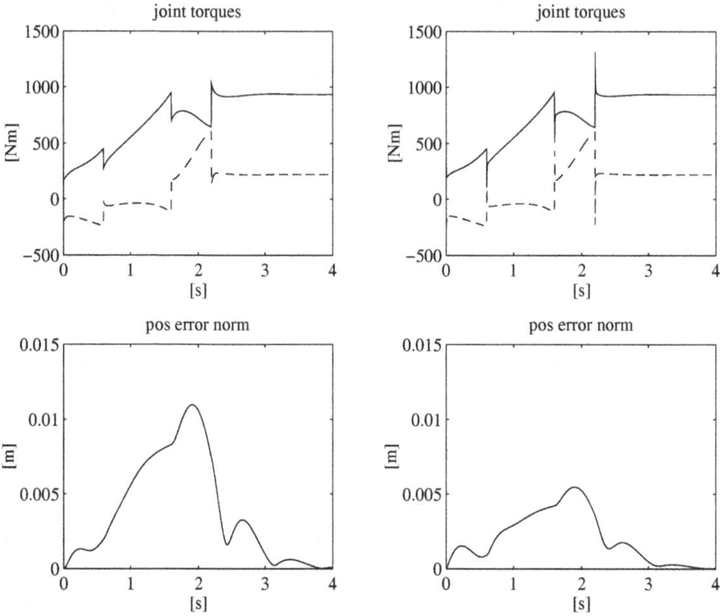

Fig. 8.32. Time history of the joint torques and of the norm of tip position error for the *fast* trajectory; *left*: with control scheme **C**, *right*: with control scheme **D**

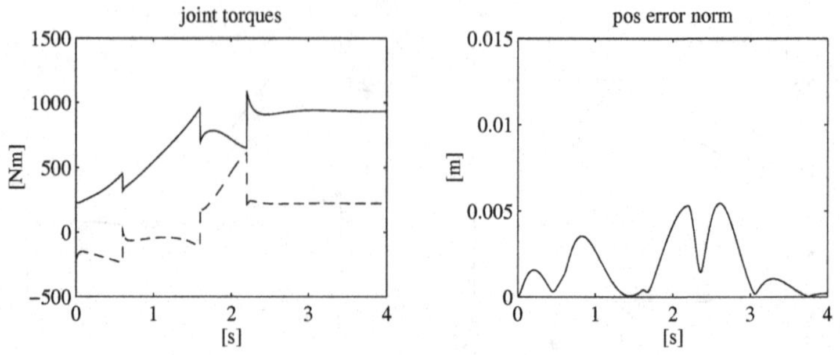

Fig. 8.33. Time history of the joint torques and of the norm of tip position error for the *fast* trajectory with control scheme **E**

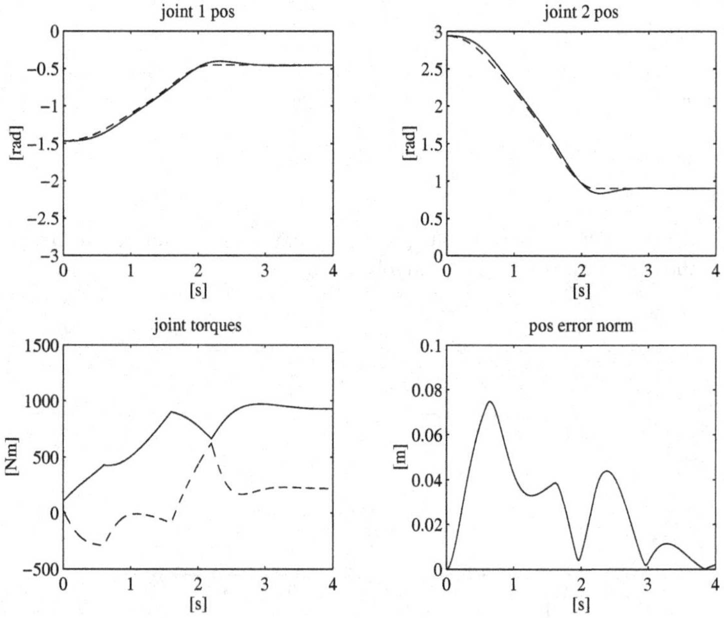

Fig. 8.34. Time history of the joint positions and torques and of the norm of tip position error for the *fast* trajectory with control scheme **F**

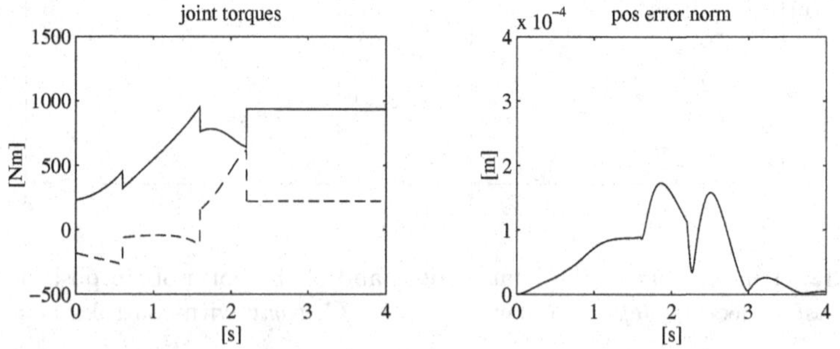

Fig. 8.35. Time history of the joint torques and of the norm of tip position error for the *fast* trajectory with control scheme **G**

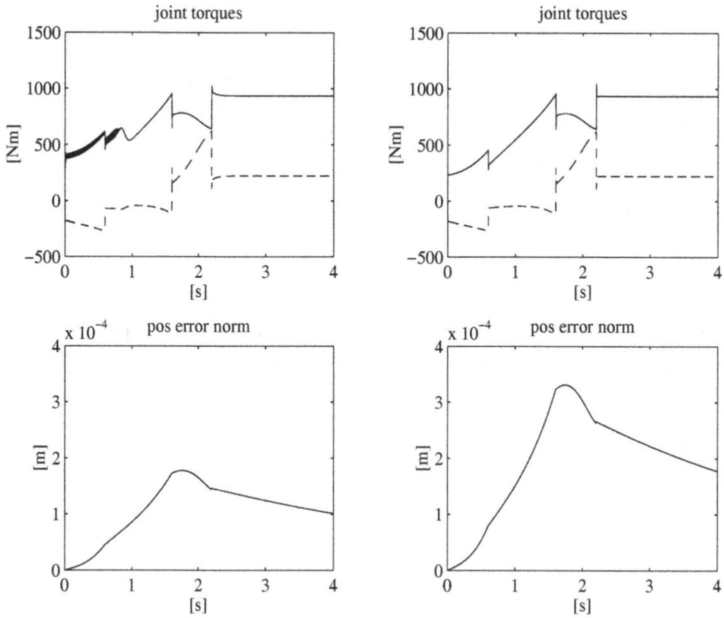

Fig. 8.36. Time history of the joint torques and of the norm of tip position error for the *fast* trajectory; *left*: with control scheme **H**, *right*: with control scheme **I**

- The dynamic behaviour of the joints is the same for schemes **A**–**E**.
- The gains of the PD actions in schemes **G**, **H**, **I** and **L** have been chosen so as to obtain the same natural frequency and damping ratios as those of schemes **A**–**E**.

The results obtained with the various control schemes are illustrated in Figs. 8.31–8.39 for the *fast* trajectory and in Figs. 8.40–8.48 for the *slow* trajectory, respectively. In the case of two quantities represented in the same plot notice that:

- For the joint trajectories, the dashed line indicates the reference trajectory obtained from the tip trajectory via inverse kinematics, while the solid line indicates the actual trajectory followed by the arm.
- For the joint torques, the solid line refers to Joint 1 while the dashed line refers to Joint 2.
- For the tip position error, the solid line indicates the error component along the horizontal axis while the dashed line indicates the error component along the vertical axis.

Finally, the representation scales have been made as uniform as possible in order to allow a more direct comparison of the results.

Regarding performance of the various control schemes for the *fast* trajectory, the obtained results lead to the following considerations.

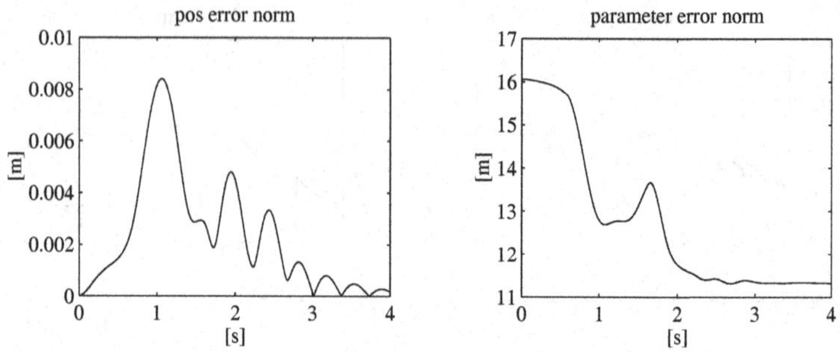

Fig. 8.37. Time history of the norm of tip position error and of the norm of parameter error vector for the *fast* trajectory with control scheme **J**

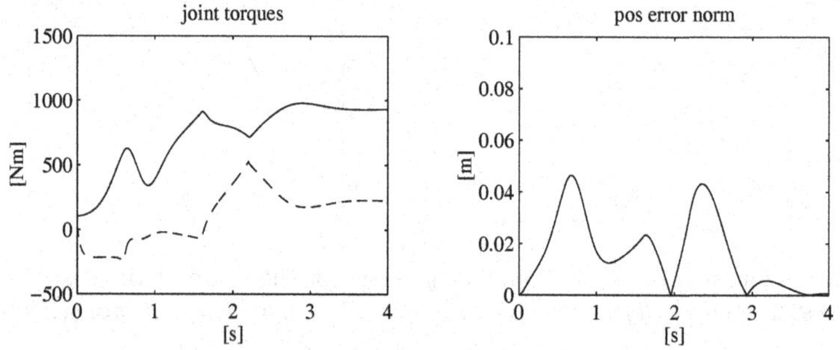

Fig. 8.38. Time history of the joint torques and of the norm of tip position error for the *fast* trajectory with control scheme **K**

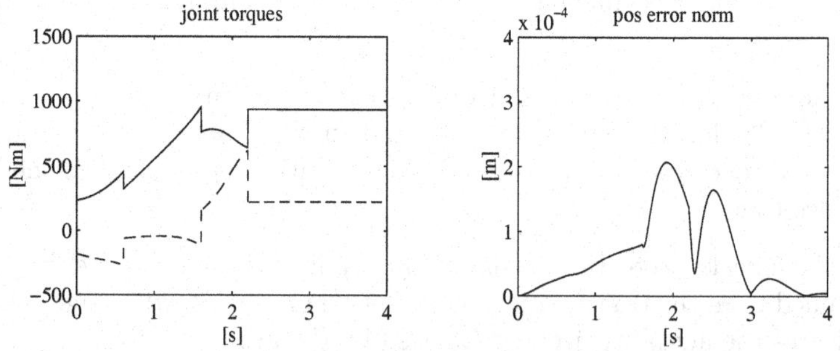

Fig. 8.39. Time history of the joint torques and of the norm of tip position error for the *fast* trajectory with control scheme **L**

Deviation of the actual joint trajectories from the desired ones shows that tracking performance of scheme **A** is quite poor (Fig. 8.31). It should be noticed, however, that the largest contribution to the error is caused by a time lag of the actual trajectory behind the desired one, while the distance

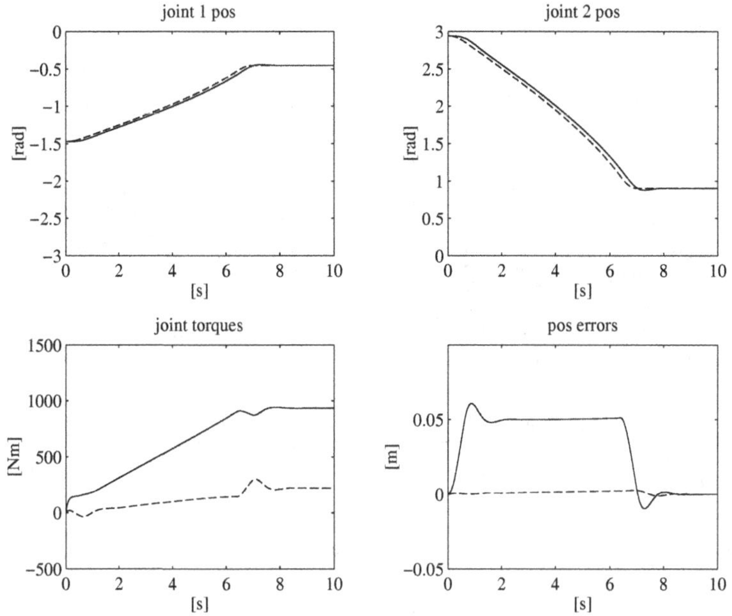

Fig. 8.40. Time history of the joint positions and torques and of the tip position errors for the *slow* trajectory with control scheme **A**

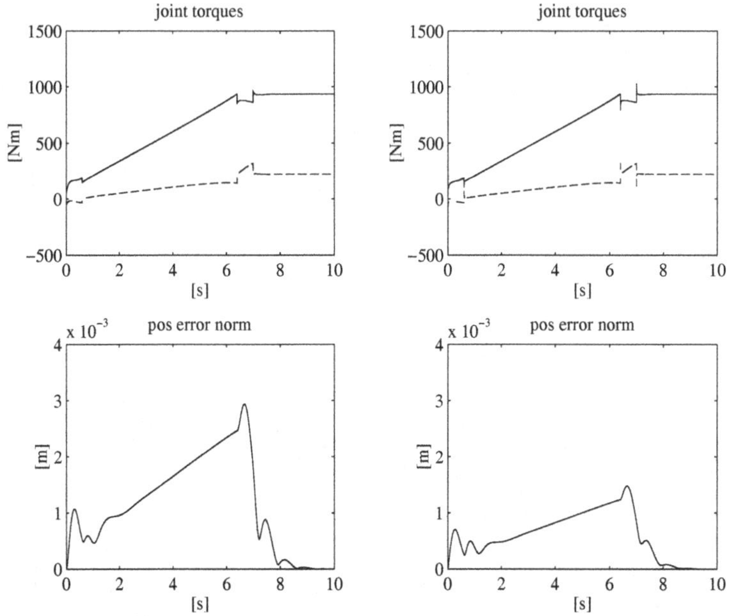

Fig. 8.41. Time history of the joint torques and of the norm of tip position error for the *slow* trajectory; *left*: with control scheme **C**, *right*: with control scheme **D**

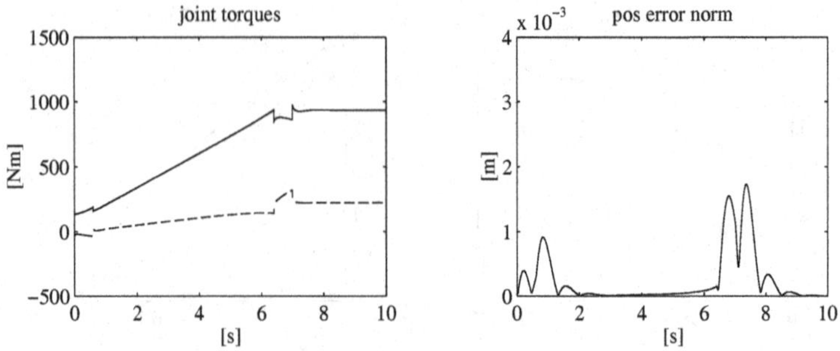

Fig. 8.42. Time history of the joint torques and of the norm of tip position error for the *slow* trajectory with control scheme **E**

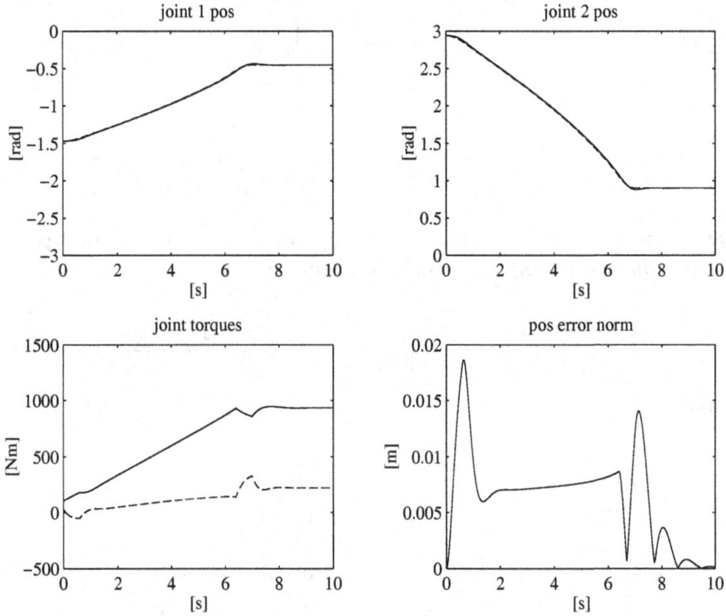

Fig. 8.43. Time history of the joint positions and torques and of the norm of tip position error for the *slow* trajectory with control scheme **F**

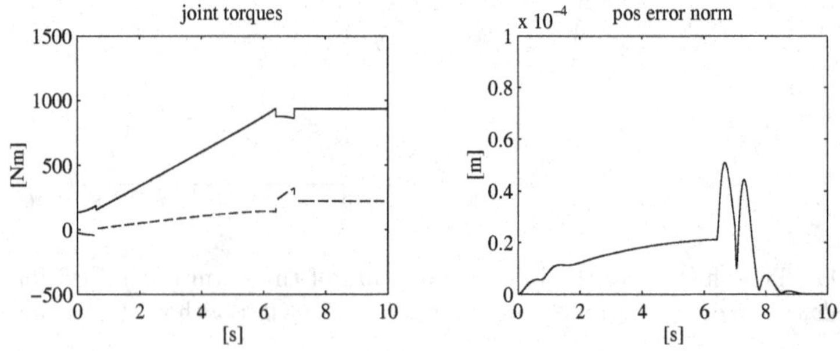

Fig. 8.44. Time history of the joint torques and of the norm of tip position error for the *slow* trajectory with control scheme **G**

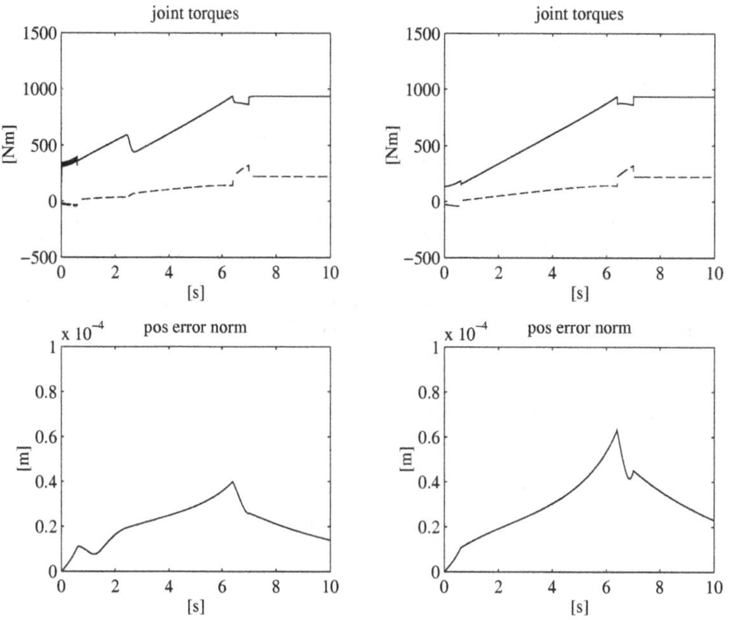

Fig. 8.45. Time history of the joint torques and of the norm of tip position error for the *slow* trajectory; *left*: with control scheme **H**, *right*: with control scheme **I**

of the tip from the geometric path is quite contained. Similar results were obtained with scheme **B**, and then they have not been reported.

With schemes **C** and **D**, an appreciable tracking accuracy improvement is observed (Fig. 8.32), with better performance for the second scheme, thanks to the outer acceleration feedback loop that allows a disturbance rejection factor twice as much as for the first scheme. Notice that the feedforward action yields a set of torques which are closer to the nominal ones required to execute the desired trajectory; the torque time history has a discontinuity in correspondence of the acceleration and deceleration fronts.

The tracking error is further decreased with scheme **E** (Fig. 8.33), by virtue of the additional nonlinear feedforward compensation.

Scheme **F** guarantees stable convergence to the final arm posture with a tracking performance which is better than that of schemes **A** and **B**, thanks to the presence of a velocity feedforward action, but worse than that of schemes **C**–**E**, in view of lack of an acceleration feedforward action (Fig. 8.34).

As would be logical to expect, the best results are observed with scheme **G** for which the tracking error is practically zero, and it is mainly due to numerical discretization of the controller (Fig. 8.35).

It is then worth comparing the performance of schemes **H** and **I** (Fig. 8.36). In fact, the choice of a small threshold value for ϵ (scheme **H**) induces high-

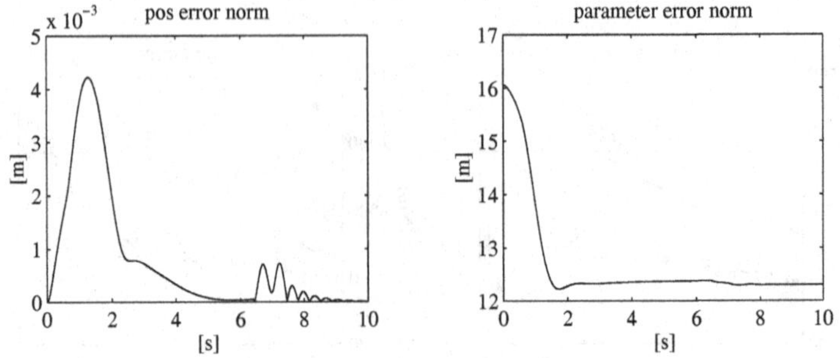

Fig. 8.46. Time history of the norm of tip position error and of the norm of parameter error vector for the *slow* trajectory with control scheme **J**

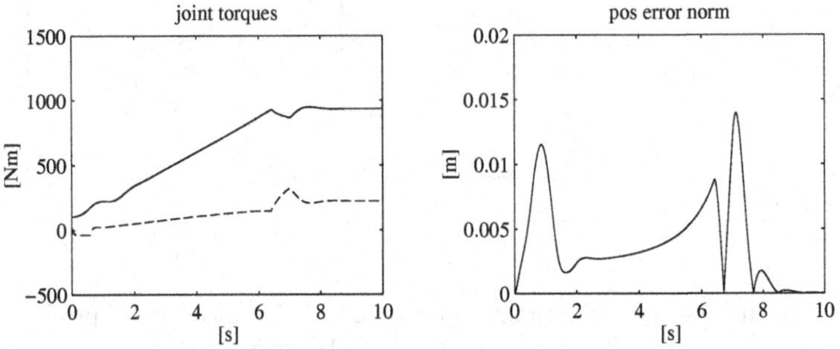

Fig. 8.47. Time history of the joint torques and of the norm of tip position error for the *slow* trajectory with control scheme **K**

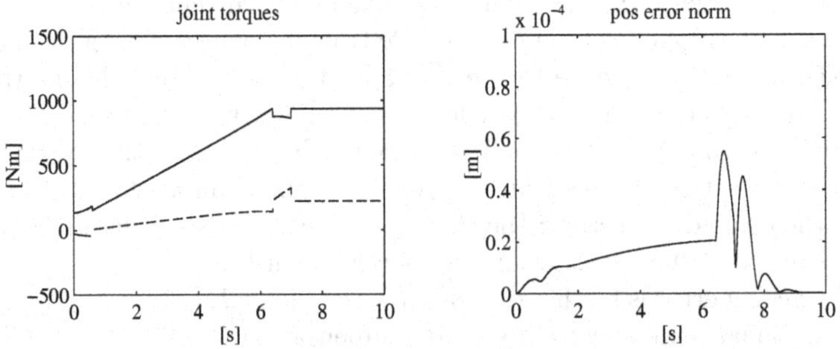

Fig. 8.48. Time history of the joint torques and of the norm of tip position error for the *slow* trajectory with control scheme **L**

frequency components in Joint 1 torque (see the thick portions of the torque plot) at the advantage of a very limited tracking error. As the threshold value is increased (scheme **I**), the torque assumes a smoother behaviour at the expense of a doubled norm of tracking error, though.

For scheme **J**, a lower tracking error than that of scheme **F** is observed, thanks to the effectiveness of the adaptive action on the parameters of the dynamic model. Nonetheless, the parameters do not converge to their nominal values, as confirmed by the time history of the norm of the parameter error vector that reaches a non-null steady-state value (Fig. 8.37).

Finally, the performance of schemes **K** and **L** is substantially comparable to that of corresponding schemes **F** and **G** (Figs. 8.38 and 8.39).

Performance of the various control schemes for the *slow* trajectory is globally better than that for the *fast* trajectory. Such improvement is particularly evident for the decentralized control schemes (Figs. 8.40–8.42), whereas the tracking error reduction for the centralized control schemes is less dramatic (Figs. 8.43–8.48), in view of the small order of magnitude of the errors already obtained for the *fast* trajectory. In any case, as regards performance of each single scheme, it is possible to make a number of remarks analogous to those previously made.

Bibliography

The independent joint control is analyzed in classical texts [180, 120, 200], and scientific articles [19, 127, 141, 101, 39]. Stability of PD control with gravity compensation is proved in [7], on the basis of the notable properties of the dynamic model in [226].

Computed torque control and inverse dynamics control were developed at the beginning of the 1970s. One of the first experimental works is [149]. Other articles on the topic are [83, 4, 117, 121, 126, 227, 29].

The main approaches of robust control are inspired to the work [50]. Among them it is worth citing [212, 84, 130, 219, 205, 216]. Robust controllers based on the high gain concept are presented in [192, 222]. A survey on robust control is [1].

One of the first approaches to adaptive control, based on the assumption of decoupled joint dynamics, is presented in [67]. The first works on adaptive control accounting for the manipultor nonlinear dynamics are [15, 167, 100], yet they exploit the notable properties of the dynamic model only to some extent. The adaptive version of inverse dynamics control is analyzed in [52, 157]. The approach based on the energy properties of the dynamic model has been proposed in [214] and further analyzed in [218]. An interesting tutorial on adaptive control is [175].

Operational space control has been proposed in [114], on the basis of the resolved acceleration control concept [143]. Inverse dynamics control schemes in the operational space are given in [30]. For the extension to redundant manipulators see [102].

Problems

8.1. With reference to the block scheme with position feedback in Fig. 5.10, find the transfer functions of the forward path, the return path, and the closed-loop system.

8.2. With reference to the block scheme with position and velocity feedback in Fig. 5.11, find the transfer functions of the forward path, the return path, and the closed-loop system.

8.3. With reference to the block scheme with position, velocity and acceleration feedback in Fig. 8.9, find the transfer functions of the forward path, the return path, and the closed-loop system.

8.4. For a single joint drive system with the data: $I = 6\,\text{kg·m}^2$, $R_a = 0.3\,\text{ohm}$, $k_t = 0.5\,\text{N·m/A}$, $k_v = 0.5\,\text{V·s/rad}$, $F_m = 0.001\,\text{N·m·s/rad}$, find the parameters of the controller with position feedback (unit transducer constant) that yield a closed-loop response with damping ratio $\zeta \geq 0.4$. Discuss disturbance rejection properties.

8.5. For the drive system of Problem 8.4, find the parameters of the controller with position and velocity feedback (unit transducer constants) that yield a closed-loop response with damping ratio $\zeta \geq 0.4$ and natural frequency $\omega_n = 20\,\text{rad/s}$. Discuss disturbance rejection properties.

8.6. For the drive system of Problem 8.4, find the parameters of the controller with position, velocity and acceleration feedback (unit transducer constants) that yield a closed-loop response with damping ratio $\zeta \geq 0.4$, natural frequency $\omega_n = 20\,\text{rad/s}$ and disturbance rejection factor $X_R = 400$. Also, design a first-order filter that allows acceleration measurement reconstruction.

8.7. Verify that the control schemes in Figs. 8.12, 8.13, 8.14 correspond to realizing (8.42), (8.43), (8.44), respectively.

8.8. Verify that the standard regulation schemes in Figs. 8.15, 8.16, 8.17 are equivalent to the schemes in Figs. 8.12, 8.13, 8.14, respectively.

8.9. Prove inequality (8.76).

8.10. For the two-link planar arm with the same data as in Sect. 8.7, design a joint control of PD type with gravity compensation. By means of a computer simulation, verify stability for the following postures $\boldsymbol{q} = [\,\pi/4 \quad -\pi/2\,]^T$ and $\boldsymbol{q} = [\,-\pi \quad -3\pi/4\,]^T$, respectively. Implement the control in discrete-time with a sampling time of 1 ms.

8.11. For the two-link planar arm with the same data as in Sect. 8.7, under the assumption of a concentrated tip payload of mass $m_L = 10\,\text{kg}$, design an independent joint control with feedforward computed torque. Perform a computer simulation of the motion of the controlled arm along the joint space rectilinear path from $\boldsymbol{q}_i = [0 \quad \pi/4]^T$ to $\boldsymbol{q}_f = [\pi/2 \quad \pi/2]^T$ with a trapezoidal velocity profile and a trajectory duration $t_f = 1\,\text{s}$. Implement the control in discrete-time with a sampling time of 1 ms.

8.12. For the two-link planar arm of Problem 8.11, design an inverse dynamics joint control. Perform a computer simulation of the motion of the controlled arm along the trajectory specified in Problem 8.11. Implement the control in discrete-time with a sampling time of 1 ms.

8.13. For the two-link planar arm of Problem 8.11, design a robust joint control. Perform a computer simulation of the motion of the controlled arm along the trajectory specified in Problem 8.11. Implement the control in discrete-time with a sampling time of 1 ms.

8.14. For the two-link planar arm of Problem 8.11, design an adaptive joint control, on the basis of a suitable parameterization of the arm dynamic model. Perform a computer simulation of the motion of the controlled arm along the trajectory specified in Problem 8.11. Implement the control in discrete-time with a sampling time of 1 ms.

8.15. For the two-link planar of Problem 8.11, design a PD control in the operational space with gravity compensation. By means of a computer simulation, verify stability for the following postures $\boldsymbol{p} = [0.5 \quad 0.5]^T$ and $\boldsymbol{p} = [0.6 \quad -0.2]^T$, respectively. Implement the control in discrete-time with a sampling time of 1 ms.

8.16. For the two-link planar arm of Problem 8.11, design an inverse dynamics control in the operational space. Perform a computer simulation of the motion of the controlled arm along the operational space rectlinear path from $\boldsymbol{p}(0) = [0.7 \quad 0.2]^T$ to $\boldsymbol{p}(1) = [0.1 \quad -0.6]^T$ with a trapezoidal velocity profile and a trajectory duration $t_f = 1\,\text{s}$. Implement the control in discrete-time with a sampling time of 1 ms.

9

Force Control

One of the fundamental requirements for the success of a manipulation task is the capacity to handle *interaction between manipulator and environment*. The quantity that describes the state of interaction more effectively is the *contact force* at the manipulator's end-effector. High values of contact force are generally undesirable since they may stress both the manipulator and the manipulated object. In this chapter, performance of operational space motion control schemes is studied first, during the interaction of a manipulator with the environment. The concepts of mechanical *compliance* and *impedance* are introduced, with special regard to the problem of integrating contact force measurements into the control strategy. Then, *force control* schemes are presented which are obtained from motion control schemes suitably modified by the closure of an outer force regulation feedback loop. For the planning of control actions to perform an interaction task, *natural constraints* set by the task geometry and *artificial constraints* set by the control strategy are established; the constraints are referred to a suitable constraint frame. The formulation is conveniently exploited to derive *hybrid force/motion control* schemes.

9.1 Manipulator Interaction with Environment

Control of interaction between a robot manipulator and the environment is crucial for successful execution of a number of practical tasks where the robot's end-effector has to manipulate an object or perform some operation on a surface. Typical examples include polishing, deburring, machining or assembly. A complete classification of possible robot tasks is practically infeasible in view of the large variety of cases that may occur, nor would such a classification be really useful to find a general strategy to *interaction control* with the environment.

During the interaction, the environment sets constraints on the geometric paths that can be followed by the end-effector. This situation is generally referred to as *constrained motion*. In such a case, the use of a purely motion control strategy for controlling interaction is a candidate to fail, as explained below.

Successful execution of an interaction task with the environment by using motion control could be obtained only if the task were accurately planned. This would, in turn, require an accurate model of both the robot manipulator (kinematics and dynamics) and the environment (geometry and mechanical features). Manipulator modelling can be achieved with enough precision, but a detailed description of the environment is difficult to obtain.

To understand the importance of task planning accuracy, it is sufficient to observe that to perform a mechanical part mating with a positional approach, the relative positioning of the parts should be guaranteed with an accuracy of an order of magnitude greater than part mechanical tolerance. Once the absolute position of one part is exactly known, the manipulator should guide the motion of the other with the same accuracy.

In practice, the planning errors may give rise to a contact force causing a deviation of the end-effector from the desired trajectory. On the other hand, the control system reacts to reduce such deviation. This ultimately leads to a build-up of the contact force until saturation of the joint actuators is reached or breakage of the parts in contact occurs.

The higher the environment stiffness and position control accuracy, the more likely a situation like the one just described can occur. This drawback can be overcome if compliant behaviour is ensured during the interaction.

From the above discussion it should be clear that the *contact force* is the quantity describing the state of interaction in the most complete fashion; to this end, the availability of force measurements is expected to provide enhanced performance for controlling interaction.

Interaction control strategies can be grouped in two categories; those performing *indirect force control* and those performing *direct force control*. The main difference between the two categories is that the former achieve force control via motion control, without explicit closure of a force feedback loop; the latter, instead, offer the possibility of controlling the contact force to a desired value, thanks to the closure of a force feedback loop. To the first category belong *compliance control* and *impedance control* which are treated next. Then, *force control* and *hybrid force/motion control* schemes will follow.

9.2 Compliance Control

For a detailed analysis of interaction between the manipulator and environment it is worth considering the behaviour of the system under a position control scheme when contact forces arise. Since these are naturally described in the operational space, it is convenient to refer to *operational space control* schemes.

Consider the manipulator dynamic model (8.7). In view of (7.42), the model can be written as

$$\boldsymbol{B}(\boldsymbol{q})\ddot{\boldsymbol{q}} + \boldsymbol{C}(\boldsymbol{q}, \dot{\boldsymbol{q}})\dot{\boldsymbol{q}} + \boldsymbol{F}\dot{\boldsymbol{q}} + \boldsymbol{g}(\boldsymbol{q}) = \boldsymbol{u} - \boldsymbol{J}^T(\boldsymbol{q})\boldsymbol{h}_e \tag{9.1}$$

where $\boldsymbol{h}_e$ is the vector of contact forces exerted by the manipulator's end-effector on the environment.[1]

It is reasonable to predict that, in the case $\boldsymbol{h}_e \neq \boldsymbol{0}$, the control scheme based on (8.110) no longer ensures that the end-effector reaches its desired pose $\boldsymbol{x}_d$. In fact, by recalling that $\widetilde{\boldsymbol{x}} = \boldsymbol{x}_d - \boldsymbol{x}_e$, where $\boldsymbol{x}_e$ denotes the end-effector pose, at the equilibrium it is

$$\boldsymbol{J}_A^T(\boldsymbol{q})\boldsymbol{K}_P\widetilde{\boldsymbol{x}} = \boldsymbol{J}^T(\boldsymbol{q})\boldsymbol{h}_e. \tag{9.2}$$

On the assumption of a full-rank Jacobian, one has

$$\widetilde{\boldsymbol{x}} = \boldsymbol{K}_P^{-1}\boldsymbol{T}_A^T(\boldsymbol{x}_e)\boldsymbol{h}_e = \boldsymbol{K}_P^{-1}\boldsymbol{h}_A \tag{9.3}$$

where $\boldsymbol{h}_A$ is the vector of equivalent forces that can be defined according to (7.128). The expression in (9.3) shows that at the equilibrium the manipulator, under a pose control action, behaves like a generalized spring in the operational space with *compliance* $\boldsymbol{K}_P^{-1}$ in respect of the equivalent force $\boldsymbol{h}_A$. By recalling the expression of the transformation matrix $\boldsymbol{T}_A$ in (3.65) and assuming matrix $\boldsymbol{K}_P$ to be diagonal, it can be recognized that linear compliance (due to force components) is independent of the configuration, whereas torsional compliance (due to moment components) does depend on the current end-effector orientation through the matrix $\boldsymbol{T}$.

On the other hand, if $\boldsymbol{h}_e \in \mathcal{N}(\boldsymbol{J}^T)$, one has $\widetilde{\boldsymbol{x}} = \boldsymbol{0}$ with $\boldsymbol{h}_e \neq \boldsymbol{0}$, namely contact forces are completely balanced by the manipulator mechanical structure; for instance, the anthropomorphic manipulator at a shoulder singularity in Fig. 3.13 does not react to any force orthogonal to the plane of the structure.

Equation (9.3) can be rewritten in the form

$$\boldsymbol{h}_A = \boldsymbol{K}_P\widetilde{\boldsymbol{x}} \tag{9.4}$$

where $\boldsymbol{K}_P$ represents a *stiffness* matrix as regards the vector of the equivalent forces $\boldsymbol{h}_A$. It is worth observing that the compliant (or stiff) behaviour of the manipulator is achieved by virtue of the control. This behaviour is termed *active compliance* whereas the term *passive compliance* denotes mechanical systems with a prevalent dynamics of elastic type.

For a better understanding of the interaction between manipulator and environment, it is necessary to analyze further the concept of passive compliance.

[1] In this chapter the term force, in general, is referred to a (6×1) vector of force and moment, unless otherwise specified.

9.2.1 Passive Compliance

Consider two elastically coupled rigid bodies R and S and two reference frames, each attached to one of the two bodies so that at equilibrium, in the absence of interaction forces and moments, the two frames coincide. Let $d\boldsymbol{x}_{r,s}$ denote an elementary displacement from the equilibrium of frame s with respect to frame r, defined as

$$d\boldsymbol{x}_{r,s} = \begin{bmatrix} d\boldsymbol{o}_{r,s} \\ \boldsymbol{\omega}_{r,s}dt \end{bmatrix} = \boldsymbol{v}_{r,s}dt \tag{9.5}$$

where $\boldsymbol{v}_{r,s} = \boldsymbol{v}_s - \boldsymbol{v}_r$ is the vector of linear and angular velocity of frame s with respect to frame r, $d\boldsymbol{o}_{r,s} = \boldsymbol{o}_s - \boldsymbol{o}_r$ is the vector corresponding to the translation of the origin $\boldsymbol{o}_s$ of frame s with respect to the origin $\boldsymbol{o}_r$ of frame r and $\boldsymbol{\omega}_{r,s}dt$, with $\boldsymbol{\omega}_{r,s} = \boldsymbol{\omega}_s - \boldsymbol{\omega}_r$, represents the vector of small rotations of frame s about the axes of frame r as in (3.106). This elementary displacement is assumed to be equivalently referred to frame r or s because, at the equilibrium, the two frames coincide; therefore, the reference frame was not explicitly denoted.

To the displacement $d\boldsymbol{x}_{r,s}$, coinciding with the deformation of the spring between R and S, it corresponds the elastic force

$$\boldsymbol{h}_s = \begin{bmatrix} \boldsymbol{f}_s \\ \boldsymbol{\mu}_s \end{bmatrix} = \begin{bmatrix} \boldsymbol{K}_f & \boldsymbol{K}_c \\ \boldsymbol{K}_c^T & \boldsymbol{K}_\mu \end{bmatrix} \begin{bmatrix} d\boldsymbol{o}_{r,s} \\ \boldsymbol{\omega}_{r,s}dt \end{bmatrix} = \boldsymbol{K}d\boldsymbol{x}_{r,s}, \tag{9.6}$$

applied by body S on the spring and referred equivalently to one of the two reference frames. In view of the action-reaction law, the force applied by R has the expression $\boldsymbol{h}_r = -\boldsymbol{h}_s = \boldsymbol{K}d\boldsymbol{x}_{s,r}$, being $d\boldsymbol{x}_{s,r} = -d\boldsymbol{x}_{r,s}$.

The (6×6) matrix $\boldsymbol{K}$ represents a *stiffness matrix*, which is symmetric and *positive semi-definite*. The (3×3) matrices $\boldsymbol{K}_f$ and $\boldsymbol{K}_\mu$ are known as *translational stiffness* and *rotational stiffness*, respectively. The (3×3) matrix $\boldsymbol{K}_c$ is known as *coupling stiffness*. An analogous decomposition can be made for the *compliance matrix* $\boldsymbol{C}$ in the mapping

$$d\boldsymbol{x}_{r,s} = \boldsymbol{C}\boldsymbol{h}_s. \tag{9.7}$$

In the real elastic systems, matrix $\boldsymbol{K}_c$ is, in general, non-symmetric. However, there are special devices, such as the RCC (*Remote Centre of Compliance*), where $\boldsymbol{K}_c$ can be symmetric or null. These are elastically compliant mechanical devices, suitably designed to achieve maximum decoupling between translation and rotation, that are interposed between the manipulator last link and the end-effector. The aim is that of introducing a *passive compliance* of desired value to facilitate the execution of assembly tasks. For instance, in a peg-in-hole insertion task, the gripper is provided with a device ensuring high stiffness along the insertion direction and high compliance along the other directions. Therefore, in the presence of unavoidable position displacements from the planned insertion trajectory, contact forces and moments arise which modify the peg position so as to facilitate insertion.

The inconvenience of such devices is their low versatility to different operating conditions and generic interaction tasks, namely, whenever a modification of the compliant mechanical hardware is required.

9.2.2 Active Compliance

The aim of *compliance control* is that of achieving a suitable *active compliance* that can be easily modified acting on the control software so as to satisfy the requirements of different interaction tasks.

Notice that the equilibrium equations in (9.3) and (9.4) show that the compliant behaviour with respect to $\boldsymbol{h}_e$ depends on the actual end-effector orientation, also for elementary displacements, so that, in practice, the selection of stiffness parameters is quite difficult. To obtain an equilibrium equation of the form (9.6), a different definition of error in the operational space must be considered.

Let O_e–$x_e y_e z_e$ and O_d–$x_d y_d z_d$ denote the end-effector frame and the desired frame respectively. The corresponding homogeneous transformation matrices are

$$\boldsymbol{T}_e = \begin{bmatrix} \boldsymbol{R}_e & \boldsymbol{o}_e \\ \boldsymbol{0}^T & 1 \end{bmatrix} \qquad \boldsymbol{T}_d = \begin{bmatrix} \boldsymbol{R}_d & \boldsymbol{o}_d \\ \boldsymbol{0}^T & 1 \end{bmatrix},$$

with obvious meaning of notation. The position and orientation displacement of the end-effector frame with respect to the desired frame can be expressed in terms of the homogeneous transformation matrix

$$\boldsymbol{T}_e^d = (\boldsymbol{T}_d)^{-1}\boldsymbol{T}_e = \begin{bmatrix} \boldsymbol{R}_e^d & \boldsymbol{o}_{d,e}^d \\ \boldsymbol{0}^T & 1 \end{bmatrix}, \tag{9.8}$$

where $\boldsymbol{R}_e^d = \boldsymbol{R}_d^T \boldsymbol{R}_e$ and $\boldsymbol{o}_{d,e}^d = \boldsymbol{R}_d^T(\boldsymbol{o}_e - \boldsymbol{o}_d)$. The new error vector in the operational space can be defined as

$$\widetilde{\boldsymbol{x}} = -\begin{bmatrix} \boldsymbol{o}_{d,e}^d \\ \boldsymbol{\phi}_{d,e} \end{bmatrix} \tag{9.9}$$

where $\boldsymbol{\phi}_{d,e}$ is the vector of Euler angles extracted from the rotation matrix $\boldsymbol{R}_e^d$. The minus sign in (9.9) depends on the fact that, for control purposes, the error is usually defined as the difference between the desired and the measured quantities.

Computing the time derivative of $\boldsymbol{o}_{d,e}^d$ and taking into account (3.10), (3.11) gives

$$\dot{\boldsymbol{o}}_{d,e}^d = \boldsymbol{R}_d^T(\dot{\boldsymbol{o}}_e - \dot{\boldsymbol{o}}_d) - \boldsymbol{S}(\boldsymbol{\omega}_d^d)\boldsymbol{R}_d^T(\boldsymbol{o}_e - \boldsymbol{o}_d). \tag{9.10}$$

On the other hand, computing the time derivative of $\boldsymbol{\phi}_{d,e}$ and taking into account (3.64), yields (see Problem 9.1)

$$\dot{\boldsymbol{\phi}}_{d,e} = \boldsymbol{T}^{-1}(\boldsymbol{\phi}_{d,e})\boldsymbol{\omega}_{d,e}^d = \boldsymbol{T}^{-1}(\boldsymbol{\phi}_{d,e})\boldsymbol{R}_d^T(\boldsymbol{\omega}_e - \boldsymbol{\omega}_d). \tag{9.11}$$

Considering that the desired quantities $\boldsymbol{o}_d$ and $\boldsymbol{R}_d$ are constant, vector $\dot{\widetilde{\boldsymbol{x}}}$ can be expressed in the form

$$\dot{\widetilde{\boldsymbol{x}}} = -\boldsymbol{T}_A^{-1}(\boldsymbol{\phi}_{d,e}) \begin{bmatrix} \boldsymbol{R}_d^T & \boldsymbol{O} \\ \boldsymbol{O} & \boldsymbol{R}_d^T \end{bmatrix} \boldsymbol{v}_e \tag{9.12}$$

being $\boldsymbol{v}_e = [\,\dot{\boldsymbol{o}}_e^T \quad \boldsymbol{\omega}_e^T\,]^T = \boldsymbol{J}(\boldsymbol{q})\dot{\boldsymbol{q}}$ the vector of linear and angular velocity of the end-effector. Therefore

$$\dot{\widetilde{\boldsymbol{x}}} = -\boldsymbol{J}_{A_d}(\boldsymbol{q}, \widetilde{\boldsymbol{x}})\dot{\boldsymbol{q}}, \tag{9.13}$$

where the matrix

$$\boldsymbol{J}_{A_d}(\boldsymbol{q}, \widetilde{\boldsymbol{x}}) = \boldsymbol{T}_A^{-1}(\boldsymbol{\phi}_{d,e}) \begin{bmatrix} \boldsymbol{R}_d^T & \boldsymbol{O} \\ \boldsymbol{O} & \boldsymbol{R}_d^T \end{bmatrix} \boldsymbol{J}(\boldsymbol{q}) \tag{9.14}$$

represents the analytic Jacobian corresponding to the definition (9.9) of error in the operational space.

The PD control with gravity compensation analogous to (8.110), with the definition (9.9) of error in the operational space, has the expression

$$\boldsymbol{u} = \boldsymbol{g}(\boldsymbol{q}) + \boldsymbol{J}_{A_d}^T(\boldsymbol{q}, \widetilde{\boldsymbol{x}})(\boldsymbol{K}_P\widetilde{\boldsymbol{x}} - \boldsymbol{K}_D\boldsymbol{J}_{A_d}(\boldsymbol{q}, \widetilde{\boldsymbol{x}})\dot{\boldsymbol{q}}). \tag{9.15}$$

Notice that, in the case where the operational space is defined only by the position components, the control laws (8.110) and (9.15) differ only because the position error (and the corresponding derivative term) is referred to the base frame in (8.110), while it is referred to the desired frame in (9.15).

In the absence of interaction, the asymptotic stability of the equilibrium pose corresponding to $\widetilde{\boldsymbol{x}} = \boldsymbol{0}$, assuming that $\boldsymbol{K}_P$ and $\boldsymbol{K}_D$ are symmetric and positive definite matrices, can be proven using the Lyapunov function

$$V(\dot{\boldsymbol{q}}, \widetilde{\boldsymbol{x}}) = \frac{1}{2}\dot{\boldsymbol{q}}^T\boldsymbol{B}(\boldsymbol{q})\dot{\boldsymbol{q}} + \frac{1}{2}\widetilde{\boldsymbol{x}}^T\boldsymbol{K}_P\widetilde{\boldsymbol{x}} > 0 \qquad \forall \dot{\boldsymbol{q}}, \widetilde{\boldsymbol{x}} \neq \boldsymbol{0},$$

as for the case of the control law (8.110).

In the presence of interaction with the environment, at the equilibrium it is

$$\boldsymbol{J}_{A_d}^T(\boldsymbol{q})\boldsymbol{K}_P\widetilde{\boldsymbol{x}} = \boldsymbol{J}^T(\boldsymbol{q})\boldsymbol{h}_e; \tag{9.16}$$

hence, assuming a full-rank Jacobian, the following equality holds:

$$\boldsymbol{h}_e^d = \boldsymbol{T}_A^{-T}(\boldsymbol{\phi}_{d,e})\boldsymbol{K}_P\widetilde{\boldsymbol{x}}. \tag{9.17}$$

The above equation, to be compared to the elastic model (9.6), must be rewritten in terms of elementary displacements. To this end, taking into account (9.12) and (9.5), it is

$$d\widetilde{\boldsymbol{x}} = \dot{\widetilde{\boldsymbol{x}}}\Big|_{\widetilde{\boldsymbol{x}}=\boldsymbol{0}} dt = \boldsymbol{T}_A^{-1}(\boldsymbol{0})(\boldsymbol{v}_d^d - \boldsymbol{v}_e^d)dt = \boldsymbol{T}_A^{-1}(\boldsymbol{0})d\boldsymbol{x}_{e,d} \tag{9.18}$$

where $d\boldsymbol{x}_{e,d}$ is the elementary displacement of the desired frame with respect to the end-effector frame about the equilibrium, referred to any of the two frames. The value of $\boldsymbol{T}_A(\boldsymbol{0})$ depends on the particular choice of Euler angles; in the following, angles XYZ are adopted, for which $\boldsymbol{T}_A(\boldsymbol{0}) = \boldsymbol{I}$ (see Problem 3.13). Therefore, rewriting (9.17) in terms of elementary displacements gives

$$\boldsymbol{h}_e = \boldsymbol{K}_P d\boldsymbol{x}_{e,d}, \tag{9.19}$$

which is formally identical to (9.6), where vectors are assumed to be referred to the desired frame or to the end-effector frame, equivalently. It follows that matrix $\boldsymbol{K}_P$ has the meaning of an active stiffness corresponding to a generalized spring acting between the end-effector frame and the desired frame. Equation (9.19) can be rewritten in the equivalent form

$$d\boldsymbol{x}_{e,d} = \boldsymbol{K}_P^{-1}\boldsymbol{h}_e, \tag{9.20}$$

showing that $\boldsymbol{K}_P^{-1}$ corresponds to an active compliance.

The selection of the elements of matrix $\boldsymbol{K}_P$ must be made taking into account geometry and mechanical features of the environment. To this end, assume that the interaction force between the end-effector and the environment derives from a generalized spring acting between the end-effector frame and a reference frame O_r–$x_r y_r z_r$ attached to the environment rest position. Considering an elementary displacement $d\boldsymbol{x}_{r,e}$ between the two reference frames, the corresponding elastic force applied by the end-effector is

$$\boldsymbol{h}_e = \boldsymbol{K} d\boldsymbol{x}_{r,e} \tag{9.21}$$

with a stiffness matrix $\boldsymbol{K}$, where vectors can be referred, equivalently, to the frame attached to the rest position of the environment or to the end-effector frame. Typically, the stiffness matrix is positive semi-definite because, in general, the interaction forces and moments belong to some particular directions, spanning $\mathcal{R}(\boldsymbol{K})$.

In view of the model (9.21), of (9.19) and of the equality

$$d\boldsymbol{x}_{r,e} = d\boldsymbol{x}_{r,d} - d\boldsymbol{x}_{e,d},$$

the following expression of the contact force at equilibrium can be found:

$$\boldsymbol{h}_e = \left(\boldsymbol{I}_6 + \boldsymbol{K}\boldsymbol{K}_P^{-1}\right)^{-1}\boldsymbol{K} d\boldsymbol{x}_{r,d}. \tag{9.22}$$

Substituting this expression into (9.20) yields

$$d\boldsymbol{x}_{e,d} = \boldsymbol{K}_P^{-1}\left(\boldsymbol{I}_6 + \boldsymbol{K}\boldsymbol{K}_P^{-1}\right)^{-1}\boldsymbol{K} d\boldsymbol{x}_{r,d}, \tag{9.23}$$

representing the pose error of the end-effector at the equilibrium.

Notice that vectors in (9.22) and (9.23) can be referred, equivalently, to the end-effector frame, to the desired frame or to the frame attached to the environment rest position; these frames coincide at equilibrium (see Problem 9.2).

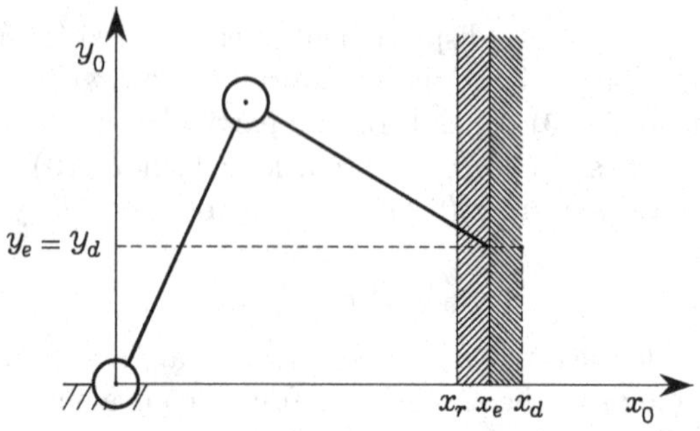

Fig. 9.1. Two-link planar arm in contact with an elastically compliant plane

The analysis of (9.23) shows that the end-effector pose error at the equilibrium depends on the environment rest position, as well as on the desired pose imposed by the control system of the manipulator. The interaction of the two systems (environment and manipulator) is influenced by the mutual weight of the respective compliance features.

In fact, it is possible to modify the active compliance $\boldsymbol{K}_P^{-1}$ so that the manipulator dominates the environment and vice versa. Such a dominance can be specified with reference to the single directions of the operational space.

For a given environment stiffness $\boldsymbol{K}$, according to the prescribed interaction task, one may choose large values of the elements of $\boldsymbol{K}_P$ for those directions along which the environment has to comply and small values of the elements of $\boldsymbol{K}_P$ for those directions along which the manipulator has to comply. As a consequence, the manipulator pose error $d\boldsymbol{x}_{e,d}$ tends to zero along the directions where the environment complies; vice versa, along the directions where the manipulator complies, the end-effector pose tends to the rest pose of the environment, namely $d\boldsymbol{x}_{e,d} \simeq d\boldsymbol{x}_{r,d}$.

Equation (9.22) gives the value of the contact force at the equilibrium. This expression reveals that, along the directions where the manipulator stiffness is much higher than the environment stiffness, the intensity of the elastic force mainly depends on the stiffness of the environment and on the displacement $d\boldsymbol{x}_{r,e}$ between the equilibrium pose of the end-effector (which practically coincides with the desired pose) and the rest pose of the environment. In the dual case that the environment stiffness is much higher than the manipulator stiffness, the intensity of the elastic force mainly depends on the manipulator stiffness and on the displacement $d\boldsymbol{x}_{e,d}$ between the desired pose and the equilibrium pose of the end-effector (which practically coincides with the rest pose of the environment).

Example 9.1

Consider the two-link planar arm whose tip is in contact with a purely frictionless elastic plane; due to the simple geometry of the problem, involving only position variables, all the quantities can be conveniently referred to the base frame. Thus, control law (8.110) will be adopted. Let $\boldsymbol{o}_r = [\,x_r \quad 0\,]^T$ denote the equilibrium position of the plane, which is assumed to be orthogonal to axis x_0 (Fig. 9.1). The environment stiffness matrix is

$$\boldsymbol{K} = \boldsymbol{K}_f = \mathrm{diag}\{k_x, 0\},$$

corresponding to the absence of interaction forces along the vertical direction ($\boldsymbol{f}_e = [\,f_x \quad 0\,]^T$). Let $\boldsymbol{o}_e = [\,x_e \quad y_e\,]^T$ be the end-effector position and $\boldsymbol{o}_d = [\,x_d \quad y_d\,]^T$ be the desired position, which is located beyond the contact plane. The proportional control action on the arm is characterized by

$$\boldsymbol{K}_P = \mathrm{diag}\{k_{Px}, k_{Py}\}.$$

The equilibrium equations for the force and position (9.22), (9.23), rewritten with $d\boldsymbol{x}_{r,d} = \boldsymbol{o}_d - \boldsymbol{o}_r$ and $d\boldsymbol{x}_{e,d} = \boldsymbol{o}_d - \boldsymbol{o}_e$, give

$$\boldsymbol{f}_e = \begin{bmatrix} \dfrac{k_{Px}k_x}{k_{Px}+k_x}(x_d - x_r) \\ 0 \end{bmatrix} \qquad \boldsymbol{o}_e = \begin{bmatrix} \dfrac{k_{Px}x_d + k_x x_r}{k_{Px}+k_x} \\ y_d \end{bmatrix}.$$

With reference to positioning accuracy, the arm tip reaches the vertical coordinate y_d, since the vertical motion direction is not constrained. As for the horizontal direction, the presence of the elastic plane imposes that the arm can move as far as it reaches the coordinate $x_e < x_d$. The value of the horizontal contact force at the equilibrium is related to the difference between x_d and x_r by an equivalent stiffness coefficient which is given by the parallel composition of the stiffness coefficients of the two interacting systems. Hence, the arm stiffness and environment stiffness influence the resulting equilibrium configuration. In the case when

$$k_{Px}/k_x \gg 1,$$

it is

$$x_e \approx x_d \qquad f_x \approx k_x(x_d - x_r)$$

and thus the arm prevails over the environment, in that the plane complies almost up to x_d and the elastic force is mainly generated by the environment (passive compliance). In the opposite case

$$k_{Px}/k_x \ll 1,$$

it is

$$x_e \approx x_r \qquad f_x \approx k_{Px}(x_d - x_r)$$

and thus the environment prevails over the arm which complies up to the equilibrium x_r, and the elastic force is mainly generated by the arm (active compliance).

To complete the analysis of manipulator compliance in contact with environment, it is worth considering the effects of a joint space position control law. With reference to (8.51), in the presence of end-effector contact forces, the equilibrium posture is determined by

$$\boldsymbol{K}_P\widetilde{\boldsymbol{q}} = \boldsymbol{J}^T(\boldsymbol{q})\boldsymbol{h}_e \tag{9.24}$$

and then

$$\widetilde{\boldsymbol{q}} = \boldsymbol{K}_P^{-1}\boldsymbol{J}^T(\boldsymbol{q})\boldsymbol{h}_e. \tag{9.25}$$

On the assumption of small displacements from the equilibrium, it is reasonable to compute the resulting operational space displacement as $d\widetilde{\boldsymbol{x}} \approx \boldsymbol{J}(\boldsymbol{q})d\widetilde{\boldsymbol{q}}$, referred to the base frame. Therefore, in view of (9.25) it is

$$d\widetilde{\boldsymbol{x}} = \boldsymbol{J}(\boldsymbol{q})\boldsymbol{K}_P^{-1}\boldsymbol{J}^T(\boldsymbol{q})\boldsymbol{h}_e, \tag{9.26}$$

corresponding to an active compliance referred to the base frame. Notice that the compliance matrix $\boldsymbol{J}(\boldsymbol{q})\boldsymbol{K}_P^{-1}\boldsymbol{J}^T(\boldsymbol{q})$ depends on the manipulator posture, both for the force and moment components. Also in this case, the occurrence of manipulator Jacobian singularities is to be analyzed apart.

9.3 Impedance Control

It is now desired to analyze the interaction of a manipulator with the environment under the action of an inverse dynamics control in the operational space. With reference to model (9.1), consider the control law (8.57)

$$\boldsymbol{u} = \boldsymbol{B}(\boldsymbol{q})\boldsymbol{y} + \boldsymbol{n}(\boldsymbol{q}, \dot{\boldsymbol{q}}),$$

with $\boldsymbol{n}$ as in (8.56). In the presence of end-effector forces, the controlled manipulator is described by

$$\ddot{\boldsymbol{q}} = \boldsymbol{y} - \boldsymbol{B}^{-1}(\boldsymbol{q})\boldsymbol{J}^T(\boldsymbol{q})\boldsymbol{h}_e \tag{9.27}$$

that reveals the existence of a nonlinear coupling term due to contact forces. Choose $\boldsymbol{y}$ in a way conceptually analogous to (8.114), as

$$\boldsymbol{y} = \boldsymbol{J}_A^{-1}(\boldsymbol{q})\boldsymbol{M}_d^{-1}\big(\boldsymbol{M}_d\ddot{\boldsymbol{x}}_d + \boldsymbol{K}_D\dot{\widetilde{\boldsymbol{x}}} + \boldsymbol{K}_P\widetilde{\boldsymbol{x}} - \boldsymbol{M}_d\dot{\boldsymbol{J}}_A(\boldsymbol{q}, \dot{\boldsymbol{q}})\dot{\boldsymbol{q}}\big) \tag{9.28}$$

where $\boldsymbol{M}_d$ is a positive definite diagonal matrix. Substituting (9.28) into (9.27) and accounting for second-order differential kinematics in the form (3.98), yields

$$\boldsymbol{M}_d\ddot{\widetilde{\boldsymbol{x}}} + \boldsymbol{K}_D\dot{\widetilde{\boldsymbol{x}}} + \boldsymbol{K}_P\widetilde{\boldsymbol{x}} = \boldsymbol{M}_d\boldsymbol{B}_A^{-1}(\boldsymbol{q})\boldsymbol{h}_A, \tag{9.29}$$

where

$$\boldsymbol{B}_A(\boldsymbol{q}) = \boldsymbol{J}_A^{-T}(\boldsymbol{q})\boldsymbol{B}(\boldsymbol{q})\boldsymbol{J}_A^{-1}(\boldsymbol{q})$$

is the inertia matrix of the manipulator in the operational space as in (7.134); this matrix is configuration-dependent and is positive definite if $\boldsymbol{J}_A$ has full rank.

The expression in (9.29) establishes a relationship through a generalized *mechanical impedance* between the vector of forces $\boldsymbol{M}_d\boldsymbol{B}_A^{-1}\boldsymbol{h}_A$ and the vector of displacements $\widetilde{\boldsymbol{x}}$ in the operational space. This impedance can be attributed to a mechanical system characterized by a mass matrix $\boldsymbol{M}_d$, a damping matrix $\boldsymbol{K}_D$, and a stiffness matrix $\boldsymbol{K}_P$, which can be used to specify the dynamic behaviour along the operational space directions.

The presence of $\boldsymbol{B}_A^{-1}$ makes the system coupled. If it is wished to keep linearity and decoupling during interaction with the environment, it is then necessary to *measure* the contact *force*; this can be achieved by means of appropriate force sensors which are usually mounted to the manipulator wrist, as discussed in Sect. 5.4.1. Choosing

$$\boldsymbol{u} = \boldsymbol{B}(\boldsymbol{q})\boldsymbol{y} + \boldsymbol{n}(\boldsymbol{q},\dot{\boldsymbol{q}}) + \boldsymbol{J}^T(\boldsymbol{q})\boldsymbol{h}_e \tag{9.30}$$

with

$$\boldsymbol{y} = \boldsymbol{J}_A^{-1}(\boldsymbol{q})\boldsymbol{M}_d^{-1}\big(\boldsymbol{M}_d\ddot{\boldsymbol{x}}_d + \boldsymbol{K}_D\dot{\widetilde{\boldsymbol{x}}} + \boldsymbol{K}_P\widetilde{\boldsymbol{x}} - \boldsymbol{M}_d\dot{\boldsymbol{J}}_A(\boldsymbol{q},\dot{\boldsymbol{q}})\dot{\boldsymbol{q}} - \boldsymbol{h}_A\big), \tag{9.31}$$

under the assumption of error-free force measurements, yields

$$\boldsymbol{M}_d\ddot{\widetilde{\boldsymbol{x}}} + \boldsymbol{K}_D\dot{\widetilde{\boldsymbol{x}}} + \boldsymbol{K}_P\widetilde{\boldsymbol{x}} = \boldsymbol{h}_A. \tag{9.32}$$

It is worth noticing that the addition of the term $\boldsymbol{J}^T\boldsymbol{h}_e$ in (9.30) exactly compensates the contact forces and then it renders the manipulator infinitely stiff as regards the external stress. In order to confer a compliant behaviour to the manipulator, the term $-\boldsymbol{J}_A^{-1}\boldsymbol{M}_d^{-1}\boldsymbol{h}_A$ has been introduced in (9.31) so that the manipulator can be characterized as a *linear impedance* with regard to the equivalent forces $\boldsymbol{h}_A$, as shown in (9.32).

The behaviour of the system in (9.32) at the equilibrium is analogous to that described by (9.4); nonetheless, compared to a control with active compliance specified by $\boldsymbol{K}_P^{-1}$, the equation in (9.32) allows a complete characterization of system dynamics through an *active impedance* specified by matrices $\boldsymbol{M}_d$, $\boldsymbol{K}_D$, $\boldsymbol{K}_P$. Also in this case, it is not difficult to recognize that, as regards $\boldsymbol{h}_e$, impedance depends on the current end-effector orientation through the matrix $\boldsymbol{T}$. Therefore, the selection of the impedance parameters becomes difficult; moreover, an inadequate behaviour may occur in the neighbourhood of representation singularities.

To avoid this problem it is sufficient to redesign the control input $\boldsymbol{y}$ as a function of the operational space error (9.9).

Under the assumption that the desired frame O_d–$x_dy_dz_d$ is time-varying, in view of (9.10), (9.11), the time derivative of (9.9) has the expression

$$\dot{\widetilde{\boldsymbol{x}}} = -\boldsymbol{J}_{A_d}(\boldsymbol{q},\widetilde{\boldsymbol{x}})\dot{\boldsymbol{q}} + \boldsymbol{b}(\widetilde{\boldsymbol{x}},\boldsymbol{R}_d,\dot{\boldsymbol{o}}_d,\boldsymbol{\omega}_d), \tag{9.33}$$

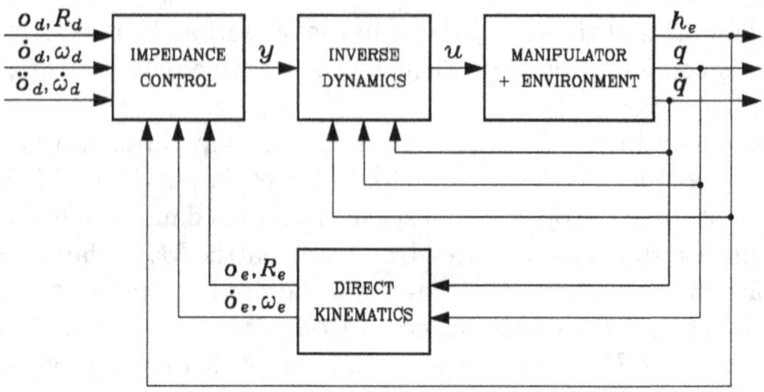

Fig. 9.2. Block scheme of impedance control

where J_{A_d} is the analytic Jacobian (9.14) and vector b is

$$b(\widetilde{x}, R_d, \dot{o}_d, \omega_d) = \begin{bmatrix} R_d^T \dot{o}_d + S(\omega_d^d) o_{d,e}^d \\ T^{-1}(\phi_{d,e}) \omega_d^d \end{bmatrix}.$$

Computing the time derivative of (9.33) yields

$$\ddot{\widetilde{x}} = -J_{A_d}\ddot{q} - \dot{J}_{A_d}\dot{q} + \dot{b}, \tag{9.34}$$

where, for simplicity, the dependence of the functions on their arguments was omitted. As a consequence, using (9.30) with

$$y = J_{A_d}^{-1} M_d^{-1} (K_D \dot{\widetilde{x}} + K_P \widetilde{x} - M_d \dot{J}_{A_d} \dot{q} + M_d \dot{b} - h_e^d), \tag{9.35}$$

yields the equation

$$M_d \ddot{\widetilde{x}} + K_D \dot{\widetilde{x}} + K_P \widetilde{x} = h_e^d, \tag{9.36}$$

where all the vectors are referred to the desired frame. This equation represents a linear impedance as regards the force vector h_e^d, independent from the manipulator configuration.

The block scheme representing impedance control is reported in Fig. 9.2.

Similar to active and passive compliance, the concept of *passive impedance* can be introduced if the interaction force h_e is generated at the contact with an environment of proper mass, damping and stiffness. In this case, the system of manipulator with environment can be regarded as a mechanical system constituted by the parallel of the two impedances, and then its dynamic behaviour is conditioned by the relative weight between them.

Example 9.2

Consider the planar arm in contact with an elastically compliant plane of the previous example. Due to the simple geometry of the problem, involving only position

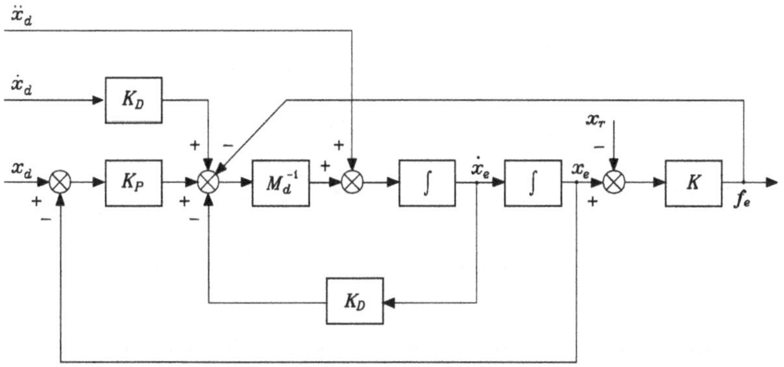

Fig. 9.3. Equivalent block scheme of a manipulator in contact with an elastic environment under impedance control

variables, all the quantities can be conveniently referred to the base frame. Thus, the impedance control law with force measurement (9.30), (9.31) will be adopted. Moreover, $\boldsymbol{x}_d = \boldsymbol{o}_d$, $\widetilde{\boldsymbol{x}} = \boldsymbol{o}_d - \boldsymbol{o}_e$, $\boldsymbol{h}_A = \boldsymbol{f}_e$ and

$$\boldsymbol{M}_d = \mathrm{diag}\{m_{dx}, m_{dy}\} \qquad \boldsymbol{K}_D = \mathrm{diag}\{k_{Dx}, k_{Dy}\} \qquad \boldsymbol{K}_P = \mathrm{diag}\{k_{Px}, k_{Py}\}.$$

The block scheme of the manipulator in contact with an elastic environment under impedance control is represented in Fig. 9.3, where $\boldsymbol{x}_e = \boldsymbol{o}_e$ and $\boldsymbol{x}_r = \boldsymbol{o}_r$.

If $\boldsymbol{x}_d$ is constant, the dynamics of the manipulator and environment system along the two directions of the operational space is described by

$$m_{dx}\ddot{x}_e + k_{Dx}\dot{x}_e + (k_{Px} + k_x)x_e = k_x x_r + k_{Px}x_d$$
$$m_{dy}\ddot{y}_e + k_{Dy}\dot{y}_e + k_{Py}y_e = k_{Py}y_d.$$

Along the vertical direction, one has an unconstrained motion whose time behaviour is determined by the following natural frequency and damping factor:

$$\omega_{ny} = \sqrt{\frac{k_{Py}}{m_{dy}}} \qquad \zeta_y = \frac{k_{Dy}}{2\sqrt{m_{dy}k_{Py}}},$$

while, along the horizontal direction, the behaviour of the contact force $f_x = k_x(x_e - x_r)$ is determined by

$$\omega_{nx} = \sqrt{\frac{k_{Px} + k_x}{m_{dx}}} \qquad \zeta_x = \frac{k_{Dx}}{2\sqrt{m_{dx}(k_{Px} + k_x)}}.$$

Below, the dynamic behaviour of the system is analyzed for two different values of environment compliance: $k_x = 10^3$ N/m and $k_x = 10^4$ N/m. The rest position of the environment is $x_r = 1$. The actual arm is the same as in Example 7.2. Apply an impedance control with force measurement of the kind (9.30), (9.31), and PD control actions equivalent to those chosen in the simulations of Sect. 8.7, namely

$$m_{dx} = m_{dy} = 100 \qquad k_{Dx} = k_{Dy} = 500 \qquad k_{Px} = k_{Py} = 2500.$$

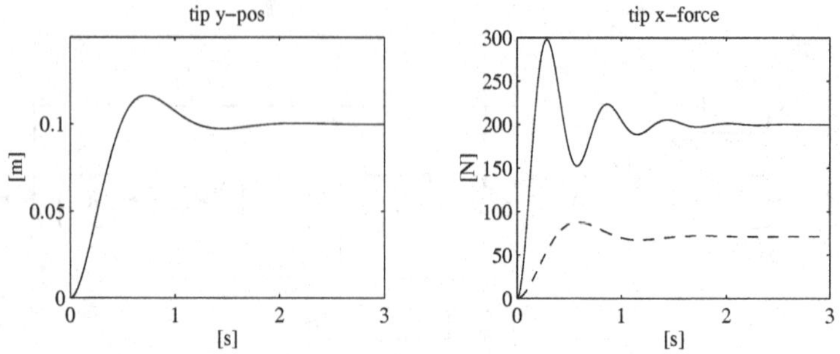

Fig. 9.4. Time history of the tip position along vertical direction and of the contact force along horizontal direction with impedance control scheme for environments of different compliance

For these values it is

$$\omega_{ny} = 5\,\text{rad/s} \qquad \zeta_y = 0.5.$$

Then, for the more compliant environment it is

$$\omega_{nx} \approx 5.9\,\text{rad/s} \qquad \zeta_x \approx 0.42$$

whereas for the less compliant environment it is

$$\omega_{nx} \approx 11.2\,\text{rad/s} \qquad \zeta_x \approx 0.22.$$

Let the arm tip be in contact with the environment at position $\boldsymbol{x}_e = [\,1 \quad 0\,]^T$; it is desired to take it to position $\boldsymbol{x}_d = [\,1.1 \quad 0.1\,]^T$.

The results in Fig. 9.4 show that motion dynamics along the vertical direction is the same in the two cases. As regards the contact force along the horizontal direction, for the more compliant environment (*dashed line*) a well-damped behaviour is obtained, whereas for the less compliant environment (*solid line*) the resulting behaviour is less damped. Further, at the equilibrium, in the first case a displacement of 7.14 cm with a contact force of 71.4 N, whereas in the second case a displacement of 2 cm with a contact force of 200 N are observed.

The selection of good impedance parameters, so as to achieve a satisfactory behaviour during the interaction, is not an easy task. Example 9.2 showed that the closed-loop dynamics along the free motion directions is different from the closed-loop dynamics along the constrained directions. In this latter case, the dynamic behaviour depends on the stiffness characteristics of the environment. The execution of a complex task, involving different types of interaction, may require different values of impedance parameters.

Notice that impedance control, in the absence of interaction or along the directions of free motion, is equivalent to an inverse dynamics position control. Therefore, for the selection of the impedance parameters, one also has to take into account the need to ensure high values to the rejection factor of the disturbances due to model uncertainties and to the approximations into the

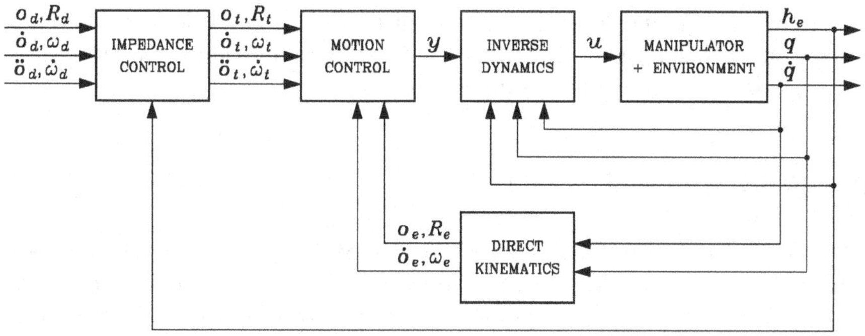

Fig. 9.5. Block scheme of admittance control

inverse dynamics computation. Such a factor increases proportionally to the gain matrix $\boldsymbol{K}_P$. Hence the closed-loop behaviour is the more degraded by disturbances, the more compliant the impedance control is made (by choosing low values for the elements of $\boldsymbol{K}_P$) to keep interaction forces limited.

A possible solution may be that of separating the motion control problem from the impedance control problem according to the control scheme represented in Fig. 9.5. The scheme is based on the concept of *compliant frame*, which is a suitable reference frame describing the ideal behaviour of the end-effector under impedance control. This frame is specified by the position of the origin $\boldsymbol{o}_t$, the rotation matrix $\boldsymbol{R}_t$, as well as by the liner and angular velocities and accelerations. These quantities can be computed by integrating the impedance equations in the form

$$\boldsymbol{M}_t\ddot{\widetilde{\boldsymbol{z}}} + \boldsymbol{K}_{Dt}\dot{\widetilde{\boldsymbol{z}}} + \boldsymbol{K}_{Pt}\widetilde{\boldsymbol{z}} = \boldsymbol{h}_e^d, \tag{9.37}$$

starting from the measurements of the force vector $\boldsymbol{h}_e$, where $\boldsymbol{M}_t$, $\boldsymbol{K}_{Dt}$, $\boldsymbol{K}_{Pt}$ are the parameters of a mechanical impedance. In the above equation, vector $\widetilde{\boldsymbol{z}}$ represents the operational space error between the desired frame and the compliant frame, as defined in (9.9), using subscript t in place of subscript e.

The kinematic variables of the compliant frame are then input to the motion control of inverse dynamics type, computed according to Eqs. (9.28), (9.30). In this way, the gains of the motion control law (9.28) can be designed so as to guarantee a high value of the disturbance rejection factor. On the other hand, the gains of the impedance control law (9.37) can be set so as to guarantee satisfactory behaviour during the interaction with the environment. Stability of the overall system can be ensured provided that the equivalent bandwidth of the motion control loop is larger than the equivalent bandwidth of the impedance control loop.

The above control scheme is also known as *admittance control* because Equation (9.37) corresponds to a mechanical admittance being used by the controller to generate the motion variables (outputs) from the force measurements (inputs). On the other hand, the control defined by Eqs. (9.31) or (9.35) and (9.30) can be interpreted as a system producing equivalent end-effector forces (outputs) form the measurements of the motion variables (inputs), thus corresponding to a mechanical impedance.

9.4 Force Control

The above schemes implement an *indirect force control*, because the interaction force can be indirectly controlled by acting on the desired pose of the end-effector assigned to the motion control system. Interaction between manipulator and environment is anyhow directly influenced by compliance of the environment and by either compliance or impedance of the manipulator.

If it is desired to control accurately the contact force, it is necessary to devise control schemes that allow the desired interaction force to be directly specified. The development of a *direct force control* system, in analogy to a motion control system, would require the adoption of a stabilizing PD control action on the force error besides the usual nonlinear compensation actions. Force measurements may be corrupted by noise, and then a derivative action may not be implemented in practice. The stabilizing action is to be provided by suitable damping of velocity terms. As a consequence, a force control system typically features a control law based not only on force measurements but also on velocity measurements, and eventually position measurements too.

The realization of a force control scheme can be entrusted to the closure of an *outer force regulation feedback loop* generating the control input for the motion control scheme the manipulator is usually endowed with. Therefore, force control schemes are presented below, which are based on the use of an inverse dynamics position control. The effectiveness of a such control scheme depends on the particular interaction cases and, in general, on the contact geometry. To this end, notice that a force control strategy is meaningful only for those directions of the operational space along which interaction forces between manipulator and environment may arise.

Below, force control schemes based on the adoption of motion control laws of inverse dynamics type are presented, assuming that the operational space is defined only by position variables. Therefore, the end-effector pose can be specified by the operational space vector $\boldsymbol{x}_e = \boldsymbol{o}_e$. Moreover, the elastic model

$$\boldsymbol{f}_e = \boldsymbol{K}(\boldsymbol{x}_e - \boldsymbol{x}_r) \tag{9.38}$$

is assumed for the environment, obtained from (9.21) with the assumption that only forces arise at the contact. In (9.38), consider $\boldsymbol{x}_r = \boldsymbol{o}_r$ and assume that the axes of the frame attached to the environment rest position are parallel to the axes of the base frame. The above assumptions allow some important features of force control to be evidenced.

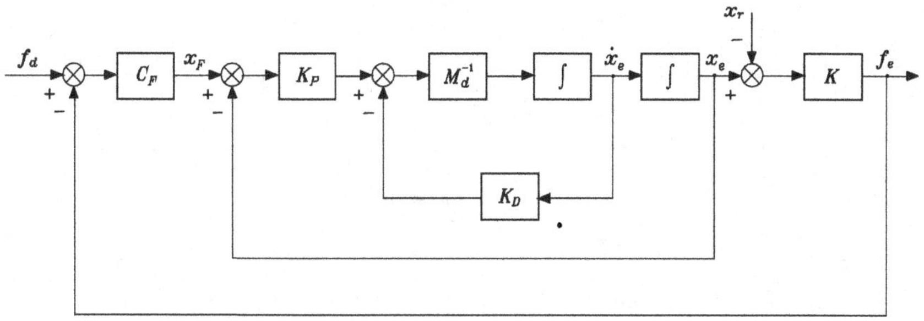

Fig. 9.6. Block scheme of force control with inner position loop

9.4.1 Force Control with Inner Position Loop

With reference to the inverse dynamics law with force measurement (9.30), choose in place of (9.31), the control

$$\boldsymbol{y} = \boldsymbol{J}^{-1}(\boldsymbol{q})\boldsymbol{M}_d^{-1}\big(-\boldsymbol{K}_D\dot{\boldsymbol{x}}_e + \boldsymbol{K}_P(\boldsymbol{x}_F - \boldsymbol{x}_e) - \boldsymbol{M}_d\dot{\boldsymbol{J}}(\boldsymbol{q},\dot{\boldsymbol{q}})\dot{\boldsymbol{q}}\big) \tag{9.39}$$

where $\boldsymbol{x}_F$ is a suitable reference to be related to a force error. Notice that the control law (9.39) does not foresee the adoption of compensating actions relative to $\dot{\boldsymbol{x}}_F$ and $\ddot{\boldsymbol{x}}_F$. Moreover, since the operational space is defined only by position variables, the analytic Jacobian coincides with the geometric Jacobian and thus $\boldsymbol{J}_A(\boldsymbol{q}) = \boldsymbol{J}(\boldsymbol{q})$.

Substituting (9.30), (9.39) into (9.1), leads, after similar algebraic manipulation as above, to the system described by

$$\boldsymbol{M}_d\ddot{\boldsymbol{x}}_e + \boldsymbol{K}_D\dot{\boldsymbol{x}}_e + \boldsymbol{K}_P\boldsymbol{x}_e = \boldsymbol{K}_P\boldsymbol{x}_F, \tag{9.40}$$

which shows how (9.30) and (9.39) perform a position control taking $\boldsymbol{x}_e$ to $\boldsymbol{x}_F$ with a dynamics specified by the choice of matrices $\boldsymbol{M}_d$, $\boldsymbol{K}_D$, $\boldsymbol{K}_P$.

Let $\boldsymbol{f}_d$ denote the desired *constant* force reference; the relation between $\boldsymbol{x}_F$ and the force error can be expressed as

$$\boldsymbol{x}_F = \boldsymbol{C}_F(\boldsymbol{f}_d - \boldsymbol{f}_e), \tag{9.41}$$

where $\boldsymbol{C}_F$ is a diagonal matrix, with the meaning of compliance, whose elements give the control actions to perform along the operational space directions of interest. The equations in (9.40), (9.41) reveal that force control is developed on the basis of a preexisting position control loop.

On the assumption of the elastically compliant environment described by (9.38), the equation in (9.40) with (9.41) becomes

$$\boldsymbol{M}_d\ddot{\boldsymbol{x}}_e + \boldsymbol{K}_D\dot{\boldsymbol{x}}_e + \boldsymbol{K}_P(\boldsymbol{I}_3 + \boldsymbol{C}_F\boldsymbol{K})\boldsymbol{x}_e = \boldsymbol{K}_P\boldsymbol{C}_F(\boldsymbol{K}\boldsymbol{x}_r + \boldsymbol{f}_d). \tag{9.42}$$

To decide about the kind of control action to specify with $\boldsymbol{C}_F$, it is worth representing (9.21), (9.40), (9.41) in terms of the block scheme in Fig. 9.6,

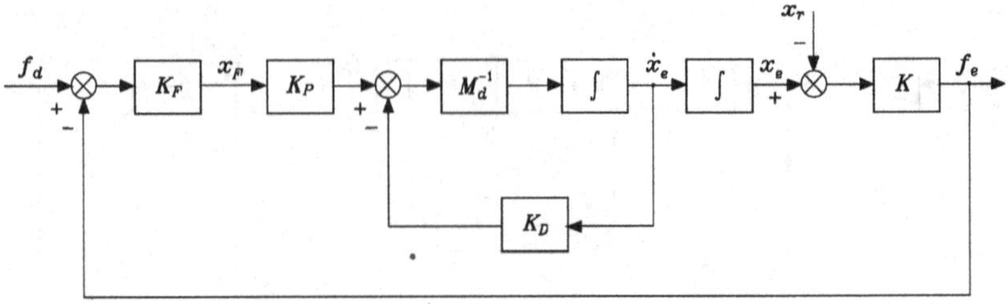

Fig. 9.7. Block scheme of force control with inner velocity loop

which is logically derived from the scheme in Fig. 9.3. This scheme suggests that if $\boldsymbol{C}_F$ has a purely proportional control action, then $\boldsymbol{f}_e$ cannot reach $\boldsymbol{f}_d$ and $\boldsymbol{x}_r$ influences the interaction force also at steady state.

If $\boldsymbol{C}_F$ also has an integral control action on the force components, then it is possible to achieve $\boldsymbol{f}_e = \boldsymbol{f}_d$ at steady state and, at the same time, to reject the effect of $\boldsymbol{x}_r$ on $\boldsymbol{f}_e$. Hence, a convenient choice for $\boldsymbol{C}_F$ is a *proportional-integral* (PI) *action*

$$\boldsymbol{C}_F = \boldsymbol{K}_F + \boldsymbol{K}_I \int^t (\cdot)\, d\varsigma. \tag{9.43}$$

The dynamic system resulting from (9.42), (9.43) is of third order, and then it is necessary to choose adequately the matrices $\boldsymbol{K}_D$, $\boldsymbol{K}_P$, $\boldsymbol{K}_F$, $\boldsymbol{K}_I$ in respect of the characteristics of the environment. Since the values of environment stiffness are typically high, the weight of the proportional and integral actions should be contained; the choice of $\boldsymbol{K}_F$ and $\boldsymbol{K}_I$ influences the stability margins and the bandwidth of the system under force control. On the assumption that a stable equilibrium is reached, it is $\boldsymbol{f}_e = \boldsymbol{f}_d$ and then

$$\boldsymbol{K}\boldsymbol{x}_e = \boldsymbol{K}\boldsymbol{x}_r + \boldsymbol{f}_d. \tag{9.44}$$

9.4.2 Force Control with Inner Velocity Loop

From the block scheme of Fig. 9.6 it can be observed that, if the position feedback loop is opened, $\boldsymbol{x}_F$ represents a velocity reference, and then an integration relationship exists between $\boldsymbol{x}_F$ and $\boldsymbol{x}_e$. This leads to recognizing that, in this case, the interaction force with the environment coincides with the desired value at steady state, even with a proportional force controller $\boldsymbol{C}_F$. In fact, choosing

$$\boldsymbol{y} = \boldsymbol{J}^{-1}(\boldsymbol{q})\boldsymbol{M}_d^{-1}(-\boldsymbol{K}_D\dot{\boldsymbol{x}}_e + \boldsymbol{K}_P\boldsymbol{x}_F - \boldsymbol{M}_d\dot{\boldsymbol{J}}(\boldsymbol{q},\dot{\boldsymbol{q}})\dot{\boldsymbol{q}}), \tag{9.45}$$

with a purely proportional control structure ($\boldsymbol{C}_F = \boldsymbol{K}_F$) on the force error yields

$$\boldsymbol{x}_F = \boldsymbol{K}_F(\boldsymbol{f}_d - \boldsymbol{f}_e) \tag{9.46}$$

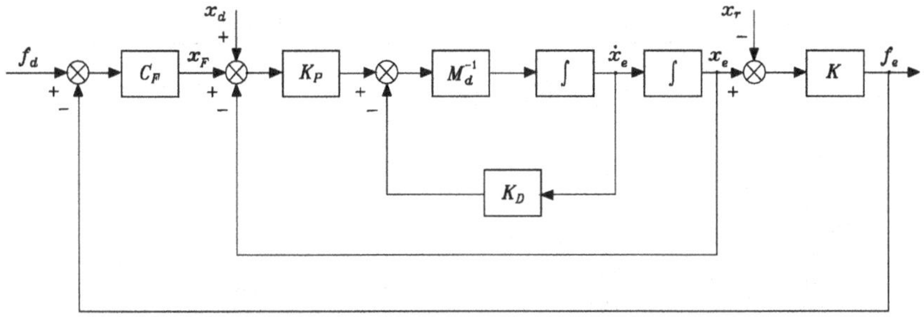

Fig. 9.8. Block scheme of parallel force/position control

and then system dynamics is described by

$$\boldsymbol{M}_d\ddot{\boldsymbol{x}}_e + \boldsymbol{K}_D\dot{\boldsymbol{x}}_e + \boldsymbol{K}_P\boldsymbol{K}_F\boldsymbol{K}\boldsymbol{x}_e = \boldsymbol{K}_P\boldsymbol{K}_F(\boldsymbol{K}\boldsymbol{x}_r + \boldsymbol{f}_d). \tag{9.47}$$

The relationship between position and contact force at the equilibrium is given by (9.44). The corresponding block scheme is reported in Fig. 9.7. It is worth emphasizing that control design is simplified, since the resulting system now is of second order;[2] it should be noticed, however, that the absence of an integral action in the force controller does not ensure reduction of the effects due to unmodelled dynamics.

9.4.3 Parallel Force/Position Control

The force control schemes presented require the force reference to be consistent with the geometric features of the environment. In fact, if $\boldsymbol{f}_d$ has components outside $\mathcal{R}(\boldsymbol{K})$, both (9.42) (in case of an integral action in $\boldsymbol{C}_F$) and (9.47) show that, along the corresponding operational space directions, the components of $\boldsymbol{f}_d$ are interpreted as velocity references which cause a drift of the end-effector position. If $\boldsymbol{f}_d$ is correctly planned along the directions outside $\mathcal{R}(\boldsymbol{K})$, the resulting motion governed by the position control action tends to take the end-effector position to zero in the case of (9.42), and the end-effector velocity to zero in the case of (9.47). Hence, the above control schemes do not allow position control even along the admissible task space directions.

If it is desired to specify a desired end-effector pose $\boldsymbol{x}_d$ as in pure position control schemes, the scheme of Fig. 9.6 can be modified by adding the reference $\boldsymbol{x}_d$ to the input where positions are summed. This corresponds to choosing

$$\boldsymbol{y} = \boldsymbol{J}^{-1}(\boldsymbol{q})\boldsymbol{M}_d^{-1}\big(-\boldsymbol{K}_D\dot{\boldsymbol{x}}_e + \boldsymbol{K}_P(\widetilde{\boldsymbol{x}} + \boldsymbol{x}_F) - \boldsymbol{M}_d\dot{\boldsymbol{J}}_A(\boldsymbol{q},\dot{\boldsymbol{q}})\dot{\boldsymbol{q}}\big) \tag{9.48}$$

where $\widetilde{\boldsymbol{x}} = \boldsymbol{x}_d - \boldsymbol{x}_e$. The resulting scheme (Fig. 9.8) is termed *parallel force/position control*, in view of the presence of a position control action

[2] The matrices $\boldsymbol{K}_P$ and $\boldsymbol{K}_F$ are not independent and one may refer to a single matrix $\boldsymbol{K}'_F = \boldsymbol{K}_P\boldsymbol{K}_F$.

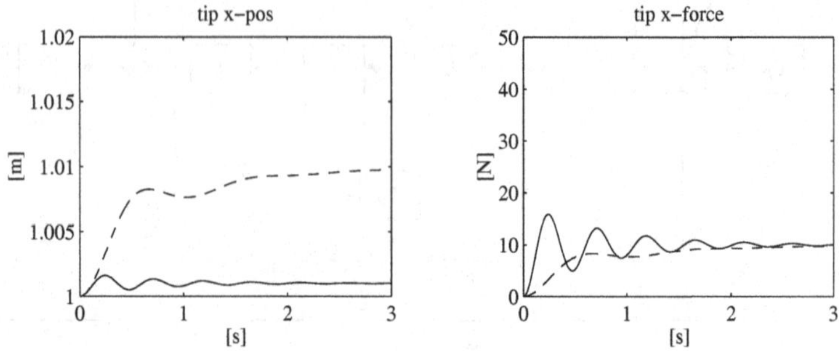

Fig. 9.9. Time history of the tip position and of the contact force along horizontal direction with force control scheme with inner position loop for two environments of different compliance

$\boldsymbol{K}_P\widetilde{\boldsymbol{x}}$ in parallel to a force control action $\boldsymbol{K}_P\boldsymbol{C}_F(\boldsymbol{f}_d - \boldsymbol{f}_e)$. It is easy to verify that, in this case, the equilibrium position satisfies the equation (see Problem 9.4)

$$\boldsymbol{x}_e = \boldsymbol{x}_d + \boldsymbol{C}_F\big(\boldsymbol{K}(\boldsymbol{x}_r - \boldsymbol{x}_e) + \boldsymbol{f}_d\big). \tag{9.49}$$

Therefore, along those directions outside $\mathcal{R}(\boldsymbol{K})$ where motion is unconstrained, the position reference $\boldsymbol{x}_d$ is reached by $\boldsymbol{x}_e$. Vice versa, along those directions in $\mathcal{R}(\boldsymbol{K})$ where motion is constrained, $\boldsymbol{x}_d$ is treated as an additional disturbance; the adoption of an integral action in $\boldsymbol{C}_F$ as for the scheme of Fig. 9.6 ensures that the force reference $\boldsymbol{f}_d$ is reached at steady state, at the expense of a position error on $\boldsymbol{x}_e$ depending on environment compliance.

Example 9.3

Consider again the planar arm in contact with the elastically compliant plane of the above examples; let the initial contact position be the same as that of Example 9.2. Performance of the various force control schemes is analyzed; as in Example 9.2, a more compliant ($k_x = 10^3$ N/m) and a less compliant ($k_x = 10^4$ N/m) environment are considered. The position control actions $\boldsymbol{M}_d$, $\boldsymbol{K}_D$, $\boldsymbol{K}_P$ are chosen as in Example 9.2; a force control action is added along the horizontal direction, namely

$$\boldsymbol{C}_F = \text{diag}\{c_{Fx}, 0\}.$$

The reference for the contact force is chosen as $\boldsymbol{f}_d = [\,10 \quad 0\,]^T$; the position reference — meaningful only for the parallel control — is taken as $\boldsymbol{x}_d = [\,1.015 \quad 0.1\,]^T$.

With regard to the scheme with inner position loop of Fig. 9.6, a PI control action c_{Fx} is chosen with parameters:

$$k_{Fx} = 0.00064 \qquad k_{Ix} = 0.0016.$$

This confers two complex poles $(-1.96, \pm j5.74)$, a real pole (-1.09), and a real zero (-2.5) to the overall system, for the more compliant environment.

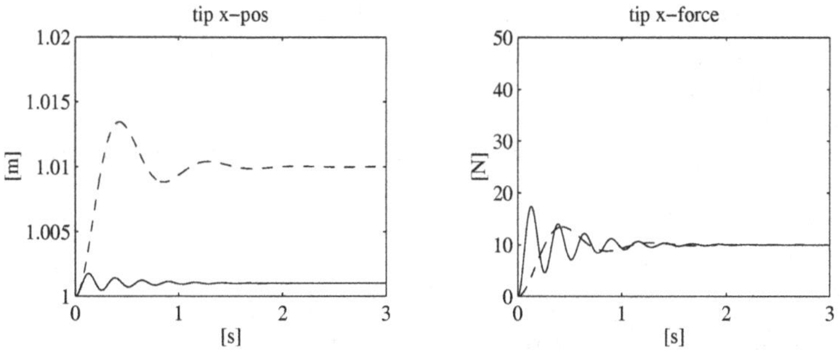

Fig. 9.10. Time history of the tip position and of the contact force along horizontal direction with force control scheme with inner velocity loop for two environments of different compliance

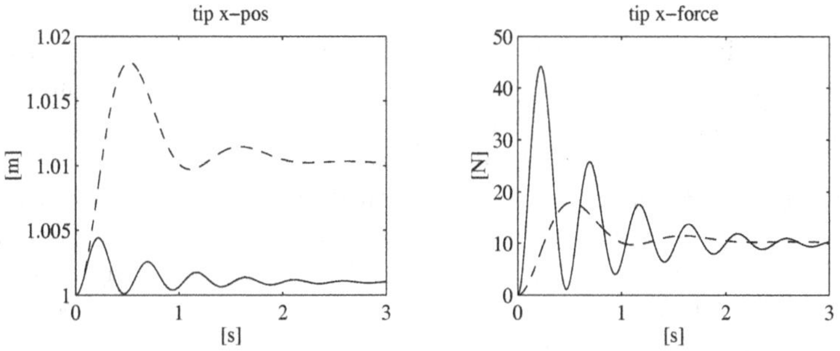

Fig. 9.11. Time history of tip position and of the contact force along horizontal direction with parallel force/position control scheme for two environments of different compliance

With regard to the scheme with inner velocity loop of Fig. 9.7, the proportional control action in c_{Fx} is

$$k_{Fx} = 0.0024$$

so that the overall system, for the more compliant environment, has two complex poles $(-2.5, \pm j7.34)$.

With regard to the parallel control scheme of Fig. 9.8, the PI control action c_{Fx} is chosen with the same parameters as for the first control scheme.

Figures 9.9, 9.10, 9.11 report the time history of the tip position and contact force along axis x_0 for the three considered schemes. A comparison between the various cases shows what follows:

- All control laws guarantee a steady-state value of contact forces equal to the desired one for both the more compliant (*dashed line*) and the less compliant (*continuous line*) environment.
- For given motion control actions ($\boldsymbol{M}_d$, $\boldsymbol{K}_D$, $\boldsymbol{K}_P$), the force control with inner velocity loop presents a faster dynamics than that of the force control with inner position loop.
- The dynamic response with the parallel control shows how the addition of a position reference along the horizontal direction degrades the transient behaviour,

but it does not influence the steady-state contact force. This effect can be justified by noting that a step position input is equivalent to a properly filtered impulse force input.

The reference position along axis y_0 is obviously reached by the arm tip according to dynamics of position control; the relative time history is not reported.

9.5 Constrained Motion

Force control schemes can be employed for the execution of a *constrained motion* as long as they suitably take into account the geometric features of the environment and the force and position references are chosen to be compatible with those features.

A real manipulation task is characterized by complex contact situations where some directions are subject to end-effector pose constraints and others are subject to interaction force constraints. During task execution, the nature of constraints may vary substantially.

The need to handle complex contact situations requires the capacity to specify and perform control of both end-effector pose and contact force. However, a fundamental aspect to be considered is that it is not possible to impose simultaneously arbitrary values of pose and force along each direction. Moreover, one should ensure that the reference trajectories for the control system be compatible with the constraints imposed by the environment during task execution.

For the above reasons, it is useful to have an analytic description of the interaction forces, which is very demanding from a modelling point of view.

A real contact situation is a naturally distributed phenomenon in which the local characteristics of the contact surfaces as well as the global dynamics of the manipulator and environment are involved. In detail:

- The environment imposes kinematic constraints on the end-effector motion, due to one or more contacts of different type; reaction forces and moments arise when the end-effector tends to violate the constraints (e.g., the case of a robot sliding a rigid tool on a frictionless rigid surface).
- The end-effector, while being subject to kinematic constraints, may also exert dynamic forces and moments on the environment, in the presence of environment dynamics (e.g., the case of a robot turning a crank, when the crank dynamics is relevant, or a robot pushing against a compliant surface).
- The contact force and moment may depend on the structural compliance of the robot, due to the finite stiffness of the joints and links of the manipulator, as well as of the wrist force/torque sensor or of the tool (e.g. an end-effector mounted on an RCC device).

- Local deformations of the contact surfaces may occur during the interaction, producing distributed contact areas; moreover, static and dynamic friction may be present in the case of non-ideally smooth contact surfaces.

The design of the interaction control is usually carried out under simplifying assumptions. The following two cases are considered:

- The robot and the environment are perfectly rigid and purely kinematics constraints are imposed by the environment.
- The robot is perfectly rigid, all the compliance of the system is localized in the environment and the contact force and moment is approximated by a linear elastic model.

In both cases, frictionless contact is assumed. It is obvious that these situations are only ideal. However, the robustness of the control should be able to cope with situations where some of the ideal assumptions are relaxed. In that case the control laws may be adapted to deal with non-ideal characteristics.

9.5.1 Rigid Environment

The kinematic constraints imposed by the environment can be represented by a set of algebraic equations that the variables describing the end-effector position and orientation must satisfy; since these variables depend on the joint variables through the direct kinematic equations, the constraint equations can also be expressed in the joint space as

$$\boldsymbol{\varphi}(\boldsymbol{q}) = \mathbf{0}. \tag{9.50}$$

Vector $\boldsymbol{\varphi}$ is an $(m \times 1)$ function, with $m < n$, where n is the number of joints of the manipulator, assumed to be nonredundant; without loss of generality, the case $n = 6$ is considered. The constraints of the form (9.50), involving only the generalized coordinates of the system, are known as *holonomic constraints*. Computing the time derivative of (9.50) yields

$$\boldsymbol{J}_{\varphi}(\boldsymbol{q})\dot{\boldsymbol{q}} = \mathbf{0}, \tag{9.51}$$

where $\boldsymbol{J}_{\varphi}(\boldsymbol{q}) = \partial\boldsymbol{\varphi}/\partial\boldsymbol{q}$ is the $(m \times 6)$ Jacobian of $\boldsymbol{\varphi}(\boldsymbol{q})$, known as *constraint Jacobian*. It is assumed that $\boldsymbol{J}_{\varphi}$ is of rank m at least locally in a neighborhood of the operating point; equivalently, the m constraint equations (9.50) are assumed to be locally independent.

In the absence of friction, the interaction forces are *reaction forces* arising when the end-effector tends to violate the constraints. These end-effector forces produce reaction torques at the joints that can be computed using the principle of virtual work, taking into account that the work of the reaction forces, by definition, should be null for all virtual displacements which satisfy the constraints. Considering the expression (3.108) of the virtual work of the joint torques $\boldsymbol{\tau}$ and that, in view of (9.51), the virtual displacement $\delta\boldsymbol{q}$ satisfy the equation

$$\boldsymbol{J}_{\varphi}(\boldsymbol{q})\delta\boldsymbol{q} = \boldsymbol{0},$$

yields

$$\boldsymbol{\tau} = \boldsymbol{J}_{\varphi}^T(\boldsymbol{q})\boldsymbol{\lambda},$$

where $\boldsymbol{\lambda}$ is a suitable $(m \times 1)$ vector. The corresponding forces applied to the end-effector are

$$\boldsymbol{h}_e = \boldsymbol{J}^{-T}(\boldsymbol{q})\boldsymbol{\tau} = \boldsymbol{S}_f(\boldsymbol{q})\boldsymbol{\lambda}, \tag{9.52}$$

assuming a nonsingular $\boldsymbol{J}$, with

$$\boldsymbol{S}_f = \boldsymbol{J}^{-T}(\boldsymbol{q})\boldsymbol{J}_{\varphi}^T(\boldsymbol{q}). \tag{9.53}$$

Notice that Eq. (9.50) corresponds to a set of *bilateral constraints*. This means that the reaction forces (9.52) act so that, during the motion, the end-effector always keeps contact with the environment, as for the case of a gripper turning a crank. However, in many applications, the interaction with the environment corresponds to *unilateral constraints*. For example, in the case of a tool sliding on a surface, the reaction forces arise only when the tool pushes against the surface and not when it tends to detach. However, Eq. (9.52) can be still applied under the assumption that the end-effector, during the motion, never loses contact with the environment.

From (9.52) it follows that $\boldsymbol{h}_e$ belongs to the m-dimensional subspace $\mathcal{R}(\boldsymbol{S}_f)$. The inverse of the linear transformation (9.52) can be computed as

$$\boldsymbol{\lambda} = \boldsymbol{S}_f^{\dagger}(\boldsymbol{q})\boldsymbol{h}_e, \tag{9.54}$$

where $\boldsymbol{S}_f^{\dagger}$ denotes a weighted pseudo-inverse of matrix $\boldsymbol{S}_f$, namely

$$\boldsymbol{S}_f^{\dagger} = (\boldsymbol{S}_f^T\boldsymbol{W}\boldsymbol{S}_f)^{-1}\boldsymbol{S}_f^T\boldsymbol{W}, \tag{9.55}$$

where $\boldsymbol{W}$ is a symmetric and positive definite weighting matrix.

Notice that, while subspace $\mathcal{R}(\boldsymbol{S}_f)$ is uniquely defined by the geometry of the contact, matrix $\boldsymbol{S}_f$ in (9.53) is not unique, because constraint equations (9.50) are not uniquely defined. Moreover, in general, the physical dimensions of the elements of vector $\boldsymbol{\lambda}$ are not homogeneous and the columns of matrix $\boldsymbol{S}_f$, as well as of matrix $\boldsymbol{S}_f^{\dagger}$, do not necessarily represent homogeneous entities. This may produce invariance problems in the transformation (9.54) if $\boldsymbol{h}_e$ represents a quantity that is subject to disturbances and, as a result, may have components outside $\mathcal{R}(\boldsymbol{S}_f)$. In particular, if a physical unit or a reference frame is changed, matrix $\boldsymbol{S}_f$ undergoes a transformation; however, the result of (9.54) with the transformed pseudo-inverse, in general, depends on the adopted physical units or reference frame! The reason is that, if $\boldsymbol{h}_e \notin \mathcal{R}(\boldsymbol{S}_f)$, the problem of computing $\boldsymbol{\lambda}$ from (9.52) does not have a solution. In this case, Eq. (9.54) represents only an approximate solution which minimizes the norm of vector $\boldsymbol{h}_e - \boldsymbol{S}_f(\boldsymbol{q})\boldsymbol{\lambda}$ weighted by matrix $\boldsymbol{W}$.[3] It is evident that the invariance of the solution can be ensured only if, in the case that a physical unit or a reference frame is changed, the weighting matrix is transformed accordingly. In the ideal case $\boldsymbol{h}_e \in \mathcal{R}(\boldsymbol{S}_f)$, the computation of the inverse of (9.52) has a unique solution, defined by (9.54), regardless the weighting matrix; hence the invariance problem does not occur.

In order to guarantee invariance, it is convenient choosing matrix $\boldsymbol{S}_f$ so that its columns represent linearly independent forces. This implies that (9.52) gives $\boldsymbol{h}_e$ as a linear combination of forces and $\boldsymbol{\lambda}$ is a dimensionless vector. Moreover, a physically consistent norm in the space of forces can be defined based on the quadratic form $\boldsymbol{h}_e^T\boldsymbol{C}\boldsymbol{h}_e$, which has the meaning of an elastic energy if $\boldsymbol{C}$ is a positive definite compliance matrix. Hence, the choice $\boldsymbol{W} = \boldsymbol{C}$ can be made for the weighting matrix and, if a physical unit or a reference frame is changed, the transformations to be applied to matrices $\boldsymbol{S}_f$ and $\boldsymbol{W}$ can be easily found on the basis of their physical meaning.

Notice that, for a given $\boldsymbol{S}_f$, the constraint Jacobian can be computed from (9.53) as $\boldsymbol{J}_\varphi(\boldsymbol{q}) = \boldsymbol{S}_f^T\boldsymbol{J}(\boldsymbol{q})$; moreover, if necessary, the constraint equations can be derived by integrating (9.51).

By using (3.4), (9.53), equality (9.51) can be rewritten in the form

$$\boldsymbol{J}_\varphi(\boldsymbol{q})\boldsymbol{J}^{-1}(\boldsymbol{q})\boldsymbol{J}(\boldsymbol{q})\dot{\boldsymbol{q}} = \boldsymbol{S}_f^T\boldsymbol{v}_e = \boldsymbol{0}, \tag{9.56}$$

which, by virtue of (9.52), is equivalent to

$$\boldsymbol{h}_e^T\boldsymbol{v}_e = 0. \tag{9.57}$$

Equation (9.57) represents the kinetostatic relationship, known as *reciprocity*, between the interaction force and moment $\boldsymbol{h}_e$ — belonging to the so-called *force controlled subspace* — which coincides with $\mathcal{R}(\boldsymbol{S}_f)$ and the end-effector linear and angular velocity $\boldsymbol{v}_e$ — belonging to the so-called *velocity controlled subspace*. The concept of reciprocity expresses the physical fact that, under the assumption of rigid and frictionless contact, the forces do not produce any work for all the end-effector displacements which satisfy the constraints. This concept is often confused with the concept of orthogonality, which is meaningless in this case because velocities and forces are non-homogeneous quantities belonging to different vector spaces.

Equations (9.56), (9.57) imply that the dimension of the velocity controlled subspace is $6-m$ whereas the dimension of the force controlled subspace is m; moreover, a $(6 \times (6-m))$ matrix $\boldsymbol{S}_v$ can be defined, which satisfies equation

$$\boldsymbol{S}_f^T(\boldsymbol{q})\boldsymbol{S}_v(\boldsymbol{q}) = \boldsymbol{O} \tag{9.58}$$

and such that $\mathcal{R}(\boldsymbol{S}_v)$ represents the velocity controlled subspace. Therefore:

$$\boldsymbol{v}_e = \boldsymbol{S}_v(\boldsymbol{q})\boldsymbol{\nu}, \tag{9.59}$$

[3] See Sect. A.7 for the computation of an approximate solution based on the left pseudo-inverse and Problem 9.5.

where $\boldsymbol{\nu}$ denotes a suitable $((6-m)\times 1)$ vector.

The inverse of the linear transformation (9.59) can be computed as

$$\boldsymbol{\nu} = \boldsymbol{S}_v^{\dagger}(\boldsymbol{q})\boldsymbol{v}_e, \tag{9.60}$$

where $\boldsymbol{S}_v^{\dagger}$ denotes a suitable weighted pseudo-inverse of matrix $\boldsymbol{S}_v$, computed as in (9.55). Notice that, as for the case of $\boldsymbol{S}_f$, although the subspace $\mathcal{R}(\boldsymbol{S}_v)$ is uniquely defined, the choice of matrix $\boldsymbol{S}_v$ itself is not unique. Moreover, about Eq. (9.60), invariance problems analogous to that considered for the case of (9.54) can be observed. In this case, it is convenient to select the matrix $\boldsymbol{S}_v$ so that its columns represent a set of independent velocities; moreover, for the computation of the pseudo-inverse, a norm in the space of velocities can be defined based on the kinetic energy of a rigid body or on the elastic energy expressed in terms of the stiffness matrix $\boldsymbol{K} = \boldsymbol{C}^{-1}$.

Matrix $\boldsymbol{S}_v$ may also have an interpretation in terms of Jacobian. In fact, due to the presence of m independent holonomic constraints (9.50), the configuration of a manipulator in contact with the environment can be locally described in terms of a $((6-m)\times 1)$ vector $\boldsymbol{r}$ of independent coordinates. From the implicit function theorem, this vector can be defined as

$$\boldsymbol{r} = \boldsymbol{\psi}(\boldsymbol{q}), \tag{9.61}$$

where $\boldsymbol{\psi}(\boldsymbol{q})$ is any $((6-m)\times 1)$ vector function such that the m components of $\boldsymbol{\phi}(\boldsymbol{q})$ and the $6-m$ components of $\boldsymbol{\psi}(\boldsymbol{q})$ are linearly independent at least locally in a neighborhood of the operating point. This means that the mapping (9.61), together with the constraint equations (9.50), is locally invertible, with inverse defined as

$$\boldsymbol{q} = \boldsymbol{\rho}(\boldsymbol{r}). \tag{9.62}$$

Equation (9.62) explicitly provides all the joint vectors $\boldsymbol{q}$ which satisfy the constraint equations (9.50), for any $\boldsymbol{r}$ arbitrary selected in a neighborhood of the operating point. Moreover, the vector $\dot{\boldsymbol{q}}$ that satisfies (9.51) can be computed as

$$\dot{\boldsymbol{q}} = \boldsymbol{J}_\rho(\boldsymbol{r})\dot{\boldsymbol{r}},$$

where $\boldsymbol{J}_\rho(\boldsymbol{r}) = \partial\boldsymbol{\rho}/\partial\boldsymbol{r}$ is a $(6\times(6-m))$ full rank Jacobian matrix. Also, the following equality holds:

$$\boldsymbol{J}_\varphi(\boldsymbol{q})\boldsymbol{J}_\rho(\boldsymbol{r}) = \boldsymbol{O},$$

which can be interpreted as a reciprocity condition between the subspace $\mathcal{R}(\boldsymbol{J}_\varphi^T)$ of the joint torques $\boldsymbol{\tau}$ corresponding to the reaction forces acting on the end-effector and the subspace $\mathcal{R}(\boldsymbol{J}_\rho)$ of the joint velocities $\dot{\boldsymbol{q}}$ which satisfy the constraints.

The above equation can be rewritten as

$$\boldsymbol{J}_\varphi(\boldsymbol{q})\boldsymbol{J}^{-1}(\boldsymbol{q})\boldsymbol{J}(\boldsymbol{q})\boldsymbol{J}_\rho(\boldsymbol{r}) = \boldsymbol{O}.$$

Hence, assuming that $\boldsymbol{J}$ is nonsingular and taking into account (9.53), (9.58), matrix $\boldsymbol{S}_v$ can be computed as

$$\boldsymbol{S}_v = \boldsymbol{J}(\boldsymbol{q})\boldsymbol{J}_\rho(\boldsymbol{r}). \tag{9.63}$$

The matrices $\boldsymbol{S}_f$, $\boldsymbol{S}_v$ and the corresponding pseudo-inverse matrices $\boldsymbol{S}_f^\dagger$, $\boldsymbol{S}_v^\dagger$ are known as *selection matrices*. These matrices play a fundamental role for task specification, since they can be used to assign the desired end-effector motion and the interaction forces and moments consistently with the constraints. Also, they are essential for control synthesis.

To this end, notice that the (6×6) matrix $\boldsymbol{P}_f = \boldsymbol{S}_f\boldsymbol{S}_f^\dagger$ projects a generic force vector $\boldsymbol{h}_e$ on the force controlled subspace $\mathcal{R}(\boldsymbol{S}_f)$. Matrix $\boldsymbol{P}_f$ is idempotent, namely $\boldsymbol{P}_f^2 = \boldsymbol{P}_f\boldsymbol{P}_f = \boldsymbol{P}_f$, and therefore is a *projection matrix*. Moreover, matrix $(\boldsymbol{I}_6 - \boldsymbol{P}_f)$ projects force vector $\boldsymbol{h}_e$ on the orthogonal complement of the force controlled subspace; also, this matrix, being idempotent, it is a projection matrix.

Similarly, it can be verified that the (6×6) matrices $\boldsymbol{P}_v = \boldsymbol{S}_v\boldsymbol{S}_v^\dagger$ and $(\boldsymbol{I}_6 - \boldsymbol{P}_v)$ are projection matrices, projecting a generic linear and angular velocity vector $\boldsymbol{v}_e$ on the controlled velocity subspace $\mathcal{R}(\boldsymbol{S}_v)$ and on its orthogonal complement.

9.5.2 Compliant Environment

In many applications, the interaction forces between the end-effector and a compliant environment can be approximated by the ideal elastic model of the form (9.21). If the stiffness matrix $\boldsymbol{K}$ is positive definite, this model corresponds to a fully constrained case and the environment deformation coincides with the elementary end-effector displacement. In general, however, the end-effector motion is only partially constrained by the environment and this situation can be modelled by introducing a suitable positive semi-definite stiffness matrix.

This kind of situation, even for a simple case, has been already considered in previous examples concerning the interaction with an elastically compliant plane. In a general case, the stiffness matrix describing the partially constrained interaction can be computed by modelling the environment as a pair of rigid bodies, S and R, connected through an ideal six-DOF spring, and assuming that the end-effector may slide on the external surface of body S.

Moreover, two reference frames are introduced, one attached to S and one attached to R. At equilibrium, corresponding to the undeformed spring, the end-effector frame is assumed to be coincident with the frames attached to S and R. The selection matrices $\boldsymbol{S}_f$ and $\boldsymbol{S}_v$ and the corresponding controlled force and velocity subspaces can be identified on the basis of the geometry of the contact between the end-effector and the environment.

Assumed frictionless contact, the interaction force applied by the end-effector on body S belongs to the force controlled subspace $\mathcal{R}(\boldsymbol{S}_f)$ and thus

$$\boldsymbol{h}_e = \boldsymbol{S}_f \boldsymbol{\lambda}, \tag{9.64}$$

where $\boldsymbol{\lambda}$ is a $(m \times 1)$ vector. Due to the presence of the generalized spring, the above force causes a deformation of the environment that can be computed as

$$d\boldsymbol{x}_{r,s} = \boldsymbol{C}\boldsymbol{h}_e, \tag{9.65}$$

where $\boldsymbol{C}$ is the compliance matrix of the spring between S and R, assumed to be nonsingular. On the other hand, the elementary displacement of the end-effector with respect to the equilibrium pose can be decomposed as

$$d\boldsymbol{x}_{r,e} = d\boldsymbol{x}_v + d\boldsymbol{x}_f, \tag{9.66}$$

where

$$d\boldsymbol{x}_v = \boldsymbol{P}_v d\boldsymbol{x}_{r,e} \tag{9.67}$$

is the component belonging to the velocity controlled subspace $\mathcal{R}(\boldsymbol{S}_v)$, where the end-effector may slide on the environment, whereas

$$d\boldsymbol{x}_f = (\boldsymbol{I}_6 - \boldsymbol{P}_v) d\boldsymbol{x}_{r,e} = (\boldsymbol{I}_6 - \boldsymbol{P}_v) d\boldsymbol{x}_{r,s} \tag{9.68}$$

is the component corresponding to the deformation of the environment. Notice that, in general, $\boldsymbol{P}_v d\boldsymbol{x}_{r,e} \neq \boldsymbol{P}_v d\boldsymbol{x}_{r,s}$.

Premultiplying both sides of (9.66) by matrix $\boldsymbol{S}_f^T$ and using (9.67), (9.68), (9.65), (9.64) yields

$$\boldsymbol{S}_f^T d\boldsymbol{x}_{r,e} = \boldsymbol{S}_f^T d\boldsymbol{x}_{r,s} = \boldsymbol{S}_f^T \boldsymbol{C} \boldsymbol{S}_f \boldsymbol{\lambda},$$

where the equality $\boldsymbol{S}_f^T \boldsymbol{P}_v = \boldsymbol{O}$ has been taken into account. The above equation can be used to compute vector $\boldsymbol{\lambda}$ which, replaced into (9.64), yields

$$\boldsymbol{h}_e = \boldsymbol{K}' d\boldsymbol{x}_{r,e}, \tag{9.69}$$

where

$$\boldsymbol{K}' = \boldsymbol{S}_f (\boldsymbol{S}_f^T \boldsymbol{C} \boldsymbol{S}_f)^{-1} \boldsymbol{S}_f^T \tag{9.70}$$

is the positive semi-definite stiffness matrix corresponding to the partially constrained elastic interaction.

Expression (9.70) is not invertible. However, using Eqs. (9.68), (9.65), the following equality can be derived:

$$d\boldsymbol{x}_f = \boldsymbol{C}' \boldsymbol{h}_e, \tag{9.71}$$

where the matrix

$$\boldsymbol{C}' = (\boldsymbol{I}_6 - \boldsymbol{P}_v) \boldsymbol{C}, \tag{9.72}$$

of rank $6 - m$, has the meaning of compliance matrix.

Notice that contact between the manipulator and the environment may be compliant along some directions and rigid along other directions. Therefore, the force control subspace can be decomposed into two distinct subspaces, one corresponding to elastic forces and the other corresponding to reaction forces. Matrices $\boldsymbol{K}'$ and $\boldsymbol{C}'$ should be modified accordingly.

9.6 Natural and Artificial Constraints

An interaction task can be assigned in terms of a desired end-effector force $\boldsymbol{h}_d$ and velocity $\boldsymbol{v}_d$. In order to be consistent with the constraints, these vectors must lie in the force and velocity controlled subspaces respectively. This can be guaranteed by specifying vectors $\boldsymbol{\lambda}_d$ and $\boldsymbol{\nu}_d$ and computing $\boldsymbol{h}_d$ and $\boldsymbol{v}_d$ as

$$\boldsymbol{h}_d = \boldsymbol{S}_f \boldsymbol{\lambda}_d, \quad \boldsymbol{v}_d = \boldsymbol{S}_v \boldsymbol{\nu}_d,$$

where $\boldsymbol{S}_f$ and $\boldsymbol{S}_v$ have to be suitably defined on the basis of the geometry of the task. Therefore vectors $\boldsymbol{\lambda}_d$ and $\boldsymbol{\nu}_d$ will be termed 'desired force' and 'desired velocity' respectively.

For many robotic tasks it is possible to define an orthogonal reference frame, eventually time-varying, where the constraints imposed by the environment can be easily identified, making task specification direct and intuitive. This reference frame O_c–$x_c y_c z_c$ is known as *constraint frame*.

Two DOFs correspond to each axis of the constraint frame: one associated with the linear velocity or to the force along the axis direction and the other associated with the angular velocity and to the moment along the axis direction.

For a given constraint frame, in the case of rigid environment and absence of friction, it can be observed that:

- Along each DOF, the environment imposes to the manipulator's end-effector either a velocity constraint — in the sense that it does not allow translation along a direction or rotation about an axis — or a force constraint — in the sense that it does not allow the application of any force along a direction or any torque about an axis; such constraints are termed *natural constraints* since they are determined directly by task geometry.
- The manipulator can control only the variables which are not subject to natural constraints; the reference values for those variables are termed *artificial constraints* since they are imposed with regard to the strategy for executing the given task.

Notice that the two sets of constraints are complementary, in that they regard different variables for each DOF. Also, they allow a complete specification of the task, since they involve all variables.

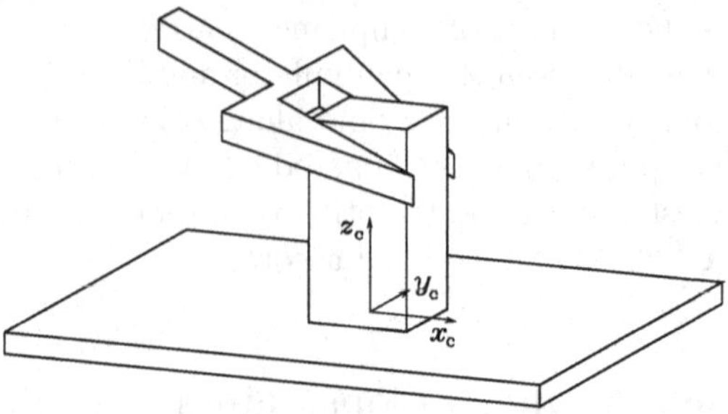

Fig. 9.12. Sliding of a prismatic object on a planar surface

In the case of compliant environment, for each DOF where interaction occurs, one can choose the variable to control, namely force or velocity, as long as the complementarity of the constraints is preserved. In case of high stiffness, it is advisable to choose the force as an artificial constraint and the velocity as a natural constraint, as for the case of rigid environment. Vice versa, in the case of low stiffness, it is convenient to make the opposite choice. Notice also that, in the presence of friction, forces and moments also arise along the DOFs corresponding to force natural constraints.

9.6.1 Analysis of Tasks

To illustrate description of an interaction task in terms of natural and artificial constraints as well as to emphasize the opportunity to use a constraint frame for task specification, in the following a number of typical case studies are analyzed.

Sliding on a planar surface

The end-effector manipulation task is the sliding of a prismatic object on a planar surface. Task geometry suggests choosing the constraint frame as attached to the contact plane with an axis orthogonal to the plane (Fig. 9.12). Alternatively, the task frame can be chosen with the same orientation but attached to the object.

Natural constraints can be determined first, assuming rigid and frictionless contact. Velocity constraints describe the impossibility to generate a linear velocity along axis z_c and angular velocities along axes x_c and y_c. Force constraints describe the impossibility to exert forces along axes x_c and y_c and a moment along axis z_c.

The artificial constraints regard the variables not subject to natural constraints. Hence, with reference to the natural constraints of force along axes x_c, y_c and moment along z_c, it is possible to specify artificial constraints for linear velocity along x_c, y_c and angular velocity along z_c. Similarly, with reference to natural constraints of linear velocity along axis z_c and angular velocity about axes x_c and y_c, it is possible to specify artificial constraints for force along z_c and moments about x_c and y_c. The set of constraints is summarized in Table 9.1.

Table 9.1. Natural and artificial constraints for the task of Fig. 9.12

Natural constraints	Artificial constraints
$\dot{o}_z^c$	f_z^c
ω_x^c	μ_x^c
ω_y^c	μ_y^c
f_x^c	$\dot{o}_x^c$
f_y^c	$\dot{o}_y^c$
μ_z^c	ω_z^c

For this task, the dimension of the force controlled subspace is $m = 3$, while the dimension of the velocity controlled subspace is $6-m = 3$. Moreover, matrices $\boldsymbol{S}_f$ and $\boldsymbol{S}_v$ can be chosen as

$$\boldsymbol{S}_f = \begin{bmatrix} 0 & 0 & 0 \\ 0 & 0 & 0 \\ 1 & 0 & 0 \\ 0 & 1 & 0 \\ 0 & 0 & 1 \\ 0 & 0 & 0 \end{bmatrix} \qquad \boldsymbol{S}_v = \begin{bmatrix} 1 & 0 & 0 \\ 0 & 1 & 0 \\ 0 & 0 & 0 \\ 0 & 0 & 0 \\ 0 & 0 & 0 \\ 0 & 0 & 1 \end{bmatrix}.$$

Notice that, if the constraint frame is chosen attached to the contact plane, matrices $\boldsymbol{S}_f$ and $\boldsymbol{S}_v$ remain constant if referred to the base frame but are time-varying if referred to the end-effector frame. Vice versa, if the constraint frame is chosen attached to the object, such matrices are constant if referred to the end-effector frame and time-varying if referred to the base frame.

In the presence of friction, non-null force and moment may also arise along the velocity controlled DOFs.

In the case of compliant plane, elastic forces and torques may be applied along the axis z_c and about the axes x_c and y_c respectively, corresponding to end-effector displacements along the same DOFs. On the basis of the expressions derived for $\boldsymbol{S}_f$ and $\boldsymbol{S}_v$, the elements of the stiffness matrix $\boldsymbol{K}'$ corresponding to the partially constrained interaction are null with the exception of those of the (3×3) block $\boldsymbol{K}'_m$ obtained selecting the rows 3, 4 and 5 of $\boldsymbol{K}'$. This block matrix can be computed as

$$\boldsymbol{K}'_m = \begin{bmatrix} c_{3,3} & c_{3,4} & c_{3,5} \\ c_{4,3} & c_{4,4} & c_{4,5} \\ c_{5,3} & c_{5,4} & c_{5,5} \end{bmatrix}^{-1},$$

where $c_{i,j} = c_{j,i}$ are the elements of the (6×6) compliant matrix $\boldsymbol{C}$.

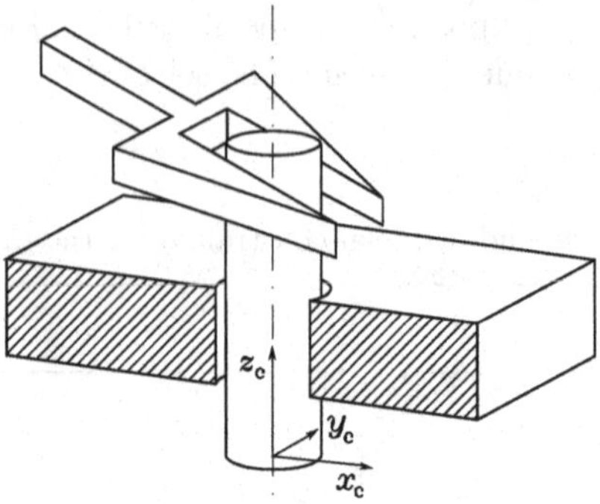

Fig. 9.13. Insertion of a cylindrical peg in a hole

Peg-in-hole

The end-effector manipulation task is the insertion of a cylindrical object (peg) in a hole. Task geometry suggests choosing the constraint frame with the origin in the centre of the hole and an axis parallel to the hole axis (Fig. 9.13). This frame can be chosen attached either to the peg or to the hole.

The natural constraints are determined by observing that it is not possible to generate arbitrary linear and angular velocities along axes x_c, y_c, nor is it possible to exert arbitrary force and moment along z_c. As a consequence, the artificial constraints can be used to specify forces and moments along x_c and y_c, as well as linear and angular velocity along z_c. Table 9.2 summarizes the constraints.

Table 9.2. Natural and artificial constraints for the task of Fig. 9.13

Natural constraints	Artificial constraints
$\dot{o}_x^c$	f_x^c
$\dot{o}_y^c$	f_y^c
ω_x^c	μ_x^c
ω_y^c	μ_y^c
f_z^c	$\dot{o}_z^c$
μ_z^c	ω_z^c

Among the variables subject to artificial constraints, $\dot{o}_z^c \neq 0$ describes insertion while the others are typically null to effectively perform the task.

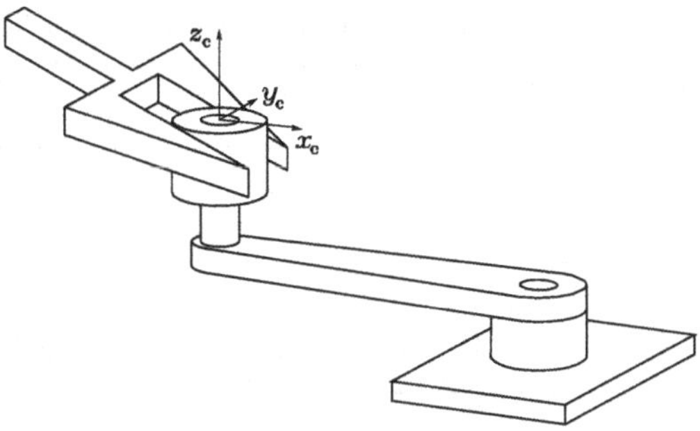

Fig. 9.14. Turning a crank

For this task, the dimension of the force controlled subspace is $m = 4$, while the dimension of the velocity controlled subspace is $6-m = 2$. Moreover, matrices $\boldsymbol{S}_f$ and $\boldsymbol{S}_v$ can be expressed as

$$\boldsymbol{S}_f = \begin{bmatrix} 1 & 0 & 0 & 0 \\ 0 & 1 & 0 & 0 \\ 0 & 0 & 0 & 0 \\ 0 & 0 & 1 & 0 \\ 0 & 0 & 0 & 1 \\ 0 & 0 & 0 & 0 \end{bmatrix} \qquad \boldsymbol{S}_v = \begin{bmatrix} 0 & 0 \\ 0 & 0 \\ 1 & 0 \\ 0 & 0 \\ 0 & 0 \\ 0 & 1 \end{bmatrix}.$$

Notice that, if the constraint frame is chosen attached to the hole, matrices $\boldsymbol{S}_f$ and $\boldsymbol{S}_v$ remain constant if referred to the base frame but are time-varying if referred to the end-effector frame. Vice versa, if the constraint frame is chosen attached to the peg, such matrices are constant if referred to the end-effector frame and time-varying if referred to the base frame.

Turning a crank

The end-effector manipulation task is the turning of a crank. Task geometry suggests choosing the constraint frame with an axis aligned with the axis of the idle handle and another axis aligned with the crank lever (Fig. 9.14). Notice that in this case the constraint frame is time-varying.

The natural constraints do not allow the generation of arbitrary linear velocities along x_c, z_c and angular velocities along x_c, y_c, nor arbitrary force along y_c and moment along z_c. As a consequence, the artificial constraints allow the specification of forces along x_c, z_c and moments along x_c, y_c, as well as a linear velocity along y_c and an angular velocity along z_c. The situation is summarized in Table 9.3.

Among the variables subject to artificial constraints, forces and moments are typically null for task execution.

Table 9.3. Natural and artificial constraints for task in Fig. 9.14

Natural constraints	Artificial constraints
$\dot{o}_x^c$	f_x^c
$\dot{o}_z^c$	f_z^c
ω_x^c	μ_x^c
ω_y^c	μ_y^c
f_y^c	$\dot{o}_y^c$
μ_z^c	ω_z^c

For this task, the dimension of the force controlled subspace is $m = 4$, while the dimension of the velocity controlled subspace is $6-m = 2$. Moreover, matrices $\boldsymbol{S}_f$ and $\boldsymbol{S}_v$ can be expressed as

$$\boldsymbol{S}_f = \begin{bmatrix} 1 & 0 & 0 & 0 \\ 0 & 0 & 0 & 0 \\ 0 & 1 & 0 & 0 \\ 0 & 0 & 1 & 0 \\ 0 & 0 & 0 & 1 \\ 0 & 0 & 0 & 0 \end{bmatrix} \qquad \boldsymbol{S}_v = \begin{bmatrix} 0 & 0 \\ 1 & 0 \\ 0 & 0 \\ 0 & 0 \\ 0 & 0 \\ 0 & 1 \end{bmatrix},$$

These matrices are constant in the constraint frame but are time-varying if referred to the base frame or to the end-effector frame, because the constraint frame moves with respect to both these frames during task execution.

9.7 Hybrid Force/Motion Control

Description of an interaction task between manipulator and environment in terms of natural constraints and artificial constraints, expressed with reference to the constraint frame, suggests a control structure that utilizes the artificial constraints to specify the objectives of the control system so that desired values can be imposed only onto those variables not subject to natural constraints. In fact, the control action should not affect those variables constrained by the environment so as to avoid conflicts between control and interaction with environment that may lead to an improper system behaviour. Such a control structure is termed *hybrid force/motion control*, since definition of artificial constraints involves both force and position or velocity variables.

For the design of hybrid control, it is useful rewriting the dynamic model of the manipulator with respect to the end-effector acceleration

$$\dot{\boldsymbol{v}}_e = \boldsymbol{J}(\boldsymbol{q})\ddot{\boldsymbol{q}} + \dot{\boldsymbol{J}}(\boldsymbol{q})\dot{\boldsymbol{q}}.$$

In particular, replacing (7.127) in the above expression yields

$$\boldsymbol{B}_e(\boldsymbol{q})\dot{\boldsymbol{v}}_e + \boldsymbol{n}_e(\boldsymbol{q}, \dot{\boldsymbol{q}}) = \boldsymbol{\gamma}_e - \boldsymbol{h}_e, \tag{9.73}$$

where

$$\begin{aligned}\boldsymbol{B}_e &= \boldsymbol{J}^{-T}\boldsymbol{B}\boldsymbol{J}^{-1}\\ \boldsymbol{n}_e &= \boldsymbol{J}^{-T}(\boldsymbol{C}\dot{\boldsymbol{q}}+\boldsymbol{g}) - \boldsymbol{B}_e\dot{\boldsymbol{J}}\dot{\boldsymbol{q}}.\end{aligned}$$

In the following, hybrid force/motion control is presented first for the case of compliant environment and then for the case of rigid environment.

9.7.1 Compliant Environment

In the case of compliant environment, on the basis of the decomposition (9.66) and of Eqs. (9.67), (9.71), (9.64), the following expression can be found

$$d\boldsymbol{x}_{r,e} = \boldsymbol{P}_v d\boldsymbol{x}_{r,e} + \boldsymbol{C}'\boldsymbol{S}_f\boldsymbol{\lambda}.$$

Computing the elementary displacements in terms of velocity, in view of (9.59) and taking into account that frame r is motionless, the end-effector velocity can be decomposed as

$$\boldsymbol{v}_e = \boldsymbol{S}_v\boldsymbol{\nu} + \boldsymbol{C}'\boldsymbol{S}_f\dot{\boldsymbol{\lambda}}, \tag{9.74}$$

where the first term belongs to the velocity control subspace and the second term belongs to its orthogonal complement. All the quantities are assumed to be referred to a common reference frame which, for simplicity, was not specified.

In the following, the base frame is chosen as the common reference frame; moreover, the contact geometry and the compliance matrix are assumed to be constant, namely $\dot{\boldsymbol{S}}_v = \boldsymbol{O}$, $\dot{\boldsymbol{S}}_f = \boldsymbol{O}$ and $\dot{\boldsymbol{C}}' = \boldsymbol{O}$. Therefore, computing the time derivative of (9.74) yields the following decomposition for the end-effector acceleration:

$$\dot{\boldsymbol{v}}_e = \boldsymbol{S}_v\dot{\boldsymbol{\nu}} + \boldsymbol{C}'\boldsymbol{S}_f\ddot{\boldsymbol{\lambda}}. \tag{9.75}$$

By adopting the inverse dynamics control law

$$\boldsymbol{\gamma}_e = \boldsymbol{B}_e(\boldsymbol{q})\boldsymbol{\alpha} + \boldsymbol{n}_e(\boldsymbol{q},\dot{\boldsymbol{q}}) + \boldsymbol{h}_e,$$

where $\boldsymbol{\alpha}$ is a new control input, in view of (9.73), the closed-loop equation is

$$\dot{\boldsymbol{v}}_e = \boldsymbol{\alpha}. \tag{9.76}$$

On the basis of the decomposition (9.75), with the choice

$$\boldsymbol{\alpha} = \boldsymbol{S}_v\boldsymbol{\alpha}_\nu + \boldsymbol{C}'\boldsymbol{S}_f\boldsymbol{f}_\lambda, \tag{9.77}$$

a complete decoupling between force control and velocity control can be achieved. In fact, replacing (9.75) and (9.77) into (9.76) and premultiplying both sides of the resulting equation once by $\boldsymbol{S}_v^{\dagger}$ and once by $\boldsymbol{S}_f^T$, the following equalities are obtained:

$$\dot{\boldsymbol{\nu}} = \boldsymbol{\alpha}_\nu \tag{9.78}$$

$$\ddot{\boldsymbol{\lambda}} = \boldsymbol{f}_\lambda. \tag{9.79}$$

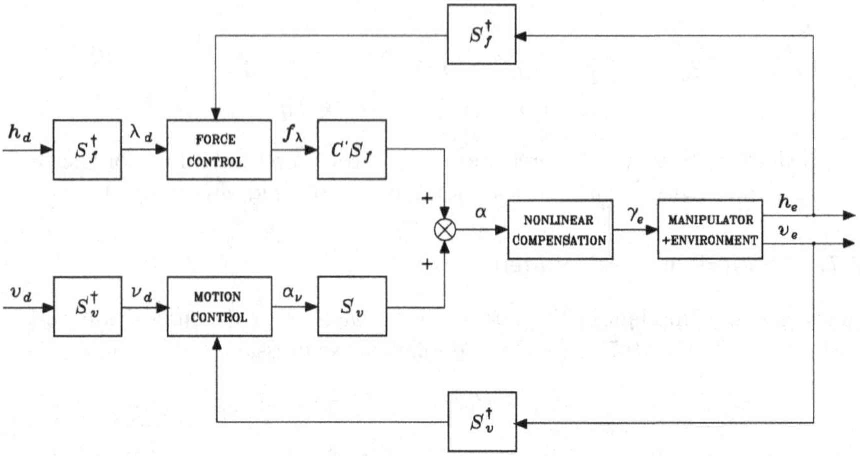

Fig. 9.15. Block scheme of a hybrid force/motion control for a compliant environment

Therefore, the task can be assigned specifying a desired force, in terms of vector $\boldsymbol{\lambda}_d(t)$, and a desired velocity, in terms of vector $\boldsymbol{\nu}_d(t)$. This control scheme is referred to as *hybrid force/velocity control*.

The desired velocity $\boldsymbol{\nu}_d$ can be achieved using the control law

$$\boldsymbol{\alpha}_\nu = \dot{\boldsymbol{\nu}}_d + \boldsymbol{K}_{P\nu}(\boldsymbol{\nu}_d - \boldsymbol{\nu}) + \boldsymbol{K}_{I\nu}\int_0^t (\boldsymbol{\nu}_d(\varsigma) - \boldsymbol{\nu}(\varsigma))d\varsigma, \tag{9.80}$$

where $\boldsymbol{K}_{P\nu}$ and $\boldsymbol{K}_{I\nu}$ are positive definite matrices. Vector $\boldsymbol{\nu}$ can be computed using (9.60), where the linear and angular velocity of the end-effector $\boldsymbol{v}_e$ is computed from joint position and velocity measurements.

The desired force $\boldsymbol{\lambda}_d$ can be achieved using the control law

$$\boldsymbol{f}_\lambda = \ddot{\boldsymbol{\lambda}}_d + \boldsymbol{K}_{D\lambda}(\dot{\boldsymbol{\lambda}}_d - \dot{\boldsymbol{\lambda}}) + \boldsymbol{K}_{P\lambda}(\boldsymbol{\lambda}_d - \boldsymbol{\lambda}), \tag{9.81}$$

where $\boldsymbol{K}_{D\lambda}$ and $\boldsymbol{K}_{P\lambda}$ are positive definite matrices. The implementation of the above control law requires the computation of vector $\boldsymbol{\lambda}$ via (9.54), using the measurements of end-effector forces and moments $\boldsymbol{h}_e$. Also, $\dot{\boldsymbol{\lambda}}$ can be computed as

$$\dot{\boldsymbol{\lambda}} = \boldsymbol{S}_f^\dagger \dot{\boldsymbol{h}}_e$$

in the ideal case that $\dot{\boldsymbol{h}}_e$ is available.

The block scheme of a hybrid force/motion control law is shown in Fig. 9.15. The output variables are assumed to be the vector of end-effector forces and moments $\boldsymbol{h}_e$ and the vector of end-effector linear and angular velocities $\boldsymbol{v}_e$.

Since force measurements are often noisy, the use of $\dot{\boldsymbol{h}}_e$ is not feasible. Hence, the feedback of $\dot{\boldsymbol{\lambda}}$ is often replaced by

$$\dot{\boldsymbol{\lambda}} = \boldsymbol{S}_f^{\dagger}\boldsymbol{K}'\boldsymbol{J}(\boldsymbol{q})\dot{\boldsymbol{q}}, \tag{9.82}$$

where $\boldsymbol{K}'$ is the positive semi-definite stiffness matrix (9.70).

If the contact geometry is known, but only an estimate of the stiffness/compliance of the environment is available, control law (9.77) can be rewritten in the form

$$\boldsymbol{\alpha} = \boldsymbol{S}_v\boldsymbol{\alpha}_\nu + \widehat{\boldsymbol{C}}'\boldsymbol{S}_f\boldsymbol{f}_\lambda,$$

where $\widehat{\boldsymbol{C}}' = (\boldsymbol{I}_6 - \boldsymbol{P}_v)\widehat{\boldsymbol{C}}$ and $\widehat{\boldsymbol{C}}$ is an estimate of $\boldsymbol{C}$.

In this case, Eq. (9.78) still holds while, in place of (9.79), the following equality can be derived:

$$\ddot{\boldsymbol{\lambda}} = \boldsymbol{L}_f\boldsymbol{f}_\lambda$$

where $\boldsymbol{L}_f = (\boldsymbol{S}_f^T\boldsymbol{C}\boldsymbol{S}_f)^{-1}\boldsymbol{S}_f^T\widehat{\boldsymbol{C}}\boldsymbol{S}_f$ is a nonsingular matrix. This implies that the force and velocity control subspaces remain decoupled and thus velocity control law (9.80) does not need to be modified.

Since matrix $\boldsymbol{L}_f$ is unknown, it is not possible to achieve the same performance of the force control as in the previous case. Also, if vector $\dot{\boldsymbol{\lambda}}$ is computed starting from velocity measurements using (9.82) with an estimate of $\boldsymbol{K}'$, only an estimate $\dot{\widehat{\boldsymbol{\lambda}}}$ is available that, in view of (9.82), (9.70), can be expressed in the form

$$\dot{\widehat{\boldsymbol{\lambda}}} = (\boldsymbol{S}_f^T\widehat{\boldsymbol{C}}\boldsymbol{S}_f)^{-1}\boldsymbol{S}_f^T\boldsymbol{J}(\boldsymbol{q})\dot{\boldsymbol{q}}.$$

Replacing (9.74) in the above equation and using (9.72) yields

$$\dot{\widehat{\boldsymbol{\lambda}}} = \boldsymbol{L}_f^{-1}\dot{\boldsymbol{\lambda}}. \tag{9.83}$$

Considering the control law

$$\boldsymbol{f}_\lambda = -k_{D\lambda}\dot{\widehat{\boldsymbol{\lambda}}} + \boldsymbol{K}_{P\lambda}(\boldsymbol{\lambda}_d - \boldsymbol{\lambda}), \tag{9.84}$$

with a constant $\boldsymbol{\lambda}_d$, the dynamics of the closed-loop system is

$$\ddot{\boldsymbol{\lambda}} + k_{D\lambda}\dot{\boldsymbol{\lambda}} + \boldsymbol{L}_f\boldsymbol{K}_{P\lambda}\boldsymbol{\lambda} = \boldsymbol{L}_f\boldsymbol{K}_{P\lambda}\boldsymbol{\lambda}_d,$$

where expression (9.83) has been used. This equation shows that the equilibrium solution $\boldsymbol{\lambda} = \boldsymbol{\lambda}_d$ is also asymptotically stable in the presence of an uncertain matrix $\boldsymbol{L}_f$, with a suitable choice of gain $k_{D\lambda}$ and of matrix $\boldsymbol{K}_{P\lambda}$.

Example 9.4

Consider a two-link planar arm in contact with a purely frictionless elastic plane; unlike the above examples, the plane is at an angle of $\pi/4$ with axis x_0 (Fig. 9.16).

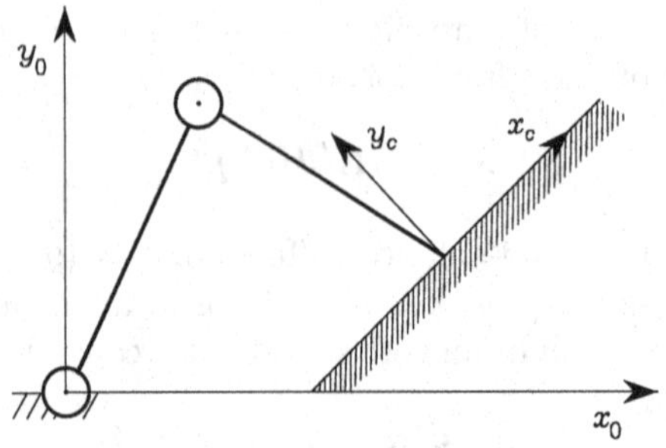

Fig. 9.16. Characterization of constraint frame for a two-link planar arm in contact with an elastically compliant plane

The natural choice of the constraint frame is that with axis x_c along the plane and axis y_c orthogonal to the plane; the task is obviously characterized by two DOFs. For the computation of the analytic model of the contact force, reference frames s and r are chosen so that, in the case of null force, they coincide with the constraint frame. In the presence of interaction, frame r remains attached to the rest position of the plane while frame s remains attached to the contact plane in the deformed position; the constraint frame is assumed to be attached to frame s. Matrices $\boldsymbol{S}_f^c$ and $\boldsymbol{S}_v^c$, referred to the constraint frame, have the form

$$\boldsymbol{S}_f^c = \begin{bmatrix} 0 \\ 1 \end{bmatrix} \qquad \boldsymbol{S}_v^c = \begin{bmatrix} 1 \\ 0 \end{bmatrix},$$

and the corresponding projection matrices are

$$\boldsymbol{P}_f^c = \begin{bmatrix} 0 & 0 \\ 0 & 1 \end{bmatrix} \qquad \boldsymbol{P}_v^c = \begin{bmatrix} 1 & 0 \\ 0 & 0 \end{bmatrix}.$$

In view of (9.70), (9.72), the stiffness and the compliance matrices, referred to the constraint frame, have the expression

$$\boldsymbol{K}'^c = \begin{bmatrix} 0 & 0 \\ 0 & c_{2,2}^{-1} \end{bmatrix} \qquad \boldsymbol{C}'^c = \begin{bmatrix} 0 & 0 \\ 0 & c_{2,2} \end{bmatrix},$$

where $c_{2,2}$ characterizes the compliance of frame s with respect to frame r along the direction orthogonal to the plane, aligned to axis y_c of the constraint frame.

It is evident that, under the assumption that the plane is compliant only along the orthogonal direction and that this direction remains fixed, then the constraint frame orientation remains constant with respect to the base frame. The corresponding rotation matrix is given by

$$\boldsymbol{R}_c = \begin{bmatrix} 1/\sqrt{2} & -1/\sqrt{2} \\ 1/\sqrt{2} & 1/\sqrt{2} \end{bmatrix}. \tag{9.85}$$

Moreover, if the task is to slide the manipulator tip along the plane, the end-effector velocity can be decomposed according to (9.74) in the form

$$\boldsymbol{v}_e^c = \boldsymbol{S}_v^c \nu + \boldsymbol{C}'^c \boldsymbol{S}_f^c \dot{\lambda}, \tag{9.86}$$

where all the quantities are referred to the constraint frame. It can be easily shown that, if $\boldsymbol{f}_e^c = [\,f_x^c \quad f_y^c\,]^T$ and $\boldsymbol{v}_e^c = [\,\dot{o}_x^c \quad \dot{o}_y^c\,]^T$, it is $\nu = \dot{o}_x^c$ and $\lambda = f_y^c$. This equation can also be referred to the base frame, where matrices

$$\boldsymbol{S}_f = \boldsymbol{R}_c \boldsymbol{S}_f^c = \begin{bmatrix} -1/\sqrt{2} \\ 1/\sqrt{2} \end{bmatrix} \qquad \boldsymbol{S}_v = \boldsymbol{R}_c \boldsymbol{S}_v^c = \begin{bmatrix} 1/\sqrt{2} \\ 1/\sqrt{2} \end{bmatrix},$$

are constant and the compliance matrix

$$\boldsymbol{C}' = \boldsymbol{R}_c \boldsymbol{C}'^c \boldsymbol{R}_c^T = c_{2,2} \begin{bmatrix} 1/2 & -1/2 \\ -1/2 & 1/2 \end{bmatrix}$$

is constant during the end-effector motion on the plane, for constant $c_{2,2}$. The adoption of an inverse dynamics control law, with the choice (9.77), leads to

$$\begin{aligned} \dot{\nu} &= \ddot{o}_x^c = \alpha_\nu \\ \ddot{\lambda} &= \ddot{f}_y^c = f_\lambda, \end{aligned}$$

showing that hybrid control achieves motion control along axis x_c and force control along axis y_c, provided that α_ν and f_λ are set according to (9.80) and (9.81) respectively.

Finally, notice that the formulation of the control law can be further simplified if the base frame is chosen parallel to the constraint frame.

9.7.2 Rigid Environment

In the case of rigid environment, the interaction force and moment can be written in the form $\boldsymbol{h}_e = \boldsymbol{S}_f \boldsymbol{\lambda}$. Vector $\boldsymbol{\lambda}$ can be eliminated from (9.73) by solving (9.73) for $\dot{\boldsymbol{v}}_e$ and substituting it into the time derivative of the equality (9.56). This yields

$$\boldsymbol{\lambda} = \boldsymbol{B}_f(\boldsymbol{q})\Big(\boldsymbol{S}_f^T \boldsymbol{B}_e^{-1}(\boldsymbol{q})(\boldsymbol{\gamma}_e - \boldsymbol{n}_e(\boldsymbol{q}, \dot{\boldsymbol{q}})) + \dot{\boldsymbol{S}}_f^T \boldsymbol{v}_e\Big), \tag{9.87}$$

where $\boldsymbol{B}_f = (\boldsymbol{S}_f^T \boldsymbol{B}_e^{-1} \boldsymbol{S}_f)^{-1}$.

Hence, the dynamic model (9.73) for the manipulator constrained by the rigid environment can be rewritten in the form

$$\boldsymbol{B}_e(\boldsymbol{q})\dot{\boldsymbol{v}}_e + \boldsymbol{S}_f \boldsymbol{B}_f(\boldsymbol{q})\dot{\boldsymbol{S}}_f^T \boldsymbol{v}_e = \boldsymbol{P}(\boldsymbol{q})(\boldsymbol{\gamma}_e - \boldsymbol{n}_e(\boldsymbol{q}, \dot{\boldsymbol{q}})), \tag{9.88}$$

with $\boldsymbol{P} = \boldsymbol{I}_6 - \boldsymbol{S}_f \boldsymbol{B}_f \boldsymbol{S}_f^T \boldsymbol{B}_e^{-1}$. Notice that $\boldsymbol{P}\boldsymbol{S}_f = \boldsymbol{O}$; moreover, this matrix is idempotent. Therefore, matrix $\boldsymbol{P}$ is a (6×6) projection matrix that filters out all the components of the end-effector forces lying in the subspace $\mathcal{R}(\boldsymbol{S}_f)$.

Equation (9.87) reveals that vector $\boldsymbol{\lambda}$ instantaneously depends on the control force $\boldsymbol{\gamma}_e$. Hence, by suitably choosing $\boldsymbol{\gamma}_e$, it is possible to control directly the m independent components of the end-effector forces that tend to violate the constraints; these components can be computed from $\boldsymbol{\lambda}$, using (9.52).

On the other hand, (9.88) represents a set of six second order differential equations whose solution, if initialized on the constraints, automatically satisfies Eq. (9.50) at all times.

The *reduced-order* dynamic model of the constrained system is described by $6-m$ independent equations that are obtained premultiplying both sides of (9.88) by matrix $\boldsymbol{S}_v^T$ and substituting the acceleration $\dot{\boldsymbol{v}}_e$ with

$$\dot{\boldsymbol{v}}_e = \boldsymbol{S}_v\dot{\boldsymbol{\nu}} + \dot{\boldsymbol{S}}_v\boldsymbol{\nu}.$$

Using the identities (9.58) and $\boldsymbol{S}_v^T\boldsymbol{P} = \boldsymbol{S}_v^T$ yields

$$\boldsymbol{B}_v(\boldsymbol{q})\dot{\boldsymbol{\nu}} = \boldsymbol{S}_v^T\left(\boldsymbol{\gamma}_e - \boldsymbol{n}_e(\boldsymbol{q},\dot{\boldsymbol{q}}) - \boldsymbol{B}_e(\boldsymbol{q})\dot{\boldsymbol{S}}_v\boldsymbol{\nu}\right), \tag{9.89}$$

where $\boldsymbol{B}_v = \boldsymbol{S}_v^T\boldsymbol{B}_e\boldsymbol{S}_v$. Moreover, expression (9.87) can be rewritten as

$$\boldsymbol{\lambda} = \boldsymbol{B}_f(\boldsymbol{q})\boldsymbol{S}_f^T\boldsymbol{B}_e^{-1}(\boldsymbol{q})\left(\boldsymbol{\gamma}_e - \boldsymbol{n}_e(\boldsymbol{q},\dot{\boldsymbol{q}}) - \boldsymbol{B}_e(\boldsymbol{q})\dot{\boldsymbol{S}}_v\boldsymbol{\nu}\right), \tag{9.90}$$

where the identity $\dot{\boldsymbol{S}}_f^T\boldsymbol{S}_v = -\boldsymbol{S}_f^T\dot{\boldsymbol{S}}_v$ has been exploited.

With reference to (9.89), consider the choice

$$\boldsymbol{\gamma}_e = \boldsymbol{B}_e(\boldsymbol{q})\boldsymbol{S}_v\boldsymbol{\alpha}_v + \boldsymbol{S}_f\boldsymbol{f}_\lambda + \boldsymbol{n}_e(\boldsymbol{q},\dot{\boldsymbol{q}}) + \boldsymbol{B}_e(\boldsymbol{q})\dot{\boldsymbol{S}}_v\boldsymbol{\nu}, \tag{9.91}$$

where $\boldsymbol{\alpha}_v$ and $\boldsymbol{f}_\lambda$ are new control inputs. By replacing (9.91) in (9.89), (9.90), the following two equations can be found:

$$\dot{\boldsymbol{\nu}} = \boldsymbol{\alpha}_\nu$$
$$\boldsymbol{\lambda} = \boldsymbol{f}_\lambda,$$

showing that the inverse dynamics control law (9.91) allows a complete decoupling between force and velocity controlled subspaces.

It is worth noticing that, for the implementation of control law (9.91), constraint equations (9.50) as well as Eq. (9.61) defining the vector of the configuration variables for the constrained system are not required, provided that matrices $\boldsymbol{S}_f$ and $\boldsymbol{S}_v$ are known. These matrices can be computed on the basis of the geometry of the environment or estimated on-line, using force and velocity measurements.

The task can easily be assigned by specifying a desired force, in terms of vector $\boldsymbol{\lambda}_d(t)$, and a desired velocity, in terms of vector $\boldsymbol{\nu}_d(t)$; the resulting scheme of *hybrid force/velocity control* is conceptually analogous to that shown in Fig. 9.15.

The desired velocity $\boldsymbol{\nu}_d$ can be achieved by setting $\boldsymbol{\alpha}_\nu$ according to (9.80), as for the case of compliant environment.

The desired force $\boldsymbol{\lambda}_d$ can be achieved by setting

$$\boldsymbol{f}_\lambda = \boldsymbol{\lambda}_d, \tag{9.92}$$

but this choice is very sensitive to disturbance forces, since it contains no force feedback. Alternative choices are

$$\boldsymbol{f}_{\lambda} = \boldsymbol{\lambda}_d + \boldsymbol{K}_{P\lambda}(\boldsymbol{\lambda}_d - \boldsymbol{\lambda}), \tag{9.93}$$

or

$$\boldsymbol{f}_{\lambda} = \boldsymbol{\lambda}_d + \boldsymbol{K}_{I\lambda}\int_0^t (\boldsymbol{\lambda}_d(\varsigma) - \boldsymbol{\lambda}(\varsigma))d\varsigma, \tag{9.94}$$

where $\boldsymbol{K}_{P\lambda}$ and $\boldsymbol{K}_{I\lambda}$ are suitable positive definite matrices. The proportional feedback is able to reduce the force error due to disturbance forces, while the integral action is able to compensate for constant bias disturbances.

The implementation of force feedback requires the computation of vector $\boldsymbol{\lambda}$ from the measurement of the end-effector force and moment $\boldsymbol{h}_e$, that can be achieved using (9.54).

When Eqs. (9.50) and (9.61) are available, matrices $\boldsymbol{S}_f$ and $\boldsymbol{S}_v$ can be computed according to (9.53) and (9.63), respectively. Moreover, a *hybrid force/position control* can be designed specifying a desired force $\boldsymbol{\lambda}_d(t)$, and a desired position $\boldsymbol{r}_d(t)$.

The force control law can be designed as above, while the desired position $\boldsymbol{r}_d$ can be achieved with the choice (see Problem 9.11)

$$\boldsymbol{\alpha}_{\nu} = \ddot{\boldsymbol{r}}_d + \boldsymbol{K}_{Dr}(\dot{\boldsymbol{r}}_d - \boldsymbol{\nu}) + \boldsymbol{K}_{Pr}(\boldsymbol{r}_d - \boldsymbol{r}), \tag{9.95}$$

where $\boldsymbol{K}_{Dr}$ and $\boldsymbol{K}_{Pr}$ are suitable positive definite matrices. Vector $\boldsymbol{r}$ can be computed from the joint position measurements using (9.61).

Bibliography

Scientific publications on force control are numerous and cover a time period of about 30 years. Review papers are [243] for the first decade, and [63] for the second decade. Recent monographs on this subject are [90, 209].

Control based on the concept of compliance was originally proposed by [165] in the joint space and [190] in the Cartesian space. The Remote Centre of Compliance concept is presented in [55] and its use for assembling operation is discussed in [242]. A reference paper for modelling of six-DOF elastic systems is [136] and their properties are analyzed in [177, 74]. The idea of impedance control was presented in [95] and a similar formulation can be found in [105]. Various schemes of impedance based on different representations of the orientation error are presented in [31, 32] and a rigorous analysis can be found in [223].

Early works on force control are described in [241]. Approaches not requiring the exact knowledge of the environment model are force control with position feedforward [65] and parallel force/position control [40, 43].

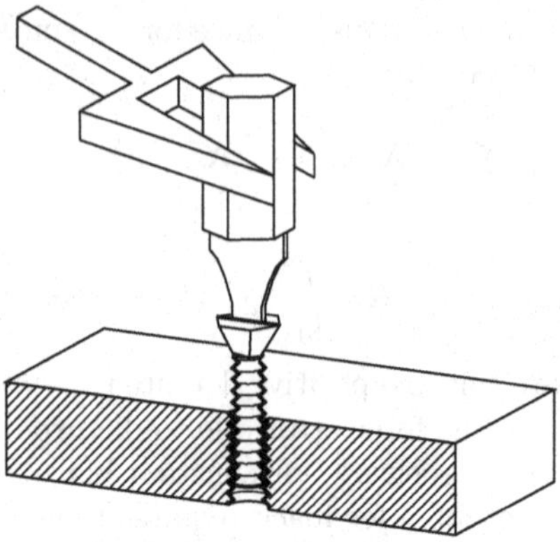

Fig. 9.17. Driving a screw in a hole

Natural and artificial constraints were introduced in [150] and further developed in [64, 27]. The concept of reciprocity of forces and velocity is discussed in [133], while invariance problems are analyzed in [66]. Models of elastic systems with semi-definite stiffness and compliance matrices are presented in [176]. The concept of hybrid force/motion control was introduced in [184] and the explicit inclusion of the manipulator dynamic model is presented in [114]. In [57] a systematic approach to modelling and control of interaction in the case of dynamic environment is introduced. Hybrid control in the presence of constraints in the Cartesian space is presented in [247, 249], while in [152] the constraints are formulated in the joint space. The use of impedance control in a hybrid framework is discussed in [5].

Adaptive versions of force/motion control schemes are proposed in [235]. The case of complex contact situations and time-varying constraints is presented in [28]. In [228, 62] the issues of controlling contact transitions is discussed, in order to overcome the instability problems evidenced in [70].

Problems

9.1. Derive expressions (9.10), (9.11).

9.2. Show that the equilibrium equations for the compliance control scheme are expressed by (9.22), (9.23).

9.3. Consider the planar arm in contact with the elastically compliant plane in Fig. 9.16. The plane forms an angle of $\pi/4$ with axis x_0 and its undeformed position intersects axis x_0 in the point of coordinates $(1, 0)$; the environment stiffness along axis y_c is $5 \cdot 10^3$ N/m. With the data of the arm in Sect. 8.7, design an impedance control. Perform a computer simulation of the interaction of the controlled manipulator along the rectilinear path from position $\boldsymbol{p}_i =$

$[1+0.1\sqrt{2} \;\; 0]^T$ to $\boldsymbol{p}_f = [1.2+0.1\sqrt{2} \;\; 0.2]^T$ with a trapezoidal velocity profile and a trajectory duration $t_f = 1\,\text{s}$. Implement the control in discrete-time with a sampling time of 1 ms.

9.4. Show that the equilibrium position for the parallel force/position control scheme satisfies (9.49).

9.5. Show that expression (9.54) with (9.55) is the solution which minimizes the norm $\|\boldsymbol{h}_e - \boldsymbol{S}_f(\boldsymbol{q})\boldsymbol{\lambda}\|$ with weighting matrix $\boldsymbol{W}$.

9.6. Show that stiffness matrix (9.70) can be expressed in the form $\boldsymbol{K}' = \boldsymbol{P}_f\boldsymbol{K}$.

9.7. For the manipulation task of driving a screw in a hole illustrated in Fig. 9.17, find the natural constraints and artificial constraints with respect to a suitably chosen constraint frame.

9.8. Show that the hybrid control scheme of Example 9.4, in the force controlled subspace, is equivalent to a force control scheme with inner velocity loop.

9.9. For the arm and environment of Example 9.4 compute the expressions of $\boldsymbol{S}_f^{\dagger}$ and $\boldsymbol{S}_v^{\dagger}$ in the constraint frame and in the base frame.

9.10. For the arm and environment of Problem 9.3, design a hybrid control in which a motion control law operates along axis x_c while a force control law operates along axis y_c; let the desired contact force along axis y_c be 50 N. Perform a computer simulation of the interaction of the controlled manipulator along a trajectory on the plane equivalent to that of Problem 9.3. Implement the control in discrete-time with a sampling time of 1 ms.

9.11. Show that control law (9.95) ensures tracking of a desired position $\boldsymbol{r}_d(t)$.

10

Visual Servoing

Vision allows a robotic system to obtain geometrical and qualitative information on the surrounding environment to be used both for motion planning and control. In particular, control based on feedback of visual measurements is termed *visual servoing*. In the first part of this chapter, some basic algorithms for image processing, aimed at extracting numerical information referred to as *image feature parameters*, are presented. These parameters, relative to images of objects present in the scene observed by a camera, can be used to estimate the pose of the camera with respect to the objects and vice versa. To this end, analytic *pose estimation methods*, based on the measurement of a certain number of points or *correspondences* are presented. Also, numerical pose estimation methods, based on the integration of the linear mapping between the camera velocity in the operational space and the time derivative of the feature parameters in the *image plane*, are introduced. In cases in which multiple images of the same scene, taken from different viewpoints, are available, additional information can be obtained using *stereo vision* techniques and *epipolar geometry*. A fundamental operation is also *camera calibration*; to this end, a calibration method based on the measurement of a certain number of correspondences is presented. Then, the two main approaches to visual servoing are introduced, namely *position-based visual servoing* and *image-based visual servoing*, as well as a scheme, termed *hybrid visual servoing*, which combines the benefits of both approaches.

10.1 Vision for Control

Vision plays a key role in a robotic system, as it can be used to obtain geometrical and qualitative information on the environment where the robot operates, without physical interaction. Such information may be employed by the control system at different levels, for the sole task planning and also for feedback control.

As an example, consider the case of a robot manipulator, equipped with a camera, which has to grasp an object using a gripper. Through vision the robot may acquire information capable of identifying the relative pose of the object with respect to the gripper. This information allows the control system to plan a trajectory leading the manipulator in an appropriate grasping configuration, computed on the basis of the pose and of the shape of the object, from which the closure of the gripper can be commanded.

The planned trajectory can be executed using a simple motion controller. In this approach, termed *look-and-move*, visual measurements are used in open loop, making the system very sensitive to uncertainties due, for instance, to poor positioning accuracy of the manipulator or to the fact that the object may have moved while the gripper reaches the grasp position.

On the other hand, in *vision-based control* or *visual servoing*, the visual measurements are fed back to the control to compute an appropriate error vector defined between the current pose of the object and the pose of the manipulator's end-effector.

A key characteristic of visual servoing, compared to motion and force control, is the fact that the controlled variables are not directly measured by the sensor, but are obtained from the measured quantities through complex elaborations, based on algorithms of *image processing* and *computational vision*.

In Sect. 5.4.3 it was shown that a monochrome camera simply provides a two-dimensional matrix of values of light intensity. From this matrix, the so-called *image feature parameters* are to be extracted in real time. The geometric relationships between one or more two-dimensional views of a scene and the corresponding 3D space are the basis of techniques of *pose estimation* of objects in the manipulator workspace or of the end-effector with respect to the surrounding objects. In this regard, of fundamental importance is the operation of *camera calibration*, which is necessary for calculating the *intrinsic parameters*, relating the quantities measured in the image plane to those referred to the camera frame, and the *extrinsic parameters*, relating the latter to quantities defined in a frame attached to the manipulator.

The vision-based control schemes can be divided into two categories, namely, those that realize *visual servoing in operational space*, also termed *position-based visual servoing*, and those that realize *visual servoing in the image space*, also known as *image-based visual servoing*. The main difference lies in the fact that the schemes of the first category use visual measurements to reconstruct the relative pose of the object with respect to the robot, or vice versa, while the schemes of the second category are based on the comparison of the feature parameters of the image of the object between the current and the desired pose. There are also schemes combining characteristics common to both categories, that can be classified as *hybrid visual servoing*.

Another aspect to be considered for vision-based control is the type of camera (colour or monochrome, resolution, fixed or variable focal length, CCD or CMOS technology). In this chapter, only the case of monochrome cameras with fixed focal length will be considered.

Equally important is the choice of the number of cameras composing the visual system and their location; this issue is briefly discussed in the following.

10.1.1 Configuration of the Visual System

A visual system may consist of only one camera, or two or more cameras. If more cameras are used to observe the same object of a scene, it is possible to retrieve information about its depth by evaluating its distance with respect to the visual system. This situation is referred to as *3D vision* or *stereo vision*, where the term *stereo* derives from the Greek and means solid. The human capability of perceiving objects in three dimensions relies on the fact that the brain receives the same images from two eyes, observing the same scene from slightly different angles.

It is clear that 3D vision can be achieved even with one camera, provided that two images of the same object, taken from two different poses, are available. If only a single image is available, the depth can be estimated on the basis of certain geometrical characteristics of the object known in advance. This means that, in many applications, mono-camera systems are often preferred to multi-camera systems, because they are cheaper and easier to calibrate, although characterized by lower accuracy.

Another feature that distinguishes visual systems for robot manipulators is the placement of cameras. For mono-camera systems there are two options: the *fixed configuration*, often referred to as *eye-to-hand*, where the camera is mounted in a fixed location, and the *mobile configuration*, or *eye-in-hand*, with the camera attached to the robot. For multi-camera systems, in addition to the mentioned solutions, it is also possible to consider the *hybrid configuration*, consisting of one or more cameras in eye-to-hand configuration, and one or more cameras in eye-in-hand configuration.

In the eye-to-hand configuration, the visual system observes the objects to be manipulated by a fixed pose with respect to the base frame of the manipulator. The advantage is that the camera field of view does not change during the execution of the task, implying that the accuracy of such measurements is, in principle, constant. However, in certain applications, such as assembly, it is difficult to prevent that the manipulator, moving in the camera field of view, occludes, in part or in whole, the view of the objects.

In the eye-in-hand configuration, the camera is placed on the manipulator and can be mounted both before and after the wrist. In the first case, the camera can observe the end-effector by a favourable pose and without occlusions caused by the manipulator arm; in the latter case, the camera is attached to the end-effector and typically observes only the object. In both situations, the camera field of view changes significantly during the motion and this produces a high variability in the accuracy of measurements. However, when the end-effector is close to the object, the accuracy becomes almost constant and is usually higher than that achievable with eye-to-hand cameras, with the advantage that occlusions are virtually absent.

Finally, hybrid configuration combines the benefits of the other two configurations, namely, ensures a good accuracy throughout the workspace, while avoiding the problems of occlusions.

A separate category is represented by robotic heads, which are typically equipped with a stereo vision system consisting of two cameras mounted on motorized mechanisms that allow for yaw motion, or *pan*, and pitch motion, or *tilt*, hence the name of *pan-tilt* cameras.

In this chapter, only schemes based on a single eye-in-hand camera will be considered. The extension of the algorithms to the case of a eye-to-hand camera, or to the case of multiple cameras, requires only minor modifications.

10.2 Image Processing

Visual information, unlike the information provided by other types of sensors, is very rich and varied and thus requires complex and computational expensive transformations before it can be used for controlling a robotic system. The objective of these transformations is the extraction of numerical information from the image, which provides a synthetic and robust description of the objects of interest in the scene, through the so-called *image feature parameters*.

To this end, two basic operations are required. The first is so-called *segmentation*, which aims at obtaining a representation suitable for the identification of measurable features of the image. The subsequent operation, termed *interpretation* is concerned with the measurement of the feature parameters of the image.

The source information is contained in a framestore, namely the two-dimensional memory array representing the spatial sample of the image. On the set of pixels the so-called *image function* is defined which, in general, is a vector function whose components represent the values of one or more physical quantities related to the pixel, in a sampled and quantized form.

For example, in the case of color images, the image function defined on a pixel of coordinates (X_I, Y_I) has three components $I_r(X_I, Y_I)$, $I_g(X_I, Y_I)$ and $I_b(X_I, Y_I)$, corresponding to the light intensity in the wavelengths of red, green and blue. For a monocrome black-and-white image, the image function is scalar and coincides with the light intensity in shades of gray $I(X_I, Y_I)$, also referred to as *gray level*. In the following, for simplicity, only monochrome images will be considered.

The number of gray levels depends on the adopted grey-scale resolution. In all cases, the gray scale is bounded by two gray levels, black and white, corresponding to the minimum and maximum measurable light intensity respectively. Most current acquisition equipments adopt a scale consisting of 256 gray levels, that can be represented by a single byte of memory.

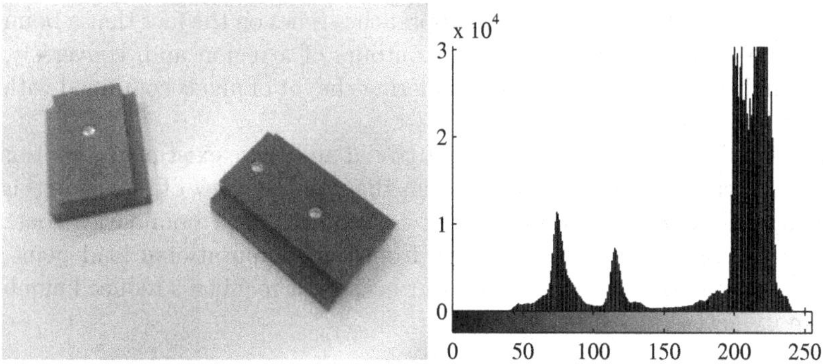

Fig. 10.1. Black-and-white image and corresponding gray-level histogram on the right

A representation of the framestore which is particularly useful for subsequent processing is the gray-level histogram, which provides the frequency of occurrence of each gray level in the image.

Where the gray levels are quantized from 0 to 255, the value $h(p)$ of the histogram at a particular gray level $p \in [0, 255]$ is the number of image pixels with gray level p. If this value is divided by the total number of pixels, the histogram is termed *normalized histogram*.

Figure 10.1 shows a black-and-white image and the corresponding gray-level histogram. Proceeding from left to right, three main peaks can be observed — from left to right — corresponding to the darkest object, the lightest object, and the background.

10.2.1 Image Segmentation

Segmentation consists of a grouping process, by which the image is divided into a certain number of groups, referred to as *segments*, so that the component of each group are similar with respect to one or more characteristics. Typically, distinct segments of the image correspond to distinct objects of the environment, or homogeneous object parts.

There are two complementary approaches to the problem of image segmentation: one is based on finding connected *regions* of the image, the other is concerned with detection of *boundaries*. The objective of region-based segmentation is that of grouping sets of pixels sharing common features into two-dimensional connected areas, with the implicit assumption that the resulting regions correspond to real-world surfaces or objects. On the other hand, boundary-based segmentation is aimed at identifying the pixels corresponding to object contours and isolating them from the rest of the image. The boundary of an object, once extracted, can be used to define the position and shape of the object itself.

The complementarity of the two approaches relies on the fact that a boundary can be achieved by isolating the contours of a region and, conversely, a region can be achieved simply by considering the set of pixels contained within a closed boundary.

The problem of segmentation is not trivial and there exist many solutions, some of which are sketched below. From the point of view of memory usage, boundary-based segmentation is more convenient, since boundaries contain a reduced number of pixels. However, from the computational load point of view, region-based segmentation is faster because it requires a reduced number of memory accesses.

Region-based segmentation

The central idea underlying region-based segmentation techniques is that of obtaining connected regions by continuous merging of initially small groups of adjacent pixels into larger ones.

Merging of two adjacent regions may happen only if the pixels belonging to these regions satisfy a common property, termed *uniformity predicate*. Often the uniformity predicate requires that the gray level of the pixels of the region belongs to a given interval.

In many applications of practical interest a thresholding approach is adopted and a light intensity scale composed of only two values (0 and 1) is considered. This operation is referred to as *binary segmentation* or image *binarization*, and corresponds to separating one or more objects present in the image from the background by comparing the gray level of each pixel with a threshold l. For light objects against a dark background, all the pixels whose gray level is greater than the threshold are considered to belong to a set S_o, corresponding to the objects, while all the other pixels are considered to belong to a set S_b corresponding to the background. It is obvious that this operation can be reversed for dark objects against a light background. When only an object is present in the image, the segmentation ends with the detection of sets S_o and S_b, representing two regions; in the presence of multiple objects, a further elaboration is required to separate the connected regions corresponding to the single objects. The image obtained assigning a light intensity equal to 0 to all the pixels of set S_o, and a light intensity equal to 1 to all the pixels of set S_b, or vice versa, is termed *binary image*.

A crucial factor for the success of binary segmentation is the choice of the threshold. A widely adopted method for selecting the threshold is based on the gray-level histogram, under the assumption that it contains clearly distinguishable minimum and maximum values, corresponding to the gray levels of the objects and of the background; the peaks of the histogram are also termed *modes*. For dark objects against a light background, the background corresponds to the mode which is located further to the right — as, for example, in the case of Fig. 10.1 — and the threshold can be chosen at the closest minimum to the left. For light objets against a dark background, the

Fig. 10.2. Binary image corresponding to image of Fig. 10.1

background corresponds to the mode which is located further to the left and the threshold should be selected accordingly. With reference to Fig. 10.1, the threshold can be set to $l = 152$. The corresponding binary image is reported in Fig. 10.2.

In practice, the gray-scale histogram is noisy and the modes are difficult to identify. Often, there is no clear separation between the gray levels of the objects and those of the background. To this end, various techniques have been developed to increase the robustness of binary segmentation, which require appropriate filtering of the image before binarization and the adoption of algorithms for automatic selection of the threshold.

Boundary-based segmentation

Boundary-based segmentation techniques usually obtain a boundary by grouping many single local edges, corresponding to local discontinuities of image gray level. In other words, local edges are sets of pixels where the light intensity changes abruptly.

The algorithms for boundary detection first derive an intermediate image based on local edges from the original gray-scale image, then they construct short-curve segments by edge linking, and finally obtain the boundaries by joining these curve segments through geometric primitives often known in advance.

Boundary-based segmentation algorithms vary in the amount of a priori knowledge they incorporate in associating or linking the edges and their effectiveness clearly depends on the quality of the intermediate image based on local edges. The more reliable the local edges in terms of their position, orientation and 'authenticity', the easier the task of the boundary detection algorithm.

Notice that edge detection is essentially a filtering process and can often be implemented via hardware, whereas boundary detection is a higher level task usually requiring more sophisticated software. Therefore, the current trend is that of using the most effective edge detector to simplify the boundary detection process. In the case of simple and well-defined shapes, boundary detection becomes straightforward and segmentation reduces to the sole edge detection.

Several edge detection techniques exist. Most of them require the calculation of the gradient or of the Laplacian of function $I(X_I, Y_I)$.

Since a local edge is defined as a transition between two regions of significantly different gray levels, it is obvious that the spatial gradient of function $I(X_I, Y_I)$, which measures the rate of change of the gray level, will have large magnitude close to these transitional boundary areas. Therefore, edge detection can be performed by grouping the pixels where the magnitude of the gradient is greater than a threshold. Moreover, the direction of the gradient vector will be the direction of maximum variation of the gray level.

Again, the choice of the value of the threshold is extremely important; in the presence of noise, the threshold is the result of a trade-off between the possibility of losing valid edges and that of detecting false edges.

For gradient computation, it suffices to evaluate the directional derivatives of function $I(X_I, Y_I)$ along two orthogonal directions. Since this function is defined on a discrete set of pixels, the derivatives are computed in an approximate way. The essential differences between gradient-based edge detection techniques are the directions used for the computation of the derivatives and the manner in which they approximate these derivatives and compute the gradient magnitude.

The most common operator for the computation of the gradient is that approximating the derivative along directions X_I and Y_I with the first differences:

$$\begin{aligned}\Delta_1 &= I(X_I + 1, Y_I) - I(X_I, Y_I)\\ \Delta_2 &= I(X_I, Y_I + 1) - I(X_I, Y_I).\end{aligned}$$

Other operators, less sensitive to noise effects are, for example, the *Roberts operator*, based on the first differences computed along the diagonals of a (2×2) square of pixels:

$$\begin{aligned}\Delta_1 &= I(X_I + 1, Y_I + 1) - I(X_I, Y_I)\\ \Delta_2 &= I(X_I, Y_I + 1) - I(X_I + 1, Y_I),\end{aligned}$$

and the *Sobel operator*, defined on a (3×3) square of pixels:

$$\begin{aligned}\Delta_1 = &\,(I(X_I + 1, Y_I - 1) + 2I(X_I + 1, Y_I) + I(X_I + 1, Y_I + 1))\\ &-(I(X_I - 1, Y_I - 1) + 2I(X_I - 1, Y_I) + I(X_I - 1, Y_I + 1))\\ \Delta_2 = &\,(I(X_I - 1, Y_I + 1) + 2I(X_I, Y_I + 1) + I(X_I + 1, Y_I + 1))\\ &-(I(X_I - 1, Y_I - 1) + 2I(X_I, Y_I - 1) + I(X_I + 1, Y_I - 1)).\end{aligned}$$

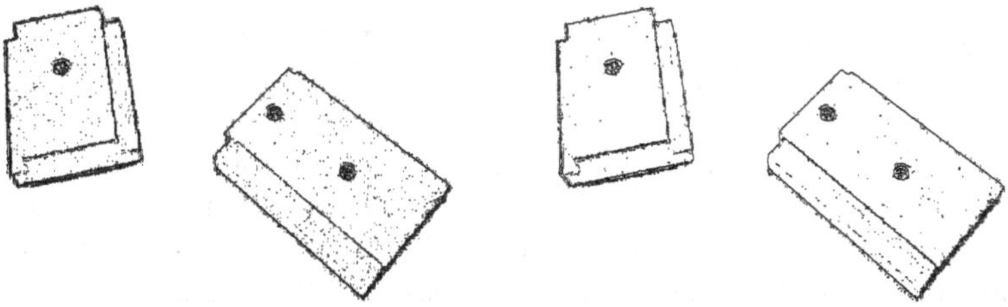

Fig. 10.3. Contours of image of Fig. 10.1 obtained using Roberts (*left*) and Sobel (*right*) operators

Then, the approximated magnitude, or norm, of gradient $G(X_I, Y_I)$ can be evaluated using one of the following two expressions:

$$G(X_I, Y_I) = \sqrt{\Delta_1^2 + \Delta_2^2}$$
$$G(X_I, Y_I) = |\Delta_1| + |\Delta_2|,$$

and direction $\theta(X_I, Y_I)$ with the relationship

$$\theta(X_I, Y_I) = \text{Atan2}\,(\Delta_2, \Delta_1)\,.$$

Figure 10.3 shows the images obtained from that of Fig. 10.1 by applying the gradient operators of Sobel and Roberts and binarization; the thresholds have been set to $l = 0.02$ and $l = 0.0146$, respectively.

An alternative edge detection method is based on the *Laplacian operator*, which requires the computation of the second derivatives of function $I(X_I, Y_I)$ along two orthogonal directions. Also in this case, suitable operators are used to discretize the computation of derivatives. One of the most common approximations is the following:

$$\begin{aligned} L(X_I, Y_I) = I(X_I, Y_I) - \frac{1}{4}\,&(I(X_I, Y_I + 1) + I(X_I, Y_I - 1) \\ &+ I(X_I + 1, Y_I) + I(X_I - 1, Y_I))\,. \end{aligned}$$

In this case, the pixels of the contour are those where the Laplacian is lower than a threshold. The reason is that the Laplacian is null at the points of maximum magnitude of the gradient. The Laplacian, unlike the gradient, does not provide directional information; moreover, being based on the calculation of second derivatives, it is more sensitive to noise than the gradient.

10.2.2 Image Interpretation

Image interpretation is the process of calculating the image feature parameters from the segments, whether they are represented in terms of boundaries or in terms of regions.

The feature parameters used in visual servoing applications sometimes require the computation of the so-called *moments*. These parameters are defined on a region $\mathcal{R}$ of the image and can be used to characterize the position, orientation and shape of the two-dimensional object corresponding to the region itself.

The general definition of moment $m_{i,j}$ of a region $\mathcal{R}$ of a framestore, with $i, j = 0, 1, 2, \ldots$, is the following:

$$m_{i,j} = \sum_{X_I, Y_I \in \mathcal{R}} I(X_I, Y_I) X_I^i Y_I^j.$$

In the case of binary images, by assuming the light intensity equal to one for all the points of region $\mathcal{R}$, and equal to zero for all the points not belonging to $\mathcal{R}$, the following simplified definition of moment is obtained:

$$m_{i,j} = \sum_{X_I, Y_I \in \mathcal{R}} X_I^i Y_I^j. \tag{10.1}$$

In view of this definition, moment $m_{0,0}$ coincides with the area of the region, computed in terms of the total number of pixels of region $\mathcal{R}$.

The quantities

$$\bar{x} = \frac{m_{1,0}}{m_{0,0}} \qquad \bar{y} = \frac{m_{0,1}}{m_{0,0}}$$

define the coordinates of the so-called *centroid* of the region. These coordinates can be used to detect uniquely the position of region $\mathcal{R}$ on the image plane.

Using an analogy from mechanics, region $\mathcal{R}$ can be seen as a two-dimensional rigid body of density equal to light intensity. Hence, moment $m_{0,0}$ corresponds to the mass of the body and the centroid corresponds to the centre of mass.

The value of moment $m_{i,j}$ in (10.1) depends on the position of region $\mathcal{R}$ in the image plane. Therefore, the so-called *central moments* are often considered, defined as

$$\mu_{i,j} = \sum_{X_I, Y_I \in \mathcal{R}} (X_I - \bar{x})^i (Y_I - \bar{y})^j,$$

which are invariant with respect to translation.

According to the mechanical analogy, it is easy to recognize that the central moments of second order $\mu_{2,0}$ and $\mu_{0,2}$ have the meaning of inertia moments with respect to axes X_I and Y_I respectively, while $\mu_{1,1}$ is an inertia product, and the matrix

$$\mathcal{I} = \begin{bmatrix} \mu_{2,0} & \mu_{1,1} \\ \mu_{1,1} & \mu_{0,2} \end{bmatrix},$$

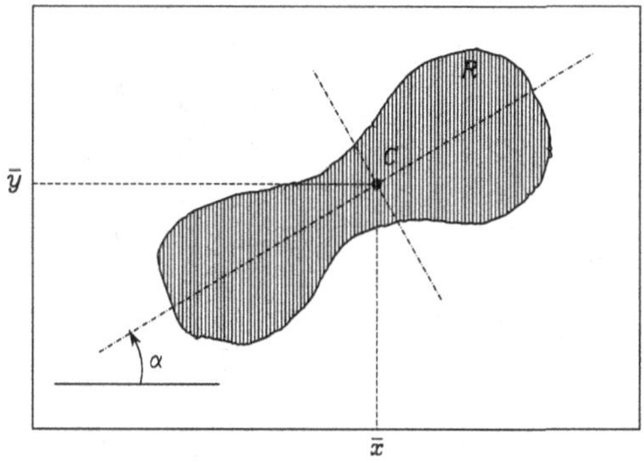

Fig. 10.4. Region of a binary image and some feature parameters

has the meaning of inertia tensor relative to the centre of mass. The eigenvalues of matrix $\mathcal{I}$ define the principal moments of inertia, termed *principal moments* of the region and the corresponding eigenvectors define the principal axes of inertia, termed *principal axes* of the region.

If region $\mathcal{R}$ is asymmetric, the principal moments of $\mathcal{I}$ are different and it is possible to characterize the orientation of $\mathcal{R}$ in terms of the angle α between the principal axis corresponding to the maximum moment and axis X. This angle can be computed with the equation (see Problem 10.1)

$$\alpha = \frac{1}{2}\tan^{-1}\left(\frac{2\mu_{1,1}}{\mu_{2,0} - \mu_{0,2}}\right). \tag{10.2}$$

As an example, in Fig. 10.4, the region of a binary image is shown; centroid C, the principal axes, and the angle α are evidenced.

Notice that the moments and the corresponding parameters can also be computed from the boundaries of the objects; moreover, these quantities are especially useful to characterize objects of generic form. Often, however, the objects present in the scene, especially those manufactured, have geometric characteristics which are useful to take into account for image interpretation.

For example, many objects have edges that, in the image plane, correspond to the intersection of linear parts of contour or to contour points of high curvature. The coordinates of these points on the image plane can be detected using algorithms robust against the noise, and therefore can be used as feature parameters of the image. They are usually termed *feature points*.

In other cases, it is possible to identify true geometric primitives such as lines or line segments, which are projections of linear edges or solids of revolution (cones, cylinders), or ellipses, obtained as projections of circles or spheres. These primitives can be characterized on the image plane in terms of a minimum set of parameters. For example, a line segment can be characterized by the coordinates of its endpoints, or alternatively, by the coordinates of its midpoint (centroid), its length (moment $m_{0,0}$) and its orientation (angle α); in both cases, the characterization of the line segment requires four parameters.

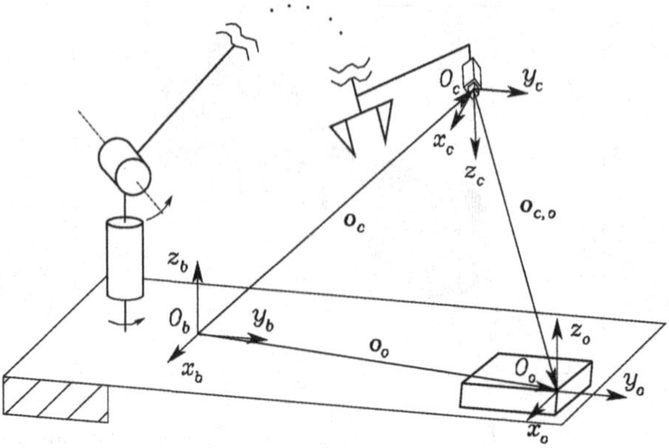

Fig. 10.5. Reference frames for an eye-in-hand camera

10.3 Pose Estimation

Visual servoing is based on the mapping between the feature parameters of an object measured in the image plane of the camera and the operational space variables defining the relative pose of the object with respect to the camera or, equivalently, of the camera with respect to the object. Often, it is sufficient to derive a differential mapping in terms of velocity. As for the computation of the inverse kinematics of a manipulator, the differential problem is easier to solve because the velocity mapping is linear; moreover, the solution to the differential problem can be used to compute the pose by using numerical integration algorithms.

The set of feature parameters of an image defines a $(k \times 1)$ vector $\boldsymbol{s}$, termed *feature vector*. In the following, to simplify notation, normalized coordinates (X, Y) defined in (5.44) will be used in place of pixel coordinates (X_I, Y_I) to define the feature vector. Since only pixel coordinates can be directly measured, the normalized coordinates should be computed from pixel coordinates using the inverse of mapping (5.45), provided that the intrinsic parameters of the camera are known.

The feature vector $\boldsymbol{s}$ of a point is defined as

$$\boldsymbol{s} = \begin{bmatrix} X \\ Y \end{bmatrix}, \tag{10.3}$$

while

$$\widetilde{\boldsymbol{s}} = \begin{bmatrix} X \\ Y \\ 1 \end{bmatrix}$$

denotes its representation in homogeneous coordinates.

10.3.1 Analytic Solution

Consider a camera, for example an eye-in-hand camera, and a reference frame O_c–$x_c y_c z_c$ attached to the camera; consider also a reference frame O_o–$x_o y_o z_o$ attached to the object, supposed to be rigid, and let $\boldsymbol{T}_o^c$ be the homogeneous transformation matrix corresponding to the relative pose of the object with respect to the camera, defined as

$$\boldsymbol{T}_o^c = \begin{bmatrix} \boldsymbol{R}_o^c & \boldsymbol{o}_{c,o}^c \\ \boldsymbol{0}^T & 1 \end{bmatrix}, \tag{10.4}$$

with $\boldsymbol{o}_{c,o}^c = \boldsymbol{o}_o^c - \boldsymbol{o}_c^c$, where $\boldsymbol{o}_c^c$ is the position vector of the origin of the camera frame with respect to the base frame, expressed in camera frame, $\boldsymbol{o}_o^c$ is the position vector of the origin of the object frame with respect to the base frame, expressed in the camera frame, and $\boldsymbol{R}_o^c$ is the rotation matrix of the object frame with respect to the camera frame (Fig. 10.5).

The problem to solve is that of computing the elements of matrix $\boldsymbol{T}_o^c$ from the measurements of object feature parameters in the camera image plane. To this end, consider n points of the object and let $\boldsymbol{r}_{o,i}^o = \boldsymbol{p}_i^o - \boldsymbol{o}_o^o$, $i = 1, \ldots, n$, denote the corresponding position vectors with respect to the object frame. These quantities are assumed to be known, for example, from a CAD model of the object. The projections of these points on the image plane have coordinates

$$\boldsymbol{s}_i = \begin{bmatrix} X_i \\ Y_i \end{bmatrix},$$

and define the feature vector

$$\boldsymbol{s} = \begin{bmatrix} \boldsymbol{s}_1 \\ \vdots \\ \boldsymbol{s}_n \end{bmatrix}. \tag{10.5}$$

The homogeneous coordinates of the points of the object with respect to the camera frame can be expressed as

$$\widetilde{\boldsymbol{r}}_{o,i}^c = \boldsymbol{T}_o^c \widetilde{\boldsymbol{r}}_{o,i}^o.$$

Therefore, using (5.44), the homogeneous coordinates of the projections of these points on the image plane are given by

$$\lambda_i \widetilde{\boldsymbol{s}}_i = \boldsymbol{\Pi} \boldsymbol{T}_o^c \widetilde{\boldsymbol{r}}_{o,i}^o, \tag{10.6}$$

with $\lambda_i > 0$.

Assume that n *correspondences* are available, namely n equations of the form (10.6) for n points of the object, whose coordinates are known both in the object frame and in the image plane. These correspondences define a system of equations to be solved for the unknown elements of matrix $\boldsymbol{T}_o^c$.

Computing the solution is a difficult task because, depending on the type and on the number of correspondences, multiple solutions may exist. This problem, in photogrammetry, is known as PnP (Perspective-n-Point) problem. In particular, it can be shown that:

- P3P problem has four solutions, in the case of three non-collinear points.
- P4P and P5P problems each have at least two solutions, in the case of non-coplanar points, while the solution is unique in the case of at least four coplanar points and no triplets of collinear points.
- PnP problem, with $n \geq 6$ non-coplanar points, has only one solution.

The analytic solution to PnP problem is rather laborious. However, the derivation becomes simpler in some particular cases as, for example, in the case of coplanar points.

Without loss of generality, assume that the plane containing the points of the object coincides with one of the three coordinate planes of the object frame, for instance, with the plane of equation $z_o = 0$; this implies that all the points of the plane have the third coordinate equal to zero. Multiplying both sides of (10.6) by the skew-symmetric matrix $\boldsymbol{S}(\widetilde{\boldsymbol{s}}_i)$, the product on the left-hand side is zero, leading to the homogeneous equation

$$\boldsymbol{S}(\widetilde{\boldsymbol{s}}_i)\boldsymbol{H}\begin{bmatrix} r_{x,i} & r_{y,i} & 1 \end{bmatrix}^T = \boldsymbol{0}, \tag{10.7}$$

where $r_{x,i}$ and $r_{y,i}$ are the two non-null components of vector $\boldsymbol{r}^o_{o,i}$ and $\boldsymbol{H}$ is the (3×3) matrix

$$\boldsymbol{H} = \begin{bmatrix} \boldsymbol{r}_1 & \boldsymbol{r}_2 & \boldsymbol{o}^c_{c,o} \end{bmatrix}, \tag{10.8}$$

$\boldsymbol{r}_1$ and $\boldsymbol{r}_2$ being the first and the second column of rotation matrix $\boldsymbol{R}^c_o$, respectively.

Vector equation (10.7), defined on homogeneous coordinates of points belonging to two planes, is know as *planar homography*; this denomination is used also for matrix $\boldsymbol{H}$.

Notice that Eq. (10.7) is linear with respect to $\boldsymbol{H}$ and, therefore, can be rewritten in the form

$$\boldsymbol{A}_i(\boldsymbol{s}_i)\boldsymbol{h} = \boldsymbol{0},$$

where $\boldsymbol{h}$ is the (9×1) column vector obtained by staking the columns of matrix $\boldsymbol{H}$, while $\boldsymbol{A}_i$ is the (3×9) matrix

$$\boldsymbol{A}_i(\boldsymbol{s}_i) = \begin{bmatrix} r_{x,i}\boldsymbol{S}(\widetilde{\boldsymbol{s}}_i) & r_{y,i}\boldsymbol{S}(\widetilde{\boldsymbol{s}}_i) & \boldsymbol{S}(\widetilde{\boldsymbol{s}}_i) \end{bmatrix}. \tag{10.9}$$

Since the rank of $\boldsymbol{S}(\cdot)$ is at most 2, then the rank of matrix $\boldsymbol{A}_i$ is also at most 2; therefore, to compute $\boldsymbol{h}$ (up to a scale factor), it is necessary to consider at least 4 equations of the form (10.9) written for 4 points of the plane, leading to the system of 12 equations with 9 unknowns

$$\begin{bmatrix} \boldsymbol{A}_1(\boldsymbol{s}_1) \\ \boldsymbol{A}_2(\boldsymbol{s}_2) \\ \boldsymbol{A}_3(\boldsymbol{s}_3) \\ \boldsymbol{A}_4(\boldsymbol{s}_4) \end{bmatrix} \boldsymbol{h} = \boldsymbol{A}(\boldsymbol{s})\boldsymbol{h} = \boldsymbol{0}, \tag{10.10}$$

with $\boldsymbol{s}$ defined in (10.5).

It can be shown that, considering a set of four points with no triplets of collinear points, matrix $\boldsymbol{A}$ has rank 8 and the system of equations in (10.10) admits a non-null solution $\zeta\boldsymbol{h}$, defined up to a scaling factor ζ (see Problem 10.2). As a result, matrix $\zeta\boldsymbol{H}$ can be computed up to a scaling factor ζ. The presented derivation is general and can be applied to any kind of planar homography defined by an equation of the form (10.7).

In view of (10.8), it is

$$\begin{aligned}\boldsymbol{r}_1 &= \zeta\boldsymbol{h}_1\\ \boldsymbol{r}_2 &= \zeta\boldsymbol{h}_2\\ \boldsymbol{o}^c_{c,o} &= \zeta\boldsymbol{h}_3\end{aligned}$$

where $\boldsymbol{h}_i$ denotes the i-th column of matrix $\boldsymbol{H}$. The absolute value of constant ζ can be computed by imposing the unit norm constraint to vectors $\boldsymbol{r}_1$ and $\boldsymbol{r}_2$:

$$|\zeta| = \frac{1}{\|\boldsymbol{h}_1\|} = \frac{1}{\|\boldsymbol{h}_2\|},$$

while the sign of ζ can be determined by choosing the solution corresponding to the object in front of the camera. Finally, the third column $\boldsymbol{r}_3$ of matrix $\boldsymbol{R}^c_o$ can be computed as

$$\boldsymbol{r}_3 = \boldsymbol{r}_1 \times \boldsymbol{r}_2.$$

Notice that, because of the noise affecting the measurements of the coordinates in the image plane, the results of this derivation are affected by errors that can be reduced by considering a number $n > 4$ of correspondences and computing the solution $\zeta\boldsymbol{h}$ to the $3n$ equations in (10.10), up to a scaling factor ζ, using least-squares techniques. This, however, does not guarantee that the resulting matrix $\boldsymbol{Q} = [\,\boldsymbol{r}_1 \quad \boldsymbol{r}_2 \quad \boldsymbol{r}_3\,]$ is a rotation matrix.

A possible solution overcoming this problem consists of computing the rotation matrix 'closest' to $\boldsymbol{Q}$ with respect to a given norm such as the matrix which minimizes the Frobenius norm[1]

$$\|\boldsymbol{R}^c_o - \boldsymbol{Q}\|_F = \left(\mathrm{Tr}\left((\boldsymbol{R}^c_o - \boldsymbol{Q})^T(\boldsymbol{R}^c_o - \boldsymbol{Q})\right)\right)^{1/2}, \tag{10.11}$$

with the constraint that $\boldsymbol{R}^c_o$ is a rotation matrix. The problem of minimizing norm (10.11) is equivalent to that of maximizing the trace of matrix $\boldsymbol{R}^{c\,T}_o\boldsymbol{Q}$. It can be shown that the solution to this problem is

$$\boldsymbol{R}^c_o = \boldsymbol{U}\begin{bmatrix}1 & 0 & 0\\ 0 & 1 & 0\\ 0 & 0 & \sigma\end{bmatrix}\boldsymbol{V}^T \tag{10.12}$$

where $\boldsymbol{U}$ and $\boldsymbol{V}^T$ are, respectively, the left and right orthogonal matrices of the singular value decomposition of $\boldsymbol{Q} = \boldsymbol{U}\boldsymbol{\Sigma}\boldsymbol{V}^T$. The choice $\sigma = \det(\boldsymbol{U}\boldsymbol{V}^T)$ ensures that the determinant of $\boldsymbol{R}^c_o$ is equal to one (see Problem 10.3).

[1] The Frobenius norm is defined in Sect. A.4.

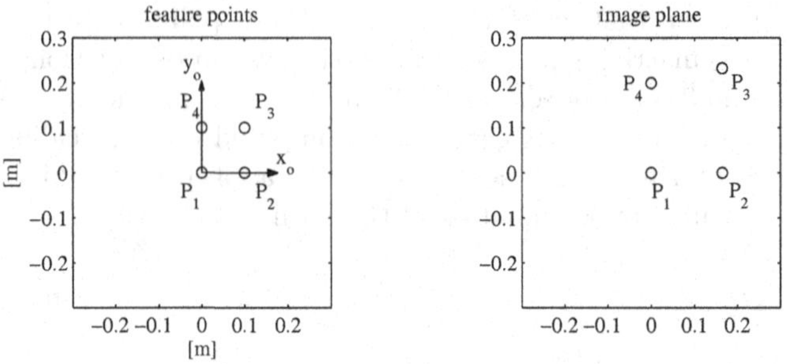

Fig. 10.6. Planar object; *left*: object frame and feature points; *right*: feature points projections on the normalized image plane of a camera in the pose of Example 10.1

Example 10.1

Consider a planar object characterized by four feature points shown in Fig. 10.6, where the object frame is represented. The feature points P_1, P_2, P_3, P_4 are the vertices of a square with side length $l = 0.1$ m. In Fig. 10.6, the images of the projections of the four points of the object on the normalized image plane of the camera are shown as well, under the assumption that the relative pose of the object frame with respect to the camera frame is characterized by rotation matrix

$$\boldsymbol{R}_o^c = \boldsymbol{R}_z(0)\boldsymbol{R}_y(\pi/4)\boldsymbol{R}_x(0) = \begin{bmatrix} 0.7071 & 0 & 0.7071 \\ 0 & 1 & 0 \\ -0.7071 & 0 & 0.7071 \end{bmatrix}$$

and position vector $\boldsymbol{o}_{c,o}^c = [0 \quad 0 \quad 0.5]^T$ m. The normalized coordinates of the four points of the object can be computed from the position vectors in the object frame $\boldsymbol{r}_{o,1}^o = [0 \quad 0 \quad 0]^T$ m, $\boldsymbol{r}_{o,2}^o = [0.1 \quad 0 \quad 0]^T$ m, $\boldsymbol{r}_{o,3}^o = [0.1 \quad 0.1 \quad 0]^T$ m, $\boldsymbol{r}_{o,4}^o = [0 \quad 0.1 \quad 0]^T$ m, using (10.6), which gives

$$\boldsymbol{s}_1 = \begin{bmatrix} 0 \\ 0 \end{bmatrix} \quad \boldsymbol{s}_2 = \begin{bmatrix} 0.1647 \\ 0 \end{bmatrix} \quad \boldsymbol{s}_3 = \begin{bmatrix} 0.1647 \\ 0.2329 \end{bmatrix} \quad \boldsymbol{s}_4 = \begin{bmatrix} 0 \\ 0.2 \end{bmatrix}.$$

To solve the inverse problem, namely that of computing matrix $\boldsymbol{T}_o^c$ from the coordinates of the four points both in the image plane and in the object frame, it is necessary to build matrix $\boldsymbol{A}(\boldsymbol{s})$ from four matrices $\boldsymbol{A}_i(\boldsymbol{s}_i)$ defined in (10.9). It is easy to verify that matrix $\boldsymbol{A}(\boldsymbol{s})$ has rank 8; moreover, a non-null solution to the system of equations in (10.10) can be computed using the singular value decomposition

$$\boldsymbol{A} = \boldsymbol{U}\boldsymbol{\Sigma}\boldsymbol{V}^T$$

and coincides with the last column of matrix $\boldsymbol{V}$, namely, with the right eigenvector corresponding to the null singular value of $\boldsymbol{A}$. From this computation, matrix

$$\zeta\boldsymbol{H} = \begin{bmatrix} -0.4714 & 0 & 0 \\ 0 & -0.6667 & 0 \\ 0.4714 & 0 & -0.3333 \end{bmatrix}$$

can be obtained. Normalization of the first column yields $|\zeta| = 1.5$. It is possible to verify that, with the choice $\zeta = -1.5$, the exact solution for $\boldsymbol{o}^c_{c,o}$ is obtained; moreover, matrix $\boldsymbol{Q}$ coincides numerically with the sought rotation matrix $\boldsymbol{R}^c_o$, without using any kind of approximate solution. This result was expected, due to the absence of measurement noise affecting image plane coordinates $\boldsymbol{s}_i$.

The above derivation is a particular case of a method, termed *direct linear transformation*, which is aimed at computing the elements of matrix $\boldsymbol{T}^c_o$ by solving a system of linear equations obtained using n correspondences relative to points in generic configuration. In detail, from equalities

$$\boldsymbol{S}(\widetilde{\boldsymbol{s}}_i)\,[\,\boldsymbol{R}^c_o \quad \boldsymbol{o}^c_{c,o}\,]\,\widetilde{\boldsymbol{r}}^o_{o,i} = \boldsymbol{0}, \tag{10.13}$$

which coincide with (10.7) in the case that points $\boldsymbol{r}^o_{o,i}$ belong to plane $z_0 = 0$, two independent linear equations with 12 unknowns are obtained, taking into account that matrix $\boldsymbol{S}(\cdot)$ is, at most, of rank 2. Therefore, n correspondences produce $2n$ equations.

It can be shown that, considering a set of 6 points not all coplanar, the matrix of coefficients of the corresponding system of 12 equations with 12 unknowns has rank 11; therefore, the solution is defined up to a scaling factor. Once this solution has been computed, the elements of rotation matrix $\boldsymbol{R}^c_o$ and of vector $\boldsymbol{o}^c_{c,o}$ can be obtained using a derivation similar to that presented above. Notice that, in practical applications, due to the presence of noise, the system of equations has rank 12 and admits only the null solution. In this case, it is necessary to consider a number $n > 6$ of correspondences and to compute the solution of the resulting system of equations, defined up to a scaling factor, using least-squares techniques.

In conclusion, the presented method permits the computation of matrix $\boldsymbol{T}^c_o$, characterizing the relative pose of the object frame with respect to the camera frame, from the projections of n points of the object on the camera image plane. To this end, it is necessary to know the position of these points with respect to the object frame, and thus the object geometry, besides the camera *intrinsic parameters*. The latter are required for computing normalized coordinates $\boldsymbol{s}_i$ from pixel coordinates.

Notice that, if it is required to compute the object pose with respect to the base frame (as usually happens in the case of eye-to-hand cameras) or to the end-effector frame (as usually happens in the case of eye-in-hand cameras), then it is also necessary to know the camera *extrinsic parameters*. In fact, in the first case, it is

$$\boldsymbol{T}^b_o = \boldsymbol{T}^b_c \boldsymbol{T}^c_o, \tag{10.14}$$

where the elements of matrix $\boldsymbol{T}^b_c$ represent the extrinsic parameters of an eye-to-hand camera; on the other hand, in the case of eye-in-hand camera, it is

$$\boldsymbol{T}^e_o = \boldsymbol{T}^e_c \boldsymbol{T}^c_o, \tag{10.15}$$

where the extrinsic parameters matrix $\boldsymbol{T}_c^e$ characterize the pose of the camera with respect to the end-effector frame.

10.3.2 Interaction Matrix

If the object is in motion with respect to the camera, the feature vector $\boldsymbol{s}$ is, in general, time-varying. Therefore, it is possible to define a $(k \times 1)$ velocity vector in the image plane $\dot{\boldsymbol{s}}$.

The motion of the object with respect to the camera is characterized by the relative velocity

$$\boldsymbol{v}_{c,o}^c = \begin{bmatrix} \dot{\boldsymbol{o}}_{c,o}^c \\ \boldsymbol{R}_c^T(\boldsymbol{\omega}_o - \boldsymbol{\omega}_c) \end{bmatrix}, \tag{10.16}$$

where $\dot{\boldsymbol{o}}_{c,o}^c$ is the time derivative of vector $\boldsymbol{o}_{c,o}^c = \boldsymbol{R}_c^T(\boldsymbol{o}_o - \boldsymbol{o}_c)$, representing the relative position of the origin of the object frame with respect to the origin of the camera frame, while $\boldsymbol{\omega}_o$ and $\boldsymbol{\omega}_c$ are the angular velocities of the object frame and camera frame, respectively.

The equation relating $\dot{\boldsymbol{s}}$ to $\boldsymbol{v}_{c,o}^c$ is

$$\dot{\boldsymbol{s}} = \boldsymbol{J}_s(\boldsymbol{s}, \boldsymbol{T}_o^c)\boldsymbol{v}_{c,o}^c, \tag{10.17}$$

where $\boldsymbol{J}_s$ is a $(k \times 6)$ matrix termed *image Jacobian*. This equation is linear but $\boldsymbol{J}_s$ depends, in general, on the current value of the feature vector $\boldsymbol{s}$ and on the relative pose of the object with respect to the camera $\boldsymbol{T}_o^c$.

It is useful to consider also the mapping between the image plane velocity $\dot{\boldsymbol{s}}$, the absolute velocity of the camera frame

$$\boldsymbol{v}_c^c = \begin{bmatrix} \boldsymbol{R}_c^T \dot{\boldsymbol{o}}_c \\ \boldsymbol{R}_c^T \boldsymbol{\omega}_c, \end{bmatrix}$$

and the absolute velocity of the object frame

$$\boldsymbol{v}_o^c = \begin{bmatrix} \boldsymbol{R}_c^T \dot{\boldsymbol{o}}_o \\ \boldsymbol{R}_c^T \boldsymbol{\omega}_o \end{bmatrix}.$$

To this end, vector $\dot{\boldsymbol{o}}_{c,o}^c$ can be expressed as

$$\dot{\boldsymbol{o}}_{c,o}^c = \boldsymbol{R}_c^T(\dot{\boldsymbol{o}}_o - \dot{\boldsymbol{o}}_c) + \boldsymbol{S}(\boldsymbol{o}_{c,o}^c)\boldsymbol{R}_c^T\boldsymbol{\omega}_c,$$

which allows equality (10.16) to be rewritten in the compact form

$$\boldsymbol{v}_{c,o}^c = \boldsymbol{v}_o^c + \boldsymbol{\Gamma}(\boldsymbol{o}_{c,o}^c)\boldsymbol{v}_c^c, \tag{10.18}$$

with

$$\boldsymbol{\Gamma}(\cdot) = \begin{bmatrix} -\boldsymbol{I} & \boldsymbol{S}(\cdot) \\ \boldsymbol{O} & -\boldsymbol{I} \end{bmatrix}.$$

Therefore, Eq. (10.17) can be rewritten in the form

$$\dot{\boldsymbol{s}} = \boldsymbol{J}_s \boldsymbol{v}_o^c + \boldsymbol{L}_s \boldsymbol{v}_c^c, \tag{10.19}$$

where the $(k \times 6)$ matrix

$$\boldsymbol{L}_s = \boldsymbol{J}_s(\boldsymbol{s}, \boldsymbol{T}_o^c)\boldsymbol{\Gamma}(\boldsymbol{o}_{c,o}^c) \tag{10.20}$$

is termed *interaction matrix*. This matrix, in view of (10.19), defines the linear mapping between the absolute velocity of the camera $\boldsymbol{v}_c^c$ and the corresponding image plane velocity $\dot{\boldsymbol{s}}$, in the case that the object is fixed with respect to the base frame ($\boldsymbol{v}_o^c = \boldsymbol{0}$).

The analytic expression of the interaction matrix is, in general, simpler than that of the image Jacobian. The latter can be computed from the interaction matrix using the equation

$$\boldsymbol{J}_s(\boldsymbol{s}, \boldsymbol{T}_o^c) = \boldsymbol{L}_s \boldsymbol{\Gamma}(-\boldsymbol{o}_{c,o}^c), \tag{10.21}$$

obtained from (10.20), with $\boldsymbol{\Gamma}^{-1}(\boldsymbol{o}_{c,o}^c) = \boldsymbol{\Gamma}(-\boldsymbol{o}_{c,o}^c)$. In the following, examples of computation of interaction matrix and image Jacobian for some of the most common cases in applications are provided.

Interaction matrix of a point

Consider a point P of the object characterized, with respect to the camera frame, by the vector of coordinates

$$\boldsymbol{r}_c^c = \boldsymbol{R}_c^T(\boldsymbol{p} - \boldsymbol{o}_c), \tag{10.22}$$

where $\boldsymbol{p}$ is the position of point P with respect to the base frame. Choose vector $\boldsymbol{s}$ of normalized coordinates (10.3) as the feature vector of the point. In view of (5.44), the following expression holds:

$$\boldsymbol{s} = \boldsymbol{s}(\boldsymbol{r}_c^c), \tag{10.23}$$

with

$$\boldsymbol{s}(\boldsymbol{r}_c^c) = \frac{1}{z_c}\begin{bmatrix} x_c \\ y_c \end{bmatrix} = \begin{bmatrix} X \\ Y \end{bmatrix}, \tag{10.24}$$

and $\boldsymbol{r}_c^c = \begin{bmatrix} x_c & y_c & z_c \end{bmatrix}^T$. Computing the time derivative of (10.23) and using (10.24) yields

$$\dot{\boldsymbol{s}} = \frac{\partial \boldsymbol{s}(\boldsymbol{r}_c^c)}{\partial \boldsymbol{r}_c^c}\dot{\boldsymbol{r}}_c^c, \tag{10.25}$$

with

$$\frac{\partial \boldsymbol{s}(\boldsymbol{r}_c^c)}{\partial \boldsymbol{r}_c^c} = \frac{1}{z_c}\begin{bmatrix} 1 & 0 & -x_c/z_c \\ 0 & 1 & -y_c/z_c \end{bmatrix} = \frac{1}{z_c}\begin{bmatrix} 1 & 0 & -X \\ 0 & 1 & -Y \end{bmatrix}.$$

To compute the interaction matrix, vector $\dot{\boldsymbol{r}}_c^c$ can be obtained from the time derivative of (10.22) under the assumption that $\boldsymbol{p}$ is constant:

$$\dot{\boldsymbol{r}}_c^c = -\boldsymbol{R}_c^T \dot{\boldsymbol{o}}_c + \boldsymbol{S}(\boldsymbol{r}_c^c)\boldsymbol{R}_c^T \boldsymbol{\omega}_c = [-\boldsymbol{I} \quad \boldsymbol{S}(\boldsymbol{r}_c^c)]\,\boldsymbol{v}_c^c. \tag{10.26}$$

Combining Eqs. (10.25), (10.26), the following expression of interaction matrix of a point can be found:

$$\boldsymbol{L}_s(\boldsymbol{s}, z_c) = \begin{bmatrix} -\dfrac{1}{z_c} & 0 & \dfrac{X}{z_c} & XY & -(1+X^2) & Y \\ 0 & -\dfrac{1}{z_c} & \dfrac{Y}{z_c} & 1+Y^2 & -XY & -X \end{bmatrix}, \tag{10.27}$$

revealing that this matrix depends on the components of vector $\boldsymbol{s}$ and the sole component z_c of vector $\boldsymbol{r}_c^c$.

The image Jacobian of a point can be computed using (10.21), (10.27) and has the expression

$$\boldsymbol{J}_s(\boldsymbol{s}, \boldsymbol{T}_o^c) = \frac{1}{z_c}\begin{bmatrix} 1 & 0 & -X & -r_{o,y}^c X & r_{o,z}^c + r_{o,x}^c X & -r_{o,y}^c \\ 0 & 1 & -Y & -(r_{o,z}^c + r_{o,y}^c Y) & r_{o,x}^c Y & r_{o,x}^c \end{bmatrix},$$

where $r_{o,x}^c$, $r_{o,y}^c$, $r_{o,z}^c$ are the components of vector $\boldsymbol{r}_o^c = \boldsymbol{r}_c^c - \boldsymbol{o}_{c,o}^c = \boldsymbol{R}_o^c \boldsymbol{r}_o^o$, $\boldsymbol{r}_o^o$ being the constant vector expressing the position of point P with respect to the object frame.

Interaction matrix of a set of points

The interaction matrix of a set of n points of the object $P_1, \dots P_n$ can be built by considering the $(2n \times 1)$ feature vector (10.5). If $\boldsymbol{L}_{s_i}(\boldsymbol{s}_i, z_{c,i})$ denotes the interaction matrix corresponding to point P_i, then the interaction matrix of the set of points will be the $(2n \times 6)$ matrix

$$\boldsymbol{L}_s(\boldsymbol{s}, \boldsymbol{z}_c) = \begin{bmatrix} \boldsymbol{L}_{s_1}(\boldsymbol{s}_1, z_{c,1}) \\ \vdots \\ \boldsymbol{L}_{s_n}(\boldsymbol{s}_n, z_{c,n}) \end{bmatrix},$$

with $\boldsymbol{z}_c = [z_{c,1} \dots z_{c,n}]^T$.

The image Jacobian of a set of points can be easily computed from the interaction matrix, using (10.21).

Interaction matrix of a line segment

A line segment is the part of the line connecting two points P_1 and P_2. The projection on the image plane is still a line segment that can be characterized in terms of the middle point coordinates $\bar{x}$, $\bar{y}$, the length L, and the angle α

formed by the line with respect to axis X. Therefore, the feature vector can be defined as

$$\boldsymbol{s} = \begin{bmatrix} \bar{x} \\ \bar{y} \\ L \\ \alpha \end{bmatrix} = \begin{bmatrix} (X_1 + X_2)/2 \\ (Y_1 + Y_2)/2 \\ \sqrt{\Delta X^2 + \Delta Y^2} \\ \tan^{-1}(\Delta Y/\Delta X) \end{bmatrix} = \boldsymbol{s}(\boldsymbol{s}_1, \boldsymbol{s}_2) \tag{10.28}$$

with $\Delta X = X_2 - X_1$, $\Delta Y = Y_2 - Y_1$ and $\boldsymbol{s}_i = [\, X_i \quad Y_i \,]^T$, $i = 1, 2$. Computing the time derivative of this equation yields

$$\begin{aligned} \dot{\boldsymbol{s}} &= \frac{\partial \boldsymbol{s}}{\partial \boldsymbol{s}_1} \dot{\boldsymbol{s}}_1 + \frac{\partial \boldsymbol{s}}{\partial \boldsymbol{s}_2} \dot{\boldsymbol{s}}_2 \\ &= \left(\frac{\partial \boldsymbol{s}}{\partial \boldsymbol{s}_1} \boldsymbol{L}_{s_1}(\boldsymbol{s}_1, z_{c,1}) + \frac{\partial \boldsymbol{s}}{\partial \boldsymbol{s}_2} \boldsymbol{L}_{s_2}(\boldsymbol{s}_2, z_{c,2}) \right) \boldsymbol{v}_c^c, \end{aligned}$$

where $\boldsymbol{L}_{s_i}$ is the interaction matrix of point P_i, under the assumption that the line segment is fixed with respect to the base frame. Therefore, the interaction matrix of a line segment is

$$\boldsymbol{L}_s(\boldsymbol{s}, \boldsymbol{z}_c) = \frac{\partial \boldsymbol{s}}{\partial \boldsymbol{s}_1} \boldsymbol{L}_{s_1}(\boldsymbol{s}_1, z_{c,1}) + \frac{\partial \boldsymbol{s}}{\partial \boldsymbol{s}_2} \boldsymbol{L}_{s_2}(\boldsymbol{s}_2, z_{c,2}),$$

with

$$\frac{\partial \boldsymbol{s}}{\partial \boldsymbol{s}_1} = \begin{bmatrix} 1/2 & 0 \\ 0 & 1/2 \\ -\Delta X/L & -\Delta Y/L \\ \Delta Y/L^2 & -\Delta X/L^2 \end{bmatrix} \qquad \frac{\partial \boldsymbol{s}}{\partial \boldsymbol{s}_2} = \begin{bmatrix} 1/2 & 0 \\ 0 & 1/2 \\ \Delta X/L & \Delta Y/L \\ -\Delta Y/L^2 & \Delta X/L^2 \end{bmatrix}.$$

Notice that vectors $\boldsymbol{s}_1$ and $\boldsymbol{s}_2$ can be computed as functions of parameters $\bar{x}$, $\bar{y}$, L, α, using (10.28). Therefore, the interaction matrix can be expressed as a function of the feature vector $\boldsymbol{s} = [\, \bar{x} \quad \bar{y} \quad L \quad \alpha \,]^T$, besides the components $z_{c,1}$ and $z_{c,2}$ of the endpoints P_1 and P_2 of the line segment.

The image Jacobian of a line segment can be easily computed from the interaction matrix using (10.21).

10.3.3 Algorithmic Solution

The interaction matrix $\boldsymbol{L}_s$ is, in general, a matrix of dimension $(k \times m)$, where k is equal to the number of feature parameters of the image and m is the dimension of velocity vector $\boldsymbol{v}_c^c$. Usually $m = 6$, but it may happen that $m < 6$, when the relative motion of the object with respect to the camera is constrained.

The image Jacobian $\boldsymbol{J}_s$ is also of dimension $(k \times m)$, being related to $\boldsymbol{L}_s$ by mapping (10.21). Since this mapping is invertible, the rank of $\boldsymbol{J}_s$ coincides with that of $\boldsymbol{L}_s$.

In the case that $\boldsymbol{L}_s$ is full rank, by using (10.17), it is possible to compute $\boldsymbol{v}^c_{c,o}$ from $\dot{\boldsymbol{s}}$.

In particular, if $k = m$, the velocity $\boldsymbol{v}^c_{c,o}$ can be obtained using the expression

$$\boldsymbol{v}^c_{c,o} = \boldsymbol{\Gamma}(\boldsymbol{o}^c_{c,o})\boldsymbol{L}_s^{-1}\dot{\boldsymbol{s}}, \tag{10.29}$$

which requires the computation of the inverse of the interaction matrix.

In the case $k > m$, the interaction matrix has more rows than columns and Eq. (10.17) can be solved using a least-squares technique, whose solution can be written in the form

$$\boldsymbol{v}^c_{c,o} = \boldsymbol{\Gamma}(\boldsymbol{o}^c_{c,o})(\boldsymbol{L}_s^T\boldsymbol{L}_s)^{-1}\boldsymbol{L}_s^T\dot{\boldsymbol{s}}, \tag{10.30}$$

where $(\boldsymbol{L}_s^T\boldsymbol{L}_s)^{-1}\boldsymbol{L}_s^T$ is the left pseudo-inverse of $\boldsymbol{L}_s$. This situation is rather frequent in applications, because it permits using interaction matrices with good condition numbers.

Finally, in the case $k < m$, the interaction matrix has more columns than rows and Eq. (10.17) admits infinite solutions. This implies that the number of parameters of the observed image is not sufficient to determine uniquely the relative motion of the object with respect to the camera. Hence, there exist relative motions of the object with respect to the camera (or vice versa) that do not produce variations of the image feature parameters. The velocities associated with these relative motions belong to the null subspace of $\boldsymbol{J}_s$, which has the same dimension of the null subspace of $\boldsymbol{L}_s$. If the problem is that of computing uniquely the relative pose of the object with respect to the camera from feature parameters in the image plane, this case has no interest.

Example 10.2

The interaction matrix of a point P is a matrix with more columns than rows of dimension (2×6) and rank 2; therefore, the null subspace has dimension 4. It can be seen immediately that this subspace contains the velocity of the camera translational motion along the visual ray projecting point P on the image plane, proportional to vector

$$\boldsymbol{v}_1 = \begin{bmatrix} X & Y & 1 & 0 & 0 & 0 \end{bmatrix}^T,$$

as well as the velocity of the camera rotational motion about this visual ray, proportional to vector

$$\boldsymbol{v}_2 = \begin{bmatrix} 0 & 0 & 0 & X & Y & 1 \end{bmatrix}^T.$$

Vectors $\boldsymbol{v}_1$ and $\boldsymbol{v}_2$ are independent and belong to a base of the null subspace of $\boldsymbol{L}_s$. The remaining base vectors are not easy to find geometrically, but can be easily computed analytically.

The pose estimation problem may be cast in a form analogous to that of inverse kinematics algorithms for robot manipulators. To this end, it is necessary to represent the relative pose of the object with respect to the camera using a minimum number of coordinates, in terms of the $(m \times 1)$ vector

$$\boldsymbol{x}_{c,o} = \begin{bmatrix} \boldsymbol{o}_{c,o}^c \\ \boldsymbol{\phi}_{c,o} \end{bmatrix}, \tag{10.31}$$

where $\boldsymbol{o}_{c,o}^c$ characterizes the position of the origin of the object frame with respect to the camera frame and $\boldsymbol{\phi}_{c,o}$ characterizes the relative orientation. If Euler angles are used to represent orientation, then $\boldsymbol{\phi}_{c,o}$ is the vector of the angles extracted from rotation matrix $\boldsymbol{R}_o^c$ and the mapping between $\boldsymbol{v}_{c,o}^c$ and $\dot{\boldsymbol{x}}_{c,o}$ is expressed by

$$\boldsymbol{v}_{c,o}^c = \begin{bmatrix} \boldsymbol{I} & \boldsymbol{O} \\ \boldsymbol{O} & \boldsymbol{T}(\boldsymbol{\phi}_{c,o}) \end{bmatrix} \dot{\boldsymbol{x}}_{c,o} = \boldsymbol{T}_A(\boldsymbol{\phi}_{c,o})\dot{\boldsymbol{x}}_{c,o}. \tag{10.32}$$

Example 10.3

Consider a camera mounted on the end-effector of the SCARA manipulator of Fig. 2.36. Choose the camera frame parallel to the end-effector frame, with axis z_c pointing downward. Assume that the camera observes a fixed planar object, parallel to the image plane, and that axis z_o of the object frame is parallel to axis z_c and points downward.

The geometry of the problem suggests that the relative position of the object with respect to the camera can be represented by a vector $\boldsymbol{o}_{c,o}^c$, whereas the relative orientation can be defined by the angle α between the object frame and the camera frame about axis z_c. Therefore, $m = 4$ and

$$\boldsymbol{x}_{c,o} = \begin{bmatrix} \boldsymbol{o}_{c,o}^c \\ \alpha \end{bmatrix}. \tag{10.33}$$

Moreover, the time derivative $\dot{\alpha}$ coincides with the component of $\boldsymbol{\omega}_{c,o}^c$ along z_c, and this is the sole non-null component of the angular velocity of the relative motion of the object frame with respect to the camera frame. Hence, in (10.32), $\boldsymbol{T}_A(\boldsymbol{\phi}_{c,o})$ is the (4×4) identity matrix.

Equation (10.17) can be rewritten in the form

$$\dot{\boldsymbol{s}} = \boldsymbol{J}_{A_s}(\boldsymbol{s}, \boldsymbol{x}_{c,o})\dot{\boldsymbol{x}}_{c,o}, \tag{10.34}$$

where the matrix

$$\boldsymbol{J}_{A_s}(\boldsymbol{s}, \boldsymbol{x}_{c,o}) = \boldsymbol{L}_s \boldsymbol{\Gamma}(-\boldsymbol{o}_{c,o}^c)\boldsymbol{T}_A(\boldsymbol{\phi}_{c,o}) \tag{10.35}$$

has a meaning analogous to that of the analytic Jacobian of a manipulator.

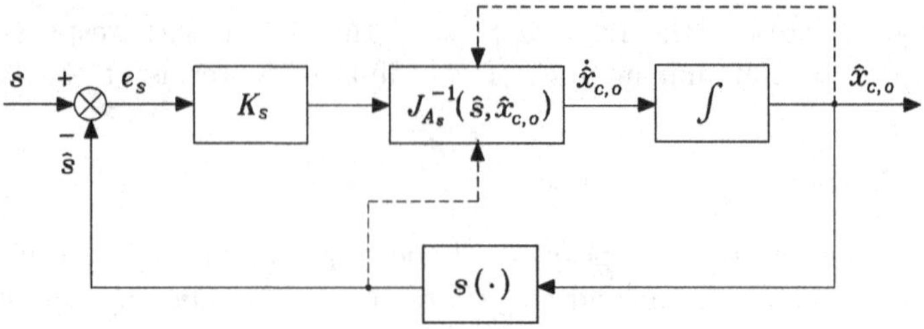

Fig. 10.7. Pose estimation algorithm based on the inverse of image Jacobian

Equation (10.34) is the starting point of a numeric integration algorithm for the computation of $\boldsymbol{x}_{c,o}$, similar to inverse kinematics algorithms. Let $\widehat{\boldsymbol{x}}_{c,o}$ denote the current estimate of vector $\boldsymbol{x}_{c,o}$ and let

$$\widehat{\boldsymbol{s}} = \boldsymbol{s}(\widehat{\boldsymbol{x}}_{c,o})$$

be the corresponding vector of image feature parameters computed from the pose specified by $\widehat{\boldsymbol{x}}_{c,o}$; the objective of this algorithm is the minimization of the error

$$\boldsymbol{e}_s = \boldsymbol{s} - \widehat{\boldsymbol{s}}. \tag{10.36}$$

Notice that, for the purpose of numerical integration, vector $\boldsymbol{s}$ is constant while the current estimate $\widehat{\boldsymbol{s}}$ depends on the current integration time. Therefore, computing the time derivative of (10.36) yields

$$\dot{\boldsymbol{e}}_s = -\dot{\widehat{\boldsymbol{s}}} = -\boldsymbol{J}_{A_s}(\widehat{\boldsymbol{s}}, \widehat{\boldsymbol{x}}_{c,o})\dot{\widehat{\boldsymbol{x}}}_{c,o}. \tag{10.37}$$

Assumed that matrix $\boldsymbol{J}_{A_s}$ is square and nonsingular, the choice

$$\dot{\widehat{\boldsymbol{x}}}_{c,o} = \boldsymbol{J}_{A_s}^{-1}(\widehat{\boldsymbol{s}}, \widehat{\boldsymbol{x}}_{c,o})\boldsymbol{K}_s\boldsymbol{e}_s \tag{10.38}$$

leads to the equivalent linear system

$$\dot{\boldsymbol{e}}_s + \boldsymbol{K}_s\boldsymbol{e}_s = \boldsymbol{0}. \tag{10.39}$$

Therefore, if $\boldsymbol{K}_s$ is a positive definite matrix (usually diagonal), system (10.39) is asymptotically stable and the error tends to zero with a convergence speed that depends on the eigenvalues of matrix $\boldsymbol{K}_s$. The convergence to zero of error $\boldsymbol{e}_s$ ensures the asymptotic convergence of the estimate $\widehat{\boldsymbol{x}}_{c,o}$ to the true value $\boldsymbol{x}_{c,o}$.

The block scheme of the pose estimation algorithm is shown in Fig. 10.7, where $\boldsymbol{s}(\cdot)$ denotes the function computing the feature vector of the 'virtual' image corresponding to the current estimate $\widehat{\boldsymbol{x}}_{c,o}$ of the object pose with respect to the camera. This algorithm can be used as an alternative to the analytic methods for pose estimation illustrated in Sect. 10.3.1. Obviously, the convergence properties depend on the choice of the image feature parameters and on the initial value of estimate $\widehat{\boldsymbol{x}}_{c,o}(0)$, which may produce instability problems related to the singularities of matrix $\boldsymbol{J}_{A_s}$.

Notice that, in view of (10.35), the singularities of matrix $\boldsymbol{J}_{A_s}$ are both the representation singularities of the orientation and those of the interaction matrix. The most critical singularities are those of the interaction matrix, since they depend on the choice of the image feature parameters.

To separate the effects of the two types of singularities, it is convenient to compute (10.38) in two steps, evaluating first

$$\widehat{\boldsymbol{v}}^c_{c,o} = \boldsymbol{\Gamma}(\boldsymbol{o}^c_{c,o})\boldsymbol{L}_s^{-1}\boldsymbol{K}_s\boldsymbol{e}_s, \tag{10.40}$$

and then

$$\dot{\widehat{\boldsymbol{x}}}_{c,o} = \boldsymbol{T}_A^{-1}(\boldsymbol{\phi}_{c,o})\widehat{\boldsymbol{v}}^c_{c,o}. \tag{10.41}$$

Assumed to work far from representation singularities, the problem of singularities of $\boldsymbol{L}_s$ can be overcome by using a number k of feature parameters greater than the minimum required m. This choice also allows a reduction of the effects of measurement noise. The resulting estimation algorithm requires the use of the left pseudo-inverse of $\boldsymbol{L}_s$ in place of the inverse, namely

$$\widehat{\boldsymbol{v}}^c_{c,o} = \boldsymbol{\Gamma}(\boldsymbol{o}^c_{c,o})(\boldsymbol{L}_s^T\boldsymbol{L}_s)^{-1}\boldsymbol{L}_s^T\boldsymbol{K}_s\boldsymbol{e}_s \tag{10.42}$$

in place of (10.40). The convergence of error (10.36) can be shown using the direct Lyapunov method based on the positive definite function [2]

$$V(\boldsymbol{e}_s) = \frac{1}{2}\boldsymbol{e}_s^T\boldsymbol{K}_s\boldsymbol{e}_s > 0 \qquad \forall \boldsymbol{e}_s \neq \boldsymbol{0}.$$

Computing the time derivative of this function, and using (10.37), (10.35), (10.41), (10.42), yields

$$\dot{V} = -\boldsymbol{e}_s^T\boldsymbol{K}_s\boldsymbol{L}_s(\boldsymbol{L}_s^T\boldsymbol{L}_s)^{-1}\boldsymbol{L}_s^T\boldsymbol{K}_s\boldsymbol{e}_s,$$

which is negative semi-definite because $\mathcal{N}(\boldsymbol{L}_s^T) \neq \emptyset$, $\boldsymbol{L}_s^T$ being a matrix with more columns than rows. Therefore, the system is stable but not asymptotically stable. This implies that the error is bounded, but in some cases the algorithm can get stuck with $\boldsymbol{e}_s \neq \boldsymbol{0}$ and $\boldsymbol{K}_s\boldsymbol{e}_s \in \mathcal{N}(\boldsymbol{L}_s^T)$.

Notice that the pose estimation methods based on inverse Jacobian are as efficient in terms of accuracy, speed of convergence and computational load, as the initial estimate $\widehat{\boldsymbol{x}}_{c,o}(0)$ is close to the true value $\boldsymbol{x}_{c,o}$. Therefore, these methods are mainly adopted for real-time 'visual tracking' applications, where the estimate on an image taken at time $\bar{t}$ is computed assuming as initial value the estimate computed on the image taken at time $\bar{t}-T$, T being the sampling time of the image (multiple of the sampling time of the numerical integration algorithm).

[2] See Sect. C.3 for the illustration of the direct Lyapunov method.

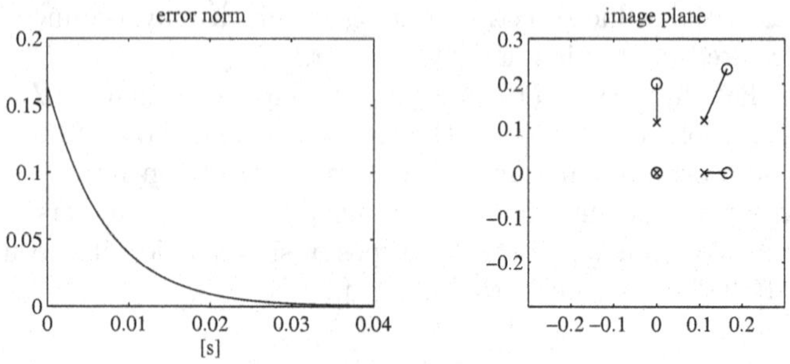

Fig. 10.8. Time history of the norm of the estimation error and corresponding paths of the feature points projections on image plane

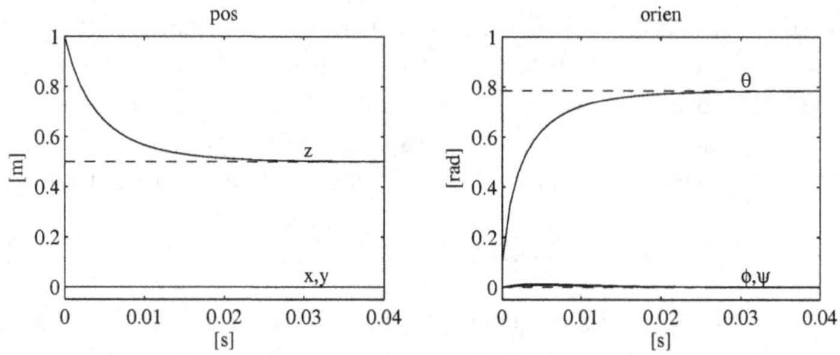

Fig. 10.9. Time history of camera pose estimate

Example 10.4

Consider the object of Example 10.1. Using the algorithmic solution, it is desired to compute the relative pose of the object frame, with respect to the camera frame, from the image plane coordinates of the projections of the four feature points of the object. The same numerical values of Example 10.1 are used.

Since the image Jacobian has dimension (6×8), it is necessary to use the algorithm based on the pseudo-inverse of the interaction matrix. This algorithm was simulated on a computer by adopting the Euler numerical integration scheme with an integration time $\Delta t = 1$ ms, matrix gain $\boldsymbol{K}_s = 160\boldsymbol{I}_8$, and initial estimate $\widehat{\boldsymbol{x}}_{c,o} = [0 \quad 0 \quad 1 \quad 0 \quad \pi/32 \quad 0]^T$.

The results in Fig. 10.8 show that the norm of the estimation error of the feature parameters $\boldsymbol{e}_s$ tends to zero asymptotically with convergence of exponential type; moreover, due to the fact that matrix gain $\boldsymbol{K}_s$ was chosen diagonal with equal elements, the paths of the projections of the feature points on the image plane (between the initial positions marked with crosses and the final positions marked with circles) are line segments.

The corresponding time histories of the components of vector $\widehat{\boldsymbol{x}}_{c,o}$ for position and orientation are reported in Fig. 10.9, together with the time histories of the components of the true value $\boldsymbol{x}_{c,o} = [0 \quad 0 \quad 0.5 \quad 0 \quad \pi/4 \quad 0]^T$ (represented with dashed lines). It can be verified that, with the chosen value of $\boldsymbol{K}_s$, the algorithm converges to the true value in about 0.03 s, corresponding to 30 iterations.

The time histories of Fig. 10.9 can be interpreted as the position and orientation trajectories of a 'virtual' camera in motion between the initial pose $\widehat{\boldsymbol{x}}_{c,o}(0)$ and the final pose $\boldsymbol{x}_{c,o}$.

Notice that, for the purpose of pose estimation, the illustrated algorithm may converge also in the case that only three feature points are used. In fact, in this case, $\boldsymbol{J}_{As}$ is a square (6×6) matrix and the convergence is ensured provided that this matrix is nonsingular. However, since P3P problem has four solutions, the algorithm may converge to a solution different from that sought, unless, as for visual tracking applications, the initial estimate $\widehat{\boldsymbol{x}}_{c,o}(0)$ is close enough to the true value $\boldsymbol{x}_{c,o}$.

10.4 Stereo Vision

The bidimensional image provided by a camera does not give any explicit information on depth, namely, the distance of the observed object from the camera. This information can be recovered in an indirect way from the geometric model of the object, assumed to be known.

On the other hand, the depth of a point can be directly computed in the case that two images of the same scene are available, taken from two different points of view. The two images can be obtained using two cameras, or sequentially, using a moving camera. These cases are referred to as *stereo vision*.

In the framework of stereo vision, two fundamental problems can be devised. The first is the *correspondence problem*, which consists of the identification of the points of the two images that are projections of the same point of the scene. These points are termed *conjugate* or *corresponding*. This problem is not easy to solve, and the solution is based on the existence of geometric constraints between two images of the same point, besides the fact that some details of the scene appear to be similar in the two images.

The second problem, illustrated below in some fundamental aspects, is that of *3D reconstruction* which, in general, consists of the computation of the relative pose of the cameras (calibrated and not) and thus, starting from this pose, the position in the 3D space of the points of the observed object.

10.4.1 Epipolar Geometry

Assume that two cameras are available, with respective reference frames, denoted as 1 and 2. Moreover, let $\boldsymbol{o}_{1,2}^1$ denote the position vector and $\boldsymbol{R}_2^1$ the rotation matrix of Frame 2 with respect to Frame 1 and let $\boldsymbol{T}_2^1$ be the corresponding homogeneous transformation matrix. The coordinates of a point P expressed in the two frames are related by equation

$$\boldsymbol{p}^1 = \boldsymbol{o}_{1,2}^1 + \boldsymbol{R}_2^1 \boldsymbol{p}^2. \tag{10.43}$$

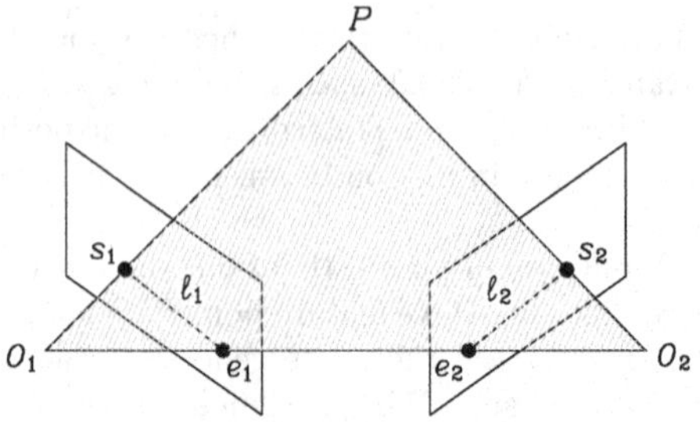

Fig. 10.10. Epipolar geometry

Let $\boldsymbol{s}_1$ and $\boldsymbol{s}_2$ be the coordinates of the projections of P on the image planes of the cameras; in view of (5.44) it is

$$\lambda_i \widetilde{\boldsymbol{s}}_i = \boldsymbol{\Pi} \widetilde{\boldsymbol{p}}^i = \boldsymbol{p}^i, \qquad i = 1, 2. \tag{10.44}$$

Substituting (10.44) into (10.43) yields

$$\lambda_1 \widetilde{\boldsymbol{s}}_1 = \boldsymbol{o}_{1,2}^1 + \lambda_2 \boldsymbol{R}_2^1 \widetilde{\boldsymbol{s}}_2. \tag{10.45}$$

Premultiplying both sides of (10.45) by $\boldsymbol{S}(\boldsymbol{o}_{1,2}^1)$ gives

$$\lambda_1 \boldsymbol{S}(\boldsymbol{o}_{1,2}^1) \widetilde{\boldsymbol{s}}_1 = \lambda_2 \boldsymbol{S}(\boldsymbol{o}_{1,2}^1) \boldsymbol{R}_2^1 \widetilde{\boldsymbol{s}}_2.$$

Hence, premultiplying both sides of the above equation by $\widetilde{\boldsymbol{s}}_1^T$, the following equality is obtained

$$\lambda_2 \widetilde{\boldsymbol{s}}_1^T \boldsymbol{S}(\boldsymbol{o}_{1,2}^1) \boldsymbol{R}_2^1 \widetilde{\boldsymbol{s}}_2 = 0,$$

which has to be satisfied for any value of scalar λ_2. Therefore, this equality is equivalent to the so-called *epipolar constraint* equation

$$\widetilde{\boldsymbol{s}}_1^T \boldsymbol{E} \widetilde{\boldsymbol{s}}_2 = 0, \tag{10.46}$$

where $\boldsymbol{E} = \boldsymbol{S}(\boldsymbol{o}_{1,2}^1)\boldsymbol{R}_2^1$ is a (3×3) matrix known as *essential matrix*. Equation (10.46) expresses in analytic form the geometric constraint existing between the projections of the same point on the image planes of the two cameras.

The geometric interpretation of the epipolar constraint can be derived from Fig. 10.10, where the projections of a point P on the image planes of the two cameras are reported as well as the respective optical centers O_1 and O_2. Notice that points O_1, O_2 and P are the vertices of a triangle whose sides O_1P and O_2P belong to the visual rays projecting point P into the points of coordinates $\boldsymbol{s}_1$ and $\boldsymbol{s}_2$ of the image plane, respectively. These rays lay along the directions of vectors $\widetilde{\boldsymbol{s}}_1$ and $\boldsymbol{R}_2^1 \widetilde{\boldsymbol{s}}_2$ respectively, expressed with respect to Frame 1. Line segment O_1O_2, termed *base line*, is represented by vector $\boldsymbol{o}_{1,2}^1$. The epipolar constraint (10.46) corresponds to imposing that vectors $\widetilde{\boldsymbol{s}}_1$, $\boldsymbol{R}_2^1 \widetilde{\boldsymbol{s}}_2$

and $\boldsymbol{o}_{1,2}^1$ are coplanar. The plane containing these vectors is termed *epipolar plane*.

Notice that line segment O_1O_2 belongs to the visual ray projecting point O_2 on the image plane of Camera 1 and, at the same time, to the ray projecting point O_1 on the image plane of Camera 2. These projections, of coordinates $\boldsymbol{e}_1$ and $\boldsymbol{e}_2$, respectively, are termed *epipoles*. Line ℓ_1, passing through the points of coordinates $\boldsymbol{s}_1$ and $\boldsymbol{e}_1$, and line ℓ_2, passing through the points of coordinates $\boldsymbol{s}_2$ and $\boldsymbol{e}_2$, are termed *epipolar lines*. The epipolar lines can also be obtained as the intersection of the epipolar plane with the image planes of the two cameras. Notice that, by varying point P, the epipolar plane describes a set of planes about the base line and the epipoles do not change.

For the purpose of computing the correspondences, the epipolar constraint can be exploited to reduce the complexity of the problem to find conjugate points. In fact, if $\boldsymbol{s}_1$ is the image of a point of the visual ray passing through O_1 and the point of coordinates $\boldsymbol{s}_1$, the corresponding conjugate point of the image plane of Camera 2 must necessarily belong to the epipolar line ℓ_2, which is known because the epipolar plane is uniquely defined by O_1, O_2 and by the point of coordinates $\boldsymbol{s}_1$. Finding correspondences, therefore, reduces to searching for a point along the epipolar line and not on the whole image plane.

In the framework of 3D reconstruction, different scenarios may arise, depending on the type of information which is available a priori.

10.4.2 Triangulation

In the case that both intrinsic and extrinsic parameters of the two cameras are known, the reconstruction problem consists of computing the position in the scene of the points projected on the two image planes using a geometric method known as *triangulation*. This method allows the computation of coordinates $\boldsymbol{p} = [\,p_x \quad p_y \quad p_z\,]^T$ of a point P with respect to the base frame, starting from normalized coordinates $\boldsymbol{s}_1 = [\,X_1 \quad Y_1\,]^T$ and $\boldsymbol{s}_2 = [\,X_2 \quad Y_2\,]^T$ of the projections of P on the image planes of the two cameras. Assume that, for simplicity, the base frame coincides with Frame 1, then $\boldsymbol{p}^1 = \boldsymbol{p}$, $\boldsymbol{o}_{1,2}^1 = \boldsymbol{o}$ and $\boldsymbol{R}_2^1 = \boldsymbol{R}$.

From (10.44), (10.45) the following equalities can be derived:

$$\boldsymbol{p} = \lambda_1 \widetilde{\boldsymbol{s}}_1 \tag{10.47}$$

$$\boldsymbol{p} = \boldsymbol{o} + \lambda_2 \boldsymbol{R} \widetilde{\boldsymbol{s}}_2, \tag{10.48}$$

where the first equality is the parametric equation of the visual ray passing through O_1 and the point of coordinates $\boldsymbol{s}_1$, while the second represents the parametric equation of the visual ray passing through O_2 and the point of coordinates $\boldsymbol{s}_2$; both equations are expressed in the base frame.

Therefore, the coordinates of point P at the intersection of the two visual rays can be computed by solving the system of two Eqs. (10.47), (10.48) with respect to $\boldsymbol{p}$. To this end, from (10.47), by computing λ_1 in the third equation and replacing its value into the other two equations, the following system is obtained:

$$\begin{bmatrix} 1 & 0 & -X_1 \\ 0 & 1 & -Y_1 \end{bmatrix} \boldsymbol{p} = \mathbf{0}. \tag{10.49}$$

Using a similar derivation for the equation obtained by premultiplying both sides of (10.48) by $\boldsymbol{R}^T$, the following system is obtained

$$\begin{bmatrix} \boldsymbol{r}_1^T - X_2 \boldsymbol{r}_3^T \\ \boldsymbol{r}_2^T - Y_2 \boldsymbol{r}_3^T \end{bmatrix} \boldsymbol{p} = \begin{bmatrix} o_x - o_z X_2 \\ o_y - o_z Y_2 \end{bmatrix}, \tag{10.50}$$

with $\boldsymbol{R} = [\boldsymbol{r}_1 \quad \boldsymbol{r}_2 \quad \boldsymbol{r}_3]$ and $\boldsymbol{R}^T \boldsymbol{o} = [o_x \quad o_y \quad o_z]^T$. Equations (10.49), (10.50) define a system of four equations and three unknowns, which is linear with respect to $\boldsymbol{p}$. Of these equations, only three are independent in the ideal case that the two visual rays intersect at point P. In practical applications, because of noise, these equations are all independent and the system has no solution; hence, suitable algorithms based on least-squares techniques have to be adopted to compute an approximate solution.

Computation of $\boldsymbol{p}$ can be greatly simplified in the case, rather frequent in applications, that the two cameras have parallel and aligned image planes, with $\boldsymbol{R} = \boldsymbol{I}$ and $\boldsymbol{R}^T \boldsymbol{o} = [b \quad 0 \quad 0]^T$, $b > 0$ being the distance between the origins of the two camera frames. This implies that the solution to the system of Eqs. (10.49), (10.50) is

$$p_x = \frac{X_1 b}{X_1 - X_2} \tag{10.51}$$

$$p_y = \frac{Y_1 b}{X_1 - X_2} = \frac{Y_2 b}{X_1 - X_2} \tag{10.52}$$

$$p_z = \frac{b}{X_1 - X_2}. \tag{10.53}$$

10.4.3 Absolute Orientation

In the case of a calibrated system of two cameras observing a rigid object of unknown shape, triangulation can be used to compute the variation of pose of the object or of the system of cameras, due to the relative motion of the system with respect to the cameras. This problem, known as *absolute orientation*, requires the measurement of the positions of the projections of a certain number of feature points of the object.

If the stereo camera system is moving and the object is fixed, let $\boldsymbol{p}_1, \ldots, \boldsymbol{p}_n$ denote the position vectors of n points of the rigid object measured at time t and $\boldsymbol{p}'_1, \ldots, \boldsymbol{p}'_n$ the position vectors of the same points measured at time t' using triangulation. These vectors are all referred to Frame 1 and, under the assumption of rigid motion, satisfy the equations

$$\boldsymbol{p}_i = \boldsymbol{o} + \boldsymbol{R}\boldsymbol{p}'_i \qquad i = 1, \ldots, n \tag{10.54}$$

where vector $\boldsymbol{o}$ and rotation matrix $\boldsymbol{R}$ define the position and orientation displacement of Frame 1 between time t and time t'. The absolute orientation problem consists of the computation of $\boldsymbol{R}$ and $\boldsymbol{o}$ from $\boldsymbol{p}_i$ and $\boldsymbol{p}'_i$.

From rigid body mechanics it is known that this problem has a unique solution in the case of three non-collinear points. In this case, nine nonlinear equations can be derived from (10.54), in terms of the nine independent parameters which characterize $\boldsymbol{o}$ and $\boldsymbol{R}$. However, since the points are obtained using triangulation, the measurements are affected by error and the system may have no solution. In this case, it is convenient to consider a number $n > 3$ of points and compute $\boldsymbol{o}$ and $\boldsymbol{R}$ as the quantities which minimize the linear quadratic function

$$\sum_{i=1}^{n} \|\boldsymbol{p}_i - \boldsymbol{o} - \boldsymbol{R}\boldsymbol{p}'_i\|^2, \tag{10.55}$$

with the constraint that $\boldsymbol{R}$ is a rotation matrix. The problem of computing $\boldsymbol{o}$ can be separated from the problem of computing $\boldsymbol{R}$ observing that the value of $\boldsymbol{o}$ which minimizes function (10.55) is (see Problem 10.6)

$$\boldsymbol{o} = \bar{\boldsymbol{p}} - \boldsymbol{R}\bar{\boldsymbol{p}}' \tag{10.56}$$

where $\bar{\boldsymbol{p}}$ and $\bar{\boldsymbol{p}}'$ are the centroids of the set of points $\{\boldsymbol{p}_i\}$ and $\{\boldsymbol{p}'_i\}$, defined as

$$\bar{\boldsymbol{p}} = \frac{1}{n}\sum_{i=1}^{n} \boldsymbol{p}_i, \qquad \bar{\boldsymbol{p}}' = \frac{1}{n}\sum_{i=1}^{n} \boldsymbol{p}'_i.$$

Hence the problem becomes that of computing the rotation matrix $\boldsymbol{R}$ which minimizes the linear quadratic function

$$\sum_{i=1}^{n} \|\bar{\boldsymbol{p}}_i - \boldsymbol{R}\bar{\boldsymbol{p}}'_i\|^2, \tag{10.57}$$

where $\bar{\boldsymbol{p}}_i = \boldsymbol{p}_i - \bar{\boldsymbol{p}}$ and $\bar{\boldsymbol{p}}'_i = \boldsymbol{p}'_i - \bar{\boldsymbol{p}}'$ are the deviations with respect to the centroids.

It can be proven that the matrix $\boldsymbol{R}$ which minimizes function (10.57) is the matrix which maximizes the trace of $\boldsymbol{R}^T\boldsymbol{K}$, with

$$\boldsymbol{K} = \sum_{i=1}^{n} \bar{\boldsymbol{p}}_i \bar{\boldsymbol{p}}'^T_i;$$

see Problem 10.7. Therefore, the solution has the form (10.12) where, for the purpose of this problem, $\boldsymbol{U}$ and $\boldsymbol{V}$ are respectively the left and right orthogonal matrices of the singular value decomposition of $\boldsymbol{K}$. Once that rotation matrix $\boldsymbol{R}$ is known, vector $\boldsymbol{o}$ can be computed using (10.56).

10.4.4 3D Reconstruction from Planar Homography

Another interesting case of application of 3D reconstruction occurs when the feature points of the observed object lie on the same plane. This geometric property represents an additional constraint between the projections of each point on the image plane of the two cameras, besides the epipolar constraint. This constraint is a *planar homography.*

Let $\boldsymbol{p}^2$ be the position vector of a point P of the object, expressed with respect to Frame 2. Moreover, let $\boldsymbol{n}^2$ denote the unit vector orthogonal to the plane containing the feature points and $d_2 > 0$ the distance of the plane from the origin of Frame 2. By virtue of simple geometric considerations, the following equation can be derived:

$$\frac{1}{d_2}\boldsymbol{n}^{2T}\boldsymbol{p}^2 = 1$$

which defines the set of points $\boldsymbol{p}^2$ belonging to the plane. In view of the above equality, Eq. (10.43) can be rewritten in the form

$$\boldsymbol{p}^1 = \boldsymbol{H}\boldsymbol{p}^2, \tag{10.58}$$

with

$$\boldsymbol{H} = \boldsymbol{R}_2^1 + \frac{1}{d_2}\boldsymbol{o}_{1,2}^1\boldsymbol{n}^{2T}. \tag{10.59}$$

Replacing (10.44) into (10.58) yields

$$\widetilde{\boldsymbol{s}}_1 = \lambda\boldsymbol{H}\widetilde{\boldsymbol{s}}_2, \tag{10.60}$$

where $\lambda = \lambda_2/\lambda_1 > 0$ is an arbitrary constant. Premultiplication of both sides of (10.60) by $\boldsymbol{S}(\widetilde{\boldsymbol{s}}_1)$ yields the equality

$$\boldsymbol{S}(\widetilde{\boldsymbol{s}}_1)\boldsymbol{H}\widetilde{\boldsymbol{s}}_2 = \boldsymbol{0}, \tag{10.61}$$

representing a planar homography defined by matrix $\boldsymbol{H}$.

Using a derivation similar to that presented in Sect. 10.3.1, it is possible to compute numerically matrix $\zeta\boldsymbol{H}$, up to a scaling factor ζ, starting from the coordinates of n points of the plane, with $n \geq 4$.

The value of the scaling factor ζ can be computed using a numerical algorithm based on expression (10.59) of matrix $\boldsymbol{H}$; once $\boldsymbol{H}$ is known, it is possible to compute quantities $\boldsymbol{R}_2^1$, $\boldsymbol{o}_{1,2}^1/d_2$ and $\boldsymbol{n}^2$ in (10.59) — actually, it can be shown that two admissible solutions exist.

This result is of a certain relevance for visual servoing applications. For example, in the case of a camera in motion with respect to the object, if Frames 1 and 2 represent the poses of the camera in two different time instants, the computation of $\boldsymbol{H}$ with decomposition (10.59) can be used to evaluate the orientation displacement of the camera frame and the position displacement of the origin, the latter defined up to a scaling factor d_2. This information can be achieved without knowing the object geometry, as long as the feature points all belong to the same plane.

Example 10.5

Consider the SCARA manipulator of Example 10.3 and the planar object of Example 10.1. Assume that the feature vector of the four points of the object, measured by the camera at time t', is

$$\boldsymbol{s}' = \begin{bmatrix} 0 & 0 & 0.2 & 0 & 0.2 & 0.2 & 0 & 0.2 \end{bmatrix}^T,$$

while the feature vector, measured by the camera at time t'', is

$$\boldsymbol{s}'' = \begin{bmatrix} -0.1667 & 0.1667 & -0.0833 & 0.0223 & 0.0610 & 0.1057 & -0.0223 & 0.2500 \end{bmatrix}^T.$$

It is desired to compute the quantities $\boldsymbol{R}_2^1$, $(1/d_2)\boldsymbol{o}_{1,2}^1$ and $\boldsymbol{n}^2$ of the planar homography (10.59).

For simplicity, assume that the orientation of the planar object is known, namely

$$\boldsymbol{n}^2 = \begin{bmatrix} 0 & 0 & 1 \end{bmatrix}^T.$$

A further simplification to the problem derives from the fact that, in this case, matrix $\boldsymbol{R}_2^1$ corresponds to a rotation about axis z of an angle β, namely $\boldsymbol{R}_2^1 = \boldsymbol{R}_z(\beta)$. Therefore, in view of (10.59), planar homography $\boldsymbol{H}$ has the symbolic expression

$$\boldsymbol{H} = \begin{bmatrix} c_\beta & -s_\beta & o_x/d_2 \\ s_\beta & c_\beta & o_y/d_2 \\ 0 & 0 & 1 + o_z/d_2 \end{bmatrix},$$

with $\boldsymbol{o}_{1,2}^1 = \begin{bmatrix} o_x & o_y & o_z \end{bmatrix}^T$.

On the other hand, starting from numerical values of $\boldsymbol{s}'$ and $\boldsymbol{s}''$ and using a derivation similar to that used in Example 10.1, the following matrix can be obtained:

$$\zeta\boldsymbol{H} = \begin{bmatrix} 0.3015 & -0.5222 & 0.1373 \\ 0.5222 & 0.3015 & 0.0368 \\ 0 & 0 & 0.5025 \end{bmatrix},$$

which is the numerical value of the planar homography, up to a scaling factor ζ.

The symbolic expression of $\boldsymbol{H}$ reveals that the first and second column of this matrix have unit norm. This property can be used to compute $|\zeta| = 1.6583$. The sign of ζ can be evaluated by imposing, for any of the feature points of the object, the constraint

$$\boldsymbol{p}^{1T}\boldsymbol{p}^1 = \boldsymbol{p}^{1T}\boldsymbol{H}\boldsymbol{p}^2 = \lambda_1\lambda_2\widetilde{\boldsymbol{s}}_1^T\boldsymbol{H}\widetilde{\boldsymbol{s}}_2 > 0.$$

Since scalars λ_i are positive, this inequality is equivalent to

$$\widetilde{\boldsymbol{s}}_1^T\boldsymbol{H}\widetilde{\boldsymbol{s}}_2 > 0,$$

hence, in this case, it is $\zeta > 0$. Therefore

$$\boldsymbol{H} = \begin{bmatrix} 0.5 & -0.8660 & 0.2277 \\ 0.8660 & 0.5 & 0.0610 \\ 0 & 0 & 0.8333 \end{bmatrix}.$$

At this point, the value of angle β can be computed from the elements h_{11} and h_{21} of matrix $\boldsymbol{H}$ using equation

$$\beta = \text{Atan2}(h_{21}, h_{11}) = \frac{\pi}{3}.$$

Finally, vector $(1/d_2)\boldsymbol{o}_{1,2}^1$ can be computed from the elements of the last row of matrix $\boldsymbol{H}$ using the equality

$$\frac{1}{d_2}\boldsymbol{o}_{1,2}^1 = \begin{bmatrix} h_{13} \\ h_{23} \\ h_{33}-1 \end{bmatrix} = \begin{bmatrix} 0.2277 \\ 0.0610 \\ -0.0667 \end{bmatrix}.$$

Notice that the derivation illustrated in Example 10.5, in the case that $\boldsymbol{n}^2$ is unknown and $\boldsymbol{R}_2^1$ is a generic rotation matrix, becomes much more complex.

10.5 Camera Calibration

An important problem for visual servoing applications is the calibration of the camera, which is the sensor providing the information fed back to the controller. Calibration consists of the estimation of the *intrinsic parameters*, characterizing matrix $\boldsymbol{\Omega}$ defined in (5.41), and of the *extrinsic parameters*, characterizing the pose of the camera frame with respect to the base frame (for eye-to-end cameras) or to the end-effector frame (for eye-in-hand cameras). Various calibration techniques exist, which are based on algorithms similar to those used for the pose estimation of an object with respect to a camera.

In particular, the solution method of a PnP problem with n coplanar points illustrated in Sect. 10.3.1 can be directly used for the computation of the extrinsic parameters of the camera, if the intrinsic parameters are known.

In fact, in view of (10.14), extrinsic parameters of an eye-to-hand camera can be computed as

$$\boldsymbol{T}_c^b = \boldsymbol{T}_o^b(\boldsymbol{T}_o^c)^{-1}, \qquad (10.62)$$

where matrix $\boldsymbol{T}_o^c$ is the output of the algorithm solving a PnP planar problem, provided that matrix $\boldsymbol{T}_o^b$, expressing the position and the orientation of the object frame with respect to the base frame, is known. Similarly, in view of (10.15), the extrinsic parameters of an eye-in-hand camera can be computed as

$$\boldsymbol{T}_c^e = \boldsymbol{T}_o^e(\boldsymbol{T}_o^c)^{-1}, \qquad (10.63)$$

provided that matrix $\boldsymbol{T}_o^e$, expressing the pose of the object frame with respect to the end-effector frame, is known.

If the intrinsic parameters are not known, the derivation of Sect. 10.3.1 has to be suitably extended and can be broken down in the three phases described below.

Phase 1

A planar homography can be computed starting from pixel coordinates

$$\boldsymbol{c}_i = \begin{bmatrix} X_{Ii} \\ Y_{Ii} \end{bmatrix}$$

in place of normalized coordinates $\boldsymbol{s}_i$. In detail, an equation formally identical to (10.7) can be obtained:

$$\boldsymbol{S}(\widetilde{\boldsymbol{c}}_i)\boldsymbol{H}'\begin{bmatrix} r_{x,i} & r_{y,i} & 1 \end{bmatrix}^T = \boldsymbol{0}, \tag{10.64}$$

where $\boldsymbol{H}'$ is the (3×3) matrix

$$\boldsymbol{H}' = \boldsymbol{\Omega}\boldsymbol{H}, \tag{10.65}$$

$\boldsymbol{\Omega}$ being the matrix of intrinsic parameters (5.41) and $\boldsymbol{H}$ the matrix defined in (10.8). Using an algorithm similar to that presented in Sect. 10.3.1, planar homography $\zeta\boldsymbol{H}'$ can be computed, up to a scaling factor ζ, from the coordinates of n points of the plane, with $n \geq 4$.

Phase 2

Matrix $\boldsymbol{\Omega}$ can be computed from the elements of matrix $\zeta\boldsymbol{H}'$. In fact, taking into account (10.65), and the definition of $\boldsymbol{H}$ in (10.8), yields

$$\zeta\begin{bmatrix} \boldsymbol{h}_1' & \boldsymbol{h}_2' & \boldsymbol{h}_3' \end{bmatrix} = \boldsymbol{\Omega}\begin{bmatrix} \boldsymbol{r}_1 & \boldsymbol{r}_2 & \boldsymbol{o}_{c,o}^c \end{bmatrix},$$

where $\boldsymbol{h}_i'$ denotes the i-th column of matrix $\boldsymbol{H}'$. Computing $\boldsymbol{r}_1$ and $\boldsymbol{r}_2$ from this equation, and imposing the orthogonality and unit norm constraints on these vectors, the following two scalar equations are obtained:

$$\begin{aligned} \boldsymbol{h}_1'^T\boldsymbol{\Omega}^{-T}\boldsymbol{\Omega}^{-1}\boldsymbol{h}_2' &= 0 \\ \boldsymbol{h}_1'^T\boldsymbol{\Omega}^{-T}\boldsymbol{\Omega}^{-1}\boldsymbol{h}_1' &= \boldsymbol{h}_2'^T\boldsymbol{\Omega}^{-T}\boldsymbol{\Omega}^{-1}\boldsymbol{h}_2' \end{aligned}$$

which, being linear, can be rewritten in the form

$$\boldsymbol{A}'\boldsymbol{b} = \boldsymbol{0}. \tag{10.66}$$

In the above equation, $\boldsymbol{A}'$ is a (2×6) matrix of coefficients depending on $\boldsymbol{h}_1'$, $\boldsymbol{h}_2'$, while $\boldsymbol{b} = \begin{bmatrix} b_{11} & b_{12} & b_{22} & b_{13} & b_{23} & b_{33} \end{bmatrix}^T$, where b_{ij} is the generic element of the symmetric matrix

$$\boldsymbol{B} = \boldsymbol{\Omega}^{-T}\boldsymbol{\Omega}^{-1} = \begin{bmatrix} 1/\alpha_x^2 & 0 & -X_0/\alpha_x^2 \\ * & 1/\alpha_y^2 & -Y_0/\alpha_y^2 \\ * & * & 1 + X_0^2/\alpha_x^2 + Y_0^2/\alpha_y^2 \end{bmatrix}. \tag{10.67}$$

By repeating Phase 1 k times, with the same plane placed each time in a different pose, $2k$ equations of the form (10.66) are obtained. These equations, in the case $k \geq 3$, have a unique solution $\gamma\boldsymbol{b}$ defined up to a scaling factor γ. From matrix $\gamma\boldsymbol{B}$, in view of (10.67), the following expressions for the intrinsic parameters can be found:

$$X_0 = -b_{13}'/b_{11}'$$

Fig. 10.11. Example of calibration plane

$$Y_0 = -b'_{23}/b'_{22}$$
$$\alpha_x = \sqrt{\gamma/b'_{11}}$$
$$\alpha_y = \sqrt{\gamma/b'_{22}},$$

where $b'_{i,j} = \gamma b_{i,j}$ and γ can be computed as

$$\gamma = b'_{13} + b'_{23} + b'_{33}.$$

Phase 3

Once the matrix $\boldsymbol{\Omega}$ of intrinsic parameters has been evaluated, it is possible to compute $\boldsymbol{H}$ from $\boldsymbol{H}'$, up to a scaling factor ζ, using (10.65) for one of the k poses of the plane. Hence, matrix $\boldsymbol{T}_o^c$ can be computed from $\boldsymbol{H}$ as for the case of the solution of a PnP problem shown in Sect. 10.3.1. Finally, using equations (10.62), (10.63), the extrinsic parameters of the camera can be evaluated.

The method illustrated above is merely conceptual, because it does not provide satisfactory solutions in the presence of measurement noise or lens distortion — especially for the intrinsic parameters; however, the accuracy of the result can be improved using models that take into account the distortion phenomena of the lenses, together with nonlinear optimization techniques.

From the experimental point of view, the calibration method described above requires the use of calibration planes where a certain number of points can be easily detected; also, the position of these points with respect to a suitable reference frame must be known with high accuracy. An example of calibration plane with a chessboard pattern is shown in Fig. 10.11.

Finally, notice that a calibration method can also be set up starting from the solution of a nonplanar PnP problem, using the direct linear transformation method.

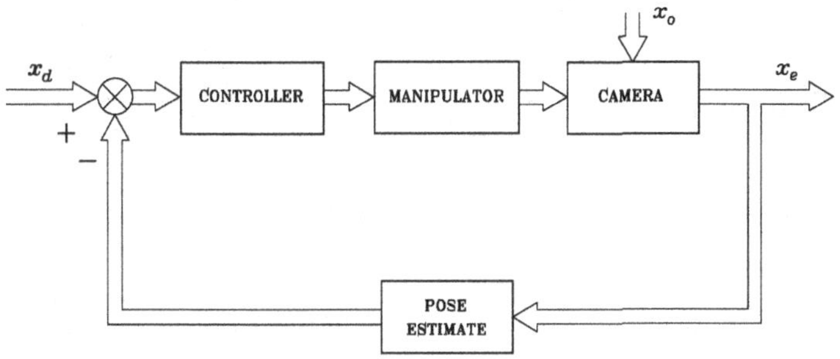

Fig. 10.12. General block scheme of position-based visual servoing

10.6 The Visual Servoing Problem

Visual measurements allow a robot to collect information on the surrounding environment. In the case of robot manipulators, such information is typically used to compute the end-effector pose with respect to an object observed by the camera. The objective of visual servoing is to ensure that the end-effector, on the basis of visual measurements elaborated in real time, reaches and keeps a (constant or time-varying) desired pose with respect to the observed object.

It is worth remarking that the direct measurements provided by the visual system are concerned with feature parameters in the image plane, while the robotic task is defined in the operational space, in terms of the relative pose of the end-effector with respect to the object. This fact naturally leads to considering two kinds of control approaches, illustrated by the block schemes of Figs. 10.12 and 10.13, namely *position-based visual servoing*, also termed *visual servoing in the operational space*, and *image-based visual servoing*, also termed *visual servoing in the image space*. In these schemes, the case of eye-in-hand camera is considered; for eye-to-hand cameras, similar schemes can be adopted.

The *position-based visual servoing* approach is conceptually similar to the operational space control illustrated in Fig. 8.2. The main difference is that feedback is based on the real-time estimation of the pose of the observed object with respect to the camera using visual measurements. Estimation can be performed analytically or using iterative numerical algorithms. Its conceptual advantage regards the possibility of acting directly on operational space variables. Therefore, the control parameters can be selected on the basis of suitable specifications imposed to the time response of the end-effector motion variables, both at steady state and during the transient. The drawback of this approach is that, due to the absence of direct control of the image features, the object may exit from the camera field of view during the transient or as a consequence of planning errors; hence, the feedback loop turns out to be open due to lack of visual measurements and instability may occur.

In the *image-space visual servoing* approach, the control action is computed on the basis of the error defined as the difference between the value of

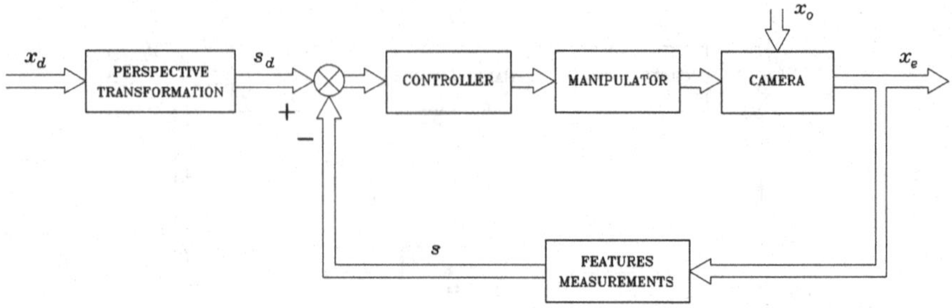

Fig. 10.13. General block scheme of image-based visual servoing

the image feature parameters in the desired configuration — computed using perspective transformation or directly measured with the camera in the desired pose — and the value of the parameters measured with the camera in the current pose. The conceptual advantage of this solution regards the fact that the real-time estimate of the pose of the object with respect to the camera is not required. Moreover, since the control acts directly in the image feature parameters, it is possible to keep the object within the camera field of view during the motion. However, due to the nonlinearity of the mapping between the image feature parameters and the operational space variables, singular configurations may occur, which cause instability or saturation of the control action. Also, the end-effector trajectories cannot be easily predicted in advance and may produce collisions with obstacles or joint limits violation.

To compare the two control strategies it is also worth considering the operating conditions. Of particular importance is the issue of camera calibration. It is easy to understand that position-based visual servoing is more sensitive to camera calibration errors compared to image-based visual servoing. In fact, for the first approach, the presence of uncertainties on calibration parameters, both intrinsic and extrinsic, produces errors on the estimate of operational space variables that may be seen as an external disturbance acting on the feedback path of the control loop, where disturbance rejection capability is low. On the other hand, in the image-based visual servoing approach, the quantities used for the computation of the control action are directly defined in the image plane and measured in pixel units; moreover, the desired value of the feature parameters is measured using the camera. This implies that the uncertainty affecting calibration parameters can be seen as a disturbance acting on the forward path of the control loop, where disturbance rejection capability is high.

A further aspect to analyze concerns knowledge of the geometric model of the object. It is evident that, for position-based visual servoing, the object geometry must be known if only one camera is used, because it is necessary for pose estimation, while it may be unknown when a stereo camera system is employed. On the other hand, image-based visual servoing does not require, in principle, knowledge of the object geometry, even for mono-camera systems.

On the above premises, in the following, the main position-based and image-based visual servoing schemes are illustrated. For both approaches, the problem of regulation to a constant set-point is presented and the object is assumed to be fixed with respect to the base frame. Without loss of generality, the case of a single calibrated camera, mounted on the manipulator's end-effector, is considered (see Fig. 10.5); moreover, the end-effector frame is chosen so as to coincide with the camera frame.

10.7 Position-based Visual Servoing

In position-based visual servoing schemes, visual measurements are used to estimate in real time the homogeneous transformation matrix $\boldsymbol{T}_o^c$, representing the relative pose of the object frame with respect to the camera frame. From matrix $\boldsymbol{T}_o^c$, the $(m \times 1)$ vector of independent coordinates $\boldsymbol{x}_{c,o}$, defined in (10.31), can be extracted.

Assumed that the object is fixed with respect to the base frame, the position-based visual servoing problem can be formulated by imposing a desired value to the relative pose of the object frame with respect to the camera frame. This quantity can be specified in terms of homogeneous transformation matrix $\boldsymbol{T}_o^d$, where superscript d denotes the desired pose of the camera frame. From this matrix, the $(m \times 1)$ operational space vector $\boldsymbol{x}_{d,o}$ can be extracted.

Matrices $\boldsymbol{T}_o^c$ and $\boldsymbol{T}_o^d$ can be used to obtain the homogeneous transformation matrix

$$\boldsymbol{T}_c^d = \boldsymbol{T}_o^d(\boldsymbol{T}_o^c)^{-1} = \begin{bmatrix} \boldsymbol{R}_c^d & \boldsymbol{o}_{d,c}^d \\ \boldsymbol{0}^T & 1 \end{bmatrix}, \tag{10.68}$$

expressing the position and orientation displacement of the camera frame in the current pose with respect to the desired pose. From this matrix, a suitable error vector in the operational space can be computed, defined as

$$\widetilde{\boldsymbol{x}} = -\begin{bmatrix} \boldsymbol{o}_{d,c}^d \\ \boldsymbol{\phi}_{d,c} \end{bmatrix}, \tag{10.69}$$

where $\boldsymbol{\phi}_{d,c}$ is the vector of the Euler angles extracted from rotation matrix $\boldsymbol{R}_c^d$. Vector $\widetilde{\boldsymbol{x}}$ does not depend on the object pose and represents the error between the desired pose and the current pose of the camera frame. It is worth observing that this vector does not coincide with the difference between $\boldsymbol{x}_{d,o}$ and $\boldsymbol{x}_{c,o}$, but it can be computed from the corresponding homogeneous transformation matrices, using (10.68), (10.69).

The control has to be designed so that the operational space error $\widetilde{\boldsymbol{x}}$ tends to zero asymptotically.

Notice that, for the choice of the set point $\boldsymbol{x}_{d,o}$, the knowledge of the object pose is not required. However, the control objective can be satisfied provided that the desired pose of the camera frame with respect to the base frame, corresponding to the homogeneous transformation matrix

$$\boldsymbol{T}_d = \boldsymbol{T}_c(\boldsymbol{T}_c^d)^{-1} = \begin{bmatrix} \boldsymbol{R}_d & \boldsymbol{o}_d \\ \boldsymbol{0}^T & 1 \end{bmatrix}, \tag{10.70}$$

belongs to the dexterous workspace of the manipulator. If the object is fixed with respect to the base frame, this matrix is constant.

10.7.1 PD Control with Gravity Compensation

Position-based visual servoing can be implemented using PD control with gravity compensation, suitably modified with respect to that used for motion control.

Computing the time derivative of (10.69), for the position part, gives

$$\dot{\boldsymbol{o}}_{d,c}^d = \dot{\boldsymbol{o}}_c^d - \dot{\boldsymbol{o}}_d^d = \boldsymbol{R}_d^T \dot{\boldsymbol{o}}_c,$$

while, for the orientation part, it gives

$$\dot{\boldsymbol{\phi}}_{d,c} = \boldsymbol{T}^{-1}(\boldsymbol{\phi}_{d,c})\boldsymbol{\omega}_{d,c}^d = \boldsymbol{T}^{-1}(\boldsymbol{\phi}_{d,c})\boldsymbol{R}_d^T \boldsymbol{\omega}_c.$$

To compute the above expressions, equalities $\dot{\boldsymbol{o}}_d^d = \boldsymbol{0}$ and $\boldsymbol{\omega}_d^d = \boldsymbol{0}$ have been taken into account, observing that $\boldsymbol{o}_d$ and $\boldsymbol{R}_d$ are constant. Therefore, $\dot{\widetilde{\boldsymbol{x}}}$ has the expression

$$\dot{\widetilde{\boldsymbol{x}}} = -\boldsymbol{T}_A^{-1}(\boldsymbol{\phi}_{d,c}) \begin{bmatrix} \boldsymbol{R}_d^T & \boldsymbol{O} \\ \boldsymbol{O} & \boldsymbol{R}_d^T \end{bmatrix} \boldsymbol{v}_c. \tag{10.71}$$

Since the end-effector frame and the camera frame coincide, the following equality holds:

$$\dot{\widetilde{\boldsymbol{x}}} = -\boldsymbol{J}_{A_d}(\boldsymbol{q}, \widetilde{\boldsymbol{x}})\dot{\boldsymbol{q}}, \tag{10.72}$$

where the matrix

$$\boldsymbol{J}_{A_d}(\boldsymbol{q}, \widetilde{\boldsymbol{x}}) = \boldsymbol{T}_A^{-1}(\boldsymbol{\phi}_{d,c}) \begin{bmatrix} \boldsymbol{R}_d^T & \boldsymbol{O} \\ \boldsymbol{O} & \boldsymbol{R}_d^T \end{bmatrix} \boldsymbol{J}(\boldsymbol{q}) \tag{10.73}$$

has the meaning of analytic Jacobian of the manipulator in the operational space, as for the Jacobian in (9.14).

Position-based visual servoing of PD type with gravity compensation has the expression

$$\boldsymbol{u} = \boldsymbol{g}(\boldsymbol{q}) + \boldsymbol{J}_{A_d}^T(\boldsymbol{q}, \widetilde{\boldsymbol{x}})(\boldsymbol{K}_P \widetilde{\boldsymbol{x}} - \boldsymbol{K}_D \boldsymbol{J}_{A_d}(\boldsymbol{q}, \widetilde{\boldsymbol{x}})\dot{\boldsymbol{q}}), \tag{10.74}$$

analogous to motion control law (8.110), but using a different definition of operational space error. The asymptotic stability of the equilibrium pose corresponding to $\widetilde{\boldsymbol{x}} = \boldsymbol{0}$, under the assumption of symmetric and positive definite matrices $\boldsymbol{K}_P$, $\boldsymbol{K}_D$, can be proven using the Lyapunov function

$$V(\dot{\boldsymbol{q}}, \widetilde{\boldsymbol{x}}) = \frac{1}{2}\dot{\boldsymbol{q}}^T \boldsymbol{B}(\boldsymbol{q})\dot{\boldsymbol{q}} + \frac{1}{2}\widetilde{\boldsymbol{x}}^T \boldsymbol{K}_P \widetilde{\boldsymbol{x}} > 0 \qquad \forall \dot{\boldsymbol{q}}, \widetilde{\boldsymbol{x}} \neq \boldsymbol{0},$$

similarly to the case of control law (8.110).

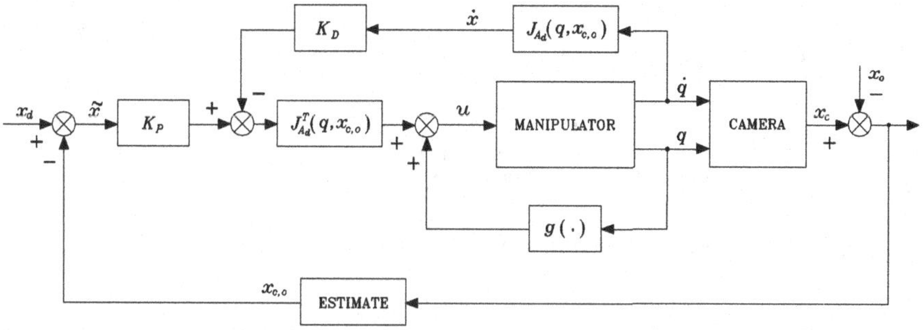

Fig. 10.14. Block scheme of position-based visual servoing of PD type with gravity compensation

Notice that, for the computation of control law (10.74), the estimation of $\boldsymbol{x}_{c,o}$ and the measurements of $\boldsymbol{q}$ and $\dot{\boldsymbol{q}}$ are required. Moreover, the derivative term can also be chosen as $-\boldsymbol{K}_D\dot{\boldsymbol{q}}$.

The block scheme of position-based visual servoing of PD type with gravity compensation is shown in Fig. 10.14. Notice that the sum block computing error $\widetilde{\boldsymbol{x}}$ and that computing the output of the controlled system have a purely conceptual meaning and do not correspond to algebraic sums.

10.7.2 Resolved-velocity Control

The information deriving from visual measurements is computed at a frequency lower or equal to the camera frame rate. This quantity, especially for CCD cameras, is at least of one order of magnitude lower than the typical frequencies used for motion control of robot manipulators. As a consequence, in the digital implementation of control law (10.74), to preserve stability of the closed-loop system, the control gains must be set to values much lower than those typically used for motion control; therefore, the performance of the closed-loop system in terms of speed of convergence and disturbance rejection capability turns out to be poor.

This problem can be avoided assuming that the manipulator is equipped with a high-gain motion controller in the joint space or in the operational space. Neglecting the effects on the tracking errors deriving from manipulator dynamics and disturbances, the controlled manipulator can be considered as an ideal positioning device. This implies that, in the case of joint space motion control, the following equality holds:

$$\boldsymbol{q}(t) \approx \boldsymbol{q}_r(t), \tag{10.75}$$

$\boldsymbol{q}_r(t)$ being the imposed reference trajectory for the joint variables.

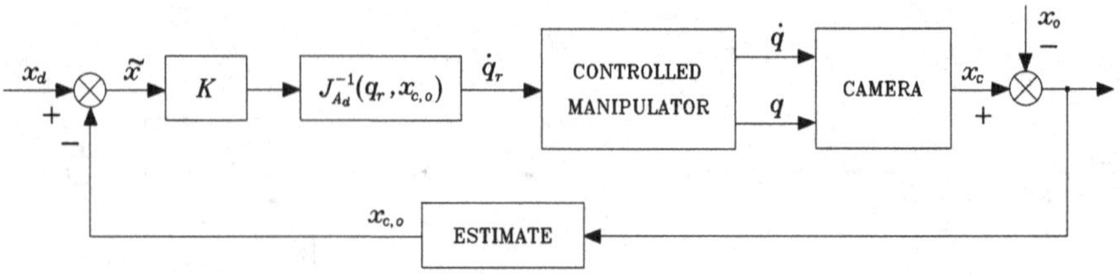

Fig. 10.15. Block scheme of resolved-velocity position-based visual servoing

Therefore, visual servoing can be achieved by computing the trajectory $\boldsymbol{q}_r(t)$ on the basis of visual measurements, so that the operational space tracking error (10.69) goes asymptotically to zero.

To this end, Eq. (10.72) suggests the following choice for the joint space reference velocity:

$$\dot{\boldsymbol{q}}_r = \boldsymbol{J}_{A_d}^{-1}(\boldsymbol{q}_r, \widetilde{\boldsymbol{x}})\boldsymbol{K}\widetilde{\boldsymbol{x}} \tag{10.76}$$

which, replaced in (10.72), by virtue of equality (10.75), yields the linear equation

$$\dot{\widetilde{\boldsymbol{x}}} + \boldsymbol{K}\widetilde{\boldsymbol{x}} = \boldsymbol{0}. \tag{10.77}$$

This equality, for a positive definite matrix $\boldsymbol{K}$, implies that the operational space error tends to zero asymptotically with a convergence of exponential type and speed depending on the eigenvalues of matrix $\boldsymbol{K}$; the larger the eigenvalues, the faster the convergence.

The above scheme is termed *resolved-velocity control* in the operational space, because it is based on the computation of velocity $\dot{\boldsymbol{q}}_r$ from the operational space error. Trajectory $\boldsymbol{q}_r(t)$ is computed from (10.76) via a simple integration.

The block scheme of resolved-velocity position-based visual servoing is reported in Fig. 10.15. Also in this case, the sum block computing the error $\widetilde{\boldsymbol{x}}$ and that computing the output of the scheme have a purely conceptual meaning and do not correspond to algebraic sums.

Notice that the choice of $\boldsymbol{K}$ influences the transient behaviour of the trajectory of the camera frame, which is the solution to the differential equation (10.77). If $\boldsymbol{K}$ is a diagonal matrix with the same gains for the positional part, the origin of the camera frame follows the line segment connecting the initial position to the desired position. On the other hand, the orientation trajectory depends on the particular choice of Euler angles and, more in general, of the orientation error. The possible choices of the orientation error are those presented in Sect. 3.7.3, with the appropriate definition of Jacobian (10.73). The possibility of knowing in advance the trajectory of the camera is important because, during the motion, the object may exit from the camera field of view, making visual measurements unavailable.

10.8 Image-based Visual Servoing

If the object is fixed with respect to the base frame, image-based visual servoing can be formulated by stipulating that the vector of the object feature parameters has a desired constant value $\boldsymbol{s}_d$ corresponding to the desired pose of the camera. Therefore, it is implicitly assumed that a desired pose $\boldsymbol{x}_{d,o}$ exists so that the camera pose belongs to the dexterous workspace of the manipulator and

$$\boldsymbol{s}_d = \boldsymbol{s}(\boldsymbol{x}_{d,o}). \tag{10.78}$$

Moreover, $\boldsymbol{x}_{d,o}$ is supposed to be unique. To this end, the feature parameters can be chosen as the coordinates of n points of the object, with $n \geq 4$ for coplanar points (and no triplets of collinear points) or $n \geq 6$ in case of non-coplanar points. Notice that, if the operational space dimension is $m < 6$, as for the case of SCARA manipulator, a reduced number of points can be used.

The interaction matrix $\boldsymbol{L}_s(\boldsymbol{s}, \boldsymbol{z}_c)$ depends on variables $\boldsymbol{s}$ and $\boldsymbol{z}_c$ with $\boldsymbol{z}_c = [\, z_{c,1} \dots z_{c,n}]^T$, $z_{c,i}$ being the third coordinate of the generic feature point of the object.

It is worth noticing that the task is assigned directly in terms of feature vector $\boldsymbol{s}_d$, while pose $\boldsymbol{x}_{d,o}$ does not need to be known. In fact, $\boldsymbol{s}_d$ can be computed by measuring the feature parameters when the object is in the desired pose with respect to the camera.

The control law must be designed so as to guarantee that the image space error

$$\boldsymbol{e}_s = \boldsymbol{s}_d - \boldsymbol{s} \tag{10.79}$$

tends asymptotically to zero.

10.8.1 PD Control with Gravity Compensation

Image-based visual servoing can be implemented using a PD control with gravity compensation defined on the basis of the image space error.

To this end, consider the following positive definite quadratic form as Lyapunov function candidate:

$$V(\dot{\boldsymbol{q}}, \boldsymbol{e}_s) = \frac{1}{2}\dot{\boldsymbol{q}}^T \boldsymbol{B}(\boldsymbol{q})\boldsymbol{q} + \frac{1}{2}\boldsymbol{e}_s^T \boldsymbol{K}_{Ps}\boldsymbol{e}_s > 0 \qquad \forall \dot{\boldsymbol{q}}, \boldsymbol{e}_s \neq \boldsymbol{0}, \tag{10.80}$$

with $\boldsymbol{K}_{Ps}$ symmetric and positive definite $(k \times k)$ matrix.

Computing the time derivative of (10.80) and taking into account the expression (8.7) of the joint space dynamic model of the manipulator and property (7.49) yields

$$\dot{V} = -\dot{\boldsymbol{q}}^T \boldsymbol{F}\dot{\boldsymbol{q}} + \dot{\boldsymbol{q}}^T(\boldsymbol{u} - \boldsymbol{g}(\boldsymbol{q})) + \dot{\boldsymbol{e}}_s^T \boldsymbol{K}_{Ps}\boldsymbol{e}_s. \tag{10.81}$$

Since $\dot{\boldsymbol{s}}_d = \boldsymbol{0}$ and the object is fixed with respect to the base frame, the following equality holds:

$$\dot{\boldsymbol{e}}_s = -\dot{\boldsymbol{s}} = -\boldsymbol{J}_L(\boldsymbol{s}, \boldsymbol{z}_c, \boldsymbol{q})\dot{\boldsymbol{q}}, \tag{10.82}$$

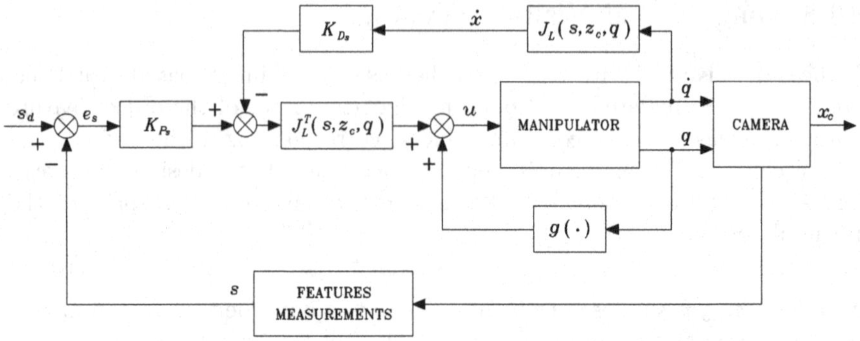

Fig. 10.16. Block scheme of image-based visual servoing of PD type with gravity compensation

where

$$J_L(s, z_c, q) = L_s(s, z_c) \begin{bmatrix} R_c^T & O \\ O & R_c^T \end{bmatrix} J(q), \tag{10.83}$$

the camera frame and the end-effector frame being coincident.

Therefore, with the choice

$$u = g(q) + J_L^T(s, z_c, q)\left(K_{Ps}e_s - K_{Ds}J_L(s, z_c, q)\dot{q}\right), \tag{10.84}$$

where K_{Ds} is a symmetric and positive definite $(k \times k)$ matrix, Eq. (10.81) becomes

$$\dot{V} = -\dot{q}^T F \dot{q} - \dot{q}^T J_L^T K_{Ds} J_L \dot{q}. \tag{10.85}$$

Control law (10.84) includes a nonlinear compensation action of gravitational forces in the joint space and a linear PD action in the image space. The last term, in view of (10.82), corresponds to a derivative action in the image space and has been added to increase damping. The resulting block scheme is reported in Fig. 10.16.

The direct measurement of $\dot{s}$ would permit the computation of the derivative term as $-K_{Ds}\dot{s}$; this measurement, however, is not available. As an alternative, the derivative term can simply be set as $-K_D\dot{q}$, with K_D symmetric and positive definite $(n \times n)$ matrix.

Equation (10.85) reveals that, for all trajectories of the system, the Lyapunov function decreases until $\dot{q} \neq 0$. Therefore the system reaches an equilibrium state, characterized by

$$J_L^T(s, z_c, q)K_{Ps}e_s = 0. \tag{10.86}$$

Equations (10.86), (10.83) show that, if the interaction matrix and the geometric Jacobian of the manipulator are full rank, then $e_s = 0$, which is the sought result.

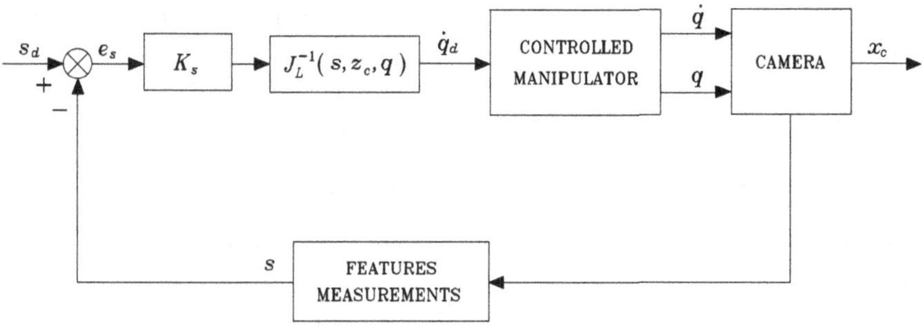

Fig. 10.17. Block scheme of resolved-velocity image-based visual servoing

Notice that control law (10.84) requires not only the measurement of $\boldsymbol{s}$ but also the computation of vector $\boldsymbol{z}_c$ which, in the image-based visual servoing philosophy, should be avoided. In some applications $\boldsymbol{z}_c$ is known with good approximation, as in the case that the relative motion of the camera with respect to the object belongs to a plane. Alternatively, estimated or constant values can be used for $\boldsymbol{z}_c$, as the value in the initial configuration or that in the desired configuration. This is equivalent to using an estimate $\widehat{\boldsymbol{L}}_s$ of the interaction matrix. In such cases, however, the stability proof becomes much more complex.

10.8.2 Resolved-velocity Control

The concept of resolved-velocity control can easily be extended to the image space. In such a case, Eq. (10.82) suggests the following choice of the reference velocity in joint space

$$\dot{\boldsymbol{q}}_r = \boldsymbol{J}_L^{-1}(\boldsymbol{s}, \boldsymbol{z}_c, \boldsymbol{q}_r)\boldsymbol{K}_s\boldsymbol{e}_s, \tag{10.87}$$

under the assumption of invertible matrix $\boldsymbol{J}_L$. This control law, replaced in (10.82), yields the linear equation

$$\dot{\boldsymbol{e}}_s + \boldsymbol{K}_s\boldsymbol{e}_s = \boldsymbol{0}. \tag{10.88}$$

Therefore, if $\boldsymbol{K}_s$ is a positive definite matrix, Eq. (10.88) is asymptotically stable and error $\boldsymbol{e}_s$ tends asymptotically to zero with convergence of exponential type and speed depending on the eigenvalues of matrix $\boldsymbol{K}_s$. The convergence to zero of the image space error $\boldsymbol{e}_s$ ensures the asymptotic convergence of $\boldsymbol{x}_{c,o}$ to the desired pose $\boldsymbol{x}_{d,o}$.

The block scheme of resolved-velocity image-based visual servoing is shown in Fig. 10.17.

Notice that this control scheme requires the computation of the inverse of matrix $\boldsymbol{J}_L$; therefore it is affected by problems related to the singularities of this matrix which, in view of (10.83), are both those of the geometric Jacobian and those of the interaction matrix. The most critical singularities are those of the interaction matrix, since they depend on the choice of the image feature parameters.

Therefore, it is convenient to compute control law (10.87) in two steps. The first step is the computation of vector

$$\boldsymbol{v}_r^c = \boldsymbol{L}_s^{-1}(\boldsymbol{s}, \boldsymbol{z}_c)\boldsymbol{K}_s \boldsymbol{e}_s. \tag{10.89}$$

The second step is the computation of the joint space reference velocity using the relationship

$$\dot{\boldsymbol{q}}_r = \boldsymbol{J}^{-1}(\boldsymbol{q}) \begin{bmatrix} \boldsymbol{R}_c & \boldsymbol{O} \\ \boldsymbol{O} & \boldsymbol{R}_c \end{bmatrix} \boldsymbol{v}_r^c. \tag{10.90}$$

Far from the kinematic singularities of the manipulator, the problem of the singularities of the interaction matrix can be overcome by using a number k of feature parameters greater than the minimum required m, similarly to the case considered in Sect. 10.3.3. The control law can be modified by using the left pseudo-inverse of interaction matrix $\boldsymbol{L}_s$ in place of the inverse, namely

$$\boldsymbol{v}_r^c = (\boldsymbol{L}_s^T \boldsymbol{L}_s)^{-1} \boldsymbol{L}_s^T \boldsymbol{K}_s \boldsymbol{e}_s \tag{10.91}$$

in place of (10.89). Stability of the closed-loop system with control law (10.90), (10.91) can be shown using the direct Lyapunov method based on the positive definite function

$$V(\boldsymbol{e}_s) = \frac{1}{2}\boldsymbol{e}_s^T \boldsymbol{K}_s \boldsymbol{e}_s > 0 \qquad \forall \boldsymbol{e}_s \neq \boldsymbol{0}.$$

Computing the time derivative of this function and taking into account (10.82), (10.83), (10.90), (10.91), yields

$$\dot{V} = -\boldsymbol{e}_s^T \boldsymbol{K}_s \boldsymbol{L}_s (\boldsymbol{L}_s^T \boldsymbol{L}_s)^{-1} \boldsymbol{L}_s^T \boldsymbol{K}_s \boldsymbol{e}_s$$

which is negative semi-definite because $\mathcal{N}(\boldsymbol{L}_s^T) \neq \emptyset$, $\boldsymbol{L}_s^T$ being a matrix with more columns than rows. Therefore, the closed-loop system is stable but not asymptotically stable. This implies that the error is bounded, but in some cases the system may reach an equilibrium with $\boldsymbol{e}_s \neq \boldsymbol{0}$ and $\boldsymbol{K}_s \boldsymbol{e}_s \in \mathcal{N}(\boldsymbol{L}_s^T)$.

Another problem connected with the implementation of control law (10.89) or (10.91) and (10.90) depends on the fact that the computation of interaction matrix $\boldsymbol{L}_s$ requires knowledge of z_c. Similar to Sect. 10.8.1, this problem can be solved by using an estimate of matrix $\widehat{\boldsymbol{L}}_s^{-1}$ (or of its pseudo-inverse). In this case, the Lypapunov method can be used to prove that the control scheme remains stable provided that matrix $\boldsymbol{L}_s \widehat{\boldsymbol{L}}_s^{-1}$ is positive definite. Notice that z_c is the only information depending on object geometry. Therefore, it can also be seen that image-based visual servoing, in the case that only one camera is used, does not require exact knowledge of object geometry.

The choice of the elements of matrix $\boldsymbol{K}_s$ influences the trajectories of the feature parameters, which are solution to differential equation (10.88). In the case of feature points, if a diagonal matrix $\boldsymbol{K}_s$ with equal elements is set, the projections of these points on the image plane will follow line segments. The corresponding camera motion, however, cannot be easily predicted, because of the nonlinearity of the mapping between image plane variables and operational space variables.

10.9 Comparison Among Various Control Schemes

In order to make a comparison between the various control schemes presented, consider the SCARA manipulator of Example 10.3 and the planar object of Example 10.1. The base frame of the SCARA manipulator of Fig. 2.36 is set with the origin at the intersection of the axis of joint 1 with the horizontal plane containing the origin of the end-effector frame when $d_3 = 0$, d_3 being the prismatic joint variable; the axis z of the base frame points downward. The operational space, of dimension $m = 4$, is characterized by vector $\boldsymbol{x}_{c,o}$ in (10.33).

With reference to the dynamic model of Problem 7.3, the same data of Example 7.2 are considered; in addition, $m_{\ell_3} = 2\,\mathrm{kg}$ and $I_{\ell_4} = 1\,\mathrm{kg{\cdot}m^2}$, while the contribution of the motors of the last two links are neglected.

For position-based visual servoing schemes, the real-time estimation of vector $\boldsymbol{x}_{c,o}$ from a suitable feature vector is required. To this end, the algorithm of Sect. 10.3.3 can be used, based on the inverse of the image Jacobian. This is a classical visual tracking application that can be accomplished using only two feature points, because the corresponding Jacobian $\boldsymbol{J}_{A_s}$ is a (4×4) matrix. The selected points are P_1 and P_2 of the object of Fig. 10.6.

The same points can also be used for image-based visual servoing, because the corresponding Jacobian $\boldsymbol{J}_L$ is a (4×4) matrix.

It is assumed that at time $t = 0$ the pose of the camera frame, with respect to the base frame, is defined by the operational space vector

$$\boldsymbol{x}_c(0) = \begin{bmatrix} 1 & 1 & 1 & \pi/4 \end{bmatrix}^T \mathrm{m}$$

and the pose of the object frame, with respect to the camera frame, is defined by the operational space vector

$$\boldsymbol{x}_{c,o}(0) = \begin{bmatrix} 0 & 0 & 0.5 & 0 \end{bmatrix}^T \mathrm{m}.$$

The desired pose of the object frame with respect to the camera frame is defined by vector

$$\boldsymbol{x}_{d,o} = \begin{bmatrix} -0.1 & 0.1 & 0.6 & -\pi/3 \end{bmatrix}^T \mathrm{m}.$$

This quantity is assumed as the initial value of the pose estimation algorithm used by position-based visual servoing schemes.

For image-based visual servoing schemes, the desired value of the feature parameters of points P_1 and P_2 of the object, in the desired pose $\boldsymbol{x}_{d,o}$, is

$$\boldsymbol{s}_d = \begin{bmatrix} -0.1667 & 0.1667 & -0.0833 & 0.0223 \end{bmatrix}^T.$$

For all the schemes, a discrete-time implementation of the controller with sampling time of 0.04 s has been adopted, corresponding to a 25 Hz frequency. This value coincides with the minimum frame rate of analog cameras and allows the use of visual measurements also in the worst case.

In the numerical simulations, the following control schemes have been utilized:

A. Position-based visual servoing of PD type with gravity compensation with the following data:

$$\boldsymbol{K}_P = \text{diag}\{500, 500, 10, 10\}$$
$$\boldsymbol{K}_D = \text{diag}\{500, 500, 10, 10\}.$$

B. Resolved-velocity position-based visual servoing with the following data:

$$\boldsymbol{K} = \text{diag}\{1, 1, 1, 2\},$$

corresponding to a time constant of 1 s for the three position variables and of 0.5 s for the orientation variable.

C. Image-based visual servoing of PD type with gravity compensation with the following data:

$$\boldsymbol{K}_{Ps} = 300\boldsymbol{I}_4 \qquad \boldsymbol{K}_{Ds} = 330\boldsymbol{I}_4.$$

D. Resolved-velocity image-based visual servoing with the following data:

$$\boldsymbol{K}_s = \boldsymbol{I}_4,$$

corresponding to a time constant of 1 s for the feature parameters.

For the simulation of resolved-velocity control schemes, the dynamics of the velocity controlled manipulator has been neglected. Therefore, a pure kinematic model has been considered, based on the analytic Jacobian of the manipulator.

For position-based control schemes, the pose estimation algorithm based on the inverse of the image Jacobian has been adopted, with integration step $\Delta t = 1$ ms and gain $\boldsymbol{K}_s = 160\boldsymbol{I}_4$. As shown in Example 10.4, this implies that the algorithm converges in a time of about 0.03 s, which is lower than the sampling time of the control, as required for a correct operation of position-based control.

For image-based control schemes, matrix $\boldsymbol{L}_s(\boldsymbol{s}, \boldsymbol{z}_c)$ has been approximated with matrix $\widehat{\boldsymbol{L}}_s = \boldsymbol{L}_s(\boldsymbol{s}, z_d)$, where z_d is the third component of vector $\boldsymbol{x}_{d,o}$.

The parameters of the various controllers have been chosen in such a way as to show the particular features of the different control laws and, at the same time, to allow a significant comparison of the performance of each scheme in response to congruent control actions. In particular, it can be observed what follows:

- The gains in schemes **A** and **C** have been tuned in simulation so as to obtain transient behaviors similar to those of schemes **B** and **D**.
- In control scheme **B** the gains of the position variables have been intentionally chosen equal to one another but different from the gain of the orientation variable to show that a desired dynamics can be imposed to each operational space variable.
- In control scheme **D** the gains have all been chosen equal to one another, since imposing different dynamics to different coordinates of the projections of the feature points on the image plane is not significant.

The results obtained with the various control schemes are illustrated in Figs. 10.18–10.25 in terms of:

- The time history of position and orientation of the camera frame with respect to the base frame and corresponding desired values (represented with dashed lines).
- The time history of feature parameters and corresponding desired values (represented with dashed lines).
- The path of feature points projections on the camera image plane, from initial positions (marked with crosses) to final positions (marked with circles).

Regarding performance of the various control schemes, the following considerations can be drawn from the obtained results.

In principle, if position-based visual servoing is adopted, a desired transient behaviour can be assigned to the operational space variables. This is only partially true for control scheme **A**, because the dynamics of the closed-loop system, for PD control with gravity compensation, is nonlinear and coupled. Therefore, the transient behaviour shown in Fig. 10.18 may be different if the manipulator starts from a different initial pose or has to reach a different desired pose. Vice versa, for control scheme **B**, the time history of operational space variables reported in Fig. 10.20 shows transient behaviours of exponential type, whose characteristics depend only on the choice of matrix $\boldsymbol{K}$.

For both schemes **A** and **B**, the trajectories of the projections of the feature points and the corresponding paths in the image plane (Figs. 10.19 and 10.21 respectively) have evolutions that cannot be predicted in advance. This implies that, although the feature points projections are inside the image plane both in the initial and in the desired configuration, they may exit from the image plane during the transient, thus causing problems of convergence to the controlled system.

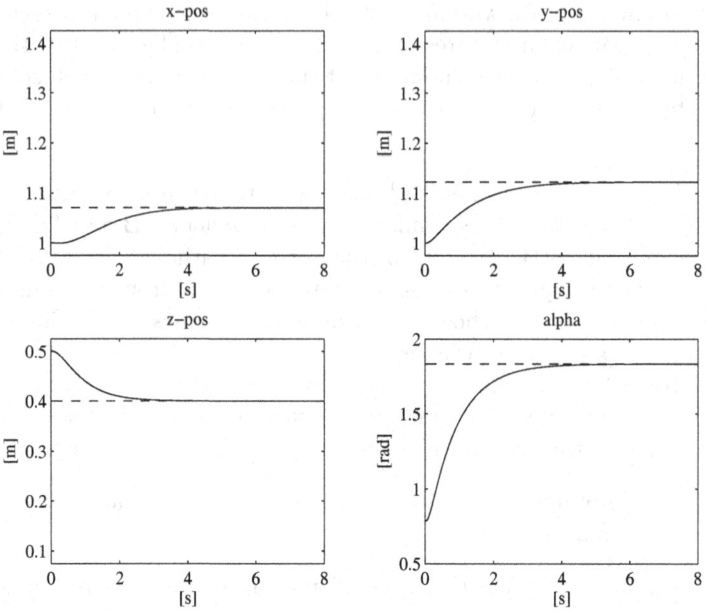

Fig. 10.18. Time history of camera frame position and orientation with control **A**

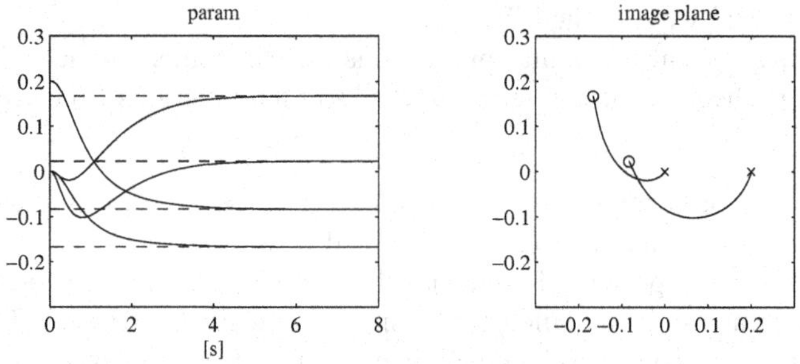

Fig. 10.19. Time history of feature parameters and corresponding path of feature points projections on image plane with control **A**

If image-based control is adopted, a desired transient behaviour can be assigned to the time histories of feature parameters and not to the operational space variables, in a dual fashion with respect to position-based control. This is confirmed by the results shown in Figs. 10.22–10.25, relative to control schemes **C** and **D**, respectively. In detail, especially in the case of control **C**, the time histories of the operational space variables are quite different from those reported in Figs. 10.18 and 10.20, despite the same initial and final configuration and a similar transient duration. This implies that the camera path, being unpredictable, may lead to joint limits violation or to collision of

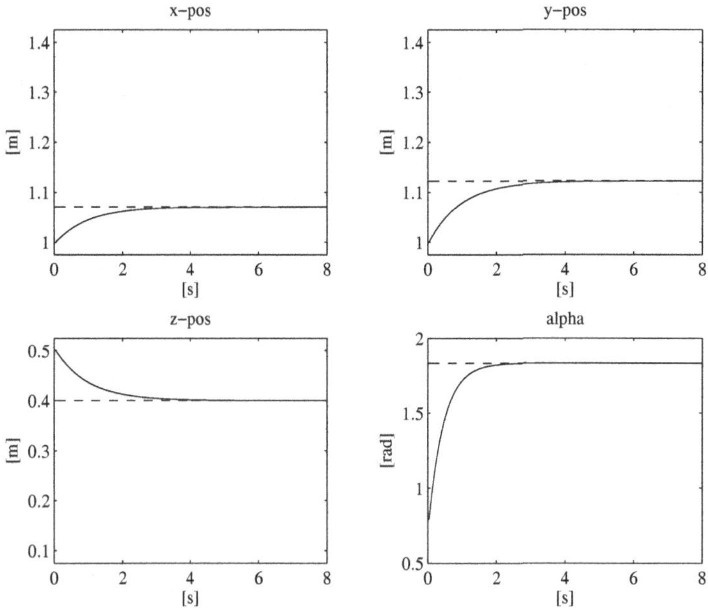

Fig. 10.20. Time history of camera frame position and orientation with control **B**

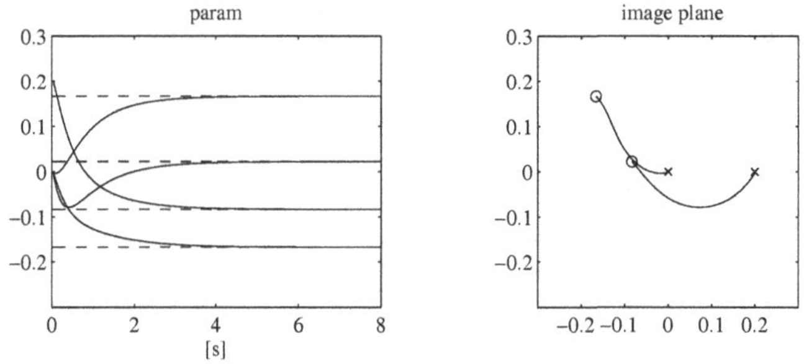

Fig. 10.21. Time history of feature parameters and corresponding path of feature points projections on image plane with control **B**

the manipulator with an obstacle. In the specific case of control scheme **C**, a 300% overshoot is present on the z component of the camera trajectory (see Fig. 10.22), which corresponds to a camera retreat movement with respect to the object of amplitude much higher than the distance reached at the end of the transient. The overshoot on the z component is present also for control scheme **D**, but is 'only' of 50% (see Fig. 10.24).

Notice that, for control scheme **C**, the presence of large displacements on some operational space variables does not correspond to significant deviations of the feature parameters with respect to their final values during the transient

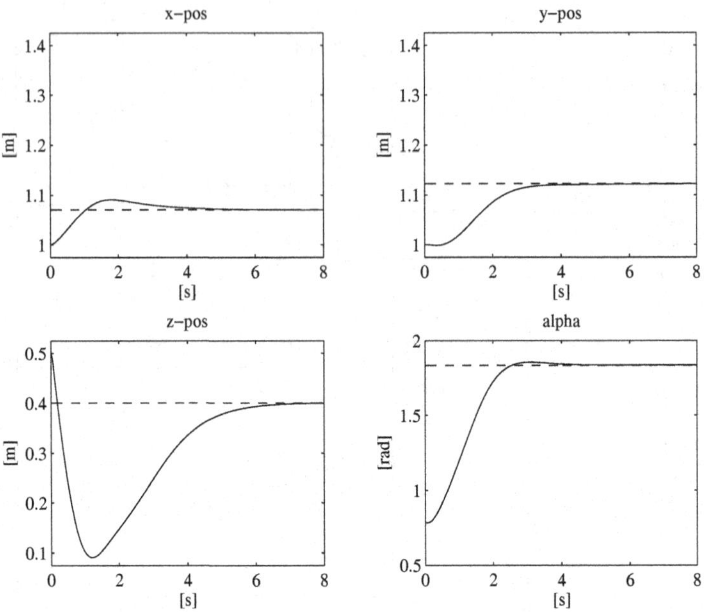

Fig. 10.22. Time history of camera frame position and orientation with control **C**

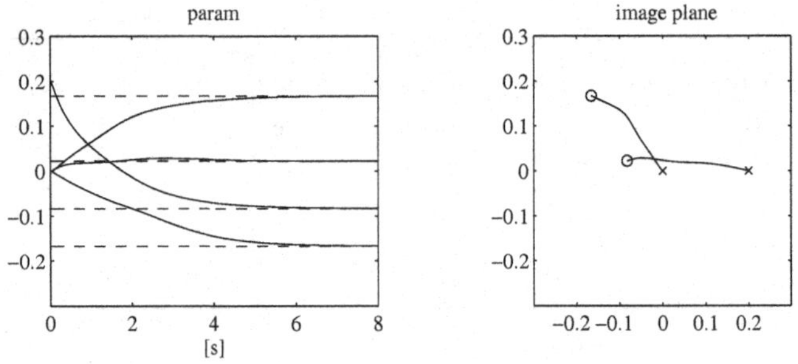

Fig. 10.23. Time history of feature parameters and corresponding path of feature points projections on image plane with control **C**

(see Fig. 10.23). Indeed, the paths of the feature points projections do not deviate much from the line segments connecting these points.

Figure 10.25, relative to control scheme **D**, reveals that the trajectories of the feature parameters are of exponential type. In this case, the transient behaviour depends only on matrix $\boldsymbol{K}_s$; choosing a diagonal matrix with equal elements implies that the paths of the feature points projections are linear. In the case at hand, in view of the approximation $\boldsymbol{L}_s(\boldsymbol{s}, \boldsymbol{z}_c) \approx \boldsymbol{L}_s(\boldsymbol{s}, z_d)$, the paths of the feature points projections shown in Fig. 10.25 are not perfectly linear.

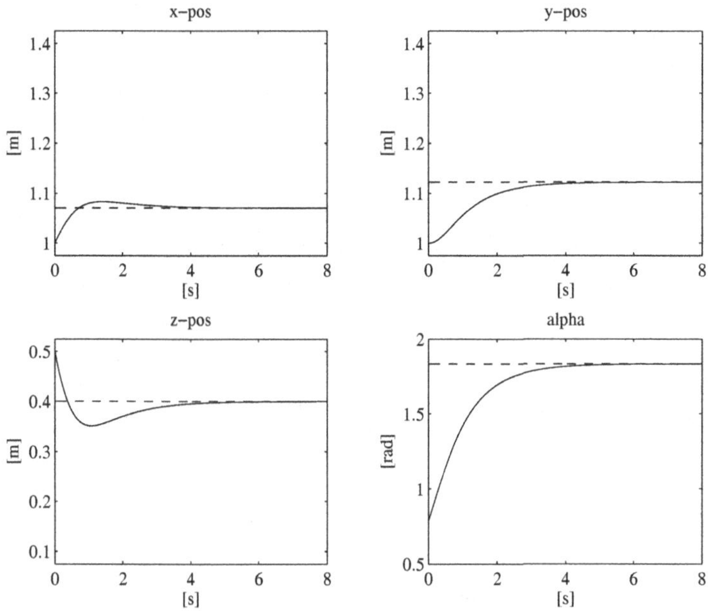

Fig. 10.24. Time history of camera frame position and orientation with control **D**

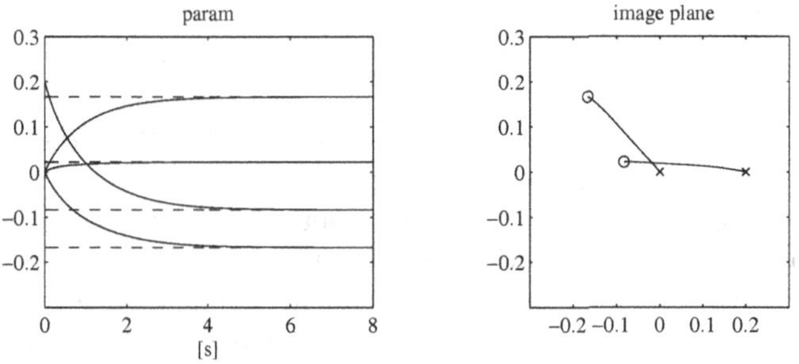

Fig. 10.25. Time history of feature parameters and corresponding path of feature points projections on image plane with control **D**

To conclude, Fig. 10.26 shows the paths of the origin of the camera frame obtained using the four control schemes. It can be observed that, with control scheme **B**, a perfectly linear path is obtained, thanks to the choice of a diagonal gain matrix $\boldsymbol{K}_s$ with equal weights for the positional part. Using control scheme **A**, the path is almost linear, because, unlike case **B**, this type of control does not guarantee a decoupled dynamics for each operational space variable. Vice versa, using control schemes **C** and **D**, the path of the origin of the camera frame is far from being linear. In both cases, the phenomenon of camera retreat with respect to the object can be observed. To this end, notice

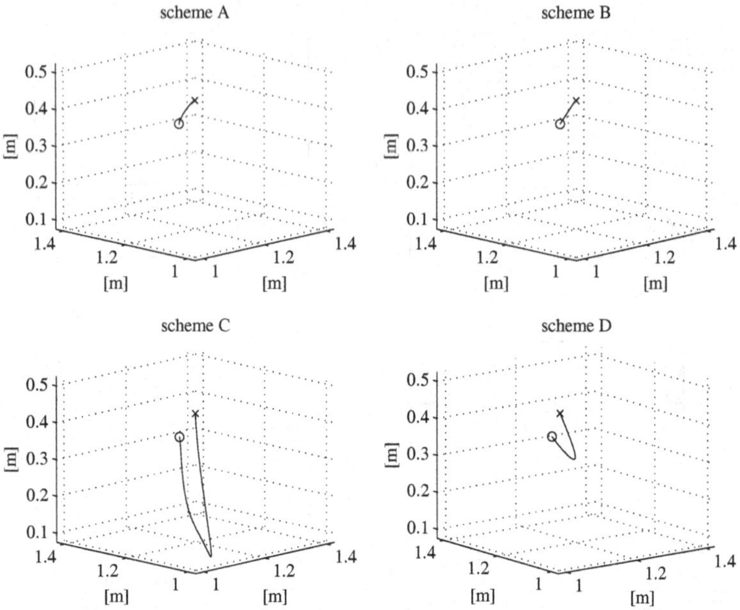

Fig. 10.26. Paths of camera frame origin in Cartesian space with the four control schemes

the axis z of the base frame of the SCARA manipulator and the axis z_c of the camera frame are aligned to the vertical direction and point downward; therefore, with respect to the reference frames of Fig. 10.26, the object is on the top and the camera points upward.

The phenomenon of camera retreat appears whenever the camera is required to perform large rotations about the optical axis; this phenomenon can be intuitively explained through a simple example. Assume that a pure rotation of the camera about the axis z_c is required and control scheme **D** is adopted. Therefore, visual servoing imposes that the feature points projections on the image plane follow rectilinear paths from the initial to the desired positions, whereas simple camera rotation would have required circular paths. The constraint on the path in the image plane implies that, during rotation, the origin of the camera frame must move, with respect to the object, first backward and then forward to reach again, asymptotically, the initial position. It can be shown that, if the desired rotation tends to π, then the distance of the camera from the object tends to ∞ and the system becomes unstable.

10.10 Hybrid Visual Servoing

An approach which combines the benefits of position-based and image-based visual servoing is *hybrid visual servoing*. The name stems from the fact that the control error is defined in the operational space for some components and in the image space for the others. This implies that a desired motion can be specified, at least partially, in the operational space so that the camera trajectory during visual servoing can be predicted in advance for some components. On the other hand, the presence of error components in the image space helps keep the image features in the camera field of view, which is a difficult task in position-based approaches.

Hybrid visual servoing requires the estimation of some operational space variables. Assume that the object has a planar surface where at least four feature points, and no triplets of collinear points, can be selected. Using the coordinates of these points in the camera image plane, both in the current and in the desired pose of the camera frame, it is possible to compute the planar homography $\boldsymbol{H}$ as described in Sect. 10.4.4. Notice that, for this computation, knowledge of the current and the desired camera pose is not required, provided that the feature vectors $\boldsymbol{s}$ and $\boldsymbol{s}_d$ are known.

In view of (10.59), assuming that Frame 1 coincides with the camera frame in the current pose and Frame 2 coincides with the camera frame in the desired pose, the following equality holds:

$$\boldsymbol{H} = \boldsymbol{R}_d^c + \frac{1}{d_d}\boldsymbol{o}_{c,d}^c \boldsymbol{n}^{dT},$$

where $\boldsymbol{R}_d^c$ is the rotation matrix between the desired orientation and the current orientation of the camera frame, $\boldsymbol{o}_{c,d}^c$ is the position vector of the origin of the camera frame in the desired pose with respect to the current pose, $\boldsymbol{n}^d$ is the unit vector normal to the plane containing the feature points, and d_d is the distance between this plane and the origin of the camera frame in the desired pose. The quantities $\boldsymbol{R}_d^c$, $\boldsymbol{n}^d$, $(1/d_d)\boldsymbol{o}_{c,d}^c$, in the current camera pose, can be computed at each sampling time from matrix $\boldsymbol{H}$.

Adopting a resolved-velocity approach, the control objective consists of computing the reference absolute velocity of the camera frame

$$\boldsymbol{v}_r^c = \begin{bmatrix} \boldsymbol{\nu}_r^c \\ \boldsymbol{\omega}_r^c \end{bmatrix}$$

from a suitably defined error vector.

To this end, the orientation error between the desired and the current camera pose can be computed from matrix $\boldsymbol{R}_d^c$, as for position-based visual servoing. If $\boldsymbol{\phi}_{c,d}$ denotes the vector of the Euler angles extracted from $\boldsymbol{R}_d^c$, the control vector $\boldsymbol{\omega}_r^c$ can be chosen as

$$\boldsymbol{\omega}_r^c = -\boldsymbol{T}(\boldsymbol{\phi}_{c,d})\boldsymbol{K}_o\boldsymbol{\phi}_{c,d}, \tag{10.92}$$

where $\boldsymbol{K}_o$ is a (3×3) matrix. With this choice, the equation of the orientation error has the form

$$\dot{\boldsymbol{\phi}}_{c,d} + \boldsymbol{K}_o\boldsymbol{\phi}_{c,d} = \boldsymbol{0}. \tag{10.93}$$

Equation (10.93), if $\boldsymbol{K}_o$ is a symmetric and positive definite matrix, implies that the orientation error tends to zero asymptotically with convergence of exponential type and speed depending on the eigenvalues of matrix $\boldsymbol{K}_o$.

The control vector $\boldsymbol{\nu}_r^c$ should be selected so that the positional part of the error between the desired and the current camera pose converges to zero. The position error could be defined as the difference of the coordinates of a point of the object in the desired camera frame $\boldsymbol{r}_d^c = [\,x_d \quad y_d \quad z_d\,]^T$ and those in the current camera frame $\boldsymbol{r}_c^c = [\,x_c \quad y_c \quad z_c\,]^T$, namely $\boldsymbol{r}_d^c - \boldsymbol{r}_c^c$. These coordinates, however, cannot be directly measured, unlike the corresponding coordinates in the image plane, defining the feature vectors $\boldsymbol{s}_{p,d} = [\,X_d \quad Y_d\,]^T = [\,x_d/z_d \quad y_d/z_d\,]^T$ and $\boldsymbol{s}_p = [\,X \quad Y\,]^T = [\,x_c/z_c \quad y_c/z_c\,]^T$.

The information deriving from the computation of homography $\boldsymbol{H}$ can be used to rewrite the ratio

$$\rho_z = z_c/z_d$$

in terms of known or measurable quantities in the form

$$\rho_z = \frac{d_c}{d_d}\frac{\boldsymbol{n}^{dT}\widetilde{\boldsymbol{s}}_{p,d}}{\boldsymbol{n}^{cT}\widetilde{\boldsymbol{s}}_p} \tag{10.94}$$

with

$$\frac{d_c}{d_d} = 1 + \boldsymbol{n}^{cT}\frac{\boldsymbol{o}_{c,d}^c}{d_d} = \det(\boldsymbol{H}), \tag{10.95}$$

and $\boldsymbol{n}^c = \boldsymbol{R}_d^c\boldsymbol{n}^d$, where vectors $\widetilde{\boldsymbol{s}}_p$ and $\widetilde{\boldsymbol{s}}_{p,d}$ denote the representations in homogeneous coordinates of $\boldsymbol{s}_p$ and $\boldsymbol{s}_{p,d}$, respectively (see Problem 10.12).

The position error, expressed in terms of known or measurable quantities, can be defined as

$$\boldsymbol{e}_p(\boldsymbol{r}_d^c, \boldsymbol{r}_c^c) = \begin{bmatrix} X_d - X \\ Y_d - Y \\ \ln\rho_z \end{bmatrix}.$$

Notice that, in view of (5.44), convergence to zero of $\boldsymbol{e}_p$ implies convergence to zero of $\boldsymbol{r}_d^c - \boldsymbol{r}_c^c$ and vice versa.

Computing the time derivative of $\boldsymbol{e}_p$ yields

$$\dot{\boldsymbol{e}}_p = \frac{\partial \boldsymbol{e}_p(\boldsymbol{r}_c^c)}{\partial \boldsymbol{r}_c^c}\dot{\boldsymbol{r}}_c^c,$$

$\boldsymbol{r}_d^c$ being constant. By taking into account (10.26) and the decomposition

$$\boldsymbol{v}_c^c = \begin{bmatrix} \boldsymbol{\nu}_c^c \\ \boldsymbol{\omega}_c^c \end{bmatrix}$$

with $\boldsymbol{\nu}_c^c = \boldsymbol{R}_c^T\dot{\boldsymbol{o}}_c$, the above expression can be rewritten in the form

$$\dot{\boldsymbol{e}}_p = -\boldsymbol{J}_p\boldsymbol{\nu}_c^c - \boldsymbol{J}_o\boldsymbol{\omega}_c^c, \tag{10.96}$$

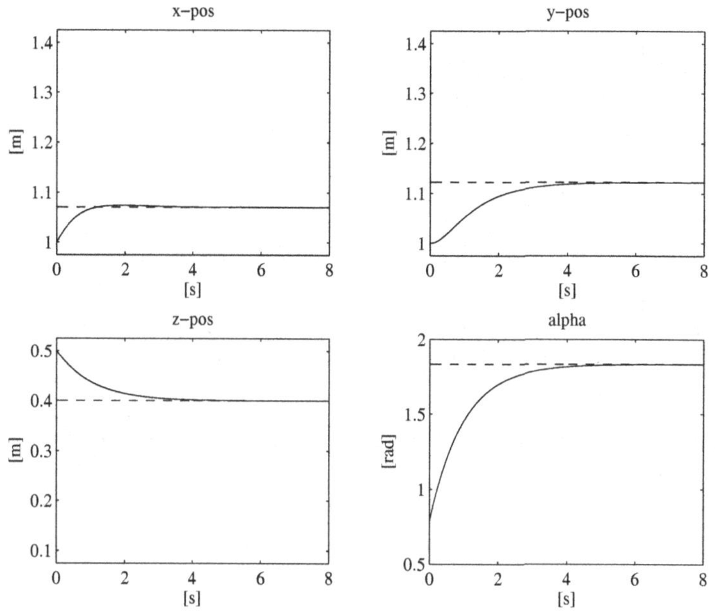

Fig. 10.27. Time history of position and orientation of camera frame with hybrid visual servoing

with (see Problem 10.13)

$$\boldsymbol{J}_p = \frac{1}{z_d \rho_z}\begin{bmatrix} -1 & 0 & X \\ 0 & -1 & Y \\ 0 & 0 & -1 \end{bmatrix}$$

and

$$\boldsymbol{J}_o = \begin{bmatrix} XY & -1-X^2 & Y \\ 1+Y^2 & -XY & -X \\ -Y & X & 0 \end{bmatrix}.$$

Equation (10.96) suggests the following choice of control vector $\boldsymbol{\nu}_r^c$

$$\boldsymbol{\nu}_r^c = \boldsymbol{J}_p^{-1}(\boldsymbol{K}_p \boldsymbol{e}_p - \boldsymbol{J}_o \boldsymbol{\omega}_r^c), \tag{10.97}$$

$\boldsymbol{J}_p$ being a nonsingular matrix.

Notice that, for the computation of $\boldsymbol{J}_p^{-1}$, knowledge of the constant quantity z_d is required.

If z_d is known, control law (10.97), in view of assumptions $\dot{\boldsymbol{o}}_c^c \approx \boldsymbol{\nu}_r^c$ and $\boldsymbol{\omega}_c^c \approx \boldsymbol{\omega}_r^c$, yields the following error equation:

$$\dot{\boldsymbol{e}}_p + \boldsymbol{K}_p \boldsymbol{e}_p = \boldsymbol{0},$$

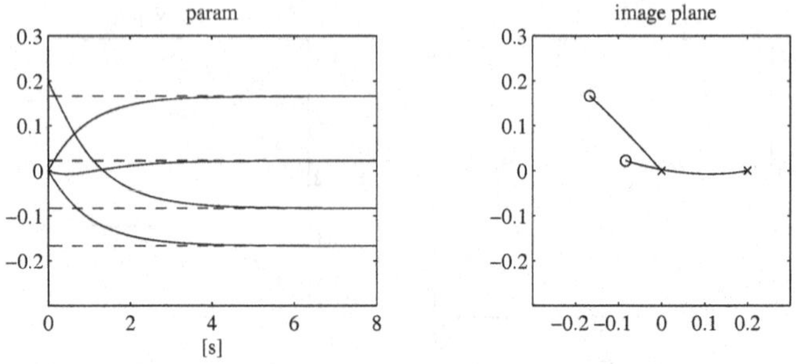

Fig. 10.28. Time history of feature parameters and corresponding path of feature points projections on image plane with hybrid visual servoing

which implies the exponential convergence of $\boldsymbol{e}_p$ to zero, provided that $\boldsymbol{K}_p$ is a positive definite matrix.

If z_d is not known, an estimate $\widehat{z}_d$ can be adopted. Therefore, in control law (10.97), matrix $\boldsymbol{J}_p^{-1}$ can be replaced by an estimate $\widehat{\boldsymbol{J}}_p^{-1}$. In view of the equality

$$\widehat{\boldsymbol{J}}_p^{-1} = \frac{\widehat{z}_d}{z_d}\boldsymbol{J}_p^{-1},$$

the following error equation is obtained:

$$\dot{\boldsymbol{e}}_p + \frac{\widehat{z}_d}{z_d}\boldsymbol{K}_p\boldsymbol{e}_p = \left(1 - \frac{\widehat{z}_d}{z_d}\right)\boldsymbol{J}_o\boldsymbol{\omega}_r^c.$$

This equation shows that the use of an estimate $\widehat{z}_d$ in place of the true value z_d implies a simple gain scaling in the error equation, and asymptotic stability is preserved. Moreover, due to the presence in the right-hand side of the above equation of a term depending on $\boldsymbol{\omega}_r^c$, the time history of $\boldsymbol{e}_p$ is influenced by the orientation error, which evolves according to (10.93).

Example 10.6

For the SCARA manipulator and the task of Sect. 10.9, consider the hybrid visual servoing law with gains

$$\boldsymbol{K}_p = \boldsymbol{I}_3 \quad k_o = 2,$$

and compute the positional part of the error with respect to point P_1. The planar homography and the corresponding parameters are estimated as in Example 10.5, using four points. The results are reported in Figs. 10.27 and 10.28, in terms of the same variables shown in Sect. 10.9. Notice that the time histories of the variables in the operational space of Fig. 10.27 are quite similar to that obtained with resolved-velocity position-based visual servoing (Fig. 10.20). On the other hand, the time histories of the feature parameters of Fig. 10.28 are substantially similar to those obtained with resolved-velocity image-based visual servoing (Fig. 10.25) except for

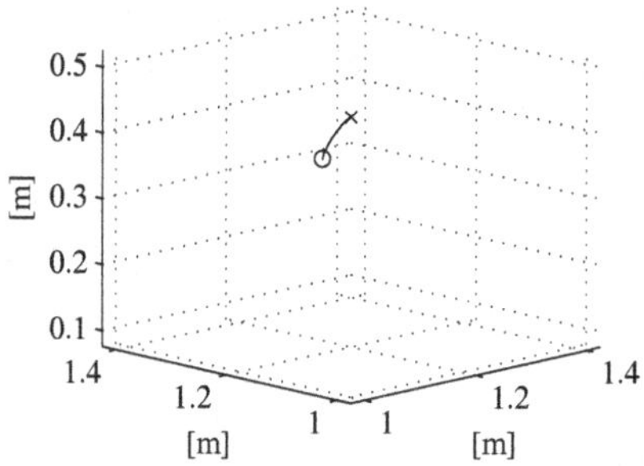

Fig. 10.29. Path of camera frame origin in Cartesian space with hybrid visual servoing

the fact that the path in the image plane of the projection of point P_1 is perfectly linear, as imposed by the control. The corresponding path of the camera frame origin, reported in Fig. 10.29, shows a substantial improvement with respect to those obtained with the image-based visual servoing schemes of Fig. 10.26.

The method illustrated above is only one of the possible visual servoing approaches based on the computation of the planar homography and on its decomposition. It is worth pointing out that knowledge of $(1/d_d)\boldsymbol{o}^c_{c,d}$ and $\boldsymbol{R}^c_d$ allows the computation of the operational space error (10.69), up to a scaling factor for the positional part; therefore, it is also possible to use the position-based visual servoing schemes presented in Sect. 10.7. On the other hand, in hybrid visual servoing approaches, different choices are possible for the error components depending on feature parameters as well as for those depending on the operational space variables.

Bibliography

The literature dealing with computational vision is quite extensive and various. Image processing is treated, e.g., in [93], while geometrical issues are considered in [75] and [146]. The concept of interaction matrix was originally proposed in [239] under a different name, while the actual name was introduced in [71]. The pose estimation algorithm based on the inverse of the image Jacobian is also termed virtual visual servoing [47]. Position-based visual servoing was proposed in [239] and, more recently, in [244] and [134]. Several papers dealing with image-based visual servoing can be found, starting from early works of [239] and [79]. A rigorous stability analysis is reported in [108] for PD control with gravity compensation and in [71] for resolved-velocity control. Hybrid visual servoing, presented in [148], is only one of advanced control schemes based on decoupling of the camera DOFs, e.g., the partitioned approach [49] and the control based on image moments [35]. Finally, visual servoing based on stereo vision is considered in [92]. An interesting review of the state of the art of visual servoing until mid 1990s is presented in [103], while a more recent review can be found in [36].

Problems

10.1. Derive Eq. (10.2).

10.2. Show that a non-null solution to (10.10) is the right eigenvector corresponding to the null singular value of matrix $\boldsymbol{A}$.

10.3. Show that (10.12) is the matrix which minimizes Frobenius norm (10.11) with the constraint that $\boldsymbol{R}_o^c$ is a rotation matrix, in the case $\sigma > 0$. [*Hint*: consider that $\mathrm{Tr}(\boldsymbol{R}_o^{cT}\boldsymbol{U}\boldsymbol{\Sigma}\boldsymbol{V}^T) = \mathrm{Tr}(\boldsymbol{V}^T\boldsymbol{R}_o^{cT}\boldsymbol{U}\boldsymbol{\Sigma})$; moreover, the absolute values of the diagonal elements of matrix $\boldsymbol{V}^T\boldsymbol{R}_o^{cT}\boldsymbol{U}$ are lower or equal to 1.]

10.4. Consider the SCARA manipulator of Example 10.3. Perform a computer implementation of the pose estimation algorithm based on the inverse of the image Jacobian considering the points P_1 and P_2 of the object of Example 10.1. Compute the homogeneous transformation matrix corresponding to the feature vector $\boldsymbol{s} = [\,-0.1667 \quad 0.1667 \quad -0.0833 \quad 0.0223\,]^T$. Assume that, in the initial pose, the axes of the camera frame are parallel to those of the object frame and the origin is at a distance of 0.5 m along the vertical axis.

10.5. Solve the previous problem using the feature parameters of the line segment P_1P_2.

10.6. Show that the value of $\boldsymbol{o}$ which minimizes function (10.55) has expression (10.56). [*Hint*: Use the equalities $\boldsymbol{p}_i = \bar{\boldsymbol{p}}_i + \bar{\boldsymbol{p}}$ and $\boldsymbol{p}'_i = \bar{\boldsymbol{p}}'_i + \bar{\boldsymbol{p}}'$ in (10.55), and the properties $\|\boldsymbol{a}+\boldsymbol{b}\|^2 = \|\boldsymbol{a}\|^2 + 2\boldsymbol{a}^T\boldsymbol{b} + \|\boldsymbol{b}\|^2$ and $\sum_{i=1}^n \bar{\boldsymbol{p}}_i = \sum_{i=1}^n \bar{\boldsymbol{p}}'_i = \boldsymbol{0}$.]

10.7. Show that the matrix $\boldsymbol{R}$ which minimizes Frobenius norm (10.57) is the matrix which maximizes the trace of $\boldsymbol{R}\boldsymbol{K}$.

10.8. For the SCARA manipulator of Example 10.3, design a position-based visual servoing scheme of PD type with gravity compensation using the measurement of the feature parameters of line segment P_1P_2 of Example 10.1. Perform a computer simulation in the same operating conditions of Sect. 10.9 and compare the results.

10.9. For the SCARA manipulator of Example 10.3, design a resolved-velocity position-based visual servoing scheme using the measurement of the feature parameters of line segment P_1P_2 of Example 10.1. Perform a computer simulation in the same operating conditions of Sect. 10.9 and compare the results.

10.10. For the SCARA manipulator of Example 10.3, design an image-based visual servoing scheme of PD type with gravity compensation using the measurement of the feature parameters of the line segment P_1P_2 of Example 10.1. Perform a computer simulation in the same operating conditions of Sect. 10.9 and compare the results.

10.11. For the SCARA manipulator of Example 10.3, design a resolved-velocity image-based visual servoing scheme using the measurement of the feature parameters of line segment P_1P_2 of Example 10.1. Perform a computer simulation in the same operating conditions of Sect. 10.9 and compare the results.

10.12. Derive the expressions (10.94), (10.95).

10.13. Derive the expressions of $\boldsymbol{J}_p$ and $\boldsymbol{J}_o$ in (10.96).

11

Mobile Robots

The previous chapters deal mainly with articulated manipulators that represent the large majority of robots used in industrial settings. However, *mobile robots* are becoming increasingly important in advanced applications, in view of their potential for autonomous intervention. This chapter presents techniques for modelling, planning and control of *wheeled* mobile robots. The structure of the kinematic constraints arising from the pure rolling of the wheels is first analyzed; it is shown that such constraints are in general *nonholonomic* and consequently reduce the local mobility of the robot. The *kinematic model* associated with the constraints is introduced to describe the instantaneous admissible motions, and conditions are given under which it can be put in *chained form*. The *dynamic model*, that relates the admissible motions to the generalized forces acting on the robot DOFs, is then derived. The peculiar nature of the kinematic model, and in particular the existence of *flat outputs*, is exploited to devise *trajectory planning* methods that guarantee that the nonholonomic constraints are satisfied. The structure of *minimum-time trajectories* is also analyzed. The *motion control* problem for mobile robots is then discussed, with reference to two basic motion tasks, i.e., *trajectory tracking* and *posture regulation*. The chapter concludes by surveying some techniques for *odometric localization* that is necessary to implement feedback control schemes.

11.1 Nonholonomic Constraints

Wheels are by far the most common mechanism to achieve locomotion in mobile robots. Any wheeled vehicle is subject to kinematic constraints that reduce in general its local mobility, while leaving intact the possibility of reaching arbitrary configurations by appropriate manoeuvres. For example, any driver knows by experience that, while it is impossible to move instantaneously a car in the direction orthogonal to its heading, it is still possible to park it arbitrarily, at least in the absence of obstacles. It is therefore important to analyze in detail the structure of these constraints.

In accordance with the terminology introduced in Sect. B.4, consider a mechanical system whose *configuration* $\boldsymbol{q} \in \mathcal{C}$ is described by a vector of *generalized coordinates*, and assume that the *configuration space* $\mathcal{C}$ (i.e., the space of all possible robot configurations) coincides[1] with $\mathbb{R}^n$. The motion of the system that is represented by the evolution of $\boldsymbol{q}$ over time may be subject to constraints that can be classified under various criteria. For example, they may be expressed as equalities or inequalities (respectively, *bilateral* or *unilateral* constraints), and they may depend explicitly on time or not (*rheonomic* or *scleronomic* constraints). In this chapter, only bilateral scleronomic constraints will be considered.

Constraints that can be put in the form

$$h_i(\boldsymbol{q}) = 0 \qquad i = 1, \ldots, k < n \tag{11.1}$$

are called *holonomic* (or *integrable*). In the following, it is assumed that the functions $h_i : \mathcal{C} \mapsto \mathbb{R}$ are of class C^∞ (*smooth*) and independent. The effect of holonomic constraints is to reduce the space of accessible configurations to a subset of $\mathcal{C}$ with dimension $n - k$. A mechanical system for which all the constraints can be expressed in the form (11.1) is called *holonomic.*

In the presence of holonomic constraints, the implicit function theorem can be used in principle to solve the equations in (11.1) by expressing k generalized coordinates as a function of the remaining $n-k$, so as to eliminate them from the formulation of the problem. However, in general this procedure is only valid locally, and may introduce singularities. A convenient alternative is to replace the original generalized coordinates with a reduced set of $n - k$ new coordinates that are directly defined on the accessible subspace, in such a way that the available DOFs are effectively characterized. The mobility of the reduced system thus obtained is completely equivalent to that of the original mechanism.

Holonomic constraints are generally the result of mechanical interconnections between the various bodies of the system. For example, prismatic and revolute joints used in robot manipulators are a typical source of such constraints, and joint variables are an example of reduced sets of coordinates in the above sense. Constraints of the form (11.1) may also arise in particular operating conditions; for example, one may mention the case of a kinematically redundant manipulator that moves while keeping the end-effector fixed at a certain pose (*self-motion*).

Constraints that involve generalized coordinates and velocities

$$a_i(\boldsymbol{q}, \dot{\boldsymbol{q}}) = 0 \qquad i = 1, \ldots, k < n$$

[1] This assumption is taken for simplicity. In the general case, the configuration space $\mathcal{C}$ may be identified with a Euclidean space only on a local basis, because its global geometric structure is more complex; this will be further discussed in Chap. 12. The material presented in this chapter is, however, still valid.

are called *kinematic*. They constrain the instantaneous admissible motion of the mechanical system by reducing the set of generalized velocities that can be attained at each configuration. Kinematic constraints are generally expressed in *Pfaffian form*, i.e., they are linear in the generalized velocities:

$$\boldsymbol{a}_i^T(\boldsymbol{q})\dot{\boldsymbol{q}} = 0 \qquad i = 1, \ldots, k < n, \tag{11.2}$$

or, in matrix form

$$\boldsymbol{A}^T(\boldsymbol{q})\dot{\boldsymbol{q}} = \boldsymbol{0}. \tag{11.3}$$

Vectors $\boldsymbol{a}_i : \mathcal{C} \mapsto \mathbb{R}^n$ are assumed to be smooth as well as linearly independent.

Clearly, the existence of k holonomic constraints (11.1) implies that of an equal number of kinematic constraints:

$$\frac{dh_i(\boldsymbol{q})}{dt} = \frac{\partial h_i(\boldsymbol{q})}{\partial \boldsymbol{q}}\dot{\boldsymbol{q}} = 0 \qquad i = 1, \ldots, k.$$

However, the converse is not true in general. A system of kinematic constraints in the form (11.3) may or may not be integrable to the form (11.1). In the negative case, the kinematic constraints are said to be *nonholonomic* (or *non-integrable*). A mechanical system that is subject to at least one such constraint is called *nonholonomic*.

Nonholonomic constraints reduce the mobility of the mechanical system in a completely different way with respect to holonomic constraints. To appreciate this fact, consider a single Pfaffian constraint

$$\boldsymbol{a}^T(\boldsymbol{q})\dot{\boldsymbol{q}} = 0. \tag{11.4}$$

If the constraint is holonomic, it can be integrated and written as

$$h(\boldsymbol{q}) = c, \tag{11.5}$$

where $\partial h / \partial \boldsymbol{q} = \gamma(\boldsymbol{q})\, \boldsymbol{a}^T(\boldsymbol{q})$, with $\gamma(\boldsymbol{q}) \neq 0$ an *integrating factor* and c an integration constant. Therefore, there is a loss of *accessibility* in the configuration space, because the motion of the mechanical system in $\mathcal{C}$ is confined to a particular *level surface* of the scalar function h. This surface, which depends on the initial configuration $\boldsymbol{q}_0$ through the value of $h(\boldsymbol{q}_0) = c$, has dimension $n - 1$.

Assume instead that the constraint (11.4) is nonholonomic. In this case, generalized velocities are indeed constrained to belong to a subspace of dimension $n-1$, i.e., the null space of matrix $\boldsymbol{a}^T(\boldsymbol{q})$. Nevertheless, the fact that the constraint is non-integrable means that there is no loss of accessibility in $\mathcal{C}$ for the system. In other words, while the number of DOFs decreases to $n - 1$ due to the constraint, the number of generalized coordinates cannot be reduced, not even locally.

The conclusion just drawn for the case of a single constraint is general. An n-dimensional mechanical system subject to k nonholonomic constraints

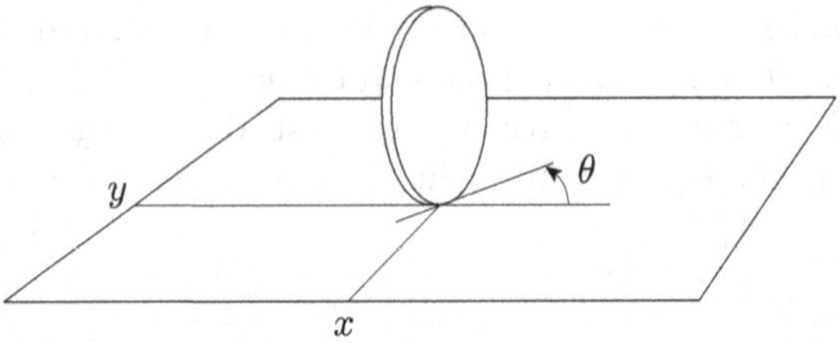

Fig. 11.1. Generalized coordinates for a disk rolling on a plane

can access its whole configuration space $\mathcal{C}$, although at any configuration its generalized velocities must belong to an $(n-k)$-dimensional subspace.

The following is a classical example of nonholonomic mechanical system, that is particularly relevant in the study of mobile robots.

Example 11.1

Consider a disk that rolls without slipping on the horizontal plane, while keeping its *sagittal plane* (i.e., the plane that contains the disk) in the vertical direction (Fig. 11.1). Its configuration is described by three[2] generalized coordinates: the Cartesian coordinates (x, y) of the contact point with the ground, measured in a fixed reference frame, and the angle θ characterizing the orientation of the disk with respect to the x axis. The configuration vector is therefore $\boldsymbol{q} = [\,x \quad y \quad \theta\,]^T$.

The *pure rolling* constraint for the disk is expressed in the Pfaffian form as

$$\dot{x}\sin\theta - \dot{y}\cos\theta = [\,\sin\theta \quad -\cos\theta \quad 0\,]\,\dot{\boldsymbol{q}} = 0, \tag{11.6}$$

and entails that, in the absence of slipping, the velocity of the contact point has zero component in the direction orthogonal to the sagittal plane. The angular velocity of the disk around the vertical axis instead is unconstrained.

Constraint (11.6) is nonholonomic, because it implies no loss of accessibility in the configuration space of the disk. To substantiate this claim, consider that the disk can be driven from any initial configuration $\boldsymbol{q}_i = [\,x_i \quad y_i \quad \theta_i\,]^T$ to any final configuration $\boldsymbol{q}_f = [\,x_f \quad y_f \quad \theta_f\,]^T$ through the following sequence of movements that do not violate constraint (11.6):

1. rotate the disk around its vertical axis so as to reach the orientation θ_v for which the *sagittal axis* (i.e., the intersection of the sagittal plane and the horizontal plane) goes through the final contact point (x_f, y_f);
2. roll the disk on the plane at a constant orientation θ_v until the contact point reaches its final position (x_f, y_f);
3. rotate again the disk around its vertical axis to change the orientation from θ_v to θ_f.

---

[2] One could add to this description an angle ϕ measuring the rotation of the disk around the horizontal axis passing through its centre. Such a coordinate is however irrelevant for the analysis presented in this chapter, and is therefore ignored in the following.

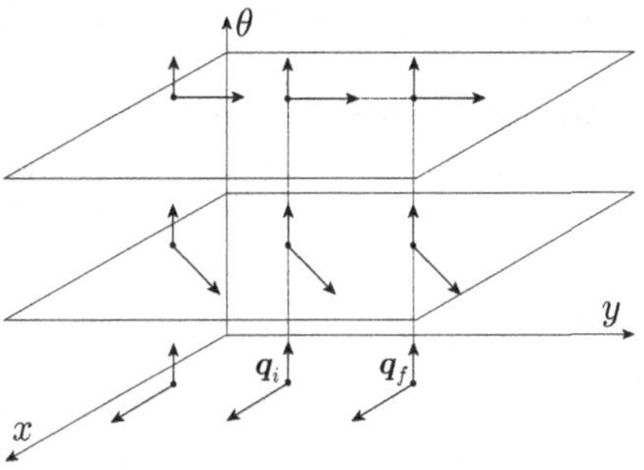

Fig. 11.2. A local representation of the configuration space for the rolling disk with an example manoeuvre that transfers the configuration from $\boldsymbol{q}_i$ to $\boldsymbol{q}_f$ (*dashed line*)

An example of this manoeuvre is shown in Fig. 11.2. Two possible directions of instantaneous motion are shown at each configuration: the first, that is aligned with the sagittal axis, moves the contact point while keeping the orientation constant (rolling); the second varies the orientation while keeping the contact point fixed (rotation around the vertical axis).

It is interesting to note that, in addition to wheeled vehicles, there exist other robotic systems that are nonholonomic in nature. For example, the pure rolling constraint also arises in manipulation problems with round-fingered robot hands. Another kind of nonholonomic behaviour is found in multibody systems that 'float' freely (i.e., without a fixed base), such as manipulators used in space operations. In fact, in the absence of external generalized forces, the conservation of the angular momentum represents a non-integrable Pfaffian constraint for the system.

11.1.1 Integrability Conditions

In the presence of Pfaffian kinematic constraints, integrability conditions can be used to decide whether the system is holonomic or nonholonomic.

Consider first the case of a *single* Pfaffian constraint:

$$\boldsymbol{a}^T(\boldsymbol{q})\dot{\boldsymbol{q}} = \sum_{j=1}^{n} a_j(\boldsymbol{q})\dot{q}_j = 0. \tag{11.7}$$

For this constraint to be integrable, there must exist a scalar function $h(\boldsymbol{q})$ and an integrating factor $\gamma(\boldsymbol{q}) \neq 0$ such that the following condition holds:

$$\gamma(\boldsymbol{q})a_j(\boldsymbol{q}) = \frac{\partial h(\boldsymbol{q})}{\partial q_j} \qquad j = 1, \ldots, n. \tag{11.8}$$

The converse is also true: if there exists an integrating factor $\gamma(\boldsymbol{q}) \neq 0$ such that $\gamma(\boldsymbol{q})\boldsymbol{a}(\boldsymbol{q})$ is the gradient of a scalar function $h(\boldsymbol{q})$, constraint (11.7) is integrable. By using Schwarz theorem on the symmetry of second derivatives, the integrability condition (11.8) may be replaced by the following system of partial differential equations:

$$\frac{\partial(\gamma a_k)}{\partial q_j} = \frac{\partial(\gamma a_j)}{\partial q_k} \qquad j,k = 1,\dots,n, \quad j \neq k, \tag{11.9}$$

that does not contain the unknown function $h(\boldsymbol{q})$. Note that condition (11.9) implies that a Pfaffian constraint with constant coefficients a_j is always holonomic.

Example 11.2

Consider the following kinematic constraint in $\mathcal{C} = \mathbb{R}^3$:

$$\dot{q}_1 + q_1\dot{q}_2 + \dot{q}_3 = 0.$$

The holonomy condition (11.9) gives

$$\begin{aligned} \frac{\partial\gamma}{\partial q_2} &= \gamma + q_1\frac{\partial\gamma}{\partial q_1} \\ \frac{\partial\gamma}{\partial q_3} &= \frac{\partial\gamma}{\partial q_1} \\ q_1\frac{\partial\gamma}{\partial q_3} &= \frac{\partial\gamma}{\partial q_2}. \end{aligned}$$

By substituting the second and third equations into the first, it is easy to conclude that the only solution is $\gamma = 0$. Therefore, the constraint is nonholonomic.

Example 11.3

Consider the pure rolling constraint (11.6). In this case, the holonomy condition (11.9) gives

$$\begin{aligned} \sin\theta\,\frac{\partial\gamma}{\partial y} &= -\cos\theta\,\frac{\partial\gamma}{\partial x} \\ \cos\theta\,\frac{\partial\gamma}{\partial\theta} &= \gamma\sin\theta \\ \sin\theta\,\frac{\partial\gamma}{\partial\theta} &= -\gamma\cos\theta. \end{aligned}$$

Squaring and adding the last two equations gives $\partial\gamma/\partial\theta = \pm\gamma$. Assume for example $\partial\gamma/\partial\theta = \gamma$. Using this in the above equations leads to

$$\begin{aligned} \gamma\cos\theta &= \gamma\sin\theta \\ \gamma\sin\theta &= -\gamma\cos\theta \end{aligned}$$

whose only solution is $\gamma = 0$. The same conclusion is reached by letting $\partial\gamma/\partial\theta = -\gamma$. This confirms that constraint (11.6) is nonholonomic.

The situation becomes more complicated when dealing with a system of $k > 1$ kinematic constraints of the form (11.3). In fact, in this case it may happen that the single constraints are not integrable if taken separately, but the whole system is integrable. In particular, if $p \leq k$ independent linear combinations of the constraints

$$\sum_{i=1}^{k} \gamma_{ji}(\boldsymbol{q})\boldsymbol{a}_i^T(\boldsymbol{q})\dot{\boldsymbol{q}} \qquad j = 1,\ldots,p$$

are integrable, there exist p independent scalar functions $h_j(\boldsymbol{q})$ such that

$$\text{span}\left\{\frac{\partial h_1(\boldsymbol{q})}{\partial \boldsymbol{q}},\ldots,\frac{\partial h_p(\boldsymbol{q})}{\partial \boldsymbol{q}}\right\} \subset \text{span}\left\{\boldsymbol{a}_1^T(\boldsymbol{q}),\ldots,\boldsymbol{a}_k^T(\boldsymbol{q})\right\} \quad \forall \boldsymbol{q} \in \mathcal{C}.$$

Therefore, the configurations that are accessible for the mechanical system belong to the $(n-p)$-dimensional subspace consisting of the particular level surfaces of the functions h_j:

$$\{\boldsymbol{q} \in \mathcal{C}\colon h_1(\boldsymbol{q}) = c_1,\ldots,h_p(\boldsymbol{q}) = c_p\}$$

on which the motion is started (see Problem 11.2). In the case $p = k$, the system of kinematic constraints (11.3) is completely integrable, and hence holonomic.

Example 11.4

Consider the system of Pfaffian constraints

$$\begin{aligned} \dot{q}_1 + q_1\dot{q}_2 + \dot{q}_3 &= 0 \\ \dot{q}_1 + \dot{q}_2 + q_1\dot{q}_3 &= 0. \end{aligned}$$

Taken separately, these constraints are found to be non-integrable (in particular, the first is the nonholonomic constraint of Example 11.2). However, subtracting the second from the first gives

$$(q_1 - 1)(\dot{q}_2 - \dot{q}_3) = 0$$

so that $\dot{q}_2 = \dot{q}_3$, because the constraints must be satisfied for any value of $\boldsymbol{q}$. The assigned system of constraints is then equivalent to

$$\begin{aligned} \dot{q}_2 &= \dot{q}_3 \\ \dot{q}_1 + (1 + q_1)\dot{q}_2 &= 0, \end{aligned}$$

which can be integrated as

$$\begin{aligned} q_2 - q_3 &= c_1 \\ \log(q_1 + 1) + q_2 &= c_2 \end{aligned}$$

with integration constants c_1, c_2.

The integrability conditions of a system of Pfaffian kinematic constraints are quite complex and derive from a fundamental result of differential geometry known as *Frobenius theorem*. However, as shown later in the chapter, it is possible to derive such conditions more directly from a different perspective.

11.2 Kinematic Model

The system of k Pfaffian constraints (11.3) entails that the admissible generalized velocities at each configuration $\boldsymbol{q}$ belong to the $(n-k)$-dimensional null space of matrix $\boldsymbol{A}^T(\boldsymbol{q})$. Denoting by $\{\boldsymbol{g}_1(\boldsymbol{q}),\ldots,\boldsymbol{g}_{n-k}(\boldsymbol{q})\}$ a basis of $\mathcal{N}(\boldsymbol{A}^T(\boldsymbol{q}))$, the admissible trajectories for the mechanical system can then be characterized as the solutions of the nonlinear dynamic system

$$\dot{\boldsymbol{q}} = \sum_{j=1}^{m} \boldsymbol{g}_j(\boldsymbol{q})u_j = \boldsymbol{G}(\boldsymbol{q})\boldsymbol{u} \qquad m = n-k, \tag{11.10}$$

where $\boldsymbol{q} \in {\rm I\!R}^n$ is the state vector and $\boldsymbol{u} = [\,u_1 \quad \ldots \quad u_m\,]^T \in {\rm I\!R}^m$ is the input vector. System (11.10) is said to be *driftless* because one has $\dot{\boldsymbol{q}} = \boldsymbol{0}$ if the input is zero.

The choice of the *input vector fields* $\boldsymbol{g}_1(\boldsymbol{q}),\ldots,\boldsymbol{g}_m(\boldsymbol{q})$ (and thus of matrix $\boldsymbol{G}(\boldsymbol{q})$) in (11.10) is not unique. Correspondingly, the components of $\boldsymbol{u}$ may have different meanings. In general, it is possible to choose the basis of $\mathcal{N}(\boldsymbol{A}^T(\boldsymbol{q}))$ in such a way that the u_js have a physical interpretation, as will be shown later for some examples of mobile robots. In any case, vector $\boldsymbol{u}$ may not be directly related to the actual control inputs, that are in general forces and/or torques. For this reason, Eq. (11.10) is referred to as the *kinematic model* of the constrained mechanical system.

The holonomy or nonholonomy of constraints (11.3) can be established by analyzing the controllability[3] properties of the associated kinematic model (11.10). In fact, two cases are possible:

1. If system (11.10) is controllable, given two arbitrary configurations $\boldsymbol{q}_i$ and $\boldsymbol{q}_f$ in $\mathcal{C}$, there exists a choice of $\boldsymbol{u}(t)$ that steers the system from $\boldsymbol{q}_i$ to $\boldsymbol{q}_f$, i.e., there exists a trajectory that joins the two configurations and satisfies the kinematic constraints (11.3). Therefore, these do not affect in any way the accessibility of $\mathcal{C}$, and they are (completely) nonholonomic.
2. If system (11.10) is not controllable, the kinematic constraints (11.3) reduce the set of accessible configurations in $\mathcal{C}$. Hence, the constraints are partially or completely integrable depending on the dimension $\nu < n$ of the accessible configuration space. In particular:

[3] Refer to Appendix D for a short survey of nonlinear controllability theory, including the necessary tools from differential geometry.

2a. If $m < \nu < n$, the loss of accessibility is not maximal, and thus constraints (11.3) are only partially integrable. The mechanical system is still nonholonomic.
2b. If $\nu = m$, the loss of accessibility is maximal, and constraints (11.3) are completely integrable. Therefore, the mechanical system is holonomic.

Note how this particular viewpoint, i.e., the equivalence between controllability and nonholonomy, was already implicitly adopted in Example 11.1, where the controllability of the kinematic system was proven *constructively*, i.e., by exhibiting a reconfiguration manoeuvre. A more systematic approach is to take advantage of the controllability conditions for nonlinear driftless systems. In particular, controllability may be verified using the *accessibility rank condition*

$$\dim \Delta_{\mathcal{A}}(\boldsymbol{q}) = n, \tag{11.11}$$

where $\Delta_{\mathcal{A}}$ is the *accessibility distribution* associated with system (11.10), i.e., the involutive closure of distribution $\Delta = \text{span}\{\boldsymbol{g}_1, \ldots, \boldsymbol{g}_m\}$. The following cases may occur:

1. If (11.11) holds, system (11.10) is controllable and the kinematic constraints (11.3) are (completely) nonholonomic.
2. If (11.11) does not hold, system (11.10) is not controllable and the kinematic constraints (11.3) are at least partially integrable. In particular, let

$$\dim \Delta_{\mathcal{A}}(\boldsymbol{q}) = \nu < n.$$

Then

2a. If $m < \nu < n$, constraints (11.3) are only partially integrable.
2b. If $\nu = m$, constraints (11.3) are completely integrable, and hence holonomic. This happens when $\Delta_{\mathcal{A}}$ coincides with $\Delta = \text{span}\{\boldsymbol{g}_1, \ldots, \boldsymbol{g}_m\}$, i.e., when the latter distribution is involutive.

It is easy to verify that, in the case of a single kinematic constraint (11.7), the integrability condition given by (11.9) is equivalent to the involutivity of $\Delta = \text{span}\{\boldsymbol{g}_1, \ldots, \boldsymbol{g}_{n-1}\}$. Another remarkable situation is met when the number of Pfaffian constraints is $k = n - 1$; in this case, the associated kinematic model (11.10) consists of a single vector field $\boldsymbol{g}$ ($m = 1$). Hence, $n - 1$ Pfaffian constraints are always integrable, because the distribution associated with a single vector field is always involutive. For example, a mechanical system with two generalized coordinates that is subject to a scalar Pfaffian constraint is always holonomic.

In the following, the kinematic models of two wheeled vehicles of particular interest will be analyzed in detail. A large part of the existing mobile robots have a kinematic model that is equivalent to one of these two.

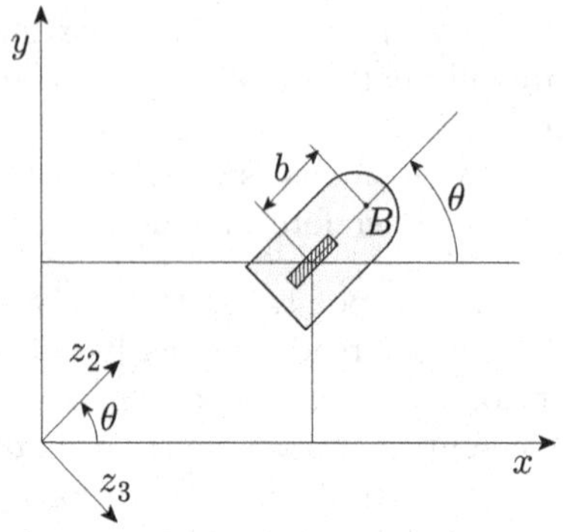

Fig. 11.3. Generalized coordinates for a unicycle

11.2.1 Unicycle

A *unicycle* is a vehicle with a single orientable wheel. Its configuration is completely described by $\boldsymbol{q} = [\, x \quad y \quad \theta \,]^T$, where (x, y) are the Cartesian coordinates of the contact point of the wheel with the ground (or equivalently, of the wheel centre) and θ is the orientation of the wheel with respect to the x axis (see Fig. 11.3).

As already seen in Example 11.1, the pure rolling constraint for the wheel is expressed as

$$\dot{x}\sin\theta - \dot{y}\cos\theta = [\, \sin\theta \quad -\cos\theta \quad 0 \,]\,\dot{\boldsymbol{q}} = 0, \tag{11.12}$$

entailing that the velocity of the contact point is zero in the direction orthogonal to the sagittal axis of the vehicle. The line passing through the contact point and having such direction is therefore called *zero motion line*. Consider the matrix

$$\boldsymbol{G}(\boldsymbol{q}) = [\, \boldsymbol{g}_1(\boldsymbol{q}) \quad \boldsymbol{g}_2(\boldsymbol{q}) \,] = \begin{bmatrix} \cos\theta & 0 \\ \sin\theta & 0 \\ 0 & 1 \end{bmatrix},$$

whose columns $\boldsymbol{g}_1(\boldsymbol{q})$ and $\boldsymbol{g}_2(\boldsymbol{q})$ are, for each $\boldsymbol{q}$, a basis of the null space of the matrix associated with the Pfaffian constraint. All the admissible generalized velocities at $\boldsymbol{q}$ are therefore obtained as a linear combination of $\boldsymbol{g}_1(\boldsymbol{q})$ and $\boldsymbol{g}_2(\boldsymbol{q})$. The kinematic model of the unicycle is then

$$\begin{bmatrix} \dot{x} \\ \dot{y} \\ \dot{\theta} \end{bmatrix} = \begin{bmatrix} \cos\theta \\ \sin\theta \\ 0 \end{bmatrix} v + \begin{bmatrix} 0 \\ 0 \\ 1 \end{bmatrix} \omega, \tag{11.13}$$

where the inputs v and ω have a clear physical interpretation. In particular, v is the *driving velocity*, i.e., the modulus[4] (with sign) of the contact point

[4] Note that v is given by the angular speed of the wheel around its horizontal axis multiplied by the wheel radius.

velocity vector, whereas the *steering velocity* ω is the wheel angular speed around the vertical axis.

The Lie bracket of the two input vector fields is

$$[\boldsymbol{g}_1, \boldsymbol{g}_2](\boldsymbol{q}) = \begin{bmatrix} \sin\theta \\ -\cos\theta \\ 0 \end{bmatrix},$$

that is always linearly independent from $\boldsymbol{g}_1(\boldsymbol{q}), \boldsymbol{g}_2(\boldsymbol{q})$. Therefore, the iterative procedure (see Sect. D.2) for building the accessibility distribution $\Delta_{\mathcal{A}}$ ends with

$$\dim \Delta_{\mathcal{A}} = \dim \Delta_2 = \dim \operatorname{span}\{\boldsymbol{g}_1, \boldsymbol{g}_2, [\boldsymbol{g}_1, \boldsymbol{g}_2]\} = 3.$$

This indicates that the unicycle is controllable with degree of nonholonomy $\kappa = 2$, and that constraint (11.12) is nonholonomic — the same conclusion reached in Example 11.3 by applying the integrability condition.

A unicycle in the strict sense (i.e., a vehicle equipped with a single wheel) is a robot with a serious problem of balance in static conditions. However, there exist vehicles that are kinematically equivalent to a unicycle but more stable from a mechanical viewpoint. Among these, the most important are the *differential drive* and the *synchro drive* vehicles, already introduced in Sect. 1.2.2.

For the differential drive mobile robot of Fig. 1.13, denote by (x, y) the Cartesian coordinates of the midpoint of the segment joining the two wheel centres, and by θ the common orientation of the fixed wheels (hence, of the vehicle body). Then, the kinematic model (11.13) of the unicycle also applies to the differential drive vehicle, provided that the driving and steering velocities v and ω are expressed as a function of the actual velocity inputs, i.e., the angular speeds ω_R and ω_L of the right and left wheel, respectively. Simple arguments (see Problem 11.6) can be used to show that there is a one-to-one correspondence between the two sets of inputs:

$$v = \frac{r\,(\omega_R + \omega_L)}{2} \qquad \omega = \frac{r\,(\omega_R - \omega_L)}{d}, \tag{11.14}$$

where r is the radius of the wheels and d is the distance between their centres.

The equivalence with the kinematic model (11.13) is even more straightforward for the *synchro drive* mobile robot of Fig. 1.14, whose control inputs are indeed the driving velocity v and the steering velocity ω, that are common to the three orientable wheels. The Cartesian coordinates (x, y) may represent in this case any point of the robot (for example, its centroid), while θ is the common orientation of the wheels. Note that, unlike a differential drive vehicle, the orientation of the body of a synchro drive vehicle never changes, unless a third actuator is added for this specific purpose.

11.2.2 Bicycle

Consider now a *bicycle*, i.e., a vehicle having an orientable wheel and a fixed wheel arranged as in Fig. 11.4. A possible choice for the generalized coordi-

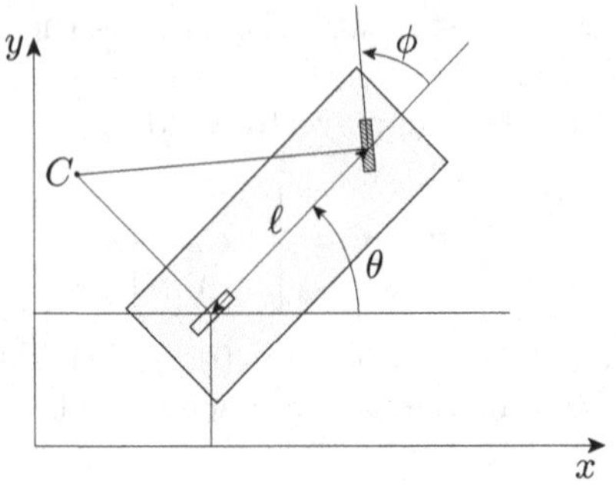

Fig. 11.4. Generalized coordinates and instantaneous centre of rotation for a bicycle

nates is $\boldsymbol{q} = [\,x \quad y \quad \theta \quad \phi\,]^T$, where (x, y) are the Cartesian coordinates of the contact point between the rear wheel and the ground (i.e., of the rear wheel centre), θ is the orientation of the vehicle with respect to the x axis, and ϕ is the steering angle of the front wheel with respect to the vehicle.

The motion of the vehicle is subject to two pure rolling constraints, one for each wheel:

$$\dot{x}_f \sin(\theta + \phi) - \dot{y}_f \cos(\theta + \phi) = 0 \tag{11.15}$$

$$\dot{x} \sin\theta - \dot{y} \cos\theta = 0, \tag{11.16}$$

where (x_f, y_f) is the Cartesian position of the centre of the front wheel. The geometric meaning of these constraints is obvious: the velocity of the centre of the front wheel is zero in the direction orthogonal to the wheel itself, while the velocity of the centre of the rear wheel is zero in the direction orthogonal to the sagittal axis of the vehicle. The zero motion lines of the two wheels meet at a point C called *instantaneous centre of rotation* (Fig. 11.4), whose position depends only on (and changes with) the configuration $\boldsymbol{q}$ of the bicycle. Each point of the vehicle body then moves instantaneously along an arc of circle with centre in C (see also Problem 11.7).

Using the rigid body constraint

$$x_f = x + \ell \cos\theta$$

$$y_f = y + \ell \sin\theta,$$

where ℓ is the distance between the wheels, constraint (11.15) can be rewritten as

$$\dot{x} \sin(\theta + \phi) - \dot{y} \cos(\theta + \phi) - \ell\,\dot{\theta} \cos\phi = 0. \tag{11.17}$$

The matrix associated with the Pfaffian constraints (11.16), (11.17) is then

$$\boldsymbol{A}^T(\boldsymbol{q}) = \begin{bmatrix} \sin\theta & -\cos\theta & 0 & 0 \\ \sin(\theta+\phi) & -\cos(\theta+\phi) & -\ell\cos\phi & 0 \end{bmatrix},$$

with constant rank $k = 2$. The dimension of its null space is $n - k = 2$, and all the admissible velocities at $\boldsymbol{q}$ may be written as a linear combination of a basis of $\mathcal{N}(\boldsymbol{A}^T(\boldsymbol{q}))$, for example

$$\begin{bmatrix} \dot{x} \\ \dot{y} \\ \dot{\theta} \\ \dot{\phi} \end{bmatrix} = \begin{bmatrix} \cos\theta\cos\phi \\ \sin\theta\cos\phi \\ \sin\phi/\ell \\ 0 \end{bmatrix} u_1 + \begin{bmatrix} 0 \\ 0 \\ 0 \\ 1 \end{bmatrix} u_2.$$

Since the front wheel is orientable, it is immediate to set $u_2 = \omega$, where ω is the *steering* velocity. The expression of u_1 depends instead on how the vehicle is driven.

If the bicycle has *front-wheel drive*, one has directly $u_1 = v$, where v is the *driving* velocity of the front wheel. The corresponding kinematic model is

$$\begin{bmatrix} \dot{x} \\ \dot{y} \\ \dot{\theta} \\ \dot{\phi} \end{bmatrix} = \begin{bmatrix} \cos\theta\cos\phi \\ \sin\theta\cos\phi \\ \sin\phi/\ell \\ 0 \end{bmatrix} v + \begin{bmatrix} 0 \\ 0 \\ 0 \\ 1 \end{bmatrix} \omega. \tag{11.18}$$

Denoting by $\boldsymbol{g}_1(\boldsymbol{q})$ and $\boldsymbol{g}_2(\boldsymbol{q})$ the two input vector fields, simple computations give

$$\boldsymbol{g}_3(\boldsymbol{q}) = [\boldsymbol{g}_1, \boldsymbol{g}_2](\boldsymbol{q}) = \begin{bmatrix} \cos\theta\sin\phi \\ \sin\theta\sin\phi \\ -\cos\phi/\ell \\ 0 \end{bmatrix} \qquad \boldsymbol{g}_4(\boldsymbol{q}) = [\boldsymbol{g}_1, \boldsymbol{g}_3](\boldsymbol{q}) = \begin{bmatrix} -\sin\theta/\ell \\ \cos\theta/\ell \\ 0 \\ 0 \end{bmatrix},$$

both linearly independent from $\boldsymbol{g}_1(\boldsymbol{q})$ and $\boldsymbol{g}_2(\boldsymbol{q})$. Hence, the iterative procedure for building the accessibility distribution $\Delta_{\mathcal{A}}$ ends with

$$\dim \Delta_{\mathcal{A}} = \dim \Delta_3 = \dim \mathrm{span}\{\boldsymbol{g}_1, \boldsymbol{g}_2, \boldsymbol{g}_3, \boldsymbol{g}_4\} = 4.$$

This means that the front-wheel drive bicycle is controllable with degree of nonholonomy $\kappa = 3$, and constraints (11.15), (11.16) are (completely) nonholonomic.

The kinematic model of a bicycle with *rear-wheel drive* can be derived by noting that in this case the first two equations must coincide with those of the unicycle model (11.13). It is then sufficient to set $u_1 = v/\cos\phi$ to obtain

$$\begin{bmatrix} \dot{x} \\ \dot{y} \\ \dot{\theta} \\ \dot{\phi} \end{bmatrix} = \begin{bmatrix} \cos\theta \\ \sin\theta \\ \tan\phi/\ell \\ 0 \end{bmatrix} v + \begin{bmatrix} 0 \\ 0 \\ 0 \\ 1 \end{bmatrix} \omega, \tag{11.19}$$

where v is the *driving* velocity of the rear wheel.[5] In this case, one has

$$\boldsymbol{g}_3(\boldsymbol{q}) = [\boldsymbol{g}_1, \boldsymbol{g}_2](\boldsymbol{q}) = \begin{bmatrix} 0 \\ 0 \\ -\dfrac{1}{\ell \cos^2 \phi} \\ 0 \end{bmatrix} \qquad \boldsymbol{g}_4(\boldsymbol{q}) = [\boldsymbol{g}_1, \boldsymbol{g}_3](\boldsymbol{q}) = \begin{bmatrix} -\dfrac{\sin\theta}{\ell \cos^2 \phi} \\ \dfrac{\cos\theta}{\ell \cos^2 \phi} \\ 0 \\ 0 \end{bmatrix},$$

again linearly independent from $\boldsymbol{g}_1(\boldsymbol{q})$ and $\boldsymbol{g}_2(\boldsymbol{q})$. Hence, the rear-wheel drive bicycle is also controllable with degree of nonholonomy $\kappa = 3$.

Like the unicycle, the bicycle is also unstable in static conditions. Kinematically equivalent vehicles that are mechanically balanced are the *tricycle* and the *car-like* robot, introduced in Sect. 1.2.2 and shown respectively in Fig. 1.15 and 1.16. In both cases, the kinematic model is given by (11.18) or by (11.19) depending on the wheel drive being on the front or the rear wheels. In particular, (x, y) are the Cartesian coordinates of the midpoint of the rear wheel axle, θ is the orientation of the vehicle, and ϕ is the steering angle.

11.3 Chained Form

The possibility of transforming the kinematic model (11.10) of a mobile robot in a canonical form is of great interest for solving planning and control problems with efficient, systematic procedures. Here, the analysis is limited to systems with two inputs, like the unicycle and bicycle models.

A $(2, n)$ *chained form* is a two-input driftless system

$$\dot{\boldsymbol{z}} = \boldsymbol{\gamma}_1(\boldsymbol{z})v_1 + \boldsymbol{\gamma}_2(\boldsymbol{z})v_2,$$

whose equations are expressed as

$$\begin{aligned} \dot{z}_1 &= v_1 \\ \dot{z}_2 &= v_2 \\ \dot{z}_3 &= z_2 v_1 \\ &\;\vdots \\ \dot{z}_n &= z_{n-1} v_1. \end{aligned} \tag{11.20}$$

Using the following notation for a 'repeated' Lie bracket:

$$\mathrm{ad}_{\boldsymbol{\gamma}_1}\boldsymbol{\gamma}_2 = [\boldsymbol{\gamma}_1, \boldsymbol{\gamma}_2] \qquad \mathrm{ad}^k_{\boldsymbol{\gamma}_1}\boldsymbol{\gamma}_2 = [\boldsymbol{\gamma}_1, \mathrm{ad}^{k-1}_{\boldsymbol{\gamma}_1}\boldsymbol{\gamma}_2],$$

[5] Note that the kinematic model (11.19) is no longer valid for $\phi = \pm\pi/2$, where the first vector field is not defined. This corresponds to the mechanical jam in which the front wheel is orthogonal to the sagittal axis of the vehicle. This singularity does not arise in the front-wheel drive bicycle (11.18), that in principle can still pivot around the rear wheel contact point in such a situation.

one has for system (11.20)

$$\gamma_1 = \begin{bmatrix} 1 \\ 0 \\ z_2 \\ z_3 \\ \vdots \\ z_{n-1} \end{bmatrix} \quad \gamma_2 = \begin{bmatrix} 0 \\ 1 \\ 0 \\ 0 \\ \vdots \\ 0 \end{bmatrix} \quad \Rightarrow \quad \mathrm{ad}^k_{\gamma_1}\gamma_2 = \begin{bmatrix} 0 \\ \vdots \\ (-1)^k \\ \vdots \\ 0 \end{bmatrix},$$

where $(-1)^k$ is the $(k+2)$-th component. This implies that the system is controllable, because the accessibility distribution

$$\Delta_{\mathcal{A}} = \mathrm{span}\,\{\gamma_1, \gamma_2, \mathrm{ad}_{\gamma_1}\gamma_2, \ldots, \mathrm{ad}^{n-2}_{\gamma_1}\gamma_2\}$$

has dimension n. In particular, the degree of nonholonomy is $\kappa = n - 1$.

There exist necessary and sufficient conditions for transforming a generic two-input driftless system

$$\dot{\boldsymbol{q}} = \boldsymbol{g}_1(\boldsymbol{q})u_1 + \boldsymbol{g}_2(\boldsymbol{q})u_2 \tag{11.21}$$

in the chained form (11.20) via coordinate and input transformations

$$\boldsymbol{z} = \boldsymbol{T}(\boldsymbol{q}) \qquad \boldsymbol{v} = \boldsymbol{\beta}(\boldsymbol{q})\boldsymbol{u}. \tag{11.22}$$

In particular, it can be shown that systems like (11.21) with dimension n not larger than 4 can *always* be put in chained form. This applies, for example, to the kinematic models of the unicycle and the bicycle.

There also exist sufficient conditions for transformability in chained form that are relevant because they are constructive. Define the distributions

$$\begin{aligned} \Delta_0 &= \mathrm{span}\,\{\boldsymbol{g}_1, \boldsymbol{g}_2, \mathrm{ad}_{\boldsymbol{g}_1}\boldsymbol{g}_2, \ldots, \mathrm{ad}^{n-2}_{\boldsymbol{g}_1}\boldsymbol{g}_2\} \\ \Delta_1 &= \mathrm{span}\,\{\boldsymbol{g}_2, \mathrm{ad}_{\boldsymbol{g}_1}\boldsymbol{g}_2, \ldots, \mathrm{ad}^{n-2}_{\boldsymbol{g}_1}\boldsymbol{g}_2\} \\ \Delta_2 &= \mathrm{span}\,\{\boldsymbol{g}_2, \mathrm{ad}_{\boldsymbol{g}_1}\boldsymbol{g}_2, \ldots, \mathrm{ad}^{n-3}_{\boldsymbol{g}_1}\boldsymbol{g}_2\}. \end{aligned}$$

Assume that, in a certain set, it is $\dim \Delta_0 = n$, Δ_1 and Δ_2 are involutive, and there exists a scalar function $h_1(\boldsymbol{q})$ whose differential $d\boldsymbol{h}_1$ satisfies

$$d\boldsymbol{h}_1 \cdot \Delta_1 = 0 \qquad d\boldsymbol{h}_1 \cdot \boldsymbol{g}_1 = 1,$$

where the symbol $\cdot$ denotes the inner product between a row vector and a column vector — in particular, $\cdot\Delta_1$ is the inner product with any vector generated by distribution Δ_1. In this case, system (11.21) can be put in the form (11.20) through the coordinate transformation[6]

$$z_1 = h_1$$

[6] This transformation makes use of the Lie derivative (see Appendix D).

$$
\begin{aligned}
z_2 &= L_{\boldsymbol{g}_1}^{n-2} h_2 \\
&\vdots \\
z_{n-1} &= L_{\boldsymbol{g}_1} h_2 \\
z_n &= h_2,
\end{aligned}
$$

where h_2 must be chosen independent of h_1 and such that $dh_2 \cdot \Delta_2 = 0$. The input transformation is given by

$$
\begin{aligned}
v_1 &= u_1 \\
v_2 &= \left(L_{\boldsymbol{g}_1}^{n-1} h_2\right) u_1 + \left(L_{\boldsymbol{g}_2} L_{\boldsymbol{g}_1}^{n-2} h_2\right) u_2.
\end{aligned}
$$

In general, the coordinate and input transformations are not unique.

Consider the kinematic model (11.13) of the unicycle. With the change of coordinates

$$
\begin{aligned}
z_1 &= \theta \\
z_2 &= x \cos\theta + y \sin\theta \\
z_3 &= x \sin\theta - y \cos\theta
\end{aligned}
\tag{11.23}
$$

and the input transformation

$$
\begin{aligned}
v &= v_2 + z_3 v_1 \\
\omega &= v_1,
\end{aligned}
\tag{11.24}
$$

one obtains the (2,3) chained form

$$
\begin{aligned}
\dot{z}_1 &= v_1 \\
\dot{z}_2 &= v_2 \\
\dot{z}_3 &= z_2 v_1.
\end{aligned}
\tag{11.25}
$$

Note that, while z_1 is simply the orientation θ, coordinates z_2 and z_3 represent the position of the unicycle in a moving reference frame whose z_2 axis is aligned with the sagittal axis of the vehicle (see Fig. 11.3).

As for mobile robots with bicycle-like kinematics, consider for example the model (11.19) corresponding to the rear-wheel drive case. Using the change of coordinates

$$
\begin{aligned}
z_1 &= x \\
z_2 &= \frac{1}{\ell} \sec^3\theta \tan\phi \\
z_3 &= \tan\theta \\
z_4 &= y
\end{aligned}
$$

and the input transformation

$$v = \frac{v_1}{\cos\theta}$$
$$\omega = -\frac{3}{\ell}v_1 \sec\theta \sin^2\phi + \frac{1}{\ell}v_2 \cos^3\theta \cos^2\phi,$$

the (2,4) chained form is obtained:

$$\begin{aligned}
\dot{z}_1 &= v_1 \\
\dot{z}_2 &= v_2 \\
\dot{z}_3 &= z_2 v_1 \\
\dot{z}_4 &= z_3 v_1.
\end{aligned}$$

This transformation is defined everywhere in the configuration space, with the exception of points where $\cos\theta = 0$. The equivalence between the two models is then subject to the condition $\theta \neq \pm k\pi/2$, with $k = 1, 2, \ldots$.

11.4 Dynamic Model

The derivation of the dynamic model of a mobile robot is similar to the manipulator case, the main difference being the presence of nonholonomic constraints on the generalized coordinates. An important consequence of nonholonomy is that exact linearization of the dynamic model via feedback is no longer possible. In the following, the Lagrange formulation is used to obtain the dynamic model of an n-dimensional mechanical system subject to $k < n$ kinematic constraints in the form (11.3), and it is shown how this model can be partially linearized via feedback.

As usual, define the Lagrangian $\mathcal{L}$ of the mechanical system as the difference between its kinetic and potential energy:

$$\mathcal{L}(\boldsymbol{q}, \dot{\boldsymbol{q}}) = \mathcal{T}(\boldsymbol{q}, \dot{\boldsymbol{q}}) - \mathcal{U}(\boldsymbol{q}) = \frac{1}{2}\dot{\boldsymbol{q}}^T \boldsymbol{B}(\boldsymbol{q})\dot{\boldsymbol{q}} - \mathcal{U}(\boldsymbol{q}), \tag{11.26}$$

where $\boldsymbol{B}(\boldsymbol{q})$ is the (symmetric and positive definite) inertia matrix of the mechanical system. The Lagrange equations are in this case

$$\frac{d}{dt}\left(\frac{\partial \mathcal{L}}{\partial \dot{\boldsymbol{q}}}\right)^T - \left(\frac{\partial \mathcal{L}}{\partial \boldsymbol{q}}\right)^T = \boldsymbol{S}(\boldsymbol{q})\boldsymbol{\tau} + \boldsymbol{A}(\boldsymbol{q})\boldsymbol{\lambda}, \tag{11.27}$$

where $\boldsymbol{S}(\boldsymbol{q})$ is an $(n \times m)$ matrix mapping the $m = n - k$ external inputs $\boldsymbol{\tau}$ to generalized forces performing work on $\boldsymbol{q}$, $\boldsymbol{A}(\boldsymbol{q})$ is the transpose of the $(k \times n)$ matrix characterizing the kinematic constraints (11.3), and $\boldsymbol{\lambda} \in \mathbb{R}^m$ is the vector of *Lagrange multipliers*. The term $\boldsymbol{A}(\boldsymbol{q})\boldsymbol{\lambda}$ represents the vector of reaction forces at the generalized coordinate level. It has been assumed that the number of available inputs matches the number of DOFs (*full actuation*), that is, in turn, equal to the number n of generalized coordinates minus the number k of constraints.

Using (11.26), (11.27), the *dynamic model* of the constrained mechanical system is expressed as

$$\boldsymbol{B}(\boldsymbol{q})\ddot{\boldsymbol{q}} + \boldsymbol{n}(\boldsymbol{q}, \dot{\boldsymbol{q}}) = \boldsymbol{S}(\boldsymbol{q})\boldsymbol{\tau} + \boldsymbol{A}(\boldsymbol{q})\boldsymbol{\lambda} \tag{11.28}$$

$$\boldsymbol{A}^T(\boldsymbol{q})\dot{\boldsymbol{q}} = \boldsymbol{0}, \tag{11.29}$$

where

$$\boldsymbol{n}(\boldsymbol{q}, \dot{\boldsymbol{q}}) = \dot{\boldsymbol{B}}(\boldsymbol{q})\dot{\boldsymbol{q}} - \frac{1}{2}\left(\frac{\partial}{\partial \boldsymbol{q}}\left(\dot{\boldsymbol{q}}^T\boldsymbol{B}(\boldsymbol{q})\dot{\boldsymbol{q}}\right)\right)^T + \left(\frac{\partial \mathcal{U}(\boldsymbol{q})}{\partial \boldsymbol{q}}\right)^T.$$

Consider now a matrix $\boldsymbol{G}(\boldsymbol{q})$ whose columns are a basis for the null space of $\boldsymbol{A}^T(\boldsymbol{q})$, so that $\boldsymbol{A}^T(\boldsymbol{q})\boldsymbol{G}(\boldsymbol{q}) = \boldsymbol{0}$. One can replace the constraint given by (11.29) with the kinematic model

$$\dot{\boldsymbol{q}} = \boldsymbol{G}(\boldsymbol{q})\boldsymbol{v} = \sum_{i=1}^{m} \boldsymbol{g}_i(\boldsymbol{q})\, v_i, \tag{11.30}$$

where $\boldsymbol{v} \in \mathbb{R}^m$ is the vector of *pseudo-velocities*;[7] for example, in the case of a unicycle the components of this vector are the driving velocity v and the steering velocity ω. Moreover, the Lagrange multipliers in (11.28) can be eliminated premultiplying both sides of the equation by $\boldsymbol{G}^T(\boldsymbol{q})$. This leads to the *reduced dynamic model*

$$\boldsymbol{G}^T(\boldsymbol{q})\left(\boldsymbol{B}(\boldsymbol{q})\ddot{\boldsymbol{q}} + \boldsymbol{n}(\boldsymbol{q}, \dot{\boldsymbol{q}})\right) = \boldsymbol{G}^T(\boldsymbol{q})\boldsymbol{S}(\boldsymbol{q})\boldsymbol{\tau}, \tag{11.31}$$

a system of m differential equations.

Differentiation of (11.30) with respect to time gives

$$\ddot{\boldsymbol{q}} = \dot{\boldsymbol{G}}(\boldsymbol{q})\boldsymbol{v} + \boldsymbol{G}(\boldsymbol{q})\dot{\boldsymbol{v}}.$$

Premultiplying this by $\boldsymbol{G}^T(\boldsymbol{q})\boldsymbol{B}(\boldsymbol{q})$ and using the reduced dynamic model (11.31), one obtains

$$\boldsymbol{M}(\boldsymbol{q})\dot{\boldsymbol{v}} + \boldsymbol{m}(\boldsymbol{q}, \boldsymbol{v}) = \boldsymbol{G}^T(\boldsymbol{q})\boldsymbol{S}(\boldsymbol{q})\boldsymbol{\tau}, \tag{11.32}$$

where

$$\boldsymbol{M}(\boldsymbol{q}) = \boldsymbol{G}^T(\boldsymbol{q})\boldsymbol{B}(\boldsymbol{q})\boldsymbol{G}(\boldsymbol{q})$$

$$\boldsymbol{m}(\boldsymbol{q}, \boldsymbol{v}) = \boldsymbol{G}^T(\boldsymbol{q})\boldsymbol{B}(\boldsymbol{q})\dot{\boldsymbol{G}}(\boldsymbol{q})\boldsymbol{v} + \boldsymbol{G}^T(\boldsymbol{q})\boldsymbol{n}(\boldsymbol{q}, \boldsymbol{G}(\boldsymbol{q})\boldsymbol{v}),$$

---

[7] In the dynamic modeling context, the use of this term emphasizes the difference between $\boldsymbol{v}$ and $\dot{\boldsymbol{q}}$, that are the actual (generalized) velocities of the mechanical system.

with $\boldsymbol{M}(\boldsymbol{q})$ positive definite and

$$\dot{\boldsymbol{G}}(\boldsymbol{q})\boldsymbol{v} = \sum_{i=1}^{m} \left(v_i \frac{\partial \boldsymbol{g}_i}{\partial \boldsymbol{q}}(\boldsymbol{q}) \right) \boldsymbol{G}(\boldsymbol{q})\boldsymbol{v}.$$

This finally leads to the *state-space reduced model*

$$\dot{\boldsymbol{q}} = \boldsymbol{G}(\boldsymbol{q})\boldsymbol{v} \tag{11.33}$$

$$\dot{\boldsymbol{v}} = \boldsymbol{M}^{-1}(\boldsymbol{q})\boldsymbol{m}(\boldsymbol{q},\boldsymbol{v}) + \boldsymbol{M}^{-1}(\boldsymbol{q})\boldsymbol{G}^T(\boldsymbol{q})\boldsymbol{S}(\boldsymbol{q})\boldsymbol{\tau}, \tag{11.34}$$

that represents in a compact form the kinematic and dynamic models of the constrained system as a set of $n+m$ differential equations.

Suppose now that

$$\det\left(\boldsymbol{G}^T(\boldsymbol{q})\boldsymbol{S}(\boldsymbol{q})\right) \neq 0,$$

an assumption on the 'control availability' that is satisfied in many cases of interest. It is then possible to perform a *partial linearization via feedback* of (11.33), (11.34) by letting

$$\boldsymbol{\tau} = \left(\boldsymbol{G}^T(\boldsymbol{q})\boldsymbol{S}(\boldsymbol{q})\right)^{-1}\left(\boldsymbol{M}(\boldsymbol{q})\boldsymbol{a} + \boldsymbol{m}(\boldsymbol{q},\boldsymbol{v})\right), \tag{11.35}$$

where $\boldsymbol{a} \in \mathbb{R}^m$ is the *pseudo-acceleration* vector. The resulting system is

$$\dot{\boldsymbol{q}} = \boldsymbol{G}(\boldsymbol{q})\boldsymbol{v} \tag{11.36}$$

$$\dot{\boldsymbol{v}} = \boldsymbol{a}. \tag{11.37}$$

Note the structure of this system: the first n equations are the kinematic model, of which the last m — that represent the inclusion of m integrators on the input channels — are a *dynamic extension*. If the system is unconstrained and fully actuated, it is $\boldsymbol{G}(\boldsymbol{q}) = \boldsymbol{S}(\boldsymbol{q}) = \boldsymbol{I}_n$; then, the feedback law (11.35) simply reduces to an inverse dynamics control analogous to (8.57), and correspondingly the closed-loop system is equivalent to n decoupled double integrators.

The implementation of the feedback control (11.35) in principle requires the measurement of $\boldsymbol{v}$, and this may not be available. However, pseudo-velocities can be computed via the kinematic model as

$$\boldsymbol{v} = \boldsymbol{G}^\dagger(\boldsymbol{q})\dot{\boldsymbol{q}} = \left(\boldsymbol{G}^T(\boldsymbol{q})\boldsymbol{G}(\boldsymbol{q})\right)^{-1}\boldsymbol{G}^T(\boldsymbol{q})\dot{\boldsymbol{q}}, \tag{11.38}$$

provided that $\boldsymbol{q}$ and $\dot{\boldsymbol{q}}$ are measured. Note that the left pseudo-inverse of $\boldsymbol{G}(\boldsymbol{q})$ has been used here.

By defining the state $\boldsymbol{x} = (\boldsymbol{q},\boldsymbol{v}) \in \mathbb{R}^{n+m}$ and the input $\boldsymbol{u} = \boldsymbol{a} \in \mathbb{R}^m$, system (11.36), (11.37) can be expressed as

$$\dot{\boldsymbol{x}} = \boldsymbol{f}(\boldsymbol{x}) + \boldsymbol{G}(\boldsymbol{x})\boldsymbol{u} = \begin{bmatrix} \boldsymbol{G}(\boldsymbol{q})\boldsymbol{v} \\ \boldsymbol{0} \end{bmatrix} + \begin{bmatrix} \boldsymbol{0} \\ \boldsymbol{I}_m \end{bmatrix}\boldsymbol{u}, \tag{11.39}$$

i.e., a nonlinear system with drift also known as the *second-order kinematic model* of the constrained mechanical system. Accordingly, Eq. (11.36) is sometimes called *first-order kinematic model*. In view of the results recalled in Appendix D, the controllability of the latter guarantees the controllability of system (11.39).

Summarizing, in nonholonomic mechanical systems — such as wheeled mobile robots — it is possible to 'cancel' the dynamic effects via nonlinear state feedback, provided that the dynamic parameters are exactly known and the complete state of the system (generalized coordinates and velocities $\boldsymbol{q}$ and $\dot{\boldsymbol{q}}$) is measured.

Under these assumptions, the control problem can be directly addressed at the (pseudo-)velocity level, i.e., by choosing $\boldsymbol{v}$ in such a way that the kinematic model

$$\dot{\boldsymbol{q}} = \boldsymbol{G}(\boldsymbol{q})\boldsymbol{v}$$

behaves as desired. From $\boldsymbol{v}$, it is possible to derive the actual control inputs at the generalized force level through (11.35). Since $\boldsymbol{a} = \dot{\boldsymbol{v}}$ appears in this equation, the pseudo-velocities $\boldsymbol{v}$ must be differentiable with respect to time.

Example 11.5

For illustration, the above procedure for deriving, reducing and partially linearizing the dynamic model is now applied to the unicycle. Let m be the mass of the unicycle, I its moment of inertia around the vertical axis through its centre, τ_1 the driving force and τ_2 the steering torque. With the kinematic constraint expressed as (11.12), the dynamic model (11.28), (11.29) takes on the form

$$\begin{bmatrix} m & 0 & 0 \\ 0 & m & 0 \\ 0 & 0 & I \end{bmatrix} \begin{bmatrix} \ddot{x} \\ \ddot{y} \\ \ddot{\theta} \end{bmatrix} = \begin{bmatrix} \cos\theta & 0 \\ \sin\theta & 0 \\ 0 & 1 \end{bmatrix} \begin{bmatrix} \tau_1 \\ \tau_2 \end{bmatrix} + \begin{bmatrix} \sin\theta \\ -\cos\theta \\ 0 \end{bmatrix} \lambda$$

$$\dot{x}\sin\theta - \dot{y}\cos\theta = 0.$$

In this case one has

$$\begin{aligned} \boldsymbol{n}(\boldsymbol{q}, \dot{\boldsymbol{q}}) &= \boldsymbol{0} \\ \boldsymbol{G}(\boldsymbol{q}) &= \boldsymbol{S}(\boldsymbol{q}) \\ \boldsymbol{G}^T(\boldsymbol{q})\boldsymbol{S}(\boldsymbol{q}) &= \boldsymbol{I} \\ \boldsymbol{G}^T(\boldsymbol{q})\boldsymbol{B}\dot{\boldsymbol{G}}(\boldsymbol{q}) &= \boldsymbol{0}, \end{aligned}$$

and thus the reduced model in state-space is obtained as

$$\begin{aligned} \dot{\boldsymbol{q}} &= \boldsymbol{G}(\boldsymbol{q})\boldsymbol{v} \\ \dot{\boldsymbol{v}} &= \boldsymbol{M}^{-1}(\boldsymbol{q})\boldsymbol{\tau} \end{aligned}$$

where

$$\boldsymbol{M}^{-1}(\boldsymbol{q}) = \begin{bmatrix} 1/m & 0 \\ 0 & 1/I \end{bmatrix}.$$

By using the input transformation

$$\boldsymbol{\tau} = \boldsymbol{M}\boldsymbol{u} = \begin{bmatrix} m & 0 \\ 0 & I \end{bmatrix} \boldsymbol{u},$$

the second-order kinematic model is obtained as

$$\dot{\xi} = \begin{bmatrix} v\cos\theta \\ v\sin\theta \\ \omega \\ 0 \\ 0 \end{bmatrix} + \begin{bmatrix} 0 \\ 0 \\ 0 \\ 1 \\ 0 \end{bmatrix} u_1 + \begin{bmatrix} 0 \\ 0 \\ 0 \\ 0 \\ 1 \end{bmatrix} u_2$$

with the state vector $\xi = [x \quad y \quad \theta \quad v \quad \omega]^T \in \mathbb{R}^5$.

11.5 Planning

As with manipulators, the problem of planning a trajectory for a mobile robot can be broken down in finding a *path* and defining a *timing law* on the path. However, if the mobile robot is subject to nonholonomic constraints, the first of these two subproblems becomes more difficult than in the case of manipulators. In fact, in addition to meeting the boundary conditions (interpolation of the assigned points and continuity of the desired degree) the path must also satisfy the nonholonomic constraints at all points.

11.5.1 Path and Timing Law

Assume that one wants to plan a trajectory $\boldsymbol{q}(t)$, for $t \in [t_i, t_f]$, that leads a mobile robot from an initial configuration $\boldsymbol{q}(t_i) = \boldsymbol{q}_i$ to a final configuration $\boldsymbol{q}(t_f) = \boldsymbol{q}_f$ in the absence of obstacles. The trajectory $\boldsymbol{q}(t)$ can be broken down into a geometric path $\boldsymbol{q}(s)$, with $d\boldsymbol{q}(s)/ds \neq 0$ for any value of s, and a timing law $s = s(t)$, with the parameter s varying between $s(t_i) = s_i$ and $s(t_f) = s_f$ in a monotonic fashion, i.e., with $\dot{s}(t) \geq 0$, for $t \in [t_i, t_f]$. A possible choice for s is the arc length along the path; in this case, it would be $s_i = 0$ and $s_f = L$, where L is the length of the path.

The above space-time separation implies that

$$\dot{\boldsymbol{q}} = \frac{d\boldsymbol{q}}{dt} = \frac{d\boldsymbol{q}}{ds}\dot{s} = \boldsymbol{q}'\dot{s},$$

where the prime symbol denotes differentiation with respect to s. The generalized velocity vector is then obtained as the product of the vector $\boldsymbol{q}'$, which is directed as the tangent to the path in configuration space, by the scalar $\dot{s}$, that varies its modulus. Note that the vector $[x' \quad y']^T \in \mathbb{R}^2$ is directed as the tangent to the Cartesian path, and has unit norm if s is the cartesian arc length (see Sect. 5.3.1).

Nonholonomic constraints of the form (11.3) can then be rewritten as

$$\boldsymbol{A}(\boldsymbol{q})\dot{\boldsymbol{q}} = \boldsymbol{A}(\boldsymbol{q})\boldsymbol{q}'\dot{s} = \boldsymbol{0}.$$

If $\dot{s}(t) > 0$, for $t \in [t_i, t_f]$, one has

$$\boldsymbol{A}(\boldsymbol{q})\boldsymbol{q}' = \boldsymbol{0}. \tag{11.40}$$

This condition, that must be verified at all points by the tangent vector on the configuration space path, characterizes the notion of *geometric* path admissibility induced by the kinematic constraint (11.3) that actually affects generalized velocities. Similar to what has been done for trajectories in Sect. 11.2, geometrically admissible paths can be explicitly defined as the solutions of the nonlinear system

$$\boldsymbol{q}' = \boldsymbol{G}(\boldsymbol{q})\widetilde{\boldsymbol{u}}, \tag{11.41}$$

where $\widetilde{\boldsymbol{u}}$ is a vector of *geometric* inputs that are related to the velocity inputs $\boldsymbol{u}$ by the relationship $\boldsymbol{u}(t) = \widetilde{\boldsymbol{u}}(s)\dot{s}(t)$. Once the geometric inputs $\widetilde{\boldsymbol{u}}(s)$ are assigned for $s \in [s_i, s_f]$, the path of the robot in configuration space is uniquely determined. The choice of a timing law $s = s(t)$, for $t \in [t_i, t_f]$, will then identify a particular trajectory along this path.

For example, in the case of a mobile robot with unicycle-like kinematics, the pure rolling constraint (11.6) entails the following condition for geometric admissibility of the path:

$$[\sin\theta \quad -\cos\theta \quad 0]\,\boldsymbol{q}' = x'\sin\theta - y'\cos\theta = 0,$$

that simply expresses the fact that the tangent to the Cartesian path must be aligned with the robot sagittal axis. As a consequence, a path whose tangent is discontinuous (e.g., a broken line) is *not* admissible, unless the unicycle is allowed to stop at discontinuity points by setting $\dot{s} = 0$ for the time necessary to rotate on the spot so as to align with the new tangent.

Geometrically admissible paths for the unicycle are the solutions of the system

$$\begin{aligned} x' &= \widetilde{v}\cos\theta \\ y' &= \widetilde{v}\sin\theta \\ \theta' &= \widetilde{\omega}, \end{aligned} \tag{11.42}$$

where $\widetilde{v}, \widetilde{\omega}$ are related to v, ω by

$$v(t) = \widetilde{v}(s)\dot{s}(t) \tag{11.43}$$

$$\omega(t) = \widetilde{\omega}(s)\dot{s}(t). \tag{11.44}$$

11.5.2 Flat Outputs

Many kinematic models of mobile robots, including the unicycle and the bicycle, exhibit a property known as *differential flatness*, that is particularly relevant in planning problems. A nonlinear dynamic system $\dot{\boldsymbol{x}} = \boldsymbol{f}(\boldsymbol{x}) + \boldsymbol{G}(\boldsymbol{x})\boldsymbol{u}$ is *differentially flat* if there exists a set of outputs $\boldsymbol{y}$, called *flat* outputs, such that the state $\boldsymbol{x}$ and the control inputs $\boldsymbol{u}$ can be expressed algebraically as a function of $\boldsymbol{y}$ and its time derivatives up to a certain order:

$$\begin{aligned}\boldsymbol{x} &= \boldsymbol{x}(\boldsymbol{y}, \dot{\boldsymbol{y}}, \ddot{\boldsymbol{y}}, \ldots, \boldsymbol{y}^{(r)})\\ \boldsymbol{u} &= \boldsymbol{u}(\boldsymbol{y}, \dot{\boldsymbol{y}}, \ddot{\boldsymbol{y}}, \ldots, \boldsymbol{y}^{(r)}).\end{aligned}$$

As a consequence, once an output trajectory is assigned for $\boldsymbol{y}$, the associated trajectory of the state $\boldsymbol{x}$ and history of control inputs $\boldsymbol{u}$ are uniquely determined.

In the unicycle and bicycle cases, the Cartesian coordinates are indeed flat outputs. In the following, this property is established for the unicycle. This can be done with reference to either the kinematic model (11.13) or the geometric model (11.42). For simplicity, refer to the latter. Its first two equations imply that, given a Cartesian path $(x(s), y(s))$, the associated state trajectory is $\boldsymbol{q}(s) = [\,x(s) \quad y(s) \quad \theta(s)\,]^T$ where

$$\theta(s) = \text{Atan2}\,(y'(s), x'(s)) + k\pi \qquad k = 0, 1. \tag{11.45}$$

The two possible choices for k account for the fact that the same Cartesian path may be followed moving forward ($k = 0$) or backward ($k = 1$). If the initial orientation of the robot is assigned, only one of the choices for k is correct. The geometric inputs that drive the robot along the Cartesian path are easily obtained from (11.42),(11.45) as

$$\widetilde{v}(s) = \pm\sqrt{(x'(s))^2 + (y'(s))^2} \tag{11.46}$$

$$\widetilde{\omega}(s) = \frac{y''(s)x'(s) - x''(s)y'(s)}{(x'(s))^2 + (y'(s))^2}. \tag{11.47}$$

These equations deserve some comments:

- The choice of the sign of $\widetilde{v}(s)$ depends on the type of motion (forward or backward).
- If $x'(\bar{s}\,) = y'(\bar{s}\,) = 0$ for some $\bar{s} \in [s_i, s_f]$, one has $\widetilde{v}(\bar{s}\,) = 0$. This happens, for example, in correspondence of cusps (motion inversions) in the Cartesian path. In these points, Eq. (11.45) does not define the orientation, that can however be derived by continuity, i.e., as the limit of its right-hand side for $s \to \bar{s}^-$. Similar arguments can be repeated for the steering velocity $\widetilde{\omega}$ given by (11.47).
- The possibility of reconstructing θ and $\widetilde{\omega}$ is lost when the Cartesian trajectory degenerates to a point, because in this case it is $x'(s) = y'(s) = 0$ identically.

It is interesting to note that for driftless dynamic systems — like the kinematic models of mobile robots — differential flatness is a necessary and sufficient condition for transformability in the chained form introduced in Sect. 11.3. In particular, it is easy to prove that the flat outputs of a $(2, n)$ chained form are z_1 and z_n, from which it is possible to compute all the other state variables as well as the associated control inputs. For example, in the case of the $(2, 3)$ chained form (11.25) it is

$$z_2 = \frac{\dot{z}_3}{\dot{z}_1} \qquad v_1 = \dot{z}_1 \qquad v_2 = \frac{\dot{z}_1 \ddot{z}_3 - \ddot{z}_1 \dot{z}_3}{\dot{z}_1^2}.$$

Note that z_2 and v_2 can be actually reconstructed only if $\dot{z}_1(t) \neq 0$, for $t \in [t_i, t_f]$.

11.5.3 Path Planning

Whenever a mobile robot admits a set of flat outputs $\boldsymbol{y}$, these may be exploited to solve planning problems efficiently. In fact, one may use any interpolation scheme to plan the path of $\boldsymbol{y}$ in such a way as to satisfy the appropriate boundary conditions. The evolution of the other configuration variables, together with the associated control inputs, can then be computed algebraically from $\boldsymbol{y}(s)$. The resulting configuration space path will *automatically* satisfy the nonholonomic constraints (11.40).

In particular, consider the problem of planning a path that leads a unicycle from an initial configuration $\boldsymbol{q}(s_i) = \boldsymbol{q}_i = [\,x_i \quad y_i \quad \theta_i\,]^T$ to a final configuration $\boldsymbol{q}(s_f) = \boldsymbol{q}_f = [\,x_f \quad y_f \quad \theta_f\,]^T$.

Planning via Cartesian polynomials

As mentioned above, the problem can be solved by interpolating the initial values x_i, y_i and the final values x_f, y_f of the flat outputs x, y. Letting $s_i = 0$ and $s_f = 1$, one may use the following cubic polynomials:

$$\begin{aligned} x(s) &= s^3 x_f - (s-1)^3 x_i + \alpha_x s^2 (s-1) + \beta_x s (s-1)^2 \\ y(s) &= s^3 y_f - (s-1)^3 y_i + \alpha_y s^2 (s-1) + \beta_y s (s-1)^2, \end{aligned}$$

that automatically satisfy the boundary conditions on x, y. The orientation at each point being related to x', y' by (11.45), it is also necessary to impose the additional boundary conditions

$$\begin{aligned} x'(0) &= k_i \cos\theta_i \qquad & x'(1) &= k_f \cos\theta_f \\ y'(0) &= k_i \sin\theta_i \qquad & y'(1) &= k_f \sin\theta_f, \end{aligned}$$

where $k_i \neq 0$, $k_f \neq 0$ are free parameters that must however have the same sign. This condition is necessary to guarantee that the unicycle arrives in $\boldsymbol{q}_f$ with the same kind of motion (forward or backward) with which it leaves $\boldsymbol{q}_i$;

in fact, since $x(s)$ and $y(s)$ are cubic polynomials, the Cartesian path does not contain motion inversions in general.

For example, by letting $k_i = k_f = k > 0$, one obtains

$$\begin{bmatrix} \alpha_x \\ \alpha_y \end{bmatrix} = \begin{bmatrix} k\cos\theta_f - 3x_f \\ k\sin\theta_f - 3y_f \end{bmatrix} \qquad \begin{bmatrix} \beta_x \\ \beta_y \end{bmatrix} = \begin{bmatrix} k\cos\theta_i + 3x_i \\ k\sin\theta_i + 3y_i \end{bmatrix}.$$

The choice of k_i and k_f has a precise influence on the obtained path. In fact, by using (11.46) it is easy to verify that

$$\widetilde{v}(0) = k_i \qquad \widetilde{v}(1) = k_f.$$

The evolution of the robot orientation along the path and the associated geometric inputs can then be computed by using Eqs. (11.45) and (11.46), (11.47), respectively.

Planning via the chained form

Another technique, which can be immediately generalized to other kinematic models of mobile robots (e.g., the bicycle), is planning the path in the chained form coordinates $\boldsymbol{z}$. To this end, it is first necessary to compute the initial and final values $\boldsymbol{z}_i$ and $\boldsymbol{z}_f$ that correspond to $\boldsymbol{q}_i$ and $\boldsymbol{q}_f$, by using the change of coordinates (11.23). It is then sufficient to interpolate the initial and final values of z_1 and z_3 (the flat outputs) with the appropriate boundary conditions on the remaining variable $z_2 = z_3'/z_1'$.

Again, it is possible to adopt a cubic polynomial to solve the problem. As an alternative, one may use polynomials of different degree for x and y in order to reduce the number of unknown coefficients to be computed. For example, under the assumption $z_{1,i} \neq z_{1,f}$, consider the following interpolation scheme:

$$\begin{aligned} z_1(s) &= z_{1,f}s - (s-1)z_{1,i} \\ z_3(s) &= s^3 z_{3,f} - (s-1)^3 z_{3,i} + \alpha_3 s^2(s-1) + \beta_3 s(s-1)^2, \end{aligned}$$

with $s \in [0,1]$. Note that $z_1'(s)$ is constant and equal to $z_{1,f} - z_{1,i} \neq 0$. The unknowns α_3, β_3 must be determined by imposing the boundary conditions on z_2:

$$\frac{z_3'(0)}{z_1'(0)} = z_{2i} \qquad \frac{z_3'(1)}{z_1'(1)} = z_{2f},$$

from which

$$\begin{aligned} \alpha_3 &= z_{2,f}(z_{1,f} - z_{1,i}) - 3z_{3,f} \\ \beta_3 &= z_{2,i}(z_{1,f} - z_{1,i}) + 3z_{3,i}. \end{aligned}$$

This scheme cannot be directly applied when $z_{1,i} = z_{1,f}$, i.e., when $\theta_i = \theta_f$. To handle this singular case, one may introduce a *via point* $\boldsymbol{q}_v = [\,x_v \quad y_v \quad \theta_v\,]^T$

such that $\theta_v \neq \theta_i$, and solve the original planning problem using two consecutive paths, the first from $\boldsymbol{q}_i$ to $\boldsymbol{q}_v$ and the second from $\boldsymbol{q}_v$ to $\boldsymbol{q}_f$. Another possibility, which avoids the introduction of the via point, is to let $z_{1,f} = z_{1,i} + 2\pi$ (i.e., to replace θ_f with $\theta_f + 2\pi$); this obviously corresponds to the same final configuration of the unicycle. With the resulting manoeuvre, the robot will reach its destination while performing a complete rotation of orientation along the path.

Once the path has been planned for the chained form, the path $\boldsymbol{q}(s)$ in the original coordinates and the associated geometric inputs $\widetilde{\boldsymbol{u}}(s)$ are reconstructed by inverting the change of coordinates (11.23) and of inputs (11.24), respectively.

Planning via parameterized inputs

A conceptually different approach to path planning consists of writing the inputs — rather than the path — in parameterized form, and computing the value of the parameters so as to drive the robot from $\boldsymbol{q}_i$ to $\boldsymbol{q}_f$. Again, it is convenient to work on the chained form, whose equations are easily integrable in closed form under appropriate inputs. For the sake of generality, refer to the $(2, n)$ chained form (11.20), whose geometric version is

$$\begin{aligned} z_1' &= \widetilde{v}_1 \\ z_2' &= \widetilde{v}_2 \\ z_3' &= z_2 \widetilde{v}_1 \\ &\vdots \\ z_n' &= z_{n-1} \widetilde{v}_1. \end{aligned}$$

Let the geometric input be chosen as

$$\widetilde{v}_1 = \mathrm{sgn}(\Delta) \tag{11.48}$$

$$\widetilde{v}_2 = c_0 + c_1 s + \ldots + c_{n-2} s^{n-2}, \tag{11.49}$$

with $\Delta = z_{1,f} - z_{1,i}$ and $s \in [s_i, s_f] = [0, |\Delta|]$. Parameters $c_0, \ldots, c_{n-2}$ must be chosen so as to give $\boldsymbol{z}(s_f) = \boldsymbol{z}_f$. It is possible to verify that such condition is expressed as a linear system of equations

$$\boldsymbol{D}(\Delta) \begin{bmatrix} c_0 \\ c_1 \\ \vdots \\ c_{n-2} \end{bmatrix} = \boldsymbol{d}(\boldsymbol{z}_i, \boldsymbol{z}_f, \Delta) \tag{11.50}$$

where matrix $\boldsymbol{D}(\Delta)$ is invertible if $\Delta \neq 0$. For example, in the case of the $(2, 3)$ chained form, one obtains

$$\boldsymbol{D} = \begin{bmatrix} |\Delta| & \frac{\Delta^2}{2} \\ \mathrm{sgn}(\Delta)\frac{\Delta^2}{2} & \frac{\Delta^3}{6} \end{bmatrix} \qquad \boldsymbol{d} = \begin{bmatrix} z_{2,f} - z_{2,i} \\ z_{3,f} - z_{3,i} - z_{2,i}\Delta \end{bmatrix}. \tag{11.51}$$

If $z_{1,i} = z_{1,f}$ a singular case is met that can be handled as before.

Both $\boldsymbol{z}(s)$ and $\widetilde{\boldsymbol{v}}(s)$ must then be converted to $\boldsymbol{q}(s)$ and $\widetilde{\boldsymbol{u}}(s)$ using the inverse coordinate and input transformations that apply to the specific case.

The above method does not make explicit use of the flat outputs, but relies on the closed-form integrability of the chained form, whose existence, as already mentioned, is equivalent to differential flatness. Note also that in this planning scheme, as in the previous two, parameter s does not represent the arc length on the path.

Other classes of parameterized inputs that can be used in place of (11.48), (11.49) are sinusoidal and piecewise constant functions.

Numerical results

For illustration, some numerical results of the planning methods so far described are now presented. The considered vehicle is a unicycle that must perform various 'parking' manoeuvres.

Two typical paths produced by the planner that uses cubic Cartesian polynomials are shown in Fig. 11.5. As already noticed, the unicycle never inverts its motion, that is forward in these two manoeuvres because $k = 5 > 0$. For $k < 0$, the manoeuvres would have been performed in backward motion with different paths.

In Fig. 11.6, the same planner is used to solve a *parallel parking* problem, in which the difference between the initial and the final configuration of the unicycle is a pure displacement in the direction orthogonal to the sagittal axis. Note how the path changes as $k_i = k_f = k$ is changed; in particular, an increase in k leads to elongated 'take-off' (from $\boldsymbol{q}_i$) and 'landing' (on $\boldsymbol{q}_f$) phases.

Figure 11.7 refers to the case in which $\boldsymbol{q}_i$ and $\boldsymbol{q}_f$ differ only for the value of θ (a pure reorientation); the unicycle leaves the initial position and follows a path that leads back to it with the correct orientation. This behaviour is to be expected, because with this planner a Cartesian path of nonzero length is needed to achieve any kind of reconfiguration.

To allow a comparison, the same planning problems have also been solved with the method based on the use of parameterized inputs in conjunction with the chained form.

Figure 11.8 shows the path obtained with this planner for the same two parking problems of Fig. 11.5. While in the first case the obtained manoeuvre is similar to the one obtained before, in the second the path contains a cusp, corresponding to a motion inversion. In fact, in view of its nature, this planner can only generate paths along which the robot orientation stays between its initial value θ_i and its final value θ_f.

In Fig. 11.9, two different solutions produced by this planner are reported for the parallel parking problem of Fig. 11.6. The singularity due to $\theta_i = \theta_f$ has been solved in two different ways: by adding a via point $\boldsymbol{q}_v$, and redefining θ_f as $\theta_f = \theta_i + 2\pi$. Note that in the latter case the path produced by the

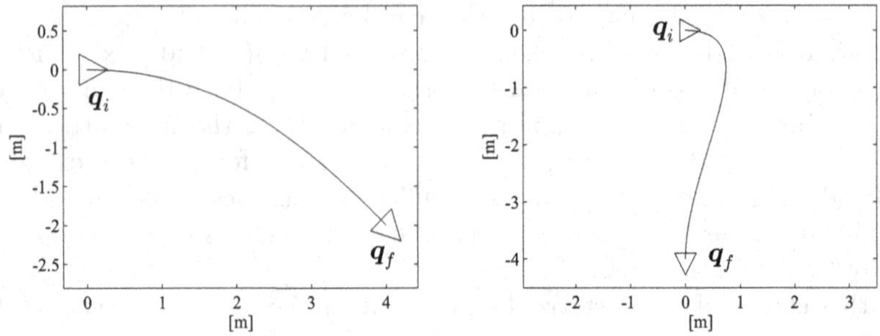

Fig. 11.5. Two parking manoeuvres planned via cubic Cartesian polynomials; in both cases $k = 5$ has been used

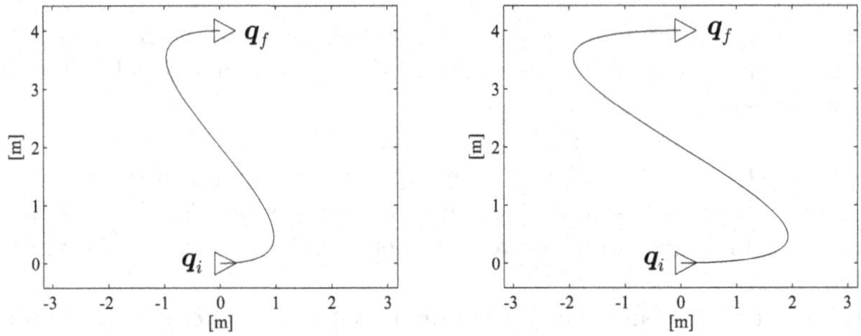

Fig. 11.6. Planning a parallel parking manoeuvre via cubic Cartesian polynomials; *left*: with $k = 10$, *right*: with $k = 20$

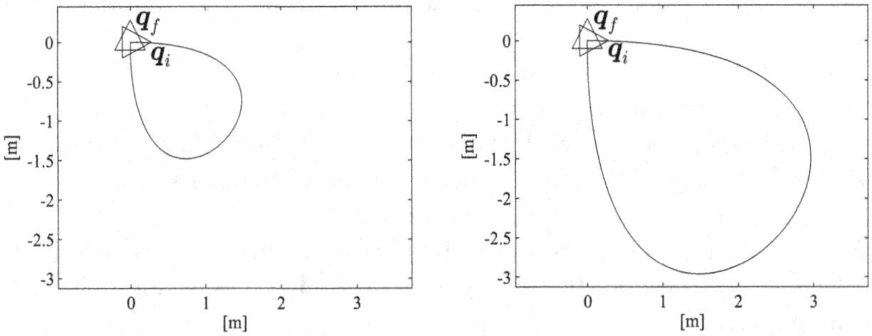

Fig. 11.7. Planning a pure reorientation manoeuvre via cubic Cartesian polynomials; *left*: with $k = 10$, *right*: with $k = 20$

planner leads the robot to the destination with a complete rotation of its orientation.

Finally, the same pure reorientation manoeuvre of Fig. 11.7 has been considered in Fig. 11.10. The path on the left has been obtained as outlined before, i.e., exploiting the transformations of coordinates (11.23) and of in-

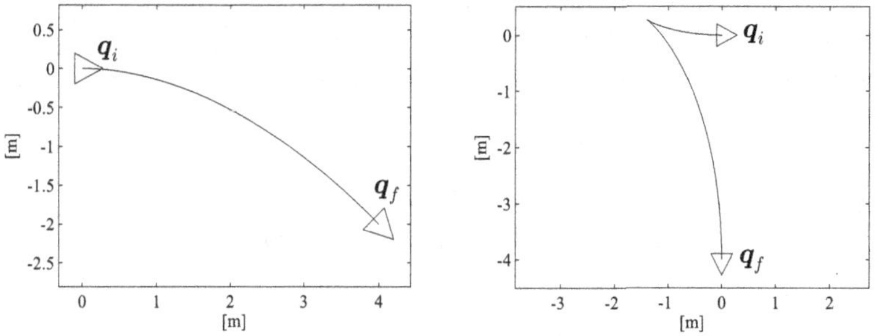

Fig. 11.8. Two parking manoeuvres planned via the chained form

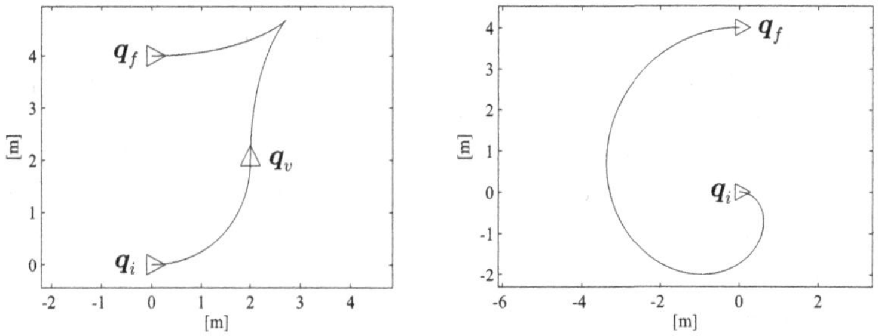

Fig. 11.9. Planning a parallel parking manoeuvre via the chained form; *left*: adding a via point $\boldsymbol{q}_v$, *right*: letting $\theta_f = \theta_i + 2\pi$

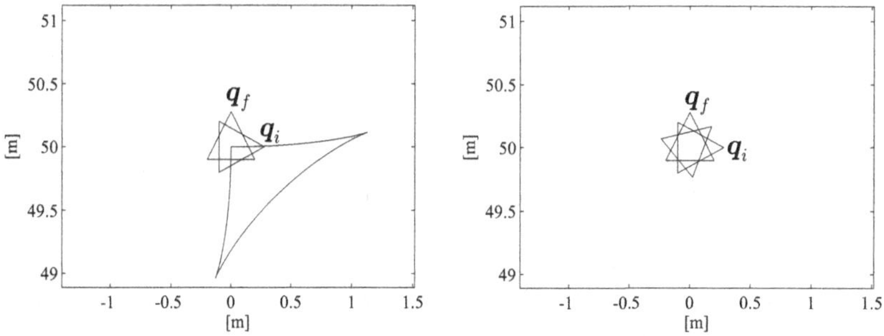

Fig. 11.10. Planning a pure reorientation manoeuvre via the chained form; *left*: with the coordinate transformation (11.23), *right*: with the coordinate transformation (11.52)

puts (11.24) to put the system in chained form, and then using the parameterized inputs (11.48), (11.49). As in the previous case, the required reorientation is realized by a Cartesian path. This is a consequence of the structure of (11.23), for which $\theta_i \neq \theta_f$ implies in general $z_{2,i} \neq z_{2,f}$ and $z_{3,i} \neq z_{3,f}$, even when $x_i = x_f$, $y_i = y_f$.

The manoeuvre on the right of Fig. 11.10, which is a rotation on the spot, was achieved by using a different change of coordinates to put the system in (2,3) chained form. In particular, the following transformation has been used

$$\begin{aligned} z_1 &= \theta - \theta_f \\ z_2 &= (x - x_i)\cos\theta + (y - y_i)\sin\theta \\ z_3 &= (x - x_i)\sin\theta - (y - y_i)\cos\theta, \end{aligned} \tag{11.52}$$

which places the origin of the (z_2, z_3) reference frame in correspondence of the initial Cartesian position of the unicycle. With this choice one has $z_{2,i} = z_{2,f}$ and $z_{3,i} = z_{3,f}$ for a pure reorientation, and thus the manoeuvre is efficiently obtained as a simple rotation.

As a matter of fact, using (11.52) in place of (11.23) is always recommended. In fact, the analysis of (11.51) shows that in general the magnitude of the coefficients of $\widetilde{v}_2$ — and therefore, the length of the obtained path — depends not only on the amount of reconfiguration required for z_2 and z_3, but also on the value of $z_{2,i}$ itself. The adoption of (11.52), which implies $z_{2,i} = 0$, makes the size of the manoeuvre invariant with respect to the Cartesian position of the unicycle.

11.5.4 Trajectory Planning

Once a path $\boldsymbol{q}(s)$, $s \in [s_i, s_f]$, has been determined, it is possible to choose a timing law $s = s(t)$ with which the robot should follow it. In this respect, considerations similar to those of Sect. 5.3.1 apply. For example, if the velocity inputs of the unicycle are subject to bounds of the form[8]

$$|v(t)| \le v_{\max} \qquad |\omega(t)| \le \omega_{\max} \qquad \forall t, \tag{11.53}$$

it is necessary to verify whether the velocities along the planned trajectory are admissible. In the negative case, it is possible to slow down the timing law via *uniform scaling*. To this end, it is convenient to rewrite the timing law by replacing t with the normalized time variable $\tau = t/T$, with $T = t_f - t_i$. From (11.43), (11.44) one has

$$v(t) = \widetilde{v}(s)\frac{ds}{d\tau}\frac{d\tau}{dt} = \widetilde{v}(s)\frac{ds}{d\tau}\frac{1}{T} \tag{11.54}$$

$$\omega(t) = \widetilde{\omega}(s)\frac{ds}{d\tau}\frac{d\tau}{dt} = \widetilde{\omega}(s)\frac{ds}{d\tau}\frac{1}{T}, \tag{11.55}$$

and therefore is is sufficient to increase T (i.e., the duration of the trajectory) to reduce uniformly v and ω, so as to stay within the given bounds.

[8] For a differential drive unicycle, the actual bounds affect the wheel angular speeds ω_L and ω_R. Through Eqs. (11.14), these bounds can be mapped to constraints on v and ω (see Problem 11.9).

It is also possible to plan directly a trajectory without separating the geometric path from the timing law. To this end, all the techniques presented before can be used with the time variable t directly in place of the path parameter s. A drawback of this approach is that the duration $t_f - t_i = s_f - s_i$ of the trajectory is fixed, and uniform scaling cannot be used to satisfy bounds on the velocity inputs. In fact, an increase (or decrease) of $t_f - t_i$ would modify the geometric path associated with the planned trajectory.

11.5.5 Optimal Trajectories

The planning techniques so far presented can be used to compute trajectories that lead the robot from an initial configuration $\boldsymbol{q}_i$ to a final configuration $\boldsymbol{q}_f$ while complying with the nonholonomic constraints and, possibly, bounds on the velocity inputs. Often, other requirements are added, such as limiting the path curvature, avoiding workspace obstacles or reducing energy consumption. In general, these are integrated in the design procedure as the optimization of a suitable cost criterion along the trajectory. For example, the previous objectives will be respectively formulated as the minimization of the maximum curvature, the maximization of the minimum distance between the robot and the obstacles, or the minimization of the total energy needed by the mobile robot to follow the path.

A simple technique for attacking the optimal planning problem consists of *over-parameterizing* the adopted interpolation scheme, so as to pursue the optimization — typically, via numerical techniques — of the cost criterion by appropriately choosing the redundant parameters. Clearly, the obtained trajectory will be optimal only with respect to the set of trajectories that can be generated by the chosen scheme, and will be a *suboptimal* solution for the original planning problem; this may or may not lead to the fulfilment of the original specifications. For example, the planning scheme based on cubic Cartesian polynomials contains two free parameters (k_i and k_f), that may be chosen so as to maximize the minimum distance along the path between the unicycle and certain obstacles. However, depending on the placement of the obstacles with respect to $\boldsymbol{q}_i$ and $\boldsymbol{q}_f$, a collision-free path (i.e., a path for which the above distance is always positive) may or may not[9] exist within the chosen family of cubic polynomials.

A more systematic approach to the problem relies on the use of *optimal control* theory. The basic problem considered in this discipline is in fact the determination of a control law that transfers a dynamic system between two assigned states so as to minimize a chosen cost functional along the trajectory. A powerful tool for solving this problem is the *Pontryagin minimum principle* that provides *necessary* conditions for optimality. By exploiting these conditions in conjunction with the analysis of the specific characteristics of the

[9] The complexity of the problem of planning collision-free motions in the presence of obstacles is such that specific solution techniques are needed; these are presented in Chap. 12.

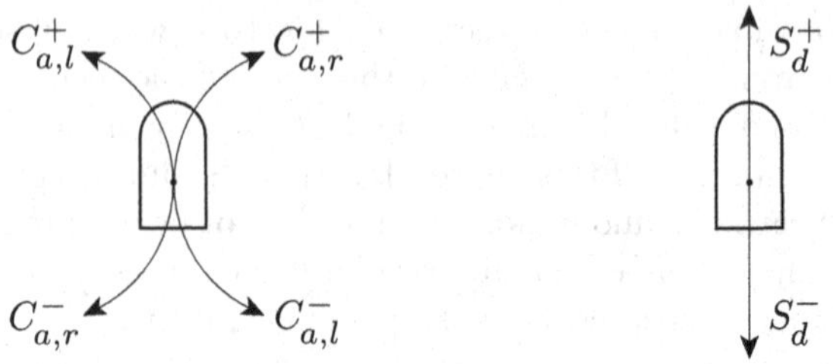

Fig. 11.11. The elementary arcs that constitute the trajectories of the sufficient family for the minimum-time planning problem for the unicycle

considered problem, it is often possible to identify a reduced set of candidate trajectories, also referred to as a *sufficient family*, among which there is certainly the desired optimal solution (if it exists).

In any case, each optimal planning problem must be formulated in the appropriate context. When minimizing curvature or avoiding static obstacles, the timing law part of the trajectory is irrelevant, and the problem can be solved by planning a path for the geometric model (11.41). If the cost criterion depends on the path as well as on the timing law, it is necessary to plan directly on the kinematic model (11.10). A particularly important example of the latter situation is met when minimum-time trajectories are sought in the presence of bounds on the velocity inputs.

Minimum-time trajectories

Consider the problem of transferring the unicycle (11.13) from the initial configuration $\boldsymbol{q}_i$ to the final configuration $\boldsymbol{q}_f$ while minimizing the functional

$$J = t_f - t_i = \int_{t_i}^{t_f} dt,$$

under the assumption that the driving and steering velocity inputs v and ω are bounded as in (11.53). By combining the conditions provided by the minimum principle with geometrical arguments, it is possible to determine a sufficient family for the solution to this problem. This family consists of trajectories obtained by concatenating *elementary arcs* of two kinds only:

- arcs of circle of variable length, covered with velocities $v(t) = \pm v_{\max}$ and $\omega(t) = \pm\omega_{\max}$ (the radius of the circle is always $v_{\max}/\omega_{\max}$);
- line segments of variable length, covered with velocities $v(t) = \pm v_{\max}$ and $\omega(t) = 0$.

These elementary arcs are shown in Fig. 11.11, where a compact notation is also defined for identifying them. In particular, C_a and S_d indicate an arc of circle of duration a and a line segment of duration d, respectively (in the particular case $v_{\max} = 1$, a and d are also the lengths of these arcs). The superscript indicates forward (+) o'r backward (−) motion, while for circular arcs the second subscript indicates a rotation in the clockwise (r) or counterclockwise (l) direction. With this notation, and considering for simplicity the case $v_{\max} = 1$ and $\omega_{\max} = 1$, the trajectories of the sufficient family (also called *Reeds–Shepp curves*) can be classified in the following nine *groups*:

$$\begin{array}{lll} \text{I} & C_a|C_b|C_e & a \geq 0, b \geq 0, e \geq 0, a+b+e \leq \pi \\ \text{II} & C_a|C_bC_e & 0 \leq a \leq b, 0 \leq e \leq b, 0 \leq b \leq \pi/2 \\ \text{III} & C_aC_b|C_e & 0 \leq a \leq b, 0 \leq e \leq b, 0 \leq b \leq \pi/2 \\ \text{IV} & C_aC_b|C_bC_e & 0 \leq a \leq b, 0 \leq e \leq b, 0 \leq b \leq \pi/2 \\ \text{V} & C_a|C_bC_b|C_e & 0 \leq a \leq b, 0 \leq e \leq b, 0 \leq b \leq \pi/2 \\ \text{VI} & C_a|C_{\pi/2}S_eC_{\pi/2}|C_b & 0 \leq a \leq \pi/2, 0 \leq b \leq \pi/2, e \geq 0 \\ \text{VII} & C_a|C_{\pi/2}S_eC_b & 0 \leq a \leq \pi, 0 \leq b \leq \pi/2, e \geq 0 \\ \text{VIII} & C_aS_eC_{\pi/2}|C_b & 0 \leq a \leq \pi/2, 0 \leq b \leq \pi, e \geq 0 \\ \text{IX} & C_aS_eC_b & 0 \leq a \leq \pi/2, 0 \leq b \leq \pi/2, e \geq 0, \end{array} \qquad (11.56)$$

where the symbol "|" between two elementary arcs indicates the presence of a cusp (motion inversion) on the path. Each group contains trajectories consisting of a *sequence* of no more than five elementary arcs; the duration of circular arcs is bounded by either $\pi/2$ or π, while the duration of line segments depends on the Cartesian distance between $\boldsymbol{q}_i$ and $\boldsymbol{q}_f$. The number of possible sequences produced by each group is finite; they are obtained by instantiating the elementary arcs depending on the direction of motion and of rotation. For example, it is easy to show that group IX generates eight sequence types, each corresponding to a trajectory entirely covered in forward or backward motion:

$$\begin{array}{c} C^+_{a,r}S^+_eC^+_{a,r},\; C^+_{a,r}S^+_eC^+_{a,l},\; C^+_{a,l}S^+_eC^+_{a,r},\; C^+_{a,l}S^+_eC^+_{a,l} \\ C^-_{a,r}S^-_eC^-_{a,r},\; C^-_{a,r}S^-_eC^-_{a,l},\; C^-_{a,l}S^-_eC^-_{a,r},\; C^-_{a,l}S^-_eC^-_{a,l}. \end{array}$$

By this kind of argument, it may be verified that the above groups generate a total number of 48 different sequences.

In practice, one may use an exhaustive algorithm to identify the minimum-time trajectory that leads the unicycle from $\boldsymbol{q}_i$ to $\boldsymbol{q}_f$:

- Determine all the trajectories belonging to the sufficient family that join $\boldsymbol{q}_i$ and $\boldsymbol{q}_f$.
- Compute the value of the cost criterion $t_f - t_i$ along these trajectories, and choose the one to which the minimum value is associated.

The first is clearly the most difficult step. Essentially, for each of the aforementioned 48 sequences it is necessary to identify — if it exists — the corresponding trajectory going from $\boldsymbol{q}_i$ to $\boldsymbol{q}_f$, and to compute the duration of the associated elementary arcs. To this end, for each sequence, it is possibly

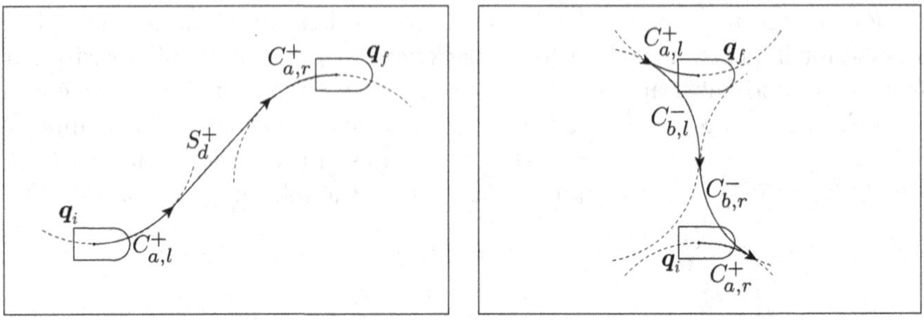

Fig. 11.12. Two examples of minimum-time trajectories for the unicycle

to express and invert in closed form the relationship between the duration of the arcs and the obtained change in configuration. By doing so, the first step can also be completed very quickly.

Figure 11.12 shows two examples of minimum-time trajectories for the unicycle. The first (on the left) belongs to group IX and contains three elementary arcs without inversions of motion. The second (on the right) is a group V trajectory consisting of four arcs of circle, along which there are two motion inversions.

11.6 Motion Control

The motion control problem for wheeled mobile robots is generally formulated with reference to the kinematic model (11.10), i.e., by assuming that the control inputs determine directly the generalized velocities $\dot{\boldsymbol{q}}$. For example, in the case of the unicycle (11.13) and of the bicycle (11.18) or (11.19) this means that the inputs are the driving and steering velocity inputs v and ω. There are essentially two reasons for taking this simplifying assumption. First, as already seen in Sect. 11.4, under suitable assumptions it is possible to cancel the dynamic effects via state feedback, so that the control problem is actually transferred to the second-order kinematic model, and from the latter to the first-order kinematic model. Second, in the majority of mobile robots it is not possible to command directly the wheel torques, because there are *low-level* control loops that are integrated in the hardware or software architecture. These loops accept as input a reference value for the wheel angular speed, that is reproduced as accurately as possible by standard regulation actions (e.g., PID controllers). In this situation, the actual inputs available for high-level controls are precisely the reference velocities.

In this section, a unicycle-like vehicle is again considered, although some of the presented control schemes may be extended to other kinds of mobile robots. Two basic control problems, illustrated in Fig. 11.13, will be considered for system (11.13):

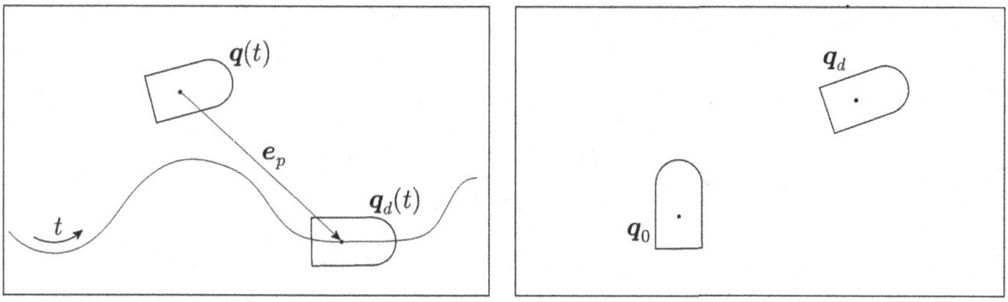

Fig. 11.13. Control problems for a unicycle; *left*: trajectory tracking, *right*: posture regulation

- *Trajectory tracking*: the robot must asymptotically track a desired Cartesian trajectory $(x_d(t), y_d(t))$, starting from an initial configuration $\boldsymbol{q}_0 = [\,x_0 \quad y_0 \quad \theta_0\,]^T$ that may or may not be 'matched' with the trajectory.
- *Posture regulation*: the robot must asymptotically reach a given posture, i.e., a desired configuration $\boldsymbol{q}_d$, starting from an initial configuration $\boldsymbol{q}_0$.

From a practical point of view, the most relevant of these problems is certainly the first. This is because, unlike industrial manipulators, mobile robots must invariably operate in unstructured workspaces containing obstacles. Clearly, forcing the robot to move along (or close to) a trajectory planned in advance considerably reduces the risk of collisions. On the other hand, a preliminary planning step is not required when posture regulation is performed, but the Cartesian trajectory along which the robot approaches $\boldsymbol{q}_d$ cannot be specified by the user.

11.6.1 Trajectory Tracking

For the tracking problem to be soluble, it is necessary that the desired Cartesian trajectory $(x_d(t), y_d(t))$ is admissible for the kinematic model (11.13), i.e., it must satisfy the equations

$$\begin{aligned}\dot{x}_d &= v_d \cos\theta_d \\ \dot{y}_d &= v_d \sin\theta_d \\ \dot{\theta}_d &= \omega_d\end{aligned} \tag{11.57}$$

for some choice of the reference inputs v_d and ω_d. For example, this is certainly true if the trajectory has been produced using one of the planning schemes of the previous section. In any case, as seen in Sect. 11.5.2, since the unicycle coordinates x and y are flat outputs ,the orientation along the desired trajectory $(x_d(t), y_d(t))$ can be computed as

$$\theta_d(t) = \text{Atan2}\,(\dot{y}_d(t), \dot{x}(t)_d) + k\pi \qquad k = 0, 1, \tag{11.58}$$

as well as the reference inputs

$$v_d(t) = \pm\sqrt{\dot{x}_d^2(t) + \dot{y}_d^2(t)} \tag{11.59}$$

$$\omega_d(t) = \frac{\ddot{y}_d(t)\dot{x}_d(t) - \ddot{x}_d(t)\dot{y}_d(t)}{\dot{x}_d^2(t) + \dot{y}_d^2(t)}. \tag{11.60}$$

Note that (11.58) and (11.59), (11.60), correspond respectively to (11.45) and (11.46), (11.47) with $s = t$. In the following, it is assumed that the value of k in (11.58) — and correspondingly, the sign of v_d in (11.59) — has been chosen.

By comparing the desired state $\boldsymbol{q}_d(t) = [\,x_d(t) \quad y_d(t) \quad \theta_d(t)\,]^T$ with the current measured state $\boldsymbol{q}(t) = [\,x(t) \quad y(t) \quad \theta(t)\,]^T$ it is possible to compute an error vector that can be fed to the controller. However, rather than using directly the difference between $\boldsymbol{q}_d$ and $\boldsymbol{q}$, it is convenient to define the tracking error as

$$\boldsymbol{e} = \begin{bmatrix} e_1 \\ e_2 \\ e_3 \end{bmatrix} = \begin{bmatrix} \cos\theta & \sin\theta & 0 \\ -\sin\theta & \cos\theta & 0 \\ 0 & 0 & 1 \end{bmatrix} \begin{bmatrix} x_d - x \\ y_d - y \\ \theta_d - \theta \end{bmatrix}.$$

The positional part of $\boldsymbol{e}$ is the Cartesian error $\boldsymbol{e}_p = [\,x_d - x \quad y_d - y\,]^T$ expressed in a reference frame aligned with the current orientation θ of the robot (see Fig. 11.13). By differentiating $\boldsymbol{e}$ with respect to time, and using (11.13) and (11.57), one easily finds

$$\begin{aligned} \dot{e}_1 &= v_d \cos e_3 - v + e_2\,\omega \\ \dot{e}_2 &= v_d \sin e_3 - e_1\,\omega \\ \dot{e}_3 &= \omega_d - \omega. \end{aligned} \tag{11.61}$$

Using the input transformation

$$v = v_d \cos e_3 - u_1 \tag{11.62}$$

$$\omega = \omega_d - u_2, \tag{11.63}$$

which is clearly invertible, the following expression is obtained for the tracking error dynamics:

$$\dot{\boldsymbol{e}} = \begin{bmatrix} 0 & \omega_d & 0 \\ -\omega_d & 0 & 0 \\ 0 & 0 & 0 \end{bmatrix} \boldsymbol{e} + \begin{bmatrix} 0 \\ \sin e_3 \\ 0 \end{bmatrix} v_d + \begin{bmatrix} 1 & -e_2 \\ 0 & e_1 \\ 0 & 1 \end{bmatrix} \begin{bmatrix} u_1 \\ u_2 \end{bmatrix}. \tag{11.64}$$

Note that the first term of this dynamics is linear, while the second and third are nonlinear. Moreover, the first and second terms are in general time-varying, due to the presence of the reference inputs $v_d(t)$ and $\omega_d(t)$.

Control based on approximate linearization

The simplest approach to designing a tracking controller consists of using the approximate linearization of the error dynamics around the reference trajectory, on which clearly $\boldsymbol{e} = \mathbf{0}$. This approximation, whose accuracy increases as the tracking error $\boldsymbol{e}$ decreases, is obtained from (11.64) simply setting $\sin e_3 = e_3$ and evaluating the input matrix on the trajectory. The result is

$$\dot{\boldsymbol{e}} = \begin{bmatrix} 0 & \omega_d & 0 \\ -\omega_d & 0 & v_d \\ 0 & 0 & 0 \end{bmatrix} \boldsymbol{e} + \begin{bmatrix} 1 & 0 \\ 0 & 0 \\ 0 & 1 \end{bmatrix} \begin{bmatrix} u_1 \\ u_2 \end{bmatrix}. \tag{11.65}$$

Note that the approximate system is still time-varying. Consider now the linear feedback

$$u_1 = -k_1 \, e_1 \tag{11.66}$$

$$u_2 = -k_2 \, e_2 - k_3 \, e_3 \tag{11.67}$$

that leads to the following closed-loop linearized dynamics:

$$\dot{\boldsymbol{e}} = \boldsymbol{A}(t) \, \boldsymbol{e} = \begin{bmatrix} -k_1 & \omega_d & 0 \\ -\omega_d & 0 & v_d \\ 0 & -k_2 & -k_3 \end{bmatrix} \boldsymbol{e}. \tag{11.68}$$

The characteristic polynomial of matrix $\boldsymbol{A}$ is

$$p(\lambda) = \lambda(\lambda + k_1)(\lambda + k_3) + \omega_d^2(\lambda + k_3) + v_d k_2(\lambda + k_1).$$

At this point, it is sufficient to let

$$k_1 = k_3 = 2\zeta a \qquad k_2 = \frac{a^2 - \omega_d^2}{v_d}, \tag{11.69}$$

with $\zeta \in (0, 1)$ and $a > 0$, to obtain

$$p(\lambda) = (\lambda + 2\zeta a)(\lambda^2 + 2\zeta a \lambda + a^2).$$

The closed-loop linearized error dynamics is then characterized by three constant eigenvalues: one real negative eigenvalue in $-2\zeta a$ and a pair of complex eigenvalues with negative real part, damping coefficient ζ and natural frequency a. However, in view of its time-varying nature, there is no guarantee that system (11.68) is asymptotically stable.

A notable exception is when v_d and ω_d are constant, as in the case of circular or rectilinear trajectories. In fact, the linearized system (11.68) is then time-invariant and therefore asymptotically stable with the choice of gains in (11.69). Hence, by using the control law (11.66), (11.67) with the same gains, the origin of the original error system (11.64) is also asymptotically stable, although this result is not guaranteed to hold globally. For sufficiently small initial errors, the unicycle will certainly converge to the desired Cartesian trajectory (either circular or rectilinear).

In general, the feedback controller (11.66), (11.67) is linear but also time-varying in view of the expression of k_2 in (11.69). The actual velocity inputs v and ω should be reconstructed from u_1 and u_2 through (11.62), (11.63). In particular, it is easy to verify that v and ω tend to coincide with the reference inputs v_d and ω_d (i.e., they reduce to a pure feedforward action) as the tracking error $\boldsymbol{e}$ vanishes.

Finally, note that k_2 in (11.69) diverges when v_d goes to zero, i.e., when the reference Cartesian trajectory tends to stop. Therefore, the above control scheme can only be used for *persistent* Cartesian trajectories, i.e., trajectories such that $|v_d(t)| \geq \bar{v} > 0, \forall t \geq 0$. This also means that motion inversions (from forward to backward motion, or vice versa) on the reference trajectory are not allowed.

Nonlinear control

Consider again the exact expression (11.64) of the tracking error dynamics, now rewritten for convenience in the 'mixed' form:

$$\begin{aligned}\dot{e}_1 &= e_2\,\omega + u_1\\ \dot{e}_2 &= v_d \sin e_3 - e_1\omega\\ \dot{e}_3 &= u_2,\end{aligned} \tag{11.70}$$

and the following nonlinear version of the control law (11.66), (11.67):

$$u_1 = -k_1(v_d, \omega_d)\, e_1 \tag{11.71}$$

$$u_2 = -k_2\, v_d \frac{\sin e_3}{e_3}\, e_2 - k_3(v_d, \omega_d)\, e_3, \tag{11.72}$$

where $k_1(\cdot,\cdot) > 0$ and $k_3(\cdot,\cdot) > 0$ are bounded functions with bounded derivatives, and $k_2 > 0$ is constant. If the reference inputs v_d and ω_d are also bounded with bounded derivatives, and they do not both converge to zero, the tracking error $\boldsymbol{e}$ converges to zero globally, i.e., for any initial condition.

A sketch is now given of the proof of this result. Consider the closed-loop error dynamics

$$\begin{aligned}\dot{e}_1 &= e_2\,\omega - k_1(v_d, \omega_d)\, e_1\\ \dot{e}_2 &= v_d \sin e_3 - e_1\,\omega\\ \dot{e}_3 &= -k_2\, v_d \frac{\sin e_3}{e_3}\, e_2 - k_3(v_d, \omega_d)\, e_3,\end{aligned} \tag{11.73}$$

and the candidate Lyapunov function

$$V = \frac{k_2}{2}\left(e_1^2 + e_2^2\right) + \frac{e_3^2}{2},$$

whose time derivative along the system trajectories

$$\dot{V} = -k_1(v_d, \omega_d)k_2 \, e_1^2 - k_3(v_d, \omega_d)e_3^2$$

is negative semi-definite. This means that V (which is bounded below) tends to a limit value, and also that the norm of $\boldsymbol{e}$ is bounded. As system (11.73) is time-varying, it is not possible to use La Salle theorem to gain further insight. However, under the above assumptions, one may verify that $\ddot{V}$ is limited, and therefore $\dot{V}$ is uniformly continuous. Hence, Barbalat lemma (see Sect. C.3) may be used to conclude that $\dot{V}$ tends to zero, i.e., that e_1 and e_3 tend to zero. From this and the system equations it is possible to prove that

$$\lim_{t \to \infty} (v_d^2 + \omega_d^2)e_2^2 = 0,$$

and thus e_2 tends to zero as well, provided that at least one of the reference inputs is persistent.

Again, the actual velocity inputs v and ω must be computed from u_1 and u_2 using (11.62), (11.63). Note that the control law (11.71), (11.72) requires the persistency of the state trajectory $\boldsymbol{q}(t)$, but not of the Cartesian trajectory. In particular, the reference velocity $v_d(t)$ can converge to zero as long as $\omega_d(t)$ does not, and vice versa. For example, this controller can be used to track a Cartesian trajectory that degenerates to a simple rotation on the spot.

Input/output linearization

A well-known systematic approach to the design of trajectory tracking controllers is based on input/output linearization via feedback (see Sect. 8.5.2). In the case of the unicycle, consider the following outputs:

$$\begin{aligned} y_1 &= x + b\cos\theta \\ y_2 &= y + b\sin\theta, \end{aligned}$$

with $b \neq 0$. They represent the Cartesian coordinates of a point B located along the sagittal axis of the unicycle at a distance $|b|$ from the contact point of the wheel with the ground (see Fig. 11.3). In particular, B is 'ahead' of the contact point if b is positive and 'behind' if it is negative.

The time derivatives of y_1 and y_2 are

$$\begin{bmatrix} \dot{y}_1 \\ \dot{y}_2 \end{bmatrix} = \begin{bmatrix} \cos\theta & -b\sin\theta \\ \sin\theta & b\cos\theta \end{bmatrix} \begin{bmatrix} v \\ \omega \end{bmatrix} = \boldsymbol{T}(\theta) \begin{bmatrix} v \\ \omega \end{bmatrix}. \tag{11.74}$$

Matrix $\boldsymbol{T}(\theta)$ has determinant b, and is therefore invertible under the assumption that $b \neq 0$. It is then sufficient to use the following input transformation

$$\begin{bmatrix} v \\ \omega \end{bmatrix} = \boldsymbol{T}^{-1}(\theta) \begin{bmatrix} u_1 \\ u_2 \end{bmatrix} = \begin{bmatrix} \cos\theta & \sin\theta \\ -\sin\theta/b & \cos\theta/b \end{bmatrix} \begin{bmatrix} u_1 \\ u_2 \end{bmatrix}$$

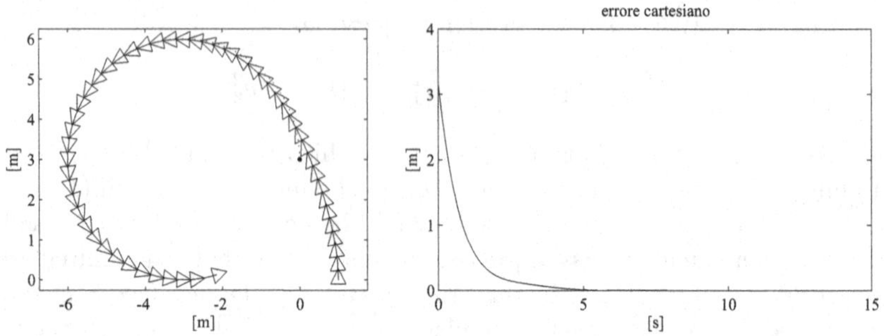

Fig. 11.14. Tracking a circular reference trajectory (*dotted*) with the controller based on approximate linearization; *left*: Cartesian motion of the unicycle, *right*: time evolution of the norm of the Cartesian error $\boldsymbol{e}_p$

to put the equations of the unicycle in the form

$$\begin{aligned} \dot{y}_1 &= u_1 \\ \dot{y}_2 &= u_2 \\ \dot{\theta} &= \frac{u_2 \cos\theta - u_1 \sin\theta}{b}. \end{aligned} \tag{11.75}$$

An input/output linearization via feedback has therefore been obtained. At this point, a simple linear controller of the form

$$u_1 = \dot{y}_{1d} + k_1(y_{1d} - y_1) \tag{11.76}$$

$$u_2 = \dot{y}_{2d} + k_2(y_{2d} - y_2), \tag{11.77}$$

with $k_1 > 0$, $k_2 > 0$, guarantees exponential convergence to zero of the Cartesian tracking error, with decoupled dynamics on its two components. Note that the orientation, whose evolution is governed by the third equation in (11.75), is not controlled. In fact, this tracking scheme does not use the orientation error; hence, it is based on the output error rather than the state error.

It should be emphasized that the reference Cartesian trajectory for point B can be arbitrary, and in particular the associated path may exhibit isolated points with discontinuous geometric tangent (like in a broken line) without requiring the robot to stop and reorient itself at those points. This is true as long as $b \neq 0$, and therefore such a possibility does not apply to the contact point between the wheel and the ground, whose velocity, as already discussed, cannot have a component in the direction orthogonal to the sagittal axis of the vehicle.

Simulations

An example of application of the trajectory tracking controller (11.66), (11.67) based on linear approximation is shown in Fig. 11.14. The desired trajectory

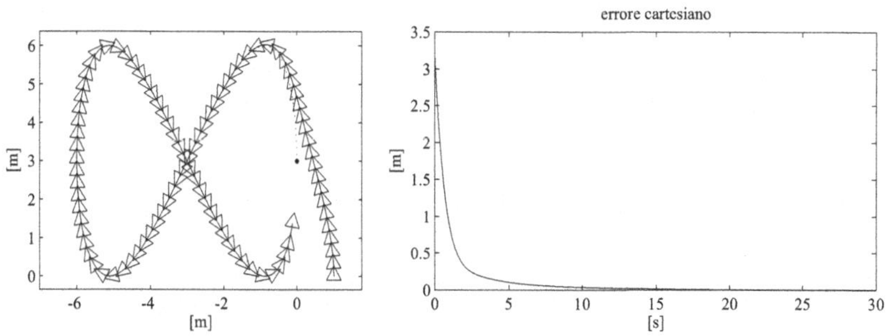

Fig. 11.15. Tracking a figure of eight-shaped reference trajectory (*dotted*) with the nonlinear controller; *left*: Cartesian motion of the unicycle, *right*: time evolution of the norm of the Cartesian error $\boldsymbol{e}_p$

in this case is the circle of centre (x_c, y_c) and radius R described by the parametric equations

$$x_d(t) = x_c + R\cos(\omega_d t)$$
$$y_d(t) = y_c + R\sin(\omega_d t),$$

with $R = 3$ m and $\omega_d = 1/3$ rad/s. Hence, the reference driving velocity on the circle is constant and equal to $v_d = R\,\omega_d = 1$ m/s. The controller gains have been chosen according to (11.69) with $\zeta = 0.7$ and $a = 1$. Note the exponential convergence to zero of the Cartesian error.

In the second simulation (Fig. 11.15) the reference trajectory is figure of eight-shaped, with centre in (x_c, y_c), and is described by the parametric equations

$$x_d(t) = x_c + R_1\sin(2\omega_d t)$$
$$y_d(t) = y_c + R_2\sin(\omega_d t),$$

with $R_1 = R_2 = 3$ m and $\omega_d = 1/15$ rad/s. The reference driving velocity $v_d(t)$ in this case varies over time and must be computed using (11.59). The results shown have been obtained with the nonlinear controller (11.71), (11.72), with k_1 and k_3 again given by (11.69) in which $\zeta = 0.7$ and $a = 1$, while k_2 has been set to 1. Also in this case, the error converges to zero very quickly.

The third Cartesian reference trajectory is a square with a side of 4 m, to be traced with constant velocity. This requirement means that the unicycle cannot stop at the vertices in order to reorient itself, and therefore the representative point (x, y) of the unicycle cannot follow the reference trajectory. As a consequence, the tracking scheme based on input/output linearization has been adopted. In particular, two simulations are presented, both obtained using the control law (11.76), (11.77) with $k_1 = k_2 = 2$. In the first, the point B that tracks the reference trajectory is located at a distance $b = 0.75$ m from the contact point between the wheel and the ground. As shown in Fig. 11.16,

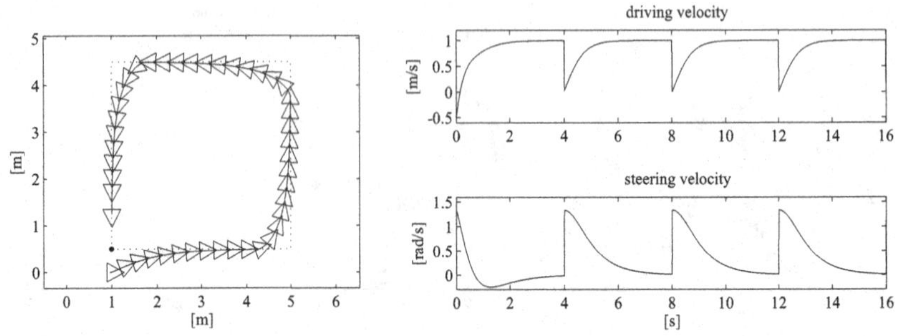

Fig. 11.16. Tracking a square reference trajectory (*dotted*) with the controller based on input/output linearization, with $b = 0.75$; *left*: Cartesian motion of the unicycle, *right*: time evolution of the velocity inputs v and ω

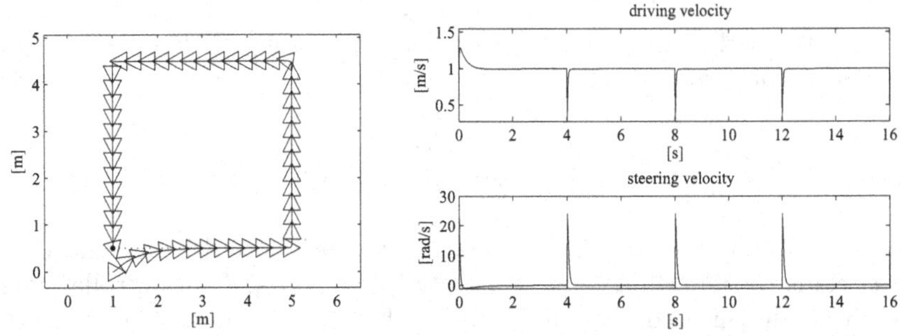

Fig. 11.17. Tracking a square reference trajectory (*dotted*) with the controller based on input/output linearization, with $b = 0.2$; *left*: Cartesian motion of the unicycle, *right*: time evolution of the velocity inputs v and ω

the unicycle actually moves on a 'smoothed' trajectory; therefore, an obstacle-free channel around the trajectory must be available to account for the area swept by the unicycle in correspondence of the square vertices.

The second simulation (Fig. 11.17) shows what can happen when b is reduced in order to achieve more accurate tracking for the unicycle representative point; in particular, $b = 0.2$ m was chosen in this case. While it is true that the unicycle tracks the square more closely with respect to the first simulation, the steering velocity is much higher in correspondence of the vertices, and this might be a problem in the presence of saturations on the velocity inputs. This situation is consistent with the fact that matrix $\boldsymbol{T}$ in (11.74) tends to become singular when b approaches zero.

11.6.2 Regulation

Consider now the problem of designing a feedback control law that drives the unicycle (11.13) to a desired configuration $\boldsymbol{q}_d$. A reasonable approach could be to plan first a trajectory that stops in $\boldsymbol{q}_d$, and then track it via feedback. However, none of the tracking schemes so far described can be used to this end. In fact, the controller based on approximate linearization and the nonlinear controller both require a persistent state trajectory. The scheme based on input/output linearization can also track non-persistent trajectories, but will drive point B to the destination rather than the representative point of the unicycle. Besides, the final orientation at the destination will not be controlled.

Actually, the difficulty of identifying feedback control laws for tracking non-persistent trajectories is structural. It is in fact possible to prove that, due to the nonholonomy of the system, the unicycle does not admit any *universal* controller, i.e., a controller that can asymptotically stabilize arbitrary state trajectories, either persistent or not. This situation is completely different from the case of manipulators, for which the scheme based on inverse dynamics is an example of universal controller. As a consequence, the posture regulation problem in nonholonomic mobile robots must be addressed using purposely designed control laws.

Cartesian regulation

Consider first the problem of designing a feedback controller for a *partial* regulation task, in which the objective is to drive the unicycle to a given Cartesian position, without specifying the final orientation. This simplified version of the regulation problem is of practical interest. For example, a mobile robot exploring an unknown environment must visit a sequence of Cartesian positions (*view points*) from where it perceives the characteristics of the surrounding area using its on-board sensors. If these are isotropically distributed on the robot (as in the case of a ring of equispaced ultrasonic sensors, a rotating laser range finder, or a panoramic camera), the orientation of the robot at the view point is irrelevant.

Without loss of generality, assume that the desired Cartesian position is the origin; the Cartesian error $\boldsymbol{e}_p$ is then simply $[-x \quad -y]^T$. Consider the following control law

$$v = -k_1(x\cos\theta + y\sin\theta) \tag{11.78}$$

$$\omega = k_2(\mathrm{Atan2}(y,x) - \theta + \pi), \tag{11.79}$$

where $k_1 > 0$, $k_2 > 0$. These two commands have an immediate geometric interpretation: the driving velocity v is proportional to the projection of the Cartesian error $\boldsymbol{e}_p$ on the sagittal axis of the unicycle, whereas the steering velocity ω is proportional to the difference between the orientation of the unicycle and that of vector $\boldsymbol{e}_p$ (*pointing error*).

Consider the following 'Lyapunov-like' function:

$$V = \frac{1}{2}(x^2 + y^2),$$

that is only positive semi-definite at the origin, because it is zero in all configurations such that $x = y = 0$, independently from the value of the orientation θ. Using the unicycle equations in (11.13) and the control inputs (11.78), (11.79) one obtains

$$\dot{V} = -k_1(x\cos\theta + y\sin\theta)^2,$$

that is negative semi-definite at the origin. This indicates that V, which is bounded below, tends to a limit value, and also that the position error $\boldsymbol{e}_p$ is bounded in norm. It is easy to verify that $\ddot{V}$ is also bounded, and thus $\dot{V}$ is uniformly continuous. Barbalat lemma[10] implies that $\dot{V}$ tends to zero. Hence

$$\lim_{t\to\infty}(x\cos\theta + y\sin\theta) = 0,$$

i.e., the projection of the Cartesian error vector $\boldsymbol{e}_p$ on the sagittal axis of the unicycle tends to vanish. This cannot happen in a point different from the origin, because the steering velocity (11.79) would then force the unicycle to rotate so as to align with $\boldsymbol{e}_p$. One may then conclude that the Cartesian error tends to zero for any initial configuration.

Posture regulation

To design a feedback controller that is able to regulate the whole configuration vector (Cartesian position and vehicle orientation) of the unicycle, it is convenient to formulate the problem in polar coordinates. It is again assumed, without loss of generality, that the desired configuration is the origin $\boldsymbol{q}_d = [0 \quad 0 \quad 0]^T$.

With reference to Fig. 11.18, let ρ be the distance between the representative point (x, y) of the unicycle and the origin of the Cartesian plane, γ the angle between vector $\boldsymbol{e}_p$ and the sagittal axis of the vehicle, and δ the angle between the same vector and the x axis. In formulae:

$$\begin{aligned}\rho &= \sqrt{x^2 + y^2}\\ \gamma &= \text{Atan2}(y, x) - \theta + \pi\\ \delta &= \gamma + \theta.\end{aligned}$$

In these coordinates, the kinematic model of the unicycle is expressed as

$$\begin{aligned}\dot{\rho} &= -v\cos\gamma\\ \dot{\gamma} &= \frac{\sin\gamma}{\rho}v - \omega\\ \dot{\delta} &= \frac{\sin\gamma}{\rho}v.\end{aligned} \qquad (11.80)$$

Note that the input vector field associated with v is singular for $\rho = 0$.

[10] La Salle theorem cannot be used because V is not positive definite at the origin.

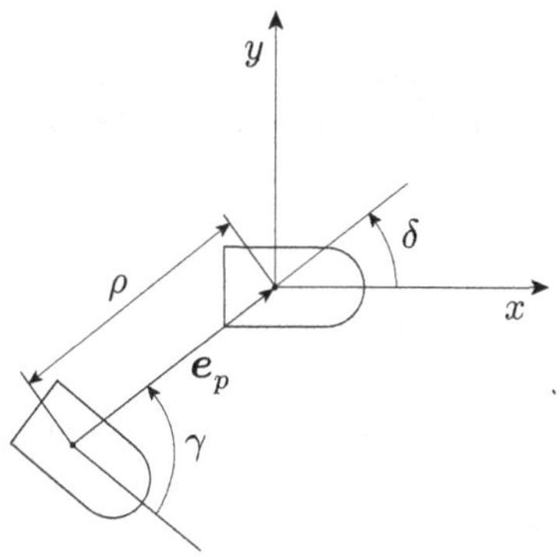

Fig. 11.18. Definition of polar coordinates for the unicycle

Define the feedback control as[11]

$$v = k_1 \rho \cos\gamma \qquad (11.81)$$

$$\omega = k_2\,\gamma + k_1\,\frac{\sin\gamma\cos\gamma}{\gamma}\,(\gamma + k_3\,\delta), \qquad (11.82)$$

with $k_1 > 0$, $k_2 > 0$. The kinematic model (11.80) under the action of the control law (11.81), (11.82) asymptotically converges to the desired configuration $[\rho \;\; \gamma \;\; \delta]^T = [0 \;\; 0 \;\; 0]^T$.

The proof of this result is based on the following Lyapunov candidate:

$$V = \frac{1}{2}\left(\rho^2 + \gamma^2 + k_3\,\delta^2\right),$$

whose time derivative along the closed-loop system trajectories

$$\dot{V} = -k_1 \cos^2\!\gamma\,\rho^2 - k_2\,\gamma^2$$

is negative semi-definite. As a consequence, V tends to a limit value and the system state is bounded. It can also be shown that $\ddot{V}$ is bounded, so that $\dot{V}$ is uniformly continuous. In view of Barbalat lemma, it can be inferred that $\dot{V}$ tends to zero, and likewise do ρ and γ. Further analysis of the closed-loop system leads to concluding that δ converges to zero as well.

Note that angles γ and δ are undefined for $x = y = 0$. They are, however, always well-defined during the motion of the unicycle, and asymptotically tend to the desired zero value.

[11] It is easy to verify that the expression (11.81) for v coincides with (11.78), the driving velocity prescribed by the Cartesian regulation scheme, except for the presence of ρ, whose effect is to modulate v according to the distance of the robot from the destination. As for the steering velocity, (11.82) differs from (11.79) for the presence of the second term, that contains the orientation error θ (through δ) in addition to the pointing error.

It should be noted that the control law (11.81), (11.82), once mapped back to the original coordinates, is discontinuous at the origin of the configuration space $\mathcal{C}$. As a matter of fact, it can be proven that any feedback law that can regulate the posture of the unicycle must be necessarily discontinuous with respect to the state and/or time-varying.[12]

Simulations

To illustrate the characteristics of the above regulation schemes, simulation results are now presented for two parking manoeuvres performed in feedback by a unicycle mobile robot.

Figure 11.19 shows the robot trajectories produced by the Cartesian regulator (11.78), (11.79), with $k = 1$ and $k_2 = 3$, for two different initial configurations. Note that the final orientation of the robot varies with the approach direction, and that the unicycle reaches the destination in forward motion, after inverting its motion at most once (like in the second manoeuvre). It is possible to prove that such behaviour is general with this controller.

The results of the application of the posture regulator (11.81), (11.82) starting from the same initial conditions are shown in Fig. 11.20. The gains have been set to $k_1 = 1$, $k_2 = 2.5$ and $k_3 = 3$. The trajectories obtained are quite similar to the previous ones, but as expected the orientation is driven to zero as well. As before, the final approach to the destination is always in forward motion, with at most one motion inversion in the transient phase.

11.7 Odometric Localization

The implementation of any feedback controller requires the availability of the robot configuration at each time instant. Unlike the case of manipulators, in which the joint encoders provide a direct measurement of the configuration, mobile robots are equipped with incremental encoders that measure the rotation of the wheels, but not directly the position and orientation of the vehicle with respect to a fixed world frame. It is therefore necessary to devise a *localization* procedure that estimates in real time the robot configuration.

Consider a unicycle moving under the action of velocity commands v and ω that are constant within each sampling interval. This assumption, which

[12] This result, which actually applies to all nonholonomic robots, derives from the application of a necessary condition (*Brockett theorem*) for the smooth stabilizability of control systems. In the particular case of a driftless system of the form (11.10), in which there are fewer inputs than states and the input vector fields are linearly independent, such a condition is violated and no control law that is continuous in the state $\boldsymbol{q}$ can asymptotically stabilize an equilibrium point. Brockett theorem does not apply to time-varying stabilizing controllers that may thus be continuous in $\boldsymbol{q}$.

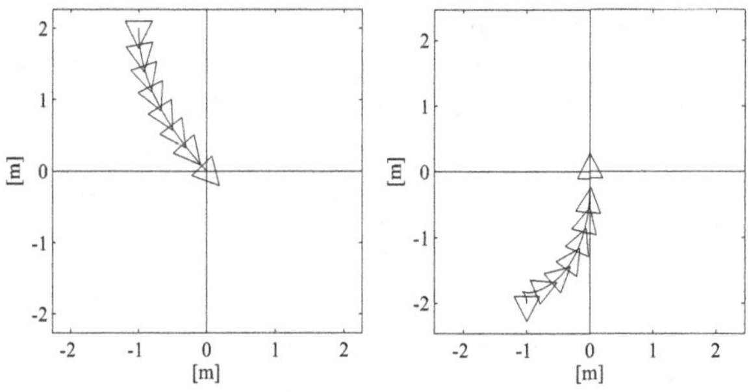

Fig. 11.19. Regulation to the origin of the Cartesian position of the unicycle with the controller (11.78), (11.79), for two different initial configurations

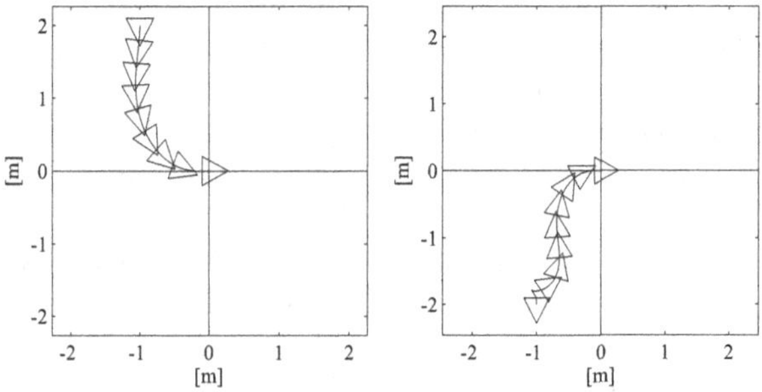

Fig. 11.20. Regulation to the origin of the posture of the unicycle with the controller (11.81), (11.82), for two different initial configurations

is generally satisfied[13] in digital control implementations, implies that during the interval the robot moves along an arc of circle of radius $R = v_k/\omega_k$, which degenerates to a line segment if $\omega_k = 0$. Assume that the robot configuration $\boldsymbol{q}(t_k) = \boldsymbol{q}_k$ at the sampling time t_k is known, together with the value of the velocity inputs v_k and ω_k applied in the interval $[t_k, t_{k+1})$. The value of the configuration variables $\boldsymbol{q}_{k+1}$ at the sampling time t_{k+1} can then be reconstructed by forward integration of the kinematic model (11.13).

A first possibility is to use the approximate formulae

$$\begin{aligned} x_{k+1} &= x_k + v_k T_s \cos\theta_k \\ y_{k+1} &= y_k + v_k T_s \sin\theta_k \\ \theta_{k+1} &= \theta_k + \omega_k T_s, \end{aligned} \tag{11.83}$$

where $T_s = t_{k+1} - t_k$ is the duration of the sampling interval. These equations, which correspond to the use of the Euler method for numerical integration

[13] In particular, this is certainly true if the velocity commands computed by the digital controller are converted to control inputs for the robot through a zero-order hold (ZOH).

of (11.13), introduce an error in the computation of x_{k+1} and y_{k+1}, that is performed as if the orientation θ_k remained constant throughout the interval. This error becomes smaller as T_s is decreased. The third formula instead is exact.

If the accuracy of the Euler method proves to be inadequate, one may use the following estimate with the same T_s:

$$\begin{aligned} x_{k+1} &= x_k + v_k T_s \cos\left(\theta_k + \frac{\omega_k T_s}{2}\right) \\ y_{k+1} &= y_k + v_k T_s \sin\left(\theta_k + \frac{\omega_k T_s}{2}\right) \\ \theta_{k+1} &= \theta_k + \omega_k T_s, \end{aligned} \tag{11.84}$$

corresponding to the adoption of the second-order Runge–Kutta integration method. Note how the first two formulae use the average value of the unicycle orientation in the sampling interval.

To obtain an exact reconstruction of $\boldsymbol{q}_{k+1}$ under the assumption of constant velocity inputs within the sampling interval one may use simple geometric arguments or exploit the transformability of the unicycle kinematic model in the chained form (11.25). As already seen, this form is easily integrable in closed form, leading to an exact expression for $\boldsymbol{z}_{k+1}$. The configuration $\boldsymbol{q}_{k+1}$ can then be computed by inverting the coordinate and input transformations (11.23) and (11.24). This procedure, which can be generalized to any mobile robot that can be put in chained form, provides the formulae:

$$\begin{aligned} x_{k+1} &= x_k + \frac{v_k}{\omega_k}(\sin\theta_{k+1} - \sin\theta_k) \\ y_{k+1} &= y_k - \frac{v_k}{\omega_k}(\cos\theta_{k+1} - \cos\theta_k) \\ \theta_{k+1} &= \theta_k + \omega_k T_s. \end{aligned} \tag{11.85}$$

Note that the first two are still defined for $\omega_k = 0$; in this case, they coincide with the corresponding formulae of Euler and Runge–Kutta methods (which are exact over line segments). In the implementation, however, it is necessary to handle this situation with a conditional instruction.

Figure 11.21 allows a comparison among the configurations $\boldsymbol{q}_{k+1}$ reconstructed via the three aforementioned integration methods. In practice, the difference is obviously much less dramatic, and tends to disappear as the duration T_s of the sampling interval is decreased.

In the previous formulae it has been assumed that the velocity inputs v_k and ω_k applied in the sampling interval are available. In view of the non-ideality of any actuation system, rather than relying on the 'commanded' values, it is convenient to reconstruct v_k and ω_k using the robot proprioceptive sensors. First of all, note that

$$v_k T_s = \Delta s \qquad \omega_k T_s = \Delta\theta \qquad \frac{v_k}{\omega_k} = \frac{\Delta s}{\Delta\theta}, \tag{11.86}$$

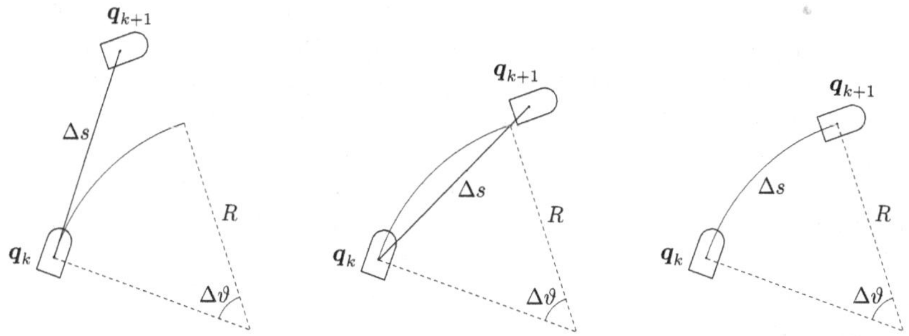

Fig. 11.21. Odometric localization for a unicycle moving along an elementary tract corresponding to an arc of circle; *left*: integration via Euler method, *centre*: integration via Runge–Kutta method, *right*: exact integration

where Δs is the length of the elementary tract travelled by the robot and $\Delta\theta$ is the total variation of orientation along the tract. For example, in the case of a differential drive unicycle, denote by $\Delta\phi_R$ and $\Delta\phi_L$ the rotation of the right and left wheel, respectively, as measured by the incremental encoders during the sampling interval. From (11.14) one easily finds

$$\Delta s = \frac{r}{2}\left(\Delta\phi_R + \Delta\phi_L\right) \qquad \Delta\theta = \frac{r}{d}\left(\Delta\phi_R - \Delta\phi_L\right)$$

that, used in (11.86), allow the implementation of all the previous formulae for the reconstruction of $\boldsymbol{q}_{k+1}$.

The forward integration of the kinematic model using the velocity commands reconstructed via the proprioceptive sensors — the encoders of the wheel actuators — is referred to as *odometric localization* or *passive localization* or *dead reckoning* — the latter is a term of uncertain etymology used in marine navigation. This method, relying on the iterated use of the previous formulae starting from an estimate of the initial configuration, provides an estimate whose accuracy cannot be better than that of $\boldsymbol{q}_0$. In any case, odometric localization — independently from the adopted integration method — is subject in practice to an error that grows over time (*drift*) and quickly becomes significant over sufficiently long paths. This error is the result of several causes, that include wheel slippage, inaccuracy in the calibration of kinematic parameters (e.g., the radius of the wheels), as well as the numerical error introduced by the integration method, if Euler or Runge–Kutta methods are used. It should also be noted that, once an odometric localization technique has been chosen, its performance also depends on the specific kinematic arrangement of the robot; for example, differential drive is usually better than synchro drive in this respect.

A more robust solution is represented by *active localization* methods. For example, this kind of approach can be adopted when the robot is equipped with proximity exteroceptive sensors (like a laser range finder) and knows a map of the workspace, either given in advance or built by the robot itself during the motion. It is then possible to correct the estimate provided by the passive localization methods by comparing the expected measures of the exteroceptive sensors with the actual readings. These techniques, which make use of tools from *Bayesian estimation theory* such as the *Extended Kalman Filter* or the *Particle Filter*, provide greater accuracy than pure odometric localization and are therefore essential in navigation tasks over long paths.

Bibliography

The literature on mobile robots is very rich and includes some comprehensive treatises, among which one of the most recent is [210].

Many books deal with the realization of mobile robots, with particular emphasis on electro-mechanical design and sensor equipment, e.g., [106, 72, 21]. For the modelling of systems subject to nonholonomic constraints, see [164]. A general classification of wheeled mobile robots based on the number, placement and type of wheels is proposed in [18], where the derivation of the kinematic and dynamic models is also presented.

Conditions for transforming driftless systems in chained form are discussed in detail in [159], while a general reference on flat outputs is [80]. The presented schemes for Cartesian trajectory tracking based on linear and nonlinear control are taken from [34], while the method for posture regulation based on polar coördinates was proposed in [2]. Complete (input/state) linearization of the unicycle kinematic model can be obtained using a dynamic feedback law, as shown for example in [174]. The non-existence of a universal controller for nonholonomic robots is proven in [135].

A detailed extension of some of the planning and control techniques described in this chapter to the case of bicycle-like kinematics is given in [58]. The design of time-varying and/or discontinuous feedback control laws for posture regulation in mobile robots was addressed in many works, including [193] and [153].

The reader interested in sensor-based robot localization and map building can refer, e.g., to [231].

Problems

11.1. Consider the mobile robot obtained by connecting N trailers to a rear-wheel drive tricycle. Each trailer is a rigid body with an axle carrying two fixed wheels, that can be assimilated to a single wheel located at the midpoint of the axle, and is hinged to the midpoint of the preceding axle through a revolute joint. Find a set of generalized coordinates for the robot.

11.2. Consider an omnidirectional mobile robot having three Mecanum wheels placed at the vertices of an equilateral triangle, each oriented in the direction orthogonal to the bisectrix of its angle. Let q_1 and q_2 be the Cartesian coordinates of the centre of the robot, q_3 the vehicle orientation, while q_4, q_5, and q_6 represent the angle of rotation of each wheel around its axis. Also, denote by r the radius of the wheels and by ℓ the distance between the centre of the robot and the centre of each wheel. This mechanical system is subject to the following Pfaffian constraints:

$$\boldsymbol{A}_1^T(\boldsymbol{q})\begin{bmatrix}\dot{q}_1\\ \dot{q}_2\\ \dot{q}_3\end{bmatrix}+\boldsymbol{A}_2^T\begin{bmatrix}\dot{q}_4\\ \dot{q}_5\\ \dot{q}_6\end{bmatrix}=0,$$

where

$$\boldsymbol{A}_1(q)=\begin{bmatrix}\frac{\sqrt{3}}{2}\cos q_3-\frac{1}{2}\sin q_3 & \sin q_3 & -\frac{1}{2}\sin q_3-\frac{\sqrt{3}}{2}\cos q_3\\ \frac{1}{2}\cos q_3+\frac{\sqrt{3}}{2}\sin q_3 & -\cos q_3 & \frac{1}{2}\cos q_3-\frac{\sqrt{3}}{2}\sin q_3\\ \ell & \ell & \ell\end{bmatrix}$$

and $\boldsymbol{A}_2 = r\,\boldsymbol{I}_3$. Show that this set of constraints is partially integrable, and that in particular the orientation of the vehicle is a linear function of the wheel rotation angles. [*Hint*: add the kinematic constraints side-by-side.]

11.3. Show that, for a single kinematic constraint in the form (11.7), the integrability condition expressed by (11.9) coincides with the involutivity of the distribution $\Delta = \text{span}\{\boldsymbol{g}_1,\dots,\boldsymbol{g}_{n-1}\}$ associated with the input vector fields of the corresponding kinematic model.

11.4. Using the controllability condition (11.11), show that a set of Pfaffian constraints that does not depend on the generalized coordinates is always integrable.

11.5. With reference to the kinematic model (11.13) of the unicycle, consider the following sequence of velocity inputs:

$$\begin{aligned}&v(t)=1 \quad \omega(t)=0,\ t\in[0,\varepsilon)\\ &v(t)=0 \quad \omega(t)=1 \ \ t\in[\varepsilon,2\varepsilon)\\ &v(t)=-1\ \omega(t)=0,\ t\in[2\varepsilon,3\varepsilon)\\ &v(t)=0 \quad \omega=-1 \quad t\in[3\varepsilon,4\varepsilon).\end{aligned}$$

By forward integration of the model equations, show that when ε is infinitesimal the displacement of the unicycle in configuration space is aligned with the Lie bracket of the input vector fields.

11.6. Prove the relationships (11.14) that allow the transformation of the velocity inputs of a differential drive vehicle to those of the equivalent unicycle.

[*Hint*: for the velocity of the midpoint of the segment joining the two wheels, it is sufficient to differentiate with respect to time the Cartesian coordinates of such a point expressed as a function of the coordinates of the wheel centres. As for ω, use the formula (B.4) for the composition of the velocities of a rigid body.]

11.7. For a generic configuration of a bicycle mobile robot, determine the Cartesian position of the instantaneous centre of rotation, and derive the expression of the angular velocity of the body as a function of the robot configuration $\boldsymbol{q}$ and of the modulus of the velocity of rear wheel centre. In the particular case of the rear-wheel drive bicycle, show that such expression is consistent with the evolution of θ predicted by the kinematic model (11.19). Moreover, compute the velocity v_P of a generic point P on the robot chassis.

11.8. Derive the kinematic model of the tricycle robot towing N trailers considered in Problem 11.1. Denote by ℓ the distance between the front wheel and rear wheel axle of the tricycle, and by ℓ_i the joint-to-joint length of the i-th trailer.

11.9. Consider a differential drive robot whose angular speed inputs — one for the right wheel and one for the left wheel — are subject to the following bounds:

$$|\omega_R(t)| \leq \omega_{RL} \qquad |\omega_L(t)| \leq \omega_{RL} \qquad \forall t,$$

that correspond to a square admissible region in the ω_R, ω_L plane. Derive the expression of the resulting constraints for the driving and steering velocity v and ω of the equivalent unicycle model. In particular, show that the admissible region in the v, ω plane is a rhombus.

11.10. Compute the expression of matrix $\boldsymbol{D}(\Delta)$ and vector $\boldsymbol{d}(\boldsymbol{z}_i, \boldsymbol{z}_f, \Delta)$ in (11.50) for the case of the (2,4) chained form.

11.11. Modify the path planning scheme based on parameterized inputs so as to obtain $s_f = 1$.

11.12. Show that the path planning schemes based on polynomials of different degree and on parameterized inputs give exactly the same result in the case of the (2,3) chained form.

11.13. Implement in a computer program the path planning method for a unicycle based on cubic Cartesian polynomials. Use the program to plan a path leading the robot from the configuration $\boldsymbol{q}_i = [0 \quad 0 \quad 0]^T$ [m,m,rad] to the configuration $\boldsymbol{q}_f = [2 \quad 1 \quad \pi/2]^T$ [m,m,rad]. Then, determine a timing law over the path so as to satisfy the following velocity bounds:

$$|v(t)| \leq 1 \text{ m/s} \qquad |\omega(t)| \leq 1 \text{ rad/s} \qquad \forall t.$$

11.14. Formulate the trajectory tracking problem for a (2,3) chained form and derive the corresponding error dynamics. Derive a feedback controller using the linear approximation around the reference trajectory.

11.15. Consider the kinematic model (11.19) of the rear-wheel drive bicycle. In analogy to the case of the unicycle, identify two outputs y_1, y_2 for which it is possible to perform a static input/output linearization. [*Hint*: consider a point P on the line passing through the centre of the front wheel and oriented as the wheel itself.]

11.16. Implement in a computer program the Cartesian regulator (11.78), (11.79), including a modification to allow the unicycle to reach the origin either in forward or in backward motion. [*Hint*: modify (11.79) so as to force the robot to orient itself as vector $\boldsymbol{e}_p$ or vector $-\boldsymbol{e}_p$, depending on which choice is the most convenient.]

11.17. Prove formulae (11.85) for the exact odometric localization of a unicycle under velocity inputs that are constant in the sampling interval.

11.18. Implement in a computer program the unicycle posture regulator based on polar coordinates, with the state feedback computed through the Runge–Kutta odometric localization method.

12

Motion Planning

The trajectory planning methods presented in Chaps. 4 and 11, respectively for manipulators and mobile robots, operate under the simplifying assumption that the workspace is empty. In the presence of obstacles, it is necessary to plan motions that enable the robot to execute the assigned task without colliding with them. This problem, referred to as *motion planning*, is the subject of this chapter. After defining a canonical version of the problem, the concept of *configuration space* is introduced in order to achieve an efficient formulation. A selection of representative planning techniques is then presented. The method based on the notion of *retraction* characterizes the connectivity of the free configuration space using a *roadmap*, i.e., a set of collision-free paths, while the *cell decomposition* method identifies a network of *channels* with the same property. The *PRM* and *bidirectional RRT* techniques are *probabilistic* in nature and rely on the randomized sampling of the configuration space and the memorization of those samples that do not cause a collision between the robot and the obstacles. The *artificial potential* method is also described as a heuristic approach particularly suited to *on-line* planning problems, where the geometry of the workspace obstacles is unknown in advance. The chapter ends with a discussion of the application of the presented planning methods to the *robot manipulator* case.

12.1 The Canonical Problem

Robotic systems are expected to perform tasks in a workspace that is often populated by physical objects, which represent an obstacle to their motion. For example, a manipulator working in a robotized cell must avoid collision with its static structures, as well as with other moving objects that may access it, such as other manipulators. Similarly, a mobile robot carrying baggage in an airport has to navigate among obstacles that may be fixed (fittings, conveyor belts, construction elements) or mobile (passengers, workers). Planning a motion amounts then to deciding which path the robot must follow in order to execute a transfer task from an initial to a final posture without colliding with the obstacles. Clearly, one would like to endow the robot with the capability of *autonomously* planning its motion, starting from a high-level description of the task provided by the user and a geometric characterization of the workspace, either made available entirely in advance (*off-line* planning) or gathered by the robot itself during the motion by means of on-board sensors (*on-line* planning).

However, developing automatic methods for motion planning is a very difficult endeavour. In fact, the spatial reasoning that humans instinctively use to move safely among obstacles has proven hard to replicate and codify in an algorithm that can be executed by a robot. To this date, motion planning is still an active topic of research, with contributions coming from different areas such as algorithm theory, computational geometry and automatic control.

To address the study of methods for motion planning, it is convenient to introduce a version of the problem that highlights its fundamental issues. The *canonical problem* of motion planning is then formulated as follows.

Consider a robot $\mathcal{B}$, which may consist of a single rigid body (mobile robot) or of a kinematic chain whose base is either fixed (standard manipulator) or mobile (mobile robot with trailers or mobile manipulator). The robot moves in a Euclidean space $\mathcal{W} = \mathbb{R}^N$, with $N = 2$ or 3, called *workspace*. Let $\mathcal{O}_1, \ldots, \mathcal{O}_p$ be the *obstacles*, i.e., fixed rigid objects in $\mathcal{W}$. It is assumed that both the geometry of $\mathcal{B}$, $\mathcal{O}_1, \ldots, \mathcal{O}_p$ and the pose of $\mathcal{O}_1, \ldots, \mathcal{O}_p$ in $\mathcal{W}$ are known. Moreover, it is supposed that $\mathcal{B}$ is *free-flying*, that is, the robot is not subject to any kinematic constraint. The motion planning problem is the following: given an initial and a final posture of $\mathcal{B}$ in $\mathcal{W}$, find if exists a *path*, i.e., a continuous sequence of postures, that drives the robot between the two postures while avoiding collisions (including contacts) between $\mathcal{B}$ and the obstacles $\mathcal{O}_1, \ldots, \mathcal{O}_p$; report a failure if such a path does not exist.

In the particular case in which the robot is a single body moving in $\mathbb{R}^2$, the canonical motion planning problem is also known as the *piano movers' problem*, as it captures the difficulties faced by movers when manoeuvring a piano (without lifting it) among obstacles. The *generalized movers' problem* is the canonical problem for a single-body robot moving in $\mathbb{R}^3$.

Clearly, some of the hypotheses of the canonical problem may not be satisfied in applications. For example, the assumption that the robot is the only object in motion in the workspace rules out the relevant case of moving obstacles (e.g., other robots). Advance knowledge of obstacle geometry and placement is another strong assumption: especially in *unstructured* environments, which are not purposely designed to accommodate robots, the robot itself is typically in charge of detecting obstacles by means of its sensors, and the planning problem must therefore be solved on-line during the motion. Moreover, as shown in Chap. 11, the free-flying robot hypothesis does not hold in nonholonomic mechanical systems, which cannot move along arbitrary paths in the workspace. Finally, manipulation and assembly problems are excluded from the canonical formulation since they invariably involve contacts between rigid bodies. As a matter of fact, all the above assumptions are introduced in order to reduce motion planning to the purely geometrical — but still quite difficult — problem of generating a collision-free path. However, many methods that successfully solve this simplified version of the problem lend themselves to an extension to more general versions.

The notion of configuration space is essential to obtain a more convenient formulation of the canonical motion planning problem, as well as to envisage approaches to its solution.

12.2 Configuration Space

A very effective scheme for motion planning is obtained by representing the robot as a mobile point in an appropriate space, where the images of the workspace obstacles are also reported. To this end, it is natural to refer to the generalized coordinates of the mechanical system, whose value identifies the *configuration* of the robot (see Sect. B.4). This associates to each posture of the latter a point in the *configuration space* $\mathcal{C}$, i.e., the set of all the configurations that the robot can assume.

Generalized coordinates of robots are essentially of two types. Cartesian coordinates are used to describe the position of selected points on the links of the kinematic chain and take value in Euclidean spaces. Angular coordinates are used to represent the orientations of the bodies; independently from the adopted representation (rotation matrices, Euler angles, quaternions), they take values in $SO(m)$ ($m = 2, 3$), the *special orthonormal group* of real $(m \times m)$ matrices with orthonormal columns and determinant equal to 1 (see Sect. 2.2). It is well known that a minimal parameterization of $SO(m)$ requires $m(m-1)/2$ parameters. The configuration space of a robot is then obtained in general as a Cartesian product of these spaces.

Some examples of configuration spaces are presented below:

- The configuration of a polygonal mobile robot in $\mathcal{W} = \mathbb{R}^2$ is described by the position of a representative point on the body (e.g., a vertex) and by the orientation of the polygon, both expressed with respect to a fixed reference frame. The configuration space $\mathcal{C}$ is then $\mathbb{R}^2 \times SO(2)$, whose dimension is $n = 3$.
- For a polyhedral mobile robot in $\mathcal{W} = \mathbb{R}^3$, the configuration space $\mathcal{C}$ is $\mathbb{R}^3 \times SO(3)$, whose dimension is $n = 6$.
- For a fixed-base planar manipulator with n revolute joints, the configuration space is a subset of $(\mathbb{R}^2 \times SO(2))^n$. The dimension of $\mathcal{C}$ equals the dimension of $(\mathbb{R}^2 \times SO(2))^n$ minus the number of constraints due to the presence of the joints, i.e., $3n-2n = n$. In fact, in a planar kinematic chain, each joint imposes two holonomic constraints on the following body.
- For a fixed-base spatial manipulator with n revolute joints, the configuration space is a subset of $(\mathbb{R}^3 \times SO(3))^n$. Since in this case each joint imposes five constraints on the following body, the dimension of $\mathcal{C}$ is $6n - 5n = n$.

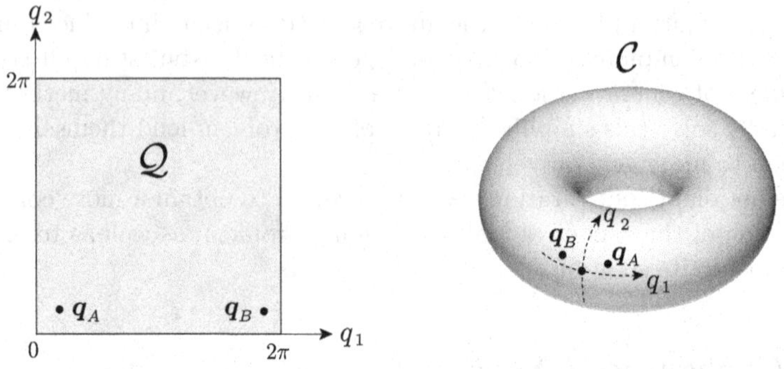

Fig. 12.1. The configuration space of a 2R manipulator; *left*: a locally valid representation as a subset of $\mathbb{R}^2$, *right*: a topologically correct representation as a two-dimensional torus

- For a unicycle-like vehicle with a trailer in $\mathbb{R}^2$, the configuration space is a subset of $(\mathbb{R}^2 \times SO(2)) \times (\mathbb{R}^2 \times SO(2))$. If the trailer is connected to the unicycle by a revolute joint, the configuration of the robot can be described by the position and orientation of the unicycle and the orientation of the trailer. The dimension of $\mathcal{C}$ is therefore $n = 4$.

If n is the dimension of $\mathcal{C}$, a configuration in $\mathcal{C}$ can be described by a vector $\boldsymbol{q} \in \mathbb{R}^n$. However, this description is only valid locally: the geometric structure of the configuration space $\mathcal{C}$ is in general more complex than that of a Euclidean space, as shown in the following example.

Example 12.1

Consider the planar manipulator with two revolute joints (2R manipulator) of Fig. 2.14. The configuration space has dimension 2, and may be locally represented by $\mathbb{R}^2$, or more precisely by its subset

$$\mathcal{Q} = \{\boldsymbol{q} = (q_1, q_2) : q_1 \in [0, 2\pi), q_2 \in [0, 2\pi)\}.$$

This guarantees that the representation is injective, i.e., that a single value of $\boldsymbol{q}$ exists for each manipulator posture. However, this representation is not topologically correct: for example, the configurations denoted as $\boldsymbol{q}_A$ and $\boldsymbol{q}_B$ in Fig. 12.1, left, which correspond to manipulator postures that are 'close' in the workspace $\mathcal{W}$, appear to be 'far' in $\mathcal{Q}$. To take this into account, one should 'fold' the square $\mathcal{Q}$ onto itself (so as to make opposite sides meet) in sequence along its two axes. This procedure generates a ring, properly called *torus*, which can be visualized as a two-dimensional surface immersed in $\mathbb{R}^3$ (Fig. 12.1, right). The correct expression of this space is $SO(2) \times SO(2)$.

When the configuration of a robot (either articulated or mobile) includes angular generalized coordinates, its configuration space is properly described as an n-dimensional *manifold*, i.e., a space in which a neighbourhood of a point can be put in correspondence with $\mathbb{R}^n$ through a continuous bijective function whose inverse is also continuous (a *homeomorphism*).

12.2.1 Distance

Having discussed the nature of the robot configuration space, it is useful to define a distance function in $\mathcal{C}$. In fact, the planning methods that will be discussed in the following make use of this notion.

Given a configuration $\boldsymbol{q}$, let $\mathcal{B}(\boldsymbol{q})$ be the subset of the workspace $\mathcal{W}$ occupied by the robot $\mathcal{B}$, and $p(\boldsymbol{q})$ be the position in $\mathcal{W}$ of a point p on $\mathcal{B}$. Intuition suggests that the distance between two configurations $\boldsymbol{q}_1$ and $\boldsymbol{q}_2$ should go to zero when the two regions $\mathcal{B}(\boldsymbol{q}_1)$ and $\mathcal{B}(\boldsymbol{q}_2)$ tend to coincide. A definition that satisfies this property is

$$d_1(\boldsymbol{q}_1, \boldsymbol{q}_2) = \max_{p \in \mathcal{B}} \|p(\boldsymbol{q}_1) - p(\boldsymbol{q}_2)\|, \tag{12.1}$$

where $\|\cdot\|$ denotes the Euclidean distance in $\mathcal{W} = \mathbb{R}^N$. In other words, the distance between two configurations in $\mathcal{C}$ is the maximum displacement in $\mathcal{W}$ they induce on a point, as the point moves all over the robot.

However, the use of function d_1 is cumbersome, because it requires characterizing the volume occupied by the robot in the two configurations and the computation of the maximum distance in $\mathcal{W}$ between corresponding points. For algorithmic purposes, the simple Euclidean norm is often chosen as a configuration space distance:

$$d_2(\boldsymbol{q}_1, \boldsymbol{q}_2) = \|\boldsymbol{q}_1 - \boldsymbol{q}_2\|. \tag{12.2}$$

Nevertheless, one must keep in mind that this definition is appropriate only when $\mathcal{C}$ is a Euclidean space. Going back to Example 12.1, it is easy to realize that, unlike $d_1(\boldsymbol{q}_A, \boldsymbol{q}_B)$, the Euclidean norm $d_2(\boldsymbol{q}_A, \boldsymbol{q}_B)$ does not represent correctly the distance on the torus. A possible solution is to modify the definition of d_2 by suitably computing differences of angular generalized coordinates (see Problem 12.2).

12.2.2 Obstacles

In order to characterize paths that represent a solution to the canonical motion planning problem — those that avoid collisions between the robot and the workspace obstacles — it is necessary to build the 'images' of the obstacles in the configuration space of the robot.

In the following, it is assumed that the obstacles are closed (i.e., they contain their boundaries) but not necessarily limited subsets of $\mathcal{W}$. Given an obstacle $\mathcal{O}_i$ ($i = 1, \ldots, p$) in $\mathcal{W}$, its image in configuration space $\mathcal{C}$ is called $\mathcal{C}$-*obstacle* and is defined as

$$\mathcal{CO}_i = \{\boldsymbol{q} \in \mathcal{C} : \mathcal{B}(\boldsymbol{q}) \cap \mathcal{O}_i \neq \emptyset\}. \tag{12.3}$$

In other words, $\mathcal{CO}_i$ is the subset of configurations that cause a collision (including simple contacts) between the robot $\mathcal{B}$ and the obstacle $\mathcal{O}_i$ in the workspace. The union of all $\mathcal{C}$-obstacles

$$\mathcal{CO} = \bigcup_{i=1}^{p} \mathcal{CO}_i \tag{12.4}$$

defines the $\mathcal{C}$-*obstacle region*, while its complement

$$\mathcal{C}_{\text{free}} = \mathcal{C} - \mathcal{CO} = \{\boldsymbol{q} \in \mathcal{C} : \mathcal{B}(\boldsymbol{q}) \cap \left(\bigcup_{i=1}^{p} \mathcal{O}_i\right) = \emptyset\} \tag{12.5}$$

is the *free configuration space*, that is, the subset of robot configurations that do not cause collision with the obstacles. A path in configuration space is called *free* if it is entirely contained in $\mathcal{C}_{\text{free}}$.

Although $\mathcal{C}$ in itself is a connected space — given two arbitrary configuration there exists a path that joins them — the free configuration space $\mathcal{C}_{\text{free}}$ may not be connected as a consequence of occlusions due to $\mathcal{C}$-obstacles. Note also that the assumption of free-flying robot in the canonical problem means that the robot can follow any path in the free configuration space $\mathcal{C}_{\text{free}}$.

It is now possible to give a more compact formulation of the canonical motion planning problem. Assume that the initial and final posture of the robot $\mathcal{B}$ in $\mathcal{W}$ are mapped to the corresponding configurations in $\mathcal{C}$, respectively called *start* configuration $\boldsymbol{q}_s$ and *goal* configuration $\boldsymbol{q}_g$. Planning a collision-free motion for the robot means then generating a safe path between $\boldsymbol{q}_s$ and $\boldsymbol{q}_g$ if they belong to the same connected component of $\mathcal{C}_{\text{free}}$, and reporting a failure otherwise.

12.2.3 Examples of Obstacles

In the following, the $\mathcal{C}$-obstacle generation procedure is presented in some representative cases. For the sake of simplicity, it is assumed that obstacles in $\mathcal{W}$ are either polygonal or polyhedral.

Example 12.2

Consider the case of a point robot $\mathcal{B}$. In this case, the configuration of the robot is described by the coordinates of point $\mathcal{B}$ in the workspace $\mathcal{W} = \mathbb{R}^N$ and the configuration space $\mathcal{C}$ is a copy of $\mathcal{W}$. Similarly, the $\mathcal{C}$-obstacles are copies of the obstacles in $\mathcal{W}$.

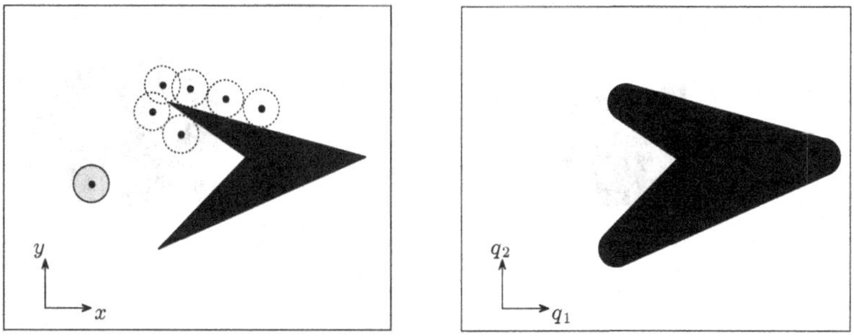

Fig. 12.2. $\mathcal{C}$-obstacles for a circular robot in $\mathbb{R}^2$; *left*: the robot $\mathcal{B}$, an obstacle $\mathcal{O}_i$ and the growing procedure for building $\mathcal{C}$-obstacles, *right*: the configuration space $\mathcal{C}$ and the $\mathcal{C}$-obstacle $\mathcal{CO}_i$

Example 12.3

If the robot is a sphere[1] in $\mathcal{W} = \mathbb{R}^N$, its configuration can be described by the Cartesian coordinates of a representative point, e.g., its centre — note that the orientation of the sphere is irrelevant for collision checking. Therefore, as in the previous example, the configuration space $\mathcal{C}$ is a copy of the workspace $\mathcal{W}$. However, the $\mathcal{C}$-obstacles are no longer simple copies of the obstacles in $\mathcal{W}$, and they must be built through a *growing* procedure. In particular, the boundary of the $\mathcal{C}$-obstacle $\mathcal{CO}_i$ is the locus of configurations that put the robot is in contact with the obstacle $\mathcal{O}_i$, and it can be obtained as the surface described by the representative point as the robot slides on the boundary of $\mathcal{O}_i$. As a consequence, to build $\mathcal{CO}_i$ it is sufficient to grow $\mathcal{O}_i$ isotropically by the radius of the robot. This procedure is shown in Fig. 12.2 for the case $N = 2$ (circular robot in $\mathbb{R}^2$); in this case, each $\mathcal{C}$-obstacle is a *generalized polygon*, i.e., a planar region whose boundary consists of line segments and/or circular arcs. If the representative point of the robot is different from the centre of the sphere, the growing procedure is not isotropic.

Example 12.4

Consider now the case of a polyhedral robot that is free to translate (with a fixed orientation) in $\mathbb{R}^N$. Its configuration can be described by the Cartesian coordinates of a representative point, for example a vertex of the polyhedron. Therefore, the configuration space $\mathcal{C}$ is again a copy of $\mathbb{R}^N$. Again, a growing procedure must be applied to the workspace obstacles to obtain their image in the configuration space. In particular, the boundary of the $\mathcal{C}$-obstacle $\mathcal{CO}_i$ is the surface described by the representative point of the robot when the robot $\mathcal{B}$ slides at a fixed orientation on the boundary of $\mathcal{O}_i$. Figure 12.3 shows this procedure for the case $N = 2$ (polygonal robot in $\mathbb{R}^2$). The resulting shape of $\mathcal{CO}_i$ depends on the position of the representative point on the robot, but in any case the $\mathcal{C}$-obstacle is itself a polyhedron. $\mathcal{CO}_i$ has in general a larger number of vertices than $\mathcal{O}_i$, and is a convex polyhedron provided that $\mathcal{B}$ and $\mathcal{O}_i$ are convex. Note also that, although the result of the growing

[1] For simplicity, the term *sphere* will be used in Euclidean spaces of arbitrary dimension n in place of *n–sphere*.

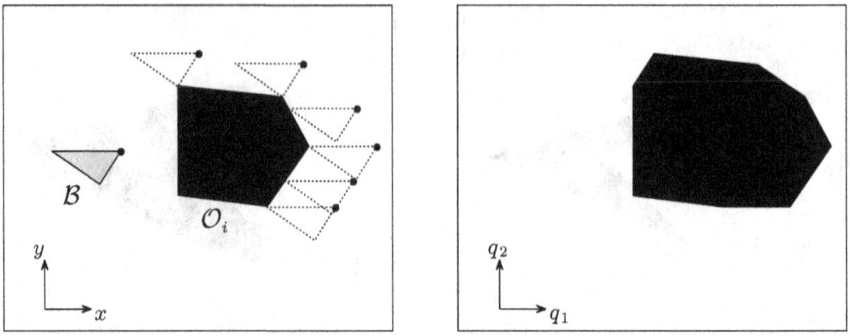

Fig. 12.3. $\mathcal{C}$-obstacles for a polygonal robot translating in $\mathrm{I\!R}^2$; *left*: the robot $\mathcal{B}$, an obstacle $\mathcal{O}_i$ and the growing procedure for building $\mathcal{C}$-obstacles, *right*: the configuration space $\mathcal{C}$ and the $\mathcal{C}$-obstacle $\mathcal{CO}_i$

procedure — and thus the shape of the $\mathcal{C}$-obstacles — depends on the choice of the representative point on the robot, all the obtained planning problems in configuration space are equivalent. In particular, the existence of a solution for any of them implied the existence of a solution for all the others. Moreover, to each path in configuration space that solves one of these problems corresponds a (different) free path in any of the others, to which the same motion of the robot in the workspace is associated.

Example 12.5

For a polyhedral robot that can translate and rotate in $\mathrm{I\!R}^N$, the dimension of the configuration space is increased with respect to the previous example, because it is also necessary to describe the orientation DOFs. For example, consider the case of a polygon that can translate and rotate in $\mathrm{I\!R}^2$. The configuration of the robot can be characterized by the Cartesian coordinates of a representative point (for example, a vertex of the polygon) and an angular coordinate θ representing the orientation of the polygon with respect to a fixed reference frame. The configuration space $\mathcal{C}$ is then $\mathrm{I\!R}^2 \times SO(2)$, which can be locally represented by $\mathrm{I\!R}^3$. To build the image in $\mathcal{C}$ of an obstacle $\mathcal{O}_i$, one should in principle repeat the procedure illustrated in Fig. 12.3 for each possible value of the robot orientation θ. The $\mathcal{C}$-obstacle $\mathcal{CO}_i$ is the volume generated by 'stacking' (in the direction of the θ axis) all the constant-orientation slices thus obtained.

Example 12.6

For a robot manipulator $\mathcal{B}$ made by rigid links $\mathcal{B}_1, \ldots, \mathcal{B}_n$ connected by joints, there exist in principle two kinds of $\mathcal{C}$-obstacles: those that represent the collision between a body $\mathcal{B}_i$ and an obstacle $\mathcal{O}_j$, and those accounting for *self-collisions*, i.e., interference between two links $\mathcal{B}_i$ and $\mathcal{B}_j$ of the kinematic chain. Even considering for simplicity only the first type, the procedure for building $\mathcal{C}$-obstacles is much more complicated than in the previous examples. In fact, to obtain the boundary of the $\mathcal{C}$-obstacle $\mathcal{CO}_i$, it is necessary to identify through appropriate inverse kinematics computations all the configurations that bring one or more links of the manipulator $\mathcal{B}$ in contact with $\mathcal{O}_i$. Figure 12.4 shows the result of the $\mathcal{C}$-obstacle building

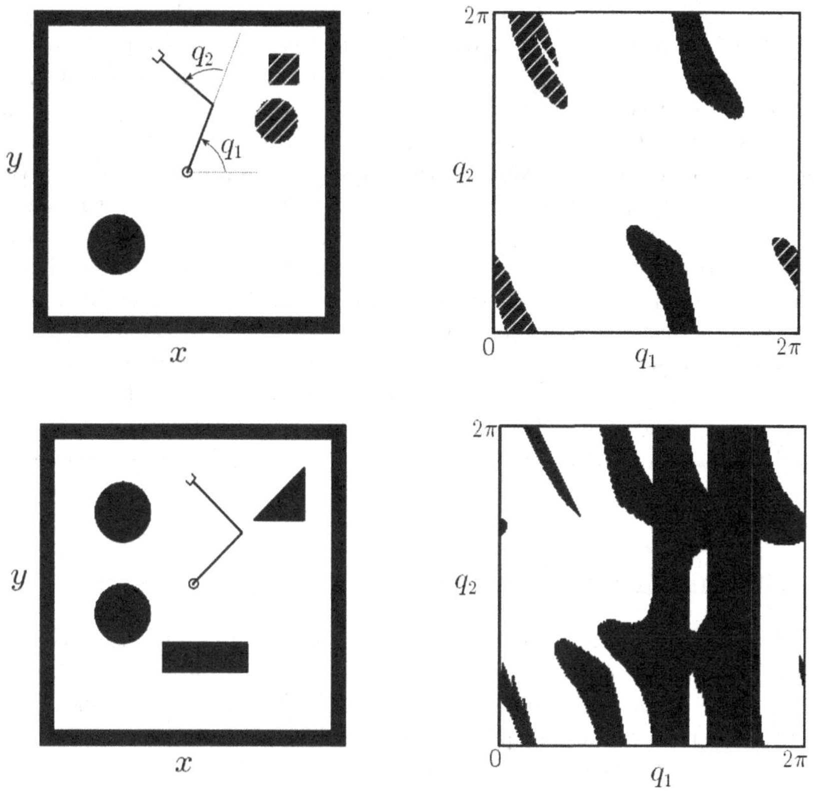

Fig. 12.4. $\mathcal{C}$-obstacles for a wire-frame 2R manipulator in two different cases; *left*: the robot and the obstacles in $\mathcal{W} = \mathbb{R}^2$, *right*: the configuration space $\mathcal{C}$ and the $\mathcal{C}$-obstacle region $\mathcal{CO}$

procedure for a wire-frame 2R manipulator in two different cases (self-collisions are not considered); note how, in spite of the simple shape of the obstacles in $\mathcal{W}$, the profile of the $\mathcal{C}$-obstacles is quite complicated. For simplicity, the configuration space has been represented as a subset (a square) of $\mathbb{R}^2$; however, to correctly visualize the $\mathcal{C}$-obstacles, one should keep in mind that the correct representation of $\mathcal{C}$ is a two-dimensional torus, so that the upper/lower and left/right sides of the square are actually coincident. Note also that in the first case (top of Fig. 12.4) the images of the two obstacles in the upper right corner of $\mathcal{W}$ merge in a single $\mathcal{C}$-obstacle, and the free configuration space $\mathcal{C}_{\text{free}}$ consists of a single connected component. In the second case (bottom of Fig. 12.4) $\mathcal{C}_{\text{free}}$ is instead partitioned into three disjoint connected components.

Whatever the nature of the robot, an *algebraic* model of the workspace obstacles is needed (e.g., derived from a CAD model of the workspace) to compute exactly the $\mathcal{C}$-obstacle region $\mathcal{CO}$. However, except for the most elementary cases, the procedures for generating $\mathcal{CO}$ are extremely complex. As a consequence, it is often convenient to resort to approximate representations of $\mathcal{CO}$. A simple (although computationally intensive) way to build such a representation is to extract configuration samples from $\mathcal{C}$ using a regular grid, compute the corresponding volume occupied by the robot via direct kinematics, and finally identify through *collision checking*[2] those samples that bring the robot to collide with obstacles. These samples can be considered as a discrete representation of $\mathcal{CO}$, whose accuracy can be improved at will by increasing the resolution of the grid in $\mathcal{C}$.

Some of the motion planning methods that will be presented in this chapter do not require an explicit computation of the $\mathcal{C}$-obstacle region. In particular, this is true for the probabilistic planners described in Sect. 12.5, and for the technique based on artificial potential fields and control points discussed in Sect. 12.7.

12.3 Planning via Retraction

The basic idea of motion planning via retraction is to represent the free configuration space by means of a *roadmap* $\mathcal{R} \subset \mathcal{C}_{\rm free}$, i.e., a network of paths that describe adequately the connectivity of $\mathcal{C}_{\rm free}$. The solution of a particular instance of a motion planning problem is then obtained by connecting (*retracting*) to the roadmap the start configuration $\boldsymbol{q}_s$ and the goal configuration $\boldsymbol{q}_g$, and finding a path on $\mathcal{R}$ between the two connection points. Depending on the type of roadmap, and on the retraction procedure, this general approach leads to different planning methods. One of these is described in the following, under the simplifying assumption that $\mathcal{C}_{\rm free}$ is a limited subset of $\mathcal{C} = {\rm I\!R}^2$ and is *polygonal*, i.e., its boundary is entirely made of line segments.[3] As the boundary of $\mathcal{C}_{\rm free}$ coincides with the boundary of $\mathcal{CO}$, this assumption implies that the $\mathcal{C}$-obstacle region is itself a polygonal subset of $\mathcal{C}$.

For each configuration $\boldsymbol{q}$ in $\mathcal{C}_{\rm free}$, let its *clearance* be defined as

$$\gamma(\boldsymbol{q}) = \min_{\boldsymbol{s}\in\partial\mathcal{C}_{\rm free}} \|\boldsymbol{q} - \boldsymbol{s}\|, \tag{12.6}$$

where $\partial\mathcal{C}_{\rm free}$ is the boundary of $\mathcal{C}_{\rm free}$. The clearance $\gamma(\boldsymbol{q})$ represents the minimum Euclidean distance between the configuration $\boldsymbol{q}$ and the $\mathcal{C}$-obstacle re-

[2] Many algorithms based on computational geometry techniques are available (in the literature, but also implemented in software packages) to test collision in ${\rm I\!R}^2$ o ${\rm I\!R}^3$. The most efficient, such as *I-Collide* and *V-Collide*, use a hierarchical representation of geometric models of bodies, and can speed up collision checking by re-using the results of previous checks in spatially similar situations.

[3] According to the definition, a polygonal subset is not necessarily connected, and may contain 'holes'.

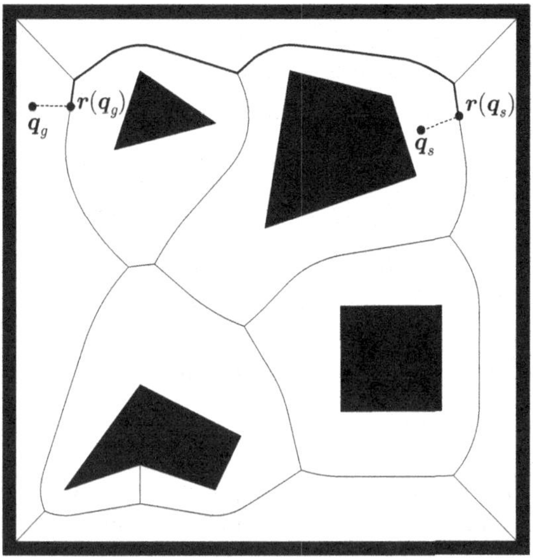

Fig. 12.5. An example of generalized Voronoi diagram and the solution of a particular instance of the planning problem, obtained by retracting q_s and q_g on the diagram. The solution path from q_s to q_g consists of the two *dashed* segments and the *thick* portion of the diagram joining them

gion. Moreover, consider the set of points on the boundary of $\mathcal{C}_{\text{free}}$ that are *neighbours* of q:

$$N(q) = \{s \in \partial\mathcal{C}_{\text{free}} : \|q - s\| = \gamma(q)\}, \tag{12.7}$$

i.e., the points on $\partial\mathcal{C}_{\text{free}}$ that determine the value of the clearance for q. With these definitions, the *generalized*[4] *Voronoi diagram* of $\mathcal{C}_{\text{free}}$ is the locus of its configurations having more than one neighbour:

$$\mathcal{V}(\mathcal{C}_{\text{free}}) = \{q \in \mathcal{C}_{\text{free}} : \text{card}(N(q)) > 1\}, \tag{12.8}$$

in which card$(\cdot)$ denotes the cardinality of a set. Figure 12.5 shows an example of generalized Voronoi diagram for a polygonal free configuration space; note how the connectivity of $\mathcal{C}_{\text{free}}$ is well captured by the diagram.

It is easy to show that $\mathcal{V}(\mathcal{C}_{\text{free}})$ is made of elementary arcs that are either rectilinear — each made of contiguous configurations whose clearance is due to the same pair of edges or vertices — or parabolic — each made of contiguous configurations whose clearance is determined by the same pair edge-vertex. Therefore, one can build an analytical expression of $\mathcal{V}(\mathcal{C}_{\text{free}})$ starting from the pair of features (side/side, side/vertex, vertex/vertex) that determine the

[4] A proper *Voronoi diagram* is obtained in the particular case in which the $\mathcal{C}$-obstacles are isolated points.

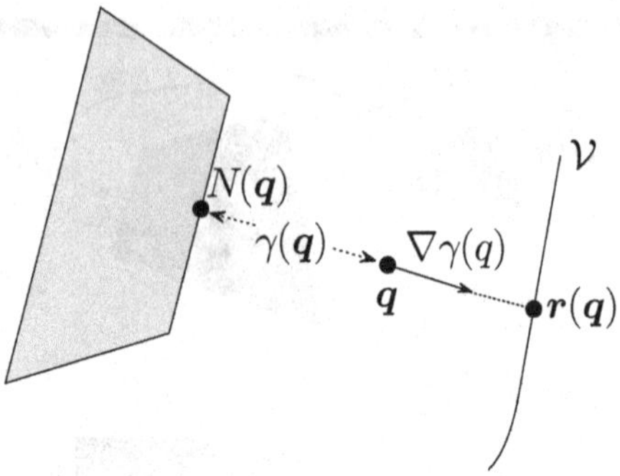

Fig. 12.6. The retraction procedure for connecting a generic configuration $\boldsymbol{q}$ in $\mathcal{C}_{\text{free}}$ to $\mathcal{V}(\mathcal{C}_{\text{free}})$

appearance of each arc. From an abstract point of view, $\mathcal{V}(\mathcal{C}_{\text{free}})$ can be considered as a *graph* having elementary arcs of the diagram as *arcs* and endpoints of arcs as *nodes*.

By construction, the generalized Voronoi diagram has the property of locally maximizing clearance, and is therefore a natural choice as a roadmap of $\mathcal{C}_{\text{free}}$ for planning motions characterized by a healthy safety margin with respect to the possibility of collisions. To use $\mathcal{V}(\mathcal{C}_{\text{free}})$ as a roadmap, a retraction procedure must be defined for connecting a generic configuration in $\mathcal{C}_{\text{free}}$ to the diagram. To this end, consider the geometric construction shown in Fig. 12.6. Since $\boldsymbol{q} \notin \mathcal{V}(\mathcal{C}_{\text{free}})$, it is $\text{card}\big(N(\boldsymbol{q})\big) = 1$, i.e., there exists a single point on the polygonal boundary of $\mathcal{CO}$ (either a vertex or a point on a side) that determines the value of the clearance $\gamma(\boldsymbol{q})$. The gradient $\nabla\gamma(\boldsymbol{q})$, which identifies the direction of steepest ascent for the clearance at the configuration $\boldsymbol{q}$, is directed as the half-line originating in $N(\boldsymbol{q})$ and passing through $\boldsymbol{q}$. The first intersection of this half-line with $\mathcal{V}(\mathcal{C}_{\text{free}})$ defines $\boldsymbol{r}(\boldsymbol{q})$, i.e., the connection point of $\boldsymbol{q}$ to the generalized Voronoi diagram. To guarantee that $\boldsymbol{r}(\cdot)$ is continuous, it is convenient to extend its domain of definition to all $\mathcal{C}_{\text{free}}$ by letting $\boldsymbol{r}(\boldsymbol{q}) = \boldsymbol{q}$ if $\boldsymbol{q} \in \mathcal{V}(\mathcal{C}_{\text{free}})$.

From a topological viewpoint, the function $\boldsymbol{r}(\cdot)$ defined above is actually an example of *retraction* of $\mathcal{C}_{\text{free}}$ on $\mathcal{V}(\mathcal{C}_{\text{free}})$, i.e., a continuous surjective map from $\mathcal{C}_{\text{free}}$ to $\mathcal{V}(\mathcal{C}_{\text{free}})$ such that its restriction to $\mathcal{V}(\mathcal{C}_{\text{free}})$ is the identity map. In view of its definition, $\boldsymbol{r}(\cdot)$ preserves the connectivity of $\mathcal{C}_{\text{free}}$, in the sense that $\boldsymbol{q}$ and $\boldsymbol{r}(\boldsymbol{q})$ — as well as the segment joining them — always lie in the same connected component of $\mathcal{C}_{\text{free}}$. This property is particularly important, because it is possible to show that, given a generic retraction ρ of $\mathcal{C}_{\text{free}}$ on a connectivity-preserving roadmap $\mathcal{R}$, there exists a free path between two configurations $\boldsymbol{q}_s$ and $\boldsymbol{q}_g$ if and only if there exists a path on $\mathcal{R}$ between $\rho(\boldsymbol{q}_s)$ and $\rho(\boldsymbol{q}_g)$. As a consequence, the problem of planning a path in $\mathcal{C}_{\text{free}}$ reduces to the problem of planning a path on its retraction $\mathcal{R} = \rho(\mathcal{C}_{\text{free}})$.

Given the start and goal configurations $\boldsymbol{q}_s$ and $\boldsymbol{q}_g$, the motion planning method via retraction goes through the following steps (see Fig. 12.5):

1. Build the generalized Voronoi diagram $\mathcal{V}(\mathcal{C}_{\rm free})$.
2. Compute the retractions $\boldsymbol{r}(\boldsymbol{q}_s)$ and $\boldsymbol{r}(\boldsymbol{q}_g)$ on $\mathcal{V}(\mathcal{C}_{\rm free})$.
3. Search $\mathcal{V}(\mathcal{C}_{\rm free})$ for a sequence of consecutive arcs such that $\boldsymbol{r}(\boldsymbol{q}_s)$ belongs to the first and $\boldsymbol{r}(\boldsymbol{q}_g)$ to the last.
4. If the search is successful, replace the first arc of the sequence with its subarc originating in $\boldsymbol{r}(\boldsymbol{q}_s)$ and the last arc of the sequence with its subarc terminating in $\boldsymbol{r}(\boldsymbol{q}_g)$, and provide as output the path consisting of the line segment joining $\boldsymbol{q}_s$ to $\boldsymbol{r}(\boldsymbol{q}_s)$, the modified arc sequence, and the line segment joining $\boldsymbol{q}_g$ to $\boldsymbol{r}(\boldsymbol{q}_g)$; otherwise, report a failure.

If simplicity of implementation is desired, the graph search[5] required at Step 3 can be performed using basic strategies such as breadth-first or depth-first search. On the other hand, if one wishes to identify the minimum-length path (among those which can be produced by the method) between $\boldsymbol{q}_s$ and $\boldsymbol{q}_g$, each arc must be labelled with a cost equal to its actual length. The minimum-cost path can then be computed with an *informed* algorithm — i.e., an algorithm using a heuristic estimate of the minimum cost path from a generic node to the goal, in this case the Euclidean distance — such as $A^\star$.

Whatever the adopted search strategy, the above motion planning method via retraction of $\mathcal{C}_{\rm free}$ on $\mathcal{V}(\mathcal{C}_{\rm free})$ is *complete*, i.e., it is guaranteed to find a solution if one exists, and to report a failure otherwise. Its time complexity is a function of the number v of vertices of the polygonal region $\mathcal{C}_{\rm free}$, and depends essentially on the construction of the generalized Voronoi diagram (Step 1), on the retraction of $\boldsymbol{q}_s$ and $\boldsymbol{q}_g$ (Step 2), and on the search on the diagram (Step 3). As for Step 1, the most efficient algorithms can build $\mathcal{V}(\mathcal{C}_{\rm free})$ in time $O(v \log v)$. The retraction procedure requires $O(v)$, mainly to compute $N(\boldsymbol{q}_s)$ and $N(\boldsymbol{q}_g)$. Finally, since it is possible to prove that $\mathcal{V}(\mathcal{C}_{\rm free})$ has $O(v)$ arcs, the complexity of breadth-first or depth-first search would be $O(v)$, whereas $A^\star$ would require $O(v \log v)$. Altogether, the time complexity of the motion planning method via retraction is $O(v \log v)$.

It should be noted that, once the generalized Voronoi diagram of $\mathcal{C}_{\rm free}$ has been computed, it can be used again to solve quickly other instances (*queries*) of the same motion planning problem, i.e., to generate collision-free paths between different start and goal configurations in the same $\mathcal{C}_{\rm free}$. For example, this is useful when a robot must repeatedly move between different postures in the same static workspace. The motion planning method based on retraction can then be considered *multiple-query*. It is also possible to extend the method to the case in which the $\mathcal{C}$-obstacles are generalized polygons.

[5] See Appendix E for a quick survey on graph search strategies and algorithm complexity.

12.4 Planning via Cell Decomposition

Assume that the free configuration space $\mathcal{C}_{\text{free}}$ can be decomposed in simply-shaped regions, called *cells*, with the following basic characteristics:

- Given two configurations belonging to the same cell, it is 'easy' to compute a collision-free path that joins them.
- Given two adjacent cells — i.e., two cells having in common a portion of their boundaries of non-zero measure — it is 'easy' to generate a collision-free path going from one cell to the other.

Starting from one such cell decomposition of $\mathcal{C}_{\text{free}}$, it is easy to build the associated *connectivity graph*. The nodes of this graph represent cells, while an arc between two nodes indicates that the two corresponding cells are adjacent. By searching the connectivity graph for a path from the cell containing the start configuration $\boldsymbol{q}_s$ to the cell containing the goal configuration $\boldsymbol{q}_g$, one obtains (if it exists) a sequence of adjacent cells, called *channel*, from which it is possible — in view of the above mentioned characteristics of cells — to extract a path that joins $\boldsymbol{q}_s$ to $\boldsymbol{q}_g$ and is entirely contained in $\mathcal{C}_{\text{free}}$.

The general approach so far outlined generates different motion planning methods, essentially depending on the type of cells used for the decomposition. In the following, two algorithms are described, respectively based on exact and approximate cell decomposition of $\mathcal{C}_{\text{free}}$. As before, it is assumed that $\mathcal{C}_{\text{free}}$ is a limited polygonal subset of $\mathcal{C} = \mathbb{R}^2$.

12.4.1 Exact Decomposition

When an exact decomposition is used, the free configuration space is partitioned in a collection of cells whose union gives exactly $\mathcal{C}_{\text{free}}$. A typical choice for cells are convex polygons. In fact, convexity guarantees that the line segments joining two configurations belonging to the same cell lies entirely in the cell itself, and therefore in $\mathcal{C}_{\text{free}}$. Moreover, it is easy to travel safely between two adjacent cells by passing through the midpoint of the segment that constitutes their common boundary. A simple way to decompose $\mathcal{C}_{\text{free}}$ in a collection of convex polygons is to use the *sweep line* algorithm, which turns out to be useful in a number of computational geometry problems. In the present setting, the application of this algorithm proceeds as follows.

Choose a line that is not parallel to any side of the boundary of $\mathcal{C}_{\text{free}}$, and let it translate ('sweep') all over $\mathcal{C}_{\text{free}}$. Whenever the line passes through one of the vertices of $\mathcal{C}_{\text{free}}$, consider the two segments (*extensions*) that originate from the vertex, lie on the line and point in opposite directions, terminating at the first intersection with $\partial\mathcal{C}_{\text{free}}$. Every extension that lies in $\mathcal{C}_{\text{free}}$ (except for its endpoints) is part of the boundary of a cell; the rest of the boundary is made of (part of) sides of $\partial\mathcal{C}_{\text{free}}$ and possibly other extensions. This procedure is illustrated in Fig. 12.7, where a vertical sweep line has been used; note that some of the vertices of $\mathcal{C}_{\text{free}}$ contribute to the decomposition with a single or no

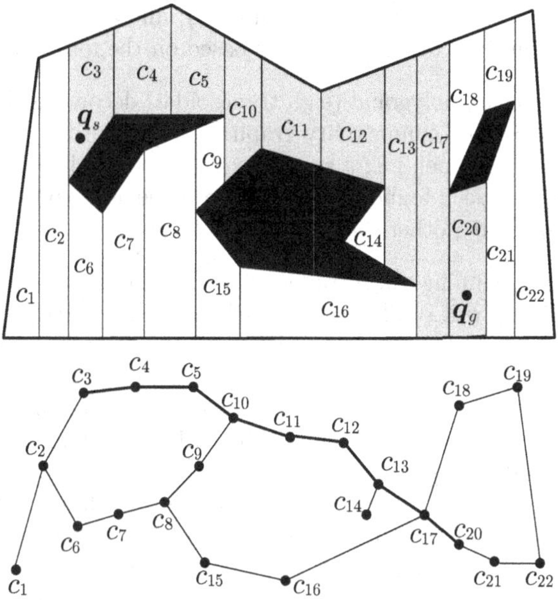

Fig. 12.7. An example of trapezoidal decomposition via the sweep line algorithm (*above*) and the associated connectivity graph (*below*). Also shown is the solution of a particular planning problem ($c_s = c_3$ and $c_g = c_{20}$), both as a channel in $\mathcal{C}_{\text{free}}$ and as a path on the graph

extension at all. The result is a special case of *convex polygonal decomposition*, that is called *trapezoidal* because its cells are trapezoids — triangular cells, if present, are regarded as degenerate trapezoids having one side of zero length.

After the decomposition of $\mathcal{C}_{\text{free}}$ has been computed, it is possible to build the associated *connectivity graph* C. This is the graph whose nodes are the cells of the decomposition, while an arc exists between two nodes if the corresponding cells are adjacent, i.e., the intersection of their boundary is a line segment of non-zero length; therefore, cells with side-vertex or vertex-vertex contacts are not adjacent. At this point, it is necessary to identify the cells c_s and c_g in the decomposition that respectively contain $\boldsymbol{q}_s$ and $\boldsymbol{q}_g$, the start and goal configurations for the considered planning problem. A graph search algorithm can then be used to find a *channel* from c_s to c_g, i.e., a path on C that joins the two corresponding nodes (see Fig. 12.7). From the channel, which is a sequence of adjacent cells, one must extract a path in $\mathcal{C}_{\text{free}}$ going from $\boldsymbol{q}_s$ to $\boldsymbol{q}_g$. Since the interior of the channel is contained in $\mathcal{C}_{\text{free}}$ and the cells are convex polygons, such extraction is straightforward. For example, one may identify the midpoints of the segments that represent the common boundaries between consecutive cells of the channel, and connect them through a broken line starting in $\boldsymbol{q}_s$ and ending in $\boldsymbol{q}_g$.

Wrapping up, given the two configurations $\boldsymbol{q}_s$ and $\boldsymbol{q}_g$, the motion planning algorithm via exact cell decomposition is based on the following steps:

1. Compute a convex polygonal (e.g., trapezoidal) decomposition of $\mathcal{C}_{\text{free}}$.
2. Build the associated connectivity graph C.
3. Search C for a channel, i.e., a sequence of adjacent cells from c_s to c_g.
4. If a channel has been found, extract and provide as output a collision-free path from $\boldsymbol{q}_s$ to $\boldsymbol{q}_g$; otherwise, report a failure.

As in motion planning via retraction, using a non-informed graph search algorithm in Step 3 will result in a channel that is not optimal, in the sense that all paths from $\boldsymbol{q}_s$ to $\boldsymbol{q}_g$ that can be extracted from the channel may be longer than necessary. To compute efficient paths, the use of $A^\star$ is advisable as a search algorithm. To this end, one should build a modified connectivity graph C' having as nodes $\boldsymbol{q}_s$, $\boldsymbol{q}_g$ and all the midpoints of adjacency segments between cells, and arcs joining nodes belonging to the same cell (note that nodes on adjacency segments belong to two cells). Each arc is then labelled with a cost equal to the distance between the nodes connected by the arc. If the heuristic function is chosen as the distance between the current node and $\boldsymbol{q}_g$, the use of $A^\star$ will produce the shortest path in C', if a solution exists.

The motion planning method based on exact cell decomposition is complete. As for its time complexity, it depends mainly on the cell decomposition and on the connectivity graph search. Using the sweep line algorithm, the decomposition procedure (including the generation of the connectivity graph) has complexity $O(v \log v)$, where v is the number of vertices of $\mathcal{C}_{\text{free}}$. Moreover, it can be shown that the connectivity graph C has $O(v)$ arcs. Hence, regardless of the adopted search strategy, the motion planning method based on exact cell decomposition has complexity $O(v \log v)$.

Note the following facts:

- Any planner based on exact cell decomposition can be considered multiple-query. In fact, once computed, the connectivity graph associated with the decomposition can be used to solve different instances of the same motion planning problem.
- The connectivity graph represents a network of channels, each implicitly containing an *infinity* of paths that traverse the channel and are topologically equivalent, i.e., differ only for a continuous deformation. Therefore, cell decomposition provides as output a structure that is more flexible than the roadmap used by retraction-based methods. This may be useful to plan paths in the channel that are also admissible with respect to possible kinematic constraints, as well as to avoid unexpected obstacles during the actual execution of the motion.
- The solution paths produced by the planning method based on exact cell decomposition are broken lines. It is however possible to smooth the path using *curve fitting* techniques. In practice, one selects a sufficient number

of intermediate points (*via points*) on the path, among which it is necessary to include $\boldsymbol{q}_s$ and $\boldsymbol{q}_g$, and then interpolates them using functions with an appropriate level of differentiability (e.g., polynomial functions of sufficiently high degree).
- The above method can be extended to the case in which $\mathcal{C}_{\text{free}}$ is a limited polyhedral subset of $\mathcal{C} = \mathbb{R}^3$. In particular, the decomposition of $\mathcal{C}_{\text{free}}$ can obtained through the *sweep plane* algorithm, which produces polyhedral cells. The common boundary between adjacent cells consists of trapezoid of non-zero area. This boundary can be safely crossed, e.g., at the barycentre of the trapezoid.

Finally, it should be mentioned that there exist in the literature methods based on exact cell decomposition that can solve essentially any motion planning problem, regardless of the dimension of $\mathcal{C}$ and of the geometry of $\mathcal{C}_{\text{free}}$, which can also be non-polyhedral. However, the complexity of these planners is prohibitive, being exponential in the dimension of $\mathcal{C}$, and their importance is therefore mainly theoretical.

12.4.2 Approximate Decomposition

In approximate decompositions of $\mathcal{C}_{\text{free}}$, disjoint cells of predefined shape are used; for example, when $\mathcal{C} = \mathbb{R}^2$ one may choose square or rectangular cells. In general, the union of all cells will represent an approximation by defect of $\mathcal{C}_{\text{free}}$. To achieve a reasonable trade-off between the accuracy of the approximation and the efficiency of the decomposition procedure, a recursive algorithm is typically used, which starts with a coarse grid whose resolution is locally increased to adapt better to the geometry of $\mathcal{C}_{\text{free}}$. As in motion planning methods based on exact decomposition, the connectivity graph associated with the obtained approximate decomposition is searched for a channel, from which a solution path can be extracted.

In the following, a motion planning method based on approximate decomposition is described for the case in which $\mathcal{C}_{\text{free}}$ is a limited polygonal subset of $\mathcal{C} = \mathbb{R}^2$. Without loss of generality, it will be assumed that the 'external' boundary of $\mathcal{C}_{\text{free}}$ is a square, and that square cells are therefore used. The decomposition algorithm (Fig. 12.8) starts by dividing initially $\mathcal{C}$ into four cells, that are classified according to the categories below:

- *free* cells, whose interior has no intersection with the $\mathcal{C}$-obstacle region;
- *occupied* cells, entirely contained in the $\mathcal{C}$-obstacle region;
- *mixed* cells, that are neither free nor occupied.

At this point, one builds the connectivity graph C associated with the current level of decomposition: this is the graph having free and mixed cells as nodes, and arcs that join nodes representing adjacent cells. Once the nodes corresponding to the cells that contain $\boldsymbol{q}_s$ and $\boldsymbol{q}_g$ have been identified, C is searched for a path between them, e.g., using the $A^\star$ algorithm. If such a path

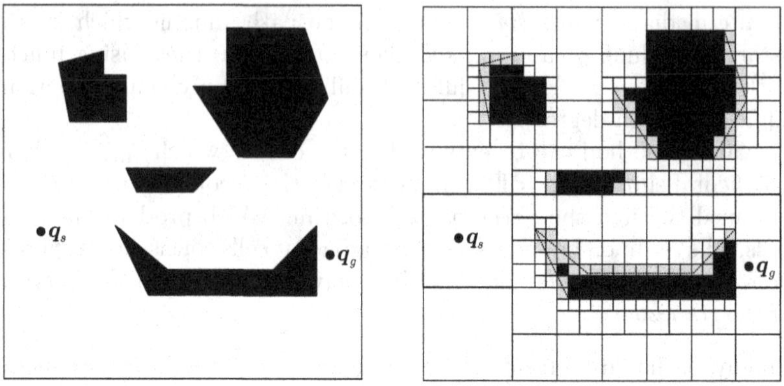

Fig. 12.8. An example of motion planning via approximate cell decomposition; *left*: the assigned problem, *right*: the solution as a free channel (*thick line*)

does not exist, a failure is reported. If the path exists, it consists of a sequence of cells that may be either all free (*free channel*) or not (*mixed channel*). In the first case, a solution to the motion planning problem has been found; in particular, a configuration space path can be easily extracted from the free channel as in the method based on exact cell decomposition. Instead, if the channel contains mixed cells, each of them is further divided into fourcells, which are then classified as free, occupied or mixed. The algorithm proceeds by iterating these steps, until a free channel going from $\boldsymbol{q}_s$ to $\boldsymbol{q}_g$ has been found or a minimum admissible size has been reached for the cells. Figure 12.8 shows an example of application of this technique. Note that, at the resolution level where the solution has been found, free and occupied cells represent an approximation by defect of $\mathcal{C}_{\text{free}}$ and $\mathcal{CO}$, respectively. The missing part of $\mathcal{C}$ is occupied by mixed cells (in gray).

At each iteration, the search algorithm on the connectivity graph can find a free channel going from $\boldsymbol{q}_s$ to $\boldsymbol{q}_g$ only if such a channel exists on the approximation of $\mathcal{C}_{\text{free}}$ (i.e., on the free cells) at the current level of decomposition. This motion planning method is therefore *resolution complete*, in the sense that a sufficient reduction of the minimum admissible size for cells guarantees that a solution is found whenever one exists.

A comparison between Figs. 12.7 and 12.8 clearly shows that, unlike what happens in exact decomposition, the boundary of a cell in an approximate decomposition does not correspond in general to a change in the spatial constraints imposed by obstacles. One of the consequences of this fact is that the implementation of a motion planning method based on approximate decomposition is remarkably simpler, as it only requires a recursive division of cells followed by a collision check between the cells and the $\mathcal{C}$-obstacle region. In particular, the first can be realized using a data structure called *quadtree*. This is a tree in which any internal node (i.e., a node that is not a leaf) has exactly four child nodes. In the cell decomposition case, the tree is rooted at

$\mathcal{C}$, i.e., the whole configuration space. Lower level nodes represent cells that are free, occupied or mixed; only the latter have children.

Another difference with respect to the method based on exact decomposition is that an approximate decomposition is intrinsically associated with a particular instance of a planning problem, as the decomposition procedure itself is guided by the search for a free channel between start and goal.

The planning method based on approximate decomposition is conceptually applicable in configuration spaces of arbitrary dimension. For example, in $\mathbb{R}^3$ it is possible to use an *octree*, a tree in which any internal node has eight children. In $\mathbb{R}^n$, the corresponding data structure is a 2^n*-tree*. Since the maximum number of leaves of a 2^n-tree is 2^{np}, where p is the depth (number of levels) of the tree, the complexity of approximate cell decomposition — and thus of the associated motion planning method — is exponential in the dimension of $\mathcal{C}$ and in the maximal resolution of the decomposition. As a consequence, this technique is effective in practice only in configuration spaces of low dimension (typically, not larger than 4).

12.5 Probabilistic Planning

Probabilistic planners represent a class of methods of remarkable efficiency, especially in problems involving high-dimensional configuration spaces. They belong to the general family of *sampling-based* methods, whose basic idea consists of determining a finite set of collision-free configurations that adequately represent the connectivity of $\mathcal{C}_{\text{free}}$, and using these configurations to build a roadmap that can be employed for solving motion planning problems. This is realized by choosing at each iteration a sample configuration and checking if it entails a collision between the robot and the workspace obstacles. If the answer is affirmative, the sample is discarded. A configuration that does not cause a collision is instead added to the current roadmap and connected if possible to other already stored configurations.

The above strategy is quite general and may lead to different planning methods depending on the specific design choices, and mainly on the criterion for selecting the samples in $\mathcal{C}$ to be checked for collision. One may proceed in a *deterministic* fashion, choosing the samples by means of a regular grid that is applied to $\mathcal{C}$. However, it is preferable to use a *randomized* approach, in which the sample configurations are chosen according to some probability distribution. In the following, two planners of this type are described.

12.5.1 PRM Method

The basic iteration of the PRM (*Probabilistic Roadmap*) method begins by generating a random sample $\boldsymbol{q}_{\text{rand}}$ of the configuration space using a uniform probability distribution in $\mathcal{C}$. Then, $\boldsymbol{q}_{\text{rand}}$ is tested for collision, by using kinematic and geometric relationships to compute the corresponding posture of

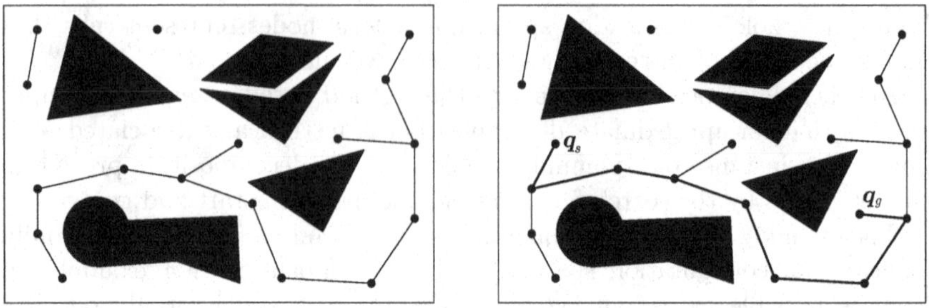

Fig. 12.9. A PRM in a two-dimensional configuration space (*left*) and its use for solving a particular planning problem (*right*)

the robot and invoking an algorithm that can detect collisions (including contacts) between the latter and the obstacles. If $\boldsymbol{q}_{\mathrm{rand}}$ does not cause collisions, it is added to the roadmap and connected (if possible) through free *local paths* to sufficiently 'near' configurations already in the roadmap. Usually, 'nearness' is defined on the basis of the Euclidean distance in $\mathcal{C}$, but it is possible to use different distance notions; for example, as mentioned in Sect. 12.2.1, one may use a configuration space distance notion induced by a distance in the workspace. The generation of a free local path between $\boldsymbol{q}_{\mathrm{rand}}$ and a near configuration $\boldsymbol{q}_{\mathrm{near}}$ is delegated to a procedure known as *local planner*. A common choice is to throw a rectilinear path in $\mathcal{C}$ between $\boldsymbol{q}_{\mathrm{rand}}$ and $\boldsymbol{q}_{\mathrm{near}}$ and test it for collision, for example by sampling the segment with sufficient resolution and checking the single samples for collision. If the local path causes a collision, it is discarded and no direct connection between $\boldsymbol{q}_{\mathrm{rand}}$ and $\boldsymbol{q}_{\mathrm{near}}$ appears in the roadmap.

The PRM incremental generation procedure stops when either a maximum number of iterations has been reached, or the number of connected components in the roadmap becomes smaller than a given threshold. At this point, one verifies whether it is possible to solve the assigned motion planning problem by connecting $\boldsymbol{q}_s$ and $\boldsymbol{q}_g$ to the *same* connected component of the PRM by free local paths. Figure 12.9 shows an example of PRM and the solution to a particular problem. Note the presence of multiple connected components of the PRM, one of which consists of a single configuration.

If a solution cannot be found, the PRM can be improved by performing more iterations of the basic procedure, or using special strategies aimed at reducing the number of its connected components. For example, a possible technique consists of trying to connect configurations that are close but belong to different components via more general (e.g., not rectilinear) local paths.

The main advantage of the PRM method is its remarkable speed in finding a solution to motion planning problems, provided that the roadmap has been sufficiently developed. In this respect, it should be noted that new instances of the same problem induce a potential enhancement of the PRM, which improves with usage both in terms of connectivity and of time efficiency. The

PRM method is therefore intrinsically multiple-query. In high-dimensional configuration spaces, the time needed by this method to compute a solution can be several orders of magnitude smaller than with the previously presented techniques. Another aspect of the method that is worth mentioning is the simplicity of implementation. In particular, note that the generation of $\mathcal{C}$-obstacles is completely eliminated.

The downside of the PRM method is that it is only *probabilistically complete*, i.e., the probability of finding a solution to the planning problem when one exists tends to 1 as the execution time tends to infinity. This means that, if no solution exists, the algorithm will run indefinitely. In practice, a maximum number of iterations is enforced so as to guarantee its termination.

A situation that is critical for the PRM method is the presence of *narrow passages* in $\mathcal{C}_{\text{free}}$, such as the one shown in the upper-right quadrant of the scene in Fig. 12.9. In fact, using a uniform distribution for generating $\boldsymbol{q}_{\text{rand}}$, the probability of placing a sample in a certain region of $\mathcal{C}_{\text{free}}$ is proportional to its volume. As a consequence, depending on its size, it may be very unlikely that a path crossing a narrow passage appears in the PRM within a reasonable time. To alleviate this problem, the method can be modified by using non-uniform probability distributions. For example, there exist strategies for generating $\boldsymbol{q}_{\text{rand}}$ that are biased towards those regions of $\mathcal{C}$ that contain fewer samples, and therefore are more likely to contain close obstacles and the associated narrow passages.

12.5.2 Bidirectional RRT Method

Single-query probabilistic methods are aimed at quickly solving a particular instance of a motion planning problem. Unlike multiple-query planners such as PRM, these techniques do not rely on the generation of a roadmap that represents exhaustively the connectivity of the free configuration space; in fact, they tend to explore only a subset of $\mathcal{C}_{\text{free}}$ that is relevant for solving the problem at hand. This results in a further reduction of the time needed to compute a solution.

An example of single-query probabilistic planner is the *bidirectional RRT* method, which makes use of a data structure called RRT (*Rapidly-exploring Random Tree*). The incremental expansion of an RRT, denoted by T in the following, relies on a simple randomized procedure to be repeated at each iteration (see Fig. 12.10). The first step is the generation of a random configuration $\boldsymbol{q}_{\text{rand}}$ according to a uniform probability distribution in $\mathcal{C}$ (as in the PRM method). Then, the configuration $\boldsymbol{q}_{\text{near}}$ in T that is closer to $\boldsymbol{q}_{\text{rand}}$ is found, and a new candidate configuration $\boldsymbol{q}_{\text{new}}$ is produced on the segment joining $\boldsymbol{q}_{\text{near}}$ to $\boldsymbol{q}_{\text{rand}}$ at a predefined distance δ from $\boldsymbol{q}_{\text{near}}$. A collision check is then run to verify that both $\boldsymbol{q}_{\text{new}}$ and the segment going from $\boldsymbol{q}_{\text{near}}$ to $\boldsymbol{q}_{\text{new}}$ belong to $\mathcal{C}_{\text{free}}$. If this is the case, T is *expanded* by incorporating $\boldsymbol{q}_{\text{new}}$ and the segment joining it to $\boldsymbol{q}_{\text{near}}$. Note that $\boldsymbol{q}_{\text{rand}}$ is not added to the tree, so

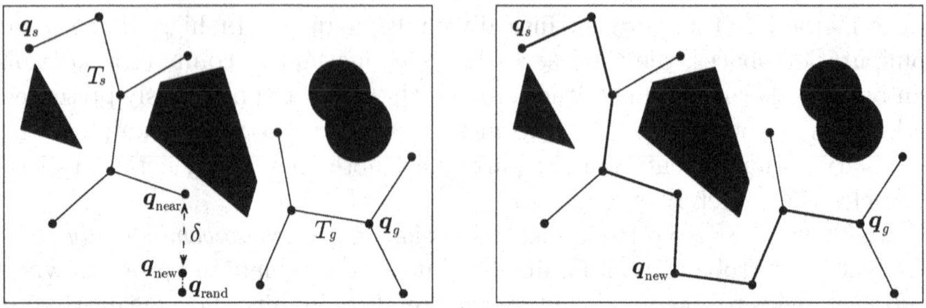

Fig. 12.10. The bidirectional RRT method in a two-dimensional configuration space *left*: the randomized mechanism for expanding a tree, *right*: the extension procedure for connecting the two trees

that it is not necessary to check whether it belongs to $\mathcal{C}_{\text{free}}$; its only function is to indicate a direction of expansion for T.

It is worth pointing out that the RRT expansion procedure, although quite simple, results in a very efficient 'exploration' of $\mathcal{C}$. In fact, it may be shown that the procedure for generating new candidate configuration is intrinsically biased towards those regions of $\mathcal{C}_{\text{free}}$ that have not been visited yet. Moreover, the probability that a generic configuration of $\mathcal{C}_{\text{free}}$ is added to the RRT tends to 1 as the execution time tends to infinity, provided that the configuration lies in the same connected component of $\mathcal{C}_{\text{free}}$ where the RRT is rooted.

To speed up the search for a free path going from $\boldsymbol{q}_s$ to $\boldsymbol{q}_g$, the bidirectional RRT method uses two trees T_s and T_g, respectively rooted at $\boldsymbol{q}_s$ and $\boldsymbol{q}_g$. At each iteration, both trees are expanded with the previously described randomized mechanism. After a certain number of expansion steps, the algorithm enters a phase where it tries to connect the two trees by *extending* each one of them towards the other. This is realized by generating a $\boldsymbol{q}_{\text{new}}$ as an expansion of T_s, and trying to connect T_g to $\boldsymbol{q}_{\text{new}}$. To this end, one may modify the above expansion procedure. In particular, once $\boldsymbol{q}_{\text{new}}$ has been generated from T_s, it acts as a $\boldsymbol{q}_{\text{rand}}$ for T_g: one finds the closest configuration $\boldsymbol{q}_{\text{near}}$ in T_g, and moves from $\boldsymbol{q}_{\text{near}}$ trying to actually *reach* $\boldsymbol{q}_{\text{rand}} = \boldsymbol{q}_{\text{new}}$, hence with a variable stepsize as opposed to a constant δ.

If the segment joining $\boldsymbol{q}_{\text{near}}$ to $\boldsymbol{q}_{\text{new}}$ is collision-free, the extension is complete and the two trees have been connected; otherwise the free portion of the segment is extracted and added to T_g together with its endpoint. At this point, T_g and T_s exchange their roles and the connection attempt is repeated. If this is not successful within a certain number of iterations, one may conclude that the two trees are still far apart and resume the expansion phase.

Like the PRM method, bidirectional RRT is probabilistically complete. A number of variations can be made on the basic scheme. For example, rather than using a constant δ for generating $\boldsymbol{q}_{\text{new}}$, one may define the stepsize as a function of the available free space, possibly going as far as $\boldsymbol{q}_{\text{rand}}$ (as in the extension procedure). This *greedy* version of the method can be much more efficient if $\mathcal{C}$ contains extensive free regions. Moreover, RRT-based methods can be adapted to robots that do not satisfy the free-flying assumption, such as robots that are subject to nonholonomic constraints.

Extension to nonholonomic robots

Consider now the motion planning problem for a nonholonomic mobile robot whose kinematic model is expressed as (11.10). As seen in the previous chapter, admissible paths in configuration space must satisfy constraint (11.40). For example, in the case of a robot with unicycle kinematics, rectilinear paths in configuration space — such as those used in the RRT expansion to move from $\boldsymbol{q}_{\text{near}}$ to $\boldsymbol{q}_{\text{new}}$ — are not admissible in general.

A simple yet general approach to the design of nonholonomic motion planning methods is to use *motion primitives*, i.e., a finite set of admissible local paths in configuration space, each produced by a specific choice of the velocity inputs in the kinematic model. Admissible paths are generated as a concatenation of motion primitives. In the case of a unicycle robot, for example, the following set of velocity inputs

$$v = \bar{v} \qquad \omega = \{-\bar{\omega}, 0, \bar{\omega}\} \qquad t \in [0, \Delta] \tag{12.9}$$

results in three[6] admissible local paths: the first and the third are respectively a left turn and a right turn along arcs of circle, while the second is a rectilinear path (Fig. 12.11, left).

The expansion of an RRT for a nonholonomic mobile robot equipped with a set of motion primitives is quite similar to the previously described procedure. The difference is that, once identified the configuration $\boldsymbol{q}_{\text{near}}$ on T that is closest to $\boldsymbol{q}_{\text{rand}}$, the new configuration $\boldsymbol{q}_{\text{new}}$ is generated by applying the motion primitives starting from $\boldsymbol{q}_{\text{near}}$ and choosing one of the produced configurations, either randomly or as the closest to $\boldsymbol{q}_{\text{rand}}$. Clearly, $\boldsymbol{q}_{\text{new}}$ and the admissible local path joining $\boldsymbol{q}_{\text{near}}$ to $\boldsymbol{q}_{\text{new}}$ are subject to a collision test. Figure 12.11, right, shows an example of RRT — more precisely, its projection on the workspace — for a unicycle equipped with the motion primitives (12.9).

Under suitable assumptions, it is possible to show that if the goal configuration $\boldsymbol{q}_g$ can be reached from the start configuration $\boldsymbol{q}_s$ through a collision-free concatenation of motion primitives,[7] the probability that $\boldsymbol{q}_g$ is added to

---

[6] Note that these particular motion primitives include neither backward motion nor rotation on the spot. A unicycle with constant positive driving velocity v and bounded steering velocity is called *Dubins car* in the literature.

[7] This hypothesis, obviously necessary, implies that the choice of motion primitives must be sufficiently 'rich' to guarantee that the set of configuration reachable through concatenation is 'dense' enough with respect to $\mathcal{C}_{\text{free}}$. For example, the unicycle with the motion primitives (12.9) with a variable time interval Δ can reach any configuration in $\mathcal{C}_{\text{free}}$, although not necessarily with a path that is entirely contained in $\mathcal{C}_{\text{free}}$. This property is instead guaranteed if the *Reeds–Shepp curves* given by (11.56) are used as motion primitives.

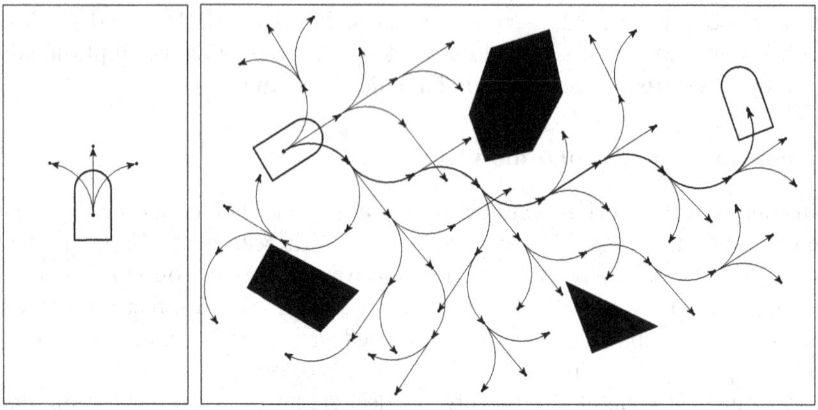

Fig. 12.11. RRT-based motion planning for a unicycle; *left*: a set of motion primitives, *right*: an example of RRT

the tree T tends to 1 as the execution time tends to infinity. To increase the efficiency of the search, also in this case it is possible to devise a bidirectional version of the method.

12.6 Planning via Artificial Potentials

All the methods so far presented are suitable for off-line motion planning, because they require a priori knowledge of the geometry and the pose of the obstacles in the robot workspace. This assumption is reasonable in many cases, e.g., when an industrial manipulator is moving in a robotized cell. However, in service robotics applications the robot must be able to plan its motion on-line, i.e., using partial information on the workspace gathered during the motion on the basis of sensor measurements.

An effective paradigm for on-line planning relies on the use of *artificial potential fields*. Essentially, the point that represents the robot in configuration space moves under the influence of a potential field U obtained as the superposition of an *attractive* potential to the goal and a *repulsive* potential from the $\mathcal{C}$-obstacle region. Planning takes place in an incremental fashion: at each robot configuration $\boldsymbol{q}$, the artificial force generated by the potential is defined as the negative gradient $-\nabla U(\boldsymbol{q})$ of the potential, which indicates the most promising direction of local motion.

12.6.1 Attractive Potential

The attractive potential is designed so as to guide the robot to the goal configuration $\boldsymbol{q}_g$. To this end, one may use a *paraboloid* with vertex in $\boldsymbol{q}_g$:

$$U_{a1}(\boldsymbol{q}) = \frac{1}{2}k_a\boldsymbol{e}^T(\boldsymbol{q})\boldsymbol{e}(\boldsymbol{q}) = \frac{1}{2}k_a\|\boldsymbol{e}(\boldsymbol{q})\|^2, \tag{12.10}$$

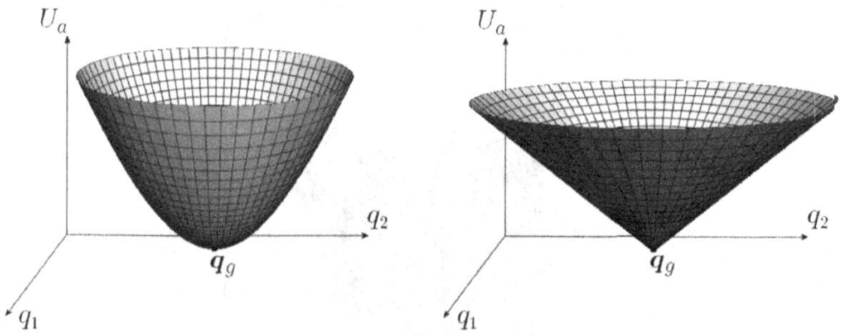

Fig. 12.12. The shape of the paraboloidic attractive potential U_{a1} (*left*) and of the conical attractive potential U_{a2} (*right*) in the case $\mathcal{C} = \mathbb{R}^2$, for $k_a = 1$

where $k_a > 0$ and $\boldsymbol{e} = \boldsymbol{q}_g - \boldsymbol{q}$ is the 'error' vector with respect to the goal configuration $\boldsymbol{q}_g$. This function is always positive and has a global minimum in $\boldsymbol{q}_g$, where it is zero. The resulting attractive force is defined as

$$\boldsymbol{f}_{a1}(\boldsymbol{q}) = -\nabla U_{a1}(\boldsymbol{q}) = k_a \boldsymbol{e}(\boldsymbol{q}). \tag{12.11}$$

Hence, $\boldsymbol{f}_{a1}$ converges linearly to zero when the robot configuration $\boldsymbol{q}$ tends to the goal configuration $\boldsymbol{q}_g$.

Alternatively, it is possible to define a *conical* attractive potential as

$$U_{a2}(\boldsymbol{q}) = k_a \|\boldsymbol{e}(\boldsymbol{q})\|. \tag{12.12}$$

Also U_{a2} is always positive, and zero in $\boldsymbol{q}_g$. The corresponding attractive force is

$$\boldsymbol{f}_{a2}(\boldsymbol{q}) = -\nabla U_{a2}(\boldsymbol{q}) = k_a \frac{\boldsymbol{e}(\boldsymbol{q})}{\|\boldsymbol{e}(\boldsymbol{q})\|}, \tag{12.13}$$

that is constant in modulus. This represents an advantage with respect to the force $\boldsymbol{f}_{a1}$ generated by the paraboloidic attractive potential, which tends to grow indefinitely as the error vector increases in norm. On the other hand, $\boldsymbol{f}_{a2}$ is indefinite in $\boldsymbol{q}_g$. Figure 12.12 shows the shape of U_{a1} and U_{a2} in the case $\mathcal{C} = \mathbb{R}^2$, with $k_a = 1$.

A choice that combines the advantages of the above two potentials is to define the attractive potential as a conical surface away from the goal and as a paraboloid in the vicinity of $\boldsymbol{q}_g$. In particular, by placing the transition between the two potentials where $\|\boldsymbol{e}(\boldsymbol{q})\| = 1$ (i.e., on the surface of the sphere of unit radius centred in $\boldsymbol{q}_g$) one obtains an attractive force that is continuous for any $\boldsymbol{q}$ (see also Problem 12.9).

12.6.2 Repulsive Potential

The repulsive potential U_r is added to the attractive potential U_a to prevent the robot from colliding with obstacles as it moves under the influence of the

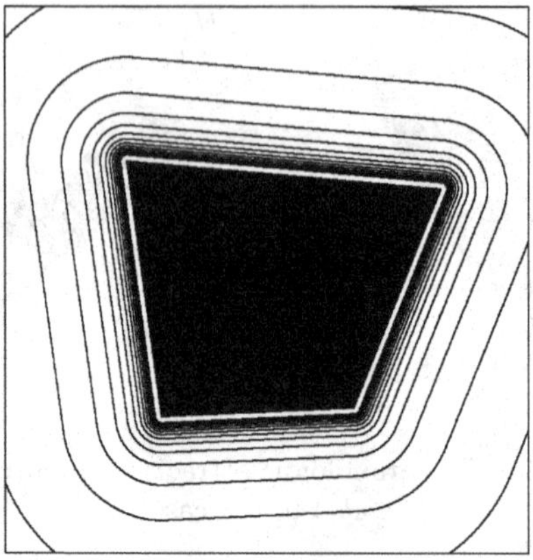

Fig. 12.13. The equipotential contours of the repulsive potential U_r in the range of influence of a polygonal $\mathcal{C}$-obstacle in $\mathcal{C} = \mathbb{R}^2$, for $k_r = 1$ and $\gamma = 2$

attractive force $\boldsymbol{f}_a$. In particular, the idea is to build a barrier potential in the vicinity of the $\mathcal{C}$-obstacle region, so as to repel the point that represents the robot in $\mathcal{C}$.

In the following, it will be assumed that the $\mathcal{C}$-obstacle region has been partitioned in convex components $\mathcal{CO}_i$, $i = 1, \ldots, p$. These components may coincide with the $\mathcal{C}$-obstacles themselves; this happens, for example, when the robot $\mathcal{B}$ is a convex polygon (polyhedron) translating with a fixed orientation in $\mathbb{R}^2$ ($\mathbb{R}^3$) among convex polygonal (polyhedral) obstacles (see Sect. 12.2.2). In the presence of non-convex $\mathcal{C}$-obstacles, however, it is necessary to perform the decomposition in convex components before building the repulsive potential.

For each convex component $\mathcal{CO}_i$, define an associated repulsive potential as

$$U_{r,i}(\boldsymbol{q}) = \begin{cases} \dfrac{k_{r,i}}{\gamma}\left(\dfrac{1}{\eta_i(\boldsymbol{q})} - \dfrac{1}{\eta_{0,i}}\right)^{\gamma} & \text{if } \eta_i(\boldsymbol{q}) \le \eta_{0,i} \\ 0 & \text{if } \eta_i(\boldsymbol{q}) > \eta_{0,i}, \end{cases} \tag{12.14}$$

where $k_{r,i} > 0$, $\eta_i(\boldsymbol{q}) = \min_{\boldsymbol{q}' \in \mathcal{CO}_i} \|\boldsymbol{q} - \boldsymbol{q}'\|$ is the distance of $\boldsymbol{q}$ from $\mathcal{CO}_i$, $\eta_{0,i}$ is the *range of influence* of $\mathcal{CO}_i$ and $\gamma = 2, 3, \ldots$. The potential $U_{r,i}$ is zero outside and positive inside the range of influence $\eta_{0,i}$ and tends to infinity as the boundary of $\mathcal{CO}_i$ is approached, more abruptly as γ is increased (a typical choice is $\gamma = 2$).

When $\mathcal{C} = \mathbb{R}^2$ and the convex component $\mathcal{CO}_i$ is polygonal, an *equipotential contour* of $U_{r,i}$ (i.e., the locus of configurations $\boldsymbol{q}$ such that $U_{r,i}$ has a certain constant value) consists of rectilinear tracts that are parallel to the sides of the polygon, connected by arcs of circle in correspondence of the vertices, as shown in Fig. 12.13. Note how the contours get closer to each other in the proximity of the $\mathcal{C}$-obstacle boundary, due to the hyperboloidic profile of the potential. When $\mathcal{C} = \mathbb{R}^3$ and the convex component $\mathcal{CO}_i$ is polyhedral, the *equipotential surfaces* of $U_{r,i}$ are copies of the faces of $\mathcal{CO}_i$, connected by patches of cylindrical surfaces in correspondence of the edges and spherical surfaces in correspondence of the vertices of $\mathcal{CO}_i$.

The repulsive force resulting from $U_{r,i}$ is

$$\boldsymbol{f}_{r,i}(\boldsymbol{q}) = -\nabla U_{r,i}(\boldsymbol{q}) = \begin{cases} \dfrac{k_{r,i}}{\eta_i^2(\boldsymbol{q})}\left(\dfrac{1}{\eta_i(\boldsymbol{q})} - \dfrac{1}{\eta_{0,i}}\right)^{\gamma-1} \nabla\eta_i(\boldsymbol{q}) & \text{if } \eta_i(\boldsymbol{q}) \le \eta_{0,i} \\ 0 & \text{if } \eta_i(\boldsymbol{q}) > \eta_{0,i}. \end{cases} \tag{12.15}$$

Denote by $\boldsymbol{q}_m$ the configuration of $\mathcal{CO}_i$ that is closer to $\boldsymbol{q}$ ($\boldsymbol{q}_m$ is uniquely determined in view of the convexity of $\mathcal{CO}_i$). The gradient vector $\nabla\eta_i(\boldsymbol{q})$, which is orthogonal to the equipotential contour (or surface) passing through $\boldsymbol{q}$, is directed as the half-line originating from $\boldsymbol{q}_m$ and passing through $\boldsymbol{q}$. If the boundary of $\mathcal{CO}_i$ is piecewise differentiable, function η_i is differentiable everywhere in $\mathcal{C}_{\text{free}}$ and $\boldsymbol{f}_{r,i}$ is continuous in the same space.[8]

The aggregate repulsive potential is obtained by adding up the individual potentials associated with the convex components of $\mathcal{CO}$:

$$U_r(\boldsymbol{q}) = \sum_{i=1}^{p} U_{r,i}(\boldsymbol{q}). \tag{12.16}$$

If $\eta_i(\boldsymbol{q}_g) \ge \eta_{0,i}$ for $i = 1, \ldots, p$ (i.e., if the goal is placed outside the range of influence of each obstacle component $\mathcal{CO}_i$), the value of the aggregate repulsive field U_r is zero in $\boldsymbol{q}_g$. In the following, it will be assumed that this is the case.

12.6.3 Total Potential

The total potential U_t is obtained by superposition of the attractive and the aggregate repulsive potentials:

$$U_t(\boldsymbol{q}) = U_a(\boldsymbol{q}) + U_r(\boldsymbol{q}). \tag{12.17}$$

This results in the force field

$$\boldsymbol{f}_t(\boldsymbol{q}) = -\nabla U_t(\boldsymbol{q}) = \boldsymbol{f}_a(\boldsymbol{q}) + \sum_{i=1}^{p} \boldsymbol{f}_{r,i}(\boldsymbol{q}). \tag{12.18}$$

[8] Note the relevance in this sense of the assumption that the component $\mathcal{CO}_i$ is convex. If it were otherwise, there would exist configurations in $\mathcal{C}_{\text{free}}$ for which $\boldsymbol{q}_m$ would not be uniquely defined. In these configurations, belonging by definition to the generalized Voronoi diagram $\mathcal{V}(\mathcal{C}_{\text{free}})$, function η_i would not be differentiable, resulting in a discontinuous repulsive force. This might induce undesired effects on the planned path (for example, oscillations).

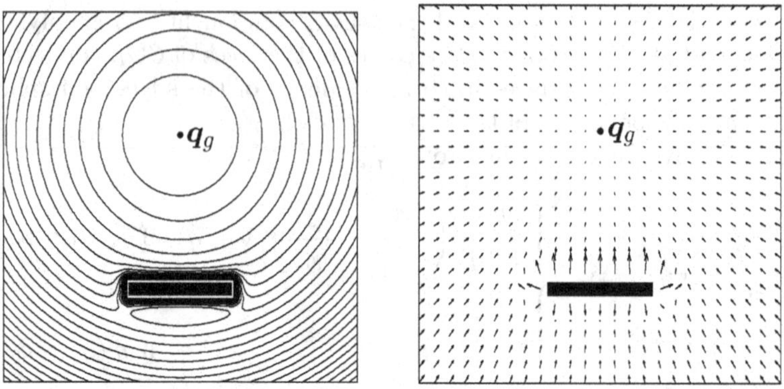

Fig. 12.14. The total potential in $\mathcal{C} = \mathrm{I\!R}^2$ obtained by superposition of a hyperboloidic attractive potential and a repulsive potential for a rectangular $\mathcal{C}$-obstacle: *left*: the equipotential contours, *right*: the resulting force field

U_t clearly has a global minimum in $\boldsymbol{q}_g$, but there may also exist some *local minima* where the force field is zero. Considering for simplicity the case $\mathcal{C} = \mathrm{I\!R}^2$, this happens in the 'shadow zone' of a $\mathcal{C}$-obstacle when the repulsive potential $U_{r,i}$ has equipotential contours with lower curvature (e.g., segments) than the attractive potential in the same area. See for example Fig. 12.14, where a local minimum is clearly present 'below' the $\mathcal{C}$-obstacle. A remarkable exception is the case (*sphere world*) in which all the convex components $\mathcal{CO}_i$ of the $\mathcal{C}$-obstacle region are spheres. In this situation, the total potential exhibits isolated saddle points (where the force field is still zero) but no local minima.

12.6.4 Planning Techniques

There are three different approaches for planning collision-free motions on the basis of a total artificial potential U_t and the associated force field $\boldsymbol{f}_t = -\nabla U_t$. They are briefly discussed below:

1. The first possibility is to let
$$\boldsymbol{\tau} = \boldsymbol{f}_t(\boldsymbol{q}), \tag{12.19}$$
hence considering $\boldsymbol{f}_t(\boldsymbol{q})$ as a vector of generalized forces that induce a motion of the robot in accordance with its dynamic model.
2. The second method regards the robot as a unit point mass moving under the influence of $\boldsymbol{f}_t(\boldsymbol{q})$, as in
$$\ddot{\boldsymbol{q}} = \boldsymbol{f}_t(\boldsymbol{q}). \tag{12.20}$$
3. The third possibility is to interpret the force field $\boldsymbol{f}_t(\boldsymbol{q})$ as a desired velocity for the robot, by letting
$$\dot{\boldsymbol{q}} = \boldsymbol{f}_t(\boldsymbol{q}). \tag{12.21}$$

In principle, one could use these three approaches for on-line as well as off-line motion planning. In the first case, (12.19) directly represents control inputs for the robot, whereas the implementation of (12.20) requires the solution of the inverse dynamics problem, i.e., the substitution of $\ddot{\boldsymbol{q}}$ in the robot dynamic model to compute the generalized forces $\boldsymbol{\tau}$ that realize such accelerations. Equation (12.21) can instead be used on-line in a kinematic control scheme, in particular to provide the reference inputs for the low-level controllers that are in charge of reproducing such generalized velocities as accurately as possible. In any case, the artificial force field $\boldsymbol{f}_t$ represents, either directly or indirectly, a true feedback control that guides the robot towards the goal, while trying at the same time to avoid collisions with the workspace obstacles that have been detected by the sensory system. To emphasize this aspect, on-line motion generation based on artificial potentials is also referred to as *reactive* planning.

In off-line motion planning, configuration space paths are generated by simulation, i.e., integrating numerically the robot dynamic model if (12.19) is used, or directly by the differential equations (12.20) and (12.21).

In general, the use of (12.19) generates smoother paths, because with this scheme the reactions to the presence of obstacles are naturally 'filtered' through the robot dynamics. On the other hand, the strategy represented by (12.21) is faster in executing the motion corrections suggested by the force field $\boldsymbol{f}_t$, and may thus be considered safer. The characteristics of scheme (12.20) are clearly intermediate between the other two. Another aspect to be considered is that using (12.21) guarantees (in the absence of local minima) the asymptotic stability of $\boldsymbol{q}_g$ (i.e., the robot reaches the goal with zero velocity), whereas this is not true for the other two motion generation strategies. To achieve asymptotic stability with (12.19) and (12.20), a damping term proportional to the robot velocity $\dot{\boldsymbol{q}}$ must be added to $\boldsymbol{f}_t$.

In view of the above discussion, it is not surprising that the most common choice is the simple numerical integration of (12.21) via the Euler method:

$$\boldsymbol{q}_{k+1} = \boldsymbol{q}_k + T\boldsymbol{f}_t(\boldsymbol{q}_k), \tag{12.22}$$

where $\boldsymbol{q}_k$ and $\boldsymbol{q}_{k+1}$ represent respectively the current and the next robot configuration, and T is the integration step. To improve the quality of the generated path, it is also possible to use a variable T, smaller when the modulus of the force field $\boldsymbol{f}_t$ is larger (in the vicinity of obstacles) or smaller (close to the destination $\boldsymbol{q}_g$). Recalling that $\boldsymbol{f}_t(\boldsymbol{q}) = -\nabla U_t(\boldsymbol{q})$, Eq. (12.22) may be easily interpreted as a numerical implementation of the *gradient method* for the minimization of $U_t(\boldsymbol{q})$, often referred to as the *algorithm of steepest descent*.

12.6.5 The Local Minima Problem

Whatever technique is used to plan motions on the basis of artificial potentials, local minima of U_t — where the total force field $\boldsymbol{f}_t$ is zero — represent a problem. For example, if (12.22) is used and the generated path enters the basin of attraction of a local minimum, the planning process is bound to stop there, without reaching the goal configuration.[9] Actually, the same problem may occur also when (12.19) or (12.20) are used if the basin of attraction of the local minimum is sufficiently large. On the other hand, as noticed previously, the total potential U_t obtained by the superposition of an attractive and a repulsive potential invariably exhibits local minima, except for very particular cases. This means that motion planning methods based on artificial potentials are not complete in general, because it may happen that the goal configuration $\boldsymbol{q}_g$ is not reached even though a solution exists.

Best-first algorithm

A simple workaround for the local minima problem is to use a *best-first* algorithm, which is formulated under the assumption that the free configuration space $\mathcal{C}_{\rm free}$ has been discretized using a regular grid. In general, the discretization procedure results in the loss of some boundary regions of $\mathcal{C}_{\rm free}$, which are not represented as free in the gridmap. Each free cell of the grid is assigned the value of the total potential U_t computed at the centroid of the cell. Planning proceeds by building a tree T rooted at the start configuration $\boldsymbol{q}_s$. At each iteration, the leaf with the minimum value of U_t is selected and its adjacent[10] cells are examined. Those that are not already in T are added as children of the considered leaf. Planning is successful when the cell containing the goal is reached (in this case, the solution path is built by tracing back the arcs of the tree from $\boldsymbol{q}_g$ to $\boldsymbol{q}_s$), whereas failure is reported when all the cells accessible from $\boldsymbol{q}_s$ have been explored without reaching $\boldsymbol{q}_g$.

The above best-first algorithm evolves as a grid-discretized version of the algorithm of steepest descent (12.22), until a cell is reached which represents a minimum for the associated total potential. If it is a local minimum, the algorithm visits ('fills') its entire basin of attraction and finally leaves it, reaching a point from which planning can continue. The resulting motion planning method is resolution complete, because a solution is found only if one exists on the gridmap that represents $\mathcal{C}_{\rm free}$ by defect. In general, increasing the grid resolution may be necessary to recover the possibility of determining a solution. In any case, the time complexity of the basin filling procedure, which is exponential in the dimension of $\mathcal{C}$ because such is the number of adjacent cells to a given one, makes the best-first algorithm applicable only in configuration spaces of low dimension (typically, not larger than 3).

---

[9] The probability of reaching a saddle point instead is extremely low; besides, any perturbation would allow the planner to exit such a point.

[10] Various definitions of adjacency may be adopted. For example, in $\mathbb{R}^2$ one may use 1-adjacency or 2-adjacency. In the first case, each cell c has four adjacent cells, while in the second they are eight. The resulting paths will obviously reflect this fact.

A more effective approach is to include in the best-first method a randomized mechanism for evading local minima. In practice, one implements a version of the best-first algorithm in which the number of iterations aimed at filling the basin of attraction of local minima is bounded. When the bound is reached, a sequence of random steps (*random walk*) is taken. This usually allows the planner to evade the basin in a much shorter time than would be needed by the complete filling procedure. As the probability of evading the basin of attraction of a local minimum approaches 1 when the number of random steps tends to infinity, this *randomized best-first* method is probabilistically complete (in addition to being resolution complete).

Navigation functions

Although the best-first algorithm (in its basic or randomized version) represents a solution to the problem of local minima, it may generate paths that are very inefficient. In fact, with this strategy the robot will still enter (and then evade) the basin of attraction of any local minimum located on its way to goal. A more radical approach is based on the use of *navigation functions*, i.e., artificial potentials that have no local minima. As already mentioned, if the $\mathcal{C}$-obstacles are spheres this property already holds for the potential U_t defined by superposition of an attractive and a repulsive field. A first possibility would then be to approximate by excess all $\mathcal{C}$-obstacles with spheres, and use the total potential U_t. Clearly, such an approximation may severely reduce the free configuration space $\mathcal{C}_{\rm free}$, and even destroy its connectedness; for example, imagine what would happen in the case depicted in Fig. 12.4.

In principle, a mathematically elegant way to define a navigation function consists of building first a differentiable homeomorphism (a *diffeomorphism*) that maps the $\mathcal{C}$-obstacle region to a collection of spheres, then generating a classical total potential in the transformed space, and finally mapping it back to the original configuration space so as to obtain a potential free of local minima. If the $\mathcal{C}$-obstacles are *star-shaped*,[11] such a diffeomorphism actually exists, and the procedure outlined above provides in fact a navigation function. Another approach is to build the potential using *harmonic functions*, that are the solutions of a particular differential equation that describes the physical process of heat transmission or fluid dynamics.

Generating a navigation function is, however, computationally cumbersome, and the associated planning methods are thus mainly of theoretical interest. A notable exception, at least in low-dimensional configuration spaces, is the *numerical navigation function*. This is a potential built on a gridmap representation of $\mathcal{C}_{\rm free}$ by assigning value 0 to the cell containing $\boldsymbol{q}_g$, value 1 to its adjacent cells, value 2 to the unvisited cells among those adjacent to

[11] A subset $S \subset \mathbb{R}^n$ is said to be star-shaped if it is homeomorphic to the closed unit sphere in $\mathbb{R}^n$ and has a point p (*centre*) such that any other point in S may be joined to p by a segment that is entirely contained in S.

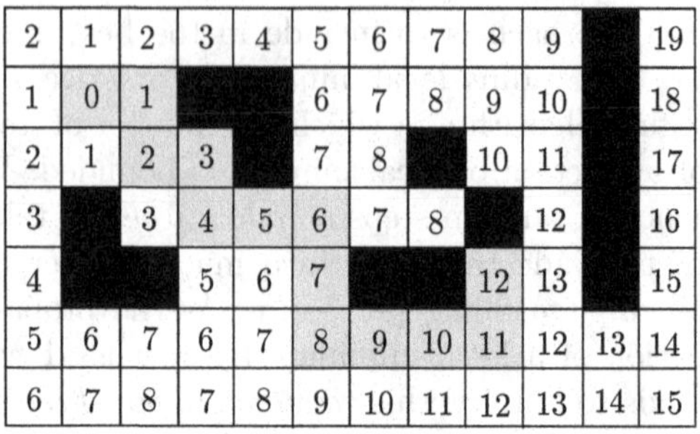

Fig. 12.15. An example of numerical navigation function in a simple two-dimensional gridmap using 1-adjacency. Cells in gray denote a particular solution path obtained by following from the start (cell 12) the steepest descent of the potential on the gridmap

cells with potential 1, and so on. To understand better the procedure, one may visualize a wavefront that originates at $\boldsymbol{q}_g$ and expands according to the adopted adjacency definition (*wavefront expansion algorithm*). It is easy to realize that the obtained potential is free of local minima, and therefore its use in conjunction with the algorithm of steepest descent provides a motion planning method that is complete in resolution (see Fig. 12.15).

Finally, it should be mentioned that the use of navigation functions is limited to off-line motion planning, because their construction — be they continuous or discrete in nature — requires the a priori knowledge of the geometry and pose of the workspace obstacles. As for on-line motion planning, in which the obstacles are gradually reconstructed via sensor measurements as the robot moves, the incremental construction of a total potential by superposition of attractive and repulsive fields represents a simple, often effective method to generate collision-free motions, even though completeness cannot be claimed. In any case, motion planning based on artificial potentials belong to the single-query category, because the attractive component of the potential depends on the goal configuration $\boldsymbol{q}_g$.

12.7 The Robot Manipulator Case

Generating collision-free movements for robot manipulators is a particularly important category of motion planning problems. In general, the computational complexity associated with this problem is substantial, due to the high dimension of the configuration space (typically $n \geq 4$) and to the presence of rotational DOFs (revolute joints).

It is sometimes possible to reduce the dimension of the configuration space $\mathcal{C}$ approximating by excess the size of the robot. For example, in a six-DOF anthropomorphic manipulator one can replace the last three links of the kinematic chain (the spherical wrist) and the end-effector with the volume they 'sweep' when the corresponding joints move across their whole available range. The dimension of the configuration space becomes three, as planning concerns only the base, shoulder and elbow joints, while the wrist can move arbitrarily. Clearly, this approximation is conservative, and hence acceptable only if the aforementioned volume is small with respect to the workspace of the manipulator.

In the presence of rotational DOFs, the other complication is the shape of the $\mathcal{C}$-obstacles, which is complex even for simple workspace obstacles due to the strong nonlinearity introduced by the manipulator inverse kinematics (recall Fig. 12.4). Apart from the intrinsic difficulty of computing $\mathcal{C}$-obstacles, their non-polyhedral shape does not allow the application of the planning methods presented in Sects. 12.3 and 12.4.

The most convenient choice for off-line planning is represented by probabilistic methods, which exhibit the best performance in high-dimensional configuration spaces. Moreover, as the computation of the $\mathcal{C}$-obstacles is not required, these planners are not affected by their shape. However, it should be noted that collision checking — which is an essential tool for probabilistic motion planning — becomes more onerous as the number of DOFs is increased.

For on-line planning, the best results are obtained by a suitable adaptation of the method based on artificial potentials. In particular, to avoid the computation of $\mathcal{C}$-obstacles and at the same time plan in a space of reduced dimension, the potential is directly built in the workspace $\mathcal{W} = \mathbb{R}^N$ rather than in the configuration space $\mathcal{C}$, and acts on a set of *control points* located on the manipulator. Among these is included a point that represents the end-effector (to which is assigned the goal of the motion planning problem) and at least one point (possibly variable in time) for each body of the linkage. While the attractive potential only influences the end-effector representative point, the repulsive potential acts on all control points. As a consequence, the artificial potentials used in this scheme are actually two: an attractive-repulsive field for the end-effector, and a repulsive field for the other control points distributed on the manipulator links.

As before, different approaches may be used to convert the force fields generated by the artificial potentials to commands for the manipulator. Denote by $\boldsymbol{p}_i(\boldsymbol{q})$, $i = 1, \ldots, P$, the coordinates in $\mathcal{W}$ of the P control points in correspondence of the configuration $\boldsymbol{q}$ of the manipulator. In particular, $\boldsymbol{p}_1, \ldots, \boldsymbol{p}_{P-1}$ are the control points located on the manipulator links, subject only to the repulsive potential U_r, while $\boldsymbol{p}_P$ is the control point for the end-effector, which is subject to the total potential $U_t = U_a + U_r$.

A first possibility is to impose to the robot joints the generalized forces which would result from the combined action of the various force fields acting on the control points in the workspace, according to

$$\boldsymbol{\tau} = -\sum_{i=1}^{P-1} \boldsymbol{J}_i^T(\boldsymbol{q})\nabla U_r(\boldsymbol{p}_i) - \boldsymbol{J}_P^T(\boldsymbol{q})\nabla U_t(\boldsymbol{p}_P), \tag{12.23}$$

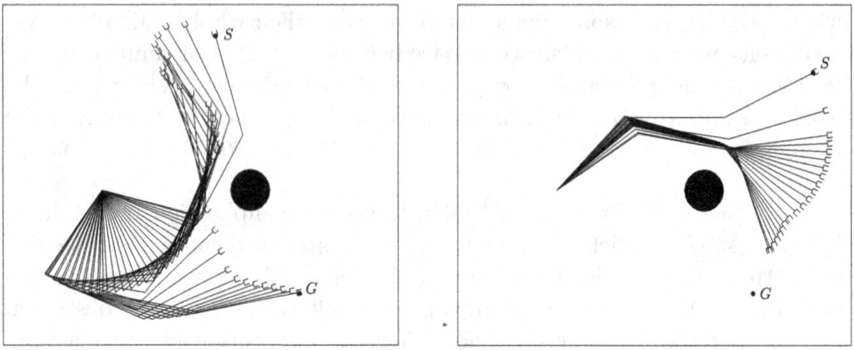

Fig. 12.16. Examples of motion planning via artificial potentials acting on control points for a planar 3R manipulator; *left*: planning is successful and leads to a collision-free motion between the start S and the goal G, *right*: a failure is reported because the manipulator is stuck at a force equilibrium

where $\boldsymbol{J}_i(\boldsymbol{q})$, $i = 1, \ldots, P$, denotes the Jacobian of the direct kinematics function associated with the control point $\boldsymbol{p}_i(\boldsymbol{q})$.

Alternatively, a purely kinematic planning scheme is obtained by letting

$$\dot{\boldsymbol{q}} = -\sum_{i=1}^{P-1} \boldsymbol{J}_i^T(\boldsymbol{q})\nabla U_r(\boldsymbol{p}_i) - \boldsymbol{J}_P^T(\boldsymbol{q})\nabla U_t(\boldsymbol{p}_P) \tag{12.24}$$

and feeding these joint velocities to the low-level control loops as reference signals. Note that Eq. (12.24) represents a gradient-based minimization step in the configuration space $\mathcal{C}$ of a combined potential defined in the workspace $\mathcal{W}$. In fact, a potential function acting on a control point in the workspace may be seen as a composite function of $\boldsymbol{q}$ through the associated direct kinematics relationship, and the Jacobian transpose $\boldsymbol{J}_i^T(\boldsymbol{q})$, $i = 1, \ldots, P$, maps a gradient in $\mathcal{W}$ to a gradient in $\mathcal{C}$. In formulae:

$$\nabla_{\boldsymbol{q}} U(\boldsymbol{p}_i) = \left(\frac{\partial U(\boldsymbol{p}_i(\boldsymbol{q}))}{\partial \boldsymbol{q}}\right)^T = \left(\frac{\partial U(\boldsymbol{p}_i)}{\partial \boldsymbol{p}_i}\frac{\partial \boldsymbol{p}_i}{\partial \boldsymbol{q}}\right)^T = \boldsymbol{J}_i^T(\boldsymbol{q})\nabla U(\boldsymbol{p}_i),$$

for $i = 1, \ldots, P$.

The above two schemes can be considered respectively the transposition of (12.19) and (12.21), and therefore they inherit the same characteristics. In particular, when (12.23) is used the motion corrections prescribed by the force fields are filtered through the dynamic model of the manipulator, and smoother movements can be expected. The kinematic scheme (12.24) instead is faster in realizing such corrections.

Finally, it should be mentioned that the use of artificial potentials that are defined in the workspace may aggravate the local minima problem. In fact, the various forces (either purely attractive or repulsive-attractive) acting on the control points may neutralize each other at the joint level, blocking the manipulator in a configuration (*force equilibrium*) where no control point is at a local minimum of the associated potential (see Fig. 12.16). As consequence, it is always advisable to use workspace potential fields in conjunction with a randomized best-first algorithm.

Bibliography

In the last three decades, the literature on motion planning has grown considerably, and this area may now be considered as a scientific discipline in itself. In the following, only a short list is given of the seminal works for the material presented in this chapter.

The systematic use of the concept of configuration space for motion planning was proposed in [138]. The method based on the retraction of the free configuration space on the generalized Voronoi diagram was originally introduced for a robot of circular shape in [170]. The technique based on trapezoidal cell decomposition is described in [122], while [197] and [33] are among the general planning methods via decomposition mentioned at the end of Sect. 12.4.1. The approach based on approximate cell decomposition was proposed in [138]. The PRM method for probabilistic planning was introduced in [107], while the RRT method with its variants is described in [125].

The use of artificial potentials for on-line motion planning was pioneered in [113]. The concept of navigation function was introduced in [185], while its numerical version on a gridmap was described in [17], together with the best-first algorithm, also in its randomized version.

For other aspects of the motion planning problem that are merely hinted at (nonholonomic motion planning) or simply ignored (managing uncertainty, mobile obstacles) in this chapter, the reader can consult many excellent books, going from the classical treatment in [122], through [123], to the most recent texts [45, 145, 124].

Problems

12.1. Describe the nature (including the dimension) of the configuration space for a mobile manipulator consisting of a unicycle-like vehicle carrying a six-DOF anthropomorphic arm, providing a choice of generalized coordinates for the system.

12.2. With reference to a 2R manipulator, modify the definition (12.2) of configuration space distance so as to take into account the fact that the manipulator posture does not change if the joint variables q_1 and q_2 are increased (or decreased) by a multiple of 2π.

12.3. Consider a polygonal robot translating at a fixed orientation in $\mathbb{R}^2$ among polygonal obstacles. Build an example showing that the same $\mathcal{C}$-obstacle region may correspond to robot and obstacles of different shapes.

12.4. With reference to the second workspace shown in Fig. 12.4, give the numerical value of three configurations of the manipulator that lie in the three connected components of $\mathcal{C}_{\text{free}}$. Moreover, sketch the manipulator posture for each of these configurations.

12.5. Discuss the basic steps of an algorithm for computing the generalized Voronoi diagram of a limited polygonal subset of $\mathbb{R}^2$. [*Hint*: a simple algorithm is obtained by considering all the possible side-side, side-vertex and vertex-vertex pairs, and generating the elementary arcs of the diagram on the basis of the intersections of the associated equidistance contours.]

12.6. For the motion planning method via exact cell decomposition, give an example in which the path obtained as a broken line joining the midpoints of the common boundary of the channel cells goes through a vertex of the $\mathcal{C}$-obstacle region, and propose an alternative procedure for extracting a free path from the channel. [*Hint*: build a situation in which the channel from c_s to c_g contains a cell for which the entrance and the exit boundary lie on the same side.]

12.7. Implement in a computer program the PRM method for a 2R planar robot moving among circular workspace obstacles. The program receives as input a geometrical description of the obstacles (centre and radius of each obstacle) as well as the start and goal configurations, and terminates after a maximum number of iterations. If a solution path has been found, it is given as output together with the associated PRM; otherwise, a failure is reported. Discuss the performance of the method with respect to the choice of configuration space distance (for example, (12.1) or (12.2)).

12.8. Implement in a computer program the RRT method for a circular-shaped unicycle moving among square obstacles, with the motion primitives defined as in (12.9). The program receives as input the start and goal configurations, in addition to a geometrical description of the obstacles (centre and side of each obstacle), and terminates after a maximum number of iterations. If a solution path has been found, it is given as output together with the associated RRT; otherwise, a failure is reported. Build a situation in which the method cannot find a solution because of the limitations inherent to the chosen motion primitives.

12.9. For the case $\mathcal{C} = \mathbb{R}^2$, build a continuously differentiable attractive potential having a paraboloidic profile inside the circle of radius ρ and a conical profile outside. [*Hint*: modify the expression (12.12) of the conical potential using a different constant k_b in place of k_a, which already characterizes the paraboloidic potential.]

12.10. For the case $\mathcal{C} = \mathbb{R}^2$, prove that the total potential U_t may exhibit a local minimum in areas where the equipotential contours of the repulsive potential U_r have lower curvature than those of the attractive potential U_a. [*Hint*: consider a polygonal $\mathcal{C}$-obstacle and use a geometric construction.]

12.11. Consider a point robot moving in a planar workspace containing three circular obstacles, respectively of radius 1, 2 and 3 and centre in $(2, 1)$, $(-1, 3)$ and $(1, -2)$. The goal is the origin of the workspace reference frame. Build the total potential U_t resulting from the superposition of the attractive potential U_a to the goal and the repulsive potential U_r from the obstacles, and derive the corresponding artificial force field. Moreover, compute the coordinates of the saddle points of U_t.

12.12. Discuss the main issues arising from the application of the artificial potential technique for planning on-line the motion of an omnidirectional circular robot. Assume that the robot is equipped with a rotating laser range finder placed at its centre that measures the distance between the sensor and the closest obstacles along each direction. If the distance is larger than the maximum measurable range R, the sensor returns R as a reading. Sketch a possible extension of the method to a unicycle-like mobile robot.

12.13. Implement in a computer program the motion planning method based on the numerical navigation function. The program receives as input a two-dimensional gridmap, in which some of the cells are labelled as 'obstacles', with the specification of the start and the goal cells. If the algorithm is successful, a sequence of free cells from the start to the goal is provided as output. With the aid of some examples, compare the average length of the solution paths obtained using 1-adjacency and 2-adjacency to build the navigation function.

Appendices

A

Linear Algebra

Since modelling and control of robot manipulators requires an extensive use of *matrices* and *vectors* as well as of matrix and vector *operations*, the goal of this appendix is to provide a brush-up of *linear algebra*.

A.1 Definitions

A *matrix* of dimensions $(m \times n)$, with m and n positive integers, is an array of elements a_{ij} arranged into m *rows* and n *columns*:

$$\boldsymbol{A} = [a_{ij}]_{\substack{i=1,\ldots,m \\ j=1,\ldots,n}} = \begin{bmatrix} a_{11} & a_{12} & \ldots & a_{1n} \\ a_{21} & a_{22} & \ldots & a_{2n} \\ \vdots & \vdots & \ddots & \vdots \\ a_{m1} & a_{m2} & \ldots & a_{mn} \end{bmatrix}. \tag{A.1}$$

If $m = n$, the matrix is said to be *square*; if $m < n$, the matrix has more columns than rows; if $m > n$ the matrix has more rows than columns. Further, if $n = 1$, the notation (A.1) is used to represent a (column) vector $\boldsymbol{a}$ of dimensions $(m \times 1)$;[1] the elements a_i are said to be vector components.

A square matrix $\boldsymbol{A}$ of dimensions $(n \times n)$ is said to be *upper triangular* if $a_{ij} = 0$ for $i > j$:

$$\boldsymbol{A} = \begin{bmatrix} a_{11} & a_{12} & \ldots & a_{1n} \\ 0 & a_{22} & \ldots & a_{2n} \\ \vdots & \vdots & \ddots & \vdots \\ 0 & 0 & \ldots & a_{nn} \end{bmatrix};$$

the matrix is said to be *lower triangular* if $a_{ij} = 0$ for $i < j$.

[1] According to standard mathematical notation, small boldface is used to denote vectors while capital boldface is used to denote matrices. Scalars are denoted by roman characters.

An $(n \times n)$ square matrix $\boldsymbol{A}$ is said to be *diagonal* if $a_{ij} = 0$ for $i \neq j$, i.e.,

$$\boldsymbol{A} = \begin{bmatrix} a_{11} & 0 & \dots & 0 \\ 0 & a_{22} & \dots & 0 \\ \vdots & \vdots & \ddots & \vdots \\ 0 & 0 & \dots & a_{nn} \end{bmatrix} = \text{diag}\{a_{11}, a_{22}, \dots, a_{nn}\}.$$

If an $(n \times n)$ diagonal matrix has all unit elements on the diagonal ($a_{ii} = 1$), the matrix is said to be *identity* and is denoted by $\boldsymbol{I}_n$.[2] A matrix is said to be *null* if all its elements are null and is denoted by $\boldsymbol{O}$. The null column vector is denoted by $\mathbf{0}$.

The *transpose* $\boldsymbol{A}^T$ of a matrix $\boldsymbol{A}$ of dimensions $(m \times n)$ is the matrix of dimensions $(n \times m)$ which is obtained from the original matrix by interchanging its rows and columns:

$$\boldsymbol{A}^T = \begin{bmatrix} a_{11} & a_{21} & \dots & a_{m1} \\ a_{12} & a_{22} & \dots & a_{m2} \\ \vdots & \vdots & \ddots & \vdots \\ a_{1n} & a_{2n} & \dots & a_{mn} \end{bmatrix}. \tag{A.2}$$

The transpose of a column vector $\boldsymbol{a}$ is the row vector $\boldsymbol{a}^T$.

An $(n \times n)$ square matrix $\boldsymbol{A}$ is said to be *symmetric* if $\boldsymbol{A}^T = \boldsymbol{A}$, and thus $a_{ij} = a_{ji}$:

$$\boldsymbol{A} = \begin{bmatrix} a_{11} & a_{12} & \dots & a_{1n} \\ a_{12} & a_{22} & \dots & a_{2n} \\ \vdots & \vdots & \ddots & \vdots \\ a_{1n} & a_{2n} & \dots & a_{nn} \end{bmatrix}.$$

An $(n \times n)$ square matrix $\boldsymbol{A}$ is said to be *skew-symmetric* if $\boldsymbol{A}^T = -\boldsymbol{A}$, and thus $a_{ij} = -a_{ji}$ for $i \neq j$ and $a_{ii} = 0$, leading to

$$\boldsymbol{A} = \begin{bmatrix} 0 & a_{12} & \dots & a_{1n} \\ -a_{12} & 0 & \dots & a_{2n} \\ \vdots & \vdots & \ddots & \vdots \\ -a_{1n} & -a_{2n} & \dots & 0 \end{bmatrix}.$$

A *partitioned* matrix is a matrix whose elements are matrices (*blocks*) of proper dimensions:

$$\boldsymbol{A} = \begin{bmatrix} \boldsymbol{A}_{11} & \boldsymbol{A}_{12} & \dots & \boldsymbol{A}_{1n} \\ \boldsymbol{A}_{21} & \boldsymbol{A}_{22} & \dots & \boldsymbol{A}_{2n} \\ \vdots & \vdots & \ddots & \vdots \\ \boldsymbol{A}_{m1} & \boldsymbol{A}_{m2} & \dots & \boldsymbol{A}_{mn} \end{bmatrix}.$$

---

[2] Subscript n is usually omitted if the dimensions are clear from the context.

A partitioned matrix may be block-triangular or block-diagonal. Special partitions of a matrix are that by columns

$$\boldsymbol{A} = [\,\boldsymbol{a}_1 \quad \boldsymbol{a}_2 \quad \dots \quad \boldsymbol{a}_n\,]$$

and that by rows

$$\boldsymbol{A} = \begin{bmatrix} \boldsymbol{a}_1^T \\ \boldsymbol{a}_2^T \\ \vdots \\ \boldsymbol{a}_m^T \end{bmatrix}.$$

Given a square matrix $\boldsymbol{A}$ of dimensions $(n \times n)$, the *algebraic complement* $\boldsymbol{A}_{(ij)}$ of element a_{ij} is the matrix of dimensions $((n-1) \times (n-1))$ which is obtained by eliminating row i and column j of matrix $\boldsymbol{A}$.

A.2 Matrix Operations

The *trace* of an $(n \times n)$ square matrix $\boldsymbol{A}$ is the sum of the elements on the diagonal:

$$\mathrm{Tr}(\boldsymbol{A}) = \sum_{i=1}^{n} a_{ii}. \tag{A.3}$$

Two matrices $\boldsymbol{A}$ and $\boldsymbol{B}$ of the same dimensions $(m \times n)$ are equal if $a_{ij} = b_{ij}$. If $\boldsymbol{A}$ and $\boldsymbol{B}$ are two matrices of the same dimensions, their *sum* is the matrix

$$\boldsymbol{C} = \boldsymbol{A} + \boldsymbol{B} \tag{A.4}$$

whose elements are given by $c_{ij} = a_{ij} + b_{ij}$. The following properties hold:

$$\begin{aligned} \boldsymbol{A} + \boldsymbol{O} &= \boldsymbol{A} \\ \boldsymbol{A} + \boldsymbol{B} &= \boldsymbol{B} + \boldsymbol{A} \\ (\boldsymbol{A} + \boldsymbol{B}) + \boldsymbol{C} &= \boldsymbol{A} + (\boldsymbol{B} + \boldsymbol{C}). \end{aligned}$$

Notice that two matrices of the same dimensions and partitioned in the same way can be summed formally by operating on the blocks in the same position and treating them like elements.

The *product of a scalar α by an $(m \times n)$ matrix* $\boldsymbol{A}$ is the matrix $\alpha\boldsymbol{A}$ whose elements are given by αa_{ij}. If $\boldsymbol{A}$ is an $(n \times n)$ diagonal matrix with all equal elements on the diagonal $(a_{ii} = a)$, it follows that $\boldsymbol{A} = a\boldsymbol{I}_n$.

If $\boldsymbol{A}$ is a square matrix, one may write

$$\boldsymbol{A} = \boldsymbol{A}_s + \boldsymbol{A}_a \tag{A.5}$$

where

$$\boldsymbol{A}_s = \frac{1}{2}(\boldsymbol{A} + \boldsymbol{A}^T) \tag{A.6}$$

is a symmetric matrix representing the *symmetric* part of $\boldsymbol{A}$, and

$$\boldsymbol{A}_a = \frac{1}{2}(\boldsymbol{A} - \boldsymbol{A}^T) \tag{A.7}$$

is a skew-symmetric matrix representing the *skew-symmetric* part of $\boldsymbol{A}$.

The row-by-column *product* of a matrix $\boldsymbol{A}$ of dimensions $(m \times p)$ by a matrix $\boldsymbol{B}$ of dimensions $(p \times n)$ is the matrix of dimensions $(m \times n)$

$$\boldsymbol{C} = \boldsymbol{A}\boldsymbol{B} \tag{A.8}$$

whose elements are given by $c_{ij} = \sum_{k=1}^{p} a_{ik}b_{kj}$. The following properties hold:

$$\begin{aligned}
\boldsymbol{A} &= \boldsymbol{A}\boldsymbol{I}_p = \boldsymbol{I}_m\boldsymbol{A}\\
\boldsymbol{A}(\boldsymbol{B}\boldsymbol{C}) &= (\boldsymbol{A}\boldsymbol{B})\boldsymbol{C}\\
\boldsymbol{A}(\boldsymbol{B}+\boldsymbol{C}) &= \boldsymbol{A}\boldsymbol{B} + \boldsymbol{A}\boldsymbol{C}\\
(\boldsymbol{A}+\boldsymbol{B})\boldsymbol{C} &= \boldsymbol{A}\boldsymbol{C} + \boldsymbol{B}\boldsymbol{C}\\
(\boldsymbol{A}\boldsymbol{B})^T &= \boldsymbol{B}^T\boldsymbol{A}^T.
\end{aligned}$$

Notice that, in general, $\boldsymbol{A}\boldsymbol{B} \neq \boldsymbol{B}\boldsymbol{A}$, and $\boldsymbol{A}\boldsymbol{B} = \boldsymbol{O}$ does not imply that $\boldsymbol{A} = \boldsymbol{O}$ or $\boldsymbol{B} = \boldsymbol{O}$; further, notice that $\boldsymbol{A}\boldsymbol{C} = \boldsymbol{B}\boldsymbol{C}$ does not imply that $\boldsymbol{A} = \boldsymbol{B}$.

If an $(m \times p)$ matrix $\boldsymbol{A}$ and a $(p \times n)$ matrix $\boldsymbol{B}$ are partitioned in such a way that the number of blocks for each row of $\boldsymbol{A}$ is equal to the number of blocks for each column of $\boldsymbol{B}$, and the blocks $\boldsymbol{A}_{ik}$ and $\boldsymbol{B}_{kj}$ have dimensions compatible with product, the matrix product $\boldsymbol{A}\boldsymbol{B}$ can be formally obtained by operating by rows and columns on the blocks of proper position and treating them like elements.

For an $(n \times n)$ *square* matrix $\boldsymbol{A}$, the *determinant* of $\boldsymbol{A}$ is the scalar given by the following expression, which holds $\forall i = 1, \ldots, n$:

$$\det(\boldsymbol{A}) = \sum_{j=1}^{n} a_{ij}(-1)^{i+j}\det(\boldsymbol{A}_{(ij)}). \tag{A.9}$$

The determinant can be computed according to any row i as in (A.9); the same result is obtained by computing it according to any column j. If $n = 1$, then $\det(a_{11}) = a_{11}$. The following property holds:

$$\det(\boldsymbol{A}) = \det(\boldsymbol{A}^T).$$

Moreover, interchanging two generic columns p and q of a matrix $\boldsymbol{A}$ yields

$$\det([\,\boldsymbol{a}_1 \ldots \boldsymbol{a}_p \ldots \boldsymbol{a}_q \ldots \boldsymbol{a}_n\,]) = -\det([\,\boldsymbol{a}_1 \ldots \boldsymbol{a}_q \ldots \boldsymbol{a}_p \ldots \boldsymbol{a}_n\,]).$$

As a consequence, if a matrix has two equal columns (rows), then its determinant is null. Also, it is $\det(\alpha\boldsymbol{A}) = \alpha^n\det(\boldsymbol{A})$.

Given an $(m \times n)$ matrix $\boldsymbol{A}$, the determinant of the square block obtained by selecting an equal number k of rows and columns is said to be k-order *minor* of matrix $\boldsymbol{A}$. The minors obtained by taking the *first* k rows and columns of $\boldsymbol{A}$ are said to be *principal* minors.

If $\boldsymbol{A}$ and $\boldsymbol{B}$ are square matrices, then

$$\det(\boldsymbol{AB}) = \det(\boldsymbol{A})\det(\boldsymbol{B}). \tag{A.10}$$

If $\boldsymbol{A}$ is an $(n \times n)$ triangular matrix (in particular diagonal), then

$$\det(\boldsymbol{A}) = \prod_{i=1}^{n} a_{ii}.$$

More generally, if $\boldsymbol{A}$ is block-triangular with m blocks $\boldsymbol{A}_{ii}$ on the diagonal, then

$$\det(\boldsymbol{A}) = \prod_{i=1}^{m} \det(\boldsymbol{A}_{ii}).$$

A square matrix $\boldsymbol{A}$ is said to be *singular* when $\det(\boldsymbol{A}) = 0$.

The *rank* $\varrho(\boldsymbol{A})$ of a matrix $\boldsymbol{A}$ of dimensions $(m \times n)$ is the maximum integer r so that at least a non-null minor of order r exists. The following properties hold:

$$\begin{aligned}
&\varrho(\boldsymbol{A}) \le \min\{m, n\} \\
&\varrho(\boldsymbol{A}) = \varrho(\boldsymbol{A}^T) \\
&\varrho(\boldsymbol{A}^T\boldsymbol{A}) = \varrho(\boldsymbol{A}) \\
&\varrho(\boldsymbol{AB}) \le \min\{\varrho(\boldsymbol{A}), \varrho(\boldsymbol{B})\}.
\end{aligned}$$

A matrix so that $\varrho(\boldsymbol{A}) = \min\{m, n\}$ is said to be *full-rank*.

The *adjoint* of a square matrix $\boldsymbol{A}$ is the matrix

$$\text{Adj}\,\boldsymbol{A} = [(-1)^{i+j}\det(\boldsymbol{A}_{(ij)})]^T_{\substack{i=1,\dots,n \\ j=1,\dots,n}}. \tag{A.11}$$

An $(n \times n)$ square matrix $\boldsymbol{A}$ is said to be *invertible* if a matrix $\boldsymbol{A}^{-1}$ exists, termed *inverse* of $\boldsymbol{A}$, so that

$$\boldsymbol{A}^{-1}\boldsymbol{A} = \boldsymbol{A}\boldsymbol{A}^{-1} = \boldsymbol{I}_n.$$

Since $\varrho(\boldsymbol{I}_n) = n$, an $(n \times n)$ square matrix $\boldsymbol{A}$ is invertible if and only if $\varrho(\boldsymbol{A}) = n$, i.e., $\det(\boldsymbol{A}) \ne 0$ (nonsingular matrix). The inverse of $\boldsymbol{A}$ can be computed as

$$\boldsymbol{A}^{-1} = \frac{1}{\det(\boldsymbol{A})}\text{Adj}\,\boldsymbol{A}. \tag{A.12}$$

The following properties hold:

$$\begin{aligned}
&(\boldsymbol{A}^{-1})^{-1} = \boldsymbol{A} \\
&(\boldsymbol{A}^T)^{-1} = (\boldsymbol{A}^{-1})^T.
\end{aligned}$$

If the inverse of a square matrix is equal to its transpose

$$\boldsymbol{A}^T = \boldsymbol{A}^{-1} \tag{A.13}$$

then the matrix is said to be *orthogonal*; in this case it is

$$\boldsymbol{A}\boldsymbol{A}^T = \boldsymbol{A}^T\boldsymbol{A} = \boldsymbol{I}. \tag{A.14}$$

A square matrix $\boldsymbol{A}$ is said *idempotent* if

$$\boldsymbol{A}\boldsymbol{A} = \boldsymbol{A}. \tag{A.15}$$

If $\boldsymbol{A}$ and $\boldsymbol{B}$ are invertible square matrices of the same dimensions, then

$$(\boldsymbol{A}\boldsymbol{B})^{-1} = \boldsymbol{B}^{-1}\boldsymbol{A}^{-1}. \tag{A.16}$$

Given n square matrices $\boldsymbol{A}_{ii}$ all invertible, the following expression holds:

$$\left(\text{diag}\{\boldsymbol{A}_{11}, \dots, \boldsymbol{A}_{nn}\}\right)^{-1} = \text{diag}\{\boldsymbol{A}_{11}^{-1}, \dots, \boldsymbol{A}_{nn}^{-1}\}.$$

where $\text{diag}\{\boldsymbol{A}_{11}, \dots, \boldsymbol{A}_{nn}\}$ denotes the block-diagonal matrix.

If $\boldsymbol{A}$ and $\boldsymbol{C}$ are invertible square matrices of proper dimensions, the following expression holds:

$$(\boldsymbol{A} + \boldsymbol{B}\boldsymbol{C}\boldsymbol{D})^{-1} = \boldsymbol{A}^{-1} - \boldsymbol{A}^{-1}\boldsymbol{B}(\boldsymbol{D}\boldsymbol{A}^{-1}\boldsymbol{B} + \boldsymbol{C}^{-1})^{-1}\boldsymbol{D}\boldsymbol{A}^{-1},$$

where the matrix $\boldsymbol{D}\boldsymbol{A}^{-1}\boldsymbol{B} + \boldsymbol{C}^{-1}$ must be invertible.

If a block-partitioned matrix is invertible, then its inverse is given by the general expression

$$\begin{bmatrix} \boldsymbol{A} & \boldsymbol{D} \\ \boldsymbol{C} & \boldsymbol{B} \end{bmatrix}^{-1} = \begin{bmatrix} \boldsymbol{A}^{-1} + \boldsymbol{E}\boldsymbol{\Delta}^{-1}\boldsymbol{F} & -\boldsymbol{E}\boldsymbol{\Delta}^{-1} \\ -\boldsymbol{\Delta}^{-1}\boldsymbol{F} & \boldsymbol{\Delta}^{-1} \end{bmatrix} \tag{A.17}$$

where $\boldsymbol{\Delta} = \boldsymbol{B} - \boldsymbol{C}\boldsymbol{A}^{-1}\boldsymbol{D}$, $\boldsymbol{E} = \boldsymbol{A}^{-1}\boldsymbol{D}$ and $\boldsymbol{F} = \boldsymbol{C}\boldsymbol{A}^{-1}$, under the assumption that the inverses of matrices $\boldsymbol{A}$ and $\boldsymbol{\Delta}$ exist. In the case of a block-triangular matrix, invertibility of the matrix requires invertibility of the blocks on the diagonal. The following expressions hold:

$$\begin{bmatrix} \boldsymbol{A} & \boldsymbol{O} \\ \boldsymbol{C} & \boldsymbol{B} \end{bmatrix}^{-1} = \begin{bmatrix} \boldsymbol{A}^{-1} & \boldsymbol{O} \\ -\boldsymbol{B}^{-1}\boldsymbol{C}\boldsymbol{A}^{-1} & \boldsymbol{B}^{-1} \end{bmatrix}$$

$$\begin{bmatrix} \boldsymbol{A} & \boldsymbol{D} \\ \boldsymbol{O} & \boldsymbol{B} \end{bmatrix}^{-1} = \begin{bmatrix} \boldsymbol{A}^{-1} & -\boldsymbol{A}^{-1}\boldsymbol{D}\boldsymbol{B}^{-1} \\ \boldsymbol{O} & \boldsymbol{B}^{-1} \end{bmatrix}.$$

The *derivative* of an $(m \times n)$ matrix $\boldsymbol{A}(t)$, whose elements $a_{ij}(t)$ are differentiable functions, is the matrix

$$\dot{\boldsymbol{A}}(t) = \frac{d}{dt}\boldsymbol{A}(t) = \left[\frac{d}{dt}a_{ij}(t)\right]_{\substack{i=1,\dots,m \\ j=1,\dots,n}}. \tag{A.18}$$

If an $(n \times n)$ square matrix $\boldsymbol{A}(t)$ is so that $\varrho(\boldsymbol{A}(t)) = n \; \forall t$ and its elements $a_{ij}(t)$ are differentiable functions, then the derivative of the *inverse* of $\boldsymbol{A}(t)$ is given by

$$\frac{d}{dt}\boldsymbol{A}^{-1}(t) = -\boldsymbol{A}^{-1}(t)\dot{\boldsymbol{A}}(t)\boldsymbol{A}^{-1}(t). \tag{A.19}$$

Given a scalar function $f(\boldsymbol{x})$, endowed with partial derivatives with respect to the elements x_i of the $(n \times 1)$ vector $\boldsymbol{x}$, the *gradient* of function f with respect to vector $\boldsymbol{x}$ is the $(n \times 1)$ column vector

$$\nabla_{\boldsymbol{x}} f(\boldsymbol{x}) = \left(\frac{\partial f(\boldsymbol{x})}{\partial \boldsymbol{x}}\right)^T = \begin{bmatrix} \dfrac{\partial f(\boldsymbol{x})}{\partial x_1} & \dfrac{\partial f(\boldsymbol{x})}{\partial x_2} & \dots & \dfrac{\partial f(\boldsymbol{x})}{\partial x_n} \end{bmatrix}^T. \tag{A.20}$$

Further, if $\boldsymbol{x}(t)$ is a differentiable function with respect to t, then

$$\dot{f}(\boldsymbol{x}) = \frac{d}{dt} f(\boldsymbol{x}(t)) = \frac{\partial f}{\partial \boldsymbol{x}}\dot{\boldsymbol{x}} = \nabla_{\boldsymbol{x}}^T f(\boldsymbol{x})\dot{\boldsymbol{x}}. \tag{A.21}$$

Given a vector function $\boldsymbol{g}(\boldsymbol{x})$ of dimensions $(m \times 1)$, whose elements g_i are differentiable with respect to the vector $\boldsymbol{x}$ of dimensions $(n \times 1)$, the Jacobian matrix (or simply *Jacobian*) of the function is defined as the $(m \times n)$ matrix

$$\boldsymbol{J}_g(\boldsymbol{x}) = \frac{\partial \boldsymbol{g}(\boldsymbol{x})}{\partial \boldsymbol{x}} = \begin{bmatrix} \dfrac{\partial g_1(\boldsymbol{x})}{\partial \boldsymbol{x}} \\ \dfrac{\partial g_2(\boldsymbol{x})}{\partial \boldsymbol{x}} \\ \vdots \\ \dfrac{\partial g_m(\boldsymbol{x})}{\partial \boldsymbol{x}} \end{bmatrix}. \tag{A.22}$$

If $\boldsymbol{x}(t)$ is a differentiable function with respect to t, then

$$\dot{\boldsymbol{g}}(\boldsymbol{x}) = \frac{d}{dt}\boldsymbol{g}(\boldsymbol{x}(t)) = \frac{\partial \boldsymbol{g}}{\partial \boldsymbol{x}}\dot{\boldsymbol{x}} = \boldsymbol{J}_g(\boldsymbol{x})\dot{\boldsymbol{x}}. \tag{A.23}$$

A.3 Vector Operations

Given n vectors $\boldsymbol{x}_i$ of dimensions $(m \times 1)$, they are said to be *linearly independent* if the expression

$$k_1\boldsymbol{x}_1 + k_2\boldsymbol{x}_2 + \ldots + k_n\boldsymbol{x}_n = \boldsymbol{0}$$

holds true only when all the constants k_i vanish. A necessary and sufficient condition for the vectors $\boldsymbol{x}_1, \boldsymbol{x}_2 \ldots, \boldsymbol{x}_n$ to be linearly independent is that the matrix

$$\boldsymbol{A} = \begin{bmatrix} \boldsymbol{x}_1 & \boldsymbol{x}_2 & \dots & \boldsymbol{x}_n \end{bmatrix}$$

has rank n; this implies that a necessary condition for linear independence is that $n \leq m$. If instead $\varrho(\boldsymbol{A}) = r < n$, then only r vectors are linearly independent and the remaining $n - r$ vectors can be expressed as a linear combination of the previous ones.

A system of vectors $\mathcal{X}$ is a *vector space* on the field of real numbers $\mathbb{R}$ if the operations of *sum of two vectors* of $\mathcal{X}$ and *product of a scalar by a vector* of $\mathcal{X}$ have values in $\mathcal{X}$ and the following properties hold:

$$
\begin{aligned}
&\boldsymbol{x} + \boldsymbol{y} = \boldsymbol{y} + \boldsymbol{x} \qquad \forall \boldsymbol{x}, \boldsymbol{y} \in \mathcal{X} \\
&(\boldsymbol{x} + \boldsymbol{y}) + \boldsymbol{z} = \boldsymbol{x} + (\boldsymbol{y} + \boldsymbol{z}) \qquad \forall \boldsymbol{x}, \boldsymbol{y}, \boldsymbol{z} \in \mathcal{X} \\
&\exists \boldsymbol{0} \in \mathcal{X} : \boldsymbol{x} + \boldsymbol{0} = \boldsymbol{x} \qquad \forall \boldsymbol{x} \in \mathcal{X} \\
&\forall \boldsymbol{x} \in \mathcal{X}, \ \exists (-\boldsymbol{x}) \in \mathcal{X} : \boldsymbol{x} + (-\boldsymbol{x}) = \boldsymbol{0} \\
&1\boldsymbol{x} = \boldsymbol{x} \qquad \forall \boldsymbol{x} \in \mathcal{X} \\
&\alpha(\beta \boldsymbol{x}) = (\alpha\beta)\boldsymbol{x} \qquad \forall \alpha, \beta \in \mathbb{R} \ \ \forall \boldsymbol{x} \in \mathcal{X} \\
&(\alpha + \beta)\boldsymbol{x} = \alpha\boldsymbol{x} + \beta\boldsymbol{x} \qquad \forall \alpha, \beta \in \mathbb{R} \ \ \forall \boldsymbol{x} \in \mathcal{X} \\
&\alpha(\boldsymbol{x} + \boldsymbol{y}) = \alpha\boldsymbol{x} + \alpha\boldsymbol{y} \qquad \forall \alpha \in \mathbb{R} \ \ \forall \boldsymbol{x}, \boldsymbol{y} \in \mathcal{X}.
\end{aligned}
$$

The *dimension* of the space $\dim(\mathcal{X})$ is the maximum number of linearly independent vectors $\boldsymbol{x}$ in the space. A set $\{\boldsymbol{x}_1, \boldsymbol{x}_2, \ldots, \boldsymbol{x}_n\}$ of linearly independent vectors is a *basis* of vector space $\mathcal{X}$, and each vector $\boldsymbol{y}$ in the space can be uniquely expressed as a linear combination of vectors from the basis

$$
\boldsymbol{y} = c_1\boldsymbol{x}_1 + c_2\boldsymbol{x}_2 + \ldots + c_n\boldsymbol{x}_n, \tag{A.24}
$$

where the constants $c_1, c_2, \ldots, c_n$ are said to be the *components* of the vector $\boldsymbol{y}$ in the basis $\{\boldsymbol{x}_1, \boldsymbol{x}_2, \ldots, \boldsymbol{x}_n\}$.

A subset $\mathcal{Y}$ of a vector space $\mathcal{X}$ is a *subspace* $\mathcal{Y} \subseteq \mathcal{X}$ if it is a vector space with the operations of vector sum and product of a scalar by a vector, i.e.,

$$
\alpha\boldsymbol{x} + \beta\boldsymbol{y} \in \mathcal{Y} \qquad \forall \alpha, \beta \in \mathbb{R} \ \ \forall \boldsymbol{x}, \boldsymbol{y} \in \mathcal{Y}.
$$

According to a geometric interpretation, a subspace is a *hyperplane* passing by the origin (null element) of $\mathcal{X}$.

The *scalar product* $<\boldsymbol{x}, \boldsymbol{y}>$ of two vectors $\boldsymbol{x}$ and $\boldsymbol{y}$ of dimensions $(m \times 1)$ is the scalar that is obtained by summing the products of the respective components in a given basis

$$
<\boldsymbol{x}, \boldsymbol{y}> = x_1 y_1 + x_2 y_2 + \ldots + x_m y_m = \boldsymbol{x}^T\boldsymbol{y} = \boldsymbol{y}^T\boldsymbol{x}. \tag{A.25}
$$

Two vectors are said to be *orthogonal* when their scalar product is null:

$$
\boldsymbol{x}^T\boldsymbol{y} = 0. \tag{A.26}
$$

The *norm* of a vector can be defined as

$$
\|\boldsymbol{x}\| = \sqrt{\boldsymbol{x}^T\boldsymbol{x}}. \tag{A.27}
$$

It is possible to show that both the *triangle inequality*

$$\|\boldsymbol{x} + \boldsymbol{y}\| \le \|\boldsymbol{x}\| + \|\boldsymbol{y}\| \tag{A.28}$$

and the *Schwarz inequality*

$$|\boldsymbol{x}^T \boldsymbol{y}| \le \|\boldsymbol{x}\|\,\|\boldsymbol{y}\|. \tag{A.29}$$

hold. A *unit vector* $\hat{\boldsymbol{x}}$ is a vector whose *norm* is unity, i.e., $\hat{\boldsymbol{x}}^T \hat{\boldsymbol{x}} = 1$. Given a vector $\boldsymbol{x}$, its unit vector is obtained by dividing each component by its norm:

$$\widehat{\boldsymbol{x}} = \frac{1}{\|\boldsymbol{x}\|}\boldsymbol{x}. \tag{A.30}$$

A typical example of vector space is the *Euclidean space* whose dimension is 3; in this case a basis is constituted by the unit vectors of a coordinate frame.

The *vector product* of two vectors $\boldsymbol{x}$ and $\boldsymbol{y}$ in the Euclidean space is the vector

$$\boldsymbol{x} \times \boldsymbol{y} = \begin{bmatrix} x_2 y_3 - x_3 y_2 \\ x_3 y_1 - x_1 y_3 \\ x_1 y_2 - x_2 y_1 \end{bmatrix}. \tag{A.31}$$

The following properties hold:

$$\begin{aligned} &\boldsymbol{x} \times \boldsymbol{x} = \boldsymbol{0} \\ &\boldsymbol{x} \times \boldsymbol{y} = -\boldsymbol{y} \times \boldsymbol{x} \\ &\boldsymbol{x} \times (\boldsymbol{y} + \boldsymbol{z}) = \boldsymbol{x} \times \boldsymbol{y} + \boldsymbol{x} \times \boldsymbol{z}. \end{aligned}$$

The vector product of two vectors $\boldsymbol{x}$ and $\boldsymbol{y}$ can be expressed also as the product of a matrix operator $\boldsymbol{S}(\boldsymbol{x})$ by the vector $\boldsymbol{y}$. In fact, by introducing the *skew-symmetric* matrix

$$\boldsymbol{S}(\boldsymbol{x}) = \begin{bmatrix} 0 & -x_3 & x_2 \\ x_3 & 0 & -x_1 \\ -x_2 & x_1 & 0 \end{bmatrix} \tag{A.32}$$

obtained with the components of vector $\boldsymbol{x}$, the vector product $\boldsymbol{x} \times \boldsymbol{y}$ is given by

$$\boldsymbol{x} \times \boldsymbol{y} = \boldsymbol{S}(\boldsymbol{x})\boldsymbol{y} = -\boldsymbol{S}(\boldsymbol{y})\boldsymbol{x} \tag{A.33}$$

as can be easily verified. Moreover, the following properties hold:

$$\begin{aligned} &\boldsymbol{S}(\boldsymbol{x})\boldsymbol{x} = \boldsymbol{S}^T(\boldsymbol{x})\boldsymbol{x} = \boldsymbol{0} \\ &\boldsymbol{S}(\alpha\boldsymbol{x} + \beta\boldsymbol{y}) = \alpha\boldsymbol{S}(\boldsymbol{x}) + \beta\boldsymbol{S}(\boldsymbol{y}). \end{aligned}$$

Given three vectors $\boldsymbol{x}$, $\boldsymbol{y}$, $\boldsymbol{z}$ in the Euclidean space, the following expressions hold for the *scalar triple products*:

$$\boldsymbol{x}^T(\boldsymbol{y} \times \boldsymbol{z}) = \boldsymbol{y}^T(\boldsymbol{z} \times \boldsymbol{x}) = \boldsymbol{z}^T(\boldsymbol{x} \times \boldsymbol{y}). \tag{A.34}$$

If any two vectors of three are equal, then the scalar triple product is null; e.g.,

$$\boldsymbol{x}^T(\boldsymbol{x} \times \boldsymbol{y}) = \boldsymbol{0}.$$

A.4 Linear Transformation

Consider a vector space $\mathcal{X}$ of dimension n and a vector space $\mathcal{Y}$ of dimension m with $m \leq n$. The *linear transformation* (or linear map) between the vectors $\boldsymbol{x} \in \mathcal{X}$ and $\boldsymbol{y} \in \mathcal{Y}$ can be defined as

$$\boldsymbol{y} = \boldsymbol{A}\boldsymbol{x} \tag{A.35}$$

in terms of the matrix $\boldsymbol{A}$ of dimensions $(m \times n)$. The *range space* (or simply range) of the transformation is the subspace

$$\mathcal{R}(\boldsymbol{A}) = \{\boldsymbol{y} : \boldsymbol{y} = \boldsymbol{A}\boldsymbol{x},\ \boldsymbol{x} \in \mathcal{X}\} \subseteq \mathcal{Y}, \tag{A.36}$$

which is the subspace generated by the linearly independent columns of matrix $\boldsymbol{A}$ taken as a basis of $\mathcal{Y}$. It is easy to recognize that

$$\varrho(\boldsymbol{A}) = \dim(\mathcal{R}(\boldsymbol{A})). \tag{A.37}$$

On the other hand, the *null space* (or simply null) of the transformation is the subspace

$$\mathcal{N}(\boldsymbol{A}) = \{\boldsymbol{x} : \boldsymbol{A}\boldsymbol{x} = \boldsymbol{0},\ \boldsymbol{x} \in \mathcal{X}\} \subseteq \mathcal{X}. \tag{A.38}$$

Given a matrix $\boldsymbol{A}$ of dimensions $(m \times n)$, the notable result holds:

$$\varrho(\boldsymbol{A}) + \dim(\mathcal{N}(\boldsymbol{A})) = n. \tag{A.39}$$

Therefore, if $\varrho(\boldsymbol{A}) = r \leq \min\{m, n\}$, then $\dim(\mathcal{R}(\boldsymbol{A})) = r$ and $\dim(\mathcal{N}(\boldsymbol{A})) = n - r$. It follows that if $m < n$, then $\mathcal{N}(\boldsymbol{A}) \neq \emptyset$ independently of the rank of $\boldsymbol{A}$; if $m = n$, then $\mathcal{N}(\boldsymbol{A}) \neq \emptyset$ only in the case of $\varrho(\boldsymbol{A}) = r < m$.

If $\boldsymbol{x} \in \mathcal{N}(\boldsymbol{A})$ and $\boldsymbol{y} \in \mathcal{R}(\boldsymbol{A}^T)$, then $\boldsymbol{y}^T\boldsymbol{x} = 0$, i.e., the vectors in the null space of $\boldsymbol{A}$ are orthogonal to each vector in the range space of the transpose of $\boldsymbol{A}$. It can be shown that the set of vectors orthogonal to each vector of the range space of $\boldsymbol{A}^T$ coincides with the null space of $\boldsymbol{A}$, whereas the set of vectors orthogonal to each vector in the null space of $\boldsymbol{A}^T$ coincides with the range space of $\boldsymbol{A}$. In symbols:

$$\mathcal{N}(\boldsymbol{A}) \equiv \mathcal{R}^{\perp}(\boldsymbol{A}^T) \qquad \mathcal{R}(\boldsymbol{A}) \equiv \mathcal{N}^{\perp}(\boldsymbol{A}^T) \tag{A.40}$$

where $\perp$ denotes the *orthogonal complement* of a subspace.

If the matrix $\boldsymbol{A}$ in (A.35) is square and idempotent, the matrix represents the *projection* of space $\mathcal{X}$ into a subspace.

A linear transformation allows the definition of the *norm* of a matrix $\boldsymbol{A}$ induced by the norm defined for a vector $\boldsymbol{x}$ as follows. In view of the property

$$\|\boldsymbol{A}\boldsymbol{x}\| \leq \|\boldsymbol{A}\|\ \|\boldsymbol{x}\|, \tag{A.41}$$

the norm of $\boldsymbol{A}$ can be defined as

$$\|\boldsymbol{A}\| = \sup_{\boldsymbol{x} \neq \boldsymbol{0}} \frac{\|\boldsymbol{A}\boldsymbol{x}\|}{\|\boldsymbol{x}\|} \tag{A.42}$$

which can also be computed as

$$\max_{\|\boldsymbol{x}\|=1} \|\boldsymbol{A}\boldsymbol{x}\|.$$

A direct consequence of (A.41) is the property

$$\|\boldsymbol{A}\boldsymbol{B}\| \leq \|\boldsymbol{A}\| \, \|\boldsymbol{B}\|. \tag{A.43}$$

A different norm of a matrix is the *Frobenius norm* defined as

$$\|\boldsymbol{A}\|_F = \left(\mathrm{Tr}(\boldsymbol{A}^T\boldsymbol{A})\right)^{1/2} \tag{A.44}$$

A.5 Eigenvalues and Eigenvectors

Consider the linear transformation on a vector $\boldsymbol{u}$ established by an $(n \times n)$ square matrix $\boldsymbol{A}$. If the vector resulting from the transformation has the same direction of $\boldsymbol{u}$ (with $\boldsymbol{u} \neq \boldsymbol{0}$), then

$$\boldsymbol{A}\boldsymbol{u} = \lambda\boldsymbol{u}. \tag{A.45}$$

The equation in (A.45) can be rewritten in matrix form as

$$(\lambda\boldsymbol{I} - \boldsymbol{A})\boldsymbol{u} = \boldsymbol{0}. \tag{A.46}$$

For the homogeneous system of equations in (A.46) to have a solution different from the trivial one $\boldsymbol{u} = \boldsymbol{0}$, it must be

$$\det(\lambda\boldsymbol{I} - \boldsymbol{A}) = 0 \tag{A.47}$$

which is termed a *characteristic equation*. Its solutions $\lambda_1, \ldots, \lambda_n$ are the *eigenvalues* of matrix $\boldsymbol{A}$; they coincide with the eigenvalues of matrix $\boldsymbol{A}^T$. On the assumption of distinct eigenvalues, the n vectors $\boldsymbol{u}_i$ satisfying the equation

$$(\lambda_i\boldsymbol{I} - \boldsymbol{A})\boldsymbol{u}_i = \boldsymbol{0} \qquad i = 1, \ldots, n \tag{A.48}$$

are said to be the *eigenvectors* associated with the eigenvalues λ_i.

The matrix $\boldsymbol{U}$ formed by the column vectors $\boldsymbol{u}_i$ is invertible and constitutes a basis in the space of dimension n. Further, the *similarity transformation* established by $\boldsymbol{U}$

$$\boldsymbol{\Lambda} = \boldsymbol{U}^{-1}\boldsymbol{A}\boldsymbol{U} \tag{A.49}$$

is so that $\boldsymbol{\Lambda} = \mathrm{diag}\{\lambda_1, \ldots, \lambda_n\}$. It follows that $\det(\boldsymbol{A}) = \prod_{i=1}^{n} \lambda_i$.

If the matrix $\boldsymbol{A}$ is *symmetric*, its eigenvalues are real and $\boldsymbol{\Lambda}$ can be written as

$$\boldsymbol{\Lambda} = \boldsymbol{U}^T\boldsymbol{A}\boldsymbol{U}; \tag{A.50}$$

hence, the eigenvector matrix $\boldsymbol{U}$ is orthogonal.

A.6 Bilinear Forms and Quadratic Forms

A *bilinear form* in the variables x_i and y_j is the scalar

$$B = \sum_{i=1}^{m}\sum_{j=1}^{n} a_{ij}x_i y_j$$

which can be written in matrix form

$$B(\boldsymbol{x},\boldsymbol{y}) = \boldsymbol{x}^T\boldsymbol{A}\boldsymbol{y} = \boldsymbol{y}^T\boldsymbol{A}^T\boldsymbol{x} \tag{A.51}$$

where $\boldsymbol{x} = [\,x_1 \quad x_2 \quad \dots \quad x_m\,]^T$, $\boldsymbol{y} = [\,y_1 \quad y_2 \quad \dots \quad y_n\,]^T$, and $\boldsymbol{A}$ is the $(m \times n)$ matrix of the coefficients a_{ij} representing the core of the form.

A special case of bilinear form is the *quadratic form*

$$Q(\boldsymbol{x}) = \boldsymbol{x}^T\boldsymbol{A}\boldsymbol{x} \tag{A.52}$$

where $\boldsymbol{A}$ is an $(n \times n)$ square matrix. Hence, for computation of (A.52), the matrix $\boldsymbol{A}$ can be replaced with its symmetric part $\boldsymbol{A}_s$ given by (A.6). It follows that if $\boldsymbol{A}$ is a *skew-symmetric* matrix, then

$$\boldsymbol{x}^T\boldsymbol{A}\boldsymbol{x} = \boldsymbol{0} \qquad \forall\boldsymbol{x}.$$

The quadratic form (A.52) is said to be *positive definite* if

$$\boldsymbol{x}^T\boldsymbol{A}\boldsymbol{x} > 0 \quad \forall\boldsymbol{x} \neq \boldsymbol{0} \qquad\qquad \boldsymbol{x}^T\boldsymbol{A}\boldsymbol{x} = 0 \quad \boldsymbol{x} = \boldsymbol{0}. \tag{A.53}$$

The matrix $\boldsymbol{A}$ core of the form is also said to be *positive definite*. Analogously, a quadratic form is said to be *negative definite* if it can be written as $-Q(\boldsymbol{x}) = -\boldsymbol{x}^T\boldsymbol{A}\boldsymbol{x}$ where $Q(\boldsymbol{x})$ is positive definite.

A necessary condition for a square matrix to be positive definite is that its elements on the diagonal are strictly positive. Further, in view of (A.50), the eigenvalues of a positive definite matrix are all positive. If the eigenvalues are not known, a necessary and sufficient condition for a symmetric matrix to be positive definite is that its principal minors are strictly positive (*Sylvester criterion*). It follows that a positive definite matrix is full-rank and thus it is always invertible.

A symmetric positive definite matrix $\boldsymbol{A}$ can always be decomposed as

$$\boldsymbol{A} = \boldsymbol{U}^T\boldsymbol{\Lambda}\boldsymbol{U} \tag{A.54}$$

where $\boldsymbol{U}$ is an orthogonal matrix of eigenvectors ($\boldsymbol{U}^T\boldsymbol{U} = \boldsymbol{I}$) and $\boldsymbol{\Lambda}$ is the diagonal matrix of the eigenvalues of $\boldsymbol{A}$.

Let $\lambda_{\min}(\boldsymbol{A})$ and $\lambda_{\max}(\boldsymbol{A})$ respectively denote the smallest and largest eigenvalues of a positive definite matrix $\boldsymbol{A}$ ($\lambda_{\min}, \lambda_{\max} > 0$). Then, the quadratic form in (A.52) satisfies the following inequality:

$$\lambda_{\min}(\boldsymbol{A})\|\boldsymbol{x}\|^2 \leq \boldsymbol{x}^T\boldsymbol{A}\boldsymbol{x} \leq \lambda_{\max}(\boldsymbol{A})\|\boldsymbol{x}\|^2. \tag{A.55}$$

An $(n \times n)$ square matrix $\boldsymbol{A}$ is said to be *positive semi-definite* if

$$\boldsymbol{x}^T \boldsymbol{A} \boldsymbol{x} \geq 0 \qquad \forall \boldsymbol{x}. \tag{A.56}$$

This definition implies that $\varrho(\boldsymbol{A}) = r < n$, and thus r eigenvalues of $\boldsymbol{A}$ are positive and $n - r$ are null. Therefore, a positive semi-definite matrix $\boldsymbol{A}$ has a null space of finite dimension, and specifically the form vanishes when $\boldsymbol{x} \in \mathcal{N}(\boldsymbol{A})$. A typical example of a positive semi-definite matrix is the matrix $\boldsymbol{A} = \boldsymbol{H}^T \boldsymbol{H}$ where $\boldsymbol{H}$ is an $(m \times n)$ matrix with $m < n$. In an analogous way, a *negative semi-definite* matrix can be defined.

Given the *bilinear form* in (A.51), the *gradient* of the form with respect to $\boldsymbol{x}$ is given by

$$\nabla_{\boldsymbol{x}} B(\boldsymbol{x}, \boldsymbol{y}) = \left(\frac{\partial B(\boldsymbol{x}, \boldsymbol{y})}{\partial \boldsymbol{x}} \right)^T = \boldsymbol{A} \boldsymbol{y}, \tag{A.57}$$

whereas the gradient of B with respect to $\boldsymbol{y}$ is given by

$$\nabla_{\boldsymbol{y}} B(\boldsymbol{x}, \boldsymbol{y}) = \left(\frac{\partial B(\boldsymbol{x}, \boldsymbol{y})}{\partial \boldsymbol{y}} \right)^T = \boldsymbol{A}^T \boldsymbol{x}. \tag{A.58}$$

Given the *quadratic form* in (A.52) with $\boldsymbol{A}$ *symmetric*, the *gradient* of the form with respect to $\boldsymbol{x}$ is given by

$$\nabla_{\boldsymbol{x}} Q(\boldsymbol{x}) = \left(\frac{\partial Q(\boldsymbol{x})}{\partial \boldsymbol{x}} \right)^T = 2\boldsymbol{A}\boldsymbol{x}. \tag{A.59}$$

Further, if $\boldsymbol{x}$ and $\boldsymbol{A}$ are differentiable functions of t, then

$$\dot{Q}(x) = \frac{d}{dt} Q(\boldsymbol{x}(t)) = 2\boldsymbol{x}^T \boldsymbol{A} \dot{\boldsymbol{x}} + \boldsymbol{x}^T \dot{\boldsymbol{A}} \boldsymbol{x}; \tag{A.60}$$

if $\boldsymbol{A}$ is constant, then the second term obviously vanishes.

A.7 Pseudo-inverse

The inverse of a matrix can be defined only when the matrix is square and nonsingular. The inverse operation can be extended to the case of non-square matrices. Consider a matrix $\boldsymbol{A}$ of dimensions $(m \times n)$ with $\varrho(\boldsymbol{A}) = \min\{m, n\}$

If $m < n$, a *right inverse* of $\boldsymbol{A}$ can be defined as the matrix $\boldsymbol{A}_r$ of dimensions $(n \times m)$ so that

$$\boldsymbol{A}\boldsymbol{A}_r = \boldsymbol{I}_m.$$

If instead $m > n$, a *left inverse* of $\boldsymbol{A}$ can be defined as the matrix $\boldsymbol{A}_l$ of dimensions $(n \times m)$ so that

$$\boldsymbol{A}_l \boldsymbol{A} = \boldsymbol{I}_n.$$

If $\boldsymbol{A}$ has more columns than rows ($m < n$) and has rank m, a special right inverse is the matrix

$$\boldsymbol{A}_r^\dagger = \boldsymbol{A}^T(\boldsymbol{A}\boldsymbol{A}^T)^{-1} \tag{A.61}$$

which is termed *right pseudo-inverse*, since $\boldsymbol{A}\boldsymbol{A}_r^\dagger = \boldsymbol{I}_m$. If $\boldsymbol{W}_r$ is an $(n \times n)$ *positive definite* matrix, a *weighted* right pseudo-inverse is given by

$$\boldsymbol{A}_r^\dagger = \boldsymbol{W}_r^{-1}\boldsymbol{A}^T(\boldsymbol{A}\boldsymbol{W}_r^{-1}\boldsymbol{A}^T)^{-1}. \tag{A.62}$$

If $\boldsymbol{A}$ has more rows than columns ($m > n$) and has rank n, a special left inverse is the matrix

$$\boldsymbol{A}_l^\dagger = (\boldsymbol{A}^T\boldsymbol{A})^{-1}\boldsymbol{A}^T \tag{A.63}$$

which is termed *left pseudo-inverse*, since $\boldsymbol{A}_l^\dagger\boldsymbol{A} = \boldsymbol{I}_n$.[3] If $\boldsymbol{W}_l$ is an $(m \times m)$ *positive definite* matrix, a *weighted* left pseudo-inverse is given by

$$\boldsymbol{A}_l^\dagger = (\boldsymbol{A}^T\boldsymbol{W}_l\boldsymbol{A})^{-1}\boldsymbol{A}^T\boldsymbol{W}_l. \tag{A.64}$$

The pseudo-inverse is very useful to invert a linear transformation $\boldsymbol{y} = \boldsymbol{A}\boldsymbol{x}$ with $\boldsymbol{A}$ a full-rank matrix. If $\boldsymbol{A}$ is a square nonsingular matrix, then obviously $\boldsymbol{x} = \boldsymbol{A}^{-1}\boldsymbol{y}$ and then $\boldsymbol{A}_l^\dagger = \boldsymbol{A}_r^\dagger = \boldsymbol{A}^{-1}$.

If $\boldsymbol{A}$ has more columns than rows ($m < n$) and has rank m, then the solution $\boldsymbol{x}$ for a given $\boldsymbol{y}$ is not unique; it can be shown that the expression

$$\boldsymbol{x} = \boldsymbol{A}^\dagger\boldsymbol{y} + (\boldsymbol{I} - \boldsymbol{A}^\dagger\boldsymbol{A})\boldsymbol{k}, \tag{A.65}$$

with $\boldsymbol{k}$ an arbitrary $(n \times 1)$ vector and $\boldsymbol{A}^\dagger$ as in (A.61), is a solution to the system of linear equations established by (A.35). The term $\boldsymbol{A}^\dagger\boldsymbol{y} \in \mathcal{N}^\perp(\boldsymbol{A}) \equiv \mathcal{R}(\boldsymbol{A}^T)$ minimizes the norm of the solution $\|\boldsymbol{x}\|$. The term $(\boldsymbol{I} - \boldsymbol{A}^\dagger\boldsymbol{A})\boldsymbol{k}$ is the projection of $\boldsymbol{k}$ in $\mathcal{N}(\boldsymbol{A})$ and is termed *homogeneous solution*; as $\boldsymbol{k}$ varies, all the solutions to the homogeneous equation system $\boldsymbol{A}\boldsymbol{x} = \boldsymbol{0}$ associated with (A.35) are generated.

On the other hand, if $\boldsymbol{A}$ has more rows than columns ($m > n$), the equation in (A.35) has no solution; it can be shown that an *approximate* solution is given by

$$\boldsymbol{x} = \boldsymbol{A}^\dagger\boldsymbol{y} \tag{A.66}$$

where $\boldsymbol{A}^\dagger$ as in (A.63) minimizes $\|\boldsymbol{y} - \boldsymbol{A}\boldsymbol{x}\|$. If instead $\boldsymbol{y} \in \mathcal{R}(\boldsymbol{A})$, then (A.66) is a real solution.

Notice that the use of the weighted (left or right) pseudo-inverses in the solution to the linear equation systems leads to analogous results where the minimized norms are weighted according to the metrics defined by matrices $\boldsymbol{W}_r$ and $\boldsymbol{W}_l$, respectively.

The results of this section can be easily extended to the case of (square or nonsquare) matrices $\boldsymbol{A}$ not having full-rank. In particular, the expression (A.66) (with the pseudo-inverse computed by means of the singular value decomposition of $\boldsymbol{A}$) gives the minimum-norm vector among all those minimizing $\|\boldsymbol{y} - \boldsymbol{A}\boldsymbol{x}\|$.

[3] Subscripts l and r are usually omitted whenever the use of a left or right pseudo-inverse is clear from the context.

A.8 Singular Value Decomposition

For a nonsquare matrix it is not possible to define eigenvalues. An extension of the eigenvalue concept can be obtained by singular values. Given a matrix $\boldsymbol{A}$ of dimensions $(m \times n)$, the matrix $\boldsymbol{A}^T\boldsymbol{A}$ has n nonnegative eigenvalues $\lambda_1 \geq \lambda_2 \geq \ldots \geq \lambda_n \geq 0$ (ordered from the largest to the smallest) which can be expressed in the form

$$\lambda_i = \sigma_i^2 \qquad \sigma_i \geq 0.$$

The scalars $\sigma_1 \geq \sigma_2 \geq \ldots \geq \sigma_n \geq 0$ are said to be the *singular values* of matrix $\boldsymbol{A}$. The *singular value decomposition* (SVD) of matrix $\boldsymbol{A}$ is given by

$$\boldsymbol{A} = \boldsymbol{U}\boldsymbol{\Sigma}\boldsymbol{V}^T \tag{A.67}$$

where $\boldsymbol{U}$ is an $(m \times m)$ orthogonal matrix

$$\boldsymbol{U} = \begin{bmatrix} \boldsymbol{u}_1 & \boldsymbol{u}_2 & \ldots & \boldsymbol{u}_m \end{bmatrix}, \tag{A.68}$$

$\boldsymbol{V}$ is an $(n \times n)$ orthogonal matrix

$$\boldsymbol{V} = \begin{bmatrix} \boldsymbol{v}_1 & \boldsymbol{v}_2 & \ldots & \boldsymbol{v}_n \end{bmatrix} \tag{A.69}$$

and $\boldsymbol{\Sigma}$ is an $(m \times n)$ matrix

$$\boldsymbol{\Sigma} = \begin{bmatrix} \boldsymbol{D} & \boldsymbol{O} \\ \boldsymbol{O} & \boldsymbol{O} \end{bmatrix} \qquad \boldsymbol{D} = \mathrm{diag}\{\sigma_1, \sigma_2, \ldots, \sigma_r\} \tag{A.70}$$

where $\sigma_1 \geq \sigma_2 \geq \ldots \geq \sigma_r > 0$. The number of non-null singular values is equal to the rank r of matrix $\boldsymbol{A}$.

The columns of $\boldsymbol{U}$ are the eigenvectors of the matrix $\boldsymbol{A}\boldsymbol{A}^T$, whereas the columns of $\boldsymbol{V}$ are the eigenvectors of the matrix $\boldsymbol{A}^T\boldsymbol{A}$. In view of the partitions of $\boldsymbol{U}$ and $\boldsymbol{V}$ in (A.68), (A.69), it is $\boldsymbol{A}\boldsymbol{v}_i = \sigma_i \boldsymbol{u}_i$, for $i = 1, \ldots, r$ and $\boldsymbol{A}\boldsymbol{v}_i = \boldsymbol{0}$, for $i = r+1, \ldots, n$.

Singular value decomposition is useful for analysis of the linear transformation $\boldsymbol{y} = \boldsymbol{A}\boldsymbol{x}$ established in (A.35). According to a geometric interpretation, the matrix $\boldsymbol{A}$ transforms the unit sphere in $\mathbb{R}^n$ defined by $\|\boldsymbol{x}\| = 1$ into the set of vectors $\boldsymbol{y} = \boldsymbol{A}\boldsymbol{x}$ which define an *ellipsoid* of dimension r in $\mathbb{R}^m$. The singular values are the lengths of the various axes of the ellipsoid. The *condition number* of the matrix

$$\kappa = \frac{\sigma_1}{\sigma_r}$$

is related to the eccentricity of the ellipsoid and provides a measure of ill-conditioning ($\kappa \gg 1$) for numerical solution of the system established by (A.35).

It is worth noticing that the numerical procedure of singular value decomposition is commonly adopted to compute the (right or left) pseudo-inverse $\boldsymbol{A}^\dagger$, even in the case of a matrix $\boldsymbol{A}$ not having full rank. In fact, from (A.67), (A.70) it is

$$\boldsymbol{A}^\dagger = \boldsymbol{V}\boldsymbol{\Sigma}^\dagger\boldsymbol{U}^T \tag{A.71}$$

with

$$\Sigma^\dagger = \begin{bmatrix} D^\dagger & O \\ O & O \end{bmatrix} \qquad D^\dagger = \mathrm{diag}\left\{\frac{1}{\sigma_1}, \frac{1}{\sigma_2}, \ldots, \frac{1}{\sigma_r}\right\}. \tag{A.72}$$

Bibliography

A reference text on linear algebra is [169]. For matrix computation see [88]. The properties of pseudo-inverse matrices are discussed in [24].

B

Rigid-body Mechanics

The goal of this appendix is to recall some fundamental concepts of *rigid body mechanics* which are preliminary to the study of manipulator *kinematics*, *statics* and *dynamics*.

B.1 Kinematics

A *rigid body* is a system characterized by the constraint that the distance between any two points is always constant.

Consider a rigid body $\mathcal{B}$ moving with respect to an orthonormal reference frame O–xyz of unit vectors $\boldsymbol{x}$, $\boldsymbol{y}$, $\boldsymbol{z}$, called *fixed frame*. The rigidity assumption allows the introduction of an orthonormal frame O'–$x'y'z'$ attached to the body, called *moving frame*, with respect to which the position of any point of $\mathcal{B}$ is independent of time. Let $\boldsymbol{x}'(t)$, $\boldsymbol{y}'(t)$, $\boldsymbol{z}'(t)$ be the unit vectors of the moving frame expressed in the fixed frame at time t.

The orientation of the moving frame O'–$x'y'z'$ at time t with respect to the fixed frame O–xyz can be expressed by means of the *orthogonal* (3×3) matrix

$$\boldsymbol{R}(t) = \begin{bmatrix} \boldsymbol{x}'^T(t)\boldsymbol{x} & \boldsymbol{y}'^T(t)\boldsymbol{x} & \boldsymbol{z}'^T(t)\boldsymbol{x} \\ \boldsymbol{x}'^T(t)\boldsymbol{y} & \boldsymbol{y}'^T(t)\boldsymbol{y} & \boldsymbol{z}'^T(t)\boldsymbol{y} \\ \boldsymbol{x}'^T(t)\boldsymbol{z} & \boldsymbol{y}'^T(t)\boldsymbol{z} & \boldsymbol{z}'^T(t)\boldsymbol{z} \end{bmatrix}, \tag{B.1}$$

which is termed *rotation matrix* defined in the orthonormal special group $SO(3)$ of the (3×3) matrices with orthonormal columns and determinant equal to 1. The columns of the matrix in (B.1) represent the components of the unit vectors of the moving frame when expressed in the fixed frame, whereas the rows represent the components of the unit vectors of the fixed frame when expressed in the moving frame.

Let $\boldsymbol{p}'$ be the *constant* position vector of a generic point P of $\mathcal{B}$ in the moving frame O'–$x'y'z'$. The motion of P with respect to the fixed frame O–xyz is described by the equation

$$\boldsymbol{p}(t) = \boldsymbol{p}_{O'}(t) + \boldsymbol{R}(t)\boldsymbol{p}', \tag{B.2}$$

where $\boldsymbol{p}_{O'}(t)$ is the position vector of origin O' of the moving frame with respect to the fixed frame.

Notice that a position vector is a *bound vector* since its line of application and point of application are both prescribed, in addition to its direction; the point of application typically coincides with the origin of a reference frame. Therefore, to transform a bound vector from a frame to another, both translation and rotation between the two frames must be taken into account.

If the positions of the points of $\mathcal{B}$ in the moving frame are known, it follows from (B.2) that the motion of each point of $\mathcal{B}$ with respect to the fixed frame is uniquely determined once the position of the origin and the orientation of the moving frame with respect to the fixed frame are specified in time. The origin of the moving frame is determined by *three* scalar functions of time. Since the orthonormality conditions impose six constraints on the nine elements of matrix $\boldsymbol{R}(t)$, the *orientation* of the moving frame depends only on *three* independent scalar functions, three being the minimum number of parameters to represent $SO(3)$.[1]

Therefore, a rigid body motion is described by arbitrarily specifying *six* scalar functions of time, which describe the body *pose* (position + orientation). The resulting rigid motions belong to the *special Euclidean group* $SE(3) = \mathbb{R}^3 \times SO(3)$.

The expression in (B.2) continues to hold if the position vector $\boldsymbol{p}_{O'}(t)$ of the origin of the moving frame is replaced with the position vector of any other point of $\mathcal{B}$, i.e.,

$$\boldsymbol{p}(t) = \boldsymbol{p}_Q(t) + \boldsymbol{R}(t)(\boldsymbol{p}' - \boldsymbol{p}'_Q) \tag{B.3}$$

where $\boldsymbol{p}_Q(t)$ and $\boldsymbol{p}'_Q$ are the position vectors of a point Q of $\mathcal{B}$ in the fixed and moving frames, respectively.

In the following, for simplicity of notation, the dependence on the time variable t will be dropped.

Differentiating (B.3) with respect to time gives the known velocity composition rule

$$\dot{\boldsymbol{p}} = \dot{\boldsymbol{p}}_Q + \boldsymbol{\omega} \times (\boldsymbol{p} - \boldsymbol{p}_Q), \tag{B.4}$$

where $\boldsymbol{\omega}$ is the *angular velocity* of rigid body $\mathcal{B}$. Notice that $\boldsymbol{\omega}$ is a *free vector* since its point of application is not prescribed. To transform a free vector from a frame to another, only rotation between the two frames must be taken into account.

By recalling the definition of the skew-symmetric operator $\boldsymbol{S}(\cdot)$ in (A.32), the expression in (B.4) can be rewritten as

$$\begin{aligned}\dot{\boldsymbol{p}} &= \dot{\boldsymbol{p}}_Q + \boldsymbol{S}(\boldsymbol{\omega})(\boldsymbol{p} - \boldsymbol{p}_Q) \\ &= \dot{\boldsymbol{p}}_Q + \boldsymbol{S}(\boldsymbol{\omega})\boldsymbol{R}(\boldsymbol{p}' - \boldsymbol{p}'_Q).\end{aligned}$$

---

[1] The minimum number of parameters represent a special orthonormal group $SO(m)$ is equal to $m(m-1)/2$.

Comparing this equation with the formal time derivative of (B.3) leads to the result

$$\dot{\boldsymbol{R}} = \boldsymbol{S}(\boldsymbol{\omega})\boldsymbol{R}. \tag{B.5}$$

In view of (B.4), the *elementary displacement* of a point P of the rigid body $\mathcal{B}$ in the time interval $(t, t+dt)$ is

$$\begin{aligned} d\boldsymbol{p} = \dot{\boldsymbol{p}}dt &= \big(\dot{\boldsymbol{p}}_Q + \boldsymbol{\omega} \times (\boldsymbol{p} - \boldsymbol{p}_Q)\big)dt \\ &= d\boldsymbol{p}_Q + \boldsymbol{\omega}dt \times (\boldsymbol{p} - \boldsymbol{p}_Q). \end{aligned} \tag{B.6}$$

Differentiating (B.4) with respect to time yields the following expression for acceleration:

$$\ddot{\boldsymbol{p}} = \ddot{\boldsymbol{p}}_Q + \dot{\boldsymbol{\omega}} \times (\boldsymbol{p} - \boldsymbol{p}_Q) + \boldsymbol{\omega} \times \big(\boldsymbol{\omega} \times (\boldsymbol{p} - \boldsymbol{p}_Q)\big). \tag{B.7}$$

B.2 Dynamics

Let ρdV be the mass of an elementary particle of a rigid body $\mathcal{B}$, where ρ denotes the density of the particle of volume dV. Also let $V_{\mathcal{B}}$ be the body volume and $m = \int_{V_{\mathcal{B}}} \rho dV$ its *total mass* assumed to be constant. If $\boldsymbol{p}$ denotes the position vector of the particle of mass ρdV in the frame O–xyz, the *centre of mass* of $\mathcal{B}$ is defined as the point C whose position vector is

$$\boldsymbol{p}_C = \frac{1}{m}\int_{V_{\mathcal{B}}} \boldsymbol{p}\rho dV. \tag{B.8}$$

In the case when $\mathcal{B}$ is the union of n distinct parts of mass $m_1, \ldots, m_n$ and centres of mass $\boldsymbol{p}_{C1} \ldots \boldsymbol{p}_{Cn}$, the centre of mass of $\mathcal{B}$ can be computed as

$$\boldsymbol{p}_C = \frac{1}{m}\sum_{i=1}^{n} m_i \boldsymbol{p}_{Ci}$$

with $m = \sum_{i=1}^{n} m_i$.

Let r be a line passing by O and $d(\boldsymbol{p})$ the distance from r of the particle of $\mathcal{B}$ of mass ρdV and position vector $\boldsymbol{p}$. The *moment of inertia* of body $\mathcal{B}$ with respect to line r is defined as the positive scalar

$$I_r = \int_{V_{\mathcal{B}}} d^2(\boldsymbol{p})\rho dV.$$

Let $\boldsymbol{r}$ denote the unit vector of line r; then, the moment of inertia of $\mathcal{B}$ with respect to line r can be expressed as

$$I_r = \boldsymbol{r}^T\left(\int_{V_{\mathcal{B}}} \boldsymbol{S}^T(\boldsymbol{p})\boldsymbol{S}(\boldsymbol{p})\rho dV\right)\boldsymbol{r} = \boldsymbol{r}^T\boldsymbol{I}_O\boldsymbol{r}, \tag{B.9}$$

where $\boldsymbol{S}(\cdot)$ is the skew-symmetric operator in (A.31), and the *symmetric, positive definite* matrix

$$\boldsymbol{I}_O = \begin{bmatrix} \int_{V_\mathcal{B}}(p_y^2+p_z^2)\rho dV & -\int_{V_\mathcal{B}} p_x p_y \rho dV & -\int_{V_\mathcal{B}} p_x p_z \rho dV \\ * & \int_{V_\mathcal{B}}(p_x^2+p_z^2)\rho dV & -\int_{V_\mathcal{B}} p_y p_z \rho dV \\ * & * & \int_{V_\mathcal{B}}(p_x^2+p_y^2)\rho dV \end{bmatrix}$$

$$= \begin{bmatrix} I_{Oxx} & -I_{Oxy} & -I_{Oxz} \\ * & I_{Oyy} & -I_{Oyz} \\ * & * & I_{Ozz} \end{bmatrix} \tag{B.10}$$

is termed *inertia tensor* of body $\mathcal{B}$ relative to pole O.[2] The (positive) elements I_{Oxx}, I_{Oyy}, I_{Ozz} are the *inertia moments* with respect to three coordinate axes of the reference frame, whereas the elements I_{Oxy}, I_{Oxz}, I_{Oyz} (of any sign) are said to be *products of inertia*.

The expression of the inertia tensor of a rigid body $\mathcal{B}$ depends both on the pole and the reference frame. If orientation of the reference frame with origin at O is changed according to a rotation matrix $\boldsymbol{R}$, the inertia tensor $\boldsymbol{I}'_O$ in the new frame is related to $\boldsymbol{I}_O$ by the relationship

$$\boldsymbol{I}_O = \boldsymbol{R}\boldsymbol{I}'_O\boldsymbol{R}^T. \tag{B.11}$$

The way an inertia tensor is transformed when the pole is changed can be inferred by the following equation, also known as *Steiner theorem* or parallel axis theorem:

$$\boldsymbol{I}_O = \boldsymbol{I}_C + m\boldsymbol{S}^T(\boldsymbol{p}_C)\boldsymbol{S}(\boldsymbol{p}_C), \tag{B.12}$$

where $\boldsymbol{I}_C$ is the inertia tensor relative to the centre of mass of $\mathcal{B}$, when expressed in a frame parallel to the frame with origin at O and with origin at the centre of mass C.

Since the inertia tensor is a symmetric positive definite matrix, there always exists a reference frame in which the inertia tensor attains a diagonal form; such a frame is said to be a *principal frame* (relative to pole O) and its coordinate axes are said to be *principal axes*. In the case when pole O coincides with the centre of mass, the frame is said to be a *central frame* and its axes are said to be *central axes*.

Notice that if the rigid body is moving with respect to the reference frame with origin at O, then the elements of the inertia tensor $\boldsymbol{I}_O$ become a function of time. With respect to a pole and a reference frame attached to the body (moving frame), instead, the elements of the inertia tensor represent six structural constants of the body which are known once the pole and reference frame have been specified.

[2] The symbol '*' has been used to avoid rewriting the symmetric elements.

Let $\dot{\boldsymbol{p}}$ be the velocity of a particle of $\mathcal{B}$ of elementary mass ρdV in frame O–xyz. The *linear momentum* of body $\mathcal{B}$ is defined as the vector

$$\boldsymbol{l} = \int_{V_\mathcal{B}} \dot{\boldsymbol{p}} \rho dV = m\dot{\boldsymbol{p}}_C. \tag{B.13}$$

Let Ω be any point in space and $\boldsymbol{p}_\Omega$ its position vector in frame O–xyz; then, the *angular momentum* of body $\mathcal{B}$ relative to pole Ω is defined as the vector

$$\boldsymbol{k}_\Omega = \int_{V_\mathcal{B}} \dot{\boldsymbol{p}} \times (\boldsymbol{p}_\Omega - \boldsymbol{p})\rho dV.$$

The pole can be either fixed or moving with respect to the reference frame. The angular momentum of a rigid body has the following notable expression:

$$\boldsymbol{k}_\Omega = \boldsymbol{I}_C \boldsymbol{\omega} + m\dot{\boldsymbol{p}}_C \times (\boldsymbol{p}_\Omega - \boldsymbol{p}_C), \tag{B.14}$$

where $\boldsymbol{I}_C$ is the inertia tensor relative to the centre of mass, when expressed in a frame parallel to the reference frame with origin at the centre of mass.

The *forces* acting on a generic system of material particles can be distinguished into *internal* forces and *external* forces.

The internal forces, exerted by one part of the system on another, have null linear and angular momentum and thus they do not influence rigid body motion.

The external forces, exerted on the system by an agency outside the system, in the case of a rigid body $\mathcal{B}$ are distinguished into *active* forces and *reaction* forces.

The active forces can be either *concentrated* forces or *body* forces. The former are applied to specific points of $\mathcal{B}$, whereas the latter act on all elementary particles of the body. An example of body force is the *gravitational force* which, for any elementary particle of mass ρdV, is equal to $\boldsymbol{g}_0 \rho dV$ where $\boldsymbol{g}_0$ is the gravity acceleration vector.

The reaction forces are those exerted because of surface contact between two or more bodies. Such forces can be distributed on the contact surfaces or they can be assumed to be concentrated.

For a rigid body $\mathcal{B}$ subject to gravitational force, as well as to active and or reaction forces $\boldsymbol{f}_1 \ldots \boldsymbol{f}_n$ concentrated at points $\boldsymbol{p}_1 \ldots \boldsymbol{p}_n$, the *resultant* of the external forces $\boldsymbol{f}$ and the *resultant moment* $\boldsymbol{\mu}_\Omega$ with respect to a pole Ω are respectively

$$\boldsymbol{f} = \int_{V_\mathcal{B}} \boldsymbol{g}_0 \rho dV + \sum_{i=1}^{n} \boldsymbol{f}_i = m\boldsymbol{g}_0 + \sum_{i=1}^{n} \boldsymbol{f}_i \tag{B.15}$$

$$\begin{aligned}\boldsymbol{\mu}_\Omega &= \int_{V_\mathcal{B}} \boldsymbol{g}_0 \times (\boldsymbol{p}_\Omega - \boldsymbol{p})\rho dV + \sum_{i=1}^{n} \boldsymbol{f}_i \times (\boldsymbol{p}_\Omega - \boldsymbol{p}_i) \\ &= m\boldsymbol{g}_0 \times (\boldsymbol{p}_\Omega - \boldsymbol{p}_C) + \sum_{i=1}^{n} \boldsymbol{f}_i \times (\boldsymbol{p}_\Omega - \boldsymbol{p}_i). \end{aligned} \tag{B.16}$$

In the case when $\boldsymbol{f}$ and $\boldsymbol{\mu}_\Omega$ are known and it is desired to compute the resultant moment with respect to a point Ω' other than Ω, the following relation holds:

$$\boldsymbol{\mu}_{\Omega'} = \boldsymbol{\mu}_\Omega + \boldsymbol{f} \times (\boldsymbol{p}_{\Omega'} - \boldsymbol{p}_\Omega). \tag{B.17}$$

Consider now a generic system of material particles subject to *external forces* of resultant $\boldsymbol{f}$ and resultant moment $\boldsymbol{\mu}_\Omega$. The motion of the system in a frame O–xyz is established by the following *fundamental principles of dynamics* (Newton laws of motion):

$$\boldsymbol{f} = \dot{\boldsymbol{l}} \tag{B.18}$$

$$\boldsymbol{\mu}_\Omega = \dot{\boldsymbol{k}}_\Omega \tag{B.19}$$

where Ω is a pole fixed or coincident with the centre of mass C of the system. These equations hold for any mechanical system and can be used even in the case of variable mass. For a system with constant mass, computing the time derivative of the momentum in (B.18) gives *Newton equations of motion* in the form

$$\boldsymbol{f} = m\ddot{\boldsymbol{p}}_C, \tag{B.20}$$

where the quantity on the right-hand side represents the *resultant of inertia forces*.

If, besides the assumption of constant mass, the assumption of rigid system holds too, the expression in (B.14) of the angular momentum with (B.19) yield *Euler equations of motion* in the form

$$\boldsymbol{\mu}_\Omega = \boldsymbol{I}_\Omega\dot{\boldsymbol{\omega}} + \boldsymbol{\omega} \times (\boldsymbol{I}_\Omega\boldsymbol{\omega}), \tag{B.21}$$

where the quantity on the right-hand side represents the *resultant moment of inertia forces*.

For a system constituted by a set of rigid bodies, the external forces obviously do not include the reaction forces exerted between the bodies belonging to the same system.

B.3 Work and Energy

Given a force $\boldsymbol{f}_i$ applied at a point of position $\boldsymbol{p}_i$ with respect to frame O–xyz, the *elementary work* of the force $\boldsymbol{f}_i$ on the displacement $d\boldsymbol{p}_i = \dot{\boldsymbol{p}}_i dt$ is defined as the scalar

$$dW_i = \boldsymbol{f}_i^T d\boldsymbol{p}_i.$$

For a rigid body $\mathcal{B}$ subject to a system of forces of resultant $\boldsymbol{f}$ and resultant moment $\boldsymbol{\mu}_Q$ with respect to any point Q of $\mathcal{B}$, the elementary work on the rigid displacement (B.6) is given by

$$dW = (\boldsymbol{f}^T\dot{\boldsymbol{p}}_Q + \boldsymbol{\mu}_Q^T\boldsymbol{\omega})dt = \boldsymbol{f}^T d\boldsymbol{p}_Q + \boldsymbol{\mu}_Q^T\boldsymbol{\omega}dt. \tag{B.22}$$

The *kinetic energy* of a body $\mathcal{B}$ is defined as the scalar quantity

$$\mathcal{T} = \frac{1}{2}\int_{V_{\mathcal{B}}} \dot{\boldsymbol{p}}^T \dot{\boldsymbol{p}} \rho dV$$

which, for a rigid body, takes on the notable expression

$$\mathcal{T} = \frac{1}{2} m \dot{\boldsymbol{p}}_C^T \dot{\boldsymbol{p}}_C + \frac{1}{2} \boldsymbol{\omega}^T \boldsymbol{I}_C \boldsymbol{\omega} \tag{B.23}$$

where $\boldsymbol{I}_C$ is the inertia tensor relative to the centre of mass expressed in a frame parallel to the reference frame with origin at the centre of mass.

A system of position forces, i.e., the forces depending only on the positions of the points of application, is said to be *conservative* if the work done by each force is independent of the trajectory described by the point of application of the force but it depends only on the initial and final positions of the point of application. In this case, the elementary work of the system of forces is equal to minus the total differential of a scalar function termed *potential energy*, i.e.,

$$dW = -d\mathcal{U}. \tag{B.24}$$

An example of a conservative system of forces on a rigid body is the gravitational force, with which is associated the potential energy

$$\mathcal{U} = -\int_{V_{\mathcal{B}}} \boldsymbol{g}_0^T \boldsymbol{p} \rho dV = -m \boldsymbol{g}_0^T \boldsymbol{p}_C. \tag{B.25}$$

B.4 Constrained Systems

Consider a system $\mathcal{B}_r$ of r rigid bodies and assume that all the elements of $\mathcal{B}_r$ can reach any position in space. In order to find uniquely the position of all the points of the system, it is necessary to assign a vector $\boldsymbol{x} = [\,x_1 \ \ \dots \ \ x_p\,]^T$ of $6r = p$ parameters, termed *configuration*. These parameters are termed *Lagrange* or *generalized coordinates* of the *unconstrained* system $\mathcal{B}_r$, and p determines the number of *degrees of freedom* (DOFs).

Any limitation on the mobility of the system $\mathcal{B}_r$ is termed *constraint*. A constraint acting on $\mathcal{B}_r$ is said to be *holonomic* if it is expressed by a system of equations

$$\boldsymbol{h}(\boldsymbol{x}, t) = \boldsymbol{0}, \tag{B.26}$$

where $\boldsymbol{h}$ is a vector of dimensions $(s \times 1)$, with $s < m$. On the other hand, a constraint in the form $\boldsymbol{h}(\boldsymbol{x}, \dot{\boldsymbol{x}}, t) = \boldsymbol{0}$ which is nonintegrable is said to be *nonholonomic*. For simplicity, only equality (or *bilateral*) constraints are considered. If the equations in (B.26) do not explicitly depend on time, the constraint is said to be *scleronomic*.

On the assumption that $\boldsymbol{h}$ has continuous and continuously differentiable components, and its Jacobian $\partial \boldsymbol{h}/\partial \boldsymbol{x}$ has full rank, the equations in (B.26) allow the elimination of s out of m coordinates of the system $\mathcal{B}_r$. With the remaining $n = m - s$ coordinates it is possible to determine uniquely the configurations of $\mathcal{B}_r$ satisfying the constraints (B.26). Such coordinates are the *Lagrange* or *generalized coordinates* and n is the number of *degrees of freedom* of the *unconstrained* system $\mathcal{B}_r$.[3]

The motion of a system $\mathcal{B}_r$ with n DOFs and holonomic equality constraints can be described by equations of the form

$$\boldsymbol{x} = \boldsymbol{x}(\boldsymbol{q}(t), t), \tag{B.27}$$

where $\boldsymbol{q}(t) = [\, q_1(t) \quad \dots \quad q_n(t)\,]^T$ is a vector of Lagrange coordinates.

The *elementary displacement* of system (B.27) relative to the interval $(t, t+dt)$ is defined as

$$d\boldsymbol{x} = \frac{\partial \boldsymbol{x}(\boldsymbol{q}, t)}{\partial \boldsymbol{q}} \dot{\boldsymbol{q}} dt + \frac{\partial \boldsymbol{x}(\boldsymbol{q}, t)}{\partial t} dt. \tag{B.28}$$

The *virtual displacement* of system (B.27) at time t, relative to an increment $\delta\boldsymbol{\lambda}$, is defined as the quantity

$$\delta\boldsymbol{x} = \frac{\partial \boldsymbol{x}(\boldsymbol{q}, t)}{\partial \boldsymbol{q}} \delta\boldsymbol{q}. \tag{B.29}$$

The difference between the elementary displacement and the virtual displacement is that the former is relative to an actual motion of the system in an interval $(t, t + dt)$ which is consistent with the constraints, while the latter is relative to an imaginary motion of the system when the constraints are made invariant and equal to those at time t.

For a system with time-invariant constraints, the equations of motion (B.27) become

$$\boldsymbol{x} = \boldsymbol{x}(\boldsymbol{q}(t)), \tag{B.30}$$

and then, by setting $\delta\boldsymbol{\lambda} = d\boldsymbol{\lambda} = \dot{\boldsymbol{\lambda}} dt$, the virtual displacements (B.29) coincide with the elementary displacements (B.28).

To the concept of virtual displacement can be associated that of *virtual work* of a system of forces, by considering a virtual displacement instead of an elementary displacement.

If external forces are distinguished into *active forces* and *reaction forces*, a direct consequence of the principles of dynamics (B.18), (B.19) applied to the system of rigid bodies $\mathcal{B}_r$ is that, for each virtual displacement, the following relation holds:

$$\delta W_m + \delta W_a + \delta W_h = 0, \tag{B.31}$$

where δW_m, δW_a, δW_h are the total virtual works done by the inertia, active, reaction forces, respectively.

---

[3] In general, the Lagrange coordinates of a constrained system have a local validity; in certain cases, such as the joint variables of a manipulator, they can have a global validity.

In the case of *frictionless* equality constraints, reaction forces are exerted orthogonally to the contact surfaces and the virtual work is always null. Hence, (B.31) reduces to

$$\delta W_m + \delta W_a = 0. \tag{B.32}$$

For a steady system, inertia forces are identically null. Then the condition for the equilibrium of system $\mathcal{B}_r$ is that the virtual work of the active forces is identically null on any virtual displacement, which gives the fundamental equation of *statics* of a constrained system

$$\delta W_a = 0 \tag{B.33}$$

known as *principle of virtual work*. Expressing (B.33) in terms of the increment $\delta\boldsymbol{\lambda}$ of generalized coordinates leads to

$$\delta W_a = \boldsymbol{\zeta}^T \delta \boldsymbol{q} = 0 \tag{B.34}$$

where $\boldsymbol{\zeta}$ denotes the $(n \times 1)$ vector of active *generalized* forces.

In the dynamic case, it is worth distinguishing active forces into *conservative* (that can be derived from a potential) and *nonconservative*. The virtual work of conservative forces is given by

$$\delta W_c = -\frac{\partial \mathcal{U}}{\partial \boldsymbol{q}} \delta \boldsymbol{q}, \tag{B.35}$$

where $\mathcal{U}(\boldsymbol{\lambda})$ is the total potential energy of the system. The work of nonconservative forces can be expressed in the form

$$\delta W_{nc} = \boldsymbol{\xi}^T \delta \boldsymbol{q}, \tag{B.36}$$

where $\boldsymbol{\xi}$ denotes the vector of nonconservative generalized forces. It follows that the vector of active generalized forces is

$$\boldsymbol{\zeta} = \boldsymbol{\xi} - \left(\frac{\partial \mathcal{U}}{\partial \boldsymbol{q}}\right)^T. \tag{B.37}$$

Moreover, the work of inertia forces can be computed from the total kinetic energy of system $\mathcal{T}$ as

$$\delta W_m = \left(\frac{\partial \mathcal{T}}{\partial \boldsymbol{q}} - \frac{d}{dt}\frac{\partial \mathcal{T}}{\partial \dot{\boldsymbol{q}}}\right) \delta \boldsymbol{q}. \tag{B.38}$$

Substituting (B.35), (B.36), (B.38) into (B.32) and observing that (B.32) holds true for any increment $\delta\boldsymbol{\lambda}$ leads to *Lagrange equations*

$$\frac{d}{dt}\left(\frac{\partial \mathcal{L}}{\partial \dot{\boldsymbol{q}}}\right)^T - \left(\frac{\partial \mathcal{L}}{\partial \boldsymbol{q}}\right)^T = \boldsymbol{\xi}, \tag{B.39}$$

where

$$\mathcal{L} = \mathcal{T} - \mathcal{U} \tag{B.40}$$

is the *Lagrangian* function of the system. The equations in (B.39) completely describe the dynamic behaviour of an n-DOF system with holonomic equality constraints.

The sum of kinetic and potential energy of a system with time-invariant constraints is termed *Hamiltonian* function

$$\mathcal{H} = \mathcal{T} + \mathcal{U}. \tag{B.41}$$

Conservation of energy dictates that the time derivative of the Hamiltonian must balance the power generated by the nonconservative forces acting on the system, i.e.,

$$\frac{d\mathcal{H}}{dt} = \boldsymbol{\xi}^T \dot{\boldsymbol{q}}. \tag{B.42}$$

In view of (B.37), (B.41), the equation in (B.42) becomes

$$\frac{d\mathcal{T}}{dt} = \boldsymbol{\zeta}^T \dot{\boldsymbol{q}}. \tag{B.43}$$

Bibliography

The fundamental concepts of rigid-body mechanics and constrained systems can be found in classical texts such as [87, 154, 224]. An authoritative reference on rigid-body system dynamics is [187].

C

Feedback Control

As a premise to the study of manipulator decentralized control and centralized control, the fundamental principles of *feedback control* of *linear systems* are recalled, and an approach to the determination of control laws for *nonlinear systems* based on the use of *Lyapunov functions* is presented.

C.1 Control of Single-input/Single-output Linear Systems

According to classical *automatic control* theory of *linear time-invariant single-input/single-output systems*, in order to servo the output $y(t)$ of a system to a reference $r(t)$, it is worth adopting a *negative feedback control* structure. This structure indeed allows the use of approximate mathematical models to describe the input/output relationship of the system to control, since negative feedback has a potential for reducing the effects of system parameter variations and nonmeasurable disturbance inputs $d(t)$ on the output.

This structure can be represented in the *domain of complex variable* s as in the block scheme of Fig. C.1, where $G(s)$, $H(s)$ and $C(s)$ are the transfer functions of the system to control, the transducer and the controller, respectively. From this scheme it is easy to derive

$$Y(s) = W(s)R(s) + W_D(s)D(s), \tag{C.1}$$

where

$$W(s) = \frac{C(s)G(s)}{1 + C(s)G(s)H(s)} \tag{C.2}$$

is the *closed-loop input/output transfer function* and

$$W_D(s) = \frac{G(s)}{1 + C(s)G(s)H(s)} \tag{C.3}$$

is the *disturbance/output transfer function.*

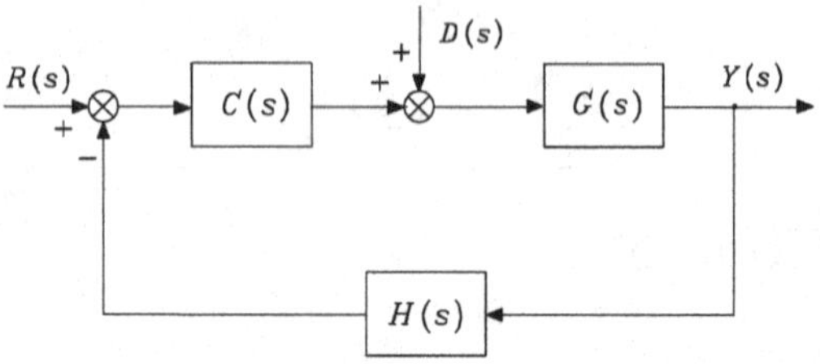

Fig. C.1. Feedback control structure

The goal of the controller design is to find a control structure $C(s)$ ensuring that the output variable $Y(s)$ tracks a reference input $R(s)$. Further, the controller should guarantee that the effects of the disturbance input $D(s)$ on the output variable are suitably reduced. The goal is then twofold, namely, *reference tracking* and *disturbance rejection.*

The basic problem for controller design consists of the determination of an action $C(s)$ which can make the system *asymptotically stable*. In the absence of positive or null real part pole/zero and zero/pole cancellation in the *open-loop* function $F(s) = C(s)G(s)H(s)$, a necessary and sufficient condition for asymptotic stability is that the *poles* of $W(s)$ and $W_D(s)$ have all *negative real parts*; such poles coincide with the zeros of the rational transfer function $1+F(s)$. Testing for this condition can be performed by resorting to stability criteria, thus avoiding computation of the function zeros.

Routh criterion allows the determination of the sign of the real parts of the zeros of the function $1+F(s)$ by constructing a table with the coefficients of the polynomial at the numerator of $1 + F(s)$ (*characteristic polynomial*).

Routh criterion is easy to apply for testing stability of a feedback system, but it does not provide a direct relationship between the open-loop function and stability of the closed-loop system. It is then worth resorting to *Nyquist criterion* which is based on the representation, in the complex plane, of the open-loop transfer function $F(s)$ evaluated in the *domain of real angular frequency* $(s = j\omega, -\infty < \omega < +\infty)$.

Drawing of Nyquist plot and computation of the number of circles made by the vector representing the complex number $1 + F(j\omega)$ when ω continuously varies from $-\infty$ to $+\infty$ allows a test on whether or not the *closed-loop* system is asymptotically stable. It is also possible to determine the number of positive, null and negative real part roots of the characteristic polynomial, similarly to application of Routh criterion. Nonetheless, Nyquist criterion is based on the plot of the open-loop transfer function, and thus it allows the determination of a direct relationship between this function and closed-loop system stability. It is then possible from an examination of the Nyquist plot to draw suggestions on the controller structure $C(s)$ which ensures closed-loop system asymptotic stability.

If the closed-loop system is asymptotically stable, the *steady-state response* to a sinusoidal input $r(t)$, with $d(t) = 0$, is sinusoidal, too. In this case, the function $W(s)$, evaluated for $s = j\omega$, is termed *frequency response function*; the frequency response function of a feedback system can be assimilated to that of a low-pass filter with the possible occurrence of a *resonance peak* inside its *bandwidth*.

As regards the transducer, this should be chosen so that its bandwidth is much greater than the feedback system bandwidth, in order to ensure a nearly instantaneous response for any value of ω inside the bandwidth of $W(j\omega)$. Therefore, setting $H(j\omega) \approx H_0$ and assuming that the *loop gain* $|C(j\omega)G(j\omega)H_0| \gg 1$ in the same bandwidth, the expression in (C.1) for $s = j\omega$ can be approximated as

$$Y(j\omega) \approx \frac{R(j\omega)}{H_0} + \frac{D(j\omega)}{C(j\omega)H_0}.$$

Assuming $R(j\omega) = H_0 Y_d(j\omega)$ leads to

$$Y(j\omega) \approx Y_d(j\omega) + \frac{D(j\omega)}{C(j\omega)H_0}; \tag{C.4}$$

i.e., the output tracks the desired output $Y_d(j\omega)$ and the frequency components of the disturbance in the bandwidth of $W(j\omega)$ produce an effect on the output which can be reduced by increasing $|C(j\omega)H_0|$. Furthermore, if the disturbance input is a constant, the steady-state output is not influenced by the disturbance as long as $C(s)$ has at least a pole at the origin.

Therefore, a feedback control system is capable of establishing a proportional relationship between the desired output and the actual output, as evidenced by (C.4). This equation, however, requires that the frequency content of the input (desired output) be inside the frequency range for which the loop gain is much greater than unity.

The previous considerations show the advantage of including a *proportional action* and an *integral action* in the controller $C(s)$, leading to the transfer function

$$C(s) = K_I \frac{1 + sT_I}{s} \tag{C.5}$$

of a *proportional-integral controller* (PI); T_I is the time constant of the integral action and the quantity $K_I T_I$ is called proportional sensitivity.

The adoption of a PI controller is effective for low-frequency response of the system, but it may involve a reduction of *stability margins* and/or a reduction of closed-loop system bandwidth. To avoid these drawbacks, a *derivative action* can be added to the proportional and integral actions, leading to the transfer function

$$C(s) = K_I \frac{1 + sT_I + s^2 T_D T_I}{s} \tag{C.6}$$

of a *proportional-integral-derivative controller* (PID); T_D denotes the time constant of the derivative action. Notice that physical realizability of (C.6) demands the introduction of a high-frequency pole which little influences the input/output relationship in the system bandwidth. The transfer function in (C.6) is characterized by the presence of two zeros which provide a stabilizing action and an enlargement of the closed-loop system bandwidth. Bandwidth enlargement implies shorter *response time* of the system, in terms of both variations of the reference signal and recovery action of the feedback system to output variations induced by the disturbance input.

The parameters of the adopted control structure should be chosen so as to satisfy requirements on the system behaviour at *steady state* and during the *transient*. Classical tools to determine such parameters are the *root locus* in the domain of the complex variable s or the *Nichols chart* in the domain of the real angular frequency ω. The two tools are conceptually equivalent. Their potential is different in that root locus allows a control law to be found which assigns the exact parameters of the closed-loop system time response, whereas Nichols chart allows a controller to be specified which confers good transient and steady-state behaviour to the system response.

A feedback system with strict requirements on the steady-state and transient behaviour, typically, has a response that can be assimilated to that of a *second-order system*. In fact, even for closed-loop functions of greater order, it is possible to identify a pair of complex conjugate poles whose real part absolute value is smaller than the real part absolute values of the other poles. Such a pair of poles is *dominant* in that its contribution to the transient response prevails over that of the other poles. It is then possible to approximate the input/output relationship with the transfer function

$$W(s) = \frac{k_W}{1 + \dfrac{2\zeta s}{\omega_n} + \dfrac{s^2}{\omega_n^2}} \tag{C.7}$$

which has to be realized by a proper choice of the controller. Regarding the values to assign to the parameters characterizing the transfer function in (C.7), the following remarks are in order. The constant k_W represents the input/output *steady-state gain*, which is equal to $1/H_0$ if $C(s)G(s)H_0$ has at least a pole at the origin. The *natural frequency* ω_n is the modulus of the complex conjugate poles, whose real part is given by $-\zeta\omega_n$ where ζ is the *damping ratio* of the pair of poles.

The influence of parameters ζ and ω_n on the closed-loop frequency response can be evaluated in terms of the resonance peak magnitude

$$M_r = \frac{1}{2\zeta\sqrt{1-\zeta^2}},$$

occurring at the resonant frequency

$$\omega_r = \omega_n\sqrt{1-2\zeta^2},$$

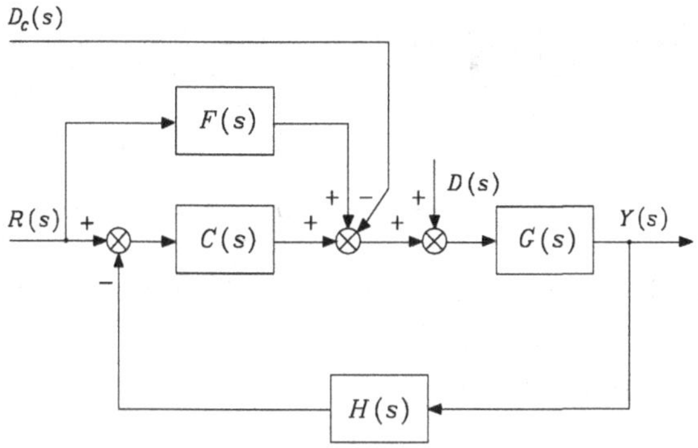

Fig. C.2. Feedback control structure with feedforward compensation

and of the 3 dB bandwidth

$$\omega_3 = \omega_n \sqrt{1 - 2\zeta^2 + \sqrt{2 - 4\zeta^2 + 4\zeta^4}}.$$

A step input is typically used to characterize the transient response in the time domain. The influence of parameters ζ and ω_n on the *step response* can be evaluated in terms of the percentage of *overshoot*

$$s\% = 100 \exp(-\pi\zeta/\sqrt{1-\zeta^2}),$$

of the *rise time*

$$t_r \approx \frac{1.8}{\omega_n}$$

and of the *settling time* within 1%

$$t_s = \frac{4.6}{\zeta\omega_n}.$$

The adoption of a *feedforward compensation* action represents a feasible solution both for tracking a time-varying reference input and for enhancing rejection of the effects of a disturbance on the output. Consider the general scheme in Fig. C.2. Let $R(s)$ denote a given input reference and $D_c(s)$ denote a computed estimate of the disturbance $D(s)$; the introduction of the feedforward action yields the input/output relationship

$$Y(s) = \left(\frac{C(s)G(s)}{1 + C(s)G(s)H(s)} + \frac{F(s)G(s)}{1 + C(s)G(s)H(s)} \right) R(s) \qquad \text{(C.8)}$$
$$+ \frac{G(s)}{1 + C(s)G(s)H(s)} \big(D(s) - D_c(s)\big).$$

By assuming that the desired output is related to the reference input by a constant factor K_d and regarding the transducer as an instantaneous system ($H(s) \approx H_0 = 1/K_d$) for the current operating conditions, the choice

$$F(s) = \frac{K_d}{G(s)} \qquad \text{(C.9)}$$

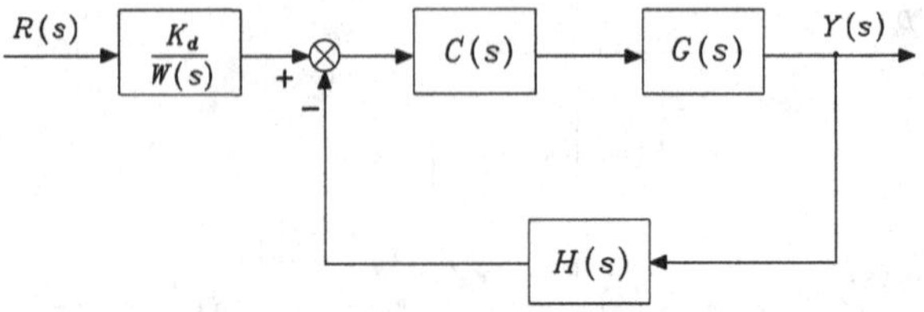

Fig. C.3. Feedback control structure with inverse model technique

yields the input/output relationship

$$Y(s) = Y_d(s) + \frac{G(s)}{1 + C(s)G(s)H_0}\big(D(s) - D_c(s)\big). \tag{C.10}$$

If $|C(j\omega)G(j\omega)H_0| \gg 1$, the effect of the disturbance on the output is further reduced by means of an accurate estimate of the disturbance.

Feedforward compensation technique may lead to a solution, termed *inverse model control*, illustrated in the scheme of Fig. C.3. It should be remarked, however, that such a solution is based on dynamics cancellation, and thus it can be employed only for a minimum-phase system, i.e., a system whose poles and zeros have all strictly negative real parts. Further, one should consider physical realizability issues as well as effects of parameter variations which prevent perfect cancellation.

C.2 Control of Nonlinear Mechanical Systems

If the system to control does not satisfy the linearity property, the control design problem becomes more complex. The fact that a *system* is qualified as *nonlinear*, whenever linearity does not hold, leads to understanding how it is not possible to resort to general techniques for control design, but it is necessary to face the problem for each class of nonlinear systems which can be defined through imposition of special properties.

On the above premise, the control design problem of nonlinear systems described by the dynamic model

$$\boldsymbol{H}(\boldsymbol{x})\ddot{\boldsymbol{x}} + \boldsymbol{h}(\boldsymbol{x}, \dot{\boldsymbol{x}}) = \boldsymbol{u} \tag{C.11}$$

is considered, where $[\,\boldsymbol{x}^T \quad \dot{\boldsymbol{x}}^T\,]^T$ denotes the $(2n \times 1)$ *state* vector of the system, $\boldsymbol{u}$ is the $(n \times 1)$ *input* vector, $\boldsymbol{H}(\boldsymbol{x})$ is an $(n \times n)$ *positive definite* (and thus invertible) matrix depending on $\boldsymbol{x}$, and $\boldsymbol{h}(\boldsymbol{x}, \dot{\boldsymbol{x}})$ is an $(n \times 1)$ vector depending on state. Several *mechanical systems* can be reduced to this class, including manipulators with rigid links and joints.

The *control* law can be found through a nonlinear compensating action obtained by choosing the following *nonlinear state feedback* law (*inverse dynamics* control):

$$\boldsymbol{u} = \widehat{\boldsymbol{H}}(\boldsymbol{x})\boldsymbol{v} + \widehat{\boldsymbol{h}}(\boldsymbol{x}, \dot{\boldsymbol{x}}) \tag{C.12}$$

where $\widehat{\boldsymbol{H}}(\boldsymbol{x})$ and $\widehat{\boldsymbol{h}}(\boldsymbol{x})$ respectively denote the *estimates* of the terms $\boldsymbol{H}(\boldsymbol{x})$ and $\boldsymbol{h}(\boldsymbol{x})$, computed on the basis of measures on the system state, and $\boldsymbol{v}$ is a new control input to be defined later. In general, it is

$$\widehat{\boldsymbol{H}}(\boldsymbol{x}) = \boldsymbol{H}(\boldsymbol{x}) + \Delta\boldsymbol{H}(\boldsymbol{x}) \tag{C.13}$$

$$\widehat{\boldsymbol{h}}(\boldsymbol{x},\dot{\boldsymbol{x}}) = \boldsymbol{h}(\boldsymbol{x},\dot{\boldsymbol{x}}) + \Delta\boldsymbol{h}(\boldsymbol{x},\dot{\boldsymbol{x}}) \tag{C.14}$$

because of the unavoidable modelling approximations or as a consequence of an intentional simplification in the compensating action. Substituting (C.12) into (C.11) and accounting for (C.13), (C.14) yields

$$\ddot{\boldsymbol{x}} = \boldsymbol{v} + \boldsymbol{z}(\boldsymbol{x},\dot{\boldsymbol{x}},\boldsymbol{v}) \tag{C.15}$$

where

$$\boldsymbol{z}(\boldsymbol{x},\dot{\boldsymbol{x}},\boldsymbol{v}) = \boldsymbol{H}^{-1}(\boldsymbol{x})\big(\Delta\boldsymbol{H}(\boldsymbol{x})\boldsymbol{v} + \Delta\boldsymbol{h}(\boldsymbol{x},\dot{\boldsymbol{x}})\big).$$

If *tracking* of a trajectory $(\boldsymbol{x}_d(t),\dot{\boldsymbol{x}}_d(t),\ddot{\boldsymbol{x}}_d(t))$ is desired, the tracking error can be defined as

$$\boldsymbol{e} = \begin{bmatrix} \boldsymbol{x}_d - \boldsymbol{x} \\ \dot{\boldsymbol{x}}_d - \dot{\boldsymbol{x}} \end{bmatrix} \tag{C.16}$$

and it is necessary to derive the error dynamics equation to study convergence of the actual state to the desired one. To this end, the choice

$$\boldsymbol{v} = \ddot{\boldsymbol{x}}_d + \boldsymbol{w}(\boldsymbol{e}), \tag{C.17}$$

substituted into (C.15), leads to the error equation

$$\dot{\boldsymbol{e}} = \boldsymbol{F}\boldsymbol{e} - \boldsymbol{G}\boldsymbol{w}(\boldsymbol{e}) - \boldsymbol{G}\boldsymbol{z}(\boldsymbol{e},\boldsymbol{x}_d,\dot{\boldsymbol{x}}_d,\ddot{\boldsymbol{x}}_d), \tag{C.18}$$

where the $(2n \times 2n)$ and $(2n \times n)$ matrices, respectively,

$$\boldsymbol{F} = \begin{bmatrix} \boldsymbol{O} & \boldsymbol{I} \\ \boldsymbol{O} & \boldsymbol{O} \end{bmatrix} \qquad \boldsymbol{G} = \begin{bmatrix} \boldsymbol{O} \\ \boldsymbol{I} \end{bmatrix}$$

follow from the error definition in (C.16). Control law design consists of finding the error function $\boldsymbol{w}(\boldsymbol{e})$ which makes (C.18) *globally asymptotically stable*,[1] i.e.,

$$\lim_{t\to\infty} \boldsymbol{e}(t) = \boldsymbol{0}.$$

In the case of *perfect* nonlinear compensation ($\boldsymbol{z}(\cdot) = \boldsymbol{0}$), the simplest choice of the control action is the *linear* one

$$\begin{aligned} \boldsymbol{w}(\boldsymbol{e}) &= -\boldsymbol{K}_P(\boldsymbol{x}_d - \boldsymbol{x}) - \boldsymbol{K}_D(\dot{\boldsymbol{x}}_d - \dot{\boldsymbol{x}}) \\ &= [\,-\boldsymbol{K}_P \quad -\boldsymbol{K}_D\,]\,\boldsymbol{e}, \end{aligned} \tag{C.19}$$

[1] *Global* asymptotic stability is invoked to remark that the equilibrium state is asymptotically stable for any perturbation.

where asymptotic stability of the error equation is ensured by choosing *positive definite* matrices $\boldsymbol{K}_P$ and $\boldsymbol{K}_D$. The error transient behaviour is determined by the eigenvalues of the matrix

$$\boldsymbol{A} = \begin{bmatrix} \boldsymbol{O} & \boldsymbol{I} \\ -\boldsymbol{K}_P & -\boldsymbol{K}_D \end{bmatrix} \tag{C.20}$$

characterizing the error dynamics

$$\dot{\boldsymbol{e}} = \boldsymbol{A}\boldsymbol{e}. \tag{C.21}$$

If compensation is *imperfect*, then $\boldsymbol{z}(\cdot)$ cannot be neglected and the error equation in (C.18) takes on the general form

$$\dot{\boldsymbol{e}} = \boldsymbol{f}(\boldsymbol{e}). \tag{C.22}$$

It may be worth choosing the control law $\boldsymbol{w}(\boldsymbol{e})$ as the sum of a nonlinear term and a linear term of the kind in (C.19); in this case, the error equation can be written as

$$\dot{\boldsymbol{e}} = \boldsymbol{A}\boldsymbol{e} + \boldsymbol{k}(\boldsymbol{e}), \tag{C.23}$$

where $\boldsymbol{A}$ is given by (C.20) and $\boldsymbol{k}(\boldsymbol{e})$ is available to make the system globally asymptotically stable. The equations in (C.22), (C.23) express nonlinear differential equations of the error. To test for stability and obtain advise on the choice of suitable control actions, one may resort to *Lyapunov direct method* illustrated below.

C.3 Lyapunov Direct Method

The philosophy of the *Lyapunov direct method* is the same as that of most methods used in control engineering to study stability, namely, testing for stability without solving the differential equations describing the dynamic system.

This method can be presented in short on the basis of the following reasoning. If it is possible to associate an energy-based description with a (linear or nonlinear) autonomous dynamic system and, for each system state with the exception of the equilibrium state, the time rate of such energy is negative, then energy decreases along any system trajectory until it attains its minimum at the equilibrium state; this argument justifies an intuitive concept of stability.

With reference to (C.22), by setting $\boldsymbol{f}(\boldsymbol{0}) = \boldsymbol{0}$, the *equilibrium state* is $\boldsymbol{e} = \boldsymbol{0}$. A scalar function $V(\boldsymbol{e})$ of the system state, continuous together with its first derivative, is defined a *Lyapunov function* if the following properties hold:

$$V(\boldsymbol{e}) > 0 \qquad \forall \boldsymbol{e} \neq \boldsymbol{0}$$

$$
\begin{aligned}
V(\boldsymbol{e}) &= 0 & \boldsymbol{e} &= \boldsymbol{0} \\
\dot{V}(\boldsymbol{e}) &< 0 & \forall \boldsymbol{e} &\neq \boldsymbol{0} \\
V(\boldsymbol{e}) &\to \infty & \|\boldsymbol{e}\| &\to \infty.
\end{aligned}
$$

The existence of such a function ensures *global asymptotic stability* of the equilibrium $\boldsymbol{e} = \boldsymbol{0}$. In practice, the equilibrium $\boldsymbol{e} = \boldsymbol{0}$ is globally asymptotically stable if a positive definite, radially unbounded function $\boldsymbol{V}(\boldsymbol{e})$ is found so that its time derivative along the system trajectories is negative definite.

If positive definiteness of $V(\boldsymbol{e})$ is realized by the adoption of a *quadratic form*, i.e.,

$$V(\boldsymbol{e}) = \boldsymbol{e}^T \boldsymbol{Q} \boldsymbol{e} \tag{C.24}$$

with $\boldsymbol{Q}$ a symmetric positive definite matrix, then in view of (C.22) it follows

$$\dot{V}(\boldsymbol{e}) = 2\boldsymbol{e}^T \boldsymbol{Q} \boldsymbol{f}(\boldsymbol{e}). \tag{C.25}$$

If $\boldsymbol{f}(\boldsymbol{e})$ is so as to render the function $\dot{V}(\boldsymbol{e})$ negative definite, the function $V(\boldsymbol{e})$ is a *Lyapunov function*, since the choice (C.24) allows system global asymptotic stability to be proved. If $\dot{V}(\boldsymbol{e})$ in (C.25) is not negative definite for the given $V(\boldsymbol{e})$, nothing can be inferred on the stability of the system, since the Lyapunov method gives only a *sufficient* condition. In such cases one should resort to different choices of $V(\boldsymbol{e})$ in order to find, if possible, a negative definite $\dot{V}(\boldsymbol{e})$.

In the case when the property of negative definiteness does not hold, but $\dot{V}(\boldsymbol{e})$ is only *negative semi-definite*

$$\dot{V}(\boldsymbol{e}) \leq 0,$$

global asymptotic stability of the equilibrium state is ensured if the only system trajectory for which $\dot{V}(\boldsymbol{e})$ is *identically* null ($\dot{V}(\boldsymbol{e}) \equiv 0$) is the equilibrium trajectory $\boldsymbol{e} \equiv \boldsymbol{0}$ (a consequence of *La Salle theorem*).

Finally, consider the stability problem of the nonlinear system in the form (C.23); under the assumption that $\boldsymbol{k}(\boldsymbol{0}) = \boldsymbol{0}$, it is easy to verify that $\boldsymbol{e} = \boldsymbol{0}$ is an equilibrium state for the system. The choice of a Lyapunov function candidate as in (C.24) leads to the following expression for its derivative:

$$\dot{V}(\boldsymbol{e}) = \boldsymbol{e}^T (\boldsymbol{A}^T \boldsymbol{Q} + \boldsymbol{Q}\boldsymbol{A})\boldsymbol{e} + 2\boldsymbol{e}^T \boldsymbol{Q}\boldsymbol{k}(\boldsymbol{e}). \tag{C.26}$$

By setting

$$\boldsymbol{A}^T \boldsymbol{Q} + \boldsymbol{Q}\boldsymbol{A} = -\boldsymbol{P}, \tag{C.27}$$

the expression in (C.26) becomes

$$\dot{V}(\boldsymbol{e}) = -\boldsymbol{e}^T \boldsymbol{P}\boldsymbol{e} + 2\boldsymbol{e}^T \boldsymbol{Q}\boldsymbol{k}(\boldsymbol{e}). \tag{C.28}$$

The matrix equation in (C.27) is said to be a *Lyapunov equation*; for any choice of a symmetric positive definite matrix $\boldsymbol{P}$, the solution matrix $\boldsymbol{Q}$ exists and is symmetric positive definite if and only if the eigenvalues of $\boldsymbol{A}$ have all negative real parts. Since matrix $\boldsymbol{A}$ in (C.20) verifies such condition, it is always possible to assign a positive definite matrix $\boldsymbol{P}$ and find a positive definite matrix solution $\boldsymbol{Q}$ to (C.27). It follows that the first term on the right-hand side of (C.28) is negative definite and the stability problem is reduced to searching a control law so that $\boldsymbol{k}(\boldsymbol{e})$ renders the total $\dot{V}(\boldsymbol{e})$ negative (semi-)definite.

It should be underlined that La Salle theorem does not hold for *time-varying* systems (also termed *non-autonomous*) in the form

$$\dot{e} = f(e, t).$$

In this case, a conceptually analogous result which might be useful is the following, typically referred to as *Barbalat lemma* — of which it is indeed a consequence. Given a scalar function $V(e,t)$ so that

1. $V(e,t)$ is lower bounded
2. $\dot{V}(e,t) \leq 0$
3. $\dot{V}(e,t)$ is *uniformly continuous*

then it is $\lim_{t\to\infty} \dot{V}(e,t) = 0$. Conditions 1 and 2 imply that $V(e,t)$ has a bounded limit for $t \to \infty$. Since it is not easy to verify the property of uniform continuity from the definition, Condition 3 is usually replaced by

3'. $\ddot{V}(e,t)$ is bounded

which is sufficient to guarantee validity of Condition 3. Barbalat lemma can obviously be used for time-invariant (autonomous) dynamic systems as an alternative to La Salle theorem, with respect to which some conditions are relaxed; in particular, $V(\boldsymbol{e})$ needs not necessarily be positive definite.

Bibliography

Linear systems analysis can be found in classical texts such as [61]. For the control of these systems see [82, 171]. For the analysis of nonlinear systems see [109]. Control of nonlinear mechanical systems is dealt with in [215].

D

Differential Geometry

The analysis of mechanical systems subject to nonholonomic constraints, such as wheeled mobile robots, requires some basic concepts of differential geometry and nonlinear controllability theory, that are briefly recalled in this appendix.

D.1 Vector Fields and Lie Brackets

For simplicity, the case of vectors $\boldsymbol{x} \in \mathbb{R}^n$ is considered. The tangent space at $\boldsymbol{x}$ (intuitively, the space of velocities of trajectories passing through $\boldsymbol{x}$) is hence denoted by $T_{\boldsymbol{x}}(\mathbb{R}^n)$. The presented notions are however valid in the more general case in which a *differentiable manifold* (i.e., a space that is locally diffeomorphic to $\mathbb{R}^n$) is considered in place of a Euclidean space.

A *vector field* $\boldsymbol{g} : \mathbb{R}^n \mapsto T_x(\mathbb{R}^n)$ is a mapping that assigns to each point $\boldsymbol{x} \in \mathbb{R}^n$ a tangent vector $\boldsymbol{g}(\boldsymbol{x}) \in T_x(\mathbb{R}^n)$. In the following it is always assumed that vector fields are *smooth*, i.e., such that the associated mappings are of class C^∞.

If the vector field $\boldsymbol{g}(\boldsymbol{x})$ is used to define a differential equation as in

$$\dot{\boldsymbol{x}} = \boldsymbol{g}(\boldsymbol{x}), \tag{D.1}$$

the *flow* $\boldsymbol{\phi}_t^{\boldsymbol{g}}(\boldsymbol{x})$ of $\boldsymbol{g}$ is the mapping that associates to each point $\boldsymbol{x}$ the value at time t of the solution of (D.1) evolving from $\boldsymbol{x}$ at time 0, or

$$\frac{d}{dt}\boldsymbol{\phi}_t^{\boldsymbol{g}}(\boldsymbol{x}) = \boldsymbol{g}(\boldsymbol{\phi}_t^{\boldsymbol{g}}(\boldsymbol{x})). \tag{D.2}$$

The family of mappings $\{\boldsymbol{\phi}_t^{\boldsymbol{g}}\}$ is a one-parameter (i.e., t) group under the composition operator

$$\boldsymbol{\phi}_{t_1}^{\boldsymbol{g}} \circ \boldsymbol{\phi}_{t_2}^{\boldsymbol{g}} = \boldsymbol{\phi}_{t_1+t_2}^{\boldsymbol{g}}.$$

For example, for time-invariant linear systems it is $\boldsymbol{g}(\boldsymbol{x}) = \boldsymbol{A}\boldsymbol{x}$ and the flow is the linear operator $\boldsymbol{\phi}_t^{\boldsymbol{g}} = e^{\boldsymbol{A}t}$.

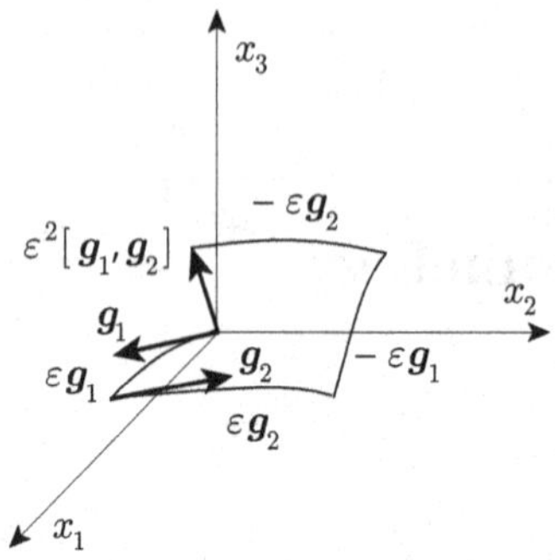

Fig. D.1. The net displacement of system (D.4) under the input sequence (D.5) is directed as the Lie bracket of the two vector fields $\boldsymbol{g}_1$ and $\boldsymbol{g}_2$

Given two vector fields $\boldsymbol{g}_1$ and $\boldsymbol{g}_2$, the composition of their flows is non-commutative in general:

$$\phi_t^{\boldsymbol{g}_1} \circ \phi_s^{\boldsymbol{g}_2} \neq \phi_s^{\boldsymbol{g}_2} \circ \phi_t^{\boldsymbol{g}_1}.$$

The vector field $[\boldsymbol{g}_1, \boldsymbol{g}_2]$ defined as

$$[\boldsymbol{g}_1, \boldsymbol{g}_2](\boldsymbol{x}) = \frac{\partial \boldsymbol{g}_2}{\partial \boldsymbol{x}} \boldsymbol{g}_1(\boldsymbol{x}) - \frac{\partial \boldsymbol{g}_1}{\partial \boldsymbol{x}} \boldsymbol{g}_2(\boldsymbol{x}) \tag{D.3}$$

is called *Lie bracket* of $\boldsymbol{g}_1$ and $\boldsymbol{g}_2$. The two vector field $\boldsymbol{g}_1$ and $\boldsymbol{g}_2$ *commute* if $[\boldsymbol{g}_1, \boldsymbol{g}_2] = 0$.

The Lie bracket operation has an interesting interpretation. Consider the driftless dynamic system

$$\dot{\boldsymbol{x}} = \boldsymbol{g}_1(\boldsymbol{x})u_1 + \boldsymbol{g}_2(\boldsymbol{x})u_2 \tag{D.4}$$

associated with the vector fields $\boldsymbol{g}_1$ and $\boldsymbol{g}_2$. If the inputs u_1 and u_2 are never active simultaneously, the solution of the differential equation (D.4) can be obtained by composing the flows of $\boldsymbol{g}_1$ and $\boldsymbol{g}_2$. In particular, consider the following input sequence:

$$u(t) = \begin{cases} u_1(t) = +1, u_2(t) = 0 & t \in [0, \varepsilon) \\ u_1(t) = 0, u_2(t) = +1 & t \in [\varepsilon, 2\varepsilon) \\ u_1(t) = -1, u_2(t) = 0 & t \in [2\varepsilon, 3\varepsilon) \\ u_1(t) = 0, u_2(t) = -1 & t \in [3\varepsilon, 4\varepsilon), \end{cases} \tag{D.5}$$

where ε is an infinitesimal time interval. The solution of (D.4) at time $t = 4\varepsilon$ can be obtained by following first the flow of $\boldsymbol{g}_1$, then of $\boldsymbol{g}_2$, then of $-\boldsymbol{g}_1$, and finally of $-\boldsymbol{g}_2$ (see Fig. D.1). By computing $\boldsymbol{x}(\varepsilon)$ through a series expansion at $\boldsymbol{x}_0 = \boldsymbol{x}(0)$ along $\boldsymbol{g}_1$, then $\boldsymbol{x}(2\varepsilon)$ as a series expansion at $\boldsymbol{x}(\varepsilon)$ along $\boldsymbol{g}_2$, and so on, one obtains

$$\begin{aligned} \boldsymbol{x}(4\varepsilon) &= \phi_\varepsilon^{-\boldsymbol{g}_2} \circ \phi_\varepsilon^{-\boldsymbol{g}_1} \circ \phi_\varepsilon^{\boldsymbol{g}_2} \circ \phi_\varepsilon^{\boldsymbol{g}_1}(\boldsymbol{x}_0) \\ &= \boldsymbol{x}_0 + \varepsilon^2 \left(\frac{\partial \boldsymbol{g}_2}{\partial \boldsymbol{x}} \boldsymbol{g}_1(\boldsymbol{x}_0) - \frac{\partial \boldsymbol{g}_1}{\partial \boldsymbol{x}} \boldsymbol{g}_2(\boldsymbol{x}_0) \right) + O(\varepsilon^3). \end{aligned}$$

If $\boldsymbol{g}_1$ and $\boldsymbol{g}_2$ commute, the net displacement resulting from the input sequence (D.5) is zero.

The above expression shows that, at each point $\boldsymbol{x}$, infinitesimal motion of the driftless system (D.4) is possible not only in the directions belonging to the linear span of $\boldsymbol{g}_1(\boldsymbol{x})$ and $\boldsymbol{g}_2(\boldsymbol{x})$, but also in the direction of their Lie bracket $[\boldsymbol{g}_1, \boldsymbol{g}_2](\boldsymbol{x})$. It can be proven that more complicated input sequences can be used to generate motion in the direction of higher-order Lie brackets, such as $[\boldsymbol{g}_1, [\boldsymbol{g}_1, \boldsymbol{g}_2]]$.

Similar constructive procedures can be given for systems with a *drift*[1] vector field, such as the following:

$$\dot{\boldsymbol{x}} = \boldsymbol{f}(\boldsymbol{x}) + \boldsymbol{g}_1(\boldsymbol{x})u_1 + \boldsymbol{g}_2(\boldsymbol{x})u_2. \tag{D.6}$$

Using appropriate input sequences, it is possible to generate motion in the direction of Lie brackets involving the vector field $\boldsymbol{f}$ as well as $\boldsymbol{g}_j$, $j = 1, 2$.

Example D.1

For a single-input linear system

$$\dot{\boldsymbol{x}} = \boldsymbol{A}\,\boldsymbol{x} + \boldsymbol{b}\,u,$$

the drift and input vector fields are $\boldsymbol{f}(\boldsymbol{x}) = \boldsymbol{A}\boldsymbol{x}$ and $\boldsymbol{g}(\boldsymbol{x}) = \boldsymbol{b}$, respectively. The following Lie brackets:

$$\begin{aligned} -[\boldsymbol{f}, \boldsymbol{g}] &= \boldsymbol{A}\boldsymbol{b} \\ [\boldsymbol{f}, [\boldsymbol{f}, \boldsymbol{g}]] &= \boldsymbol{A}^2\boldsymbol{b} \\ -[\boldsymbol{f}, [\boldsymbol{f}, [\boldsymbol{f}, \boldsymbol{g}]]] &= \boldsymbol{A}^3\boldsymbol{b} \\ &\vdots \end{aligned}$$

represent well-known directions in which it is possible to move the system.

The *Lie derivative* of the scalar function $\alpha : \mathbb{R}^n \mapsto \mathbb{R}$ along vector field $\boldsymbol{g}$ is defined as

$$L_{\boldsymbol{g}}\,\alpha(\boldsymbol{x}) = \frac{\partial \alpha}{\partial \boldsymbol{x}}\,\boldsymbol{g}(\boldsymbol{x}). \tag{D.7}$$

The following properties of Lie brackets are useful in computation:

$$\begin{array}{ll} [\boldsymbol{f}, \boldsymbol{g}] = -[\boldsymbol{g}, \boldsymbol{f}] & \text{(skew-symmetry)} \\ [\boldsymbol{f}, [\boldsymbol{g}, \boldsymbol{h}]] + [\boldsymbol{h}, [\boldsymbol{f}, \boldsymbol{g}]] + [\boldsymbol{g}, [\boldsymbol{h}, \boldsymbol{f}]] = 0 & \text{(Jacobi identity)} \\ [\alpha\boldsymbol{f}, \beta\boldsymbol{g}] = \alpha\beta[\boldsymbol{f}, \boldsymbol{g}] + \alpha(L_{\boldsymbol{f}}\beta)\boldsymbol{g} - \beta(L_{\boldsymbol{g}}\alpha)\boldsymbol{f} & \text{(chain rule)} \end{array}$$

---

[1] This term emphasizes how the presence of $\boldsymbol{f}$ will in general force the system to move ($\dot{\boldsymbol{x}} \neq \boldsymbol{0}$) even in the absence of inputs.

with $\alpha, \beta : \mathbb{R}^n \mapsto \mathbb{R}$. The vector space $\mathcal{V}(\mathbb{R}^n)$ of smooth vector fields on $\mathbb{R}^n$, equipped with the Lie bracket operation, is a *Lie algebra*.

The *distribution* Δ associated with the m vector fields $\{\boldsymbol{g}_1, \ldots, \boldsymbol{g}_m\}$ is the mapping that assigns to each point $\boldsymbol{x} \in \mathbb{R}^n$ the subspace of $T_x(\mathbb{R}^n)$ defined as

$$\Delta(\boldsymbol{x}) = \mathrm{span}\{\boldsymbol{g}_1(\boldsymbol{x}), \ldots, \boldsymbol{g}_m(\boldsymbol{x})\}. \tag{D.8}$$

Often, a shorthand notation is used:

$$\Delta = \mathrm{span}\{\boldsymbol{g}_1, \ldots, \boldsymbol{g}_m\}.$$

The distribution Δ is *nonsingular* if $\dim \Delta(\boldsymbol{x}) = r$, with r constant for all $\boldsymbol{x}$. In this case, r is called the *dimension* of the distribution. Moreover, Δ is called *involutive* if it is closed under the Lie bracket operation:

$$[\boldsymbol{g}_i, \boldsymbol{g}_j] \in \Delta \qquad \forall \boldsymbol{g}_i, \boldsymbol{g}_j \in \Delta.$$

The *involutive closure* $\bar{\Delta}$ of a distribution Δ is its closure under the Lie bracket operation. Hence, Δ is involutive if and only if $\bar{\Delta} = \Delta$. Note that the distribution $\Delta = \mathrm{span}\{\boldsymbol{g}\}$ associated with a single vector field is always involutive, because $[\boldsymbol{g}, \boldsymbol{g}](\boldsymbol{x}) = \boldsymbol{0}$.

Example D.2

The distribution

$$\Delta = \mathrm{span}\{\boldsymbol{g}_1, \boldsymbol{g}_2\} = \mathrm{span}\left\{ \begin{bmatrix} \cos x_3 \\ \sin x_3 \\ 0 \end{bmatrix}, \begin{pmatrix} 0 \\ 0 \\ 1 \end{pmatrix} \right\}$$

is nonsingular and has dimension 2. It is not involutive, because the Lie bracket

$$[\boldsymbol{g}_1, \boldsymbol{g}_2](\boldsymbol{x}) = \begin{bmatrix} \sin x_3 \\ -\cos x_3 \\ 0 \end{bmatrix}$$

is always linearly independent of $\boldsymbol{g}_1(\boldsymbol{x})$ and $\boldsymbol{g}_2(\boldsymbol{x})$. Its involutive closure is therefore

$$\bar{\Delta} = \mathrm{span}\{\boldsymbol{g}_1, \boldsymbol{g}_2, [\boldsymbol{g}_1, \boldsymbol{g}_2]\}.$$

D.2 Nonlinear Controllability

Consider a nonlinear dynamic system of the form

$$\dot{\boldsymbol{x}} = \boldsymbol{f}(\boldsymbol{x}) + \sum_{j=1}^{m} \boldsymbol{g}_j(\boldsymbol{x})u_j, \tag{D.9}$$

that is called *affine* in the inputs u_j. The state $\boldsymbol{x}$ takes values in $\mathbb{R}^n$, while each component u_j of the control input $\boldsymbol{u} \in \mathbb{R}^m$ takes values in the class $\mathcal{U}$ of piecewise-constant functions.

Denote by $\boldsymbol{x}(t, 0, \boldsymbol{x}_0, \boldsymbol{u})$ the solution of (D.9) at time $t \geq 0$, corresponding to an input $\boldsymbol{u}: [0, t] \to \mathcal{U}$ and an initial condition $\boldsymbol{x}(0) = \boldsymbol{x}_0$. Such a solution exists and is unique provided that the drift vector field $\boldsymbol{f}$ and the input vector fields $\boldsymbol{g}_j$ are of class C^∞. System (D.9) is said to be *controllable* if, for any choice of $\boldsymbol{x}_1, \boldsymbol{x}_2$ in $\mathbb{R}^n$, there exists a time instant T and an input $\boldsymbol{u}: [0, T] \to \mathcal{U}$ such that $\boldsymbol{x}(T, 0, \boldsymbol{x}_1, \boldsymbol{u}) = \boldsymbol{x}_2$.

The *accessibility algebra* $\mathcal{A}$ of system (D.9) is the smallest subalgebra of $\mathcal{V}(\mathbb{R}^n)$ that contains $\boldsymbol{f}, \boldsymbol{g}_1, \ldots, \boldsymbol{g}_m$. By definition, all the Lie brackets that can be generated using these vector fields belong to $\mathcal{A}$. The *accessibility distribution* $\Delta_{\mathcal{A}}$ of system (D.9) is defined as

$$\Delta_{\mathcal{A}} = \operatorname{span}\{\boldsymbol{v} | \boldsymbol{v} \in \mathcal{A}\}. \tag{D.10}$$

In other words, $\Delta_{\mathcal{A}}$ is the involutive closure of $\Delta = \operatorname{span}\{\boldsymbol{f}, \boldsymbol{g}_1, \ldots, \boldsymbol{g}_m\}$.

The computation of $\Delta_{\mathcal{A}}$ may be organized as an iterative procedure

$$\Delta_{\mathcal{A}} = \operatorname{span}\left\{\boldsymbol{v} | \boldsymbol{v} \in \Delta_i, \forall i \geq 1\right\},$$

with

$$\begin{aligned}
\Delta_1 &= \Delta = \operatorname{span}\{\boldsymbol{f}, \boldsymbol{g}_1, \ldots, \boldsymbol{g}_m\} \\
\Delta_i &= \Delta_{i-1} + \operatorname{span}\{[\boldsymbol{g}, \boldsymbol{v}] | \, \boldsymbol{g} \in \Delta_1, \boldsymbol{v} \in \Delta_{i-1}\}, \quad i \geq 2.
\end{aligned}$$

This procedure stops after κ steps, where κ is the smallest integer such that $\Delta_{\kappa+1} = \Delta_\kappa = \Delta_{\mathcal{A}}$. This number is called the *nonholonomy degree* of the system and is related to the 'level' of Lie brackets that must be included in $\Delta_{\mathcal{A}}$. Since $\dim \Delta_{\mathcal{A}} \leq n$, it is $\kappa \leq n - m$ necessarily.

If system (D.9) is driftless

$$\dot{\boldsymbol{x}} = \sum_{i=1}^{m} \boldsymbol{g}_i(\boldsymbol{x})u_i, \tag{D.11}$$

the accessibility distribution $\Delta_{\mathcal{A}}$ associated with vector fields $\boldsymbol{g}_1, \ldots, \boldsymbol{g}_m$ characterizes its controllability. In particular, system (D.11) is controllable if and only if the following *accessibility rank condition* holds:

$$\dim \Delta_{\mathcal{A}}(\boldsymbol{x}) = n. \tag{D.12}$$

Note that for driftless systems the iterative procedure for building $\Delta_{\mathcal{A}}$ starts with $\Delta_1 = \Delta = \text{span}\{\boldsymbol{g}_1, \dots, \boldsymbol{g}_m\}$, and therefore $\kappa \leq n - m + 1$.

For systems in the general form (D.9), condition (D.12) is only necessary for controllability. There are, however, two notable exceptions:

- If system (D.11) is controllable, the system with drift obtained by performing a *dynamic extension* of (D.11)

$$\dot{\boldsymbol{x}} = \sum_{i=1}^{m} \boldsymbol{g}_i(\boldsymbol{x}) v_i \tag{D.13}$$

$$\dot{v}_i = u_i, \qquad i = 1, \dots, m, \tag{D.14}$$

  i.e., by adding an integrator on each input channel, is also controllable.
- For a linear system

$$\dot{\boldsymbol{x}} = \boldsymbol{A}\boldsymbol{x} + \sum_{j=1}^{m} \boldsymbol{b}_j u_j = \boldsymbol{A}\boldsymbol{x} + \boldsymbol{B}\boldsymbol{u}$$

  (D.12) becomes

$$\varrho([\boldsymbol{B} \quad \boldsymbol{A}\boldsymbol{B} \quad \boldsymbol{A}^2\boldsymbol{B} \quad \dots \quad \boldsymbol{A}^{n-1}\boldsymbol{B}]) = n, \tag{D.15}$$

  i.e., the well-known necessary and sufficient condition for controllability due to Kalman.

Bibliography

The concepts briefly recalled in this appendix can be studied in detail in various tests of differential geometry [94, 20] and nonlinear control theory [104, 168, 195].

E

Graph Search Algorithms

This appendix summarizes some basic concepts on algorithm complexity and graph search techniques that are useful in the study of motion planning.

E.1 Complexity

A major criterion for assessing the efficiency of an algorithm A is its *running time*, i.e., the time needed for executing the algorithm in a computational model capturing the most relevant characteristics of an actual elaboration system. In practice, one is interested in estimating the running time as a function of a single parameter n characterizing the *size* of the input within a specific class of instances of the problem. In motion planning, this parameter may be the dimension of the configuration space, or the number of vertices of the free configuration space (if it is a polygonal subset).

In *worst-case analysis*, $t(n)$ denotes the maximum running time of A in correspondence of input instances of size n. Other kinds of analyses (e.g., average-case) are possible but they are less critical or general, requiring a statistical knowledge of the input distribution that may not be available.

The exact functional expression of $t(n)$ depends on the implementation of the algorithm, and is of little practical interest because the running time in the adopted computational model is only an approximation of the actual one. More significant is the *asymptotic behaviour* of $t(n)$, i.e., the rate of growth of $t(n)$ with n. Denote by $O(f(n))$ the set of real functions $g(n)$ such that

$$c_1 f(n) \leq g(n) \leq c_2 f(n) \qquad \forall n \geq n_0,$$

with c_1, c_2 and n_0 positive constants. If the worst-case running time of A is $O(f(n))$, i.e., if $t(n) \in O(f(n))$, the *time complexity* of A is said to be $O(f(n))$.

A very important class is represented by algorithms whose worst-case running time is asymptotically polynomial in the size of the input. In particular, if $t(n) \in O(n^p)$, for some $p \geq 0$, the algorithm is said to have *polynomial* time complexity. If the asymptotic behaviour of the worst-case running time is not polynomial, the time complexity of the algorithm is *exponential*. Note that here 'exponential' actually means 'not bounded by any polynomial function'.

The asymptotic behaviour of an algorithm with exponential time complexity is such that in the worst case it can only be applied to problems of 'small' size. However, there exist algorithms of exponential complexity that are very efficient on average, i.e., for the most frequent classes of input. A well known example is the simplex algorithm for solving linear programming problems. Similarly, there are algorithms with polynomial time complexity which are inefficient in practice because c_1, c_2 or p are 'large'.

The above concepts can be extended to inputs whose size is characterized by more than one parameter, or to performance criteria different from running time. For example, the memory space required by an algorithm is another important measure. The *space complexity* of an algorithm is said to be $O(f(n))$ if the memory space required for its execution is a function in $O(f(n))$.

E.2 Breadth-first and Depth-first Search

Let $G = (N, A)$ be a graph consisting of a set N of nodes and a set A of arcs, with cardinality n and a respectively. It is assumed that G is represented by an *adjacency list*: to each node N_i is associated a list of nodes that are connected to N_i by an arc. Consider the problem of searching G to find a path from a start node N_s to a goal node N_g. The simplest graph search strategies are *breadth-first search* (BFS) and *depth-first search* (DFS). These are briefly described in the following with reference to an iterative implementation.

Breadth-first search makes use of a *queue* — i.e., a FIFO (First In First Out) data structure — of nodes called OPEN. Initially, OPEN contains only the start node N_s, which is marked *visited*. All the other nodes in G are marked *unvisited*. At each iteration the first node in OPEN is extracted, and all its *unvisited* adjacent nodes are marked *visited* and inserted in OPEN. The search terminates when either N_g is inserted in OPEN or OPEN is empty (failure). During the search, the algorithm maintains the *BFS tree*, which contains only those arcs that have led to discovering *unvisited* nodes. This tree contains one and only one path connecting the start node to each visited node, and hence also a solution path from N_s to N_g, if it exists.

In depth-first search, OPEN is a *stack*, i.e., a LIFO (Last In First Out) data structure. Like in the breadth-first case, it contains initially only the start node N_s marked *visited*. When a node N_j is inserted in OPEN, the node N_i which has determined its insertion is memorized. At each iteration, the first node in OPEN is extracted. If it is *unvisited*, it is marked *visited* and the arc connecting N_i to N_j is inserted in the *DFS tree*. All *unvisited* nodes that are adjacent to N_j are inserted in OPEN. The search terminates when either N_g is inserted in OPEN or OPEN is empty (failure). Like in the BFS, the DFS tree contains the solution path from N_s to N_g, if it exists.

Both breadth-first and depth-first search have time complexity $O(a)$. Note that BFS and DFS are actually *traversal* strategies, because they do not use any information about the goal node; the graph is simply traversed until N_g is marked *visited*. Both the algorithms are *complete*, i.e., they find a solution path if it exists and report failure otherwise.

E.3 $A^\star$ Algorithm

In many applications, the arcs of G are labelled with positive numbers called *weights*. As a consequence, one may define the *cost* of a path on G as the sum of the weights of its arcs. Consider the problem of connecting N_s to N_g on G through a path of minimum cost, simply called *minimum path*. In motion planning problems, for example, the nodes generally represent points in configuration space, and it is then natural to define the weight of an arc as the length of the path that it represents. The minimum path is obviously interesting because it is the shortest among those joining N_s to N_g on G.

A widely used strategy for determining the minimum path on a graph is the $A^\star$ algorithm. $A^\star$ visits the nodes of G iteratively starting from N_s, storing only the current minimum paths from N_s to the visited nodes in a tree T. The algorithm employs a cost function $f(N_i)$ for each node N_i visited during the search. This function, which is an *estimate* of the cost of the minimum path that connects N_s to N_g passing through N_i, is computed as

$$f(N_i) = g(N_i) + h(N_i),$$

where $g(N_i)$ is the cost of the path from N_s to N_i as stored in the current tree T, and $h(N_i)$ is a *heuristic* estimate of the cost $h^\star(N_i)$ of the minimum path between N_i and N_g. While the value of $g(N_i)$ is uniquely determined by the search, any choice of $h(\cdot)$ such that

$$\forall N_i \in N : 0 \leq h(N_i) \leq h^\star(N_i) \tag{E.1}$$

is admissible. Condition (E.1) means that $h(\cdot)$ must not 'overestimate' the cost of the minimum path from N_i to N_g.

In the following, a pseudocode description of $A^\star$ is given. For its understanding, some preliminary remarks are needed:

- all the nodes are initially *unvisited*, except N_s which is *visited*;
- at the beginning, T contains only N_s;
- OPEN is a list of nodes that initially contains only N_s;
- N_{best} is the node in OPEN with the minimum value of f (in particular, it is the first node if OPEN is sorted by increasing values of f);
- ADJ(N_i) is the adjacency list of N_i;
- $c(N_i, N_j)$ is the weight of the arc connecting N_i to N_j.

$A^\star$ algorithm

repeat
  find and extract N_{best} from OPEN
  if $N_{\text{best}} = N_g$ **then** exit
  for each node N_i in ADJ(N_{best}) **do**
    if N_i is *unvisited* **then**
      add N_i to T with a pointer toward N_{best}
      insert N_i in OPEN; mark N_i *visited*
    else if $g(N_{\text{best}}) + c(N_{\text{best}}, N_i) < g(N_i)$ **then**
      redirect the pointer of N_i in T toward N_{best}
      if N_i is not in OPEN **then**
        insert N_i in OPEN
      else update $f(N_i)$
      end if
    end if
until OPEN is empty

Under condition (E.1), the $A^\star$ algorithm is complete. In particular, if the algorithm terminates with an empty OPEN, there exists no path in G from N_s to N_g (failure); otherwise, the tree T contains the minimum path from N_s to N_g, which can be reconstructed by backtracking from N_g to N_s.

The $A^\star$ algorithm with the particular (admissible) choice $h(N_i) = 0$, for each node N_i, is equivalent to the *Dijkstra algorithm*. If the nodes in G represent points in a Euclidean space, an admissible heuristic is the Euclidean distance between N and N_g. In fact, the length of the minimum path between N_i and N_g is bounded below by the Euclidean distance.

The extraction of a node from OPEN and the visit of its adjacent nodes is called node *expansion*. Given two admissible heuristic functions h_1 and h_2 such that $h_2(N_i) \geq h_1(N_i)$, for each node N_i in G, it is possible to prove that each node in G expanded by $A^\star$ using h_2 is also expanded using h_1. This means that $A^\star$ equipped with the heuristic h_2 is *at least* as efficient as $A^\star$ equipped with the heuristic h_1; h_2 is said to be *more informed* than h_1.

The $A^\star$ algorithm can be implemented with time complexity $O(a \log n)$.

Bibliography

The notions briefly recalled in this appendix are explained in detail in various texts on algorithm theory and artificial intelligence, such as [51, 189, 202].

References

1. C. Abdallah, D. Dawson, P. Dorato, M. Jamshidi, "Survey of robust control for rigid robots," *IEEE Control Systems Magazine*, vol. 11, no. 2, pp. 24–30, 1991.
2. M. Aicardi, G. Casalino, A. Bicchi, A. Balestrino, "Closed loop steering of unicycle-like vehicles via Lyapunov techniques," *IEEE Robotics and Automation Magazine*, vol. 2, no. 1, pp. 27–35, 1995.
3. J.S. Albus, H.G. McCain, R. Lumia, *NASA/NBS Standard Reference Model for Telerobot Control System Architecture (NASREM)*, NBS tech. note 1235, Gaithersburg, MD, 1987.
4. C.H. An, C.G. Atkeson, J.M. Hollerbach, *Model-Based Control of a Robot Manipulator*, MIT Press, Cambridge, MA, 1988.
5. R.J. Anderson, M.W. Spong, "Hybrid impedance control of robotic manipulators," *IEEE Journal of Robotics and Automation*, vol. 4, pp. 549–556, 1988.
6. J. Angeles, *Spatial Kinematic Chains: Analysis, Synthesis, Optimization*, Springer-Verlag, Berlin, 1982.
7. S. Arimoto, F. Miyazaki, "Stability and robustness of PID feedback control for robot manipulators of sensory capability," in *Robotics Research: The First International Symposium*, M. Brady, R. Paul (Eds.), MIT Press, Cambridge, MA, pp. 783–799, 1984.
8. R.C. Arkin, *Behavior-Based Robotics*, MIT Press, Cambridge, MA, 1998.
9. B. Armstrong-Hélouvry, *Control of Machines with Friction*, Kluwer, Boston, MA, 1991.
10. H. Asada, J.-J.E. Slotine, *Robot Analysis and Control*, Wiley, New York, 1986.
11. H. Asada, K. Youcef-Toumi, "Analysis and design of a direct-drive arm with a five-bar-link parallel drive mechanism," *ASME Journal of Dynamic Systems, Measurement, and Control*, vol. 106, pp. 225–230, 1984.
12. H. Asada, K. Youcef-Toumi, *Direct-Drive Robots*, MIT Press, Cambridge, MA, 1987.
13. C.G. Atkeson, C.H. An, J.M. Hollerbach, "Estimation of inertial parameters of manipulator loads and links," *International Journal of Robotics Research*, vol. 5, no. 3, pp. 101–119, 1986.
14. J. Baillieul, "Kinematic programming alternatives for redundant manipulators," *Proc. 1985 IEEE International Conference on Robotics and Automation*, St. Louis, MO, pp. 722–728, 1985.

15. A. Balestrino, G. De Maria, L. Sciavicco, "An adaptive model following control for robotic manipulators," *ASME Journal of Dynamic Systems, Measurement, and Control*, vol. 105, pp. 143–151, 1983.
16. A. Balestrino, G. De Maria, L. Sciavicco, B. Siciliano, "An algorithmic approach to coordinate transformation for robotic manipulators," *Advanced Robotics*, vol. 2, pp. 327–344, 1988.
17. J. Barraquand, J.-C. Latombe, "Robot motion planning: A distributed representation approach," *International Journal of Robotics Research*, vol. 10, pp. 628–649, 1991.
18. G. Bastin, G. Campion, B. D'Andréa-Novel, "Structural properties and classification of kinematic and dynamic models of wheeled mobile robots," *IEEE Transactions on Robotics and Automation*, vol. 12, pp. 47–62, 1996.
19. A.K. Bejczy, *Robot Arm Dynamics and Control*, memo. TM 33-669, Jet Propulsion Laboratory, California Institute of Technology, 1974.
20. W.M. Boothby, *An Introduction to Differentiable Manifolds and Riemannian Geometry*, Academic Press, Orlando, FL, 1986.
21. J. Borenstein, H.R. Everett, L. Feng, *Navigating Mobile Robots: Systems and Techniques*, A K Peters, Wellesley, MA, 1996.
22. B.K.K. Bose, *Modern Power Electronics and AC Drives*, Prentice-Hall, Englewood Cliffs, NJ, 2001.
23. O. Bottema, B. Roth, *Theoretical Kinematics*, North Holland, Amsterdam, 1979.
24. T.L. Boullion, P.L. Odell, *Generalized Inverse Matrices*, Wiley, New York, 1971.
25. M. Brady, "Artificial intelligence and robotics," *Artificial Intelligence*, vol. 26, pp. 79–121, 1985.
26. M. Brady, J.M. Hollerbach, T.L. Johnson, T. Lozano-Pérez, M.T. Mason, (Eds.), *Robot Motion: Planning and Control*, MIT Press, Cambridge, MA, 1982.
27. H. Bruyninckx, J. De Schutter, "Specification of force-controlled actions in the "task frame formalism" — A synthesis," *IEEE Transactions on Robotics and Automation*, vol. 12, pp. 581–589, 1996.
28. H. Bruyninckx, S. Dumey, S. Dutré, J. De Schutter, "Kinematic models for model-based compliant motion in the presence of uncertainty," *International Journal of Robotics Research*, vol. 14, pp. 465–482, 1995.
29. F. Caccavale, P. Chiacchio, "Identification of dynamic parameters and feedforward control for a conventional industrial manipulator," *Control Engineering Practice*, vol. 2, pp. 1039–1050, 1994.
30. F. Caccavale, C. Natale, B. Siciliano, L. Villani, "Resolved-acceleration control of robot manipulators: A critical review with experiments," *Robotica*, vol. 16, pp. 565–573, 1998.
31. F. Caccavale, C. Natale, B. Siciliano, L. Villani, "Six-DOF impedance control based on angle/axis representations," *IEEE Transactions on Robotics and Automation*, vol. 15, pp. 289–300, 1999.
32. F. Caccavale, C. Natale, B. Siciliano, L. Villani, "Robot impedance control with nondiagonal stiffness," *IEEE Transactions on Automatic Control*, vol. 44, pp. 1943–1946, 1999.
33. J.F. Canny, *The Complexity of Robot Motion Planning*, MIT Press, Cambridge, MA, 1988.

34. C. Canudas de Wit, H. Khennouf, C. Samson, O.J. Sørdalen, "Nonlinear control design for mobile robots," in *Recent Trends in Mobile Robots*, Y.F. Zheng, (Ed.), pp. 121–156, World Scientific Publisher, Singapore, 1993.
35. F. Chaumette, "Image moments: A general and useful set of features for visual servoing," *IEEE Transactions on Robotics and Automation*, vol. 21, pp. 1116-1127, 2005.
36. F. Chaumette, S. Hutchinson, "Visual servo control. Part I: Basic approaches," *IEEE Robotics and Automation Magazine*, vol. 13, no. 4, pp. 82–90, 2006.
37. P. Chiacchio, S. Chiaverini, L. Sciavicco, B. Siciliano, "Closed-loop inverse kinematics schemes for constrained redundant manipulators with task space augmentation and task priority strategy," *International Journal of Robotics Research*, vol. 10, pp. 410–425, 1991.
38. P. Chiacchio, S. Chiaverini, L. Sciavicco, B. Siciliano, "Influence of gravity on the manipulability ellipsoid for robot arms," *ASME Journal of Dynamic Systems, Measurement, and Control*, vol. 114, pp. 723–727, 1992.
39. P. Chiacchio, F. Pierrot, L. Sciavicco, B. Siciliano, "Robust design of independent joint controllers with experimentation on a high-speed parallel robot," *IEEE Transactions on Industrial Electronics*, vol. 40, pp. 393–403, 1993.
40. S. Chiaverini, L. Sciavicco, "The parallel approach to force/position control of robotic manipulators," *IEEE Transactions on Robotics and Automation*, vol. 4, pp. 361–373, 1993.
41. S. Chiaverini, B. Siciliano, "The unit quaternion: A useful tool for inverse kinematics of robot manipulators," *Systems Analysis Modelling Simulation*, vol. 35, pp. 45–60, 1999.
42. S. Chiaverini, B. Siciliano, O. Egeland, "Review of the damped least-squares inverse kinematics with experiments on an industrial robot manipulator," *IEEE Transactions on Control Systems Technology*, vol. 2, pp. 123–134, 1994.
43. S. Chiaverini, B. Siciliano, L. Villani, "Force/position regulation of compliant robot manipulators," *IEEE Transactions on Automatic Control*, vol. 39, pp. 647–652, 1994.
44. S.L. Chiu, "Task compatibility of manipulator postures," *International Journal of Robotics Research*, vol. 7, no. 5, pp. 13–21, 1988.
45. H. Choset, K.M. Lynch, S. Hutchinson, G. Kantor, W. Burgard, L.E. Kavraki, S. Thrun, *Principles of Robot Motion: Theory, Algorithms, and Implementations*, MIT Press, Cambridge, MA, 2005.
46. J.C.K. Chou, "Quaternion kinematic and dynamic differential equations. *IEEE Transactions on Robotics and Automation*, vol. 8, pp. 53–64, 1992.
47. A.I. Comport, E. Marchand, M. Pressigout, F. Chaumette, "Real-time markerless tracking for augmented reality: The virtual visual servoing framework," *IEEE Transactions on Visualization and Computer Graphics*, vol. 12, pp. 615–628, 2006.
48. P.I. Corke, *Visual Control of Robots: High-Performance Visual Servoing*, Research Studies Press, Taunton, UK, 1996.
49. P. Corke, S. Hutchinson, "A new partitioned approach to image-based visual servo control," *IEEE Transactions on Robotics and Automation*, vol. 17, pp. 507-515, 2001.
50. M. Corless, G. Leitmann, "Continuous state feedback guaranteeing uniform ultimate boundedness for uncertain dynamic systems," *IEEE Transactions on Automatic Control*, vol. 26, pp. 1139–1144, 1981.

51. T.H. Cormen, C.E. Leiserson, R.L. Rivest, C. Stein, *Introduction to Algorithms*, 2nd ed., MIT Press, Cambridge, MA, 2001.
52. J.J. Craig, *Adaptive Control of Mechanical Manipulators*, Addison-Wesley, Reading, MA, 1988.
53. J.J. Craig, *Introduction to Robotics: Mechanics and Control*, 3rd ed., Pearson Prentice Hall, Upper Saddle River, NJ, 2004.
54. C. De Boor, *A Practical Guide to Splines*, Springer-Verlag, New York, 1978.
55. T.L. De Fazio, D.S. Seltzer, D.E. Whitney, "The instrumented Remote Center of Compliance," *Industrial Robot*, vol. 11, pp. 238–242, 1984.
56. A. De Luca, *A Spline Generator for Robot Arms*, tech. rep. RAL 68, Rensselaer Polytechnic Institute, Department of Electrical, Computer, and Systems Engineering, 1986.
57. A. De Luca, C. Manes, "Modeling robots in contact with a dynamic environment," *IEEE Transactions on Robotics and Automation*, vol. 10, pp. 542–548, 1994.
58. A. De Luca, G. Oriolo, C. Samson, "Feedback control of a nonholonomic car-like robot," in *Robot Motion Planning and Control*, J.-P. Laumond, (Ed.), Springer-Verlag, Berlin, Germany, 1998.
59. A. De Luca, G. Oriolo, B. Siciliano, "Robot redundancy resolution at the acceleration level," *Laboratory Robotics and Automation*, vol. 4, pp. 97–106, 1992.
60. J. Denavit, R.S. Hartenberg, "A kinematic notation for lower-pair mechanisms based on matrices," *ASME Journal of Applied Mechanics*, vol. 22, pp. 215–221, 1955.
61. P.M. DeRusso, R.J. Roy, C.M. Close, A.A. Desrochers, *State Variables for Engineers*, 2nd ed., Wiley, New York, 1998.
62. J. De Schutter, H. Bruyninckx, S. Dutré, J. De Geeter, J. Katupitiya, S. Demey, T. Lefebvre, "Estimating first-order geometric parameters and monitoring contact transitions during force-controlled compliant motions," *International Journal of Robotics Research*, vol. 18, pp. 1161–1184, 1999.
63. J. De Schutter, H. Bruyninckx, W.-H. Zhu, M.W. Spong, "Force control: A bird's eye view," in *Control Problems in Robotics and Automation*, B. Siciliano, K.P. Valavanis, (Ed.), pp. 1–17, Springer-Verlag, London, UK, 1998.
64. J. De Schutter, H. Van Brussel, "Compliant robot motion I. A formalism for specifying compliant motion tasks," *International Journal of Robotics Research*, vol. 7, no. 4, pp. 3–17, 1988.
65. J. De Schutter, H. Van Brussel, "Compliant robot motion II. A control approach based on external control loops," *International Journal of Robotics Research*, vol. 7, no. 4, pp. 18–33, 1988.
66. K.L. Doty, C. Melchiorri, C. Bonivento, "A theory of generalized inverses applied to robotics," *International Journal of Robotics Research*, vol. 12, pp. 1–19, 1993.
67. S. Dubowsky, D.T. DesForges, "The application of model referenced adaptive control to robotic manipulators," *ASME Journal of Dynamic Systems, Measurement, and Control*, vol. 101, pp. 193–200, 1979.
68. C. Edwards, L. Galloway, "A single-point calibration technique for a six-degree–of–freedom articulated arm," *International Journal of Robotics Research*, vol. 13, pp. 189–199, 1994.
69. O. Egeland, "Task-space tracking with redundant manipulators," *IEEE Journal of Robotics and Automation*, vol. 3, pp. 471–475, 1987.

70. S.D. Eppinger, W.P. Seering, "Introduction to dynamic models for robot force control," *IEEE Control Systems Magazine*, vol. 7, no. 2, pp. 48–52, 1987.
71. B. Espiau, F. Chaumette, P. Rives, "A new approach to visual servoing in robotics," *IEEE Transactions on Robotics and Automation*, vol. 8, pp. 313–326, 1992.
72. H.R. Everett, *Sensors for Mobile Robots: Theory and Application*, AK Peters, Wellesley, MA, 1995.
73. G.E. Farin, *Curves and Surfaces for CAGD: A Practical Guide*, 5th ed., Morgan Kaufmann Publishers, San Francisco, CA, 2001.
74. E.D. Fasse, P.C. Breedveld, "Modelling of elastically coupled bodies: Parts I–II", *ASME Journal of Dynamic Systems, Measurement, and Control*, vol. 120, pp. 496–506, 1998.
75. O. Faugeras, *Three-Dimensional Computer Vision: A Geometric Viewpoint*, MIT Press, Boston, MA, 1993.
76. R. Featherstone, "Position and velocity transformations between robot end-effector coordinates and joint angles," *International Journal of Robotics Research*, vol. 2, no. 2, pp. 35–45, 1983.
77. R. Featherstone, *Robot Dynamics Algorithms*, Kluwer, Boston, MA, 1987.
78. R. Featherstone, O. Khatib, "Load independence of the dynamically consistent inverse of the Jacobian matrix," *International Journal of Robotics Research*, vol. 16, pp. 168–170, 1997.
79. J. Feddema, O. Mitchell, "Vision-guided servoing with feature-based trajectory generation," *IEEE Transactions on Robotics and Automation*, vol. 5, pp. 691–700, 1989.
80. M. Fliess, J. Lévine, P. Martin, P. Rouchon, "Flatness and defect of nonlinear systems: Introductory theory and examples," *International Journal of Control*, vol. 61, pp. 1327–1361, 1995.
81. J. Fraden, *Handbook of Modern Sensors: Physics, Designs, and Applications*, Springer, New York, 2004.
82. G.F. Franklin, J.D. Powell, A. Emami-Naeini, *Feedback Control of Dynamic Systems*, 5th ed., Prentice-Hall, Lebanon, IN, 2005.
83. E. Freund, "Fast nonlinear control with arbitrary pole-placement for industrial robots and manipulators," *International Journal of Robotics Research*, vol. 1, no. 1, pp. 65–78, 1982.
84. L.-C. Fu, T.-L. Liao, "Globally stable robust tracking of nonlinear systems using variable structure control with an application to a robotic manipulator," *IEEE Transactions on Automatic Control*, vol. 35, pp. 1345–1350, 1990.
85. M. Gautier, W. Khalil, "Direct calculation of minimum set of inertial parameters of serial robots," *IEEE Transactions on Robotics and Automation*, vol. 6, pp. 368–373, 1990.
86. A.A. Goldenberg, B. Benhabib, R.G. Fenton, "A complete generalized solution to the inverse kinematics of robots," *IEEE Journal of Robotics and Automation*, vol. 1, pp. 14–20, 1985.
87. H. Goldstein, C.P. Poole, J.L. Safko, *Classical Mechanics*, 3rd ed., Addison-Wesley, Reading, MA, 2002.
88. G.H. Golub, C.F. Van Loan, *Matrix Computations*, 3rd ed., The Johns Hopkins University Press, Baltimore, MD, 1996.
89. M.C. Good, L.M. Sweet, K.L. Strobel, "Dynamic models for control system design of integrated robot and drive systems," *ASME Journal of Dynamic Systems, Measurement, and Control*, vol. 107, pp. 53–59, 1985.

90. D.M. Gorinevski, A.M. Formalsky, A.Yu. Schneider, *Force Control of Robotics Systems*, CRC Press, Boca Raton, FL, 1997.
91. W.A. Gruver, B.I. Soroka, J.J. Craig, T.L. Turner, "Industrial robot programming languages: A comparative evaluation," *IEEE Transactions on Systems, Man, and Cybernetics*, vol. 14, pp. 565–570, 1984.
92. G. Hager, W. Chang, A. Morse, "Robot feedback control based on stereo vision: Towards calibration-free hand-eye coordination," *IEEE Control Systems Magazine*, vol. 15, no. 1, pp. 30–39, 1995.
93. R.M. Haralick, L.G. Shapiro, *Computer and Robot Vision*, vols. 1 & 2, Addison-Wesley, Reading, MA, 1993.
94. S. Helgason, *Differential Geometry and Symmetric Spaces*, Academic Press, New York, NY, 1962.
95. N. Hogan, "Impedance control: An approach to manipulation: Part I — Theory," *ASME Journal of Dynamic Systems, Measurement, and Control*, vol. 107, pp. 1–7, 1985.
96. J.M. Hollerbach, "A recursive Lagrangian formulation of manipulator dynamics and a comparative study of dynamics formulation complexity," *IEEE Transactions on Systems, Man, and Cybernetics*, vol. 10, pp. 730–736, 1980.
97. J.M. Hollerbach, "Dynamic scaling of manipulator trajectories," *ASME Journal of Dynamic Systems, Measurement, and Control*, vol. 106, pp. 102–106, 1984.
98. J.M. Hollerbach, "A survey of kinematic calibration," in *The Robotics Review 1*, O. Khatib, J.J. Craig, and T. Lozano-Pérez (Eds.), MIT Press, Cambridge, MA, pp. 207–242, 1989.
99. J.M. Hollerbach, G. Sahar, "Wrist-partitioned inverse kinematic accelerations and manipulator dynamics," *International Journal of Robotics Research*, vol. 2, no. 4, pp. 61–76, 1983.
100. R. Horowitz, M. Tomizuka, "An adaptive control scheme for mechanical manipulators — Compensation of nonlinearity and decoupling control," *ASME Journal of Dynamic Systems, Measurement, and Control*, vol. 108, pp. 127–135, 1986.
101. T.C.S. Hsia, T.A. Lasky, Z. Guo, "Robust independent joint controller design for industrial robot manipulators," *IEEE Transactions on Industrial Electronics*, vol. 38, pp. 21–25, 1991.
102. P. Hsu, J. Hauser, S. Sastry, "Dynamic control of redundant manipulators," *Journal of Robotic Systems*, vol. 6, pp. 133–148, 1989.
103. S. Hutchinson, G. Hager, P. Corke, "A tutorial on visual servo control," *IEEE Transactions on Robotics and Automation*, vol. 12, pp. 651–670, 1996.
104. A. Isidori, *Nonlinear Control Systems*, 3rd ed., Springer-Verlag, London, UK, 1995.
105. H. Kazerooni, P.K. Houpt, T.B. Sheridan, "Robust compliant motion of manipulators, Part I: The fundamental concepts of compliant motion," *IEEE Journal of Robotics and Automation*, vol. 2, pp. 83–92, 1986.
106. J.L. Jones, A.M. Flynn, *Mobile Robots: Inspiration to Implementation*, AK Peters, Wellesley, MA, 1993.
107. L.E. Kavraki, P. Svestka, J.-C. Latombe, M.H. Overmars, "Probabilistic roadmaps for path planning in high-dimensional configuration spaces," *IEEE Transactions on Robotics and Automation*, vol. 12, pp. 566–580, 1996.

108. R. Kelly, R. Carelli, O. Nasisi, B. Kuchen, F. Reyes, "Stable visual servoing of camera-in-hand robotic systems," *IEEE/ASME Transactions on Mechatronics*, vol. 5, pp. 39–48, 2000.
109. H.K. Khalil, *Nonlinear Systems*, Prentice-Hall, Englewood Cliffs, NJ, 2002.
110. W. Khalil, F. Bennis, "Symbolic calculation of the base inertial parameters of closed-loop robots," *International Journal of Robotics Research*, vol. 14, pp. 112–128, 1995.
111. W. Khalil, E. Dombre, *Modeling, Identification and Control of Robots*, Hermes Penton Ltd, London, 2002.
112. W. Khalil, J.F. Kleinfinger, "Minimum operations and minimum parameters of the dynamic model of tree structure robots," *IEEE Journal of Robotics and Automation*, vol. 3, pp. 517–526, 1987.
113. O. Khatib, "Real-time obstacle avoidance for manipulators and mobile robots," *International Journal of Robotics Research*, vol. 5, no. 1, pp. 90–98, 1986.
114. O. Khatib, "A unified approach to motion and force control of robot manipulators: The operational space formulation," *IEEE Journal of Robotics and Automation*, vol. 3, pp. 43–53, 1987.
115. P.K. Khosla, "Categorization of parameters in the dynamic robot model," *IEEE Transactions on Robotics and Automation*, vol. 5, pp. 261–268, 1989.
116. P.K. Khosla, T. Kanade, "Parameter identification of robot dynamics," in *Proceedings of 24th IEEE Conference on Decision and Control*, Fort Lauderdale, FL, pp. 1754–1760, 1985.
117. P.K. Khosla, T. Kanade, "Experimental evaluation of nonlinear feedback and feedforward control schemes for manipulators," *International Journal of Robotics Research*, vol. 7, no. 1, pp. 18–28, 1988.
118. C.A. Klein, C.H. Huang, "Review of pseudoinverse control for use with kinematically redundant manipulators," *IEEE Transactions on Systems, Man, and Cybernetics*, vol. 13, pp. 245–250, 1983.
119. D.E. Koditschek, "Natural motion for robot arms," *Proc. 23th IEEE Conference on Decision and Control*, Las Vegas, NV, pp. 733–735, 1984.
120. A.J. Koivo, *Fundamentals for Control of Robotic Manipulators*, Wiley, New York, 1989.
121. K. Kreutz, "On manipulator control by exact linearization," *IEEE Transactions on Automatic Control*, vol. 34, pp. 763–767, 1989.
122. J.-C. Latombe, *Robot Motion Planning*, Kluwer, Boston, MA, 1991.
123. J.-P. Laumond, (Ed.), *Robot Motion Planning and Control*, Springer-Verlag, Berlin, 1998.
124. S.M. LaValle, *Planning Algorithms*, Cambridge University Press, New York, 2006.
125. S.M. LaValle, J.J. Kuffner, "Rapidly-exploring random trees: Progress and prospects," in *New Directions in Algorithmic and Computational Robotics*, B.R. Donald, K. Lynch, D. Rus, (Eds.), AK Peters, Wellesley, MA, pp. 293–308, 2001.
126. M.B. Leahy, G.N. Saridis, "Compensation of industrial manipulator dynamics," *International Journal of Robotics Research*, vol. 8, no. 4, pp. 73–84, 1989.
127. C.S.G. Lee, "Robot kinematics, dynamics and control," *IEEE Computer*, vol. 15, no. 12, pp. 62–80, 1982.
128. W. Leonhard, *Control of Electrical Drives*, Springer-Verlag, New York, 2001.

129. A. Liégeois, "Automatic supervisory control of the configuration and behavior of multibody mechanisms," *IEEE Transactions on Systems, Man, and Cybernetics*, vol. 7, pp. 868–871, 1977.
130. K.Y. Lim, M. Eslami, "Robust adaptive controller designs for robot manipulator systems," *IEEE Journal of Robotics and Automation*, vol. 3, pp. 54–66, 1987.
131. C.S. Lin, P.R. Chang, J.Y.S. Luh, "Formulation and optimization of cubic polynomial joint trajectories for industrial robots," *IEEE Transactions on Automatic Control*, vol. 28, pp. 1066–1073, 1983.
132. S.K. Lin, "Singularity of a nonlinear feedback control scheme for robots," *IEEE Transactions on Systems, Man, and Cybernetics*, vol. 19, pp. 134–139, 1989.
133. H. Lipkin, J. Duffy, "Hybrid twist and wrench control for a robotic manipulator," *ASME Journal of Mechanism, Transmissions, and Automation Design*, vol. 110, pp. 138–144, 1988.
134. V. Lippiello, B. Siciliano, L. Villani, "Position-based visual servoing in industrial multirobot cells using a hybrid camera configuration," *IEEE Transactions on Robotics and Automation*, vol. 23, pp. 73–86, 2007.
135. D.A. Lizárraga, "Obstructions to the existence of universal stabilizers for smooth control systems," *Mathematics of Control, Signals, and Systems*, vol. 16, pp. 255–277, 2004.
136. J. Lončarić, "Normal forms of stiffness and compliance matrices," *IEEE Journal of Robotics and Automation*, vol. 3, pp. 567–572, 1987.
137. T. Lozano-Pérez, "Automatic planning of manipulator transfer movements," *IEEE Transactions on Systems, Man, and Cybernetics*, vol. 11, pp. 681–698, 1981.
138. T. Lozano-Pérez, "Spatial planning: A configuration space approach," *IEEE Transactions on Computing*, vol. 32, pp. 108–120, 1983.
139. T. Lozano-Pérez, "Robot programming," *Proceedings IEEE*, vol. 71, pp. 821–841, 1983.
140. T. Lozano-Pérez, M.T. Mason, R.H. Taylor, "Automatic synthesis of fine-motion strategies for robots," *International Journal of Robotics Research*, vol. 3, no. 1, pp. 3–24, 1984.
141. J.Y.S. Luh, "Conventional controller design for industrial robots: A tutorial," *IEEE Transactions on Systems, Man, and Cybernetics*, vol. 13, pp. 298–316, 1983.
142. J.Y.S. Luh, M.W. Walker, R.P.C. Paul, "On-line computational scheme for mechanical manipulators," *ASME Journal of Dynamic Systems, Measurement, and Control*, vol. 102, pp. 69–76, 1980.
143. J.Y.S. Luh, M.W. Walker, R.P.C. Paul, "Resolved-acceleration control of mechanical manipulators," *IEEE Transactions on Automatic Control*, vol. 25, pp. 468–474, 1980.
144. J.Y.S. Luh, Y.-F. Zheng, "Computation of input generalized forces for robots with closed kinematic chain mechanisms," *IEEE Journal of Robotics and Automation* vol. 1, pp. 95–103, 1985.
145. V.J. Lumelsky, *Sensing, Intelligence, Motion: How Robots and Humans Move in an Unstructured World*, Wiley, Hoboken, NJ, 2006.
146. Y. Ma, S. Soatto, J. Kosecka, S. Sastry, *An Invitation to 3-D Vision: From Images to Geometric Models*, Springer, New York, 2003.

147. A.A. Maciejewski, C.A. Klein, "Obstacle avoidance for kinematically redundant manipulators in dynamically varying environments," *International Journal of Robotics Research*, vol. 4, no. 3, pp. 109–117, 1985.
148. E. Malis, F. Chaumette, S. Boudet, "2-1/2D visual servoing," *IEEE Transactions on Robotics and Automation*, vol. 15, pp. 238–250, 1999.
149. B.R. Markiewicz, *Analysis of the Computed Torque Drive Method and Comparison with Conventional Position Servo for a Computer-Controlled Manipulator*, memo. TM 33-601, JPL, Pasadena, CA, 1973.
150. M.T. Mason, "Compliance and force control for computer controlled manipulators," *IEEE Transactions on Systems, Man, and Cybernetics*, vol. 6, pp. 418–432, 1981.
151. J.M. McCarthy, *An Introduction to Theoretical Kinematics*, MIT Press, Cambridge, MA, 1990.
152. N.H. McClamroch, D. Wang, "Feedback stabilization and tracking of constrained robots," *IEEE Transactions on Automatic Control*, vol. 33, pp. 419–426, 1988.
153. R.T. M'Closkey, R.M. Murray, "Exponential stabilization of driftless nonlinear control systems using homogeneous feedback," *IEEE Transactions on Automatic Control*, vol. 42, pp. 614–628, 1997.
154. L. Meirovitch, *Dynamics and Control of Structures*, Wiley, New York, 1990.
155. C. Melchiorri, *Traiettorie per Azionamenti Elettrici*, Progetto Leonardo, Bologna, I, 2000.
156. N. Manring, *Hydraulic Control Systems*, Wiley, New York, 2005.
157. R. Middleton, G.C. Goodwin, "Adaptive computed torque control for rigid link manipulators," *Systems & Control Letters*, vol. 10, pp. 9–16, 1988.
158. R.R. Murphy, *Introduction to AI Robotics*, MIT Press, Cambridge, MA, 2000.
159. R.M. Murray, Z. Li, S.S. Sastry, *A Mathematical Introduction to Robotic Manipulation*, CRC Press, Boca Raton, CA. 1994.
160. Y. Nakamura, *Advanced Robotics: Redundancy and Optimization*, Addison-Wesley, Reading, MA, 1991.
161. Y. Nakamura, H. Hanafusa, "Inverse kinematic solutions with singularity robustness for robot manipulator control," *ASME Journal of Dynamic Systems, Measurement, and Control*, vol. 108, pp. 163–171, 1986.
162. Y. Nakamura, H. Hanafusa, "Optimal redundancy control of robot manipulators," *International Journal of Robotics Research*, vol. 6, no. 1, pp. 32–42, 1987.
163. Y. Nakamura, H. Hanafusa, T. Yoshikawa, "Task-priority based redundancy control of robot manipulators," *International Journal of Robotics Research*, vol. 6, no. 2, pp. 3–15, 1987.
164. J.I. Neimark, F.A. Fufaev, *Dynamics of Nonholonomic Systems*, American Mathematical Society, Providence, RI, 1972.
165. I. Nevins, D.E. Whitney, "The force vector assembler concept," *Proc. First CISM-IFToMM Symposium on Theory and Practice of Robots and Manipulators*, Udine, I, 1973.
166. F. Nicolò, J. Katende, "A robust MRAC for industrial robots," *Proc. 2nd IASTED International Symposium on Robotics and Automation*, pp. 162–171, Lugano, Switzerland, 1983.
167. S. Nicosia, P. Tomei, "Model reference adaptive control algorithms for industrial robots," *Automatica*, vol. 20, pp. 635–644, 1984.

168. H. Nijmeijer, A. van de Schaft, *Nonlinear Dynamical Control Systems*, Springer-Verlag, Berlin, Germany, 1990.
169. B. Noble, *Applied Linear Algebra*, 3rd ed., Prentice-Hall, Englewood Cliffs, NJ, 1987.
170. C. O'Dúnlaing, C.K. Yap, "A retraction method for planning the motion of a disc," *Journal of Algorithms*, vol. 6, pp. 104–111, 1982.
171. K. Ogata, *Modern Control Engineering*, 4th ed., Prentice-Hall, Englewood Cliffs, NJ, 2002.
172. D.E. Orin, R.B. McGhee, M. Vukobratović, G. Hartoch, "Kinematic and kinetic analysis of open-chain linkages utilizing Newton–Euler methods," *Mathematical Biosciences* vol. 43, pp. 107–130, 1979.
173. D.E. Orin, W.W. Schrader, "Efficient computation of the Jacobian for robot manipulators," *International Journal of Robotics Research*, vol. 3, no. 4, pp. 66–75, 1984.
174. G. Oriolo, A. De Luca, M. Vendittelli, "WMR control via dynamic feedback linearization: Design, implementation and experimental validation," *IEEE Transactions on Control Systems Technology*, vol. 10, pp. 835–852, 2002.
175. R. Ortega, M.W. Spong, "Adaptive motion control of rigid robots: A tutorial," *Automatica*, vol. 25, pp. 877–888, 1989.
176. T. Patterson, H. Lipkin, "Duality of constrained elastic manipulation," *Proc. 1991 IEEE International Conference on Robotics and Automation*, pp. 2820–2825, Sacramento, CA, 1991.
177. T. Patterson, H. Lipkin, "Structure of robot compliance," *ASME Journal of Mechanical Design*, vol. 115, pp. 576–580, 1993.
178. R.P. Paul, *Modelling, Trajectory Calculation, and Servoing of a Computer Controlled Arm*, memo. AIM 177, Stanford Artificial Intelligence Laboratory, 1972.
179. R.P. Paul, "Manipulator Cartesian path control," *IEEE Transactions on Systems, Man, and Cybernetics*, vol. 9, pp. 702–711, 1979.
180. R.P. Paul, *Robot Manipulators: Mathematics, Programming, and Control*, MIT Press, Cambridge, MA, 1981.
181. R.P. Paul, B.E. Shimano, G. Mayer, "Kinematic control equations for simple manipulators," *IEEE Transactions on Systems, Man, and Cybernetics*, vol. 11, pp. 449–455, 1981.
182. R.P. Paul, H. Zhang, "Computationally efficient kinematics for manipulators with spherical wrists based on the homogeneous transformation representation," *International Journal of Robotics Research*, vol. 5, no. 2, pp. 32–44, 1986.
183. D.L. Pieper, *The Kinematics of Manipulators Under Computer Control* memo. AIM 72, Stanford Artificial Intelligence Laboratory, 1968.
184. M.H. Raibert, J.J. Craig, "Hybrid position/force control of manipulators," *ASME Journal of Dynamic Systems, Measurement, and Control*, vol. 103, pp. 126–133, 1981.
185. E. Rimon, D.E. Koditschek, "The construction of analytic diffeomorphisms for exact robot navigation on star worlds," *Proc. 1989 IEEE International Conference on Robotics and Automation*, Scottsdale, AZ, pp. 21–26, 1989.
186. E.I. Rivin, *Mechanical Design of Robots*, McGraw-Hill, New York, 1987.
187. R.E. Roberson, R. Schwertassek, *Dynamics of Multibody Systems*, Springer-Verlag, Berlin, Germany, 1988.
188. Z. Roth, B.W. Mooring, B. Ravani, "An overview of robot calibration," *IEEE Journal of Robotics and Automation*, vol. 3, pp. 377–386, 1987.

189. S. Russell, P. Norvig, *Artificial Intelligence: A Modern Approach*, 2nd ed., Prentice Hall, Englewood Cliffs, NJ, 2003.
190. J.K. Salisbury, "Active stiffness control of a manipulator in Cartesian coordinates," *Proc. 19th IEEE Conference on Decision and Control*, pp. 95–100, Albuquerque, NM, 1980.
191. J.K. Salisbury, J.J. Craig, "Articulated hands: Force control and kinematic issues," *International Journal of Robotics Research*, vol. 1, no. 1, pp. 4–17, 1982.
192. C. Samson, "Robust control of a class of nonlinear systems and applications to robotics," *International Journal of Adaptive Control and Signal Processing*, vol. 1, pp. 49–68, 1987.
193. C. Samson, "Time-varying feedback stabilization of car-like wheeled mobile robots," *International Journal of Robotics Research*, vol. 12, no. 1, pp. 55–64, 1993.
194. C. Samson, M. Le Borgne, B. Espiau, *Robot Control: The Task Function Approach*, Clarendon Press, Oxford, UK, 1991.
195. S. Sastry, *Nonlinear Systems: Analysis, Stability and Control*, Springer-Verlag, Berlin, Germany, 1999.
196. V.D. Scheinman, *Design of a Computer Controlled Manipulator*, memo. AIM 92, Stanford Artificial Intelligence Laboratory, 1969.
197. J.T. Schwartz, M. Sharir, "On the 'piano movers' problem: II. General techniques for computing topological properties of real algebraic manifolds," *Advances in Applied Mathematics*, vol. 4, pp. 298–351, 1983.
198. L. Sciavicco, B. Siciliano, "Coordinate transformation: A solution algorithm for one class of robots," *IEEE Transactions on Systems, Man, and Cybernetics*, vol. 16, pp. 550–559, 1986.
199. L. Sciavicco, B. Siciliano, "A solution algorithm to the inverse kinematic problem for redundant manipulators," *IEEE Journal of Robotics and Automation*, vol. 4, pp. 403–410, 1988.
200. L. Sciavicco, B. Siciliano, *Modelling and Control of Robot Manipulators*, 2nd ed., Springer, London, UK, 2000.
201. L. Sciavicco, B. Siciliano, L. Villani, "Lagrange and Newton–Euler dynamic modeling of a gear-driven rigid robot manipulator with inclusion of motor inertia effects," *Advanced Robotics*, vol. 10, pp. 317–334, 1996.
202. R. Sedgewick, *Algorithms*, 2nd ed., Addison-Wesley, Reading, MA, 1988.
203. H. Seraji, "Configuration control of redundant manipulators: Theory and implementation," *IEEE Transactions on Robotics and Automation*, vol. 5, pp. 472–490, 1989.
204. S.W. Shepperd, S.W., "Quaternion from rotation matrix," *AIAA Journal of Guidance and Control*, vol. 1, pp. 223–224, 1978.
205. R. Shoureshi, M.E. Momot, M.D. Roesler, "Robust control for manipulators with uncertain dynamics," *Automatica*, vol. 26, pp. 353–359, 1990.
206. B. Siciliano, "Kinematic control of redundant robot manipulators: A tutorial," *Journal of Intelligent and Robotic Systems*, vol. 3, pp. 201–212, 1990.
207. B. Siciliano, "A closed-loop inverse kinematic scheme for on-line joint based robot control," *Robotica*, vol. 8, pp. 231–243, 1990.
208. B. Siciliano, J.-J.E. Slotine, "A general framework for managing multiple tasks in highly redundant robotic systems," *Proc. 5th International Conference on Advanced Robotics*, Pisa, I, pp. 1211–1216, 1991.

209. B. Siciliano, L. Villani, *Robot Force Control*, Kluwer, Boston, MA, 2000.
210. R. Siegwart, I.R. Nourbakhsh, *Introduction to Autonomous Mobile Robots*, MIT Press, Cambridge, MA, 2004.
211. D.B. Silver, "On the equivalence of Lagrangian and Newton–Euler dynamics for manipulators," *International Journal of Robotics Research*, vol. 1, no. 2, pp. 60–70, 1982.
212. J.-J.E. Slotine, "The robust control of robot manipulators," *International Jorunal of Robotics Research*, vol. 4, no. 2, pp. 49–64, 1985.
213. J.-J.E. Slotine, "Putting physics in control — The example of robotics," *IEEE Control Systems Magazine*, vol. 8, no. 6, pp. 12–18, 1988.
214. J.-J.E. Slotine, W. Li, "On the adaptive control of robot manipulators," *International Journal of Robotics Research*, vol. 6, no. 3, pp. 49–59, 1987.
215. J.-J.E. Slotine, W. Li, *Applied Nonlinear Control*, Prentice-Hall, Englewood Cliffs, NJ, 1991.
216. M.W. Spong, "On the robust control of robot manipulators," *IEEE Transactions on Automatic Control*, vol. 37, pp. 1782–1786, 1992.
217. M.W. Spong, S. Hutchinson, M. Vidyasagar, *Robot Modeling and Control*, Wiley, New York, 2006.
218. M.W. Spong, R. Ortega, R. Kelly, "Comments on "Adaptive manipulator control: A case study"," *IEEE Transactions on Automatic Control*, vol. 35, pp. 761–762, 1990.
219. M.W. Spong, M. Vidyasagar, "Robust linear compensator design for nonlinear robotic control," *IEEE Journal of Robotics and Automation*, vol. 3, pp. 345–351, 1987.
220. SRI International, *Robot Design Handbook*, G.B. Andeen, (Ed.), McGraw-Hill, New York, 1988.
221. Y. Stepanenko, M. Vukobratović, "Dynamics of articulated open-chain active mechanisms," *Mathematical Biosciences*, vol. 28, pp. 137–170, 1976.
222. Y. Stepanenko, J. Yuan, "Robust adaptive control of a class of nonlinear mechanical systems with unbounded and fast varying uncertainties," *Automatica*, vol. 28, pp. 265–276, 1992.
223. S. Stramigioli, *Modeling and IPC Control of Interactive Mechanical Systems — A Coordinate Free Approach*, Springer, London, UK, 2001.
224. K.R. Symon, *Mechanics*, 3rd ed., Addison-Wesley, Reading, MA, 1971.
225. K. Takase, R. Paul, E. Berg, "A structured approach to robot programming and teaching," *IEEE Transactions on Systems, Man, and Cybernetics*, vol. 11, pp. 274–289, 1981.
226. M. Takegaki, S. Arimoto, "A new feedback method for dynamic control of manipulators," *ASME Journal of Dynamic Systems, Measurement, and Control*, vol. 102, pp. 119–125, 1981.
227. T.-J. Tarn, A.K. Bejczy, X. Yun, Z. Li, "Effect of motor dynamics on nonlinear feedback robot arm control," *IEEE Transactions on Robotics and Automation*, vol. 7, pp. 114–122, 1991.
228. T.-J. Tarn, Y. Wu, N. Xi, A. Isidori, "Force regulation and contact transition control," *IEEE Control Systems Magazine*, vol. 16, no. 1, pp. 32–40, 1996.
229. R.H. Taylor, "Planning and execution of straight line manipulator trajectories," *IBM Journal of Research and Development*, vol. 23, pp. 424–436, 1979.
230. R.H. Taylor, D.D. Grossman, "An integrated robot system architecture," *Proceedings IEEE*, vol. 71, pp. 842–856, 1983.

231. S. Thrun, W. Burgard, D. Fox, *Probabilistic Robotics*, MIT Press, Cambridge, MA, 2005.
232. L.W. Tsai, A.P. Morgan, "Solving the kinematics of the most general six- and five-degree-of-freedom manipulators by continuation methods," *ASME Journal of Mechanisms, Transmission, and Automation in Design*, vol. 107, pp. 189–200, 1985.
233. R. Tsai, "A versatile camera calibration technique for high accuracy 3-D machine vision metrology using off-the-shelf TV cameras and lenses," *IEEE Transactions on Robotics and Automation*, vol. 3, pp. 323–344, 1987.
234. J.J.Uicker, "Dynamic force analysis of spatial linkages," *ASME Journal of Applied Mechanics*, vol. 34, pp. 418–424, 1967.
235. L. Villani, C. Canudas de Wit, B. Brogliato, "An exponentially stable adaptive control for force and position tracking of robot manipulators," *IEEE Transactions on Automatic Control*, vol. 44, pp. 798–802, 1999.
236. M. Vukobratović, "Dynamics of active articulated mechanisms and synthesis of artificial motion," *Mechanism and Machine Theory*, vol. 13, pp. 1–56, 1978.
237. M.W. Walker, D.E. Orin, "Efficient dynamic computer simulation of robotic mechanisms," *ASME Journal of Dynamic Systems, Measurement, and Control*, vol. 104, pp. 205–211, 1982.
238. C.W. Wampler, "Manipulator inverse kinematic solutions based on damped least-squares solutions," *IEEE Transactions on Systems, Man, and Cybernetics*, vol. 16, pp. 93–101, 1986.
239. L. Weiss, A. Sanderson, C. Neuman, "Dynamic sensor-based control of robots with visual feedback," *IEEE Journal of Robotics and Automation*, vol. 3, pp. 404–417, 1987.
240. D.E. Whitney, "Resolved motion rate control of manipulators and human prostheses," *IEEE Transactions on Man-Machine Systems*, vol. 10, pp. 47–53, 1969.
241. D.E. Whitney, "Force feedback control of manipulator fine motions," *ASME Journal of Dynamic Systems, Measurement, and Control*, vol. 99, pp. 91–97, 1977.
242. D.E. Whitney, "Quasi-static assembly of compliantly supported rigid parts," *ASME Journal of Dynamic Systems, Measurement, and Control*, vol. 104, pp. 65–77, 1982.
243. D.E. Whitney, "Historical perspective and state of the art in robot force control," *International Journal of Robotics Research*, vol. 6, no. 1, pp. 3–14, 1987.
244. W. Wilson, C. Hulls, G. Bell, "Relative end-effector control using Cartesian position based visual servoing," *IEEE Transactions on Robotics and Automation*, vol. 12, pp. 684–696, 1996.
245. T. Yoshikawa, "Manipulability of robotic mechanisms," *International Journal of Robotics Research*, vol. 4, no. 2, pp. 3–9, 1985.
246. T. Yoshikawa, "Dynamic manipulability ellipsoid of robot manipulators," *Journal of Robotic Systems*, vol. 2, pp. 113–124, 1985.
247. T. Yoshikawa, "Dynamic hybrid position/force control of robot manipulators — Description of hand constraints and calculation of joint driving force," *IEEE Journal of Robotics and Automation*, vol. 3, pp. 386–392, 1987.
248. T. Yoshikawa, *Foundations of Robotics*, MIT Press, Boston, MA, 1990.
249. T. Yoshikawa, T. Sugie, N. Tanaka, "Dynamic hybrid position/force control of robot manipulators — Controller design and experiment," *IEEE Journal of Robotics and Automation*, vol. 4, pp. 699–705, 1988.

250. J.S.-C. Yuan, "Closed-loop manipulator control using quaternion feedback," *IEEE Journal of Robotics and Automation*, vol. 4, pp. 434–440, 1988.

Index

acceleration
  feedback, 317
  gravity, 255, 583
  joint, 141, 256
  link, 285
accessibility
  loss, 471, 476
  rank condition, 477, 603
accuracy, 87
actuator, 3, 191
algorithm
  $A^\star$, 607
  best-first, 552
  complete, 535
  complexity, 605
  inverse kinematics, 132, 143
  pose estimation, 427
  probabilistically complete, 543
  randomized best-first, 553
  resolution complete, 540
  search, 606
  steepest descent, 551
  sweep line, 536
  sweep plane, 539
  wavefront expansion, 554
angle
  and axis, 52, 139, 187
  Euler, 48
architecture
  control, 233, 237
  functional, 233
  hardware, 242
arm
  anthropomorphic, 73, 96, 114
  anthropomorphic with spherical wrist, 77
  parallelogram, 70
  singularity, 119
  spherical, 72, 95
  three-link planar, 69, 91, 113
automation
  flexible, 17
  industrial, 24
  programmable, 16
  rigid, 16
axis
  and angle, 52
  central, 582
  joint, 62
  principal, 582

Barbalat
  lemma, 507, 512, 513, 598
bicycle
  chained-form transformation, 485
  flat outputs, 491
  front-wheel drive, 481
  rear-wheel drive, 481

calibration
  camera, 229, 440
  kinematic, 88
  matrix, 217
camera
  calibration, 440
  eye-in-hand, 409
  eye-to-hand, 409

fixed configuration, 409
hybrid configuration, 409
mobile configuration, 409
pan-tilt, 410
cell decomposition
approximate, 539
exact, 536
chained form, 482
flat outputs, 492
transformation, 483
Christoffel
symbols, 258
collision checking, 532
compensation
decentralized feedforward, 319
feedforward, 593
feedforward computed torque, 324
gravity, 328, 345, 368, 446, 449
compliance
active, 367
control, 364, 367
matrix, 366
passive, 366
configuration, 470, 525, 585
configuration space
2R manipulator, 526
as a manifold, 527
distance, 527
free, 528
free path, 528
obstacles, 527
connectivity graph, 536, 537
constraint
artificial, 391
bilateral, 386, 585
epipolar, 434
frame, 391
holonomic, 385, 470, 585
Jacobian, 385
kinematic, 471
natural, 391
nonholonomic, 469, 585
Pfaffian, 471
pure rolling, 472
scleronomic, 585
unilateral, 386
control
adaptive, 338
admittance, 377
architecture, 233, 237
centralized, 327
comparison among schemes, 349, 453
compliance, 364, 367
decentralized, 309
force, 378
force with inner position loop, 379
force with inner velocity loop, 380
hybrid force/motion, 396
hybrid force/position, 403
hybrid force/velocity, 398, 402
impedance, 372
independent joint, 311
interaction, 363
inverse dynamics, 330, 347, 372, 487, 594
inverse model, 594
Jacobian inverse, 344
Jacobian transpose, 345
joint space, 305
kinematic, 134
linear systems, 589
motion, 303
operational space, 343, 364
parallel force/position, 381
PD with gravity compensation, 328, 345, 368
PI, 311, 322, 380, 591
PID, 322, 591
$PIDD^2$, 322
points, 555
position, 206, 312, 314, 317
resolved-velocity, 448
robust, 333
system, 3
unit vector, 337
velocity, 134, 314, 317, 502
vision-based, 408
voltage, 199
controllability
and nonholonomy, 477
condition, 477
system, 603
coordinate
generalized, 247, 296, 585
homogeneous, 56, 418
Lagrange, 585
transformation, 56

degree
nonholonomy, 603
of freedom, 4, 585
Denavit–Hartenberg
convention, 61
parameters, 63, 69, 71, 72, 74, 75, 78, 79
differential flatness, 491
displacement
elementary, 366, 368, 581, 586
virtual, 385, 586
distribution
accessibility, 603
dimension, 602
involutive, 602
involutive closure, 602
disturbance
compensation, 325
rejection, 207, 376, 590
drive
electric, 198
hydraulic, 202
with gear, 204
dynamic extension, 487
dynamic model
constrained mechanical system, 486
joint space, 257
linearity in the parameters, 259
notable properties, 257
operational space, 296
parallelogram arm, 277
parameter identification, 280
reduced order, 402
skew-symmetry of matrix $\dot{B} - 2C$, 257
two-link Cartesian arm, 264
two-link planar arm, 265
dynamics
direct, 298
fundamental principles, 584
inverse, 298, 330, 347

encoder
absolute, 210
incremental, 212, 517
end-effector
force, 147
frame, 59
orientation, 187
pose, 58, 184
position, 184
energy
conservation, 588
conservation principle, 259
kinetic, 249
potential, 255, 585
environment
compliant, 389, 397
interaction, 363
programming, 238
rigid, 385, 401
structured, 15
unstructured, 25
epipolar
geometry, 433
line, 435
error
estimation, 430
force, 378
joint space, 328
operational space, 132, 345, 367, 445
orientation, 137
position, 137
tracking, 324
estimation
pose, 427
Euler
angles, 48, 137, 187

feedback
nonlinear, 594
position, 312
position and velocity, 314
position, velocity and acceleration, 317
flat outputs, 491
force
active, 583, 586
centrifugal, 256
conservative, 585, 587
contact, 364
control, 378
controlled subspace, 387
Coriolis, 257
elementary work, 584
end-effector, 147
error, 378
external, 583, 584

generalized, 248, 587
gravity, 255, 583
internal, 583
nonconservative, 587
reaction, 385, 583, 586
resultant, 583
transformation, 151
form
bilinear, 574
negative definite, 574
positive definite, 574
quadratic, 574, 597
frame
attached, 40
base, 59
central, 582
compliant, 377
constraint, 391
current, 46
fixed, 46, 579
moving, 579
principal, 582
rotation, 40
friction
Coulomb, 257
electric, 200
viscous, 257
Frobenius
norm, 421
theorem, 476
function
gradient, 569
Hamiltonian, 588
Lagrangian, 588
Lyapunov, 596

gear
reduction ratio, 205, 306
generator
torque-controlled, 200, 309
velocity-controlled, 200, 309
graph search, 606
$A^\star$, 607
breadth-first, 606
depth-first, 606
gravity
acceleration, 255, 583
compensation, 328, 345, 368, 446, 449
force, 255, 583
Hamilton
principle of conservation of energy, 259
homography
planar, 420, 438

identification
dynamic parameters, 280
kinematic parameters, 88
image
binary, 412
centroid, 416
feature parameters, 410
interpretation, 416
Jacobian, 424
moment, 416
processing, 410
segmentation, 411
impedance
active, 373
control, 372
mechanical, 373
passive, 374
inertia
first moment, 262
matrix, 254
moment, 262, 581
product, 582
tensor, 251, 582
integrability
multiple kinematic constraints, 475, 477
single kinematic constraint, 473
interaction
control, 363
environment, 363
matrix, 424
inverse kinematics
algorithm, 132
anthropomorphic arm, 96
comparison among algorithms, 143
manipulator with spherical wrist, 94
second-order algorithm, 141
spherical arm, 95
spherical wrist, 99
three-link planar arm, 91

Jacobian
analytical, 128

anthropomorphic arm, 114
computation, 111
constraint, 385
damped least-squares, 127
geometric, 105
image, 424
inverse, 133, 344
pseudo-inverse, 133
Stanford manipulator, 115
three-link planar arm, 113
transpose, 134, 345

joint
acceleration, 141, 256
actuating system, 191
axis, 62
prismatic, 4
revolute, 4
space, 84
torque, 147, 248
variable, 58, 248

kinematic chain
closed, 4, 65, 151
open, 4, 60

kinematics
anthropomorphic arm, 73
anthropomorphic arm with spherical wrist, 77
differential, 105
direct, 58
DLR manipulator, 79
humanoid manipulator, 81
inverse, 90
inverse differential, 123
parallelogram arm, 70
spherical arm, 72, 95
spherical wrist, 75
Stanford manipulator, 76
three-link planar arm, 69

kineto-statics duality, 148

La Salle
theorem, 507, 597

Lagrange
coordinates, 585
equations, 587
formulation, 247, 292
function, 588
multipliers, 124, 485

level
action, 235
gray, 410
hierarchical, 234
primitive, 236
servo, 236
task, 235

Lie
bracket, 600
derivative, 601

link
acceleration, 285
centre of mass, 249
inertia, 251
velocity, 108

local
minima, 550, 551
planner, 542

Lyapunov
direct method, 596
equation, 597
function, 135, 328, 335, 340, 341, 345, 368, 431, 446, 449, 452, 506, 513, 596

manipulability
dynamic, 299
ellipsoid, 152
measure, 126, 153

manipulability ellipsoid
dynamic, 299
force, 156
velocity, 153

manipulator
anthropomorphic, 8
Cartesian, 4
cylindrical, 5
DLR, 79
end-effector, 4
humanoid, 81
joint, 58
joints, 4
link, 58
links, 4
mechanical structure, 4
mobile, 14
parallel, 9
posture, 58
redundant, 4, 87, 124, 134, 142, 296

SCARA, 7
spherical, 6
Stanford, 76, 115
with spherical wrist, 94
wrist, 4
matrix
adjoint, 567
algebraic complement, 565
block-partitioned, 564
calibration, 217, 229
compliance, 366
condition number, 577
damped least-squares, 127
damped least-squares inverse, 282
derivative, 568
determinant, 566
diagonal, 564
eigenvalues, 573
eigenvectors, 573
essential, 434
homogeneous transformation, 56
idempotent, 568
identity, 564
inertia, 254
interaction, 424
inverse, 567
Jacobian, 569
left pseudo-inverse, 90, 281, 386, 428, 431, 452, 576
minor, 566
negative definite, 574
negative semi-definite, 575
norm, 572
null, 564
operations, 565
orthogonal, 568, 579
positive definite, 255, 574, 582
positive semi-definite, 575
product, 566
product of scalar by, 565
projection, 389, 572
right pseudo-inverse, 125, 299, 576
rotation, 40, 579
selection, 389
singular value decomposition, 577
skew-symmetric, 257, 564
square, 563
stiffness, 366
sum, 565
symmetric, 251, 255, 564
trace, 565
transpose, 564
triangular, 563
mobile robot
car-like, 13, 482
control, 502
differential drive, 12, 479
dynamic model, 486
kinematic model, 476
legged, 11
mechanical structure, 10
omnidirectional, 13
path planning, 492
planning, 489
second-order kinematic model, 488
synchro drive, 12, 479
trajectory planning, 498
tricycle-like, 12, 482
wheeled, 10, 469
moment
image, 416
inertia, 262, 581
inertia first, 262
resultant, 583
motion
constrained, 363, 384
control, 303
equations, 255
internal, 296
planning, 523
point-to-point, 163
primitives, 545
through a sequence of points, 168
motion planning
canonical problem, 523
multiple-query, 535
off-line, 524
on-line, 524
probabilistic, 541
query, 535
reactive, 551
sampling-based, 541
single-query, 543
via artificial potentials, 546
via cell decomposition, 536
via retraction, 532
motor
electric, 193

hydraulic, 193
pneumatic, 193

navigation function, 553
Newton–Euler
equations, 584
formulation, 282, 292
recursive algorithm, 286
nonholonomy, 469

octree, 541
odometric localization, 514
operational
space, 84, 445
operator
Laplacian, 415
Roberts, 414
Sobel, 414
orientation
absolute, 436
end-effector, 187
error, 137
minimal representation, 49
rigid body, 40
trajectory, 187

parameters
Denavit–Hartenberg, 63
dynamic, 259
extrinsic, 229, 440
intrinsic, 229, 440
uncertainty, 332, 444
path
circular, 183
geometrically admissible, 490
minimum, 607
primitive, 181
rectilinear, 182
plane
epipolar, 435
osculating, 181
points
feature, 417
path, 169
via, 186, 539
virtual, 173
polynomial
cubic, 164, 169
interpolating, 169
sequence, 170, 172, 175
Pontryagin
minimum principle, 499
pose
estimation, 418
regulation, 345
rigid body, 39
position
control, 206, 312
end-effector, 184
feedback, 312, 314, 317
rigid body, 39
trajectory, 184
transducer, 210
posture
manipulator, 58
regulation, 328, 503, 512
potential
artificial, 546
attractive, 546
repulsive, 547
total, 549
power
amplifier, 197
supply, 198
principle
conservation of energy, 259
virtual work, 147, 385, 587
PRM (Probabilistic Roadmap), 541
programming
environment, 238
language, 238
object-oriented, 242
robot-oriented, 241
teaching-by-showing, 240

quadtree, 540

range
sensor, 219
reciprocity, 387
redundancy
kinematic, 121
analysis, 121
kinematic, 87
resolution, 123, 298
Reeds–Shepp
curves, 501
regulation

Cartesian, 511
discontinuous and/or time-varying, 514
pose, 345
posture, 328, 503, 512
Remote Centre of Compliance (RCC), 366
resolver, 213
retraction, 534
rigid body
angular momentum, 583
angular velocity, 580
inertia moment, 581
inertia product, 582
inertia tensor, 582
kinematics, 579
linear momentum, 583
mass, 581
orientation, 40
pose, 39, 580
position, 39
potential energy, 585
roadmap, 532
robot
applications, 18
field, 26
industrial, 17
manipulator, 4
mobile, 10
origin, 1
service, 27
robotics
advanced, 25
definition, 2
fundamental laws, 2
industrial, 15
rotation
elementary, 41
instantaneous centre, 480
matrix, 40, 579
vector, 44
rotation matrix
composition, 45
derivative, 106
RRT (Rapidly-exploring Random Tree), 543

segmentation
binary, 412
image, 411
sensor
exteroceptive, 3, 215, 517
laser, 222
proprioceptive, 3, 209, 516
range, 219
shaft torque, 216
sonar, 219
vision, 225
wrist force, 216
servomotor
brushless DC, 194
electric, 193
hydraulic, 195
permanent-magnet DC, 194
simulation
force control, 382
hybrid visual servoing, 464
impedance control, 376
inverse dynamics, 269
inverse kinematics algorithms, 143
motion control schemes, 349
pose estimation, 432
regulation for mobile robots, 514
trajectory tracking for mobile robots, 508
visual control schemes, 453
visual servoing, 453
singularity
arm, 119
classification, 116
decoupling, 117
kinematic, 116, 127
representation, 130
wrist, 119
space
configuration, 470
joint, 83, 84, 162
null, 122, 149
operational, 83, 84, 296, 343
projection, 572
range, 122, 149, 572
vector, 570
work, 85
special group
Euclidean, 57, 580
orthonormal, 41, 49, 579
stability, 133, 135, 141, 328, 368, 446, 447, 452, 590, 595, 596

statics, 147, 587
Steiner
  theorem, 260, 582
stiffness
  matrix, 366

tachometer, 214
torque
  actuating, 257
  computed, 324
  controlled generator, 200
  driving, 199, 203
  friction, 257
  joint, 147, 248
  limit, 294
  reaction, 199
  sensor, 216
tracking
  error, 504
  reference, 590
  trajectory, 503, 595
  via input/output linearization, 507
  via linear control, 505
  via nonlinear control, 506
trajectory
  dynamic scaling, 294
  joint space, 162
  operational space, 179
  orientation, 187
  planning, 161, 179
  position, 184
  tracking, 503
transducer
  position, 210
  velocity, 214
transformation
  coordinate, 56
  force, 151
  homogeneous, 56
  linear, 572
  matrix, 56
  perspective, 227
  similarity, 573
  velocity, 149
transmission, 192
triangulation, 435

unicycle
  chained-form transformation, 484
  dynamic model, 488
  flat outputs, 491
  kinematic model, 478
  minimum-time trajectories, 500
  optimal trajectories, 499
  second-order kinematic model, 489
unit quaternion, 54, 140
unit vector
  approach, 59
  binormal, 181
  control, 337
  normal, 59, 181
  sliding, 59
  tangent, 181

vector
  basis, 570
  bound, 580
  column, 563
  components, 570
  feature, 418
  field, 599
  homogeneous representation, 56
  linear independence, 569
  norm, 570
  null, 564
  operations, 569
  product, 571
  product of scalar by, 570
  representation, 42
  rotation, 44
  scalar product, 570
  scalar triple product, 571
  space, 570
  subspace, 570
  sum, 570
  unit, 571
velocity
  controlled generator, 200
  controlled subspace, 387
  feedback, 314, 317
  link, 108
  transducer, 214
  transformation, 149
  trapezoidal profile, 165
  triangular profile, 167
vision
  sensor, 225
  stereo, 409, 433

visual servoing
hybrid, 460
image-based, 449
PD with gravity compensation, 446, 449
position-based, 445
resolved-velocity, 447, 451
Voronoi
generalized diagram, 533

wheel
caster, 11
fixed, 11
Mecanum, 13
steerable, 11
work
elementary, 584
virtual, 147, 385, 586
workspace, 4, 14
wrist
force sensor, 216
singularity, 119
spherical, 75, 99